Trends in Optical Fibre Metrology and Standards

NATO ASI Series

Advanced Science Institutes Series

A Series presenting the results of activities sponsored by the NATO Science Committee, which aims at the dissemination of advanced scientific and technological knowledge, with a view to strengthening links between scientific communities.

The Series is published by an international board of publishers in conjunction with the NATO Scientific Affairs Division

A	Life Sciences	Plenum Publishing Corporation
B	Physics	London and New York
C	Mathematical and Physical Sciences	Kluwer Academic Publishers
D	Behavioural and Social Sciences	Dordrecht, Boston and London
E	Applied Sciences	
F	Computer and Systems Sciences	Springer-Verlag
G	Ecological Sciences	Berlin, Heidelberg, New York, London,
H	Cell Biology	Paris and Tokyo
I	Global Environmental Change	

PARTNERSHIP SUB-SERIES

1.	Disarmament Technologies	Kluwer Academic Publishers
2.	Environment	Springer-Verlag
3.	High Technology	Kluwer Academic Publishers
4.	Science and Technology Policy	Kluwer Academic Publishers
5.	Computer Networking	Kluwer Academic Publishers

The Partnership Sub-Series incorporates activities undertaken in collaboration with NATO's Cooperation Partners, the countries of the CIS and Central and Eastern Europe, in Priority Areas of concern to those countries.

NATO-PCO-DATA BASE

The electronic index to the NATO ASI Series provides full bibliographical references (with keywords and/or abstracts) to more than 30000 contributions from international scientists published in all sections of the NATO ASI Series.
Access to the NATO-PCO-DATA BASE is possible in two ways:

– via online FILE 128 (NATO-PCO-DATA BASE) hosted by ESRIN,
Via Galileo Galilei, I-00044 Frascati, Italy.

– via CD-ROM "NATO-PCO-DATA BASE" with user-friendly retrieval software in English, French and German (© WTV GmbH and DATAWARE Technologies Inc. 1989).

The CD-ROM can be ordered through any member of the Board of Publishers or through NATO-PCO, Overijse, Belgium.

Series E: Applied Sciences - Vol. 285

Trends in Optical Fibre Metrology and Standards

edited by

Olivério D. D. Soares
CETO - Centre de Ciências e Tecnologias Ópticas,
Universidade do Porto,
Porto,
Portugal

Springer-Science+Business Media, B.V.

Proceedings of the NATO Advanced Study Institute on
Trends in Optical Fibre Metrology and Standards
Viana do Castelo, Portugal
June 27–July 8, 1994

Library of Congress Cataloging-in-Publication Data

```
Trends in optical fibre metrology and standards / edited by O.D.D.
Soares.
      p.    cm. -- (NATO ASI series. Series E, Applied sciences ; vol.
285)
      "Published in cooperation with NATO Scientific and Environmental
Affairs Division.
      Includes index.
      ISBN 978-94-010-4020-4      ISBN 978-94-011-0035-9 (eBook)
      DOI 10.1007/978-94-011-0035-9
      1. Optical fibers--Measurement. 2. Optical fibers--Standards.
I. Soares, O. D. D. (Olivério D. D.)  II. North Atlantic Treaty
Organization.  Scientific and Environmental Affairs Division.
III. Series: NATO ASI series.  Series E, Applied sciences ; no. 285.
TA1800.T73  1995
621.36'92'0287--dc20                                          95-5544
```

ISBN 978-94-010-4020-4

Printed on acid-free paper

All Rights Reserved
© 1995 Springer Science+Business Media Dordrecht
Originally published by Kluwer Academic Publishers in 1995
Softcover reprint of the hardcover 1st edition 1995
No part of the material protected by this copyright notice may be reproduced or utilized in any form or by any means, electronic or mechanical, including photocopying, recording or by any information storage and retrieval system, without written permission from the copyright owner.

TABLE OF CONTENTS

Preface _____ XI

Institute Programme _____ XV

Participants _____ XIX

Sponsors and Co-sponsors _____ XXVII

Committees _____ XXVIII

Opening Message
Ministro da Indústria e Energia
Eng° Luis Mira Amaral _____ XXIX

I INTRODUCTION

1.1 A Historical Survey of Optical Signals and Optical Fibers
Silvério P. Almeida and Olivério D.D. Soares _____ 3

1.2 Trends in Optical Fibre Metrology and Standards
Olivério D.D. Soares and Silvério P. Almeida _____ 19

II OPTICAL FIBRES AND MATERIALS

2.1 Polarised Light Evolution in Optical Fibres
J. Pelayo, F. Villuendas _____ 47

2.2 Nonlinear Optical Fibres
Raman Kashyap _____ 69

2.3 Exotic Fibres
F. Gauthier _____ 103

2.4 Fibers and Fiber Lasers for the Mid-Infrared
U.B. Unrau _____ 113

2.5 Optical Fibres in Nuclear Radiation Environments
F. Berghmans, O. Deparis, S. Coenen, M. Decréton and P. Jucker _____ 131

III Optical Fibre Components

3.1 Connectors and Splices
J. Dubard 159

3.2 Optical Attenuators and Couplers Characterization
C. Le Men 175

3.3 Wavelength Division Multiplexers and the Measurement of the Channel Wavelengths
J.P. Laude 193

3.4 Active Fibre Characterisation with Passive Measurements
J.M.S. Anacleto, Modesto Morais, Gaspar Rego, O.D.D. Soares 209

IV Optical Sources

4.1 Wavelength Tunable Laser Diodes and their Applications
M.C. Amann 217

4.2 Active Components Characterization
C. Le Men 241

V Optical Fibre Amplifiers

5.1 Optical Amplifiers
Ivan Andonovic 263

5.2 Analysis of Erbium-Doped Silica-Fibre Characterisation Techniques
M.A. Rebolledo 305

5.3 Optical Fibre Amplifiers Standardisation
P. Di Vita 327

5.4 Optical Amplification Modelling for Er-Doped Fibre
Joaquim M.S. Anacleto and O.D.D. Soares 343

VI Optical Fibre Characterisation, Calibration and Standards

6.1 Fibre Characterization and Measurements
C. Le Men 353

6.2 Mechanical and Environmental Testing
J. Dubard, P. Fournier, P. Blanchard, F. Gauthier _____ 399

6.3 Fibre Ribbon Measurements
Lauri Oksanen _____ 415

6.4 Calibration Artefacts, Traceability and Accuracy
A.J. Barlow _____ 429

VII Optical Fibre Instrumentation, Measurement and Metrology

7.1 Photodetector Calibration
Antonio Corrons _____ 441

7.2 Power Meter Calibration
Pedro Corredera Guillén _____ 453

7.3 OTDR Calibration
Francis Gauthier and Lionel Ducos _____ 469

7.4 Optical Spectrum Analyzer Calibration
J. Dubard and C. Le Men _____ 489

7.5 EDFA Testing
Eric Malzahn _____ 511

7.6 Real-time Heterodyne Fibre Polarimetry by Means of Jones and Stokes Vector Detection
R. Calvani, R. Caponi, F. Cisternino _____ 527

7.7 Application of Er-Doped Fiber Amplifiers to Optical Measurement Techniques
Haruo Okamura and Katsumi Iwatsuki _____ 541

VIII Optical Communication Systems

8.1 Optical Fiber Communication Technology and System Overview
Ira Jacobs _____ 567

8.2 Modern Optical Communication Systems
Govind P. Agrawal _____ 593

8.3 An Optical Coherent Transmission System Based on Polarization Modulation and Heterodyne Detection at 155Mbit/s Bitrate and 1.55 µm Wavelength
R. Calvani, R. Caponi, F. Delpiano and G. Marone _____ 617

IX OPTICAL SENSORS

9.1 Fiber Sensor Review
D.A. Jackson _____ 629

9.2 Fibre Optic Sensors
M.R.H. Voet, Y. Verbandt, L. Boschmans, H. Thienpont and F. Berghmans _____ 647

9.3 Fiber Optic Sensors for Environmental Monitoring
A.G. Mignani, M. Brenci, A. Mencaglia _____ 691

X FUTURE TRENDS

10.1 Nonlinear Optics: Theory and Applications
U. L. Österberg _____ 711

10.2 Advanced Optical Fibre Optoelectronics
B. Culshaw _____ 727

10.3 Plastic Optical Fibres (POF)
Demetri Kalymnios _____ 747

XI FURTHER POSTERS

11.1 Refractive Index of Planar Microlenses
S. Ríos, E. Acosta, M. Oikawa and K. Iga _____ 773

11.2 Fibre Interferometer Set for Phase and Polarisation Measurement of Passive Fibre Elements
M. Szustakowski, J.L. Jaroszewicz _____ 781

11.3 FMCW-Lidar with Tunable Twin-Guide Laser Diode
A. Dieckman, M.C. Amann _____ 791

11.4 Fibre-Optic White-Light Interferometer Insensitive to Polarization Fading
L.A. Ferreira, J.L. Santos, F. Farahi _____ 803

11.5 QND Measurement of the Photon Number using Kerr Effect in
Optical Fibres
V. Sochor, I. Paulička _____ 809

11.6 Laser Transitions Characterization by Spectral and Thermal
Dependences of the Transient Oscillations in the Fiber Lasers
Oleg G. Okhotnikov and José R. Salcedo _____ 813

11.7 Nonlinear M-Lines Spectroscopy of DMOP-PPV Polymer Planar
Optical Waveguides: Quasi-Permanent and Fast Electronic Refractive
Index Changes
Francesco Michelotti _____ 825

11.8 Plastic Optical Fibre (POF) Displacement Sensor
N. Ioannides, D. Kalymnios and I.W. Rogers _____ 827

11.9 A New Reflective Plastic Optical Fibre Displacement Transducer
S. Hadjiloucas, L.S. Karatzas, D.A. Keating and M.J. Usher _____ 829

11.10 White Light Interferometer Using Linear CCD Array with Signal
Processing
Stephen R. Taplin, Adrian GH. Podoleanu, David J. Webb
and David A. Jackson _____ 831

11.11 Radiation-to-guided Field Coupling in Ti:LiNbO$_3$ Intensity
Modulators
M. Marciniak, M. Szustakowski _____ 833

11.12 White Light Fiber Optic Interferometry a Technique for Monitoring
Tissue Properties
Scott Shukes, Faramarz Farahi, M. Yasin Akhtar Raja
and Robert Splinter _____ 833

11.13 Holographic Interferometry of Immersed Systems in Electrochemistry
K. Habib _____ 834

11.14 Pressure Effects of Optical Frequency in External Cavity
Diode Lasers
Hannu Talvitie, Johan Åman, Hanne Ludvigsen,
Antti Pietiläinen, Leslie Pendrill and Erkki Ikonen _____ 834

11.15 Methods of Measurement for Angle-Polished Optical Fiber Launching
Efficients
Darrin P. Clement and Ulf Österberg _____ 835

11.16 The Application of a Fiber Optic Gyroscope for Studies of
Extinction Ratio of Fiber Optic Polarizer
J.L. Jaroszewicz and M. Szustakowski _____ 835

11.17 Blocks and Architecture for a Multichannel Fibre Optic Correlator
Adrian Gh. Podoleanu, Ryan K. Harding, David A. Jackson _____ 836

11.18 Calibration of Fibre Optical Measurement Equipment at SP
R. Andersson, M. Holmsten and L. Liedquist _____ 836

11.19 Fibre Application in Absolute Ranging Interferometry with Computer
Generated Holograms (CGH)
L. Wang, M. Deininger, K. Gerstner, T. Tschudi _____ 837

11.20 Monitoring Through-the-Thickness Behaviour of Composites Using
Fiber Optic Sensors
W. E. Wolfe and B. C. Foos _____ 837

11.21 Fibre-Optic Fabry-Perot Sensor for Vibration and Profile Measurements
I. Paulička, V. Sochor, J. Stulpa _____ 838

11.22 Application of OTDR in the Mining Industry
V. Kumar and Dinesh Chandra _____ 838

11.23 Parametric Amplification by Four-Wave Mixing for Distributed
Optical Fibre Sensors
J. Zhang, V.A. Handerek and A.J. Rogers _____ 839

11.24 Intracavity Laser Diode Pumped and Modulated Er^{3+} Doped Fiber
Lasers
Oleg G. Okhotnikov and José R. Salcedo _____ 839

11.25 Mode-Locking of A 2.7 µm Erbium-Doped Fluoride Fiber Laser
Christian Frerichs _____ 840

AUTHOR INDEX _____

SUBJECT INDEX _____

PREFACE

Fibre Optics has gained prominence in: telecommunications, data transmission and distribution, cable television networks, sensing and control, light probing and instrumentation.

The 1990's shows an increased expansion of optical fibre networks which respond to the rapid growth on a world scale of long distance trunk lines combined with a family of emerging optical based services in which fibre-to-the-home will have the greatest impact. There is already evidence that optical communications are moving toward higher bit-rates, wavelength transparency and irrelevance of signal formats.

The rate of change in fibre optics and the emergence of new services will be a mere consequence of economics. The actual increasing of cost and the demand for high-date-rates or large bandwidth per transmission channels, and the lack of available space in the congested conduits in urban areas, strongly favour the technological change to fibre optics. The recognised advantages of fibre optic technologies and the unchallenged potential to respond to future needs requires the inclusion of fibre optics networking into new installations.

Concomitantly, current progress in the field of optical fibres (optical fibre amplifiers, optical fibre switching, WDM, fibre gratings, etc.) unfold major technical advances and greater flexibility in the designs and engineering of networks, optical fibre components and instrumentation. The explosion of growth in fibre sensors, fibre probes and the myriad of fibre based components shows that we are only using a fraction of optical fibre potential.

The optical fibre market expansion implies a simultaneous accretion of measurement techniques, particularly aimed at the three levels of testing, i.e. laboratory, factory and in the field, to encompass the world character of trade.

The global market and production delocalisation cannot function effectively unless specifications are described by quantities that can be universally measured by common agreed procedures. Interoperability and interconnectivity relies on the development of standards. Standardisation bodies at both national and international levels are active in setting specific standards and reference/alternative testing methods for fibre optics.

The fibre optics world-wide communications market is estimated to reach (USA $) 8,3 billion by the year 1998. The full exploitation of this projected enormous potential of fibre optics in scientific, engineering and the myriad of applications requires a precise characterisation of optical fibres and their associated systems and components, via effective and accurate measurements supported by reliable measuring standards and good practices in measurements and calibrations. In addition the assurance of universal metrological traceability requests permanent intercomparison practices both at vertical and horizontal level.

The ISO 9000 compliance for adequate quality control of products and services already specifies the requirements for the test, measurement and inspection equipment (ISO 10012-1) which stress the calibration imperatives.

Optical fibre metrology had in recent times a significant evolution to respond to the paradigm shift in telecommunications, to the explosion of applications, with the tightening of specifications and exigencies in performance: automation, portability, real-time surveillance and remote mode testing.

Finally, given that 30% of the investment in fibre optic network installations concerns metrology and related activities, there were sufficient reasons to organise the Advance Study Institute (ASI) on "Trends in Optical Fibre Metrology and Standards".

The paramount importance of providing a progressive and comprehensive presentation of current issues and trends concerning the optical fibre metrology and standards was recognised two years ago within EUROMET (European Federation of Metrological Institutes) were the design of the event was enunciated.

The programme of the ASI considered the science and methodology of optical fibre metrology and standards up to the components and instrumentation level. Equipment was also made available for demonstrations and manipulation by the ASI participants.

Despite the inevitable limitations, a broad and reasonably coherent coverage of the field was achieved, with significant contributions from the attendees some of which presented posters (30 in total) supported with written material of high interest, part of which is included in the proceedings.

The topics covered during the eleven working days include: optical fibres and materials (polarisation, non-linear effects, plastic fibres, fibre lasers); optical fibre components (connectors and splicers, WDM); optical sources (tuneable lasers); optical fibre amplifiers (Er-Dopped amplifiers, amplifier standardisation); optical fibre characterisation; calibration and standards (fibre characterisation, ribbon fibres, calibration artefacts); instrumentation (photodetectors, power meters, OTDRs, EDFAs, Er-Dopped fibres in metrology); optical communications systems (fibre communication technology, modern optical systems, coherent transmission systems); optical sensors (fibre sensors, environmental sensors); future trends (fibres evolution, solitons, non-linear effects, fibre optoelectronics).

Lack of time prevented one from dealing with other crucial aspects of optical fibre metrology such as: automation, remote metrologic systems, software validation, safety and legal aspects of metrology, as well as a family of recently developed fibre components.

The profusivity of material could not be matched by further enlargement of the ASI. The ASI was designed for a lecturers staff of around 10 to 15 lecturers assembled from academia and industry to produce cross-fertilisation and providing views from different backgrounds and orientations. Due to the interest of the topics and timeliness of the event we obtained 27 world leading experts who expressed their willingness to participate. The number of attendees was programmed for around 60. The actual number of participants was curtailed to 100 who came from four continents. These were selected, from about the 200 requests received.

The organisation of an ASI of these proportions and which was scheduled over a period of two weeks could only be met by laborious efforts of a responsive secretariat which the editor and the organising committee thanks for their dedication and the excellence of a job well done.

Only the financial assistance provided by the Scientific and Environmental Division of NATO, and of the Measurements and Testing Division of the European Commission could make it possible to engineer the ASI to the proportion it reached. Therefore the funding Institutions are thanked profusely for their support and faith in the project.

The indispensable sponsorship endorsed by the FLAD - Fundação Luso Americana para o Desenvolvimento, JNICT - Junta Nacional de Investigação Científica e Tecnológica, IPQ - Instituto Português da Qualidade, NSF - National Science Foundation (USA) is thanked with high appreciation.

The friendship and continued support given by VIEIRA de CASTRO & Filhos, Lda. is also deeply acknowledged.

The valuable cooperation and support of further co-sponsoring companies and organisations is also thanked with gratitude.

The extensive support and engagement by local Institutions, the Governo Civil of Viana do Castelo, the Camara Municipal of Viana do Castelo, Região de Turismo do Alto Minho, and in particular the enrichment brought by an ennobling and charming cultural and social programme, product of the efforts and dedication of the Local Hosting Committee is thanked with great reconnaissance.

The honouring presence of the Minister for Industry and Energy, as a strong sign of encouragement to the fostering of cooperation with Portuguese Industry and academia, is acknowledge with high appreciation.

It is hoped that researchers, production and field engineers concerned with design, manufacturing, installation and maintenance of networks and systems and instrumentation involved in the area of fibre optics whether in measurements, characterisation of materials, devices, systems, networks, standards or oriented R&D will find the proceedings a tool of direct interest to their professional activities and for the development of new directions of industrial and applied research. Furthermore, the book will serve as an excellent up-to-date source of bibliographical material.

The materialisation of the proceedings is of course, the result of considerable efforts by authors for which they deserve the credit and are cordially thanked and acknowledged for the excellence of their contributions.

The editor is also indebted to Mrs. Nel de Boer of KLUWER Academic Publishers for the devoted work to the publication of the proceedings.

The success of the ASI was the direct result of the excellence of the work of all the active members, lecturers and participants but it is the time ahead that shall demonstrate the relevance of the impulse, the benefits of full endeavour and the strength of the outcome toward the traced challenges.

Olivério D.D. Soares

CETO - Centro de Ciências e Tecnologias Opticas
Universidade do Porto, Portugal
August 1994

INSTITUTE PROGRAMME

xvii

TRENDS in OPTICAL FIBRE METROLOGY and STANDARDS
VIANA do CASTELO - Hotel do Parque**** - PORTUGAL
Cultural and Congress Center SANTIAGO da BARRA CASTLE

1st Week	1st Day 27th June-Monday	2nd Day 28th June-Tuesday	3rd Day 29th June-Wednesday	4th Day 30th June-Thursday	5th Day 1st July-Friday	6th Day 2nd July-Saturday	7th Day 3rd July-Sunday
9-10	REGISTRATION	Fibre Optical Metrology & Standards S.P.Almeida / O.Soares	Fibre Characterisation and Measurement C. le Men	Passive Comp. Characterisation C. le Men	Active Comp. Characterisation C. le Men	PAUSE for INFORMAL GROUP DISCUSSIONS	EXCURSION
10-11		Optical Fibre Communication Technology Ira Jacobs	Fibre Characterisation and Measurement C. le Men	Tunable Lasers and Applications M.C. Amann	WDM Technologies and Applications J.P. Laude		
Coffee Break							
11.30 - 12.30		Optical Fibre Communication Technology Ira Jacobs	Ribon Fibers Measurements L. Oksanen	Tunable Lasers and Applications M.C. Amann	WDM Technologies and Applications J.P. Laude		
Lunch							
Further Activities	16. OPENING CEREMONY	16. Fibre Characterisation and Measurement C. le Men	15-17 Heterodyne Polarimetric Measurements and Communications R. Calvani		16h30 Optical Amplifiers Ivan Andonovic		
17-18	Modern Optical Communications System G.P. Agrawal	Measurements and Testing Framework Programme 4 P. Salieri (BCR)	Mechanical and Environmental Testing F. Gauthier	Optical Fibre Connectors & Splices F. Gauthier			
Coffee Break							
18½ / 19½	Trends in Optical Fibres S. Mustafa	Fibre Characterisation and Measurement C. le Men	Mechanical and Environmental Testing F. Gauthier	Optical Fibre Connectors & Splices F. Gauthier	19-19h30 Effects Laser RIM on Optical Amplifier Noise Figure I. Jacobs		
19½ / 20½	Welcome Dinner	Posters presentation and discussion	Posters presentation and discussion	Posters presentation and discussion	Posters presentation and discussion		
Dinner							

TRENDS in OPTICAL FIBRE METROLOGY and STANDARDS
VIANA do CASTELO - Hotel do Parque**** - PORTUGAL
Cultural and Congress Center SANTIAGO da BARRA CASTLE

2nd Week	8th Day 4th July - Monday	9th Day 5th July - Tuesday	10th Day 6th July - Wednesday	11th Day 7th July - Thursday	12th Day 8th July - Friday	13th Day 9th July - Saturday
9-10	Optical Fibre Amplifiers Standardisation P. di Vita	Fibre Sensors Metrology Marc Voet	Photodetector Calibration A. Corrons	Er Doped Fibres Characterisation M. Rebolledo	Polarisation in Fibres J. Pelayo	
10-11	Plastic Fibers K. Kalymnios	OTDR Calibration F. Gauthier	Calibration Artifacts, Traceability and Accuracy A. Barlow	Er Doped Fibres Characterisation M. Rebolledo	Application Non-Linear Effects U. Österberg	
Coffee Break						
11.30 - 12.30	Plastic Fibers K. Kalymnios	OTDR Calibration F. Gauthier	Glass Fibres and Fibre Lasers Mid-IR U.B. Unrau	Polarisation in Fibres J. Pelayo	Application Non-Linear Effects U. Österberg	
Lunch						
Further Activities	16 HP EDFA - Testing Eric Malzahn		16h30 Analog Communications on Fibres I. Jacobs	16 Non-linear Fiber Optics R. Kashyap	Exotic Fibers F. Gauthier	
17-18	Fibre Sensors Review D. Jackson	Power Meter Calibration P. Corredera	AFA - Instrumentation H. Okamura			
Coffee Break						
18½ / 19½	Fibre Sensors Metrology Marc Voet	Optical Spectrum Analyser C. le Men	AFA - Instrumentation H. Okamura	Solitons in Optical Fibres G.P. Agrawal	Advanced Fibre Optoelectronics B. Culshaw	
19½ / 20½	Posters presentation and discussion	Posters presentation and discussion	Posters presentation and discussion	Posters presentation and discussion	CLOSING of SCHOOL	
Dinner				Farwell Gala Dinner		

PARTICIPANTS

Lecturers

AGRAWAL, PROF. GOVIND P.
The Institute of Optics
University of Rochester
206 WalmoBuilding
Rochester NY 14627
USA
Telf: 1-716-275-48 46
Fax: 1-716-244 49 36
E-mail: gpa@optics.rochester.edu

ALMEIDA, PROF. SILVÉRIO P.
University of North Carolina
at Charlotte
Charlotte, N.C. 28223
USA
Telf: 704-547 20 40
Fax: 704-547 31 60

AMANN, PROF. MARKUS-CHRISTIAN
Institute of Technical Electronics
University of Kassel
Heinrich-Plett-Str. 40
34109 Kassel
Germany
Telf: 00-49-561-804 44 85
Fax: 00-49-561-804 41 36

ANDONOVIC, PROF. IVAN
Strathclyde University
Dept. Electronic & Electrical
Royal College Building
204 George Street
Glasgow G1 1XW
Scotland
Telf: 44-41-552 44 00
Fax: 44-41-552 24 87
Telex: 7742 UNSLIB G

BARLOW, DR. ARTHUR
EG&G Fiber Optics
Sorbus House
Mulberry Business Park
Wokingham
Berkshire RG11 2GY
United Kingdom
Telf: 44-734-77 30 03
Fax: 44-734-77 34 93
Telex: 847164

CALVANI, DR. RICARDO
CSELT
Via G. Reins Romoldi, 274
10148 Torino
Italy
Telf: 39-11-22 85 111
Fax: 39-11-22 85 085
Telex: 220539 cselt

CORREDERA, DR. PEDRO G.
Instituto de Optica "Daza de Valdés"
Serrano, 121
28006 Madrid
Spain
Telf: 34-1-561 68 00
Fax: 34-1-564 55 57
Telex: 42182 CSICE E

CORRONS, PROF. ANTONIO
Instituto de Optica (CSIC)
c/ Serrano, 121
28006 Madrid
Spain
Telf: 34-1-561 68 00
Fax: 34-1-564 55 57
Telex: 42182 CSICE E

CULSHAW, PROF. BRIAN
Strathclyde University
Optoelectronics Division
Dept. Electronic & Electrical
Royal College Building
204 George Street
Glasgow G1 1XW
Scotland
Telf: 44-41-552 44 00
Fax: 44-41-552 24 87
Telex: 77472 UNSLIB G

DUBARD, DR. J.
LCIE
33, Av. General Leclerc
F 92266 Fotenay-aux-Roses
France
Telf: 33-1-40-95 60 60
Fax: 33-1-40 95 60 50
Telex: labelec 634 147 F

GAUTHIER, DR. F.
LCIE
33, Av. General Leclerc
F 92266 Fotenay-aux-Roses
France
Telf: 33-1-40-95 60 60
Fax: 33-1-40 95 60 50
Telex: labelec 634 147 F

JACKSON, PROF. DAVID A.
Applied Optics Group
University of Kent
Canterbury
Kent CT2 7NR
United Kingdom
Telf: 44-227-475 423
Fax: 44-227-764 000
E-mail: daj1@ukc.ac.uk

JACOBS, PROF. IRA
Virginia Polyt. Inst Bradley
Elect. Eng. Dept.
Fiber & Elect. Opt. Div.
Blacksburg, VA 24061
USA
Telf: 1-703-231-56 20
Fax: 1-703-231 33 62
E-mail: ijacobs@vtm1.cc.vt.edu

KALYMNIOS, PROF. DEMETRI
School of Electrom
Univ. North London
166 - 220 Holloway Road
London N7 8DB
UK
Telf: 44-71-753 51 26
Fax: 44-71-753 7002

KASHYAP, DR. RAMAN
BT Laboratories
DTN12, B55/131
Marthesham Heath
Ipswich, IP5 7RE
United Kingdom
Telf: 44-473-64 53 63
Fax: 44-473-64 68 85

LAUDE, DR. JEAN PIERRE
ISA / JOBIN - YVON
16-18, Rue du Canal
91163 Longjumeau Cedex
France
Telf: 33-1-64 54 13 00
Fax: 33-1-69 09 07 21
Telex: 602882 F

MALZAHN, DR. ERIC
Hewlett Packard
Herrenberger Strasse 130
71034 Böblingen
Germany
Telf: 49-7031-14 4662
Fax: 49-7031-14 7023
Telex: labelec 634 147 F
E-mail:
Eric_Malzahn@hpgrmy desk.hp.com

MEN, DR. C. LE
LCIE
33, Av. General Leclerc
F 92266 Fotenay-aux-Roses
France
Telf: 33-1-40-95 62 50
Fax: 33-1-40 95 60 50
Telex: labelec 634 147 F
E-mail: flyer@email.teaser.com

MUSTAFA, DR. SYED AHMED
CORNING
Telecommunication Products
Corning NY 14831
USA
Telf: 1-607-974 48 02
Fax: 1-607-974 70 41
E-mail:

OKAMURA, DR. HARUO
NTT
Transmission Systems Labs
1-2356 Take, Yokosuka-shi
Kanagawa 238-03
Japan
Telf: 81-468 59 32 19
Fax: 81-468 59 33 96
E-mail: okamura@nttsd.ntt.jp

OKSANEN, DR. LAURI
Nokia Cables
PO Box 77, Virkatie 8
01511 Vantaa
Finland
Telf: 358-0-68 251
Fax: 358-0-870 13 59

ÖSTERBERG, PROF. ULF
Thayer School of Engineering
Dartmouth College
8000 Cummings Hall
Hanover
NH 03755-8000
USA
Telf: 1-603-646-34 86
Fax: 1-603-646 38 56
E-mail:
Ulf.Österberg@dartmouth.edu

PELAYO, PROF. J.
University of Zaragoza
Ciudad Universitaria
50009 Zaragoza
Spain
Telf: 34-76-56 07 35
Fax: 34-76-56 52 00

REBOLLEDO, PROF. J.
University of Zaragoza
Ciudad Universitaria
50009 Zaragoza
Spain
Telf: 34-76-56 07 35
Fax: 34-76-56 52 00
E-mail: rebolledo@msf.unizar.es

SOARES, PROF. OLIVÉRIO D.D.
CETO - Centro de Ciências e Tecnologias Opticas
Fac. Ciências - Univ. do Porto
Praça Gomes Teixeira
4000 Porto
Portugal
Telf: 351-2-310 290
Fax: 351-2-319 267
E-mail: ceto@fis1.fc.up.pt

UNRAU, PROF. ING. U.
Institut für Hochfrequenztechnik
Technische Universität
Braunschweig
PO Box 3329
D-38023 Braunschweig
Germany
Telf: 49-531-391 24 58
Fax: 49-531-391 58 41
Telex: 952526 tubsw-d

VITA, DR. PIETRO DI
CSELT
Via G. Reins Romoldi, 274
10148 Torino
Italy
Telf: 39-11-22 85 278
Fax: 39-11-22 85 520
Telex: 220539 cselt i

VOET, DR. MARC R.H.
Identity E.E.I.G.
Inspralaan 75
B-2400 Mol
Belgium
Telf: 32-14-58 11 91
Fax: 32-14-59 15 14

Attendees

Anacleto, Dr. Joaquim
CETO
Faculdade de Ciências
Universidade do Porto
Praça Gomes Teixeira
4000 Porto - Portugal
Telf: 351-2-310290
Fax: 351-2-319267
E-mail: ceto@fis1.fc.up.pt

Araújo, Dr. Francisco M.C.
INESC
Rua José Falcão, 110
4000 Porto - Portugal
Telf: 351-2-2094301
Fax: 351-2-2008487
Telex: 23023 INESC P

Bao Varela, Dr. Maria del Carmen
Lab. de Óptica - Dept. Física Aplicada
Facultade de Fisica
Campus Universitario
15706 Santiago de Compostela
Spain
Telf: 34-81-521984
Fax: 34-81-21984

Barcelos, Dr. Sérgio
University of Southampton
Optoelectronics Research
Group
Southampton SO17 1BJ
UK
Telf: 44-703-59 31 72
Fax: 44-703-59 31 49
E-mail: s.barcelos@ieee.org

Beguin, Dr. Claude
Swiss Telecom
PTTResearch & Development
Fibre Optic Metrology FE 144
Ostermundigenstrasse 93
CH - 3000 Bern 29
Switzerland
Telf: 41-31-338 31 86
Fax: 41-31-338 57 47
Telex: 911 031 vpttch

Berghmans, Dr. Francis
SCK.CEN, DEPT BR3
Boeretang 200 - B-2400 MOL
Belgium
Telf: 32-14 33 26 37
Fax: 32-14 31 19 93
E-mail: fberghma@vnet3.vub.ac.be

Boccardi, Dr. Pasquale
Via LVII Strada a denominarsi, 6
70059 Trani
Italy
Telf: 39-883 44 969

Brinkmann, Dr. Sven
Hindenburgstr. 14
D-91054 Erlangen
Germany

Cervasio, Dr. Alberto
S.V. Giagumona, 90
07040 (Ottava) - Sassari
Italy
Telf: 39-79-390709

Clement, Dr. Darrin P.
Thayer School of Engineering
Dartmouth College
8000 Cummings Hall
Hanover, New Hampshire 03755
USA
Telf: 1-603-646 1466
Fax: 1-603-646 3856
E-mail:darrin.clement@dartmouth.edu

Cloninger, Dr. Todd L.
1725 Ratchford Drive
Dallas, N.C. 28034-9555
USA
Telf: 1-704-547 25 23
Fax: 1-704-547 31 60

Cortes, Dr. Santiago David Armando Reyes
Universidade da Beira Interior
Departamento de Física
R. Marquês d'Ávila e Bolama
6200 Covilhã - Portugal
Telf: 351-75-25141/4
Fax: 351-75-26198
Telex: 53733 UBI P

Costa, Dr. Manuel Filipe P.C.
Universidade do Minho
Dept. de Fisica
Largo do Paço
4719 Braga Codex - Portugal
Telf: 351-53-60 43 27
Fax: 351-53-60 43 39

Cui, Dr. Guoqi
AILUN
Via della Resistenza, 39
08-100 Nuoro - Italy
Telf: 39-784-203409
Fax: 39-784-203158

Dias, Dr. Ireneu Manuel Silva
INESC
Rua José Falcão, 110
4000 Porto
Portugal
Telf: 351-2-2094300
Fax: 351-2-2008487
Telex: 23023 INESC P

Dieckmann, Dr. Andreas
SIEMENS AG
ZFE ST KM 42
Oho-Hahn-Ring 6
81739 Munich - Germany
Telf: 49-89-63644486
Fax: 49-89-6363832

Dopazo, Dr. Juan Félix Roman
Begonias 31 - 1° D
La Coruna
Spain
Telf: 34-81-293311 / 256410

Ducos, Dr. Lionel
LCIE
33, Av. General Leclerc
F 92266 Fotenay-aux-Roses
France
Telf: 33-1-40 95 62 50
Fax: 33-1-40 95 60 50

Ferreira, Dr. Ana Cristina Bento
TELECOM
Av. Fontes Pereira de Melo, 40
1089 Lisboa Codex - Portugal
Telf: 351-1-540020
Fax: 351-1-523614

Ferreira, Dr. João M.C.
INESC
Rua José Falcão, 110
4000 Porto - Portugal
Telf: 351-2-209 4000
Fax: 351-2-318 692
Telex: 23023 INESC P

Ferreira, Dr. Luis Alberto
INESC
Rua José Falcão, 110
4000 Porto - Portugal
Telf: 351-2-2094300
Fax: 351-2-2008487
Telex: 230023 INESC P

Figueira, Dr. Ana Rita L. C.
CETO
Faculdade de Ciências
Universidade do Porto
Praça Gomes Teixeira
4000 Porto - Portugal
Telf: 351-2-310290
Fax: 351-2-319267
E-mail: ceto@fis1.fc.up.pt

Fiore, Dr. Marina
C.so L. Kossuth, 71
10132 Torino
Italy

Fonseca, Dr. António Fernando Figueiredo
TELECOM
Av. Fontes Pereira de Melo, 40
1089 Lisboa Codex - Portugal
Telf: 351-1-3504094
Fax: 351-1-523614

Foos, Dr. Bryan C.
Research Scientist
WJ/FIVEC Bldg 45
Wright-Patterson Air Force Base
334 Aberdeen Avenue
Oakwood - OHIO
USA
Telf: 1-513-2553021
Fax: 1-513-2551633
E-mail: foosbc@fivmailgw.flight.wpafb-af.mil

Frerichs, Dr. Christian
Institut für Hochfrequenztechnik
Technishe Universität Braunschweig
PO Box 3329
38023 Braunschweig - Germany
Telf: 49-531-391 24 23
Fax: 49-531-391 58 41
E-mail: C.Frerichs@tu.bs.de

Fuchs, Dr. Richard
European Patent Office
Gitschinerstrasse 103
Berlin D-10958 - Germany
Telf: 49-30-25901-201
Fax: 49-30-25901-845

Gray, Dr. George R.
University of Utah
Dept. of EE
3280 Merril Eng. Bldg.
Salt Lake City, UT 84112
USA
Telf: 1-801-585 6157
Fax: 1-801-581 5281
E-mail: gray@ee.utah.edu

Guerrero, Dr. Hèctor
Dept. Optica
Facultad de Ciencias Fisicas
Universidade Complutense
Ciudad Universitaria s/n
28040 Madrid - Spain
Telf: 34-1-394 4402/3
Fax: 34-1-394 4683
Telex:47273 FFUC
E-mail: w653@emducm11

Habib, Dr. Khalid
Materials Application Department
KISR
PO BOX 24885 SAFAT
13109 Kuwait
Kuwait
Telf: 965-4830-432
Fax: 965-57237-19

Ioannides, Dr. Nicos
109 Uxbridge Road
Hanwell, London W7 3ST
United Kingdom
Telf: 44-71-607 27 89 / 607 25 96
Fax: 44-71-753 70 02

Iodice, Dr. Mario
IRECE - CNR
Via Diocleziano, 328
80124 Napoli
Italy
Telf: 39-81-5707999
Fax: 39-81-5705734
E-mail: irece::iodice
E-mail: iodice@dimes.tudelft.nl

Karafolas, Dr. Nikos
University of Strathclyde
Optoelectronics Group
204 George Street
G1 1XW Glasgow
Scotland
Telf: 44-41-5524400
Fax: 44-41-5522487

Karoutis, Dr. Athanase D.
Univ. Creete, School of Health Sciences
Dept. Medicine, Sector Neurology
and Sense Organs
PO BOX 1352
711 10 Heraklion Crete - Greece
Telf: 30-81-54 20 70
Fax: 30-81-54 21 16

Konstantaki, Dr. Maria
Strathclyde University
Dept. E.E. Engineering
Royal College Building
204 George Street
Glasgow G1 1XW
Scotland
Telf: 44-41-552400
Fax: 44-41-5522487
E-mail: m.konstantaki@uk.ac.strath

Kumar, Dr. Virendra
Dept. of Electronics
and Instrumentation
Indian School of Mines
Dhanbad 826 004
India
Telf: 326-822273
Fax: 326-832 040
Telex: 0629-214

Lago, Dr. Maria Elena López
E.U. Optica
Univ. Santiago de Compostela
Campus Sur
15706 Santiago de Compostela
Galicia - Spain
Telf: 34-81-52 19 84 / 56 31 00
Fax: 34-81-52 19 84

Liedquist, Dr. Leif
Physical Measurements
Physics & Electronics
Sveriges Provnings-och
Forskningsinstitut
Statens Provningsanstalt
BOX 857 S-501 15 BORAS
Sweden
Telf: 46-33 16 50 00
Fax: 46 33 13 83 81

Mangia, Dr. Maria
Via Iglesias, n.9
08-100 Nuoro
Italy
Telf: 39-784-204187 / 202024

Marciniak, Dr. Marian
Military Academy of
Telecommunications
05-131 Zegrze 300/11
Poland
Telf: 48-2-6883536
Fax: 48-2-6883413
E-mail:wsowl.frodo.nask.org.pl

Martins, Dr. Maria Raquel V.
CETO
Faculdade de Ciências
Universidade do Porto
Praça Gomes Teixeira
4000 Porto - Portugal
Telf: 351-2-310290
Fax: 351-2-319267
E-mail: ceto@fis1.fc.up.pt

Marttila, Dr. Pekka
Nokia Cables
PO BOX 77
01511 - VANTAA
Finland
Telf: 358-0-68251
Fax: 358-0-8701359
Telex: 123475

Michelotti, Dr. Francesco
Univ. di Roma "La Sapienza"
Dept. di Energetica
Via A. Scarpa, 16
I-00161 Roma - Italy
Telf: 39-6-49 76 65 62
Fax: 39-6-44 24 01 83
E-mail: bertol88@itcaspur.caspur.it

Mignani, Dr. Anna Grazia
IROE - CNR
Via Panciatichi 64
50127 Firenze
Italy
Telf: 39-55-423 1 / 4235262
Fax: 39-55-4379569

Morais, Dr. Modesto Cerqueira
IEP
Rua de S. Gens, 3717
Senhora da Hora
4450 Matosinhos - Portugal
Telf: 351-2-9529675
Fax: 351-2-9530594

Navarrete, Dr. Mª Cruz
Universidade Complutense de Madrid
Dept. Optica - Fac. Ciencias Fisicas
Ciudad Universitaria s/n
28040 Madrid - Spain
Telf: 34-1-39 44403
Fax: 34-13944683

Okhotnikov, Dr. Oleg G.
INESC
Rua José Falcão, 110
4000 Porto - Portugal
Telf: 351-2-2094300
Fax: 351-2-2008487
Telex: 230023 INESC P

Olszak, Dr. Artur
Warsaw University of Technology
Optical Engineering Division
Dept. Precision Mechanics
8 Chodkiewicza St.
02-525 Warsaw
Poland
Telf: 48-22-499871
Fax: 48-22-490 392
E-mail: olszak@mp.pw.edu.pl

Otero, Dr. José Lazaro Gato
E.U. Optica
Dept. Fisica Aplicada
Facultad de Fisica
Univ. Santiago de Compostela
Campus Universitario
15706 Santiago de Compostela
Galicia - Spain
Telf: 34-81-52 19 84
Fax: 34-81-52 19 84

Pervan, Dr. Ogus
Turkish Atomic Energy
Authority
Ankara Nuclear Research and
Training Center
Saray, 06105 - Ankara
Turkey
Telf: 312-8154300/8154390
Fax: 312-8154307

Pieraccini, Dr. Massimiliano
IROE - CNR
Via Panciatichi 64
50127 Firenze - Italy
Telf: 39-55-4235214
Fax: 39-55-4223889
E-mail: brenci@iroe.iroe.fi.cnr.it

Podoleanu, Dr. Adrian
University of Kent
at Canterbury
Physics Laboratory
Kent CT2 7NR
United Kingdom
Telf: 44-227-764000
Fax:44-227-475423
E-mail: physics-lab@ukc.ac.uk

Pousa, Dr. José Marcelino
CET
Rua Engº José Ferreira Pinto Basto
3800 Aveiro - Portugal
Telf: 351-381831
Fax: 351-24723
Telex: 37371

Ravet, Dr. Fabien
Faculte Polytechnique de Mons
Electromagnétisme et
Télécommunications
Boulevard Dolez, 31
7000 Mons
Belgium
Telf: 32-65-374191
Fax: 32-65-37 41 99
E-mail: ravet@fpms.fpms.ac.be

Rego, Dr. Gaspar Mendes do
Fiopos - Barroselas
4905 Barroselas - Portugal
Telf: 058-972215

Ribeiro, Dr. A.B. Lobo
INESC
Rua José Falcão, 110
4000 Porto - Portugal
Telf: 351-2-2094300
Fax: 351-2-2008487
Telex: 230023 INESC P

Rodriguez, Dr. Jorge
U.P.C. ETSI
Campus Nord UPC
Edifici D-3
C/ Sor Eulàlia Anzizu s/n
E - 08034 Barcelona - Spain
Telf: 34-3-401 6795
Fax: 34-3-401 7232

Rodriguez, Dr. Susana Rios
E.U. Optica e Optometria
Dept. Fisica Aplicada
Facultad de Fisica
Univ. Santiago de Compostela
Campus Sur
15706 Santiago de Compostela
Galicia - Spain
Telf: 34-81-52 19 84
Fax: 34-81-52 19 84

Romolini, Dr. Andrea
IROE - CNR
Via Panciatichi 64
50127 Firenze - Italy
Telf: 39-55-4235233
Fax: 39-55-4223889
E-mail: falciai@iroe.iroe.fi.cnr.it

Santos, Dr. Fernando M. Ferreira
INEGI
Fac. Engenharia do Porto
Rua dos Bragas
4099 Porto Codex - Portugal
Telf: 351-2-2007505

Santos, Dr. José Luis C.
INESC
Rua José Falcão, 110
4000 Porto - Portugal
Telf: 351-2-2094300
Fax: 351-2-2008487
Telex: 230023 INESC P

Sartori, Dr. Giovanni
AILUN
Via della Resistenza, 39
08-100 Nuoro - Italy
Telf: 39-784-20 34 09
Fax: 39-784-20 31 58

Seker, Dr. Selim
Bogaziçi University
Dept. of Electrical-Electronic
P.K.2 80815 Bebek
Istanbul - Turkey
Telf: 212-2631540
Fax: 212-2575030

Serra, Dr. Giovanni
Via F.M.Brundu, 1
07040 Codrongianus (SS)
Italy
Telf: 39-79-435203

Sharer, Dr. Deborah
103 South Fork Road
Indian Trail - NC 28079
USA
Telf: 1-704-547 23 02

Shukes, Dr. Scott
1409 Richmond Place
Charlotte, N.C. 28209
USA
Telf: 1-704-527 0821

Sillas, Dr. Hadjiloucas
University of Reading
Dept. Cybernetics - Instrum.
Measurement Research Group
Whiteknights,
PO BOX 225
Reading - Berks RG6 2AY
UK
Telf: 44-734-318219
Fax: 44-734-318220
E-mail: shrhadji@uk.ac.reading

Sochor, Dr. Vaclav
Fac. of Nucl. and Physics Eng.
Csech Techn. Univ.
V Holesovickach 2
180 00 Prague 8
Czech Republic
Telf: 422-85762285
Fax: 422-66414818
E-mail: sochor@troja.fjfi.cvut.cz

Sousa, Dr. Fernando J. Pelicano
Portugal TELECOM
DCRS / RGT
Rua Tomás Ribeiro, 2, 3º DCRS
1000 Lisboa - Portugal
Telf: 351-1-540020
Fax: 351-1-526110
Telex: 60256 DIRS P

Sporea, Dr. Dan G.
Bucharest
P.O. BOX 31-53
Romania
Fax: 40-1-312 11 54
E-mail: sporea@roifa - Bitnet
sporea@ifa.ro - Internet

Szustakowski, Dr. Mieezislaw
Institut of Technical Physics
2 Kahskiego St.
01-489 Warsaw
Poland
Telf: 48-22-36 93 53
Fax: 48-22-36 22 54
Telex:812 535 WAT pl

Talvitie, Dr. Hannu
Helsinki Univ. of Technology
Fac. Electrical Engineering
Otakaari 5 A
SF - 02 150 Espoo
Finland
Telf: 358-0-451 23 36
Fax: 358-0-460 224
E-mail: hannu.talvitie@hut.fi

Taplin, Dr. Stephen R.
University of Kent
Room 116 - Physics Dept.
Applied Optics Group
Canterbury - Kent CT2 7NR
United Kingdom
Telf: 44-227-764000
Fax: 44-227-475423
E-mail: srt1@ukc.ac.uk

Teixeira, Dr. José Manuel Feliz
CETO
Faculdade de Ciências
Universidade do Porto
Praça Gomes Teixeira
4000 Porto - Portugal
Telf: 351-2-310290
Fax: 351-2-319267
E-mail: ceto@fis1.fc.up.pt

Traian, Dr. Dumitrica
Rudului 129
Ploiesti 2000
Romania
Telf: 40-44-140371

Varga, Dr. Jozsef
PKI Telecommunications Inst.
Budapest IX
Zombori u. 1
H-1456 P.O.B. 2 - Hungary
Telf: 36-1-147 1560
Fax: 36-1-127 5075

Velasco, Dr. Mª Lourdes Pedraza
Escuela Universitária de
Enfermeria Fisioterapia Y Podologia
3es Piso Fac. De Medicina
Univ. Complutense de Madrid
Ciudad Universitaria
Madrid 28040 - Spain
Telf: 34-1-3941525
Fax: 34-1-3941539

Vinhais, Dr. Carlos Alberto A.
CETO
Faculdade de Ciências
Universidade do Porto
Praça Gomes Teixeira
4000 Porto - Portugal
Telf: 351-2-310290
Fax: 351-2-319267
E-mail: ceto@fis1.fc.up.pt

Vorropoulos, Dr. G.
European Patent Office
Gitschinerstrasse 103
Berlin
D-10958 - Germany
Telf: 49-30-25901614
Fax: 49-30-25901845

Wang, Dr. Lingli
Institute of Applied Physics
Technishe Hochschule
Darmstadt, Hochschulstr. 6
D-6100 Darmstadt - Germany
Telf: 49-6151-16 30 17
Fax: 49-6151-16 41 23
E-mail: lingli@trudel.iap.physik.th-darmst

Wolfe, Dr. William
Ohio State University
470 Hitchcock Hall
2070 Neil Avenue
Columbus, OH 43210-1275
USA
Telf: 1-614-292 7338
Fax: 1-614-292 3780
E-mail: wolfe.10@osn.edu

Zhang, Dr. Jian
Dept. Electronic Electrical
Engineering
Kings College of London
Strand
London WC2R 2LS
UK
Telf: 44-71-873 2371
Fax: 44-71-836 4781
E-mail: zdee486@bay.cc.kcl.ac.uk

SPONSORS

Scientific and Environmental Affairs Division - NATO
Commission of the European Union (M &T Division)
CETO - Centro de Ciências e Tecnologias Opticas
The European Institute for Advanced Studies in Optics
FLAD - Fundação Luso-Americana para o Desenvolvimento
JNICT - Junta Nacional de Investigação Científica e Tecnológica
IPQ - Instituto Português da Qualidade
NSF - National Science Foundation
Governo Civil de Viana do Castelo
Câmara Municipal de Viana do Castelo
RTAM - Região de Turismo do Alto Minho
IEP - Instituto Electrotécnico Português
MARCONI - Companhia Portuguesa Rádio Marconi, SA
IVP - Instituto do Vinho do Porto
Vieira de Castro & Filhos, Lda.
CLUB TOUR - Agência de Viagens

CO-SPONSORS

EOS - European Optical Society
HP - Hewlett Packard Portugal
DECADA - Equipamentos de Electrónica e Científicos, S.A.
BA - Fábrica de Vidros Barbosa & Almeida, SA
Comissão de Viticultura da Região dos Vinhos Verdes
CANON - Copicanola
Hotel do Parque - Viana do Castelo
M. T. Brandão, Lda.
Câmara Municipal de Ponte da Barca
PERCON-Computadores
Adega Cooperativa de Monção
BCI - Banco de Comércio e Indústria
SOPETE - Casino da Póvoa de Varzim

COMMITTEES

Organising Committee

PROF. OLIVÉRIO D.D. SOARES (Chairman)
CETO - CENTRO DE CIÊNCIAS E TECNOLOGIAS ÓPTICAS
Lab. Fisica - Fac. Ciências - Univ. Porto
Praça Gomes Teixeira
4000 Porto - Portugal

PROF. S. P. ALMEIDA
Department of Physics
University of North Carolina at Charlotte
Charlotte, N.C. 28223
USA

DR. RICARDO CALVANI
CSELT
Via G. Reins Romoldi, 274
10148 Torino
Italy

Local Hosting Committee

Mr. Roleira Meirinho
Governo Civil de Viana do Castelo

Dr. Defensor Moura
Câmara Municipal de Viana do Castelo

Dr. Francisco Sampaio
RTAM - Região de Turismo do Alto Minho

Secretariat and Technical Committee

José Sousa Fernandes
Fernanda Campos
Luis Vilaça

O
MINISTRO da INDÚSTRIA
e
ENERGIA

Excelentíssimo Senhor Governador Civil de Viana do Castelo

Excelentíssimo Senhor Presidente da Câmara Municipal de Viana do Castelo

Excelentíssimo Senhor Presidente da Região de Turismo do Alto Minho

Excelentíssimo Senhor Presidente do CETO - Centro de Ciências e Tecnologias Opticas

Ladies and Gentlemen

The time is now upon us to leave the past and join the challenge of the future before it becomes too late. As individuals, as enterprises and as a country, we must continuously upgrade our educational and vocational training, replace obsolete equipment, and develop effective management methods. Portugal is now on the brink of great opportunities to modernise itself and to prepare for the 21st Century. It must not after in its pursuit of the challenges which lie ahead.

Recent advances in telecommunications, information processing and electronic industries represent the driving forces behind the present day industrial revolution. In order to keep abreast of these rapidly developing areas, Portugal must develop new and stronger partnerships between its industries and universities. We must take advantage of the position universities are in to educate and train the much needed students and industrial personnel in areas of research and applications which are vital to the future of the country's development. The Portuguese Ministry for Industry and Energy recognises these needs and has taken a deep interest in the Advanced Institute on "Trends in Optical Fibre Metrology and Standards".

xxx

The Advanced Institute on "Trends in Optical Fibre Metrology and Standards" provides a thorough coverage of fibre technology and an insight into opportunities offered to various industries. The theme of the Advanced Institute "Optical Fibre" is central to industries associated with communication, data transfer, sensors, instrumentation; and is a key element in many optoelectronic components.

The telecommunication network represents one of the largest human systems ever developed in the world. It is also one of the most complex in that it handles in excess of 1,000 billion calls each year. This, in addition to, an immense amount of transmitted data. Its very existence represents an important element of personal freedom, progress and economic globalization. Every nations' defence, trade, commerce, industry and social practices are inextricably dependent on the extent and quality of their communication and data transfer networking.

Optical networks, namely, optical fibre based links, offer the potential of gigabit switched bandwidth. Recently developed commercial optical amplifiers provide yet another enormous step forward in transmission and data processing thereby lowering the cost of both terrestrial and submarine transmission. Other important industries which make use of fibre optics include: aerospace, computing, broadcasting, consumer electronics and instrumentation. Also, a diverse range of activities broadly defined as the information technology sector make use of fibre optics.

European market experts project for Portugal, an expanded use of about 100,00 km of fibre optics cabling by the year 1995. The estimated cost for this expansion is of the order of 21 billion dollars. A fraction of 30% of the installation costs goes towards characterisation and metrological uses.

The Portuguese Ministry for Industry and Energy has recognised the need to modernise and support the electronics and information technology industries. In particular, this has recently led to the implementation of an integrated program for information and electronics technologies - PITIE, whose main goal is the setting up of a motivating environment for the development of that industry. Also the strategic program for the dynamic development of the Portuguese industry - PEDIP I was set up to help integrate industrial and entrepeneurship linkages.

The Ministry for Industry and Energy, through PEDIP I, has supported technological innovations in three portuguese optical fibre cable manufacturers. It has provided adequate funding for the instalment of optical fibre metrological facilities both in the manufacturing industries and at related technological centers and laboratories. In addition it has supported, under the supervision of the Portuguese Institute for Quality - IPQ, the Laboratório Primário de Optica - LPO, as part of the Central Laboratory for Metrology - LCM. The LPO which is currently under construction at Vila da Feira, is part of the basic infrastructure for optical metrology. It will provide calibrations, the realisation and dissemination of fibre optical standards. At international level, it will participate in metrological inter-comparisons. In particular, within the Federation of the European National Metrological Institutes - EUROMET and the Bureau Communitaire de Reference - BCR, for adequate traceability. The LCM and CETO - Centro de Ciências e Tecnologias Opticas will participate in the scientific and technological programmes of LPO. They will also provide the necessary interface to the industrial users.

Recently, a new strategical program, PEDIP II was formed. PEDIP-II was developed to insure a continued modernisation and strengthening of the portuguese industry. It will lend support to the entrepreneurs' goal to guarantee quality control and sustain its competitiveness within the global economy. In addition, it will support efforts, as those of CETO, to motivate portuguese companies to develop innovative technologies and manufacture of high-tech products and to promote the establishment of foreign companies in Portugal in the field of optoelectronics and related areas.

In conclusion, the Ministry for Industry and Energy welcomes the Advanced Institute on "Trends in Optical Fibre Metrology and Standards" for its foreseeable contribution in promoting contacts and ultimately the internationalisation of related industrial activities, and for stimulating innovating ideas in the latest technologies.

Personally, I would like to welcome all the participants and wish them a very pleasant stay and fruitful outcome.

Eng° Luis Mira Amaral
VIANA do CASTELO
27 de Junho 1994

I   Introduction

A HISTORICAL SURVEY of OPTICAL SIGNALS and OPTICAL FIBERS

Silvério P. Almeida
University of North Carolina at Charlotte
Department of Physics
Charlotte, North Carolina 28223

Olivério D. D. Soares
Universidade do Porto
Centro de Ciências e Technologias Opticas
4000 Porto, Portugal

A historical survey of optical fibers is presented in this paper. It begins with a perspective on the orgin of light signals, their use in communications by early forms of life and then by humanity. The survey covers the evolution of various aspects of fiber optic advances and applications. The time period for these discussions ranges from the beginning of time to the 20th Century.

1. Let there be light: "fiat lux"

1.1 ORIGIN OF LIGHT AND USE AS A SIGNAL

Mankind would like to believe that the use of optical light signals originated during the past few millennia. However, the 3-million year history of humanity is a relatively short period of time when compared to the 4.6-billion year evolution of the earth and our solar system. One can imagine that the first light sources in the universe began with its origin about 10-20-billion years ago in the theory of the Big Bang. During this initial period, electromagnetism became one of the fundamental constituent forces in the universe. Hence, one of the first evolutionary steps in mankind's development of vision took place. From the Big Bang explosion arose the creation of billions of galaxies each in turn composed of billions of stars. Each star is a powerful source of light. The spiral galaxy (Figure 1) in which we are located has about 300-billion stars in it; the sun is only one of these sources of light. The planet earth can be considered as having electromagnetic lightwave signals impinging on it from many of these stellar sources.

Figure 1. Spiral Galaxy stars as light signals.

About 500-million years ago, during the Cambrian period of evolution of ocean life, the sea was abundant in ocean life much of which was luminescent. Today, we can venture to depths ranging from about 800 to 4000 meters to find a sea full of luminescent animals. Light signals flashing from ocean forms such as the lantern fish and stomiatoids have been detected on a photomultiplier to range from 1-160 flashes per minute. Lantern fish (Figure 2) have large light organs on their head much like a miner's head-lamp as well as along their bodies. It is believed that these light organs are used to help the lantern fish recognize one another from the many bioluminescent fish species.

Figure 2. Lantern fish light organs shown as two rows along the bottom side. Some organs are also located on its head.

Later during the evolution of time, about 200-million years ago, the insect population on earth had many species which communicated via light signals. A modern example of these are the fireflies (Figure 3). Light emitted by these insects is unique in being cold. The female firefly lands in the grass and emits light signals to a male flying overhead. The male firefly emits his own light signals down to the female. Then, depending on the exchange of light signals, he either lands next to her or keeps on flying. The number of flashes and duration vary among fireflies implying that there is some sort of light signal code being emitted.

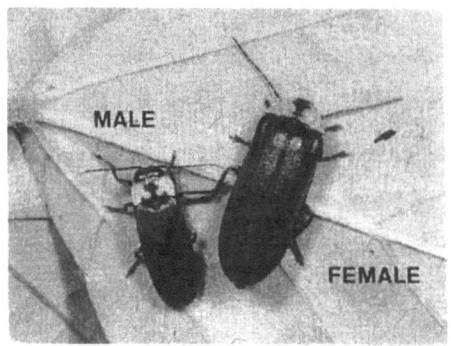
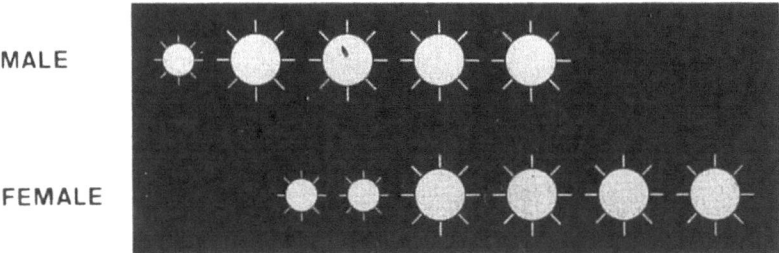

Figure 3. Fireflies, male and female, with their light signals shown below them.

1.1.1. *Mankind's early communication: 1,000 BC-19th Century*

During the Renaissance period (~1400 -1600) to the 19th century, light as a source of communication graduated from the animal kingdom to mankind beginning with the unspoken word of art. Artists began to openly communicate their creative thoughts by images and light reflections on their canvas paintings. Some artists have expanded upon the natural sources of sunlight to communicate through the beauty of stained glass windows. Still others have based masterpieces upon the lighted message of God. These nuances observed in frescos and stained glass windows in cathedrals and churches are yet another example of mankind's use of light to convey a message. In the case of paintings, stars have always maintained their importance in the artistic arena as seen in Van Gogh's painting of the "Starry Night" (1889).

Man has also used light as a more immediate avenue of communication. Early use of light signals by fire can be found to be documented from about 1,000 BC by the Greeks to the 1800s in Europe and most likely throughout the world. Fire and smoke signals were used to send signal warnings of impending enemy attacks. Distances as great as 200 km could be covered by having a series of signal relay stations. Galileo and an assistant (~1600s), used lantern signals sent between them (less than a kilometer) to try and measure the speed of light. While useful at this time, the low data rate communication achieved by signals from fire, smoke lanterns and semaphore would not usually suffice at a later period of time.

At the turn of the 18th century, the low data rate achieved by signals from fire, smoke and semaphore was overshadowed by the introduction of new methods in communication systems. Around 1880, the pursuit of communication via light signals was carried out by Alexander Graham Bell. Bell and his assistant, Sumner Tainter, developed the "photophone"[1], (Figure 4) . This optical communication invention was able to transmit a human voice about 200 meters on a light beam of sun. Bell was so excited about his invention that he almost named one of his daughters "photophone". Later, in 1880 he wrote to his father, "I have heard articulate speech produced by sunlight!.....We may talk by light to any visible distance without conducting wire" [2].

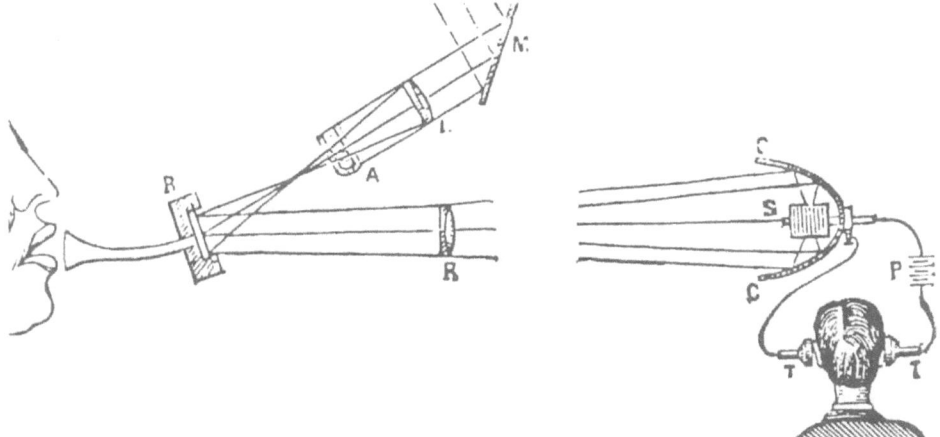

Figure 4. Diagram of Alexander Graham Bell's "Photophone". A reflected beam of sunlight, voice modulated, passes through lens R to the receiver.

In the following years, an increasingly larger portion of the electromagnetic spectrum was utilized for conveying information from one place to another. The important feature of electrical systems is that data can be transferred over the communication channel by superimposing the information signal onto a sinusoidally varying electromagnetic wave which is known as the carrier. Since the amount of information that can be transmitted is directly related to the frequency range over which the carrier wave operates, increasing the carrier frequency theoretically increases the available transmission bandwidth and, consequently, provides a larger information capacity.

2. The 20th Century

2.1 COMMUNICATIONS: HIGH DATA RATES

An important portion of the electromagnetic spectrum encompasses the optical region. Great interest in communicating at optical frequencies was created in 1960 with the advent of the laser [3], which made available a coherent optical source. Since optical frequencies are on the order of 5×10^{14} Hz, the laser has a theoretical information capacity exceeding that of microwave systems by a factor of 10^5, which is approximately equal to 10 million TV channels (Figure 5). With the potential of such wideband transmission capabilities in mind, a number of experiments using atmospheric optical channels were carried out in the early 1960s. These experiments showed the feasibility of modulating a coherent optical carrier wave at very high frequencies. However, the high installation expense, the tremendous cost of developing all the necessary components, impose severe limitations on this method. The atmospheric channel limitations due to rain, fog, snow, and dust also make such extremely high-speed systems economically unattractive in view of present demand for communication channel capacity. In 1970, Kapron, Keck, and Maurer [4] of the Corning Glass fabricated a silica fiber having a 20 dB/km attenuation. At this attenuation, repeater spacings for optical fiber links become comparable to those of copper system, thereby making lightwave technology an engineering reality. In the next two and half decades researchers worked intensively to reduce the attenuation close to its theoretical limit of 0.14 dB/km at a wavelength of 1550nm.

Carrier	Frequency Band	~Telephone Channels	~ TV Channels
UHF	(300 - 3,000) MHz	10,000	10
Microwave	$(3 \times 10^9 - 10^{12})$ Hz	100,000	100
Optical	$(5 \times 10^{13} - 10^{15})$ Hz	10^8	10^5

Figure 5. Optical versus non-optical frequency carriers.

The development and application of optical fiber systems grew from the combination of semiconductor technology, which provided necessary light sources, photodetectors and optical waveguide technology. The result was an information-transmission link that had certain inherent advantages over previous communication systems. Namely: (a) The principal material of which optical fibers are made (silica) is inexpensive and it is available almost everywhere. (b) The optical fiber cable systems transmit data over a longer distances, thereby decreasing the number of repeaters for these spans. (c) Silica is a dielectric material, therefore, silica optical fibers are optical waveguides with immunity to electromagnetic interference. (d) Fiber creates no arcing or sparking, since glass is an electrical insulator; this makes the use of fibers also attractive in electrically hazardous environment. (e) The optical signal is well confined within the waveguide, therefore high degree of data security is obtained.

The major elements of an optical fiber transmission link (Figure 6) are the transmitter, cable, and receiver. Additional elements include fiber and cable splices, repeaters, couplers and optical amplifiers. Optical amplifiers reduce the number of repeaters in the link, and fiber amplifiers are specially attractive because of their compatibility with the transmitting fiber.

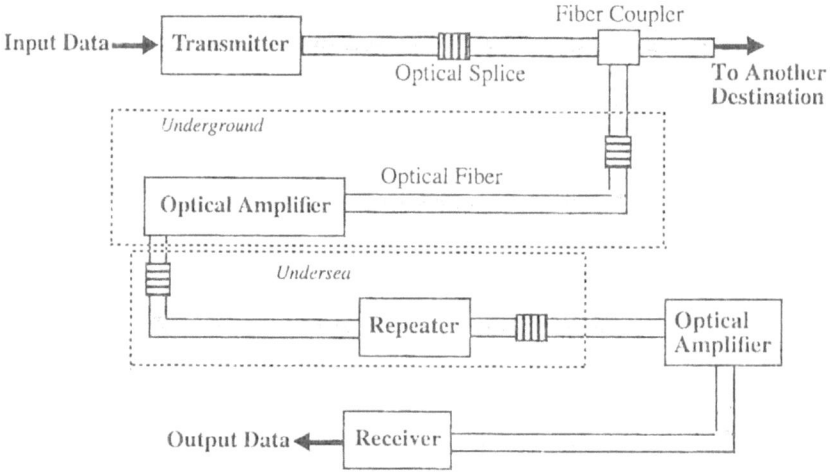

Figure 6. Schematic of an optical fiber transmission link.

Rapid growth of fiber optic communications technology has brought forward an increased need for guided-wave optical devices and components (passive and active) which permit coupling, multiplexing and demultiplexing, switching and in general, reconditioning of light signals propagating in optical waveguides. Because of the strong potential benefits of fiber optic components in the field of lightwave technology, all-fiber components associated with these optical functions have received considerable attention. Devices constructed from optical fibers could have advantages over conventional integrated optics such as low insertion loss and their compatibility with fiber optic transmission lines [5]. In addition, optical fibers can not only be utilized for their transmission characteristics but also as non-linear devices [6,7]. The design of fibers, especially for non-linear optical applications, represents a potentially rich field [8]. Such fibers could incorporate low loss, tight field confinement, polarization preservation, special dopants to enhance Raman gain and special index profiles. They provide, for instance, efficient frequency conversion means for generating tunable optical sources for practical applications such as the testing of optical components and devices.

2.1.1 Optical Devices: Nonlinear Effect

Another significant aspect is the use of non-linear optical effects in order to provide useful special optical device functions such as direct optical amplification, optical gating and switching, optical pulse shaping and short pulse generation, which may find interesting new applications in future optical signal transmission, signal processing or optical sensing systems. At present, the potential applications of fiber non-linearities far out-weigh their adverse effects in high bandwidth lightwave systems. The non-linear relationship between the electrical polarization strength P and the electrical field strength E in a dielectric optical fiber induces the non-linearity of the refractive index. The third order susceptibility $\chi^{(3)}$ is responsible for the optical Kerr effect. The second order susceptibility $\chi^{(2)}$ vanishes in principle in fibers due to the inversion symmetry of fused silica material. The physical process underlying the appearance of optical solitons in the Kerr effect leads to the self-phase modulation of strong power light pulses, propagating over a long silica fiber. It can be characterized by an intensity dependence of the refractive index [9]. If I is the beam intensity and n_0 the linear refractive index:

$$n(I) = n_0 + n_2 I = n_0 + n_2 E^2(t)$$

A pulse with the intensity envelope $E^2(t)$ will induce a non-linear refractive index variation:

$$\Delta n(t) = n_2 E^2(t)$$

Consequently, the self-phase modulation $\Delta\Phi(t)$ of the wave packet propagating along a fiber length L is expressed by

$$\Delta\Phi(t) = \frac{2\pi L}{\lambda} \Delta n(t) = \frac{2\pi L n_2}{\lambda} E^2(t)$$

The non-linearity of the refractive index of a fiber has a number of useful device and system applications, including: optical pulse compression, soliton transmission, soliton laser and non-linear pulse shaping [10]. Another important application is optical switching where the induced birefringence caused by pico-second pulses in the normal dispersion regime, induces the switching between two counter rotating, circularly polarized, linearly coupled modes.

Other important uses of fibers with non-linear properties are in development of fiber lasers and fiber amplifiers. The development of rare-earth doped fibers has been one of the most important steps in optical fibers in the last few years. Pulse fiber lasers in either Q-switched [11] or mode-locked [12] form can produce several 10s of watts of peak power when pumped by inexpensive semiconductor lasers. Pulsed fiber lasers have developed rapidly over the past few years and commercial instruments based upon them are now available. Recent development of passively mode-locked short-pulse fiber laser offers a possibility of an inexpensive source of femto-second pulses. Both Nd^{3+} and Er^{3+} doped fibers in either ring or linear cavities have been used to develop tunable and narrow linewidth (~kHz) lasers. Upconversion fiber lasers using fluoride glass fibers have now been developed where two photons at a longer wavelength than the lasing wavelength, usually in infra-red, are absorbed to produce a photon of laser output at a shorter wavelength. This was first shown in fiber in a Ho^{3+} doped fluoride fiber, and later a Pr^{3+} doped fiber was used to produce wavelength from 491 nm to 910 nm. Such lasers have many potential applications in bio-medical and environmental sensors.

3. Fiber Optic Sensors: Classifications

Progress in the development of optical fibers and fiber components have led to new field of applications, most noticeably the area of fiber optic sensors [13]. In fiber sensors, the parameter to be measured changes at least one of the properties of light, intensity, frequency, phase or polarization state. Technical advantages of fiber optic sensors can be listed as: i) Silica optical fibers can be used to power remote sensors and actuators. This can eliminate need for local electrical power supply at sensor site. ii) Optical fiber sensors emit no RF signal and are not affected by external RF noise. iii) They have electromagnetic interference immunity. iv) Dielectric nature of optical fibers precludes conduction of hazardous high voltages. v) Small size and weight of optical fiber make sensors of this type useful in many applications where these properties are critical such as aerospace and medical fields. vi) Total internal reflection and surface plasmon effects can be used to design sensitive surfaces selectively. vii) Fiber optic sensors can be multiplexed to form a sensing system.

Fiber optic sensor systems can be classified into three groups: (1) Intrinsic sensors, (2) Extrinsic sensors and (3) Evanescent sensors.

3.1 INTRINISIC SENSORS

Intrinsic sensors are those sensors where the fiber itself can act as the responsive element. Such devices can range from highly sensitive interferometric systems, to less sensitive systems where loss mechanism within the fiber can provide sufficient backscatter signal for detection, caused by measurand, as in the case of time-domain reflectometry. Technologies for applying specialized coating to a fiber to enhance its sensitivity to external effects have also been developed. Examples are aluminum coatings for temperature sensors, metallic glass or nickel coatings for magnetic sensors [14,15], and special gas or chemical species absorbing materials [16,17]. The strain applied to the fiber by these coatings can be detected. Intrinsic sensors have the advantage that radiation is not lost from the fiber. There are, therefore, possibilities for measurement systems such that a single continuous loop of fiber can be interrogated remotely to detect changes at specific or distributed points on the loop [18]. It is possible to multiplex a single fiber in this way and thereby avoid numerous couplers in the system.

The most sensitive intrinsic fiber sensors are the interferometric sensors [19]. The effect of pressure and strain in fiber arm of an interferometer produces a phase change in the interferometric output. This type of sensors have been used for highly sensitive underwater array of hydrophones [20]. Fiber optic Sagnac interferometer has been used for gyroscope application [21]. Fiber optic gyroscopes are inexpensive, can be designed to be insensitive to gravity, they are insensitive to temperature, can measure and withstand high rotation rates, can measure rotation under high acceleration and vibration. Fiber optic interferometric sensors can be used to measure magnetic field, temperature, acceleration and many other physical effects.

Polarimetric sensors are another class of intrinsic senors where the change in the state of polarization of optical beam can be affected by external signal. They can be made from polarization preserving fibers or strain/wound monomode fiber which have been used for magnetic and electric field measurements.

Fiber grating sensors are the most exciting development of recent years, where the in-fiber photorefractive gratings can be used for many sensing applications such as strain, temperature, pressure, etc. [22,23] With this technique miniature fiber sensors for smart structure have now become a reality.

The most commonly used intrinsic fiber optic sensor involving multimode fiber is the microbend sensor. Various proposal have been put forward, ranging from potentially very cheap devices for use with OTDR distributed systems to highly sensitive hydrophones. They are typically made of a length of multimode fiber laid between grooved or serrated plates. On compression, the fiber is forced into the grooves and the critical angle for total internal reflection within the core is exceeded. Light is lost to the cladding. Sensors can be constructed to measure the amount of light entering the cladding, or the loss of light from the core, or both.

3.2 EXTRINSIC SENSORS

Extrinsic optical fiber sensors are those wherein optical fibers are used to transmit radiation to and from the point or region to be sensed. The radiation is released from the transmitter fiber and modulated externally by some induced or environmental change. Such change may be caused by beam interruption by a pressure release switch, back-reflection from a moving reflector (as a function of angular, axial or lateral positioning), absorption by a gas (spectroscopy), or perturbing effects from environmentally sensitive birefringent elements. In some applications, the fiber can also have certain advantages as small, remote, well-defined, single or multiple light sources having a well defined geometry or arrangement such as for surface inspection, and laser Doppler velocimetry.

3.3 EVANESCENT SENSORS

Evanescent sensors rely on the measurable loss of guidance from an optical waveguide as a means of detecting external changes. Examples are level detection (when the effective refractive index of surrounding material around an optical fiber or exposed optical element changes), detection of specific gravity, and detection of chemical reactions, chemical species, and pH levels.

4. Conclusion

Fiber-optic communication transmission rates have risen in an exponential fashion since the deployment of fiber optics in the 1970s [24]. In particular, the bit rate-distance product (BL) has increased dramatically to a present value of about BL~ 200 Gb/s-km at a wavelength of 1550 nm. This rate can be further increased by the use of erbium-doped fiber amplifiers (EDFAs) to data trasmission rates of about BL~15 Tb/s-km, that is about ~10 Gb/s over a distance of 1500 km. Research and development of advanced fiber-optic communication systems are now underway which utilize state-of-the-art fiber-optics to transmit fiber solitons. It is expected that these advanced future systems will be capable of transmission rates of about 15Gb/s over a distance of 24,000 km, i.e. BL ~ 360 Tb/s -km.

The use of fiber optics in medicine, sensors and metrology continues to increase at a rapid pace. In medicine, optical fibers are being used to deliver a laser beam for surgery as well as for skin disorders, i.e. the removal of moles and the treatment of skin cancer is now becoming more wide spread [25]. Fiber optic sensors are also playing an important role in monitoring air pollution, sensing stress changes in structures [26], and for many other sensor applications [27].

A crucial development in the future of fiber optics is in metrology and the setting of standards for adoption in fiber couplers and many other fiber components [28-30]. The very fast growing area of fiber optics must adhere to a universally adopted set of fiber optic component standards if it is to have a non-chaotic growth pattern.

Overall, the area of fiber optics is proving to be a very exciting one which has been well accepted worldwide. Applications for the use of optical fibers is exceeding all expectations. The NATO Advanced Institute on Trends in Optical Fibre Metrology and Standards held in Viana do Castelo, Portugal serves as an important source for the distribution of the latest developments in this exciting field.

5. References

1. D.L. Hutt, K.J. Snell and P.A. Belanger (1993). *Optics and Photonics News,* **4**, No.6, p.20.
2. F.M. Mims (Feb.1982) "The first century of lightwave comm.," *IFOC*, p.10.
3. T.H. Maiman (1960) *Nature,* **187**, p.493.
4. F.P. Kapron, D.B. Keck, and R.D. Maurer (1970) *Appl. Phys. Lett.* **17**, p.423.
5. S.E. Miller, and I.P. Kaminow (1988) *Optical Fiber Telecommunication* **II**, Academic Press, N.Y.
6. C. Lin, (1986) *J. Lightwave Tech.,* **LT-4**, p.1103.

7. D. Cotter (1987) *Opt. Quantum Electron,* **19**, p.1.
8. W.A. Gambling and S.B. Poole (1988) *Optical Fiber Sensors,* **II**, J. Dakin and B. Culshaw (eds), Artech House, Ch.8, p.249.
9. R.H. Stolen and C. Lin (1978) *Phy. Rev. A,* **17**, p.1448.
10. A. Ankiewicz (1988) *Opt. Quantum Electron.,* **20**, p.329.
11. I.P. Alcock, A.C. Tropper, A.I. Ferguson, and D.C. Hanna (1986) *Electron.Lett.,* **22**, p.84.
12. I.P. Alcock, A.I.Ferguson, D.C. Hanna, and A.C. Tropper (1986) *Electron. Lett.,* 22, p.268.
13. A. Harmer and A. Scheggi (1988) *"Optical Fiber Sensors",* **II**, B. Culshaw and J. Dakin (eds), Artech House, Inc., Ch.16, p.599.
14. A. Yariv and H. Winsor 1980), *Opt. Lett.,* **5**, p.51.
15. K.P. Koo, F. Bucholtz and A. Dandridge (1986) *Conf. Proc. OFS-4,* Tokyo,p. 77.
16. F. Farahi, P.A. Leilabady, J.D.C. Jones and D.A. Jackson 1987) *J. Phys. E: Sci. Instrum.* **20**, p.432.
17. F.Farahi, P.A. Leilabady, J.D.C.Jones and D.A.Jackson (1987) J. *Phys.E: Sci..Instr.* **20**, p.435.
18. A.H. Hartog and D.N. Payne (1982) *Proc. IEE Colloq Optical Fiber Sensors,* London,
19. D.A. Jackson and J.D.C. Jones (1988) *"Optical Fiber Sensors",* **II**, B. Culshaw and J. Dakin (eds), Artech House, Inc., Ch. 10, p.329.
20. J.A. Bucaro, H.D. Dardy, E.F. Carome (1977) J. *Acoustical Society of America,* **62**, p.1302.
21. R.A. Bergh, H.C. Lefevre, and H.J. Shaw (1981) *Optics Lett.,* **6**, p.502.
22. K.O. Hill, Y. Fujii, D.C. Johnson, and B.S. Kawasaki (1978) *Appl. Phys. Lett.,* **32**, p.647.
23. G. Meltz, W.W. Morey, W.H. Glenn. (1989) *Opt. Lett.,* **14**, p.823.
24. G. Agrawal, (1994) *NATO Advanced Summer Institute on "Trends in Optical Fibre Metrology and Standards", Viana do Castelo, Portugal; "Modern Optical Communication Systems".*

25. J. Hecht (1994) *Laser Focus World, June, "Lasers in Medicine"*.
26. S.M. Melle, (1994) *Photonics Spectra, April; "Today's Sensors for Tomorrow's Structures"*.
27. A.G. Mignani, M. Brenci, A. Mencaglia (1994) *NATO Advanced Summer Institute on "Trends in Optical Fibre Metrology and Standards", Viana do Castelo, Portugal; "Fiber Optic Sensors for Environmental Monitoring"*.
28. F. Gauthier, (1994) *NATO Advanced Summer Institute on "Trends in Optical Fibre Metrology and Standards", Viana do Castelo, Portugal; "Optical Connectors & Splicers; OTDR Calibration; and Mechanical and Environmental Testing"*.
29. C. le Men, (1994) *NATO Advanced Summer Institute on "Trends in Optical Fibre Metrology and Standards",Viana do Castelo, Portugal; Fiber Characterization and Measurement; Active and Passive Comp. Characterization"*.
30. O.D.D. Soares and S.P. Almeida, (1994) *NATO Advanced Summer Institute on " Trends in Optical Fibre Metrology and Standards", Viana do Castelo, Portugal; "Trends in Optical Fibre Metrology and Standards"*.

Aknowledgements

The authors gratefully acknowledge the valuable input they received from Faramarz Farahi, Cathy Kaufman (Optical Society of America).

Trends in Optical Fibre Metrology and Standards

Olivério D. D. Soares
CETO - Centro de Ciências e Tecnologias Opticas
Faculdade de Ciências
Universidade do Porto
4000 Porto, Portugal

Silvério P. Almeida
Department of Physics
University of North Caroline - Charlotte
Charlotte, North Caroline 28223
USA

Abstract

Trends on optical fibre technologies look very promising in terms of monetary market and technical advances. The maturing of the present and foreseen potential will rely on an adequate response in optical fibre measurements, metrology, standards and instrumentation. Key aspects of this evolutionary cycle, in which an important share of the investment goes in metrological activity are described stressing the exigencies brought by exploiting technical breakthroughs.

Contents

1. Optical Fibre Measurements, Metrology, Standards, and Instrumentation
2. World Market of Fibre Optics
3. Optical Fibre Metrology and Standards
    3.1 Global Market Metrology and Trends
    3.2 Measurement Procedures
    3.3 Standardised Recommendations
    3.4 Sensing, Control and Instrumentation
    3.5 Fibre Probes
4. The Trends
    4.1 Non-invasive Measurements
    4.2 Remote Fibre Testing Systems
    4.3 Uniformity
    4.4 Passive Components
    4.5 Standardisation
    4.6 Dispersion Measurements
    4.7 Polymer Optical Fibres (POF)
    4.8 Laser Sources Components
    4.9 Repeater Spacing Enlargement
    4.10 Optical Amplifier Metrology
    4.11 Coherent Communication Systems
    4.12 Fibre Optic Instrumentation Trends
    4.13 Improved Availability of Transfer Standards
    4.14 Intelligent Measurements and Standardisation
    4.15 All-Optical Network
5. Conclusion
6. References

1. Optical Fibre Measurements, Metrology, Standards and Instrumentation

Fibre optical metrology has matured into a broad field, composed of measurements, characterisation and standards concerning telecommunications, data transfer, sensors, light probes, components and instrumentation, Fig. 1. The ultimate goal of fibre optic metrology is to consolidate the continued fast expansion of the market of optical fibres, derived components and systems progressively turned into real commodities.

The relevance of optical fibre metrology is increasing with the globalisation of the market. The transnational character of optical fibre links, Fig.2, serves well to exemplify the metrological requests: generally accepted specifications, agreed measurement procedures and universality of standards. Conversely metrology [1] works as a key aspect supporting the technological development and market expansion while responding to inherent implications of: globalisation of the economy, delocalisation of production, safety assurance, consumer protection, environment protection and instrumentation development.

The global market imposes a universally accepted metrology. Inherent procedures and standards have to be established and agreed upon. Concomitanly a supporting infrastructure has to be put into place to assure mutual accreditation, Fig. 3. This metrological infrastructure then becomes a tool for the enhancement of competitiveness, Fig.4.

The measurement procedures cover parameters concerning the source, transmitting channel, receptor, inserted components and the overall system performances, Fig.5. For optical fibre communication systems the metrology activity is rated at 30% of the investment on the installation.

There are a variety of measurement procedures to respond to the diversity of optical fibre based components and systems in tune with the universal needs of users: researchers, manufacturers, assemblers, installers, and operators.

The requests may vary considerably. The installer of a transmission link will be interested on: overall loss, attenuation per sections, position and through loss of splices, return loss of connections, through loss of connections, link length and location of breaks and faults. The link user will be more concerned with regularly checking for preventive maintenance to evaluate changes in the link performance. Regular measurements will then be compared to the reference measurements (made during link installation) and an alarm signal should be provided before transmission degradation reaches a critical level. Therefore, the link operator will prefer a remote automatic metrological system.

The degree of accuracy and complexity of measurement procedures will vary according to the application. Nonetheless the need to cover dispersion in reproducibility will emphasise the opportunity to establish reference and alternative test methods that could generally be accepted.

On the other hand the accuracy and repeatability of the measurements should become independent of the operator. Instrumentation then permanently improves in design and automation, Fig.6. Though large and fragmented the instrumentation market responds with growth.

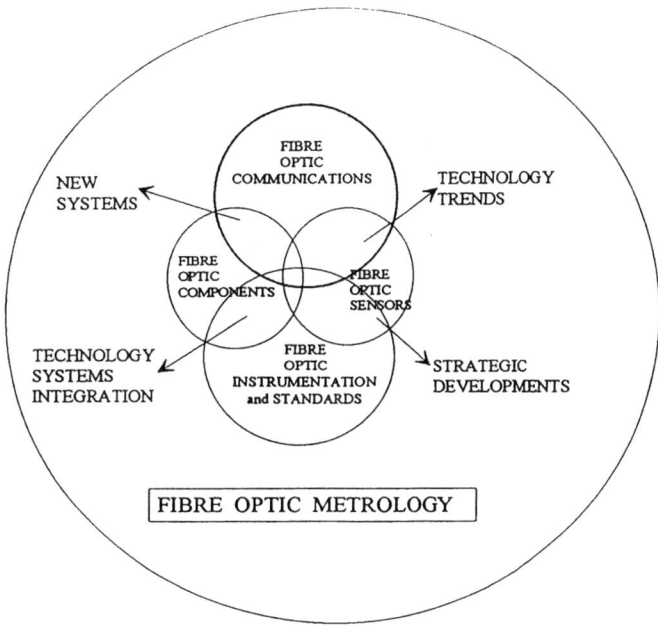

Fig. 1: Fibre Optic Metrology as a broad strategic discipline

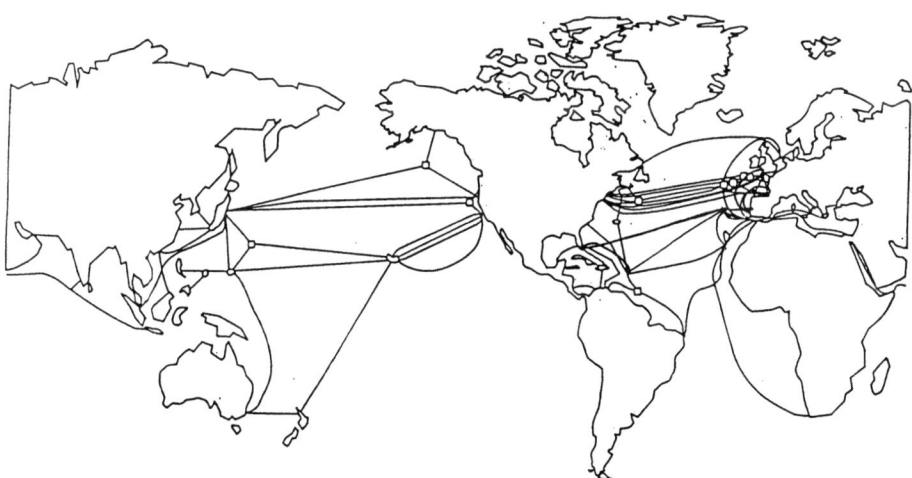

Fig. 2: Optical fibre links exemplify the transnational character of optical fibre metrology.

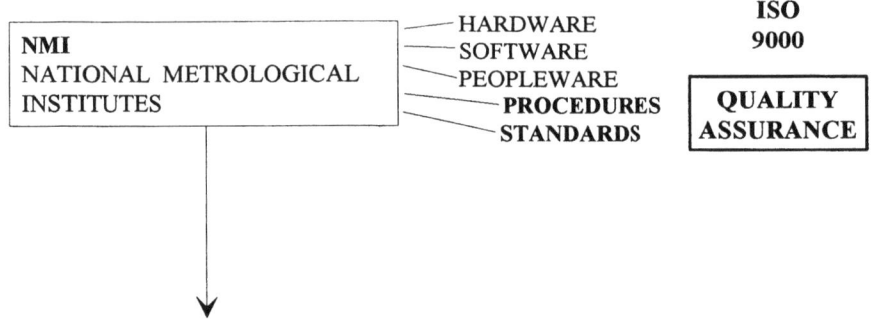

Fig. 3: Global market metrology prerequisits

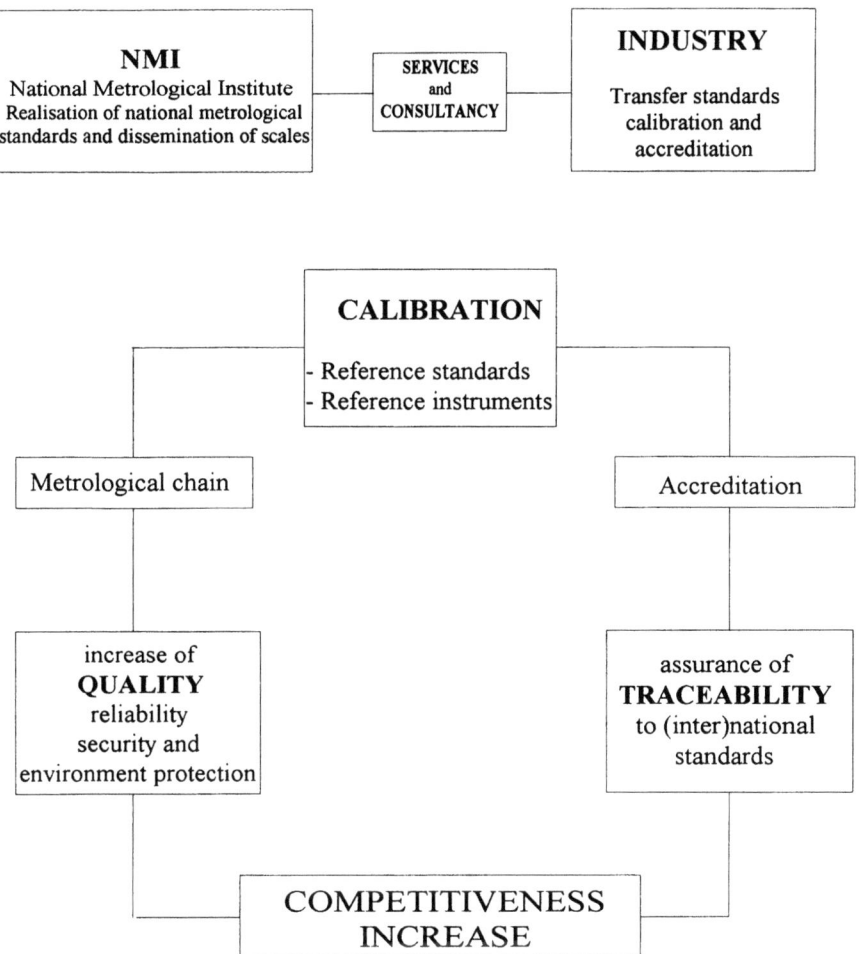

Fig. 4: Metrological Strategy

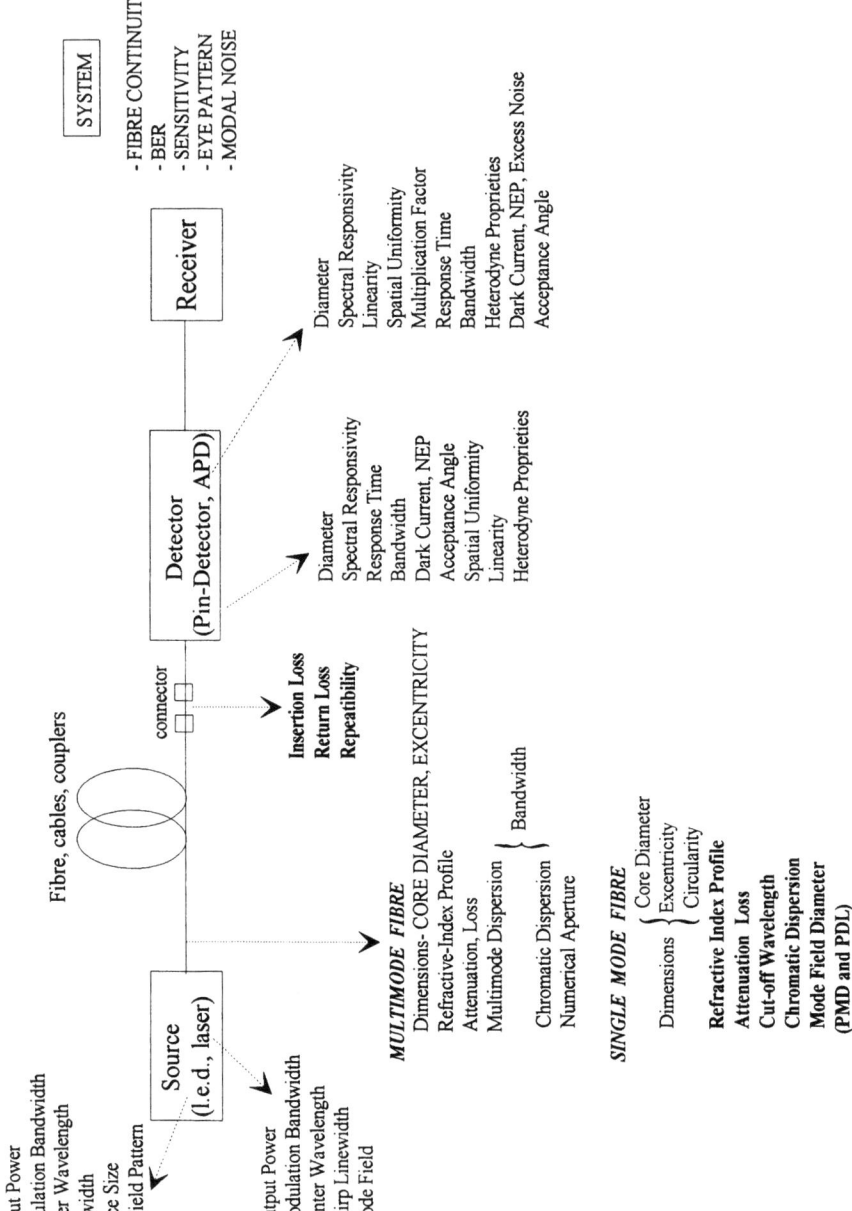

Fig. 5: Metrologic main aspects to be considered on a conventional optical fibre communication link

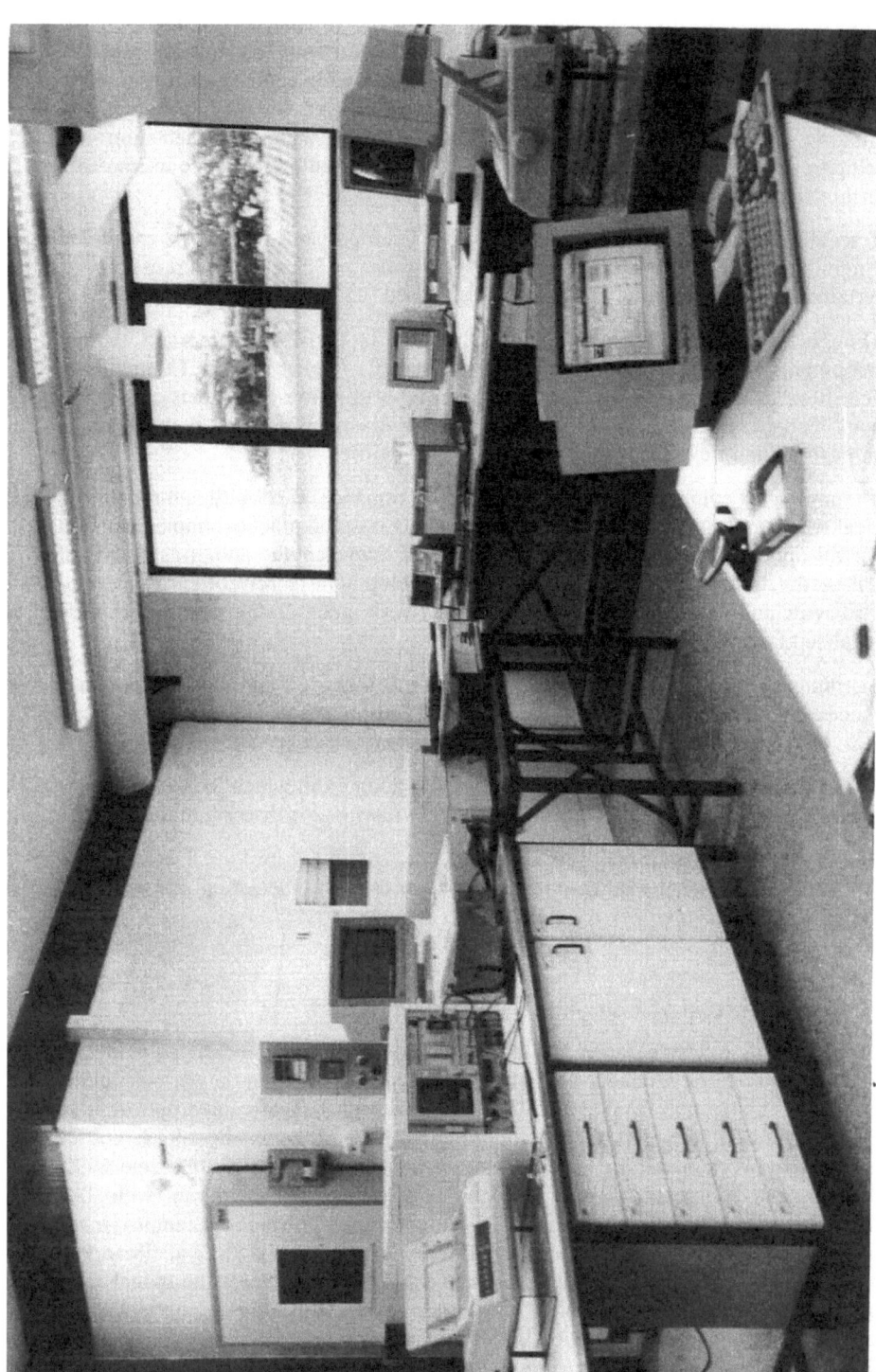

Fig. 6: Optical fibre metrological laboratory at the IEP - Instituto Electrotécnico Português. (Porto, Portugal)

The relevance of instrumentation exceeds the mere support of metrological practices today. The massive introduction of automation has transformed the instrumentation into the core source for real-time decision eventually in an automatic sequences. Therefore higher exigencies are being specified for instrumentation, its calibration and for software validation. The accuracy with instrumentation has develop to a point where there are metrological chains based on reference instruments to anchor the traceability.

Traceability and the results of concurrent inter-comparisons have to be expressed by numbers of universally accepted significance. Therefore, expression of uncertainties should preferably follow some standard [2].

Though optical fibre is a ripped technology [3] there are important current developments with great technological and foreseeable market impact. The trends in optical fibre technologies point to a large scale hybridisation of systems combining various technological developments. For example plastic fibres are likely to see a great push in their development by the graded index fibres.

The myriad of components will improve in performance, flexibility and reliability. Optical fibre amplifiers will become a key component in the practical implementation of an all-optical network, transparent to multiple wavelengths and irrespective of signal modulation. Software will continue its development to control hardware, and optical switching will realise a reconfigurable network incorporating rerouting in case of localised failures.

Instrumentation will be developed with direct inclusion of new components, advancement of software towards higher degree of automation and remote monitoring modes, directed towards an intelligent network system.

Compliance with standards will be increasingly adopted in which ISO 9000 clearly indicates the trend towards universality to match with the globalisation of the market.

This evolutionary cycle will rely in a large increase in metrological activity and progressive development of measurement procedures, instrumentation and generally accepted standards.

2. World Market of Fibre Optics

In view of its relevance, there are abundant estimates and forecasts of monetary markets for fibre optic products. One of the prognostications from the various market organisations states that by the year 1993, 12,2 million Km of cabled optical fibre (USA $3,7 billion) were installed world-wide by telecom companies: PTT, CATV and private network operators, Fig. 7. This represents an increase of 10% from 1992, with a projected increase of 15% through 1998, to reach $7,6 billion, with 32,6 million Km of installed fibre [4], Fig.8. Multimode fibre applications (campus, intra-building cabling of PCs, datacom networks) is expected to range 18% of fiberoptics cable market by 1998 corresponding to around USA $1,4 billion. The global fibre optic market (optical fibre, cables, transmitters, receivers and connectors) was estimated in 1993 as USA $5 billion with projected increase of 11% spread by rapid escalation on demand by Eastern Europe, Latin America and Far East [5, 6].

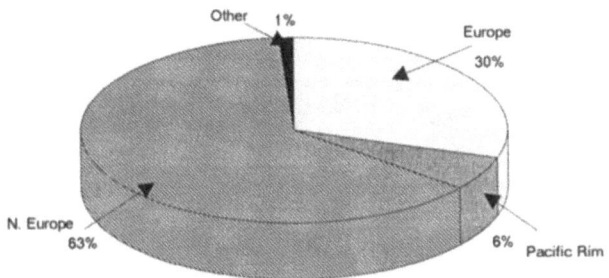

Fig. 7 - Worldwide suppliers of fibre optics in 1992 (source KMI) [4]

Total Fiber Optic World Market Revenue Forecasts
1988 - 1998

Year	Revenues ($Million	Revenue Growth Rate (%
1988	2 660,2	11,7
1989	2 970,3	13,5
1990	3 371,1	13,9
1991	3 840,4	13,7
1992	4 366,9	13,5
1993	4 958,9	12,8
1994	5 591,0	11,6
1995	6 240,7	10,5
1996	6 896,7	9,9
1997	7 578,6	9,9
1998	8 329,5	9,9

Compound Annual revenue Rate Growth (1991-1998): 11,7%

Fig. 8 - Global Optical Fibre Market (Source Market Intelligence Research Corp.) [4]

The fibre optics market in Europe is expected to reach USA $2,3 billion by 1995 [7] with an average annual growth of 17%. Where test and measurement instruments for fiber optics were rated at about USA $112 million for 1991 and expected to reach USA $300 million by 1997. The demand for fiberoptic components is expected to reach US $ 5,81 billion by the year 2000 [7]. Distribution among countries is uneven but the establishment of the unique internal market will foster better and more uniform prices. Also, a rapid expansion of emerging technologies (Fiber-to-the-home, SDH, ISDN, LAN, WAN, MAN) will take place responding to present and future market opportunities driven by new and expanded corporate cosumer requirements.

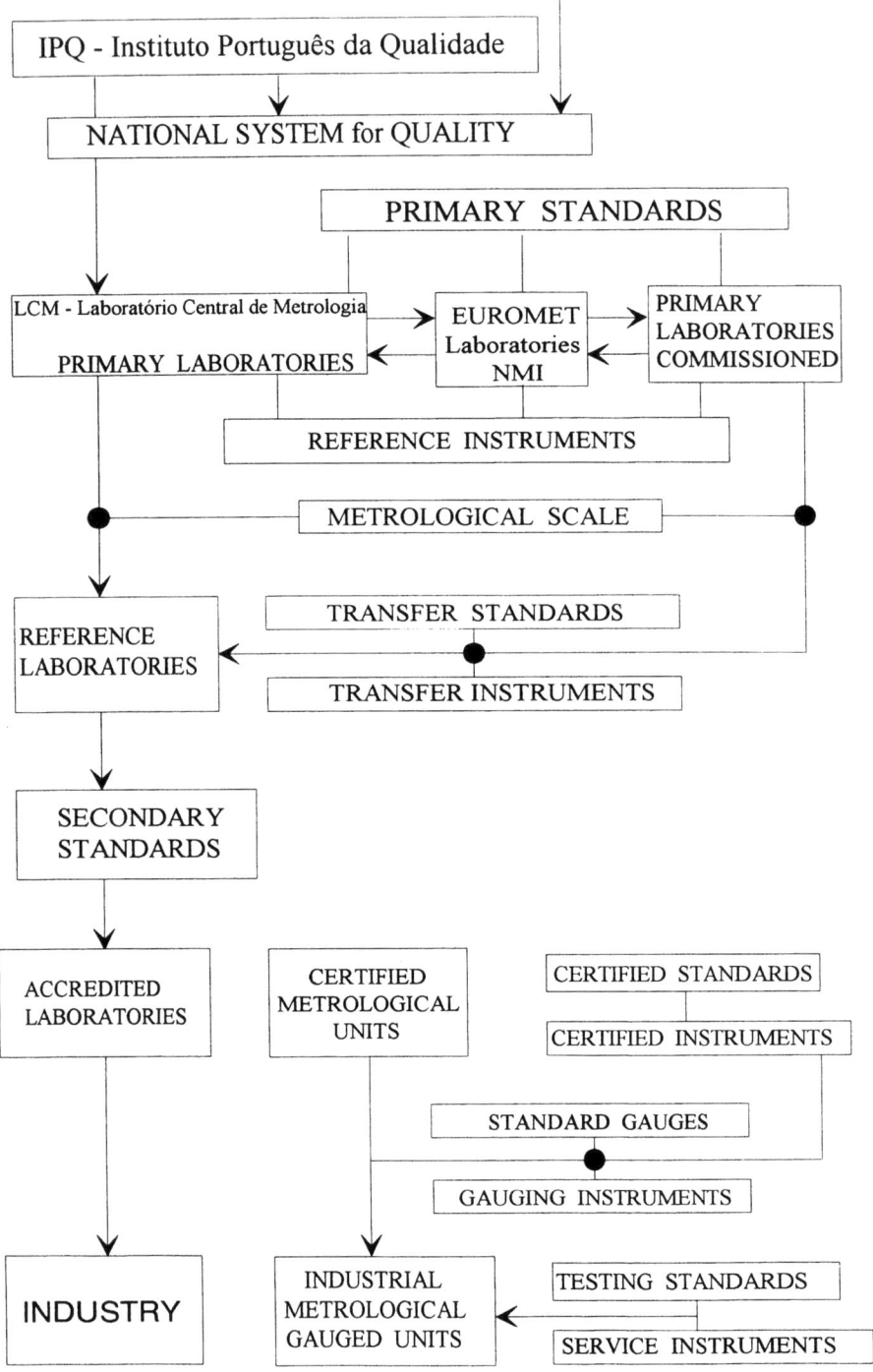

Fig. 9: Functional Structuring of Portuguese Metrological System [8]

3. Optical Fibre Metrology and Standards

3.1 Global Market Metrology and Trends

Optical fibres have already percolated from communication systems, to sensors and instrumentation. Each length of optical fibre must meet a specified performance and reliability before being incorporated into any of the concerned systems. Thus, the crucial importance in characterising performance parameters of optical fibres and assembled systems has led to a major effort by international standard organisations nationally and world-wide to develop agreements on accurate measurement procedures and test methods in view of the globalisation of both the transnational and intercontinental markets' character of information highways.

Verification of performance, however can only be mutually certified through demonstration of traceability of measurements along the metrological chain up to the primary standards (ISO 9000 and ISO 10012-1). Intermediate steps may be anchored in derived and transfer standards particularly those used for dissemination of traceable measurements. Traceability should also be combined with the expression of uncertainty [2]. Fidelity of the metrological practice is confirmed by the realisation of intercomparison (RRT - round robin test) both at a vertical and horizontal level.

National Metrological Institutes (NMI) in every country have consistently proceeded to establish the necessary infrastructure to respond to the needs of scientific, legal and industrial metrology. The ever growing needs in metrology have led to a networking type of structuring at the national level, Fig.9, and to international cooperation by memorandum of understanding (MOU) such as the one that originated EUROMET (European Federation of NMI).

One of the major problems encountered in the practical utilisation of fibre optic metrology systems is the lack of global compatibility among components and subsystems. To ensure full compatibility, fibre optic metrology practices should be supported by agree-upon procedures for standardisation.

At the international level a number of organisations are involved in standardising fibre optic measurements [9]. The EIA - Electronic Industries Association, in USA, has been early active in influencing the standardisation efforts in other countries. The EIA procedures were first published as FOTP - Fiber Optic Test Procedures. Most of them, now more than 200, have become part of Recommended Standard RS 455. In the USA, both the National Institute of Standards, and Technology (NIST) and the Department of Defence (DOD) have also contributed to setting fibre optic standards.

The work by the International Electro-Technical Committee (IETC) is done by the Technical Committee (TC) which is devoted to optical fibres and comprehending six Working Groups (WG):

86/WG 1 - Fibre Optic Terminology and Symbology
86/WG 2 - Fibre Optic Safety Aspects
86/WG 3 - Changed to SC 86C
86/WG 4 - Fibre Optic Test Equipment Calibration
86/WG 5 - Changed to SC 86C
86/WG 6 - Active Fibres

The subcommittees, SG86 have several WG:

SC 86 A - Fibres and Cables
    86 A/WG1 - Optical Fibres and Cables for Short Distance Links
    86 A/WG2 - Optical Fibres for Long Distance Links
    86 A/WG3 - Optical Fibre Cables and Installation Methods

SC 86 B - Fibre Optic Interconnection Devices and Passive Components
    86 B/WG1 - Interconnecting Devices for Optical Fibres and Cables
    86 B/WG2 - Fibre Optic Passive Components
    86 B/WG3 - Quality Assessment Procedures for Optical Fibre Interconnections

SC 86 C - Fibre Optic System Specification:
    86 Ct/WG1 - (ex 86/WG3) Fibre Optic Subsystems
    86 C/WG2 - (ex 86/WG5) Fibre Optic Sensors

Included among the most relevant IEC documentations are:

793 - Optical Fibres:
    Part 1: Generic Specifications
    Part 2: Product Specifications

794 - Optical Fibre Cables
    Part 1: Generic Specifications

874.1 - Connectors for Optical Fibres and Cables
    Part 1: Generic Specifications

812 - IEEC Standard Definitions of Terms Relating to Fibre Optics

869-1 - Fibre Optic Attenuators
    Part 1: Generic Specifications

874-1 - Connectors for Optical Fibres and Cables
    Part 1: Generic Specifications

875-1 - Fibre Optic Branching Devices
    Part 1: Generic Specifications

876-1 - Fibre Optic Switches
    Part 1: Generic Specifications

The work of the Consultive Committee for International Telegraph and Telephone (CCITT) is organised into the following Study Groups:

    Study Group VI - Optical Cables

    Study Group VII - Tests and Characterisation Procedures on Fibres

CCITT has published the following recommendations:

    G 650 - Definitions and Test Methods for the Relevant Parameters of Single - Mode Fibres

    G 651 - Characteristics of a 50/ 125 μm Multimode Graded - index Optical Fibre Cable

G 652 - Characteristics of a Single Mode Fibre Cable

G 653 - Characteristics of a Dispersion-shifted Single-Mode Optical Fibre Cable

G 654 - Characteristics of a 1550 μm Wavelength Loss Minimised Single-Mode Optical Fibre Cable

Within Europe, the European Commission Cooperation on Science and Technology (COST), is organising interlaboratory work mainly inter-comparisons with important relevance to:

COST 217- Optical Measurement Techniques for Advanced Fibre Systems

COST 239- Ultra-high Capacity Optical Transmission Network

COST 240- Techniques for Modelling and Measuring Advanced Photonic Telecommunication Components

COST 241- Characterisation of Advanced Fibres for the New Photonic Network

This is related to standardisation activity by ETSI - European Telecommunication Standards Institute (STC TM1 - WG1) and CECC - CENELEC (Comité Européen de Normalisation Electrotechnique) Electronic Components Committee (WG 28).

3.2 Measurement Procedures

The measurement procedures [9, 10] are different depending on:

i) **Laboratory Measurements**, mainly addressed towards research, development, study of proprieties, adjustments or improvements, in which accuracy and sensitivity are primarily required. They are, in essence, of a particular nature according to the situation.

ii) **Factory Measurements**, tuned towards quality control acceptance procedures, requiring maximum automation and repeatability (as they are performed in large numbers). Standardisation is crucial so that the reference test methods (RTM) and the alternative test methods (ATM) can be specified. RTM provides a measurement of a particular characteristic, strictly according to the definition (gives usually the highest reproducibility, and lower uncertainty). An ATM is, in general, more suitable for practical use and corresponds to a derived process from the definition.

iii) **Field Measurements**, targeted to link and system testing, maintenance test, requesting essentially portable instrumentation for practical use.

The optical fibre application determines the most relevant parameters and the accuracy required on its measurement. The information is transmitted via light source modulation, propagation and detection. Fibre Optic measurement techniques basically address two major areas: functional testing (e.g. fibre continuity) and performance testing (i.e. loss and dispersion measurements) [9].

The performance testing aims to characterise the fibre, in particular:

i) Transmission characteristic
ii) Geometrical and optical characteristics
iii) Mechanic characteristics

Most of the current measurements are performed on fibres by the fibre manufacturer. This information is essential for the optical communication system designer. There are however progressively increasing requirements for field measurements in order to assess fibre network installation (i.e. event localisation and characterisation), and to evaluate overall system performance and subsequent ageing and maintenance.

Geometrical and mechanical proprieties are of constituitive nature and of major relevance to system and component engineering design. Optical proprieties are essentially related to transmission and sensing potential of the optical channels. The optical properties involve intrinsic physical properties such as refractive index profile and propagation field characterisation, namely mode field diameter, dispersion, spectral attenuation, cut-off wavelength, bandwidth, and polarisation.

Measurement of the mechanical characteristics (tensile strength, durability, corrosion sensitivity) are of particular interest to optical cable manufacturers. Geometrical dimensional measurements aim to describe the optical fibre cross-section. Geometrical/optical characteristics for multimode fibre are size (core and cladding diameter), numerical aperture, and refractive index profile. For single mode optical fibres and sensors, effective cut-off wavelength of the second order mode and mode field diameter should also be determined.

The fibre transmission characteristics of greater interest are dispersion for multimode fibre, and related bandwidth evaluation. For single mode optical fibres and sensors, coherent related properties [12] are most important (time dispersion, modal noise, polarisation mode dispersion).

Attenuation is due to intrinsic properties of the material such as absorption, scattering and additional factors, namely macro-and microbending. Scattering loss varies with λ^{-4}, Rayleigh's law. While absorption loss presents peaks that are mostly reduced. The transmission windows are 0,85 µm, 1,3 µm and 1,55 µm where absorption losses are relative low.

Microbending losses result from leakage of radiation by curvature. Microbending results from the statistical distribution of microcurvature from the roughness of the fibre coating and the cable support. Macro - and microbending losses are being reduced to negligible values with tighter confinement of the fundamental mode (i.e. mode field diameter).

Attenuation is measured directly by optical power decay between sections, and indirectly by monitoring the backscattering power decay of a travelling pulse (OTDR) [13].

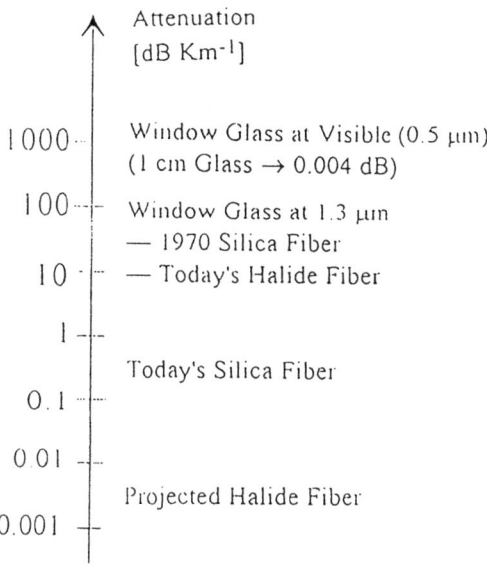

Fig. 10 Comparison of optical attenuation in different materials.

Time dispersion for single-mode fibre results from wavelength dependence of the group delay per unit of propagation length in relation to the fundamental mode. This chromatic dispersion has a material and waveguide contribution. Material dispersion is dominant far from the cut-off wavelength as the mode is tightly confined but near the cut-off waveguide dispersion may become important. Dispersion is non-linear with a dominant first order derivative and a second order derivative to be considered at 1550 nm. Polarisation mode dispersion is a marginal contribution, resulting from the small differences in phase constants of the two polarisation states in the fundamental mode. Multi-mode fibres present a dispersion several orders of magnitude greater than single mode fibre so direct measurement is done both in the time or frequency domain. Single mode fibres have their chromatic dispersion measured at selected wavelengths evaluating the propagation delay in the fibre. Polarisation dispersion is inferred from selective excitation or detection of two polarisation states in the fundamental mode.

Polarisation loss and depolarisation is relevant in transmission and detection systems that are polarisation sensitive (heterodyne and homodyne detection). The polarisation loss results in power fluctuations due to coherent combination of the two polarisation states. Depolarisation results in a non-defined polarisation state.

Mode distribution is a major concern for significant measurements to be obtained. For multimode optical fibre measurements independence of launching conditions and of mode distribution it is recommended that equilibrium mode distribution (EMD) be realised, i.e. the mode distribution after a reasonable long length of uniform propagation. The condition can be achieved on a short length of multimode fibre by power modal distribution (dummy fibre, mode scramblers). For single mode ideally a unique mode should be present so that mode filtering is used in measurements. Nevertheless some residual modal noise is present.

Modal noise appears in the presence of multimode propagation resulting from a pair of polarisation states in the fundamental mode, or the spurious presence of a second-order mode (highly attenuated under cut-off condition). Modal noise is pernicious in direct detection of intensity modulation. It is generated by localised events that favour mode coupling (i.e. connectors, splices). Modal noise characterisation is performed in the field when the transmission system or the sensor is assembled at the installation site.

Subsequent contributions by various lecturers will discuss the importance of the various quantities and the methods of measurement (RTMs and ATMs) whether single mode or multimode fibres are under consideration.

3.3 Standardised Recommendations

Standardised specifications, tolerances and testing procedures are available for industrial oriented products according to recommendations issued by EIA, IEC CCITT, ETSI and the CECC among other bodies. They cover optical fibre, optical fibre cables, optical fibre devices and components. Sensors are less covered but the trend in spreading their use will, hopefully, increase the amount of their coverage.

CCITT recommends (G652) for single mode fibres for operation at
$\lambda = 1,3$ *μm the following:*
MFD 9 to 10 ± 10% ; cladding diameter 125 ± 2 μm; Concentricity error between mode field and cladding $\lambda < 1$ μm; Cladding non circularity < 2% .
Cut-off wavelength for jacketed fibre with one primary coating

1,1 μm $< \lambda_c <$ 1,28 μm
Cut-off wavelength for cabled fibre $\lambda_{cc} <$ 1,27 μm
Attenuation < 1 dB Km^{-1}, (0,4 dB Km^{-1} achieved)
Zero dispersion wavelength 1300 $< \lambda_0 <$ 1324 nm
Maximum Dispersion Slope: $S_{Max} = 0,093$ ps nm^{-2} Km^{-1}

For the single mode fibre operation wavelength $\lambda = 1,55$ μm, for dispersion shifted fibres (G653):

MFD 7,0 μm to 8,3 μm (deviation smaller than 10% from nominal value)

Loss increase for 100 turns loosely wound 37,5 mm radius at

$\lambda = 1,55$ μm < 0,5 dB;

Dispersion slope between $\lambda = 1525$ and 1575 nm (< 3,5 ps nm^{-1} Km^{-1})
and for **loss-minimised single mode fibres**:

Attenuation < 0,25 dB Km^{-1} (values less than 0,2 dB Km^{-1} were obtained)

Chromatic dispersion 20 ps nm^{-1} Km^{-1}

Cut-off wavelength λ_c $\lambda \cong 1530$ nm

3.4 Sensing, Control and Instrumentation

The field of sensors started with the use of fibres and components developed for communications (low-loss fibres, couplers, WDM, LED, Laser Sources). Presently, specific technologies continue to be developed as well as specific fibre (low/high birefringence, polarisation maintaining fibres) and components (fibre polarisers, modulators, super-luminescent diodes).

Fibre based sensors, and instrumentation components require a comprehensive knowledge of their characteristics as they constitute the basis for the application and also due to their cross sensitivity to a number of environmental parameters. Metrological activity is abundant in this area but standardisation has been slow while waiting for the market forces.

Fibre-optics based measuring systems are going to experience a substantial and sustained development, promoted by very important fields of application: industrial sensors, robotic systems, biomedicine and environment.

3.5 Fibre Probes

Fibre optics provide means for measurement portability so that a range of fibre probes have been developed to enjoy the freedom to bring the instrument to the experiment rather than having to bring the experiment to the instrument. This tendency is likely to experience significant growth and broader use with the development of exotic type of fibres. Applications range from a very ample diversity from light delivery to fibre-optic refractometry [11].

4. The Trends

The future like in any area is uncertain and it might not be prudent to attempt predictions of events. However, one may venture to speculate on evolving trends based on the emerging needs that pose new demands on fibre metrology and standards [14].

Optical fibres have evolved with specific needs:

- R & D Laboratories
- Factories (fibre, cable)
- Network (communication, data, trunk and local)
- Components (passive, active, sensing)

The network evolution has been from trunk links with initial digital bit rates in the order of 560 Mb s^{-1} to standardised synchronous digital hierarchy (SDH) with bitrates of 156 Mb s^{-1}, 622 Mb s^{-1}, 2,5 Gb s^{-1}, 10 Gb s^{-1} in coming years, and aiming to reach 25 Gb s^{-1} and 48 Gb s^{-1} (dispersion control required). The network branching is also getting closer to the user (LAN and FITL) by passive systems (lower loss required).

Signal modulation is evolving from digital to compatibility of digital plus analogic (mainly AM, TV signals) to encompass a multitude of applications over a single fibre (ISDN and information highways) (lower return loss required).

The search for higher transmission capacity brought further technological developments such as channel multiplexing over a single fibre by wavelength division multiplexing (WDM). Thus overcoming limitations from time-division multiplexing (wavelength line width control required). Pure optical amplification will present technological advantages both increasing repeater spacing and revolutionising the instrumentation design (active fibre characterisation required). Also, coherent communications system may become viable with optical amplification and control of non-linear effects (non-linear characterisation required). Further foreseeable requirements include non-invasive measurements, remote testing systems, passive components characterisation and stable transfer standards among the ever expanding requests.

4.1 Non-Invasive Measurements

Fibre measurements are, in general, the end-connection type requiring preparation of fibre ends and the use of connectorized fibres at times or pig-tails fused spliced. Non-invasive techniques represent a major benefit once they do not affect mechanical continuity of the fibre nor the optical channel in fibre traffic. Clip-on techniques have been developed [15] operating by macrobending or microbending the fibre either to inject or extract the optical signal. Applications include live-fibre identifiers [16], local loss measurements [17] fibre optic talk sets, and specific uses in loop environment [18].

4.2 Remote Fibre Testing Systems (RFTS)

Fibre networks are investments projected to work for 24 hours a day so that reliability is a major concern. Preventive maintenance is envisaged on critical branches and in the event of trouble response must be quick and service restored in a very short time. One solution for monitoring and maintaining the optical fibre network, even in the verge of a shortage of skilled personnel, is a remote fibre testing system.

The introduction of optical fibre switches permits one to insert remotely the testing instrument, usually an OTDR, that assess periodically the status by comparison with a recorded reference trace (monitoring loss increase, outright breaks). Computer automatic allows to signal abnormalities and to establish a surveillance sequence either automatic or under manual control. Standardisation of operating software, and corresponding integrability on existing software in use by network operators is still needed to disseminate the use of RFTS, coupled with a more attractive price per channel.

4.3 Uniformity

Backscattering and back reflections could affect performance by re-entering the transmitter or affecting the multisignal discrimination of the receiver. Both optical continuous wave reflectometers (OCWR) and OTDR may assess the integrated sum of the fibre's continuous backscattering and discrete reflections. Analogue transmission is now imposing a return loss per components lower than ~ 50 dB. Older cable networks have been remeasured in search of acceptability of actual return loss. Further, fibre backscattering uniformity, mainly related to uniformity of mode-field diameter and cut-off wavelength has emerged. Fibre uniformity will open up the need, expressed by the standard, for bi-directional attenuation measurements.

4.4 Passive Components

Most optical fibre systems consists of a combination of fibres integrated with fibre components. As fibre optics technology proves viable for the subscriber loop and other optical networks, there is a growing demand for a variety of fibre optic components such as couplers, splitters, wavelength multiplexers, wavelength filters, attenuators, switches, isolators, etc. These components need to be high performance, reliable and low cost.

Progressive use of broader and complex components are bringing new challenges to fibre metrology. The metrological characterisation of the components is the first task. However, new problems emerge from the change of the bi-directional response of the network. The introduction of optical splitters (in passive optical networks, i.e. branching out to several customers) presents new difficulties in detection and localisation of events beyond the splitter due to reduction of effective dynamic range and the increase of background backscattering. Therefore, improvements on discrete reflection attenuation resolution is needed. For systems with WDMs the measurements interpretation will require a correct loss characterisation for WDM path losses.

Fibre gratings and further new components will place additional demands on metrology. Conversely optical systems which incorporate fibre gratings for complex instrumentation can be more cost effective while presenting a higher level of measurement efficiency and accuracy.

4.5 Standardisation

Establishment of standardised components for fibre-optic metrology will greatly enhance the utilisation of these techniques in industrial practice by improving the confidence of users and suppliers. Accurate and standardised fibre characterisation, well developed engineering designs, and low costs will ultimately decide the degree of success in field applications of fibre optic metrology systems.

Measurement standardisation has not kept pace with the rapid progress in fibre and components. A larger effort is likely to come through in the interest of helping market growth, product interchangeability and compatibility as well as global production delocalisation.

4.6 Dispersion Measurement

Higher transmission rates demand in-situ, more precise dispersion measurements. The zero-dispersion wavelength may need to be known in systems with fibre amplifiers. The dispersion compensation techniques (chirped lasers, dispersion compensating fibre, electronic pre-distortion, Fabry-Perot etalon, spectral inversion) for dispersion unshifted fibre at 1550 nm requires the value of chromatic dispersion of the link to be measured at the installation site. Polarisation mode dispersion is also of interest but the method of measurement has to be improved in the principle and adapted to field measurements [19].

4.7 Polymer Optical Fibres (POF)

Recently there has been considerable interest in the development of polymer optical fibres (POF).

Graded-index POF [20] is expected to present low loss combined with both high bandwidth, and larger diameters to cope with the many junctions and connections in short-range communications (LAN, interconnectors, domestic passive optical network).

4.8 Laser Sources and Components

There has been a considerable evolution in new laser sources and associated components such as: Laser Tuning Sources, Fibre Laser Gratings and Optical Fibre Amplifiers. Optical Fibre Amplifiers have already been developed to a network component stage and are reaching the phase of instrumentation components. Fibre Laser Gratings will introduce important changes in the design of sensing components and wavelength multiplexing. Laser Tuning Sources are likely to provide great improvements on metrological methods and on the development of artifacts calibration techniques [21].

- **MODULARISED**
- **MINIATURISED**
- **STANDARDISED**
- **PORTABILITY (LIGHTER, LOWER POWER CONSUMPTION, AUTONOMY)**
- **RUGGED**
- **SIMPLER** — **OPERATION / MAINTENANCE / REPAIR**
- **SMARTER** — **QUICK OPERATION / AUTOMATIC / DEEP MEMORY WITH TRANSPORTABILITY**
- **CHEAPER (DISSEMINATION OF USES)**

Fig. 11 - Instrumentation Trends

4.9 Repeater Spacing Enlargement

The possibility of a substantial increase in repeater spacing at high transmission rates, will decrease the cost of system components, installation, operation, and maintenance. For a system without transmission - line length restriction, it would be possible to install repeaters in substations instead of manholes. Immediate advantages will result with the availability of power supply thereby eliminating the need for power transmission along the fibre cable, improvement of system reliability and less expensive maintenance. Toward this aim, two major avenues are being explored: fibre amplifiers and coherent communication systems. New challenges are faced by metrology on systems incorporating active fibres and for coherent systems.

4.10 Optical Amplifier Metrology

Optical fibre amplifiers present numerous advantages (amplification of signals at multiple wavelengths whatever the signal modulation format, high efficiency, high output power, wide band, low noise, high coupling efficiency to fibres, preservation of optical phase wavelength and polarisation).

The use of optical fibre amplifiers will certainly progress rapidly both in optical fibre networks and instrumentation. This will require further efforts in metrology. The need for link transparency brings an example.

The optical amplifiers usually are mounted at the ends of optical isolators to block spurious reflections. The transparency of the network becomes inadequate for direct application of an OTDR. Various methods have been studied to overcome this problem of in line optical fibre amplifiers (modulation of the pumping laser, replacement of isolators by circulators, optical bypass at 1310 nm). A cost effective, standardised technique needs to be found in order to gain general acceptability.

The extensive characterisation of optical fibre amplifiers is in progress (gain, noise) and standardisation is following (see reference [25] on this volume) while optical fibre amplifiers are expected also to lead to new measurement and sensing techniques.

4.11 Coherent Communication Systems

Advances in single mode fibres and single spectral mode lasers (1,2 to 1,7 μm) stimulated interest in coherent communication systems. A zero dispersion single mode fibre is achieved by compensating material dispersion with waveguide dispersion at a desired wavelength. High spectral purity of the laser can be obtained with an external cavity. Coherent systems enable one to exploit the wide bandpass of low-loss silica window combining high-bit-rate with wavelength division multiplexing. The ultimate restriction comes from the non-linear effects due to self-phase modulation. These effects have to be fully characterised before the potential for solitons transmission can be commercially exploited. [26, 27]

4.12 Fibre-Optic Instrumentation Trends

The tendency for instrumentation is likely to stress field operation, Fig. 11. Portability would be a key aspect for factory testing and telecommunication industry (installation, maintenance and repair).

The instrument should have high fidelity whether in use at indoor or outdoor environments. Reducing the price will promote proliferation of its use. Low power consumption will mean autonomy (matching the working hours schedule) so that LCD rather than CRT will be the option. Simplicity of operation matching eventually the handling by one person without high skilled specialisation will aid in the dissemination of metrological practices.

The operation cycle should be quick, automatic with availability of deep memory combined with transportability to a remote centralised data processing unit. Fibre Optical Amplifier (AFA - Active Fibre Amplifier namely EDFA) in view of their inherent characteristics: high efficiency, high output power, wideband, low noise, insensitivity to polarisation, transparency irrespective of signal format, low crosstalk, high coupling efficiency to fibre, negligible non-linearity, and applicability to high-speed and/or multi-wavelength signal amplification have extremely attractive potential to be applied in measurement techniques [22].

4.13 Improved Availability of Transfer Standards

Stable transfer standards development with generalised acceptance through international standardisation is expected to permit the calibration of test equipment used in the laboratory and in the field. This will provide the necessary traceability for the inevitable inter-comparisons within the global market.

4.14 Intelligent Measurements and Standardisation [23]

There is a general trend to move from hardware-oriented to software-oriented technologies. Software is becoming the fundamental part of innovative instrumentation and productivity in scientific and technological processes. The technical elaboration of measurement protocols and quality assurance handbooks according to ISO 9000 include measurement data management. Those requirements would lead to exigencies of software accreditation and a major increase in metrological activities. This will result in lower priced instrumentation, better automation and portability, and compatibility of modularised software.

Further, miniaturisation combined with improved optoelectronics reliability are likely to generate a great number of different plug-in boards for PCs, making a diversity of optical fibre measurements readily available to industry and for in field measurements with a greater flexibility for up-dating and to adopt to the foreseeable technological breakthroughs.

4.15 All - Optical Network

The age of broadband networks i.e. higher bit rates, increased flexibility and services demands greater transmission capacity on the individual network links combined with better switching at the network nodes.

The fluctuation range of transmission requires a real time reconfigurable network, overcoming any occasional fault with the establishment of alternative paths and the consequent burden of switching and network management. The response to such a level of performance is an all-optical cross-connected [24], transparent to bit-rate, wavelength and signal format.

For an all-optical network the individual subcomponents within the node could limit the system performance. The switch must have low loss, high switching speed, low cross-talk, small size and scalability.

Development of optical amplifiers may lead to the elimination of repeaters with the penalty that crosstalk and noise will have to be corrected by other means or alternatively minimised and the network size shortened in correspondence to the number of channels and bit-rates.

The optical transparency, the reconfigurability and the high bit-rates combined with mixed formatted signals will certainly bring new challenges to metrology.

5. Conclusion

Fibre applications have matured in a fast growing field. Optical testing methods, instrumentation, and calibration artefacts, have to respond in consonance.

The theoretical and experimental challenges in the different applications (laboratory, factory and in the field) are at a high pace of evolution branching out in new applications with renewed demands for optical fibre metrology, standardisation, and instrumentation.

This TRENDS in OPTICAL FIBRE METROLOGY and STANDARDS, INSTITUTE hopefully will help to create a stimulating environment for the encouragement of increased efforts to grasp the opportunities to promote high scientific and technological research and future developments. The resulting book from the Institute may prove of interest to persons involved in the specification, design, installation, operation and maintenance of optical fibre networks in their progressively growing multiple uses.

6. References

1. O.D.D. Soares   Optical Metrology
   Martinus Nijhoff, Dordrecht (1987)

2. ISO/TAG4/WG3 Guide to the Expression of Uncertainty in
   Measurement (1991)

3. J. M. Senior   Optical Fibre Communications
   Prentice Hall, NY (1992)

4. World Markets for Cabled Fibres
   Laser & Optronics (March 1994)

5. R.B. Linscott   US Expected to Held Its Own
   In Global Photonics Spectra Fiber Optic Boom
   Photonics Spectra, (March 1994)

6. R.B. Linscott — In the Field of Fiber Communications All the Forecasters are Thinking Big
Photonics Spectra (Feb. 1993)

7. Fiberoptics in Europe — Market Projections for Europe
Fiberoptic Product News 9 (1991)

8. O.D.D. Soares (Ed.) — Colorimetria
CETO, Porto (1994)

9. B. Costa — Test and Measurement Techniques for Optical Fibres
CSELT, Torino (1992)

10. O.D.D. Soares (Ed.) — Fiber Optic Metrology and Standards
SPIE, Vol 1504, Bellingham (1992)

11. T. Takeo, H. Hattori — Skin Hydration State Estimation Using a Fiber-Optic Refractometer
Appl. Opt. 19 (1994), 4267

12. C. Hentschel — Fiber Optics Handbook
Hewlett-Packard (1989)

13. G. Cancellieri (Ed.) — Single-Mode Optical Fiber Measurement: Characterisation and Sensing
Artech House, Boston (1993)

14. F.P. Kapron — Measurement Needs for Evolving Fibre Networks
OFMC'93 - 2nd Optical Fibre Measurement Conference Technical Digest, Torino (1993), 23-28

15. Generic Requirements for Fibre Optic Clip-On Test Sets
TA-NWT-001009, (1993), Bellcore

16. Generic Criteria for Optical Fibre Identifiers
TR-NWT-000764, (1990), Bellcore

17. Generic Requirements for Splice Verification Sets
TR-NWT-001196, (1991), Bellcore

18. N. Lewis, P. Keeble, D. Fergunson — Testing Strategies for Modern Fibre Network Architectures Tech. Digest Symposium on Optical Fibre Measurement, Boulder CO, (1988), 15-18

19. Y. Namihira, J. Maedz — Polarisation Mode Dispersion in Optical Fibres
Electronics lett. (1992), 145-150

20. T. Ishgure, E. Nihei, Y. Koike — Graded-index Polymer Optical Fibre for High Speed Data Communication
Appl. Opt. 33 (1994), 4261

21. C. Hentshel Characterising Fibre-Optic Components
with a Tunable Laser
HP Lightwave Seminar (1992)

22. H. Okamura Application of Er-doped Fibre Amplifiers to
    K. Iwatsuki Optical Measurement Techniques
This volume

23. D. Hoffmann Intelligent Measurement - New Solutions for
Old Problems
Measurements 13 (1994), 23-37

24. R. Dettmer Travelling Light: an all optical transparent network for
telelecommunications
IEE Review (July 1994), 159-162

25. P. Di Vita Optical Fibre Amplifiers Standardisation
This volume

26. I. Jacobs Optical Fiber Communication Technology and
System Overview
This volume

27. G. P. Agrawal Modern Optical Communications Systems
This volume

II Optical Fibres and Materials

POLARIZED LIGHT EVOLUTION IN OPTICAL FIBRES

From Basic Concepts to Polarization Characteristics Measurement Techniques

J. PELAYO, F. VILLUENDAS
Departamento Física Aplicada. Universidad de Zaragoza
Facultad de Ciencias. Ciudad Universitaria
50009-Zaragoza, SPAIN

1. Introduction

There are several reasons for studying evolution of polarized light in optical fibres. Apart from reasons based on pure scientific knowledge, we can mention some specific examples, with a practical point of view in mind. On the one hand, principles of operation of many optical fibre systems, in particular interferometric sensing devices and coherent transmission systems, rely on the maintaining of a given polarization state in the light transmitted and detected. On the other hand, polarization mode dispersion represents a limitation concerning ultimate achievable performances of high bit-rate, optical fibre communications systems with direct (incoherent) detection.

In relation with these examples, two kinds of problems concerning polarization properties of optical fibre systems must be clearly identified from the start. They can be stated as follows.

A) Analysis of performance characteristics of the so-called polarization-maintaining fibres, that is, special fibres with high birefringence produced in the fabrication process in order to achieve the best possible polarization-holding characteristics.

B) Analysis of the effects that residual birefringence produce on the propagation of polarized light in nominally symmetrical fibres.

The analysis of A) may be done using a simple description of strictly monochromatic waves propagating in a fibre with locally well-defined principal axes of symmetry, that may rotate along the propagation distance. This analysis, developed in Section 2 below, leads to the key concept of principal states of polarization, which plays a central role in the explanation of optical fibre polarization properties.

The main features of principal states of polarization, defined as first-order wavelength independent polarization states, are discussed in Section 3. These features allow a complete description of the evolution of polarized light in fibres, even for polychromatic beams, assuming that a "weak coupling" condition between local polarization eigenstates is fulfilled, as it actually occurs in polarization maintaining fibres.

Concerning point B), however, there are important aspects of the behaviour of polarized light propagation that can be experimentally observed in conventional single-mode fibres, and cannot be explained with the same simple model applicable to polarization-maintaining fibres. Global principal states of polarization, although remaining formally well-defined in the limit of strictly monochromatic waves, lose most of its physical meaning for standard fibres, because of its instability with respect to wavelength changes inside the signal spectrum.

Notwithstanding, polarization properties of standard fibres can be conveniently analysed in terms of a concatenation of fibre trunks, with locally stable principal states of polarization, and a "strong coupling" between polarization states at the junctions between consecutive trunks. This model, presented in Section 4, predicts the spreading in the time domain of the optical power launched in an ideal pulse as the signal propagates along the fibre. Temporal delays between different fractions of the total power launched in a single pulse provide a clear description of polarization mode dispersion effects in the fibre.

The definition of a suitable parameter for characterising polarization mode dispersion penalties on optical fibre transmission systems is the main subject of Section 5.

The same pulse broadening phenomenon in the time domain can also be interpreted in a frequency domain description as the result of differences in phase shifts between principal states changing at different wavelengths in the spectral range of the signal. These equivalent descriptions in the time and frequency domains represent the fundamentals of the two main measurement techniques of polarization mode dispersion in optical fibres. As it is discussed in Section 6, interferometric and Stokes parameter measurement methods are shown to be in close relation to time and frequency domain descriptions, respectively.

2. Evolution of Polarization in High-Birefringence Fibres.

We initially assume that an infinitesimal segment of the fiber, between points z and $z+dz$ along the propagation direction, acts as a simple linear retarder introducing a phase delay $\Delta\beta(z)dz$ between electric fields components E in the directions of local symmetry axes. With a Jones matrix formalism [1,2],

$$\boldsymbol{E} + d\boldsymbol{E} = \begin{bmatrix} 1 + j(\beta + \Delta\beta/2)dz & 0 \\ 0 & 1 + j(\beta - \Delta\beta/2)dz \end{bmatrix} \boldsymbol{E} \qquad (1)$$

where β represents the mean propagation constant.

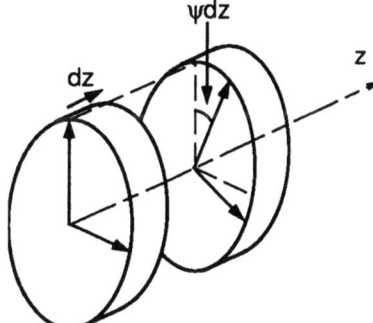

Figure 1. Rotation of local symmetry axes along the propagation direction.

In order to recursively apply the transformation given by (1) in successive dz-elements, we have to keep the electric field vector referred to local symmetry axes. Then, an infinitesimal rotation, by an angle $\psi(z)dz$ (fig. 1), must also be taken into account to give a full description of the differential change of vector $\boldsymbol{E}$:

$$\begin{aligned} \boldsymbol{E}' &= \begin{bmatrix} 1 & \psi dz \\ -\psi dz & 1 \end{bmatrix} \begin{bmatrix} 1 + j(\beta + \Delta\beta/2)dz & 0 \\ 0 & 1 + j(\beta - \Delta\beta/2)dz \end{bmatrix} \boldsymbol{E} \\ &= \begin{bmatrix} 1 + j(\beta + \Delta\beta/2)dz & \psi dz \\ -\psi dz & 1 + j(\beta - \Delta\beta/2)dz \end{bmatrix} \boldsymbol{E}. \end{aligned} \qquad (2)$$

where $\boldsymbol{E}'$ gives field components referred to local symmetry axes.

The angle $\psi(z)dz$ is defined by the relative orientations of polarization eigenstates in consecutive infinitesimal segments of the fibre. It may include the effects of both a purely geometrical twisting of the fibre symmetry axes and any kind of material optical activity in the fibre [3].

The differential relation (1) can be integrated from $z=0$ to $z=L$, to get

$$E'_L = \exp(j\overline{\beta}L)\underline{U}(L)E_o, \qquad (3)$$

with

$$\overline{\beta} = \frac{1}{L}\int_0^L \beta(z)dz, \quad \underline{U}(L) = \exp\left\{\begin{bmatrix} jI_\beta & I_\psi \\ -I_\psi & -jI_\beta \end{bmatrix}\right\}$$

and

$$I_\beta = \int_0^L dz \, \frac{\Delta\beta(z)}{2}, \quad I_\psi = \int_0^L dz \, \psi(z), \quad \Gamma = \sqrt{I_\beta^2 + I_\psi^2}.$$

Finally, the exponential form of the matrix $\underline{U}(L)$ can be evaluated by standard mathematical techniques (see ref. [3]) to give

$$\underline{U}(L) = \begin{bmatrix} \cos\Gamma + j(I_\beta/\Gamma)\sin\Gamma & (I_\psi/\Gamma)\sin\Gamma \\ -(I_\psi/\Gamma)\sin\Gamma & \cos\Gamma - j(I_\beta/\Gamma)\sin\Gamma \end{bmatrix}. \qquad (4)$$

Equation (3) gives a general description of the propagation of a monochromatic light beam along a singlemode fiber. The relation between the matrix elements of the unitary transformation $\underline{U}(L)$ and the local characteristics of the fiber $\Delta\beta(z)$ and $\psi(z)$, given by (4), is obtained with the only condition of continuity in the evolution of the field components along the fiber. The precise meaning of this condition (in fact, not very restrictive) is that no abrupt changes in the propagation conditions along the fibre, in particular, in the orientation of linear birefringence axes, take place in a small fraction of a wavelength period, so that $\Delta\beta, \psi << \beta$. The continuity condition also implies that the local birefringence $\Delta\beta(z)$ must be considered as a function that can take either positive or negative values, in order to allow the interpretation of a possible interchange between fast and slow axes for phase propagation as a change in the sign of $\Delta\beta(z)$, and not as a sudden $\pi/2$-rotation of the axes.

The extension of the formalism to include circular birefringence (optical activity) effects in the fibre is straightforward. It just requires to consider $\psi(z)$ as a complex function, preserving the unitary character of matrix $\underline{U}(L)$. The formalism could also be extended to describe attenuation effects, as well as linear and circular dichroism in the fiber, by allowing everyone of the parameters $\beta(z)$, $\Delta\beta(z)$ and $\psi(z)$, and the corresponding integrals, to take complex values, although, in this case, $\underline{U}(L)$ would not represent any more a unitary transformation. However, we will neglect all these latter effects in the following analysis, under the assumption that they do not play a very significant role in most cases of interest.

The analysis of any experimentally significant situation, concerning propagation of polychromatic and/or modulated polarized light beams, can be made by considering the dependence of matrix $\underline{U}(L)$ on the angular frequency ω. With the input field amplitude expressed in the form

$$E(t) = A(t)\, E_o \exp(j\omega_o t), \tag{5}$$

the scalar complex function $A(t)$ may describe both modulation of the signal as well as random fluctuations in a polychromatic field around the central frequency ω_o. Transformation of the spectral components $E(\omega)$ of the input field (5) by the propagation matrix $\underline{U}(\omega,L)$ allows the reconstruction, coming back to the time domain, of the output field after propagation through the fiber.

Let us consider then the general form of the dependence of matrix $\underline{U}$ on ω. To this purpose, it is convenient to write down $\underline{U}(\omega,L)$ in the form

$$\underline{U}(\omega,L) = \begin{bmatrix} 1 & 0 \\ 0 & 1 \end{bmatrix} \cos \Gamma + \underline{M}(\omega,L) \sin \Gamma, \tag{6}$$

with

$$\underline{M}(\omega,L) = \begin{bmatrix} ji_\beta & i_\psi \\ -i_\psi & -ji_\beta \end{bmatrix}_\omega$$

and i_β, i_ψ giving the ratios (I_β/Γ), (I_ψ/Γ) at frequency ω. From (6), we can analyse separately a fast dependence of the output polarization state on ω, due to changes in the phase argument $\Gamma(\omega)$, and a slow dependence, due to changes in amplitude coefficients, i_β, i_ψ, in the second term of the expression (6) of $\underline{U}(\omega,L)$.

In general, for long propagation distances, the accumulated phase difference Γ will amount to a large number of radians at the end of the fiber length L, so that a relatively small variation $\Delta\Gamma(\omega) \approx 1 \ll \Gamma(\omega_o)$ will produce a considerable change in matrix $\underline{U}$ through the sine and cosine functions, while the contribution to that change of the matrix elements in the second term of (6) would remain negligible (of order $\Delta\Gamma/\Gamma \ll 1$). It is in this case, that is, when the approximation

$$\underline{U}(\omega,L) \cong \begin{bmatrix} 1 & 0 \\ 0 & 1 \end{bmatrix} \cos \Gamma(\omega) + \underline{M}(\omega_o,L) \sin \Gamma(\omega) \qquad (7)$$

is justified, where the concept of principal states of polarization (PSP), as defined in a general way in the following section, finds its full significance, from a physical point of view. They are corresponding input-output states with polarization vectors which are approximately wavelength independent at both the input and the output ends of the fiber. From (7), PSP are obtained as eigenvectors

$$\varepsilon_{\pm} = \sqrt{\frac{1 \mp i_\beta}{2}} \begin{bmatrix} \pm j i_\psi \\ -(1 \mp i_\beta) \end{bmatrix}$$

of the matrix in the second term, with eigenvalues $\pm j$,

$$\underline{M}(\omega_o,L)\varepsilon_{\pm} = \pm j \varepsilon_{\pm}. \qquad (8)$$

The corresponding phase factors after propagation through the fiber are given by $\exp(j\overline{\beta}L \pm j\Gamma)$, so that the time delay between optical pulses in these polarization states is $\Delta\tau = 2\, d\Gamma/d\omega$.

PSP must not be confused with ω-dependent polarization eigenstates that would be obtained as eigenvectors of the whole matrix $\underline{U}(L)$, including phase factors $\cos\Gamma(\omega)$ and $\sin\Gamma(\omega)$, and with a very different physical meaning.

The application of the PSP model to high-birefringence (polarization maintaining) fibers, for which $\Gamma \approx I_\beta \gg I_\psi$, is straight-forward. In this case, PSP are obtained directly from (6) as vectors $\begin{bmatrix} 1 \\ 0 \end{bmatrix}$ and $\begin{bmatrix} 0 \\ 1 \end{bmatrix}$, referred to local symmetry axes, that are directed along principal axes of birefringence at both the input and the output ends, in fact at any point, of the fiber. The concept of PSP gives account of the polarization maintaining effect in high-birefringence fibre, which actually is observed for linear polarizations along principal axes of birefringence in the fibre. Notwithstanding, the model also provides the means for a quantitative analysis of polarization crosstalk effects related to off-diagonal matrix elements i_ψ.

3. General Definition and Properties of PSP

Principal states of polarization are defined in a general way [4] as polarization states of the light such that they do not change for first-order infinitesimal variations of the optical frequency ω.

PSP properties can be explicitly derived from the formal definition that follows. Let us express field vectors as

$$\boldsymbol{E} = \exp(j\phi)\, |E|\, \varepsilon,$$

where ε is a unit vector giving the polarization of the light.

Changes of ε with ω are given by

$$\frac{d}{d\omega}\varepsilon = \exp(-j\phi)|E|^{-1}\left(\frac{d}{d\omega}\boldsymbol{E} - j\dot{\phi}\boldsymbol{E}\right), \tag{9}$$

with a dot notation for derivatives with respect to ω, and assuming that the modulus of the field amplitude $|E|$ does not depend on ω.

From equation (3), we get the derivative of the output field,

$$\frac{d}{d\omega}\boldsymbol{E}'_L = \exp(j\beta L)\left\{j\overline{\overline{\beta}}L\underline{U}(L) + \underline{\dot{U}}(L)\right\}\boldsymbol{E}_o,$$

that, substituted into (9), gives

$$\frac{d}{d\omega}\varepsilon'_L = \exp(-j\phi + j\beta L)|E_L|^{-1}\left\{\underline{\dot{U}}(L) - j(\dot{\phi} - \overline{\overline{\beta}}L)\underline{U}(L)\right\}\boldsymbol{E}_o.$$

Thus, PSP, at the input, are defined by the eigenvalue equation

$$\underline{U}^{-1}(L)\underline{\dot{U}}(L)\varepsilon_o = j(\dot{\phi} - \overline{\overline{\beta}}L)\varepsilon_o. \tag{10}$$

The precise meaning of PSP can be conveniently explained considering a first-order expansion of matrix $\underline{U}(\omega, L)$ around $\omega = \omega_0$,

$$\underline{U}(\omega, L) = \underline{U}(\omega_o, L) + \underline{\dot{U}}(\omega_o, L)(\omega - \omega_o).$$

Taking into account the unitary character of the matrix $\underline{U}(L)$, it can be easily shown that eigenvalues of equation (10) must take opposite values of the form $\pm j(\dot{\phi} - \overline{\beta}L)$, corresponding to eigenvectors $\varepsilon_{\pm}$. Thus, the output field may be written in terms of PSP at the input as

$$E'_L = \exp(j\beta L)\underline{U}(\omega_o, L)\sum_{\pm} a_{\pm}\left\{1 \pm j(\dot{\phi}_L - \overline{\beta}L)(\omega - \omega_o)\right\}(\varepsilon_{\pm})_o,$$

or, in terms of PSP at the output, neglecting second-order terms in $(\omega-\omega_o)$,

$$E'_L = \sum_{\pm} a_{\pm} \exp\left\{j\beta L \pm j(\Delta\tau/2)(\omega - \omega_o)\right\}(\varepsilon_{\pm})'_L, \qquad (11)$$

where $a_{\pm}$ represent projection coefficients of the input polarization over PSP, and $\Delta\tau = 2(\dot{\phi}_L - \overline{\beta}L)$ gives the total time delay between both PSP terms at the output.

Another interesting point of view is provided by a description of the output polarization evolution with ω, in the Poincaré sphere [2].

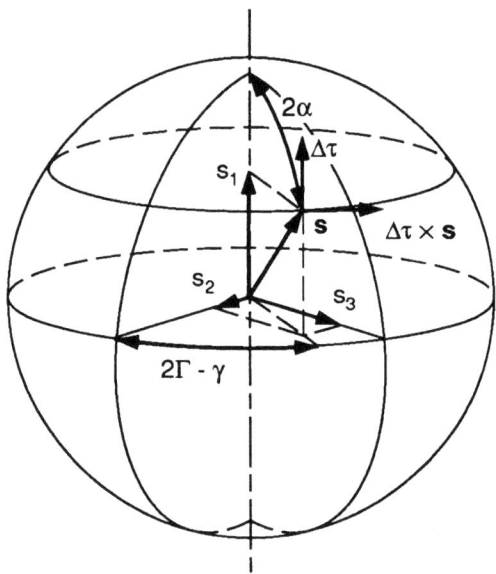

Figure 2. Evolution of polarization state in the Poincaré sphere.

Representing the output polarization, given by equation (11), in the form

$$|\varepsilon(\omega)\rangle = \cos\alpha\exp\{j\Gamma(\omega)\}|\varepsilon_+\rangle + \exp(j\gamma)\sin\alpha\exp\{-j\Gamma(\omega)\}|\varepsilon_-\rangle,$$

where α and γ define the projection coefficients on the basis provided by PSP (see fig. 2), we can easily obtain Stokes parameters of the output polarization as

$$S_0 = \langle \varepsilon | \{ |\varepsilon_+\rangle\langle\varepsilon_+| + |\varepsilon_-\rangle\langle\varepsilon_-| \} |\varepsilon\rangle = 1,$$
$$S_1 = \langle \varepsilon | \{ |\varepsilon_+\rangle\langle\varepsilon_+| - |\varepsilon_-\rangle\langle\varepsilon_-| \} |\varepsilon\rangle = \cos(2\alpha),$$
$$S_2 = \langle \varepsilon | \{ |\varepsilon_+\rangle\langle\varepsilon_-| + |\varepsilon_-\rangle\langle\varepsilon_+| \} |\varepsilon\rangle = \sin(2\alpha)\cos(2\Gamma - \gamma),$$
$$S_3 = \langle \varepsilon | \{ |\varepsilon_+\rangle\langle\varepsilon_-| - |\varepsilon_-\rangle\langle\varepsilon_+| \} |\varepsilon\rangle = \sin(2\alpha)\sin(2\Gamma - \gamma).$$

Then, the evolution of the Stokes parameter vector $s = (S_1, S_2, S_3)/S_0$ is given by the simple equation

$$\frac{d}{d\omega} s = \Delta\tau \times s, \qquad (12)$$

that represents a rotation in the Poincaré sphere around the axis defined by the orientation of PSP, with a velocity given by $|\Delta\tau| = \Delta\tau$ [5].

Considering the explicit form of matrix $\underline{M}$ in equation (6), it is very easy to show that, at a given optical frequency $\omega = \omega_o$, the following equation holds

$$\underline{U}^{-1}(\omega_o, L)\underline{\dot{U}}(\omega_o, L) = \underline{M}(\omega_o, L),$$

so that PSP definitions given by eigenvalue equations (8) and (10) are exactly coincident.

In spite of the important clarification provided by the PSP concept, concerning polarization evolution in optical fibres, there are some limitations affecting the applicability of this concept in practical situations. They come from the fact that PSP, as defined by equation (10), are actually ω-dependent!

If we require first-order independence on ω of output polarization states, in general, we get a condition on input states that makes them ω-dependent. Of course, the PSP concept remains well defined for narrow enough spectral ranges of the light.

PSP dependence on optical frequency can be better analysed from the definition given by equation (8). This definition associates changes in PSP to relatively slow changes (in comparison with fast phase changes) of the $\underline{M}$-matrix elements. PSP stability with ω is also related by equation (8) to $\underline{M}$-matrix off-diagonal terms magnitude relative to that of diagonal terms. For large enough diagonal elements, there is no actual dependence of PSP orientation on ω, although there may be some dependence of the time delay $\Delta\tau$ between PSP.

This fact gives rise to the distinction between different polarization coupling regimes that can take place in a given fibre line. A weak coupling regime, for small off-diagonal (coupling) elements, $|i_\psi| << |i_\beta|$ (as in polarization-maintaining fibres), implies stable PSP oriented along principal symmetry axes. On the contrary, a strong coupling regime, for $|i_\psi| \approx |i_\beta|$ (as in standard, nominally symmetric, fibres), implies unstable PSP with optical frequency.

4. Polarization evolution under strong coupling conditions

As it has been pointed out in some recent papers on polarization mode dispersion [6,7], the evolution of polarization in optical fibre transmission lines and optical fibre devices can be understood in terms of the propagation through a concatenation of several high birefringence fibre trunks. Every one of these trunks is determined by a fibre length in which wavelength independent principal states of polarization can be defined. Then, a polarization maintaining trunk introduces a delay between both principal states, while the relative orientation of consecutive trunks accounts for polarization mode coupling that takes place only at the junctions between trunks. The number of trunks to be taken into account and their relative orientations would be determined by the actual conditions of the specific fibre line under consideration.

Polychromatic light propagation through this concatenated system can be described either in the frequency or in the time domain, indistinctly. Both descriptions are completely equivalent, and are connected by Fourier transform relations. Descriptions in both domains are presented below, starting with the time domain description that can give a more intuitive insight into polarization mode dispersion effects.

So, the field amplitude of a general polychromatic beam with polarization state $|\varepsilon_o\rangle$ at the input of the fibre, in the time domain, can be expressed as

$$E(0,t) = A(t)\exp(j\omega_o t)|\varepsilon_o\rangle,$$

where $A(t)$ is the time dependent complex amplitude and ω_o is the central optical frequency of the light. The same field can be represented in the frequency domain as follows

$$E(0,\omega) = B(\omega)|\varepsilon_o\rangle$$

where $B(\omega)$ is the complex amplitude of the spectral component of optical frequency ω.

Field amplitudes $A(t)$ and $B(\omega)$ are connected by Fourier transform relations in the form

$$A(t)\exp(j\omega_o t) = \frac{1}{2\pi}\int_{-\infty}^{\infty} B(\omega)\exp(-j\omega t)d\omega$$

$$B(\omega) = \int_{-\infty}^{\infty} A(t)\exp(j\omega_o t)\exp(j\omega t)dt$$

and the polarization evolution in the fibre line would be determined by the concatenated influence on these amplitudes of different fibre trunks in which we can subdivide the whole line.

4.1. TIME DOMAIN DESCRIPTION

The jth polarization maintaining trunk, with principal polarization states $|-_j^i\rangle$ and $|+_j^i\rangle$, and $|-_j^o\rangle$ and $|+_j^o\rangle$ at the input and at the output of the fibre trunk, introduces delays $\tau_j - \Delta\tau_j/2$ and $\tau_j + \Delta\tau_j/2$ for the fast and slow principal states, respectively.

The propagation through this trunk can be expressed as the convolution of Dirac delta functions with the time dependent complex amplitudes launched in every one of the input principal states. Then, if $A_j^-(t)$ and $A_j^+(t)$, and $A_{j-1}^-(t)$ and $A_{j-1}^+(t)$ represent the complex amplitudes in every principal state at the output of the j and j-1 trunks (fig. 3), the influence of the jth trunk can be expressed in a matrix form as follows

$$\begin{bmatrix} A_j^-(t) \\ A_j^+(t) \end{bmatrix} = \underline{T}_{j,j-1} * \begin{bmatrix} A_{j-1}^-(t) \\ A_{j-1}^+(t) \end{bmatrix},$$

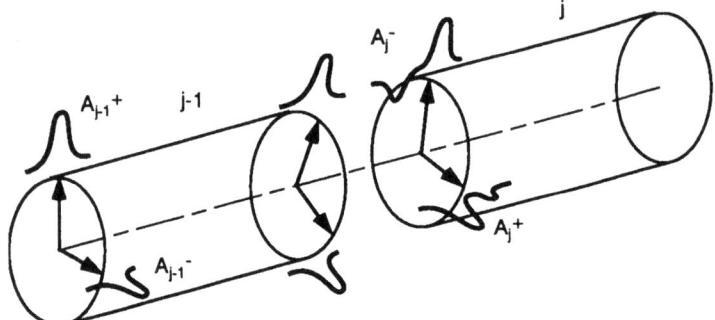

Figure 3. Schematic representation of field amplitude evolution in consecutive fibre trunks.

where

$$\underline{T}_{j,j-1} = \begin{bmatrix} \langle -_j^i | -_{j-1}^o \rangle \delta\left[t - \left(\tau_j - \frac{\Delta\tau_j}{2}\right)\right] & \langle -_j^i | +_{j-1}^o \rangle \delta\left[t - \left(\tau_j - \frac{\Delta\tau_j}{2}\right)\right] \\ \langle +_j^i | -_{j-1}^o \rangle \delta\left[t - \left(\tau_j + \frac{\Delta\tau_j}{2}\right)\right] & \langle +_j^i | +_{j-1}^o \rangle \delta\left[t - \left(\tau_j + \frac{\Delta\tau_j}{2}\right)\right] \end{bmatrix}$$

is the propagation matrix along the jth trunk, and * represent a convolution product between matrices. This formalism can be applied to every trunk so that complex amplitudes at the output of the concatenated system can be obtained from input complex amplitudes in the form

$$\begin{bmatrix} A_n^-(t) \\ A_n^+(t) \end{bmatrix} = \prod_{j=2}^{n} * \underline{T}_{j,j-1} * \begin{bmatrix} \delta\left(t - \left(\tau_1 - \frac{\Delta\tau_1}{2}\right)\right) & 0 \\ 0 & \delta\left(t - \left(\tau_1 + \frac{\Delta\tau_1}{2}\right)\right) \end{bmatrix} * \begin{bmatrix} \langle -_1^i | \varepsilon_o \rangle A(t) \\ \langle +_1^i | \varepsilon_o \rangle A(t) \end{bmatrix}$$

where $\prod_{j=1}^{n} *$ represents the convolution of propagation matrices corresponding to every trunk and n is the number of trunks in the system.

The relation between amplitudes at the input and at the output of the system only includes the convolution of Dirac delta functions corresponding to the introduced delays, with weighting factors given by projections between principal polarization states in consecutive trunks. The convolution of two Dirac delta functions results on another Dirac delta function evaluated at the sum of delays. Therefore, the final relationship between field amplitudes only includes weighted Dirac delta functions evaluated at all possible additions and subtractions of relative delays introduced in every trunk. This fact can be summarised by means of two complex functions, $F(t)$ and $G(t)$, that will determine the time evolution of the complex field amplitude at the output of the system, in a form that can be expressed as

$$E(L,t) = F(t)*A(t) \, e^{j\omega_o t}\left|-{}_n^o\right> + G(t)*A(t) \, e^{j\omega_o t}\left|+{}_n^o\right>.$$

The shape of the functions $F(t)$ and $G(t)$ will be determined by the polarization characteristics of the fibre or device considered. In particular, if a weak polarization mode coupling takes place in the fibre line, $F(t)$ and $G(t)$ are simply given by delayed Dirac delta functions $F(t) \propto \delta\left[t-\left(\tau_I - \frac{\Delta\tau_I}{2}\right)\right]$ and $G(t) \propto \delta\left[t-\left(\tau_I + \frac{\Delta\tau_I}{2}\right)\right]$, describing the effect of propagation in the case of a polarization maintaining fibre. On the contrary, in a highly perturbed low birefringence fibre a strong random polarization coupling occurs, and both functions would be described by complex distributions with similar Gaussian envelopes. These are the two opposite extreme cases of polarization evolution, so that, in a general fibre line including polarization dependent devices, complicated functions resulting from biased distributions of Dirac delta terms would describe the behaviour of the polarized light in the fibre.

4.2. FREQUENCY DOMAIN DESCRIPTION

In the frequency domain, the jth fibre trunk introduces phase changes $\omega\left(\tau_j - \frac{\Delta\tau_j}{2}\right)$ and $\omega\left(\tau_j + \frac{\Delta\tau_j}{2}\right)$ for field components of the light beam with optical frequency ω launched in every principal state. Then, using the same notation for the principal states than in the time domain description, the influence of the jth trunk can be also expressed in matrix form as follows

$$\begin{bmatrix} B_j^-(\omega) \\ B_j^+(\omega) \end{bmatrix} = \underline{U}_{j,j-1} \begin{bmatrix} B_{j-1}^-(\omega) \\ B_{j-1}^+(\omega) \end{bmatrix}$$

where

$$\underline{U}_{j,j-1} = \begin{bmatrix} \langle -{}_j^i | -{}_{j-1}^o \rangle e^{j\left[\omega\left(\tau_j - \frac{\Delta\tau_j}{2}\right)\right]} & \langle -{}_j^i | +{}_{j-1}^o \rangle e^{j\left[\omega\left(\tau_j - \frac{\Delta\tau_j}{2}\right)\right]} \\ \langle +{}_j^i | -{}_{j-1}^o \rangle e^{j\left[\omega\left(\tau_j + \frac{\Delta\tau_j}{2}\right)\right]} & \langle +{}_j^i | +{}_{j-1}^o \rangle e^{j\left[\omega\left(\tau_j + \frac{\Delta\tau_j}{2}\right)\right]} \end{bmatrix}$$

is the transmission matrix (frequency dependent Jones matrix) that relates the frequency dependent amplitudes at both sides of the trunk. This formalism can also be applied to every trunk so that the frequency dependent complex amplitudes at the output and at the input of the concatenated system are related in the form

$$\begin{bmatrix} B_n^-(\omega) \\ B_n^+(\omega) \end{bmatrix} = \prod_{j=1}^{n} \underline{U}_{j,j-1} \begin{bmatrix} e^{j\omega\left(\tau_1 - \frac{\Delta\tau_1}{2}\right)} & 0 \\ 0 & e^{j\omega\left(\tau_1 + \frac{\Delta\tau_1}{2}\right)} \end{bmatrix} \begin{bmatrix} \langle -{}_1^i | \varepsilon_o \rangle B(\omega) \\ \langle +{}_1^i | \varepsilon_o \rangle B(\omega) \end{bmatrix}.$$

The frequency dependent amplitudes at the input and at the output of the system are related by products of weighted complex exponential functions with phase factors at all possible additions and subtractions of phase changes introduced in every trunk. Then, as in the time domain, two functions $f(\omega)$ and $g(\omega)$ can be defined in order to express the field amplitudes at the output of the fibre line,

$$E(L,\omega) = f(\omega)B(\omega) |-{}_n^o\rangle + g(\omega)B(\omega)|+{}_n^o\rangle$$

that are related also with the functions $F(t)$ and $G(t)$, used in the time domain description, by Fourier transforms:

$$F(t) * A(t) e^{i\omega_o t} = \frac{1}{2\pi} \int_{-\infty}^{\infty} f(\omega)B(\omega)e^{-j\omega t} d\omega$$

$$G(t) * A(t) e^{i\omega_o t} = \frac{1}{2\pi} \int_{-\infty}^{\infty} g(\omega)B(\omega)e^{-j\omega t} d\omega$$

Obviously, the shape of the functions $f(\omega)$ and $g(\omega)$ will also be determined by the polarization characteristics of the fibre line considered. However, the matrix that relates the frequency dependent amplitudes is a unitary matrix, since only includes products of rotations matrices corresponding to projections on principal states for consecutive trunks and the propagation matrices along every trunk. Accordingly, for every optical frequency, that is to say, for strictly monochromatic light, two global principal states of polarization can be found as the eigenvectors of the corresponding transmission matrix. Then, in these global principal states of polarization, $f(\omega)$ and $g(\omega)$ only express the introduced change of phase between the two polarizations. Obviously, the global principal states of polarization, as well as, the corresponding phase factor will be dependent on the optical frequency of the light, except for the case that the system only includes a polarization maintaining trunk in which the principal states of polarization can be considered as wavelength independent.

5. Polarization mode dispersion

Polarization mode dispersion (PMD) accounts for the degradation that a transmitted signal undergoes in the propagation through a single-mode fibre as a consequence of different group velocities for different polarizations. These different group velocities cause a spreading of an impulse introduced into the fibre or a demodulation of a modulated transmitted signal. Therefore, signal degradation can only be considered for polychromatic signals, either due to the finite spectral width of the source or to the temporal evolution of the signal.

The characterisation of PMD can be derived from the output power response, $\Phi(t)$, to an impulse introduced at the input of the fibre with polarization state $|\varepsilon_o\rangle$,

$$\Phi(t) = \frac{|F(t)|^2 + |G(t)|^2}{\int_{-\infty}^{\infty} \left(|F(t)|^2 + |G(t)|^2\right) dt},$$

so that, a suitable PMD parameter can be defined as the temporal width of the impulse response [8,9].

There are several ways to characterise the temporal width of the impulse response, either as a Full-Width-at-Half-Maximum (FWHM) time interval, τ, or, better, as the root mean square (r.m.s.) temporal width, σ, defined as

$$\sigma = \left(\langle t^2 \rangle - \langle t \rangle^2\right)^{\frac{1}{2}} = \int_{-\infty}^{\infty} t^2 \Phi(t) dt - \int_{-\infty}^{\infty} t \Phi(t) dt$$

The relation between τ and σ, depends on the specific shape of $\Phi(t)$, and for the hypothetical case of an impulse response with a Gaussian distribution, $\tau = 2.35\sigma$.

Because the optical fibre system can be regarded as an overall linear system, we may use Fourier transformation to convert the impulse response into an equivalent frequency response $H(v)$, in the form

$$H(v) = \int_{-\infty}^{\infty} \Phi(t)e^{-i2\pi vt}dt$$

In general, $H(v)$, is a complex function that represents the variation of both amplitude and phase with the modulation frequency. For a frequency response function, it is usual to define an optical bandwidth of the fibre as the modulation frequency (Δv) at which $|H(v)|$ reaches a -3dB value.

The relation between Δv, τ and σ depends on the shape of the impulse response, for the Gaussian case, $\Delta v = \dfrac{0.441}{\tau} = \dfrac{0.188}{\sigma}$. However, specific penalties over the transmitted signal will be, in general, dependent on the kind of transmission system. For digital systems, the maximum bit rate, B, is a more convenient parameter than Δv, τ or σ. As demonstrated in ref. [10], for a wide range of impulse response shapes, B should not exceed $0.25/\sigma$. As a practical rule of thumb, the approximate relations

$$\Delta v \cong B \cong \frac{1}{4\sigma} \cong \frac{1}{2\tau},$$

may be used.

Polarization mode dispersion, both in terms of FWHM, τ, or r.m.s. width, σ, or, alternatively, polarization mode bandwidth, B, can be used to estimate the practical penalties of polarization spreading on the transmitted signal.

The analysis carried out represents the PMD of a single-mode fibre for a determined input polarization state $|\varepsilon_o\rangle$, and accordingly, the characteristic PMD parameter of the fibre will be determined as the worst PMD value obtained for any arbitrary input polarization state. This worst possible condition has to be considered since the input polarization state is uncontrolled in the transmission system.

Another question that requires a careful discussion is that concerning length-dependence of PMD values [11]. In a fibre system in which low polarization mode coupling takes place and, consequently, wavelength independent principal states can be defined, the PMD will increase linearly with the length of the fibre. Then, a length-normalized PMD coefficient (PMD_L) can be defined as

$$PMD_L = \frac{PMD}{L}$$

where *PMD* is the total PMD value and *L* is the fibre length. On the contrary, for a strong random coupling regime, as in the case of standard cabled telecommunication fibres, the dependence is given by the square root of the length. Then, the length-normalized PMD coefficient takes the form

$$PMD_L = \frac{PMD}{\sqrt{L}}$$

In any intermediate case, that may correspond to systems including both standard cabled telecommunication fibres and polarization sensitive components, a length dependence with a $\frac{1}{2} \leq \gamma \leq 1$ exponential factor will better describe the PMD evolution

$$PMD_L = \frac{PMD}{L^\gamma}$$

Obviously, in every case, any of the PMD parameter definitions discussed above can be used to account for the total PMD value of the system.

6. Measurements Techniques

Polarization mode dispersion has been established from the impulse response of the fibre; then the direct technique for PMD characterisation will be constituted by the measurement of the temporal response to a very short pulse introduced into the fibre, or by the measurement of a system transfer function from demodulation produced in modulated signals at variable frequencies transmitted through the fibre. However, measurements of actual PMD values for standard telecommunication fibres would require extremely short pulses and/or very high modulation frequencies, hardly attainable with conventional optoelectronics components. Then alternative techniques have been developed for PMD measurements [12].

Two main PMD measurement techniques have been developed which are in close relation with the analysis of polarization evolution that has been carried out in both time domain, with polychromatic light, and frequency domain, with monochromatic light. The required time resolution is achieved by means of wide spectrum interferometry in time domain measurements, while the rate of evolution of output polarization state in the Poincaré sphere is used for delay determination between principal states of polarization, in the frequency domain. In the next sections, the characteristics of measurements procedures in both kind of techniques are discussed.

6.1. PMD MEASUREMENT BY INTERFEROMETRIC TECHNIQUES

In the time domain, the required time resolution is achieved by interferometry of wide spectrum. Then, the required measurement set-up can derive from the basic configuration depicted in figure 4.

The light source is a LED with a linear polarizer to select a definite input polarization state. The detection chain is constituted by a low speed PIN detector, preamplifier and lock-in amplifier. A mirror mounted on a step motor displacer provides the variation of the optical path to obtain the interferometric signal. In order to get the cross-correlation signal between orthogonal polarizations, it is necessary to include some polarization coupling device in the measurement system. This function can be accomplished by means of an output polarization analyser in a bulk Michelson interferometer or by the own polarization evolution in the optical fibres of an all-fibre arrangement [13].

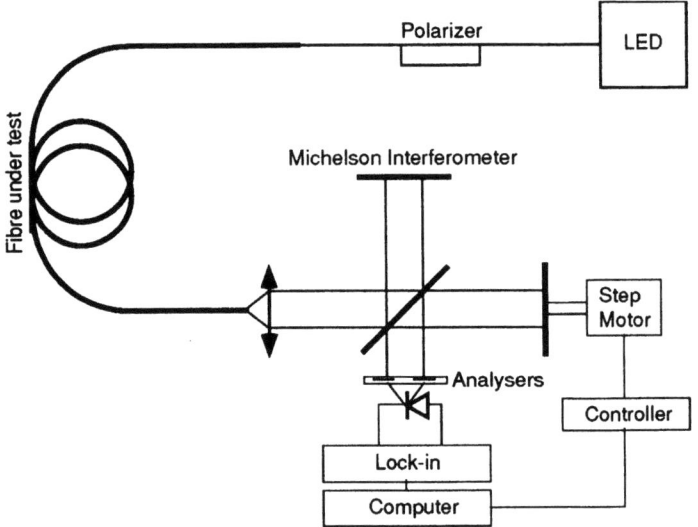

Figure 4. Experimental set-up for interferometric PMD measurements.

PMD values are determined from the temporal width of interferograms obtained for different input polarization states, looking for the worst possible case, that is, the most spreading interferogram. However there are some contributions to the interferometric signals that disturb the obtained results. In first place, interferograms consist of the convolution of the source spectrum with the impulse response of the fibre, and, in second place, an extra auto-correlation term, that should be taken off from the interferograms, appears at delay zero, corresponding to unpolarized light in the output beam. The usual way to avoid these disturbances is to choose a light source with power spectrum much narrower than the fibre impulse response, allowing the elimination of the extra term directly by software in the analysis of the final signal. Notwithstanding, both elimination of disturbing contributions to the interferometric signals and the look for the worst input polarization state can be achieved by means of a complete polarization analysis at the output of the interferometer [14].

The complete polarization analysis is performed by obtaining interferograms corresponding to a complete set of output analysers, composed by linear polarizers at 0°, 90°, 45°, and 135°, and right and left circular polarizers. In this way, the temporal dependent Stokes parameters of the light at the output of the interferometer are determined by

$$S_1 = I_o - I_{90}, \quad S_2 = I_{45} - I_{135}, \quad S_3 = I_{cr} - I_{cl},$$
$$S_0 = \sqrt{S_1^2 + S_2^2 + S_3^2}$$

where I_j is the interferogram corresponding to the j analyser. Then, S_0 represents the response of the fibre for the polarization state selected at the input, without the unpolarized extra term, while the widest S_j provides the interferogram for the worst possible input polarization state.

The PMD of the system can be obtained as the temporal width (σ or τ) of the widest S_j interferogram. Besides, a Polarization Mode Transfer function, in which the contribution of the source spectrum is eliminated, can be obtained as

$$H_j(\Delta v) = 10 \log \frac{FT_{S_j}(\Delta v)}{FT_s(\Delta v)},$$

where $FT_{S_j}(\Delta v)$ is the normalized Fourier transform of the generalised Stokes parameter $S_j(\tau)$ and $FT_s(\Delta v)$ the normalized Fourier transform of the light source spectrum. $H_j(\Delta v)$ represents the relative demodulation of a signal with modulation frequency Δv due to polarization dispersion of the system for the worst possible input polarization state. Then, a Polarization Mode Bandwidth is determined by the modulation frequency at which the transfer function reaches a -3 dB value.

The main disadvantage of interferometric techniques is the lack of information about the actual output polarization state that would result for any particular optical frequency in the source. However, the required set-up is not expensive and usually available in a standard fibre optics laboratory. Another advantage derives from the flexibility of the measurement procedure to obtain PMD values in different ranges, requiring only larger or smaller scanning ranges of the reference optical path in the interferometer, without any special requirements about spectral characteristics of the source. The analysis of the interferometric signals in terms of generalised Stokes parameters avoids all the troubles that can disturb the measurements results, without increasing significantly the experimental effort.

6.2. PMD MEASUREMENT BY POLARIMETRIC TECHNIQUES

This method provides PMD values as an average of time delays between principal states of polarization derived from the measured velocity of output polarization state evolution in the Poincaré sphere for varying optical frequencies [15].

The required polarimetric set-up is based on the arrangement depicted in figure 5.

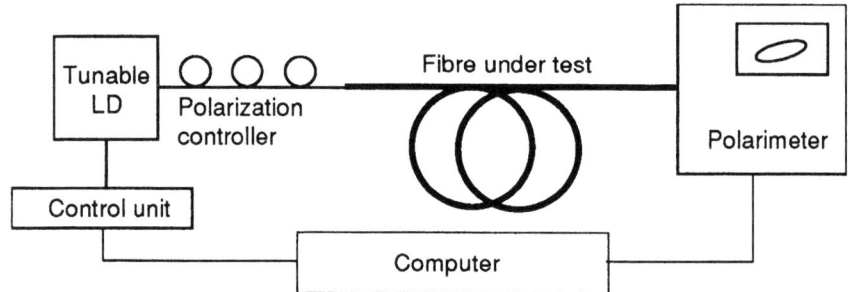

Figure 5. Experimental set-up for polarimetric PMD measurements.

A tunable semiconductor laser may be used as light source. The spectral width of the laser must be small enough to keep stable principal states inside the laser spectrum as well as to obtain completely polarized light at the output of the system. On the other hand, the tunability range of the laser must provide enough statistical information on the time delay distribution with respect to optical frequency. The detection chain consist of a polarimetric system that determines the state of polarization (SOP) of the output light beam. The delay between principal states at different wavelengths are obtained from the variation of the monitored SOP in the Poincaré sphere.

A polarization controller selects the input polarization state, so that the PMD value may be determined as the width of the delay distribution obtained for the worst possible case of input polarization state.

The advantages of this PMD measuring procedure derive from the fact that the actual output SOP and its wavelength dependence are directly determined. The main disadvantages are related to the tough requirements about spectral width and tunability range of the laser source. The time coherence of the laser has to be much higher than the PMD delays in the fibre under test, while the required tunability range must be much higher than the stability interval of principal states. In particular, a spectral width smaller than 0.1 nm and tunability range higher than 10 nm should be required for PMD delays of about 10 ps, that are commonly found in long distances (tens of km) of cabled standard fibres. For lower PMD values, that may be found in uncabled fibres, fibre devices or short cabled fibre segments, both spectral width and tunability range should be increased in about two orders of magnitude.

7. References

1. Born, M. and Wolf, E. (1993) *Principles of Optics. 6º Ed.* Pergamon Press.
2. Azzam, R.M.A. and Bashara, N.M. (1977) *Ellipsometry and Polarized Light*, North-Holland Elsevier, Amsterdam.
3. Tsao, C. (1992) *Optical Fibre Waveguide Anaysis.*, Oxford University Press, Oxford.
4. Poole, C.D., Wagner, R.E. (1986) Phenomenological approach to polarisation dispersion in long single-mode fibres, *Electron. Lett.* **22**, 1029-1030.
5. Andresciani, D., Curti, F., Matera, F. and Daino, B. (1987) Measurement of the group delay difference between the principal states of polarization on a low-birefringence terrestrial fiber cable, *Opt.Lett.*, **12**, 844-846.
6. Gisin, N., Passy, R., Bishoff, J. C. and Perny, B. (1993) Experimental investigations of the statistical properties of polarization mode dispersion in single mode fibers, *IEEE Photon. Technol. Lett.*, **5**, 819-821.
7. Gisin, N. and Pellaux, J.P. (1992) Polarization mode dispersion: time versus frequency domains, *Opt. Communications*, **89**, 316-323.

8. Blasco, P., Pelayo, J. and Villuendas, F. (1994) Uncertainty in PMD measurements by interferometric techniques. Analysis by means a transfer function approach. *EFOC'94 Proc.*, Heidelberg.
9. Gisin, N. (in press), Definition of Polarization Mode Dispersion.
10. Gowar, J. (1993) *Optical Communication Systems, 2º Ed.*, Prentice Hall.
11. Poole, C.D. (1988), Statistical treatment of polarization dispersion in single-mode fibre, *Opt.Lett.*, **13**, 687-689.
12. *2nd Optical Fibre Measurement Conference* (1993), Session 8: Polarisation properties of fibres, 163-193. Torino. And references therein
13. Gisin, N., Von der Weid, J.P. and Pellaux, J.P. (1991), Polarization mode dispersion of short and long single-mode fibers *IEEE J. Lightwave Technol.*, **9**, 821-827.
14. Pelayo, J., López, F. and Villuendas, F. (1989) Analysis of polarization characteristics of low and high-birefringent fibres by "interferopolarimetry". *ECOC'89 Proc.*, **1**, 510-513, Gothenburg.
15. Bergano, N.S., Poole, C.D. and Wagner, R.E. (1987), Investigation of polarization dispersion in long lengths of single-mode fiber using multilongitudinal mode lasers, *IEEE J. Lightwave Technol.*, **5**, 1618-1622.

NONLINEAR OPTICAL FIBRES

RAMAN KASHYAP
BT Laboratories
Martlesham Heath
Ipswich IP5 7RE
United Kingdom

1. Introduction

The dielectric optical waveguide has been in existence for nearly 40 years[1]. Early research was focused at fibre bundles for image transmission, but even after pioneering papers by Snitzer on the weakly guiding phenomenon[2] it was several years later that glass optical fibres were recognised for their transmission potential[3]. Since then, a revolution has occurred: silica based optical fibre waveguides with transmission loss at a wavelength of 1550nm of less than 0.2dB/km have been routinely manufactured and long distance optical fibre transmission at gigabit rate has had a profound impact on communications. Along with this major advance, it has become apparent that low loss optical fibres are an ideal candidate for the investigation of a host of linear and non-linear phenomena at modest optical powers. These include Rayleigh, Mie and Raman scattering[4-6], optical Kerr effect[7] all optical fibre amplification and fibre lasers[8] using rare-earth dopants such as erbium and neodymium in the silica host, all optical switching[9], soliton transmission[10], as well as the discovery of new phenomena such as photosensitivity[11] and optical damage at low optical powers[12]. Nonlinear optics[13-14] has been known since the invention of the laser, however, low loss optical fibres have made it possible to observe almost all non-linear effects; surprisingly, even material symmetry forbidden phenomena such as efficient second-harmonic generation, associated with crystalline media[11].

This chapter will briefly introduce the subject of non-linear optics from a theoretical standpoint, guided modes of the non-linear optical fibre, some important non-linear effects in optical fibres, measurement techniques, and conclude with examples of non-linear fibre experiments such as harmonic generation, optical damage and polarisation coupling. It is assumed that much of waveguiding optics and mode propagation will be familiar to the reader and that this set of lectures will introduce the subject of nonlinear optics from a phenomenological point of view, followed by a description of special nonlinear optical effects in optical fibres. Soliton propagation is specifically excluded, since it will be covered elsewhere.

2. Theory of Non-linear optics

2.1 POLARISATION RESPONSE OF TRANSPARENT DIELECTRIC MEDIA

The material response of a dielectric in the presence of an applied electric field (optical and/or static) can be described as

$$D = \varepsilon_0 E + P \qquad (1)$$

where, the polarisation response of the material is

$$P = \varepsilon_0 \chi E \qquad (2)$$

The induced polarisation, P is linearly proportional to the applied field if the applied field is much smaller than the atomic Coulombic field (3×10^{10} V m^{-1}). With a higher field intensity, such as from a focused laser beam, the polarisation response is no longer linear and must be modified to include higher order terms in a series expansion as

$$P_i(\omega_0) = \varepsilon_0 [\chi_{ij}^{(1)}(-\omega_0;\omega_1) \cdot E_j^{\omega_1} + \chi_{ijk}^{(2)}(-\omega_0;\omega_1,\omega_2) : E_j^{\omega_1} E_k^{\omega_2}$$
$$+ \chi_{ijkl}^{(3)}(-\omega_0;\omega_1,\omega_2,\omega_3) \vdots E_j^{\omega_1} E_k^{\omega_2} E_l^{\omega_3} + \ldots] \qquad (3)$$

The susceptibility tensors $\chi^{(n)}$ are of rank (n+1) and the ., : and $\vdots$ indicate tensorial dot products. The polarisation directions of the applied fields are indicated by the subscripts, j, k and l, while the polarisation of the generated field is i. In Equation (3), up to three interacting fields have been shown to express the third order non-linear interaction via $\chi^{(3)}$. According to convention, the three angular frequencies ω_1, ω_2, and ω_3 mix to generate ω_0.

$\chi_{ij}^{(1)}(-\omega_0;\omega_1)$ is a second rank tensor and describes the linear susceptibility, related to the material permittivity ε_{ij} by

$$\varepsilon_{ij} = \left[1 + \chi_{ij}^{(1)}(-\omega_0;\omega_1) \right] \qquad (4)$$

$\chi_{ijk}^{(2)}(-\omega_0;\omega_1,\omega_2)$ is the second order susceptibility and is present in materials that lack inversion symmetry. Usually crystalline materials and poled glasses or polymers possess a non-zero $\chi^{(2)}$. The linear electo-optic effect (Pockels), sum and difference frequency generation, optical rectification and in the degenerate frequency case, second-harmonic generation (SHG), are the result of such a non linearity. This tensor has 27 elements, some of which are identical depending on the crystal symmetry and whether or not Kleinman symmetry applies[16]. Kleinman symmetry proposes that in the absence of strong resonances, the dispersion of the non-linear susceptibility is insignificant and remains unchanged with the interchange of the polarisation indices, ijk. Therefore,

$$\chi^{(2)}_{ijk}(-\omega_0;\omega_1,\omega_2) = \chi^{(2)}_{ikj}(-\omega_0;\omega_1,\omega_2) = \chi^{(2)}_{jik}(-\omega_0;\omega_1,\omega_2)$$
$$= \chi^{(2)}_{jki}(-\omega_0;\omega_1,\omega_2) = \chi^{(2)}_{kij}(-\omega_0;\omega_1,\omega_2) = \chi^{(2)}_{kji}(-\omega_0;\omega_1,\omega_2) \quad (5)$$

Kleinman symmetry reduces the number of independent tensor elements significantly. It should also be noted that the second-order non linearity can only exist in materials without inversion symmetry since the induced polarisation must follow the sign of the applied electric field (excluding a constant phase difference). This may be seen in the following inequality

$$-P_i^{(2)}(\omega_0) \neq \varepsilon_0 \chi^{(2)}_{ijk}(-\omega_0;\omega_1,\omega_2):(-E_j^{\omega_1})(-E_k^{\omega_2}) \quad (6)$$

A change in the sign of the applied fields will not change the sign of the induced polarisation in a material with inversion symmetry, since the polarisation wave follows the driving wave. Therefore,

$$\chi^{(2)}_{ijk}(-\omega_0;\omega_1,\omega_2) = 0 \quad (7)$$

For the third-order non-linear polarisability, there is no such restriction. All materials possess a third order non-linear susceptibility and therefore is non-zero in non-symmetric media. $\chi^{(3)}_{ijkl}(-\omega_0;\omega_1,\omega_2,\omega_3)$ is a fourth rank tensor and in general has 81 independent non-zero elements, but in isotropic materials, these reduce to 21, of which only 3 are independent. This reduction in the number of tensor elements is fortuitous, otherwise it would be nearly impossible to deal with the large number of polarisation contributions in any third order non-linear interaction. $P_i^{(3)}(\omega_0)$ from Equation (3) is defined as

$$P_i^{(3)}(\omega_0) = \varepsilon_0 \chi^{(3)}_{ijkl}(-\omega_0;\omega_1,\omega_2,\omega_3):E_j^{\omega_1} E_k^{\omega_2} E_l^{\omega_3} \quad (8)$$

and is responsible for a large number of important parametric phenomena (ones in which the final state of the molecule/atom interacting with the optical fields, remains unchanged). These include many of the observable effects in optical fibres and are listed in Table 1, while Table 2 lists important non-parametric effects, such as stimulated Raman and Brillouin scattering and two photon absorption.

TABLE 1
Parametric Third Order Non-linear Effects

Non-linear Effect	Description
Third Harmonic Generation	$\chi^{(3)}_{ijkl}(-3\omega_0;\omega_0,\omega_0,\omega_0)$
Sum and Difference Frequency Generation	$\chi^{(3)}_{ijkl}(-\omega_0;\pm\omega_1,\pm\omega_2,\pm\omega_3)$
Optical Kerr Effect	$\chi^{(3)}_{ijkl}(-\omega_0;\omega_1,-\omega_1,\omega_0)$
Self-Focusing and Self-Phase Modulation	$\chi^{(3)}_{ijkl}(-\omega_0;\omega_0,\omega_0,\omega_0)$
DC Kerr Effect	$\chi^{(3)}_{ijkl}(-\omega_0;\omega_0,0,0)$
DC Induced Second-Harmonic Generation	$\chi^{(3)}_{ijkl}(-2\omega_0;\omega_0,\omega_0,0)$

TABLE 2
Non-Parametric Third Order Non-linear effects

Stimulated Raman Scattering	$\chi^{(3)}(-\omega_s;\omega_p,-\omega_p,\omega_s)$
Stimulated Brillouin Scattering	$\chi^{(3)}(-\omega_0;\omega_p,-\omega_p,\omega_0)$
Two Photon Absorption	$\chi^{(3)}(-\omega_0;\omega_0,-\omega_0,\omega_0)$

2.2 NON-LINEAR POLARISATION

There are several accepted methods of dealing with the description of induced polarisation and the applied field. Unfortunately, these have led to confusion since factors of 1/2 used in the complex notation for real fields can cause differences in the magnitude of the non-linear susceptibility[17-19]. In the following description, complex notation has been implied or used explicitly for consistency, but special attention will be paid to handling these factors.

The applied electric field in the frequency domain is the Fourier integral of the applied electric field impulse

$$E(r,t) = \int_{-\infty}^{+\infty} e^{-i\omega t} E(r,\omega) d\omega \qquad (9)$$

and

$$E^*(r,\omega) = E(r,-\omega) \quad (10)$$

$$P(r,t) = \int_{-\infty}^{+\infty} e^{-i\omega t} P(r,\omega) d\omega \quad (11)$$

$$P^*(r,\omega) = P(r,-\omega) \quad (13)$$

The non-linear susceptibility is defined as a summation over all frequency and permutation of the polarisation directions as

$$P_i^{(n)}(r,\omega) = \varepsilon_o \sum_{j_1}^{j_n} \int_{-\infty}^{+\infty} \chi_{ij_1\cdots j_n}^{(n)}(r,-\omega_0;\omega_1,\ldots,\omega_n) E_{j_1}(r,\omega_1)\ldots E_{j_n}(r,\omega_n) \times \delta(\omega-\omega_0) \quad (14)$$

To ensure that the polarisation in space and time is real, the following condition must apply

$$\chi_{ij_1\cdots j_n}^{(n)*}(-\omega_0;\omega_1,\ldots,\omega_n) = \chi_{ij_1\cdots j_n}^{(n)}(-\omega_0^*;-\omega_1^*,\ldots,-\omega_n^*) \quad (15)$$

In Equation (15) the subscripts $i, j_1, \ldots, j_n$ refer to the laboratory frame Cartesian co-ordinates, x, y or z, indicating the polarisation direction. The Dirac delta function $\delta(\omega-\omega_0) = \infty$ for a zero argument and is zero otherwise, ensuring energy conservation within the integral. The Fourier components of the polarisation in Equation (9,11) is a Taylor series expansion in the fields at all frequencies, ω_0 generated by the summation of the individual contribution of each term of the susceptibility,

$$P(r,\omega) = \sum_{n=1}^{\infty} P^{(n)}(r,\omega) \quad (16)$$

There is a permutation symmetry in the non-linear susceptibility and is known as *intrinsic* permutation symmetry. This may be explained by remembering that since physical laws are invariant with respect to a translation in time, the non-linear susceptibility is invariant under all permutations of the pairs $j_1\omega_1\ldots j_n\omega_n$ in the applied field. Commuting the position of all the frequency terms of the applied field does not affect the susceptibility provided that the polarisation direction subscripts are also similarly commuted. This applies to all materials and is a general theorem. Additionally, in a loss-less medium (such as an optical fibre) in which no frequencies or their combinations are resonant with material transitions, an *overall* permutation symmetry prevails. Under this symmetry, the susceptibility is invariant with all permutations of the pairs $-\omega_0 i; \omega_1 j_1, \ldots, \omega_n j_n$ of the applied fields and the generated polarisation fields (pair: frequency and polarisation). Note that Kleinman symmetry is restrictive and applies only to the commutation of the *polarisation* indices in the susceptibility away from resonances.

It is useful to apply the symmetry properties using the applied electric field, $E(r,t)$ in Equation (9) for the important case when the applied field is a summation of monochromatic fields defined in complex notation,

$$E(r,t) = 1/2 \sum_{\omega \geq 0} \left[E^\omega(r) e^{-i\omega t} + E^{-\omega}(r) e^{i\omega t} \right] \qquad (17)$$

and since the field $E(r,t)$ is always real,

$$E^{-\omega}(r) = E^{*\omega}(r) \qquad (18)$$

The induced polarisation can be described in a similar manner

$$P(r,t) = 1/2 \sum_{\omega \geq 0} \left[P^\omega(r) e^{-i\omega t} + P^{-\omega}(r) e^{i\omega t} \right] \qquad (19)$$

with

$$P^{-\omega}(r) = P^{*\omega}(r) \qquad (20)$$

Using Equation (17) in Equation (9) and Equation (11) in Equation (14) we get

$$P_i^{(n)}(\omega_0) = \varepsilon_o \sum_{j_1}^{j_n'} \int_{-\infty}^{+\infty} K(-\omega_0; \omega_1, \ldots, \omega_n) \chi_{ij_1 \ldots j_n}^{(n)}(-\omega_0; \omega_1, \ldots, \omega_n) \\ \times E_{j_1}(\omega_1) \ldots E_{j_n}(\omega_n) \qquad (21)$$

The K-factors are discussed below. The prime in Equation (21) indicates summation over all the distinct sets of frequencies $\omega_1 \ldots \omega_n$ capable of generating ω_0 since

$$\omega_0 = \omega_1 + \ldots + \omega_n \qquad (22)$$

In the most general case when all the interacting frequencies are different, the choice of the order in which the frequency combinations may be grouped remains. Invoking permutation symmetry, it can be seen that the terms of each *distinguishable* permutation are identical. This simplifies the description of the interaction, since only one term need be written. The sum of the terms for the distinguishable permutations need to be included in the K-factor. Additionally, Equations (17-19) indicate factors of 1/2 for both the applied electric field vectors and the induced polarisation. Thus for the case when $\omega_1, \ldots, \omega_n$ are all different and for non-zero $\omega_0, \omega_1, \ldots, \omega_n$ the K-factor is

$$K = 2^{1-n} \frac{n!}{p!} \qquad (23)$$

where p is the frequency degeneracy factor ($p = 2$, if $\omega_1 = \omega_2$, and $p = 3$ if $\omega_1 = \omega_2 = \omega_3$ etc.). If there are m zero frequencies and p identical frequencies out of n, Equation (23) may be modified to be completely general:

$$K = 2^{m+l-n} \frac{n!}{p!} \qquad (24)$$

where $l = 0$ if $\omega_0 = 0$ and $l = 1$ if $\omega_0 \neq 0$. In other words, the K-factor comprises of three contributions: the first is the result of the degeneracy in frequencies of the fields; the second is the result of the permutations of the frequency indices, while the third is as a result of the 1/2 in the complex notation used in describing the driving electric and polarisation fields.

Symmetry properties of the third order nonlinear susceptibility of an isotropic medium leads to the following relationship for all pairs of ij from the set $ijkl$,

$$\chi^{(3)}_{iijj} = \chi^{(3)}_{ijij} = \chi^{(3)}_{ijji} = \frac{1}{3}\chi^{(3)}_{iiii} = \frac{1}{3}\chi^{(3)}_{jjjj} = \frac{1}{3}\chi^{(3)}_{kkkk} \qquad (25)$$

while for all materials

$$\left[\chi^{(3)}_{iijj} + \chi^{(3)}_{ijij}\right] + \chi^{(3)}_{ijji} = \chi^{(3)}_{iiii} = \chi^{(3)}_{jjjj} = \chi^{(3)}_{kkkk} \qquad (26)$$

which gives the result for isotropic media[20]

$$\left[\chi^{(3)}_{iijj} + \chi^{(3)}_{ijij}\right] : \chi^{(3)}_{ijji} : \chi^{(3)}_{jjjj} = 2:1:3 \qquad (27)$$

These relationships shown in Equations (25-27) simplify the analysis of third-order processes in isotropic materials such as optical fibres.

3.1 NONLINEAR WAVE PROPAGATION

Nonlinear interactions in optical fibres is based on optical fibre modes. In this section wave-propagation will be discussed relating the propagating modes to the non-linear interaction and the generation of the polarisation waves.

The starting point for wave-guide propagation are Maxwell's equations

$$\nabla \times \boldsymbol{E} = -\frac{\partial \boldsymbol{B}}{\partial t} \tag{28}$$

$$\nabla \times \boldsymbol{B} = \frac{\partial \boldsymbol{D}}{\partial t} + \boldsymbol{J} \tag{29}$$

where D is the electric displacement vector and related to the polarisation by

$$\boldsymbol{D} = \varepsilon_0 \boldsymbol{E} + \boldsymbol{P} \tag{30}$$

Optical fibre can be assumed to be a near perfect dielectric so that J is zero.

Combining Equations(28-29) by taking the curl of Equation (28) and neglecting free charges we arrive at the wave-equation in homogeneous media

$$\nabla^2 \cdot \boldsymbol{E} = -\mu_0 \frac{\partial^2}{\partial t^2}[\varepsilon_0 \boldsymbol{E} + \boldsymbol{P}] \tag{31}$$

The induced polarisation may be described as a combination of the linear and nonlinear part

$$\boldsymbol{P} = \varepsilon_0 \chi_{ij}^{(1)} \cdot \boldsymbol{E} + \boldsymbol{P}^{NL} \tag{32}$$

so that

$$\nabla^2 \cdot \boldsymbol{E} + \mu_0 \varepsilon_0 \varepsilon_r \frac{\partial^2}{\partial t^2} \boldsymbol{E} = -\mu_0 \frac{\partial^2}{\partial t^2} \boldsymbol{P}^{NL} \tag{33}$$

where all the nonlinear polarisation terms are lumped in $\boldsymbol{P}^{NL}$ and the relative permittivity ε_r is $1 + \chi^{(1)}$. Equation (33) describes wave propagation in a charge free non-linear medium.

Equation (33) may also be transformed into the frequency domain by using the Fourier transforms described in Equations (9-13). This is useful for treating frequency mixing phenomena such as second-harmonic generation and four-wave mixing. In this case it should be noted that the time derivative in Equation (33) may be conveniently replaced by $-i\omega$, so that the wave-equation becomes

$$\nabla^2 \cdot \mathbf{E} + \mu_0 \omega^2 \varepsilon_0 \varepsilon_r \mathbf{E} = -\mu_0 \omega^2 \mathbf{P}^{NL} \qquad (34)$$

In optical waveguides, light propagates in modes and the field distribution must be calculated by first assuming that the nonlinear polarisation is zero.

The fundamental mode field is described as[21]

$$E_z(r,\omega) = \frac{1}{2}\left[A_i(\omega)f(\frac{r}{a})e^{-i\beta z} + cc\right] \qquad (35)$$

where cc is the complex conjugate. Here $A_i(\omega)$ is the normalisation constant and $f(\frac{r}{a})$ is the radial field distribution. β, is the propagation constant and Bessel functions satisfy the mode field distribution in terms of the well known normalised parameters u and w, as

$$f\left(\frac{r}{a}\right) = J_0\left(\frac{ur}{a}\right) \quad \text{for } r < a \qquad (36)$$

and

$$f\left(\frac{r}{a}\right) = K_0\left(\frac{wr}{a}\right) \quad \text{for } r > a \qquad (37)$$

Generally, the mode shape may be approximated by a Gaussian distribution. An effective area for the mode may be defined from the mode-field diameter, w_0 as

$$A_{eff} = \pi w_0^2 \qquad (38)$$

This parameter is of great importance in calculating the intensity of the mode in the fibre and has to be estimated accurately.

The nonlinear part of Equation (34) can then be included as a perturbation by assuming that the transverse field distribution is unaffected by the non-linearity. The solutions to the optical fibre modes may be incorporated into the nonlinear wave Equation (33). Variable separation can be used in Equations(33-34) to solve the time and space dependent evolution of the input pulse amplitude. To do this it is necessary to use the slowly varying envelope approximation (SVEA) which follows directly from the assumption that the amplitude of the pulse does not vary within a period of a few optical cycles or with propagation distance of a few wavelengths. This condition may be conveniently described in the frequency domain as

$$\left|\frac{\partial^2}{\partial z^2}E(\omega)\right| << \left|k\frac{\partial}{\partial z}E(\omega)\right| \tag{39}$$

while in the time domain the equivalent form is

$$\left|\frac{\partial^2}{\partial t^2}E(\omega)\right| << \left|\omega\frac{\partial}{\partial t}E(\omega)\right| \tag{40}$$

An identical definition as Equations(19) for the nonlinear polarisation can be used in the wave-equation as

$$P^{NL}(r,t) = 1/2\sum_{\omega\geq 0}\hat{u}_i\left[P^{NL}(r)e^{-i\omega t} + P^{NL-}(r)e^{i\omega t}\right] \tag{41}$$

where $\hat{u}_i$ is a unit vector in the i direction where P^{NL} includes a time dependent envelope function.

4. Nonlinear refractive index

4.1 THE OPTICAL KERR EFFECT

The optical Kerr effect is the optical equivalent of the DC Kerr effect[22] and its manifestation as polarisation rotation in liquids was first reported by Maker et al [23]. They observed strong polarisation rotation of an input optical beam focused into a liquid as a function of intensity. This was explained in terms of an induced birefringence as a result of induced molecular orientation of polar molecules. Since, the subject has attained noteriety by the vast number of publications in related topics such as Self-Phase-Modulation (SPM), Cross-Phase-Modulation (XPM), Optical Kerr gate, polarisation instabilities and self-switching, all of which have been reported in optical fibres[24-32]. In this section, a brief description of some of these phenomena will be presented to acquire a physical insight. It is not possible to do justice to this area in a short treatise, however, effects will be discussed in view of techniques used for the observation and also in the measurement of the fundamental nonlinear coefficient. Finally, some recent results on the influence of the dopant concentration on the nonlinearity of the fibre will also be presented.

The nonlinear susceptibility is complex and in general has a real and imaginary part. The real part, as will be seen later gives rise to parametric process (loss-less) while the imaginary part is responsible for non-parametric processess such as stimulated Raman scattering. It should be remembered that susceptibility represents the dielectric response of a material to an applied field via the permittivity [See Equation(4)] (and hence refractive index). Since the real part of the refractive index gives rise to the phase velocity and the imaginary part is responsible for the transmission loss, it becomes immediately apparent that the real part of the *non-linear* susceptibility gives rise to loss-less phenomena, while the imaginary part indicates lossy interactions.

For many of the problems associated with optical fibres, nonlinear refractive index effects are dominant. This field dependent nonlinearity may be described using the real part of the susceptibility as

$$P^{NL}(r,t) = \frac{3}{4}\varepsilon_0 \mathrm{Re}\,\chi^{(3)}_{iiii}|E_i(r,t)|^2 E_i(r,t) \qquad (42)$$

where the nonlinear index takes the form

$$n = n_0 + n_2 |E|^2 \qquad (43)$$

so that the nonlinear permittivity

$$\varepsilon^{NL} = \frac{3}{4}\mathrm{Re}\,\chi^{(3)}_{iiii}(-\omega_0;\omega_0,\omega_0,\omega_0)|E_i(r,t)|^2 \qquad (44)$$

since the refractive index, n is related to the permittivity as $n^2 = \varepsilon_r$, any perturbation may be calculated as

$$\varepsilon_r + \varepsilon^{NL} = (n + \Delta n)^2 = n^2 + 2n\Delta n + \Delta n^2 \qquad (45)$$

so that

$$\varepsilon^{NL} = 2n\Delta n \qquad (46)$$

Therefore, comparing Equations (44) and (45)

$$n_2 = \frac{3}{8n}\mathrm{Re}\,\chi^{(3)}_{iiii}(-\omega_0;\omega_0,-\omega_0,\omega_0) \qquad m^2\,V^{-2} \qquad (47)$$

Another relationship which is useful for many experiments is the *intensity* dependence of the refractive index. Remembering that the intensity of an optical field is

$$I_\omega = \frac{1}{2}\varepsilon_0 c n_0 |E|^2 \quad \text{W m}^{-2} \tag{48}$$

it follows that the intensity dependent refractive index is

$$n_2^I = \frac{2n_2}{\varepsilon_0 c n_0} \quad \text{m}^2 \text{ W}^{-1} \tag{49}$$

This may be related to the nonlinear susceptibility as

$$n_2^I = \frac{3\operatorname{Re}\chi_{iiii}^{(3)}(-\omega_0;\omega_0,-\omega_0,\omega_0)}{4\varepsilon_0 c n_0^2} \quad \text{m}^2 \text{ W}^{-1} \tag{50}$$

The intensity of the optical field in the fibre is calculated from the effective area of the mode, A_{eff}

$$I_\omega = \frac{P}{A_{\text{eff}}} \quad \text{W m}^{-2} \tag{51}$$

4.2 SELF-PHASE-MODULATION

When an intense optical pulse propagates in a transparent dielectric such as glass fibre, a polarisation wave is generated which can substantially disturb the electronic cloud distribution within the medium. As a result, the axis of oscillation of the polarisation can cause the local permittivity to be intensity, and therefore time dependent. This may have a profound effect on the propagation of the pulse within the fibre, since the pulse can impress a time dependent phase change directly on itself leading to Self-Phase-Modulation. The electric field intensity dependent refractive index change is described as

$$n = n_0 + n_2 |E|^2 \tag{52}$$

where n is the perturbed index of the fibre n_0 is the unperturbed index and n_2 is the nonlinear field dependent refractive index.

When E has a spatial and temporal variation, it gives rise to spatial and temporal refractive index changes. For a radially symmetric field distribution (such as a Gaussian distribution) the change in the radial index distribution can cause catastrophic effects. The effect of such a nonlinear refraction depends on the sign of the nonlinearity, n_2. For a negative sign of the nonlinearity, such as one with a thermal origin, the field may be diffracted, the material having a de-focussing effect. For positive nonlinearities, such as those with a non-resonant electronic origin, positive lensing occurs causing severe damage, since the radial field may collapse onto itself in a short distance, a process known as "self-focussing". These effects usually require extremely large intensities in bulk glass. Fortunately, the effect is not observed in single-mode waveguides, since guidance balances diffraction and increases in the refractive index of the guiding region merely stabilises the mode-diameter to a diffraction limited value.

A nonlinear thermally driven phenomena can also occur in optical fibres causing destuction of kilometer lengths of optical fibres at relatively low optical powers *even in single-mode fibres* (see Section 6). This phenomenon, called "Self-Propelled Self-Focussing" or the "Fibre Fuse" is quite new and highly unusual, with the observation being confined to optical waveguides. This phenomenon will be discussed in Section 7.

A Gaussian pulse propagating in a nonlinear medium will induce a time dependent refractive index change as

$$\Delta n = n_2 \left| E e^{-\frac{t^2}{2t_0^2}} \right|^2 \quad (53)$$

This refractive index change induces a time varying phase-shift across the pulse as it propagates along the fibre of length L, $\Delta\phi = \Delta n L$

The change in the instantaneous angular frequency, ω is therefore,

$$\partial\omega = \frac{\partial\Delta\phi}{\partial t} = -\frac{2t}{t_0^2} n_2 L \left| E e^{-\frac{t^2}{2t_0^2}} \right|^2 \quad (54)$$

Figure 1 shows the Gaussian pulse as a function of time and the associated frequency chirp across it for an arbitary length L. The effect of the frequency chirp may be seen in Figure 2, in which it is clear that the rising edge of the pulse sees a "red" shift while the trailing edge experiences a "blue" shift. If this pulse propagates in a fibre with a negative group-velocity dispersion, i. e in which the blue wavelengths travel faster than the red, then under certain conditions, a balancing of the dispersion occurs due to the nonlinear chirp, giving rise to the beautiful effect of solitary waves or solitons.

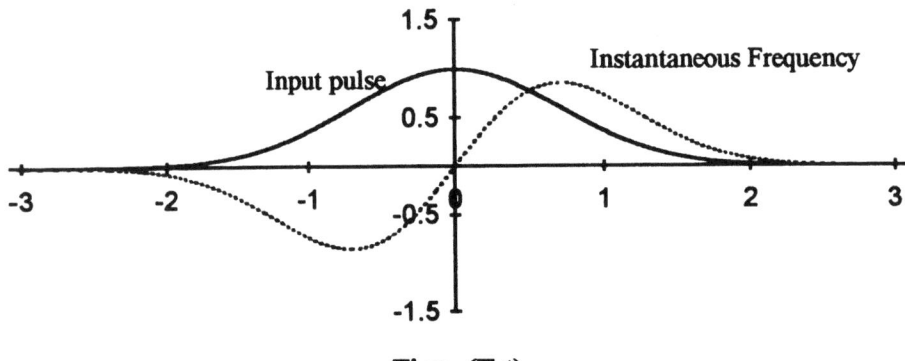

Figure 1. The optical pulse has a time dependent frequency change induced by the the intensity dependent refractive index.

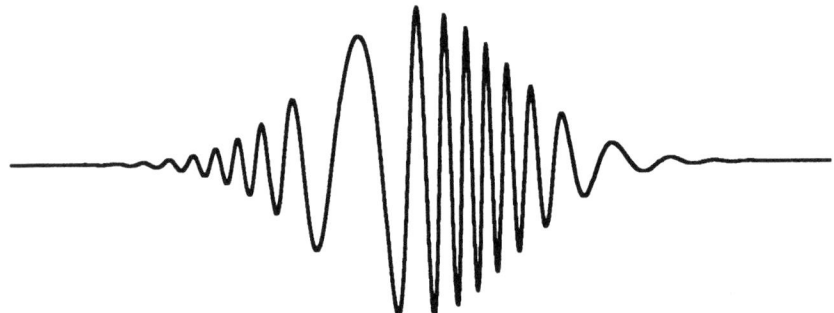

Figure 2. The frequency chirp induced in the pulse due to self-phase modulation.

The maximum length over which the nonlinear interaction accumulates is limited by the linear attenuation, α, and is given by

$$L_{\mathit{eff}} = \frac{1}{\alpha}\left[1 - e^{-\alpha L}\right] \tag{55}$$

A nonlinear coefficient can be defined by using the definition of intensity [Equation(51)] and the nonlinear index [Equation (49)] so that[33]

$$\gamma = \frac{n_2 \omega_0}{cA_{eff}} \qquad (56)$$

This definition is a type of normalisation based on the fact that the only variables in the nonlinear propagation are the phase and the input power, since the transverse field perturbation has been ignored. It has the units of m^{-1} when multiplied by power. A useful parameter is therefore the nonlinear length which is a measure of the significance of the nonlinear effect over a given length when compared to

$$L_{NL} = \frac{1}{\gamma P} \qquad (57)$$

The maximum phase-shift experienced by a pulse occurs at the peak as seen in Figure 2 and may be calculated from

$$\phi_{max} = \frac{L_{eff}}{L_{NL}} = \gamma L_{eff} P \qquad (58)$$

It is immediately evident from Equation (58) that when the nonlinear length is equal to the effective length of the fibre, the maximum phase change is unity. For typical fibres, the effective length can be several kilometers at 1.55microns, but a useful figure to remember is the phase change of approximately π radians for a propagation length of one kilometer with one watt of launched power.

Self phase modulation in optical fibres is thus easy to observe with modest optical powers and has been used for a variety of applications such as pulse-compression to generate ultra-short pulses, switching and pulse broadening.

The nonlinear propagation equation can be derived from the field definition of Equation (35) and solutions to the linear propagation Equations (36-37) in the nonlinear wave-Equation (34). Remembering that there is a spectral distribution in an optical pulse, the propagation constant can be expanded as a Taylor series

$$\beta = \beta_0 + (\omega - \omega_0)\beta' + \frac{1}{2}(\omega - \omega_0)\beta'' + \frac{1}{6}(\omega - \omega_0)\beta'''$$
$$(59)$$

where the primes indicate diffrentiation with respect to ω. Fourier transforming the field to arrive at the temporal evolution equation for nonlinear propagation, it should be remembered that in the quasi-monochromatic description, the operator $i(\partial/\partial t)$ replaces $\omega - \omega_0$. This is valid for most nonlinear processes down to around 100fs.

For the purposes of the discussion in this section, the magnitude of the nonlinearity will be restricted so that radial changes to the field may be safely ignored. The propagation equation may be described as

$$\frac{\partial A_i}{\partial z} + \frac{1}{v_g}\frac{\partial A_i}{\partial t} - \frac{i}{2v_g^2}\frac{\partial v_g}{\partial \omega}\frac{\partial^2 A_i}{\partial t^2} + \frac{\alpha}{2} = i\gamma |A_i|^2 A_i + \frac{\beta'''}{6}\frac{\partial^3 A_i}{\partial t^3}$$
$$-a_1 \frac{\partial(|A_i|^2 A_i)}{\partial t} - a_2 A_i \frac{\partial |A_i|^2}{\partial t}$$

(60)

where A_i has been defined in Equation (35), the second term on the LHS is due to linear dispersion, the third term represents group velocity dispersion and the fourth term is the linear loss. On the right hand side, the first term is due to the nonlinearity, while the last three terms are responsible for higher order effects which become significant with ultra-fast pulse propagation. β''' is the third order dispersion coefficient. The third term on the RHS arises as a result of a violation of the slowly varying approximation and is due to the derivative of the nonlinear polarisation and is the cause of such effects as self-stepeening of the pulse, while a_2 has contributions from relaxation effects due to a delayed response of the nonlinearity. This includes the Raman gain (the imaginary part of $\chi^{(3)}$), and the electronic ultra-fast response (the real part of $\chi^{(3)}$), and results in self-frequency shift.

a_1 is defined as

$$a_1 = \frac{2\gamma}{\omega_0}$$ (61)

while

$$a_2 = i\gamma \tau_r$$ (62)

Both a_1 and a_2 have units of time per watt per metre.

Usually it is sufficient to ignore the last three terms in Equation (60) for many observed nonlinear phenomena in optical fibres. A further two transformations, one to the stationary frame and the other, a normalisation of the input intensity are useful by making changes in variables as

$$\tau = \frac{t - \beta' z}{T_0}$$ (63)

and

$$A(z, \tau) = \sqrt{P} \, e^{-\alpha z/2} \, U(z, \tau) \qquad (64)$$

where $U(z,)$ is the normalised pulse amplitude. Substituting these into Equation (60) (ignoring the last three terms) gives the propagation equation generally used in analysing nonlinear effects

$$\frac{\partial U}{\partial z} + i \frac{sgn(\beta'')}{2L_D} \frac{\partial^2 U}{\partial \tau^2} + i \frac{e^{-\alpha z}}{L_{NL}} |U|^2 U = 0 \qquad (65)$$

The dispersion length L_D is a ratio of the square of the normalised pulse-width and the magnitude of the group velocity dispersion, and is a measure of the effect of dispersion on pulse propagation,

$$L_D = \frac{T_0^2}{|\beta''|} \qquad (66)$$

The sign of the dispersion determines the effect the changing spectral content of the pulse has on the output pulse width in Equation (65). Both L_D and L_{NL} are scaling parameters determining the relative importance of dispersive and nonlinear regimes (listed in Table 3).

Table 3 Summary of the regimes of pulse propagation in optical fibres.

Fibre Length	Effect	Typical Values at 1550nm:		
		T_0 (ps)	L (km)	P (W)
$L \ll L_D, L_{NL}$	Loss Limited Non-Dispersive Transmission,	100	50	1e-4
$L \geq L_D, L \ll L_{NL}$	GVD Linear Dispersion	1	0.05	1
$L \ll L_D, L \geq L_{NL}$	SPM, Spectral Broadening, Pulse Shaping	>100	0.1	1
$L > L_D, L > L_{NL}$	$-ve\ \beta''$, Solitons	10	0.1	10
$L > L_D, L > L_{NL}$	$+ve\ \beta''$, Pulse Compression and Dark Solitons	10	0.1	10

The interplay between linear dispersion and the nonlinearity may be represented graphically in Figure 3. This Figure shows the nonlinear length as a function of peak input power and also the linear dispersion length as a function of the input pulse width. The crossing point at which the nonlinear length is equal to the dispersion length is at ten kilometers for an input pulses width of 10 ps and an input power of 10mW, with B = 10ps2 km-1.

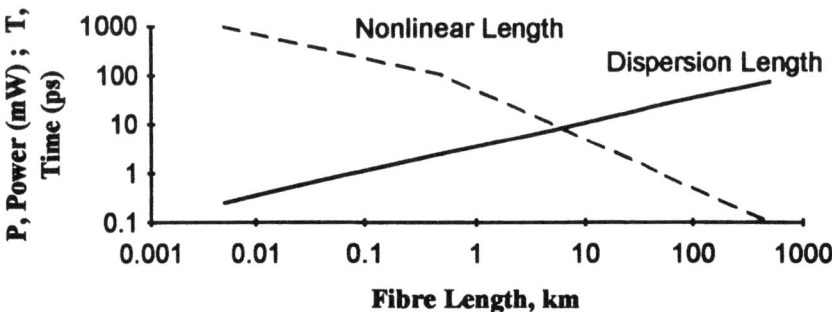

Figure 3. The dependence of nonlinear length as function of input power for $\gamma = 20$ W^{-1} km^{-1}, and dispersion length dependence on pulse width for $|\beta''|$ of ~10ps^2 km^{-1}. This distance is also approximately the soliton period. There are a series of curves for each value of the group velocity dispersion, β'' and also for each value of γ.

4.3 INDUCED BIREFRINGENCE: CROSS PHASE MODULATION (XPM)

Section 4.1 described how the optical field changes the refractive index when a single linearly polarised field is present. To deal with different frequencies and polarisations, the nonlinear polarisation $\chi^{(3)}{}_{iiii}$ may be replaced by other terms as described in Equation (21). Of the many possibile combinations of polarisations and frequencies in isotropic or birefringent fibres, three combinations are usually considered to be useful: i) degenerate frequencies with identical polarisations, as has been described already in Sections 4.1 and 4.2; ii) different frequencies with identical polarisations and iii) degenerate frequencies with different polarisations. In this section the last two conditions will be discussed.

4.3.1 XPM: Identical Frequencies
An optical wave of elliptical polarisation can be described as

$$E(\mathbf{r},t) = 1/2[E_x(\omega) + E_y(\omega)] + cc \qquad (67)$$

where $x = i$ and $y = j$. When this wave travels in an isotropic medium, each polarised component can affect the polarisation of the other through the cross terms in $\chi^{(3)}$. Fourier transforming Equation (21) and after tedious algebra, collecting terms we arrive at the following relationship for the induced refractive index changes for each component as

$$P_i = 3/4\varepsilon_0 [\chi^{(3)}_{iiii} |E_i|^2 + 2\chi^{(3)}_{iijj} |E_j|^2]E_i \qquad (68)$$

In the above equation, j may be interchanged for i to arrive at the equation for the orthogonal polarisation. Using the symmetry properties described in Equations (25-27), and after using Equation (47), Equation (68) above may be simplified to give

$$\Delta n_i = n_2 \{ |E_i|^2 + 2/3 | E_j |^2 \} \qquad (69)$$

since the terms describing the induced index change for the i direction are proportional to the field E_i, and similarly for E_j, as per Equation (42). The subscript i may be interchanged for j to arrive at the equation for the orthogonal index change.

In weakly birefringent fibres, the induced index change can be comparable to the intrinsic birefringence. This can cause strong instabilities in the state of polarisation at the output of the fibre. Cancellation of the birefringence is possible if the dominant index change is along the fast axis[29]. The wave polarised along the fast axis slows down so that the state of the output polarisation is intensity dependent if the fibre is of the order of a beat length.

The coupled mode equations describing XPM can be shown to be[34]

$$\frac{dc_+}{dz} = i\kappa c_- + i\beta |c_-|^2 c_+ \qquad (70)$$

$$\frac{dc_-}{dz} = i\kappa c_+ + i\beta |c_+|^2 c_- \qquad (71)$$

where $\beta = \frac{2}{3}\gamma$ and $\kappa = \pi/L_b$. L_b is the beat length over which coupling occurs linearly in a birefringent fibre.

These will be discussed in the section on applications.

4.3.2 XPM: Non-Degenerate Frequencies

In the case when the frequencies are non-degenerate (ω_i and ω_j), a similar analysis as in the last section results in the following equation for the induced index change for identical polarisations for each frequency (i, j)

$$\Delta n(\omega_i) = n_2 \{ |E(\omega_i)|^2 + 2 |E(\omega_j)|^2 \} \qquad (72)$$

Notice here that XPM in the case of non-degeneracy is more effective since the second term has a coefficient of 2 rather than 2/3. Another further difference is in the walk-off between pulses in the case of non-degenaracy since each frequency has a different group velocity. The effect of walk-off often limits the length over which XPM effect may be accumulated.

Cross phase modulation is an important effect which has been used in a variety of applications including all optical switching in loop mirrors, Kerr shutters and can also be a source of cross-talk in wavelength division systems[26-27, 35-39].

5. Stimulated Raman Scattering

Stimulated Raman scattering is easily observable in optical fibres[5, 40-43]. It give rise to a number of effects such as stimulated multiple Stokes generation, Raman fibre laser, Raman amplification, Raman scattering induced soliton self-frequency, modulation instability, to name a few. In this section, a brief introduction to the subject will be presented along with the basic formulae for the calculation of Raman gain and thresholds.

Raman scattering is due to the absorption of energy from a pump photon in a medium to a vibrational state in the optical branch of the phonon of the molecule. It is a very weak non-parametric effect and is described by the energy conservation equation

$$\omega_s = \omega_p - \omega_v \qquad (73)$$

where ω_v is the frequency of the Raman active molecular vibration, ω_p is the frequency of the pump photon and ω_s is the generated Stokes photon. This is a linear scattering phenomenon.

Stimulated Raman scattering occurs since a Stokes photon stimulates the generation of another as described by the energy conservation equation

$$\omega_s = \omega_p - \omega_p + \omega_s \qquad (74)$$

This may be understood as a two step process as follows

A. generation of a Stokes photon as per Equation (73)

B. generation of a phonon by mixing of a Stokes photon and the pump photon as

$$\omega_v = \omega_p - \omega_s \qquad (75)$$

Combining Equations (73) and (75) directly gives Equation (74).

Momentum is automatically conserved since

$$k_s = k_p - k_p + k_s \qquad (76)$$

so that the process is self phase-matching. The subscripts refer to the pump and Stokes waves. However, the group velocities in an optical fibre are usually different for the pump and the Stokes photons and therefore the interaction length is limited by the difference in the group velocities of the two waves in the pulsed regime. A walk-off length can be defined over which the interaction takes place

$$L_{walk\text{-}off} = T_0 / |[1/v_{pg} - 1/v_{sg}]| \qquad (77)$$

The Raman gain spectrum of germania doped silica fibres is large (approximately 40 THz) and has been measured by Stolen [40]. The Fourier transform of the spectrum give the Raman response time of the third order nonlinearity of silica fibres and has been estimated by Stolen et al[44] Blow and Wood[45], and measured by Grudinin et al[46] et al to be around 2-4fs.

Stimulated Raman gain results from the imaginary part of $\chi^{(3)}$. Allowing $\chi^{(3)}$ to be complex, Equation (60) may be modified by replacing the attenuation coefficient α, by adding a gain term at the Stokes wavelength and a loss term for the pump wave as

$$\alpha = \alpha_p + g_p |A_s|^2 A_p \qquad (78)$$

for the pump wave and

$$\alpha = \alpha_s + g_s |A_p|^2 A_s \qquad (79)$$

for the Stokes wave. The resulting pair of coupled mode equations are identical to Equation (60) except for the RHS in the case of the evolution of the pump field, A_p being replaced by

$$RHS = i\gamma_p(|A_p|^2 + 2|A_s|^2)A_p - \frac{g_p}{2}|A_s|^2 A_p \qquad (80)$$

and for the Stokes field, A_s by

$$RHS = i\gamma_s(|A_s|^2 + 2|A_p|^2)A_s - \frac{g_s}{2}|A_p|^2 A_s \qquad (81)$$

with the gain coefficients

$$g_s = g_R / A_e \qquad (82)$$

and

$$g_p = g_s (\omega_p/\omega_s) \qquad (83)$$

g_R is the Raman gain coefficient. measured from the spontaneous scattering cross-section by Stolen [40]. The set of coupled mode Equations (80-81) include the effects of SPM and XPM as well as the group velocity dispersion term as before. The interplay between Raman scattering and GVD can results in interesting effect such as the soliton self-frequency shift, with the central wavelength of the soliton shifting to longer wavelengths for picosecond pulses. This may be understood by remembering that for pulses <1ps, the soliton spectrum overlaps the soliton induced Raman spectrum. The longer wavelengths within the soliton spectrum experience a larger gain pumped by the blue part of the soliton, since the peak of the Raman gain is approximately 14THz on the longer wavelength side of the central wavelength of the soliton. This asymmetric gain "pulls" the soliton to longer wavelengths, causing the soliton to slow down, since the pulse propagates in the anomalous dispersion regime.

The threshold for stimulated Raman gain is defined as the pump power required for the Stokes power to become equal the pump power at the output of the fibre and may be calculated from

$$P_{TH} \approx 16 A_{eff} / (L_{eff} g_R) \qquad (84)$$

6. Second Harmonic Generation in Optical Fibres

The raw material for glass optical fibres, quartz is a crystalline material and posesses a second-order nonlinearity, $\chi^{(2)}$. Glass fibre, on the other hand is amorphous with a centre of inversion and consequently has a zero $\chi^{(2)}$. However, it is possible to frequency double in optical fibres as a result of two nonlinear effects. Firstly, the bulk quadrupolar nonlinearity which exists in all materials does allow frequency doubling in optical fibres provided the interaction is between modes of the appropriate symmetry. Secondly, another type of nonlinearity in waveguides is due to the core-cladding interface. There is a chemical potential across the interface and also a gradient of the optical field, giving rise to a second order nonlinearity. This also results in an interaction between any fundamental frequency mode to an odd symmetry second-harmonic mode[47]. These nonlinearities are weak and require intense pump powers, or phase-matching[48] for the frequency doubled mode to be observed at reasonable pump powers.

With fibres designed for phase-matched interaction between of the appropriate modes[49], the second-harmonic modes are readily visible with a 20-30mW of CW power at 1.064μm, as may be seen in Figures 4a-c The mode combinations are indicated in the Figure. Phase-matching is restricted to relatively short lengths due to random fluctuations in the core dimension and refractive index difference and the efficiencies are low owing to the poor overlap between the modes [47].

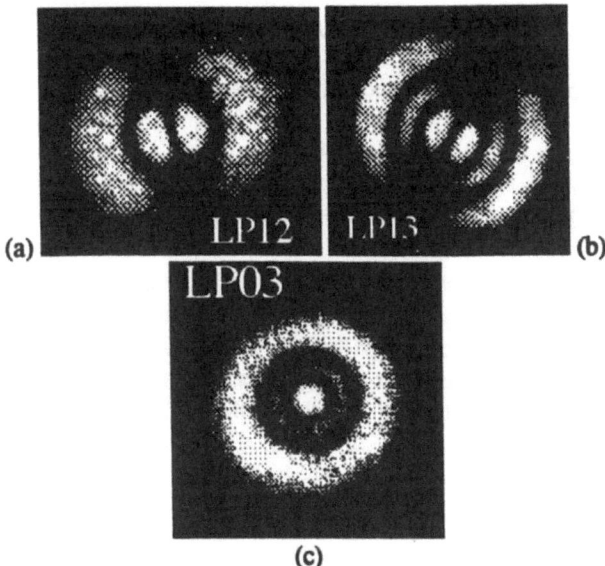

Figure 4. Experimentally observed frequency doubled modes. (a) shows the phase-matched interaction between $LP_{01}(\omega) \to LP_{12}(2\omega)$ modes. (b) is for $LP_{01}(\omega) \to LP_{13}(2\omega)$ modes. (c) is for the phase-matched interaction $LP_{01}(\omega) + LP_{11}(\omega) \to LP_{03}(2\omega)$ modes[49].

Therefore, it was totally unexpected when it was found that *efficient* second-harmonic generation was possible in optical fibres[15]. Since, the subject has been debated at length, and it has been shown that the effective nonlinearity is due to $\chi^{(3)}$ as [50] a result of Electric Field Induced Second-Harmonic Generation (EFISH)[48]. With intense pump powers photo-excitation of defects at 240nm causes self-organisation (self-phase-matching) through interference between photons of ω and 2ω [51]. An electric field grating is written into the fibre with a period which cancels the momentum mismatch between the intense fundamental field and the weak seeding second-harmonic fields as

$$1/\Lambda = 2(n^{\omega}_{eff} - n^{2\omega}_{eff})/\lambda = \Delta\beta/\lambda \qquad (85)$$

The preparation process was shown to accelerate when a weak seed at the second-harmonic wavelength is present with the fundamental field, although preparation also takes place without the seed. Conversion efficiencies as high as 13% have been reported [52]. The picture is somewhat complicated by the fact that in order to create a large enough nonlinearity, the large internal field can only be explained by invoking the photo-voltaic effect[53-54] which relates a nonlinear current density to the optical field in a smilar fashion as for the induced polarisation

$$j_{dc} = \beta E^2_{\omega} E^*_{2\omega} \exp(-i\Delta\beta z) \qquad (86)$$

where β is the photovoltaic constant and with the conductivity, σ

$$E_{dc} = \sigma \, j_{dc} \qquad (87)$$

Finally, the induced spatially periodic second order nonlinearity is

$$\chi^{(2)} = \chi^{(3)} E_{dc} \qquad (88)$$

A measured internal field as large as 10^6 Vm^{-1} has been reported[50].

Recently, it has been possible to prepare optical fibres by preferentially exciting defects with uv radiation while preparing the fibre for second harmonic generation using CW radiation at the fundamental and second-harmonic wavelengths[55-57]. The subject continues to attract considerable interest owing to its similarity of the process to photo-sensitive reflection grating formation in fibres[11, 58-59].

7. Self Propelled "Self-Focusing" Damage In Waveguides

Catastrophic optical damage in bulk materials occurs when a critical threshold power is exceeded. Self-focusing, the result of lensing due to the nonlinear refractive index, n_2, is an input power dependent phenomenon (rather than intensity) and depends on two critical powers

$$P_2 = \varepsilon_0 \pi (1.22 \, \lambda_0)^2 c / (64 \, n_2) \qquad (89)$$

and

$$P_1 = 0.273 P_2 \qquad (90)$$

where λ_0 is the free space wavelength.
For a collimated beam in bulk material when $P > P_2$, self-focusing occurs at a distance z_f which can be calculated from[60-61]

$$z_f = 0.369 k a^2 / [\,(P/P_2) - 0.858\,] \qquad (91)$$

The damage is catastrophic with filament formation, cavitation and with emission of plasma like radiation. Multiple focal points occur when $P > 2P_2$, while weak self-focusing occurs with $P_1 < P < P_2$ for the on axis part of the beam but diffraction counteracts focusing and the intensity drops as a function of z. The critical power P_1 may be calculated as 0.25 MW for silica. The observed damage filament moves away from the source.

It was surprising when it was discovered that catastrophic damage could occur over kilometres of optical fibres with CW powers as low as 1W[12,62] with the damage point moving towards the source! The similarity of the damage tracks, the cavitation, the visible emission and the self-propulsion prompted the name "Self-Propelled Self-Focusing".

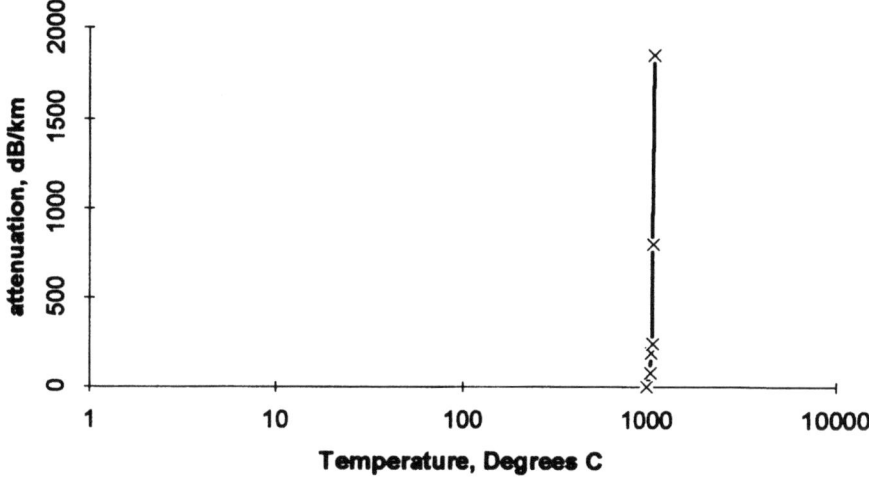

Figure 5. Measured absorption in optical fibres as a function of temperature[62].

The reason for the damage mechanism has been pin-pointed to the measured dramatic increase in the broad-band intrinsic absorption above 1100°C in silica based optical fibres (shown in Figure 5). The result of the increasing absorption with temperature is further absorption of light propagating in the fibre. Simulations have shown that this thermal runaway effect leads to the formation of a "thermal profile" shown in Figure 6 which travels towards the source (since it becomes the driving force behind the phenomenon) very much like a solitary wave, unchanged with propagation distance. Calculations also show that the temperatures in the core rapidly exceed several thousand degrees.

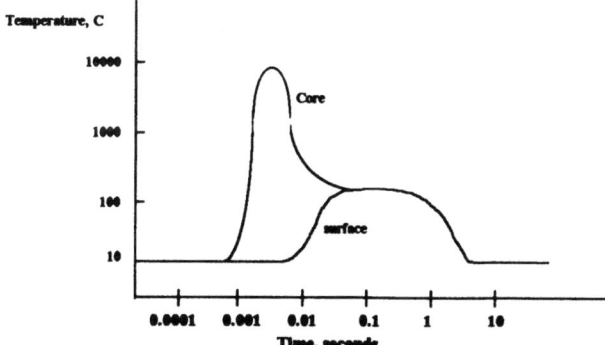

Figure 6. The moving constant temperature profile of the core and cladding. Time is measured from the occurance of the damage. Within around a ms, the core temperature rises to ~10,000 °C while heat dissipation causes the cladding temperature to rise some time later. As the temperature increases, the damage point moves closer to the source and the temperature profile reaches a steady state. This temperature profile moves at the velocity of propagation of the damage.

The elevated temperature causes a plasma like emission (Shown in Figure 7) from the damage point and leaves a trail of regular shaped cavities. (shown in Figure 8), containing oxygen. Apart from being an eerie phenomenon, it is peculiar to waveguides alone, since the power is confined by guidance to a small volume, and has the potential of destroying kilometres of optical fibres at unprecedented power levels and with a calculated threshold of only 100mW!

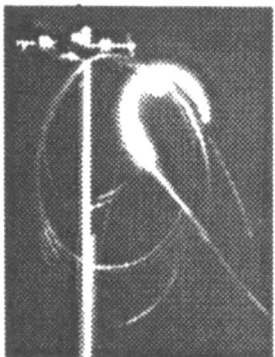

Figure 7. Visible plasma-like emission from an optical fibre coiled on a retort stand, under-going SPSF damage. The length of the illuminated region is a result of the time of the exposure for the photograph.

For an input power of W watts, the velocity of propagation of the damage can be estimated as[12]

$$v = W/(VN\rho cT) \qquad (92)$$

where T is the temperature rise, V is the volume of the damage centre, N is the number of damage centres per metre, ρ is the density of the glass, and c is the specific heat. With a temperature rise of ~6000°K, $V = 2.4 \times 10^{-16}$ m^{-3}, $N = 10^5$ m^{-1}, the velocity of propagation of the damage for an input power of 2W in a standard fibre, the velocity is calculated to be 1.8 m s^{-1} in close agreement with measured values.

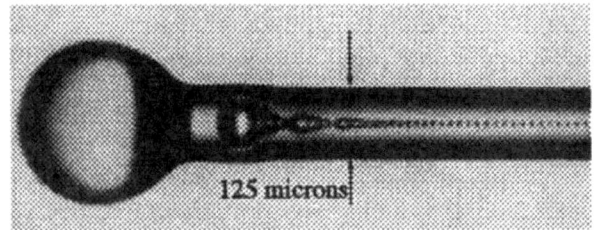

Figure 8. Photograph of an optical fibre damaged by SPSF.

7. Measurement Techniques & Applications

This section will introduce some of the techniques employed in the measurement of the nonlinear refractive index, and describe experiments on all optical switching and polarisation coupling via XPM.

7.1. KERR EFFECT

Self-Phase-Modulation (SPM) in optical fibres was reported by Stolen and Ashkin *et al*[7] using a pump wavelength of 514nm. The principle of the experiment is based on the observation of the output spectrum of a pulse as a function of intensity with $L_d \gg L_{NL}$. The spectrum is the Fourier transform of the chirped pulse shown in Figure 2. The spectrum undergoes continuous spectral broadening with a direct relationship between the number of peaks observed and the maximum phase change ϕ_{max} at the peak of the output pulse. The measured output spectrum is shown in Figure 9[63]. The bandwidth of the spectrum has a linear relationship with the peak input power until the onset of stimulated Raman scattering leading to asymmetry in the spectrum.

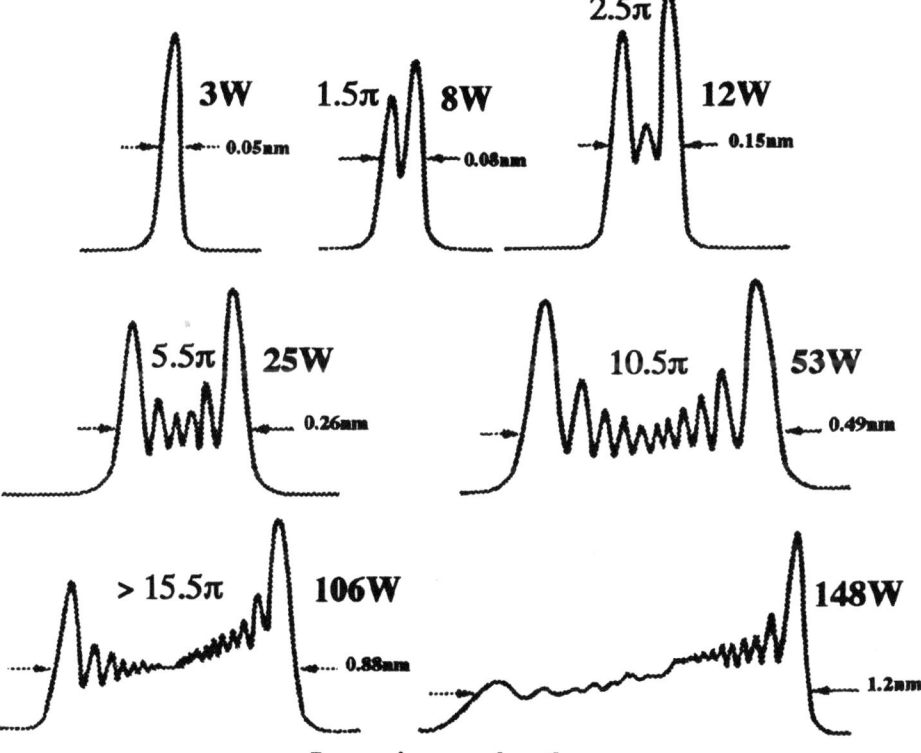

Increasing wavelength →

Figure 9. Measured spectra from a 150m long length of fibre for input powers shown with the curves. The maximum phase-shift (approximate) is also indicated. Spectra is broadened as a result of SPM. The bottom two curves show the onset of stimulated Raman scattering[63].

7.2 MEASUREMENT OF n_2

There are several methods of measuring the nonlinear refractive index of optical fibres. Traditionally, the measurement of SPM has been used to estimate the effective nonlinearity n_2 / A_{eff}[7]. By estimating the the effective mode area, it is possible to calculate n_2. Owing to measurement uncertainties, the results tend to be accurate to approximately 15-20%. Recently, the mode areas have been computed accurately and it has been shown[64] that the value of n_2 has been measured with an an uncertainty of ±5% by reducing systematic errors. The value for pure silica core is 2.36 × 10^{-16} cm^2 W^{-1}. The germanium dopant concentration shows an increase in the value to around 2.63 × 10^{-16} cm^2 W^{-1}, with the nonlinear refractive index of germania being estimated to be ~ ×3 that of pure silica. High germania fibres have large core-cladding refractive index differences so that for singlemode operation, the core diameter has to be small, increasing the effective nonlinearity, n_2 / A_{eff}.

The important parameter in designing nonlinear devices, is the change in the mode index rather than the *core refractive index* change. This is readily seen in Figure 10, in which the normalised mode-overlap integral change is shown as a function of the v-value of the fibre. The transverse distribution of the mode integrated over the change in the refractive index determines the change in the effective mode-index. The change in the normalised overlap, ignoring the cladding nonlinearity is

$$Change\ in\ the\ normalised\ mode\text{-}overlap\ integral = \frac{\Delta n_{eff} \lambda^2}{n_2 p} \quad (93)$$

It may be seen that at a v-value of approximately 2, the mode index change is maximised, and changes in the v-value have little effect. This curve is appropriate for a core nonlinearity > the cladding nonlinearity. Operating at a v-value of ~2 maximises the nonlinear effect.

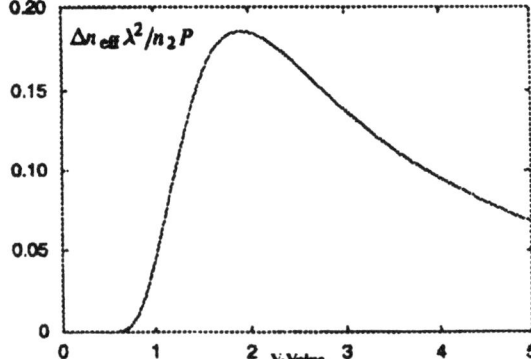

Figure 10. The induced mode-index change as a function of the fibre v-value[65].

Interferometric methods have also been used for measuring the nonlinear refractive index. in optical fibres[65-67]. These are based on a pump-probe scheme, in which a high-intensity pump induces a phase change in a low intensity probe beam, or undergoes a larger phase change. Owing to the sensitivity of the measurements, the interferometers used are generally self-balancing [65-69], to cancel out the effect of differential paths.

7.2 POLARISATION INSTABILITIES

When an optical mode propagates in a weakly birefringent fibre predominantly polarised along the fast axis, the polarisation of the mode undergoes an evolution from linear to elliptical, returning to linear once again at a distance of a beat length. If the intensity of the mode is increased, the refractive index along the fast axis increases more than the index along the slow axis, owing to the greater component along the fast axis[29]. The beat length

$$L_b = \lambda / [n^{slow}_{eff} - n^{fast}_{eff}(I)] \qquad (94)$$

is intensity dependent and goes to infinity as

$$n^{slow}_{eff} - n^{fast}_{eff}(I) \rightarrow 0 \qquad (95)$$

Under this condition, the state of polarisation at the output changes dramatically. The effect is known as polarisation instability and has been observed in optical fibres[31-32]. Experiments are difficult to perform in optical fibres since weak birefringence is required as well as a reasonable propagation length of order a beat length. Since the nonlinearity in optical fibres is small, high peak powers are required to observe the instability and polarisation rotation. However, the situation is made better by using a near isotropic waveguide with a liquid core which is highly nonlinear, eg nitrobenzene. The nonlinearity of nitrobenzene is ~100 time that of silica fibres, but has a decay time of ~35ps (re-orientation relaxation time of the molecule).

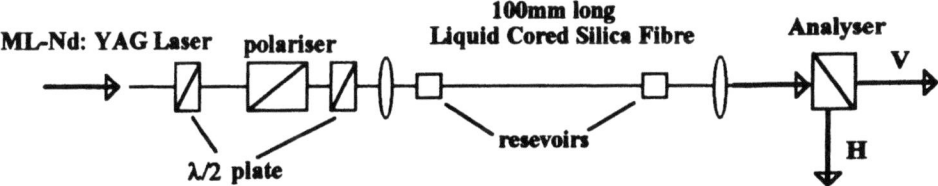

Figure 11. Experimental arrangement for observing nonlinear polarisation instabilities in weakly birefringent fibres[34]

Simple experiments were reported recently[34] using this technique showing a number of interesting features. The experimental arrangement is shown in Figure 11. Rotation of the input polarisation shows highly nonlinear output polarisation states as the intensity is increased. Figure 12 shows the transmission of the vertical and horizontal states of polarisation through an analyser at the output of a weakly birefringent, 10 cm long singlemode nitrobenzene-cored fibre capillary waveguide. At certain input polarisation angles, rapid changes in the output state occurs for slight changes in the input polarisation. Figure 13 shows the normalised transmission of the fibre for different input polarisation angles (relative to vertical polarisation), showing strong multiple coupling between the polarisations as set out by the coupled mode Equations (70-71).

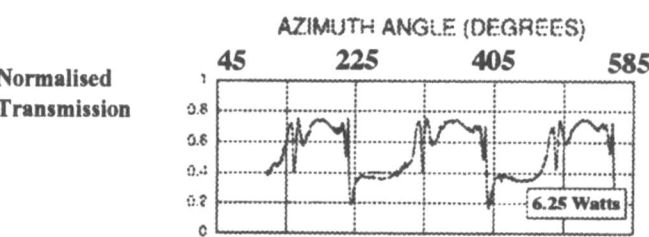

Figure 12. Output power as a function of input polarisation angle for an input power of 6.25 watts. The output polarisation changes rapidly with slight changes in the input polarisation, demonstrating a highly nonlinear dependence[34,70]. In the linear regime, the output changes as $\sin^2(\theta)$.

The output state power for a given input polarisation state may be computed using Equations(70-71). The solutions show strong polarisation coupling for certain input states. For comparison with Figure 13(60°), the results of calculations are shown in Figure 14 for two cases: for CW operation, in which the nonlinear coupling is strong, and, for pulsed operation, in which the output response is the result of integration over the pulse. The estimated phase change for the 10cm long liquid cored guide is approximately 12π for the case when the input polarisation of 60° at a maximum power of approximately 12W. Each cycle is equivalent to a phase-shift of 2π. This value is thought to be the largest for a short singlemode waveguide. The device was estimated to be ~1 beat length long. Although the energy required to induce a 2π phase shift was ~ 400pJ, the compactness of the fibre device allows observation of nonlinear coupling without interference from environmental effects.

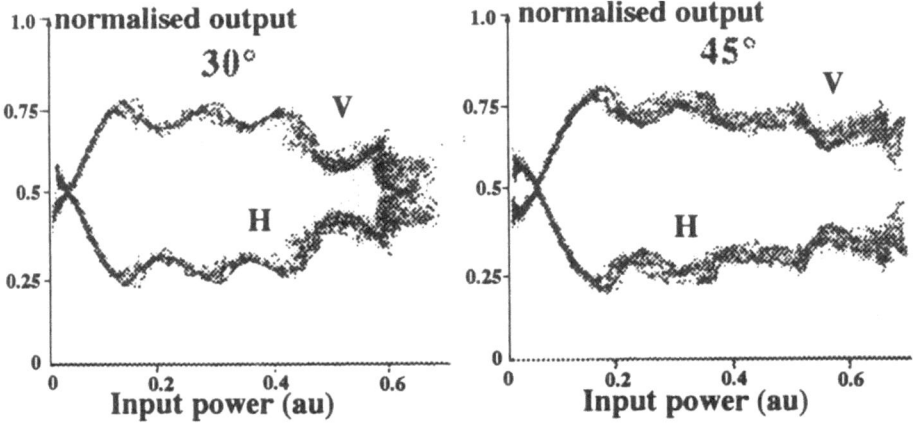

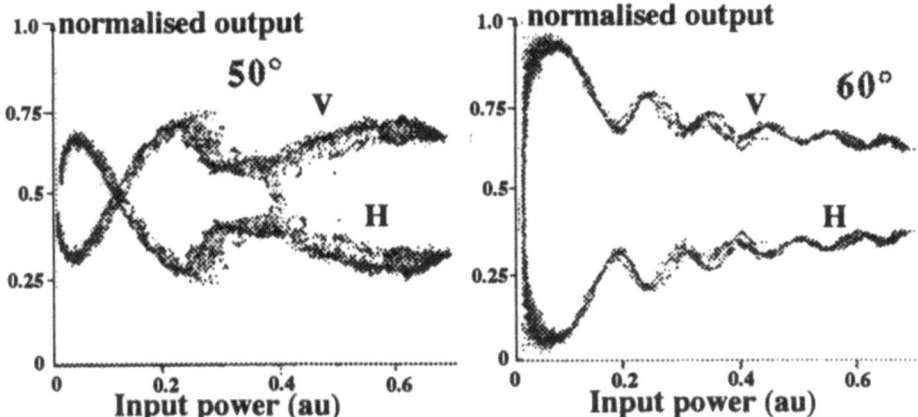

Figure 13. The curve show the normalised transmittance of the V and H polarisations as a function of launched peak power for different input polarisation angles and fixed output analyser angle. As the input polarisation changes, the normalised transmittance shows changes in the coupling between the orthogonal polarisation modes as a function of power[34,70].

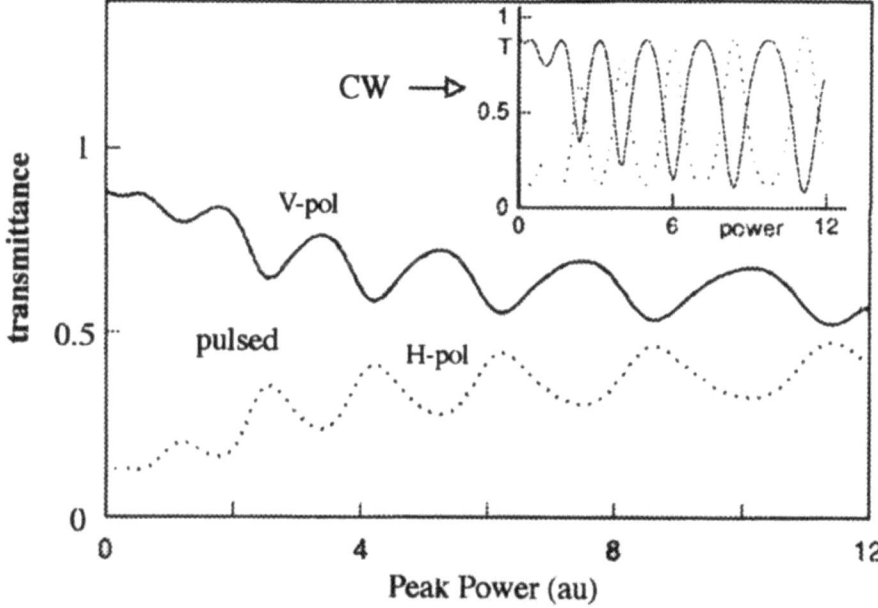

Figure 14. The above trace is the equivalent calculated response using Equations (70-71) of the experimentally determined trace shown in Figure 13 (60°). The CW response is much sharper since almost complete switching is possible using identical fibre parameters. In the pulsed regime, the coupling is an integrated response over the whole pulse[34].

8. Other Applications

There are many application of nonlinear fibres which have not been covered in this paper. Other lectures in this course cover effects such as nonlinear pulse-compression, soliton generation and transmission, soliton self-frequency shift, stimulated Brillouin scattering/ lasers, the loop mirror and the amplifying loop mirror. The subject is vast and continues to grow. The reader is directed to further references in Ref 71 and 33.

9. Conclusion

Optical fibres have excellent linear and nonlinear properties. They can be used for the investigation of a large number of physical phenomenon, as well as for application in all optical switching, amplification and soliton transmission. The low nonlinearity of optical fibres is proving to be extremely useful in a variety of ways. There are a wide variety of applications of nonlinear effects not covered; however, this chapter has explored some of the novel aspects of nonlinear optical fibres from a physical understanding of the nonlinearity to the mechanisms which form the basis of several nonlinear applications.

I gratefully acknowledge Richard Wyatt for his constructive criticism on the manuscript.

10. References:

1. Kapany N S, *J Opt Soc Am 49*, 779, 1959, and references therein.
2. Snitzer E, *J Opt Soc Am 51*, 491, 1961
3. Kao K C and Hockham G H, *Proc IEEE 113*, 1151, 1966.
4. Ainslie B J and Day C R, *J Lightwave Technol.*, LT-4, 967, 1986.
5. Stolen R H, Ippen E P and Tynes A R, *Appl Phys Lett 20*, 62, 1962.
6. Ippen E P and Stolen R H, *Appl Phys Lett 21*, 539, 1972.
7. Stolen R H and Ashkin A, *Appl Phys Lett 22*, 294, 1973.
8. Desurvire E,*Optics and Photonics News 2*, 6, 1991.
9. Blow K J, and Smith K , *BT Technol J 11*(2), 1993.
10. Hasegawa A and Tappert F, *Appl Phys Lett 23*, 142, 1973.
11. Hill K, O, Fujii Y, Johnson D C and Kawasaki B S, *Appl Phys Lett 32*, 647, 1978.
12. Kashyap R and Blow K J, *Electorn Lett 24*, 47, 1988.
13. Franken P A, Hill A E, Peters C W and Weinreich G, *Phys Rev Lett 7*, 118, 1961.
14. Bloembergen N, *Nonlinear Optics* (W A benjamin, Inc, New York, 1965).
15. Osterberg U and Margulis W, *Opts Lett 11*, 516, 1986.
16. Kleinman D A, *Phys Rev 12*, 1977, 1962.
17. Flytzanis C, *in Quantum Electronics Vol I*, ed H Rabin and C L Tang, Academic Press, 1975.
18. Butcher P N and Cotter D, *The Elements of Nonlinear Optics*, Cambridge Studies in Modern Optics: 9, Press Syndicate of the University of Cambridge, 1990.

19. Hanna D C, Yuratitch and Cotter D, in *Nonlinear Optics of Free Atoms and Molecules*, Springer Series in Optical Sciences Vol 17, Springer Verlag, 1979.
20. Kielich S, *Chem Phys Letts 2*, 569, 1968.
21. Marcuse D, in *Theory of Dielectric Optical Waveguides*, Academic Press, 1974.
22. Kielich S, in *Nonlinear Optics*, Molecular Electro-Optics: Part I - Theory and Method, Ed. O'Konski C T, Chapter 12, Marcel Dekker Inc 1976.
23. Maker P D, Terhune R W and Savage C M, *Phys Rev Lett 12*, 507, 1964.
24. Dziedzic J M, Stolen R H and Ashkin A, *Appl Opts 20*, 1403, 1981.
25. Trillo S, Wabnitz S, Finlayson N, Banyai W C, Seaton C T and Stegeman G I, *Appl Phys Lett 53*, 837, 1988.
26. Morioka T and Saruwatari M, *IEEE J Select Areas Comm 6*, 1186, 1988.
27. Trillo S, Wabnitz S, Stolen R H, Assanto G, Seaton C T and Stegeman G I, *Appl Phys Lett 49*, 1224, 1986.
28. Finlayson N, Nayar B K and Doran N J, *Electron Lett 27*, 1209, 1991.
29. Winful H G, *Opts Lett 11*, 33, 1986.
30. Daino B, Gregori G and Wabnitz S, *Opts Lett 11*, 43, 1986.
31. Feldman S F, Weinberger D A and Winful H G, *Opts Lett 15*, 311, 1990.
32. Ferro P, Haelterman M, Trillo S, Wabnitz S and Daino B, *Electron Lett 27*, 1408, 1991.
33. Agrawal G P, *Nonlinear Fibre Optics*, Quantum Electronics - Principles and Applications, Ed. Liao P F and Kelley P L, Academic Press, 1989.
34. Kashyap R and Finlayson N, *Opts Lett 17*, 407, 1992.
35. Nayar B K and Vanherzeele H, *IEEE Photon Technol Lett 2*, 603, 1990, and references therein.
36. Kitayama K, Kimura Y and Sasaki S, *Appl Phys Lett 46*, 623, 1985.
37. Winful H G, *Appl Phys Lett 47*, 213, 1985.
38. Cotter D and Hill A M, *Electron Lett 20*, 185, 1984.
39. Olsson, Hegarty J, Logan R A, Johnson L F, Walker K L, Cohen L G, Kasper B L and Campbell J C, *Electron Lett 21*, 105, 1985.
40. Stolen R H and Ippen E P, *Appl Phys Lett 22*, 276, 1973.
41. Davey S T, Williams D L, Ainslie B J, Rothwell W J M and Wakefield B, *IEE Proc 136* Pt J, 301, 1989.
42. Desurvire E, Papuchon M, Pocholle J P, Raffy J and Ostrowsky D B, *Electron Lett 19*, 751, 1983.
43. Jain R K, Lin C, Stolen R H, Pleibel W and Kaiser P, *Appl Phys Lett 30*, 162, 1977.
44. Stolen R H, Gordon J P, Tomlinson W J and Haus H, *J Opt Soc Am B*, 6, 1159, 1989.
45. Blow K J and Wood D, *IEEE J Quant Electron 25*, 2665, 1989.
46. Grudinin A B, Dianov E M, Korobin D V, Prokhorov A M, Serkin V N and Khaidarov, *JETP 34*, 62, 1972.
47. Terhune R W and Weinberger D A, *J Opt Soc Am B 4*, 661, 1987.
48. Kashyap R, *J Opt Soc Am B 6*, 313, 1989.
49. Kashyap R, Davey S T and Williams D L, "First Observation of phase-matched intrinsic second-harmonic generation in optical fibres", in OSA Annual Meeting

Technical Digest 1990, Vol 15 of the OSA Technical Digest Series (Optical Society of America, Washington D.C., 1990), pages 105-106.
50. Kamal A, Stock M L, Sapak A, Thomas C H, WeinbergerD A, Frenkel M Y, Nees J, Ozaki K, Valdmanis J, in Optical Soc Am Annual Meeting Technical Digest, (OSA, Washington D C, Paper PD24 (1990).
51. Stolen R H and Tom H W K, *Opts Lett 12*, 585, 1987.
52. Fairy M C, Fermann M E and Russell P St J, in Proc of Topical Meeting on nonlinear Guided Wave Phenomenon: Physics and Applications, pp246-249, 1989.
53. Barnova N B and Zeldovich B Ya, *JEPT Lett 45*, 717, 1987.
54. Anderson D Z, Mizrahi V and Sipe J E, *Opts Lett 16*, 796, 1989.
55. Kashyap R, Borgonjen E and Campbell R C, *SPIE 2044*, 202, 1993.
56. Margulis W, Carvalho I C S, Lesche B, *SPIE 1516*, 60, 1991.
57. Lawandy N M, *Opts Lett 14*, 571, 1992.
58. Kashyap R, Armitage J R, Wyatt R, Davey S T and Williams D L, *Electron Lett* 26(11), 730, 1990.
59. Meltz G, Morey W W and Glenn W H, *Opts Lett 14*(15), 823, 1989.
60. Dawes E L and Marburger,J H, *Phys Rev Letts 27*, 905, 1971.
61. Reintjes J F, in "Nonlinear optical parametric processes in liquids and gasses", *Quantum Electronics: Principles and Applications*, Chapter 5, pp 333-334, Academic Press Inc, 1984.
62. Kashyap R, "Self propelled self-focusing damage in optical fibres", in Proc of the international Conference on Lasers '87, Dec 7-11 1987, Ed Duarte F J, STS Press.
63. Kean P N, Smith K and Sibbett W, *IEE Proc Vol 134*, Pt J, 163, 1987.
64. Kim K S, Stolen R H, Reed W A and Quoi K W, *Opts Lett 19*, 257, 1994.
65. Manning R J, Kashyap R, Oliver S N and Cotter D, *in Proc of the Topical Meeting on Integrated Photonics Research*, 1993.
66. Nayar B K, Finlayson N, Doran N J, Davey S T, Williams D L and Arkwright J W, *Opts Lett 16*, 408, 1991.
67. Doran N J, Forrester D S and Nayer B K, *Electron Lett 25*, 267, 1989.
68. Stegemen G I and Stolen R H, *J Opt Soc Am B 6*, 267, 1989.
69. Meier J E and Heinlein, in Proceedings of *The Second Optical FibresMeasurements Conference* '93, IIC, Turin, Sept 21-22 1993, pp125-128.
70. Kashyap R and Finlayson N, in Proc of Nonlinear Guided Wave Phenomena, 1991, (Optical Society of America, Washington D C, 1991), *Vol 15*, pp74-77.
71. Kashyap R, in Proceedings of The Second *Optical FibresMeasurements Conference* '93, IIC, Turin, Sept 21-22 1993, pp119-124.

EXOTIC FIBRES
Multitude of application--specific optical fibres

F. GAUTHIER
Laboratoire Central des Industries Électriques
33, av. du Général Leclerc
92260 Fontenay aux roses
FRANCE

1. Summary

Defining exotic fibres as a starting point, then fibre history is reviewed. Development reasons and parameters are given. What makes fibres "exotic" is exposed, such as manufacturing processes, materials and dopants, structure, end termination, protective materials, dimensions, assembly, etc... Some typical examples of fibres and applications are given . Emphasis is given to commercialy available solutions.

2. Definition

Etymologically, the word exotic comes from exoticus (latin) and exôtikos (greek) which means "foreign, extraneous, not belonging to". Exotic fibres will be fibres different from the ones referred as "classical fibres", say the three common types of "Telecommunication fibres" with core/clad/coating diameters of 50/125/250, 62.5/125/250, 8/125/250 microns, with UV-acrylate coating. This is a starting point to show to the users, scientists, engineers or others, the wide variety of optical fibres available on the market, and demonstrate their huge potential of designs and applications. Every exotic fibre is an application-specific optical fibre designed or chosen to increase the performance of the application in which it is included.

3. Fibre tale

Mineral glass fibres have been made by mother nature since about 4. 10^9 years, due to volcanism. One present example is given by the permanently active Kilauea volcano. In the 15-m high lava fountains of the halemaumau lava lake, Pele hair, lava tears, are formed by gazeous blowing. They consist of a small volcanic bomb, droplet of a few millimetres in diameter, sometimes formed around crystals of olivine or pyroxene, followed by a hair made of a basaltic glassy matrix.

What human being was the first to make a fibre, rather say a small rod or cord made of a relatively homogeneous material, probably a silicate glass because of its drawing properties: egyptian glassmakers ? According to W.M. Flinders Petrie, the earliest known glaze would date from about 12,000 years BC, and the earliest relatively pure glass from about 7000. A few glazes appear about 5500 years ago. Pearls and drawn objects are known by 1500 years BC. The technique was to use a sand core wrapped with a glass cord. Drawing the glass was understood then, and the limit to obtain fibres was only a technological one: the capability to realize higher temperature furnaces to get a lower viscosity of the material to be drawn.

Alexandria (100 BC), Rome, Murano craftmen used decoration with smaller and smaller rods and fibres. The "industrial man" invented spinning and blowing machines to make rowing and glass wool. The "man of atom" invented the waveguide model and improved materials purification, leading to the "super transparency race" since the seventies.

The optical fibres are born thanks to the technological knowledge of the drawing of silicate glasses, and developped with the drawing of pure and doped silica glasses. Since twenty years, fibre making has been extended using various materials with respect to their specific purifying and forming processes.

4. Parameters leading to the development of optical fibres

Four main parameters can be considered: the glassy state, the glass forming oxides transparency in the "visible-near infrared" range, the "man of atom" understanding of the materials structure, the human society need for information exchange through practical, confined, "communication highways".

4.1. THE GLASSY STATE

Among a number of peculiar properties, glasses and specialy silicate glasses do not present a melting point but a large temperature dependent transition domain between the solid state and the liquid state. They have a viscous melting, illustrated by their viscosity-temperature curves (See reference 1, and chapter 26, U.Bunrau, this volume). This feature allows them to be elongated and drawn-down into a fibre. The glass scientists define the annealing point, the softening point, the working point. The lower the slope of the viscosity curve is in the vicinity of the working point, the easier it is to draw the glass into a fibre. Practically, the glass can be drawn from a viscous melt, or from a solid rod or preform.

4.2. THE GLASS FORMING OXIDES TRANSPARENCY IN THE "VISIBLE-NEAR INFRARED" RANGE

Fortunately, among the first glasses millenary known by mankind, are the silicon oxides based glasses, some of them being translucent or relatively transparent under low thickness. Their technological recent development lead to the pure silica glass, a highly transparent one in the visible range. Its exceptionnal properties, such as high temperature stability, chemical durability, low thermal expansion and related high thermal-shock resistance, coupled with a broad optical transmission range made it the favorite candidate for the first optical fibres.

4.3. THE UNDERSTANDING OF THE MATERIALS STRUCTURE

The modern materials science has given us a new tool to better understand the structure of the glassy state. By creating models, Zachariassen and Warren (See reference 2 and 3) opened in the thirties a new approach leading to the modern concept of glass. By the sixties, glass science emerged and the basic sciences were profitably applied to a better knowledge of glass in terms of atomic structure and composition. Fast technological improvements completed this evolution. Glasses could be elaborated, for example, by chemical vapor deposition, starting with purified raw materials, under a controlled atmosphere. The possibility to easily substitute the silicon atom at the center of the silicon-oxygen tetrahedron lead to the silica-doped (germanate) glass with a higher index of refraction than the silica glass, with other compatible properties. The complementary possibility to stuff the silicon oxide lattice with phosphorus or other elements gave to the waveguide designers all the compatible materials they need to combine them for their refractive indices values to get an efficient optical waveguide, yet much more transparent in the infra-red region than in the known visible.

4.4. THE NEED FOR PRACTICAL, SPACIALLY CONFINED, "COMMUNICATION HIGHWAYS"

The first economically developped fibres have been the telecommunication ones, because the huge investment for the development was justified by the predicted pay-off in their use in the telecommunication systems. Our modern evolved societies call for an extraordinary expansion of human relations and exchange of any kind of information. To upgrade our communications systems, optical fibres are well suited of course because of their high bandwith capability and many other qualities, but in the short term future because optical cables can physically replace the copper cables in the existing ground ducts, without imposing to create a new infrastructure.

5. Parameters leading to the variety of fibres

Ready-to use fibres may be very different from one to another due to:
-The numerous manufacturing processes used for the elaboration of their different raw materials, their purification and their forming ;
-The variety of waveguide materials and dopants susceptible to enter into the composition of the core, the optical cladding, and the cladding;
-The designed structure: refractive index profile and refractive index difference, composition distribution, shape and dimensions;
- The protective materials for coating and buffer;
- The presence of a specific end termination, protection or assembling.

These parameters can be used as design factors which directly will determine the specific properties of a given fibre, and correlatively will pre-determine it to a specific application.
The wide variety of applications first comes from the multiple ways to use an existing type of fibre, and secondly from the variety of fibres that can be custom-manufactured nowadays.

5.1. THE MANUFACTURING PROCESSES

The fibre can be obtained from a viscous melt by glass spinning. Mineral borosilicate glasses are molten at high temperature (1300 degrees or more) using the double crucible technique. Organic glasses can be drawn from a prepolymer reactor. A fibre can also be obtained from a solid cylindrical rod heated up to its working point to become viscous to allow the elongation of the material into a cone, down to a fibre diameter ranging from a few microns up to a few millimetres. These techniques refer to the preform and rod-in tube drawing (See chapter 26, U.B. Unrau, this volume). To obtain these preforms, many processes can be used, such as the vapor deposition ones, called "chemical vapor deposition" (CVD), "modified chemical vapor deposition" (MCVD), "vapor axial deposition" (VAD), "outer vapor phase oxidation" (OVPO), "outer vapor deposition" (OVD), and others.... Techniques using a plasma torch, radio frequencies, vapor deposition, liquid impregnation are used alone or combined. Molecular stuffing of a silica glass with erbium can be obtained by solution doping of an already existing porous layer deposited by a CVD technique inside a silica glass tubing. After adequate drying and vitrifying treatments, the successive layers of different composition constituting the future fibre are built-up towards the center of the tube. The tube is collapsed by viscous flow into a preform rod.
Fluoride preforms can be made either by pressure molding or by direct filling with the core melt of a cladding tube obtained by rotational casting (See chapter 26, this volume).
Some types of fibres can be extruded, starting from granulates of a raw material like polymethylmethacrylate, using a heating screw-extruder, or starting from a preform, using a piston extruder, as for polycrystalline materials.
Crystalline growth is sometimes used to get a monocrystalline fibre, out of composition such as sapphire.

5.2. CORE AND CLADDING MATERIALS AND DOPANTS

Materials and dopants of the core and optical cladding of the fibres are the main parameters to characterize an exotic fibre. For commodity, we distinguish the glass fibres -mineral or organic-, the crystalline fibres -monocrystalline or polycrystalline-, and the liquid core fibres.
The first glasses to be used to draw fibres from the melt have been the multicomponent silicates, with slightly different compositions for the core and the cladding; Fluoride-based glasses are used in the same way, or with a plastic cladding. ZBLAN glasses (a combination of zirconium, barium, lanthanum, aluminum, sodium fluorides) have been

recently doped with rare-earth to make fibre amplifiers. Chalcogenide glasses, like the commercially available Ge-As-Se, are employed with or without a plastic clad. As_2S_3 is optically cladded with an $As_xS_{(1-x)}$ composition. Unfortunately, their components are highly toxic.

Telecommunication fibres are made of "cvd-type" materials. The core can be made of silica, doped with germanium, phosphorus, fluorine, aluminum, erbium, and the optical cladding of pure or also doped silica. Do not confuse the deposited optical cladding wih the final cladding, which includes in the CVD process the silica glass coming from the tube used for the deposition. The outer diameter of the cladding is then 125 microns.

To minimize losses, the core can be made of pure silica, and the cladding of doped silica. Germanate glasses are structurally identical and are interresting for their better transparency in the near- infrared.

The "Plastic-clad silica" (PCS), and the "Hard-clad silica" (HCS) have a pure silica core and a cladding of silicone for the former, of a hard polymer composition for the later. They can be designed with a low OH content, for visible and near-infrared transmission, or with a high OH content, for UV transmission improvement.

Organic glasses have been used for about twenty years to make step-index fibres, Polymethymethacrylate, polystyrene and polycarbonate are used for the core, associated with appropriate cladding compositions (See chapter 17, D. Kalymnios, this volume). Deuterated materials, evaluated for their spectrally improved transparency, have been dropped because they are too costly to process. Research is in progress to process polymer fibres with a graded-index profile (See reference 4). Polystyrene is widely used to make scintillating and fluorescent fibres used as sensors in the research for the physics of particles.

Polycrystalline fibres can be made out of AgCl, AgBr, KRS13 (AgBr-AgCl), KRS5 (TlBr-Tll), KRS6 (TlBr-TlCl), ZnSe, and monocrystalline fibres of CsBr, CsI, or sapphire. Chloride, bromide, iodide fibres have sericus limits of use, like plastic deformation or high hygroscopy.

Other exotic structures to be mentionned here include liquid-core fibres or light guides, and hollow dielectric or metallic waveguide.

5.3. THE DESIGNED STRUCTURE

The refractive index profile is caracteristic of each application-specific optical fibre and is one of the key design factor for it. It is directly related to the compositions and concentrations of the dopants used. The cross-section internal structure of each fibre is specific too, as illustrated with the "panda" fibre, the bow-tie fibre, the elliptical or rectangular core fibres, dual-core fibre, D-shaped fibre, elliptical stress element fibre and others (see chapter 2, S.A. Mustafa, and chapter 35, B. Culshaw, this volume). Connecting two identical polarization-maintaining fibres is difficult, and intermating two different types is almost impossible to achieve.

Fibres now exist in a very large range of dimensions. The outer diameter, and at the same time optical cladding diameter, may be as small as a few microns for each fibre among the thousands used in some image guide, or as large as two to three millimetres for a customized pure silica core fibre. Core/clad/coating ratios can respectively vary depending upon design requirements and properties needed.

5.4. COATING AND BUFFER MATERIALS

The most common materials used for protection and mechanical reinforcement are the polymers, some metals, and carbon.
Polymers intensively used in the cabling industry have been adapted to fibres, like the classical UV-acrylate family, with operating temperatures of about 80°C. Selecting the appropriate coating may extend the range of performance of a fibre. For example, silicone may be used to reduce the sensitivity of the h-parameter of some polarization-maintaining fibres. Polyimide, largely known under the nylon brand name, is used for its wide operating temperature range (from -180 to 380°C) with reduced thickness due to a good Young's modulus. Polyurethane is prefered for its improved hardness. Dual coating associates qualities of different polymers, like silicone and acrylate. Tefzel and Hytrel (trade marks) are reinforcing materials.
Metals are also used to protect the fibre. Aluminum used for hermeticity and high operating temperature (450°C) is replaced by carbon for water-resistant critical applications due to the presence of pinholes in the layer. Electroless and electrolytic nickel layers, and electrolytic gold layers of a few microns are plated on pigtails depending upon specific applications to allow soldering of components.

5.5 END TERMINATION

End termination is an essential preparation step of any fibre to use it efficiently. The most common one is a planar face prepared by cleaving, with the score, pull and break technique, using many commercially available tools. This clean pristine glass face can be included in a connector, using the same technique with the help of specific tools. Polishing can be required to obtain a controlled perpendicularity, or a controlled tilt angle, or a curved end-face. Connectorized pig-tails or cables are considered exotic because connectors types are numerous, and because the compatibility of connectors of the same type made from different manufacturers is not obvious (See reference 5), that is they are not fully intermateable (see chapter 12, this volume). Polishing is also used to grind-down the cladding to allow lateral coupling between fibres, to make loops, couplers or sensors. Chemical etching with hydrofluoric acid is employed to locally thin or remove the cladding of silicate fibres.
The fibre can become loosy, by making millimetric loops or by contacting the end-face with an index-matching material. Thin-film coating can be deposited on the face, for total reflection to make a mirror, or for anti-reflection to enhance light launching with core materials of high refractive index.
Intra-core grating can be obtained directly into the fibre, opening a new field of applications (see chapter 31, R. Kashyap, this volume).
The melting property of the very stable silica and silicates glasses is again of the greatest importance. It allows the fusion splicing, even between fibres with different core/clad ratios, and the forming and shaping of the fibre end into a half-sphere, or a

sphere acting as a lens. Spheres from 600 to 2000 microns in outer diameter are made on the core of pig-tails of hard clad silica fibres to improve coupling for power applications. Stretching of the fibre is also done. Tapers or double-tapers of a few meters are made with reduction ratios up to 5, to allow coupling to fibre cores from 50 to 1500 microns .

The user may also need a discrete component to be definitively mounted at the end of the fibre, such as microlens, selfoc lens (see chapter 12, OTDR calibration, Gauthier et al, this volume)

5.6 PROTECTION AND ASSEMBLING

Each individual fibre can be used bare, "as-drawn", or protected. Fibres can be assembled into varied cables, ribbons (see chapter 7, L. Oksanen, this volume), incoherent bundles for light transport, or coherent bundlles for image guiding. Each individual fibre can be highly protected, using a tight or a loose buffer, or a metallic tubing.

6. Selecting the most appropriate fibre for an application

The users requirements for application-specific optical fibres are becoming more and more stringent. Fortunately, the design parameters offered by the manufacturers are numerous, and there is a growing tendency to use customized fibres . The user should precisely identify the main operating parameters he really needs for his application to help the fibre designer to select the appropriate factors for manufacturing the exotic fibre required. Some of them are the effective loss at the operating wavelengths, under the real environment conditions, the modal and polarization regime, the effective bandwidth, launching and transport power capability, dimensions, mechanical characteristics, resistance to ageing in the field environmental conditions. For example, reduced cladding may have accumulated advantages of lowering bending attenuation, reducing volume, and minimizing the applied tensile stress on the fibre with the consequence of improving its mechanical durability and reliability; Relative refractive index difference may be increased to lower sensitivity to bending; Selecting the appropriate coating may extend the range of performance of the fibre.

Keep in mind that almost any already existing type of fibre can be optimized for a given application, for a given operating wavelength, made to a special diameter, and protected with the appropriate coating and reinforcing materials.

7. Examples of fibres and applications

In the telecommunication silica-doped fibre family, let us consider the dispersion-shifted singlemode fibre main characteristics:

Mode field diameter /clad /coating:	8 /125 /250 microns
Core cladding concentricity error:	1 micron
Group index:	1.470
Attenuation at 1310 nm:	0.4 dB / km
at 1550 nm:	0.2 dB / km
Cut-off wavelength:	1050-1350 nm
Nul dispersion wavelength:	1555 +/- 30 nm
Chromatic dispersion: (1540-1570 nm)	2.5 ps /nm km
Proof test (on line, 1 s):	0.9 %
Weibull slope:	61
Weibull fatigue coefficient:	19
Length:	25 km

The application is the long distance transmission at high bit rate. How exotic is an erbium-doped fibre used in fibre amplifiers, compared to it?

Mode field diameter at 1550 nm/clad /coating:	5 /125 /250 microns
Cut-off wavelength:	850 nm
Length:	minimum 10 m

Beside these different values, to usefully characterize this fibre, other distinctive parameters are needed, such as:

NA:	0.29 +/- 0.02
Absorption coefficient at 1532 nm:	4 dB/m +/- 1 dB/m
Absorption coefficient at 1480 nm:	1.2 dB/m +/- 0.3 dB/m
Base-line loss at 1200 nm	lower than 10 dB/km
etc...	

Obviously, it is a completely new fibre, derived from the same technology. This example concretise the difficulty to describe the world of exotic fibres and applications. It is proposed here to correlate the major types of fibres with their main fields of applications.

Silica-doped fibres are the basis of the ultra low-loss, high bandwidth telecommunication fibres. New developments are on their way with fibre-amplifiers and fibre-lasers.

Derived from them are the polarization-maintaining fibres, with their eventually different operating wavelengths (0.48, 0.63 micron), higher attenuation, beat-length values (1 to 5 mm), crosstalk dependence on length and temperature. Specific coatings are used, like polyimide and silicone. Applications are mainly for sensors (see chapter 19, D.A. Jackson, chapter 35, B. Culshaw, this volume) like fibre optic gyroscopes, hydrophones, vibrographs, or to set laser doppler velocimeters, interferometers, displacement gauge or magnetic field monitors. For a new-coming telecom generation, they could support coherent transmissions.

The Plastic-clad silica and hard-clad silica fibres have a large core, (200 to 3000 microns), may reach NAs of 0.4, and come with different buffers. Pistoning can become a problem with silicone. The close "doped silica clad/ pure silica core" fibre type has a lower attenuation and can come with an aluminum buffer. High OH silica provides radiation resistance and improved UV transmission. Low OH silica provides a better spectral transmission range. Their common applications deal with laser delivery

systems, illumination systems, sensing systems, short-haul video or data communications. Custom fibre sizes, coatings and buffers are available on request.
Fluoride fibres can be all glass or plastic clad, with a high NA of 0.6 for the later. Cores are in the range 100-300 microns. Operating temperature may be limited by the coating, Attenuation is of the order of 0.05dB/ m at operating wavelengths in the infrared range from 2 to 3 microns. Products include fibres, cables, incoherent and coherent bundles, and applications deal with temperature sensors, spectroscopy, imaging, medical.
Chalcogenide fibres have the same applications, with a transmission range further extended in the infrared up to about 7 microns.
The silver halide fibre has a remarquable transmission in the infrared, less than 1 dB/ m in the range 8-14 microns, but it has severe draw-backs and present a limited lifetime.
Polymer fibres have large diameters (125 to 3000 microns). Note the very thin optical cladding (10 to 20 microns for a 1 mm fibre). They are limited by temperature, attenuation and bandwidth. They might be suitable for many low-cost applications, with the advantage of the easiness of the connection: short distance data links, illumination devices, displays, sensors, specially scintillating and fluorescent sensors for nuclear research.
Flexible liquid core fibre and lightguide are the last types evaluated here. Quartz and polymer tubing can be drawn-down to a wide range of outer diameters and thicknesses. They can be filled with various liquids of the required index of refraction to adapt to specific transmission requirements (operating wavelength, NA). Flexible liquid-core lightguides are available with an active core from 2 to 8 mm, NAs from 0.47 to 0.76, with a protective tube of 6 to 12.5 mm, an active length from a few centimetres to 20 metres, a minimum bending radius of 30 to 100 mm, and an end window of quartz glass. The main application is a wide variety of light transmission ranges, from UV to the infrared, depending upon the liquid core composition: UV curing, intense white light illumination, cold white light endoscope illumination, medical endoscopy.

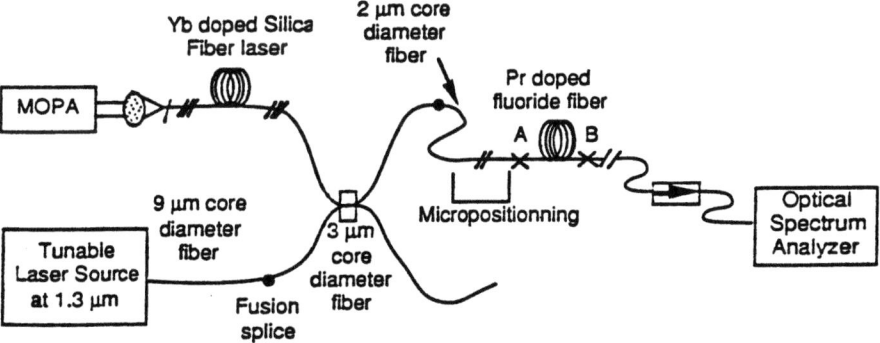

Figure 1: Exotic fibres of a 1.3-micron praseodymium-doped fibre amplifier pumped by an ytterbium-doped fibre laser

The last following example (see Figure 1) [6] shows the powerful association of five different exotic fibres to create a 1.3-micron praseodymium-doped fibre amplifier, pumped by an ytterbium-doped fibre laser. The set-up includes a 1.8-micron core praseodymium-doped fluoride fibre (12-m length, background attenuation = 0.05 dB/m, cut-off wavelength = 0.99 micron, doping concentration = 500 ppm), is fed by micropositionning from a 2-micron core silica fibre, fusion-spliced to the 3-micron core fibre output port of the all-fibre WDM. An ytterbium-doped silica fibre with intracore gratings, pumped by a Master oscillator power amplifier (MOPA), is coupled to one 3-micron core fibre port of the all-fibre WDM. A 9-micron core silica fibre, fed by a 1.3 micron tunable laser source is fusion-spliced to the other 3-micron core fibre port of the all-fibre WDM.

8. Conclusion

The huge potentiality of optical fibres is illustrated by the already existing numerous exotic fibres. Aside from a few specific large fibre-consuming applications, there is a tendency to develop custom fibres, as more and more manufacturers propose to design new fibres. Scientists, engineers, technicians and industrials are strongly encouraged to use exotic fibres to solve their problems or develop new concepts. There are market limitations for volume reasons, but there is no doubt they will be pushed away by the growing demand.

9. References

1. Doremus, R. H (1973) *Glass science*, John Wiley and sons, New York
2. Zachariasen, W. H. (1932) J. Am. Chem. Soc.,54, 3841
3. Warren B.E. and co-workers, (1934) J. Am. Ceram. Soc., 17, 249
4. Ishigure T., Nihei E., and Koike Y.(1994) Applied Optics, 33, 19, 4261-4266
5. CECC standard 86000
6. E. Taufflieb, D. Joulin, H. Lefevre, J.Y. Allain, J.F. Bayon, and M. Monerie, (1994), OFC'94, paper PD4

FIBERS AND FIBER LASERS FOR THE MID-INFRARED

UDO B. UNRAU
Technische Universität Braunschweig
Institut für Hochfrequenztechnik
P.O. Box 3329, D-38023 Braunschweig
Germany

Abstract

A compact overview of glasses and optical fibers for the mid-infrared is given, their properties and fabrication as well as special infrared characterization procedures are described. Emphasis is laid on halide glasses, especially on heavy-metal fluoride glasses. This is followed by a description of the spectroscopic properties of such fibers, rare-earth doped as laser hosts, and state-of-the-art examples for fiber lasers emitting at 2.7, 2.9 and 3.45 µm.

1. Introduction

The vast majority of optical fibers are silica-based and are used for applications in telecommunications at wavelengths between 0.8 µm and 1.6 µm. There are, however, many fiber applications in the mid-infrared beyond 2 µm, where silica becomes rapidly opaque. For example, absorption by water is very high between 2.7 - 3.0 µm, a wavelength range well suited for laser ablation of biological tissues. An atmospheric transmission window extends from 3 µm to 5 µm, where LIDAR operation and spectroscopic monitoring of environmental pollution are feasible, since many organic compounds have resonance absorptions in this region.

In the first sections of the following, glasses and optical fibers for transmission in the mid-infrared are considered, their properties and fabrication processes shortly described and, for the fibers, special IR characterization procedures presented. Since fabrication of halide fibers is the most mature technology in this class of materials, special emphasis is given to these, especially to heavy-metal fluoride glass (HMFG) fibers. In the following part spectroscopic properties of rare-earth doped HMFG are discussed, and it is shown that they are excellent laser hosts. The last section deals with the still quite esoteric field of fluoride fiber lasers emitting at wavelengths above 2.5 µm, a field which may suddenly become the focus of many interests, if output powers can be raised to about 1 W CW.

2. Glasses and fibers for the mid-infrared

2.1. OPTICAL LOSSES IN GLASSES

If the attenuation coefficient for intrinsic losses of solids is depicted vs. wavelength, λ, on a double logarithmic scale, a curve in the form of the letter V is obtained. On the short wavelength side, loss is determined by superposition of the Urbach edge due to electronic transitions and Rayleigh scattering due to local changes in the optical density of the material extending over dimensions small compared to λ. In general, the contribution of electronic transitions to the total loss is already small in the visible and can be neglected in the mid-infrared. The attenuation coefficient of the remaining Rayleigh scattering, α_{RS}, is given by

$$\alpha_{RS} = A_{RS}\, \lambda^{-4} \tag{2.1}$$

where A_{RS} is the Rayleigh scattering coefficient of the glass.

For infrared applications, the branch of the V-curve on the right hand side is of major importance. This stems from multiphonon interactions i.e. quantized vibrational absorptions. The detailed theory, involving complete description of anharmonic overtones due to non-linear bonding forces, is quite complicated; an overview can be found in [1]. Here, a more intuitive approach is chosen which, although strictly-speaking limited to ionic/covalent materials, depicts the principle admirably. Loss is due to interaction of optical waves with resonant vibrations of the crystal/glass lattice. One must thus first of all define the vibrational modes of the constituent atoms and molecules. In a simple diatomic molecule with two vibrating point masses, m_1 und m_2, the fundamental vibration frequency, ν_0, is given by

$$\nu_0 = \frac{1}{2\pi}\sqrt{\frac{\kappa}{m}} \tag{2.2}$$

where κ is the bond strength as the restoring force and $m^{-1} = m_1^{-1} + m_2^{-1}$ holds for the reduced mass, m. If now a linear chain is made up of such diatomic molecules, serving as a model for a simple crystal, two vibrational modes are possible: one with high frequency and one with low frequency vibrating molecules. Only the high frequency mode can interact with optical waves and is called the optical mode in contrast to the low frequency, acoustical mode. The highest vibration frequency of the optical mode is given by eq. (2.2) with κ replaced by 2κ. In 3-dimensional diatomic crystals, 3 optical and 3 acoustical modes propagate, and in more complicated crystals and glasses many more modes occur.

The highest frequency vibrational modes contribute most to the IR edge absorption. Quantum mechanically, energy in the different modes can be quantized as optical phonons and ω_0, the temperature dependent characteristic frequency of a glass, is an average optical phonon frequency which is close to the highest vibrational frequency. It can be derived from reflectivity measurements on a material sample [1]. Far above ω_0, the slope of the IR absorption coefficient vs. normalized frequency ω/ω_0 is essentially exponential. Minimum total loss occurs at the frequency where this infrared absorption tail intersects the Rayleigh scattering curve. In order to obtain good infrared transmissivity, it becomes clear from eq. (2.2) that glasses with low phonon energies made of heavy molecules with low bond strength must be used.

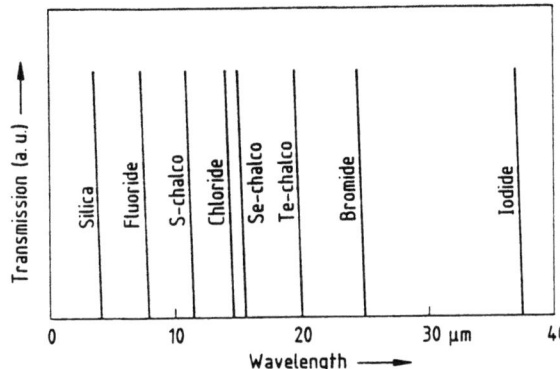

Figure 2.1: IR-absorption edge of different glass groups (schematic)

Among the halides (column VII in the periodic table of elements), the masses increase and the bond strengths to metals, M-X, generally decrease in the order X = F, Cl, Br, I. In Fig. 2.1, the infrared edge of the transmission through thin sheets of several glasses is schematically depicted vs. the wavelength. It can clearly be seen how the wavelength cut-off increases with increasing atomic mass, i.e. from fluoride to iodide glasses. Since the M-X bond is weaker than the M-O bond, fluoride glasses have a larger cut-off wavelength than oxide glasses, of which silica is the only one shown in Fig. 2.1. Chalcogenide glasses are made from the column VI elements S, Se, Te (usually with Ga, Ge or As), and the position of the infrared edge in Fig. 2.1 follows their respective masses. Note that all of these elements are hazardous to health and that Se and As are extremely toxic.

There are of course several drawbacks in using glasses with a large IR cut-off wavelength. The lower the bond strength, the lower is the chemical durability and mechanical strength. Fluorides are weakly and chlorides quite substantially hygroscopic, e.g. NaCl. Fluoride glasses have less than 1/3 of the mechanical strength of silica; iodide glasses may already deform at room temperature under their own weight. The glass transformation temperature decreases from $T_g > 1000\,°C$ in silica to less than 200°C in halide glasses. Furthermore, halide glasses show greater ionicity than oxide glasses, which results in a stronger crystallization tendency below the melting point.

On the other hand, infrared materials usually offer very low intrinsic loss minima as shown in Fig. 2.2. With a minimum loss of $\alpha \approx 0.1$ dB/km in silica fibers the theoretical limit has been reached. From this it can be seen that silica glass technology is very mature. By changing from Si as the glass former to the heavier Ge, the slightly more IR transparent Germanate glasses are obtained, but their technology is not as developed as that of silica.

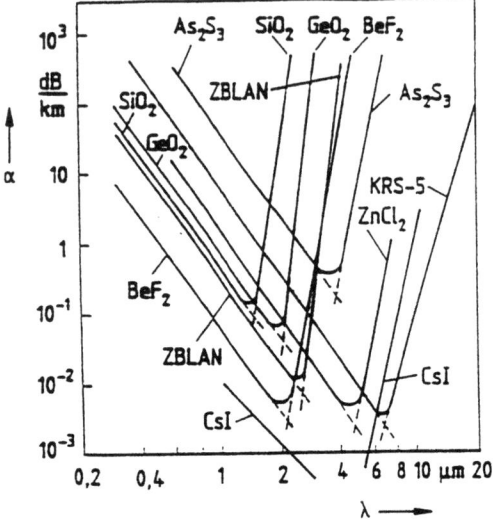

Figure 2.2: Intrinsic losses in infrared materials (after [2,3])

The next group of interesting glasses are the halide glasses, where BeF_2 and the mixed fluorozirconate ZBLAN represent the fluoride glasses. BeF_2 was the first fluoride glass but is very toxic and not in use. ZBLAN is a heavy metal fluoride glass composed of ZrF_4, BaF_2, LaF_3, AlF_3 and NaF, the most stable fluoride glass discovered to date and that mostly used for fiber drawing. It is also the glass with the most mature technology: fiber losses have been brought down to 0.45 ± 0.15 dB/km @ 2.55 μm but are still more than one order of magnitude beyond their theoretical limit of approximately 0.01 dB/km. In Fig. 2.2, $ZnCl_2$ and CsI have even lower losses but their drawbacks have already been discussed. KRS-5 is a mixed crystal of TlBr-TlI and will not be further discussed here. As an example for the chalcogenides, As_2S_3 shows that intrinsic losses are typically higher than in halide glasses.

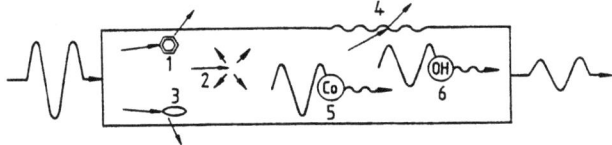

Figure 2.3: Sources of extrinsic losses in glasses (after [4])

In addition to intrinsic losses, extrinsic losses introduced by the fabrication process have to be taken into account, the origins of which are sketched in Fig. 2.3. Firstly, there are losses due to scattering from microcrystals (1), local phase separations (2), bubbles & small metal particles (3) or surface roughness (4). The wavelength dependence of the scattered power depends on the relative refractive index, the dimensions and the form of the scattering centers. If the former are small compared to λ, Rayleigh scatter - with $\sim \lambda^{-4}$-dependence - is observed; for larger values, Mie scatter ($\sim \lambda^{-2}$) or wavelength independent scatter may occur. Secondly, there are absorptions due to impurities, of which transition metal ions (5), e.g. V, Cr, Mn, Fe, Co, Ni, and OH$^-$-ions (6) are the most prominent. Transition metals have uncompletely filled inner shells where electronic transitions are possible which are strongly influenced by the glass matrix. Absorptions due to these transitions can have very detrimental effects on the fabrication of low-loss fluoride glasses, e.g. 3.6 ppb of

Fe or 3.2 ppb of Co will already cause 0.1 dB/km extra loss. OH⁻-ions, rapidly incorporated into glasses from humidity in the processing atmosphere, have their fundamental resonance in fluoride glasses near 2.9 μm, close to the predicted loss minimum of ZBLAN. In order to reduce such impurity electronic and other optical absorptions, raw materials for IR glasses must be carefully purified and processed in a very dry or reactive atmosphere.

2.2. GLASS AND PREFORM FABRICATION

2.2.1. *Fluoride glasses*

In order to ensure processing of the raw materials in a dry atmosphere, a glove box with less than 1 ppm water content in the atmosphere is used in the case of fluoride glasses. In our institute the highly purified fluorides are at first dried at 160 °C overnight before storing them in the glove box in the dry atmosphere. Then they are weighed to obtain the proper composition (in the case of ZBLAN: 53 ZrF_4, 20 BaF_2, 4 LaF_3, 3 AlF_3, 20 NAF in mol%) and placed in a platinum crucible. Melting is also performed under a dry nitrogen atmosphere at 850 °C for 1-2 hours while for the last hour nitrogen is slowly bubbling through the melt in order to homogenize it.

For fabrication of the cladding of a fiber preform, the temperature is reduced to 650 °C, and the melt is poured into a tubular mould of polished brass as shown in Fig. 2.4. The tube is held slanted to avoid the creation of bubbles. The mould is then spun at a high revolution rate while the temperature is reduced. Thus, a glass tube is formed which serves as a preform cladding. Finally, the tube cools down from 260 °C to room temperature under strict temperature control, i.e. it is tempered to remove residual stresses.

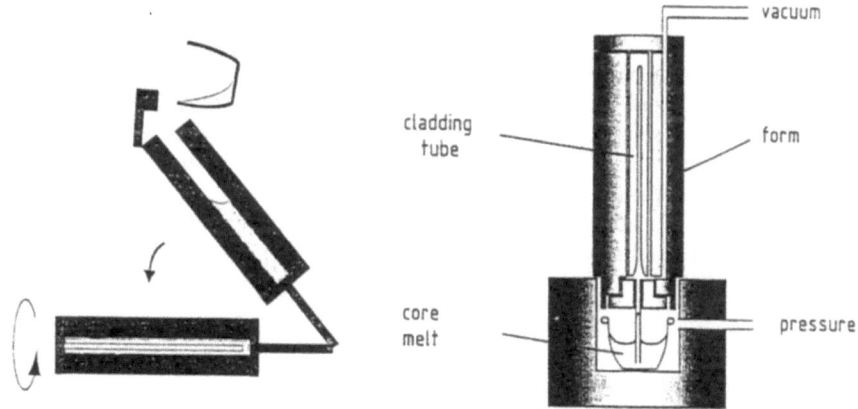

Figure 2.4: Rotational casting of the preform cladding

Figure 2.5: Pressure moulding of a fiber preform

One way to make a fiber preform is the pressure moulding procedure as used in our own institute and shown in Fig. 2.5. The cladding tube is mounted upside down in a heated mould such that there is a connection to the core melt via a thin platinum tube. The entire mould can be evacuated. At first the system is evacuated both fro: above and below. The lower vacuum pipe is then closed and a valve is opened to allo pressured nitrogen to enter the lower chamber. The melt is pressed through th platinum tube into the cladding tube; the vacuum in the tube is necessary to avoid th creation of bubbles between the melt and the inner glass wall. Finally, the mould cool down, and the preform is removed. Pressure moulding allows for the fabrication o preforms with thin cores. For fabrication of a core rod, the cladding tube can bɛ replaced by a thinner mould.

An alternative method of preform fabrication, which leads to very low-loss fiber preforms, involves rotation casting for the cladding and direct pouring of the core melt into the cladding after a short cooling period. Both of these process steps are however performed under reduced pressure [5]. However, no further details have been published since most details of glass and fiber fabrication processes are typically kept proprietary.

The quality of the preform is usually checked in a dark room by transmitting white or He-Ne laser light through it. Extrinsic scattering of the types 1 - 4 in Fig. 2.3 can easily be detected this way. More delicate is the quantitative measurement of the homogeneity of a glass with the Christiansen-Shelyubskii method [6]. The sample is ground to grains (0.3 - 0.5 mm) and placed together with an immersion fluid into a cuvette which is temperature controlled. Monochromatic light is transmitted through the cuvette while varying the temperature. This changes the refractive index difference between glass and fluid. The transmittivity of the cuvette is measured and depends due to diffraction on temperature and homogeneity. Homogeneous samples yield high transmissions extending only over small temperature ranges, the latter being a measure of the homogeneity. Furthermore, from the measured temperature for maximum transmission and knowledge of the temperature dependence of the refractive index of the immersion fluid, the average refractive index of the glass can be determined.

2.2.2. Chalcogenide glasses

Raw materials of high purity (Se: 5N [= 99.999%], As: 5N - 9N, Te: 6N, Ge: 50 Ω/m) are weighed and mixed in a glove box. Melting and refining is normally performed in a sealed silica ampoule in a rocking furnace at 900 - 1000°C for 5 - 16 hours. After cooling, the glass can be broken to grains and further purified by distillation through a fritted silica disk in a sealed ampoule. The distilled part of it is sealed and tempered. Finally, the glass is melted again and processed by the rotational casting method to a cladding tube or otherwise to a core rod [7].

2.3 FIBER DRAWING AND CHARACTERIZATION

2.3.1 *Fiber drawing*

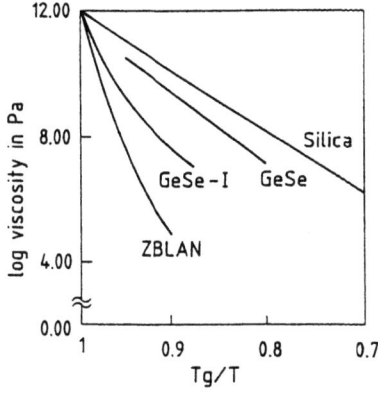

Figure 2.6: Temperature dependence of the viscosity of several IR-glasses

The principle of the fiber drawing process does not differ much from that for silica fibers, except for the lower temperature range and a very tight process control. The reason becomes clear from Fig. 2.6 which shows the temperature dependence of the viscosity of different IR glasses above the glass transformation temperature T_g where the fiber drawing process takes place. It is seen that especially with ZBLAN the viscosity changes dramatically over several orders of magnitude within a few degrees whereas for silica the change is much more smooth. The chalcogenide GeSe glass behaves similar to silica while the hybrid chalco-halide glass, GeSe-I, shows an intermediate viscosity decrease. In order to minimize crystallization during drawing, it is generally aimed at drawing with the lowest temperature possible, which on the other hand brings the risk of a fiber break.

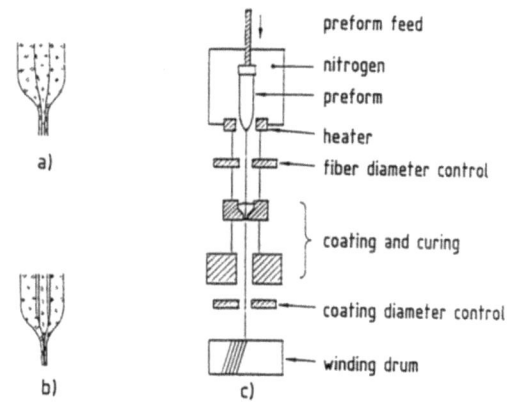

Figure 2.7: The fiber drawing process
a) diameter reduction of a preform
b) rod-in-tube fiber drawing
c) fiber drawing tower (schematic)

Fig. 2.7(a-c) shows different aspects of the fiber drawing process. Due to volume contraction of the melt, it is impossible to manufacture preforms with a very thin core such as are required for single-mode fibers. Therefore, for fabrication of such fibers, an intermediate process step has to be introduced in which a core/cladding preform is pre-drawn (collimated) to a smaller diameter as shown in Fig. 2.7(a). At our institute this is done in a separate, small drawing machine which allows for linear drawing of a preform to a length of 1.4 m. Before this is done, the preform is carefully ground, polished and etched to a diameter of ≈ 13 mm and placed in a Teflon® tube in order to protect the surface from humidity during collimation and subsequent storage. This coating is easily stripped before the next step: a rod-in-tube drawing procedure as shown in Fig. 2.7(b). Here, the thin preform rod is introduced into a closely fitting cladding tube and both are drawn together to a fiber. In the case of chalcogenide glasses a core rod is inserted into a cladding tube.

The drawing tower is shown in Fig. 2.7 c. The assembled preform is clamped to a feed mechanism which is synchronized to the fiber winding drum. The tip of the preform is brought forward into the heating zone in the center of an induction-heated graphite ring where the glass is heated until a drop - pulling fiber behind it - falls down and is attatched to the winding drum. Now the rotation speed of the drum and the temperature are varied until winding load, speed and fiber diameter reach predetermined values. The fiber is on-line coated with an acrylate.

2.3.2 IR fiber loss measurements

If possible, IR fibers are characterized in the same way as silica fibers, since fiber measurement equipment up to a wavelength of 1.6 μm is readily available. Difficulties arise for measurements further into the infrared or with fibers nearly opaque below 1.6 μm. Due to their critical fiber manufacturing technology, IR fibers have to be checked for scattering centers first before making any more detailed measurements. ZBLAN fibers, being transparent in the visible, are checked by injecting white light or low-power He-Ne laser light into the fiber. Non-guided scattered light leaves the fiber sideways through the cladding and can be seen with naked eye. Care must be taken to observe under all polar angles since many scattering centers have very distinct angular characteristics; in the case of laser excitation, looking at the fiber endface should be carefully avoided.

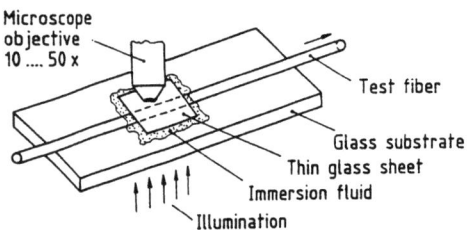

Figure 2.8: Microscope inspection for fiber defects

For a closer look at single scattering centers, a microscope inspection as shown in Fig. 2.8 is needed. The fiber under test is drawn through immersion fluid between two glass sheets (usually microscope slide and cover sheet) and inspected with a microscope while illuminated from below. If the thin fiber coating is sufficiently transparent, it is not necessary to remove it. The check can reveal, for example, whether small crystals or bubbles cause the scattering, and where they occur in the fiber. This allows conclusions to be drawn regarding the nature of difficulties with the technological processes. If the fibers to be tested are opaque in the visible, an IR source for illumination and an IR camera for inspection must be used.

As with silica fibers, the real loss measurement should preferably be done with the cut-back method. Since with each measurement, depending on the expected fiber loss, from 5 cm to several meters of fiber will be lost and since IR fibers are very expensive, broadband loss measurements in a single step are desirable. This can be done by using the external path of a Fourier Transform Infrared spectrometer (FTIR-S) in an arrangement as in Fig. 2.9. (The interior of an FTIR-S is shown in Fig. 2.10). A collimated beam of radiation leaves the FTIR-S and is focused onto the fiber end by a special focusing mirror set-up. Mirrors are used because lenses show a variation of

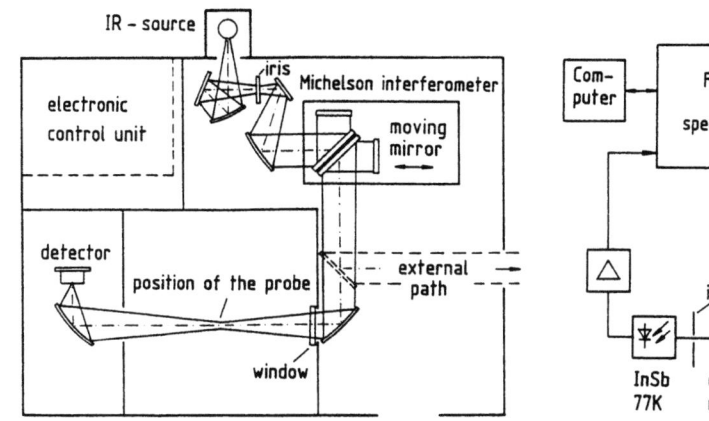
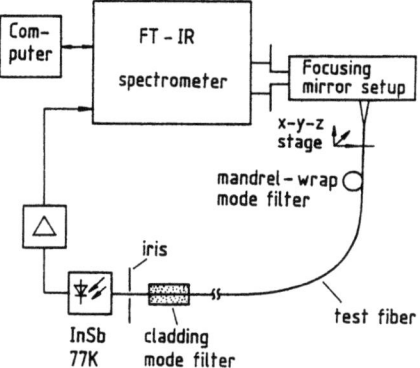

Figure 2.9: Fiber measurement with a FTIR spectrometer

Figure 2.10: Fourier Transform Infrared Spectrometer

cal length over the broad bandwidth used in an FTIR-S. A short distance after the input end, cladding modes in the fiber are stripped by a mandrel-wrap filter. Remaining or newly created cladding modes are stripped a short distance before the fiber end in an oil bath and radiation modes blocked by an iris before the fiber end radiates onto a cooled InSb detector. The detector signal is fed back to the spectrometer where the fiber transmission curve is computed.

Inside the FTIR-S (see Fig. 2.10) the radiation of an IR-source (for the mid-infrared a halogen lamp suffices) is focused by a set of mirrors and spatially filtered by an iris. The diverging beam is collimated by further mirrors and enters a Michelson interferometer in which one mirror is movable in a very precisely controlled fashion over a distance x. Both beam parts reflected by the mirrors are spatially coherent and interfere with a path difference of $2x$. The output beam is internally focused inside a sample chamber, where a bulk sample can be inserted into the focus and then guided to a detector. The detector measures the intensity which is recorded over x, digitalized and Fourier transformed.

To show the principle, the two extremes of a spectrum will be considered. A monochromatic source emitting at $\lambda = \lambda_0$ yields an intensity function (interferogram)

$$I(x) = S(\nu_0) * \cos(2\pi\nu_0 x) \qquad (2.3)$$

where $\nu_0 = 1/\lambda_0$ is the wave number and $S(\nu_0)$ the spectral intensity at ν_0. Its Fourier transform is a spectral line at ν_0. On the other hand, white light causes a spike at $I(x_0)$ and is Fourier transformed to a continuous spectrum $S(\nu)$ = const. Eq. (2.3) is used for calibrating a FTIR-S. A reference He-Ne laser causes a sinusoidal interferogram

$I(x)$, and therefore it is easy to sample the 'real' interferogram at the zeros of the sinusoidal reference with high precision, which results in a wave number accuracy down to 0.01 cm^{-1}. In the FTIR-S of Fig. 2.10 the source is a halogen lamp. Its interferogram is measured with an empty sample chamber and Fourier transformed to the reference spectrum $R(\nu)$. Then the sample, e.g. a bulk glass, is introduced and its spectrum $S(\nu)$ is determined. It is similar to $R(\nu)$ but shows less intensity at wavenumbers where the sample absorbs. The desired transmission spectrum is obtained by division, $T(\nu) = S(\nu)/R(\nu)$.

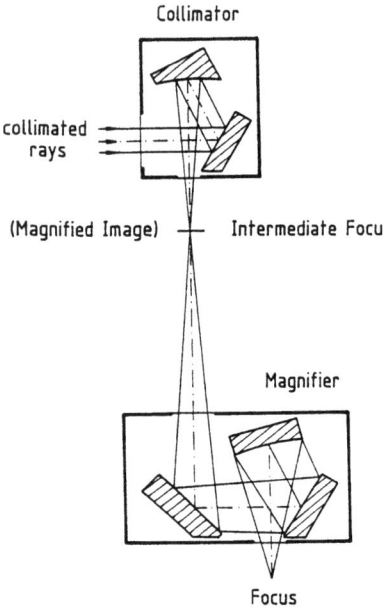

Figure 2.11: Focusing mirror set-up with low astigmatism [8]

For fiber measurements, the external exit in Fig. 2.10 is used. The collimated beam is focused with two special 26° off-axis paraboloids together with several plane mirrors as shown in Fig. 2.11. The whole set-up, consisting of a focusing mirror (or collimator in the reverse direction) and a demagnifier 2:1 (magnifier) was manufactured by the Fraunhofer-Institut für Physikalische Meßtechnik in Freiburg / Germany. It ensures a smaller and much more precise focus than 90° off-axis paraboloids. For single-mode fiber measurements it is essential that the already very low launching efficiency of a halogen lamp is optimized. Thus, the rather complicated set-up in Fig. 2.11, with several mirrors, yields a higher launching efficiency than a single 90° off-axis paraboloid. Firstly, the transmission spectrum of a long fiber is recorded, then the fiber is cut back without touching the input coupling arrangement, and the transmission spectrum for the short fiber is recorded. The spectral loss curve of the fiber under test can be computed from these spectra in the usual way. In conclusion, the FTIR-S set-up allows for quick broadband loss measurements in the infrared.

2.3.3 Fiber loss data

We will now look at fiber losses reached in different laboratories. Fig. 2.12 shows the loss curves for several halide glass fibers. A ZBLAN fiber, doped with Pb in the core to raise the refractive index and with hafnium replacing zirconium in the cladding to decrease the refractive index, reached 0.45 ± 0.15 dB/km @ $\lambda = 2.35$ μm [9]. The loss was measured on 60 m of fiber. The position of the minimum varied among different fibers between 2.3 μm and 2.55 μm; it depends on the relative loss contributions of the OH$^-$ vibration and wavelength dependent extrinsic scatter. An AlF$_3$-based composite fiber is the next best with $\approx$ 10 dB/km minimum loss near 2 μm. It should be noted that the technology of AlF$_3$ fibers is extremely difficult. Both fibers show a very steep increase in attenuation at longer wavelengths. There have been many attempts to reach high transmission at longer wavelengths with hybrid halide fibers. Fig. 2.12 shows two examples.

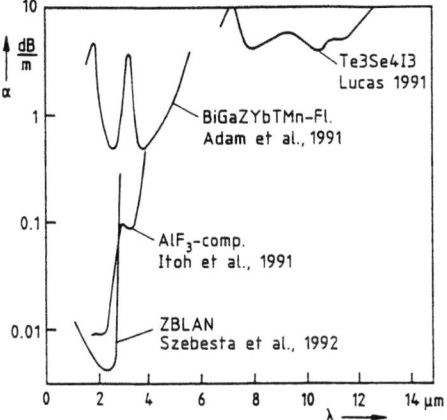

Figure 2.12: Loss spectra of fibers made from halide glasses

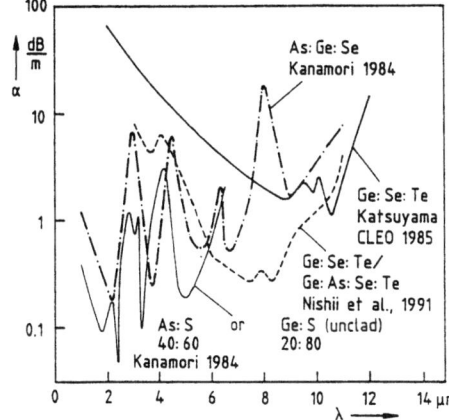

Figure 2.13: Loss spectra of fibers made from chalcogenide glasses

Fig. 2.13 shows similar curves for fibers made from chalcogenide glasses which are often made for transmission of CO-laser radiation, which can be tuned from 4.8 μm to 8.4 μm, and CO$_2$-laser radiation at 10.6 μm. These fibers are almost all both toxic and opaque in the visible. The relatively non-toxic sulphide fibers, with low loss near 2.55 μm, cannot yet compete with ZBLAN.

Finally, it should be noted that polycrystalline silver halide fibers are also on the market with losses < 1 dB/m from 5 to > 13 μm [10]. They are, however, beyond the scope of this article which deals with glass fibers only.

3. Mid-infrared fiber lasers

Fluoride glasses based on the ZBLAN glass compostion have turned out to be excellent hosts for laser-active lanthanoides, the rare-earth (RE) metals. Since fluoride fibers are transparent from approx. 400 nm to > 5 μm, many fiber lasers and amplifiers have been realized in this wavelength range, up to 3.5 μm so far as shown in Fig. 3.1. Up to approx. 2.3 μm, doped silica fibers can also be used for most laser transitions. Therefore, we shall restrict ourselves in the following to lasers emitting above 2.5 μm, which still is the domain of fluoride fibers. As shown in Fig. 3.1 there are fiber lasers emitting at 2.75 μm (Er^{3+}), 2.9 μm (Ho^{3+}) and 3.5 μm (Er^{3+}). Before presenting the laser properties it should firstly be discussed why fluoride fibers are so well suited as

IR laser hosts and, secondly, the spectroscopic properties of Er^{3+} and Ho^{3+} in ZBLAN should be explained.

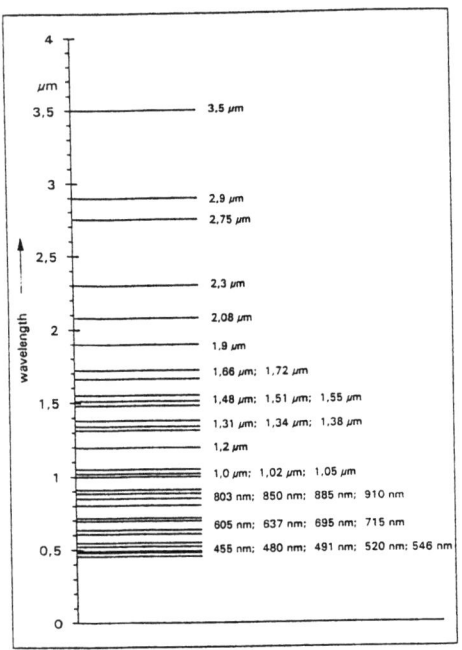

Figure 3.1: Emission wavelengths of ZBLAN fiber lasers [11]

3.1. RARE-EARTH DOPED FLUORIDE FIBERS

3.1.1. *Fluoride glasses as laser hosts*

The composition of ZBLAN as given in section 2.2.1 shows that it already contains 4 mol% LaF_3 which acts mainly as a network former. It is easy to replace La^{3+} ions by RE^{3+} ions and to dope with quite high concentrations without disturbing the glass. In contrast to its role in ZBLAN, La^{3+} in silica plays the role of a network modifier which breaks up the network [12] such that high dopant levels are fundamentally deleterious to the glass.

Another very important property of ZBLAN is its low phonon frequency, ω_0, as introduced in section 2.1. In ZBLAN the corresponding fundamental phonon energy is ≈ 500 cm^{-1} compared to 1100 cm^{-1} in silica. As the probability for non-radiative decay decreases with the number of phonons necessary to transfer energy to a lower energy level in an ion, fluoride glasses exhibit weaker non-radiative transitions than silica because more phonons have to be generated for the same process. The non-radiative transition rate Γ^{nr} across the energy gap ΔE to the next lower level follows the empirical rule

$$\Gamma^{nr} = C\ e^{-2\alpha \Delta E} \qquad (3.1)$$

where C and α are material constants; for ZBLAN $C = 2 \cdot 10^5$ s^{-1} and $\alpha = 0.0021$ cm hold at room temperature [13]. Due to the low multiphonon rate, the quantum efficiency in fluoride glasses is higher than in most other glasses. Therefore, fluoride glasses have more fluorescent transitions than other laser hosts.

For the total radiative lifetime, τ_j, in a given state j, the expression

$$\tau_j^{-1} = \Gamma_j^r + \Gamma_j^{nr} \qquad (3.2)$$

holds which is the sum of all radiative and non-radiative decay rates from that level j. Many levels in fluoride glasses have a longer lifetime than in other glasses. This leads to lower thresholds for lasing transitions. A drawback, however, is that in fluoride glasses several metastable level often exist in addition to the lasing level, which complicates the theoretical treatment. If the lower level has a longer lifetime than that of the upper, CW operation is generally inhibited if no additional processes empty the lower level.

If more than 10 phonons are needed for a transition from an excited state, the lifetime is independent of temperature and the multiphonon rate vanishes. If less than 10 phonons are needed, the measured lifetime becomes temperature dependent. If ≤ 4 phonons are required, multiphonon transitions dominate [14]. Therefore, an energy gap to be bridged by 4 phonons can be considered the minimum for the appearance of fluorescence. In ZBLAN this leads to $\Delta E = 2000$ cm^{-1} ($\lambda = 5$ μm) compared to $\Delta E = 4400$ cm^{-1} ($\lambda = 2.27$ μm) in silica. Together with sufficiently low attenuation in the mid-infrared up to approximately 4 μm, it becomes clear that ZBLAN is a very good fiber laser host, at least up to 4 μm. Actually, it is also a better laser host than silica below 2 μm, but the advantages of mature silica fiber component technology are usually more important than the modest gain in fluoride fiber laser performance.

3.1.2. Spectroscopic properties of Er^{3+} and Ho^{3+} in ZBLAN

According to Fig. 3.1, only erbium and holmium ions so far have been used to realize mid-infrared glass fiber lasers. Glasses have by definition an irregular structure. Each ion site in a glass has its own absorption and emission spectra as well as its own decay rates. The measurable absorption and emission spectra and decay rates are, however, averaged over all sites. Together with the many radiative and energy transfer processes possible in doped ZBLAN, it is therefore in general quite difficult to make quantitative calculations of IR laser behavior and to measure lifetimes accurately.

Only radiative decay processes can be measured directly [15], the multiphonon decay rate Γ^{nr} must be determined by indirect methods, e.g. via the Judd-Ofelt theory. This theory allows for estimation of the total radiative lifetime τ_j, any difference between τ_j^{-1} and a measured Γ_j^r can be, according to eq. (3.2), attributed to Γ_j^{nr}.

Fig. 3.2 shows the part of the absorption spectrum of Er^{3+} in ZBLAN which can be covered by laser diodes for pumping a fiber laser. It provides therefore information where laser diode pumping of an erbium-doped laser is possible, e.g. 650 nm, 795 nm, 980 nm and ≈ 1.5 μm. The last pump wavelength can not be used for pumping mid-IR lasers, as it will be shown further down. 650 nm is a very important pump wavelength, but high power laser diodes are not yet available commercially.

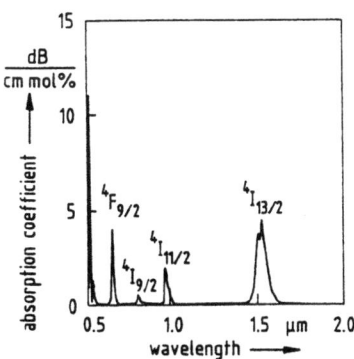

Figure 3.2: Absorption spectrum of Er^{3+}- doped ZBLAN glass [after 14]

Emission spectra depend on the pumping wavelength. Below 550 nm many densely spaced strong absorption bands exist. Therefore, in order to get a quite complete picture of fluorescent transitions, pumping with an argon laser operation in 'all lines' mode is a good way. Figs. 3.3 and 3.4 show such fluorescence spectra.

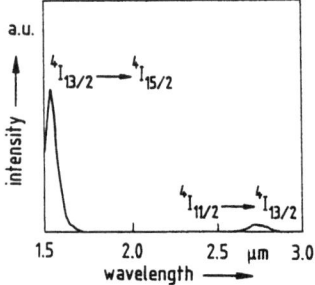
Figure 3.3: Fluorescence spectrum of Er^{3+}-doped ZBLAN glass [after 14]

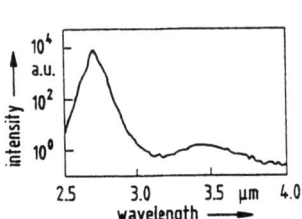
Figure 3.4: Mid-IR fluorescences of Er^{3+} in ZBLAN glass [11]

While Fig. 3.3 gives an overview from 1.5 μm to the mid-infrared and shows how weak the fluorescence near 2.7 μm is in comparison to the other emission, Fig. 3.4 shows a blow-up of the fluorescence spectra beyond 2.5 μm and reveals an even weaker fluorescence near 3.5 μm. Note that in Fig. 3.4 the intensity scale is logarithmic: on a linear scale the fluorescence near 3.5 μm would be barely visible, and in fact, it has been overlooked for many years. It is really surprising and a good demonstration of the advantages of ZBLAN fiber lasers that on the basis of such a weak fluorescence a fiber laser operating at room temperature could be built, as it will be discussed further down.

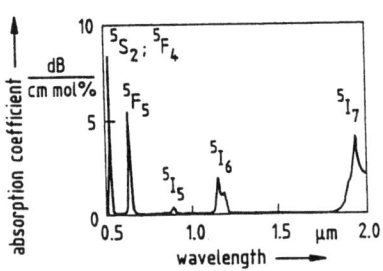
Figure 3.5: Absorption spectrum of Ho^{3+} in ZBLAN glass [after 14]

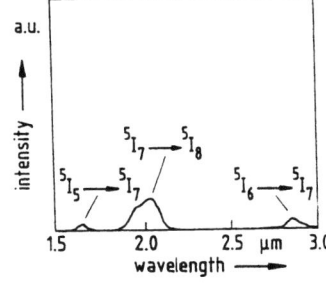
Figure 3.6: Fluorescence spectrum of Ho^{3+}-doped ZBLAN glass [after 14]

In Fig. 3.5 the absorption spectrum of holmium is shown, where 640 nm, 750 nm and 890 nm are the most interesting absorption bands for mid-IR laser operation. Unfortunately, the absorption at 750 nm is extremely weak and can not be detected in Fig. 3.5.

Fig. 3.6 shows the fluorescence spectrum of holmium if pumped with all lines of an argon laser. The fluorescence at 2.0 μm is quite important since quite high power fiber lasers have been realized near 2.0 μm as well in silica as in ZBLAN fibers. The mid-IR fluorescence at 2.9 μm is the basis for ZBLAN fiber lasers at the heart of the water absorption peak and therefore of interest for medical applications if higher powers can be reached.

3.2 ERBIUM-DOPED FIBER LASERS IN THE MID-INFRARED

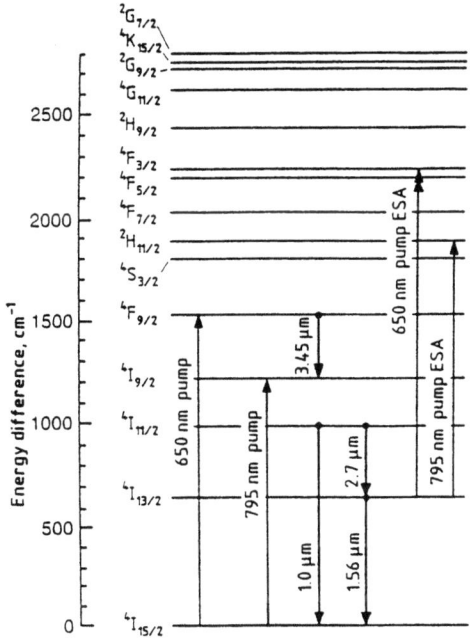

Figure 3.7: Energy level diagram for Er^{3+} in ZBLAN glass with pump and energy transfer mechanisms for mid-IR emissions

Fig. 3.7 shows the energy level diagram of Er^{3+} in ZBLAN. The 2.7 μm emission occurs between the $^4I_{11/2}$ and the $^4I_{13/2}$ levels. This transition is expected to be self-terminating and to inhibit CW operation. The lifetimes as measured with 650 nm pump are around 9 ms for $^4I_{11/2}$ and around 7 ms for $^4I_{13/2}$. Nevertheless, CW lasing was observed while pumping at 650 nm, 795 or 980 nm. This was mainly due to pump excited state absorption (ESA) processes which empty the lower laser level and reduce its lifetime. In addition, direct fluorescent transitions from the laser levels to the ground level are possible. Strong fluorescence at 1.0 μm can be observed, even during operation at 2.7 μm.

An effective way to reduce the lifetime of the lower laser level is by weakly co-doping the fiber with Pr^{3+} [16]. This enables nonradiative, resonant energy transfer from $^4I_{13/2}$ in Er^{3+} to $^3F_3$ and $^3F_4$ in Pr^{3+}. In contrast to Er^{3+}, many other levels are found in Pr^{3+} between $^3F_3$ and the ground level which allow a quick nonradiative decay.

The 3.45 μm emission occurs between the $^4F_{9/2}$ and the $^4I_{9/2}$ levels. It is pumped at 650 nm directly into the upper lasing level. In contrast to the multiphonon transition rate of 456 s^{-1}, the radiative transition rate is only 2 s^{-1} - indeed a very weak fluorescence. The measured lifetimes at room temperature are 0.59 ms for $^4F_{9/2}$ and 0.47 ms for $^4I_{9/2}$ [11]. Therefore, CW operation is possible since nonradiative are involved. At 77 K the nonradiative processes are reduced, and the lifetime of the lower laser level is slightly larger than that of the upper. Nevertheless, CW operation is again possible due to the beneficial effect of pump ESA between $^4I_{9/2}$ and $^2K_{15/2}$.

Fig. 3.8 shows the best fiber laser characteristics at 2.7 μm published so far. The highest CW output power of more than 10 mW at room temperature is obtained with 35 cm of fiber forming a laser resonator without mirrors, e.g. reflectivities of 4% at the fiber - air interfaces. (In the mean time the output power has been improved to nearly 30 mW by the author.) A much lower lasing threshold is reached by using an output mirror with 96% reflectivity. Fig. 3.8 also shows that pumping at 980 nm, directly into the upper laser level, is also possible. It is, however, not very attractive since pump ESA from the upper laser level to $^4F_{7/2}$ draws much energy from the laser. After nonradiative decay to the thermally coupled $^2H_{11/2}/^4S_{3/2}$ levels green fluorescence at 550 nm to ground level can be observed.

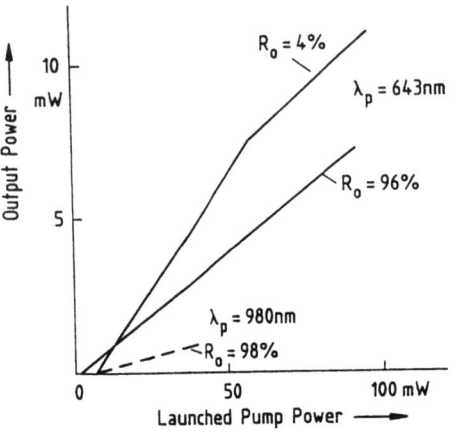

Figure 3.8: Output characteristics of $Er^{3+}:Pr^{3+}$-codoped fluoride fiber lasers emitting at 2.71 μm [after 17]
(fiber: 5000ppm Er^{3+}:300ppm Pr^{3+}; 10/125/0.4)

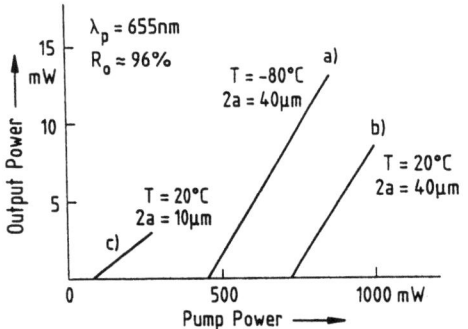

Figure 3.9: Output characteristics of Er^{3+}-doped fluoride fiber lasers emitting at 3.45 μm [after 11]
(fiber: a,b: 1 mol% Er^{3+}; 40/125/0.2
c: ditto; 10/125/0.4)

Much more modest are the results with fiber lasers emitting at 3.45 μm. Cooled to -80 °C an output power of 14 mW was measured on a 12 cm long multimode fiber with 96% reflectivity of the output mirror (curve a). At room temperature only 8 mW of pulsed output power was achieved (curve b). Lasing could be observed up to 40°C which is quite unusual for this wavelength.

In order to raise the pump intensity in the core, experiments on a small core, high NA fiber were also done. Despite of a good pump coupling efficiency reached by use of an achromatic lens only 3 mW of output power could be reached at room temperature with a 12 cm long fiber. It is assumed that due to the high dopant concentration the small core of 10 μm diameter is heated to a temperature at which nonradiative processes dominate too much.

In both cases the fibers were multimode at the signal wavelength. It was however assured through careful aligning that only the fundamental transverse mode was excited in the laser cavity.

3.2 HOLIUM-DOPED FIBER LASER IN THE MID-INFRARED

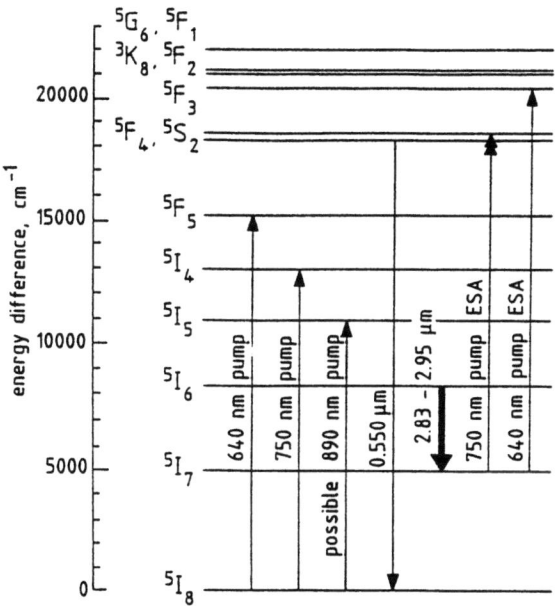

Figure 3.10: Energy level diagram for Ho^{3+} in ZBLAN glass with pump and energy transfer mechanisms for 2.9 μm emission

Fig. 3.10 shows the energy level diagram for 2.9 μm emission by Ho^{3+} in ZBLAN. Three pump wavelengths at 640, 750 and 890 nm have been considered so far [16], of which 640 nm is the most attractive because of the high absorption cross section, see Fig. 3.5. Pumping at 890 nm allowed only for single-pulse operation after switching-on. This is to be expected since the lifetime of the $^5I_7$ level is longer than that of the $^5I_6$ level. At 640 and 750 nm pump ESA processes empty the lower laser level and allow CW operation. They cause a green 550 nm fluorescence from $^5F_4/^5S_2$ similar to that observed in erbium and described in the previous section.

Fig. 3.11 shows different output characteristics for this multimode fiber laser. Due to the low dopant concentration a longer fiber than in the previous section has been used to ensure complete pump radiation absorption. The highest output power of 13 mW was reached with a 2.8 m long fiber and a mirror of 90% reflectivity. It can be seen from Fig. 3.11 that the 10 m fiber laser was too long: gain decreased due to pump absorption along the fiber, and finally signal absorption dominated the gain and lead to reduced output power. Nevertheless, without mirrors (4% reflectivity at the fiber ends) still 12 mW of output power could be reached on expense of a higher threshold.

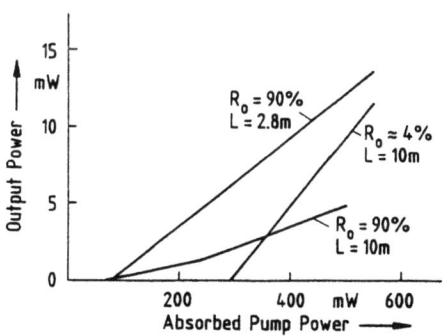

Figure 3.11: Output characteristic of an Ho^{3+}-doped fluoride fiber laser emitting at 2.9 μm (fiber: 1000ppm Ho^{3+}; 30/125/0.15) [after 16]

With an external reflection grating it was possible to tune the laser emission in the 2.83 - 2.96 μm range.

4. Final remarks

It was the aim of this contribution to give a very compact overview on fibers and fiber lasers for the mid-infrared. Since all of the described processes are much more complicated than explained here, the interested reader is strongly encouraged to consult the cited literature for deeper insights.

The author wishes to acknowledge helpful discussions with G. F. West on the glass and fiber parts and with Ch. Frerichs and J. Schneider on the fiber laser part.

References

1. Bendow, B. (1991) Transparency of Bulk Halide Glasses, in I.D. Aggarwal and G. Lu (eds), *Fluoride Glass Fiber Optics*, chapter 3, Academic Press, Inc., pp. 85-140
2. Parker, J.M. and France, P.W. (1990) Properties of Fluoride Glasses, in P.W. France et al., *Fluoride Glass Optical Fibres*, chapter 2, Blackie, Glasgow and London, pp 32-74
3. Tran, D.C. (1988) Infrared Transmitting Fluoride Glasses and Fibers. Conf. OFC'88, Tutorial, New Orleans
4. Takahashi, S. and Iwasaki, H. (1991) Preform and Fiber Fabrication, in I.D. Aggarwal and G. Lu (eds), *Fluoride Glass Fiber Optics*, chapter 5, Academic Press, Inc., pp. 213-233
5. Carter, S.F., Moore, M.W., Szebesta, D., Williams, J.R., Ranson, D. and France, P.W. (1990) Low Loss Fluoride Fibre by Reduced Pressure Casting, Electron. Lett. **26**, 2115-2117
6. Högerl, K. and Frischat, G.H. (1993) Monitoring the Glass Melting Process by Using the Christiansen-Shelyubskii Method, Ceramic Trans. **29**, :55-64
7. Nishii, J., Morimoto, S., Inagawa, I., Iizuka, R., Yamashita, T. and Yamagishi, T. (1992) Recent Advances and Trends in Chalcogenide Glass Fiber Technology - A Review, J. Non-Cryst. Solids **140**, 199-208
8. Riedel, W.J. and Knothe, M. (1991) Optics for Tunable Diode Laser Spectrometers, Proc. SPIE **1433** "Measurement of Atmospheric Gases", 179-189
9. Szebesta, D., Davey, S.T., Williams, J.R. and Moore, M.W. (1993) OH Absorption in the Low Loss Window of ZBLAN(P) Glass Fibre, J. Non-Cryst. Solids **161**, 18-22
10. CeramOptec (1991) Optran MIR Silver Halide-Fiber. CeramOptec Sales Information
11. Többen, H. (1993) Neue Faserlaser für das nahe und mittlere Infrarot (New Fiber Lasers for the Near and Middle Infrared), Dr.-Ing. thesis, TU Braunschweig, Fakultät für Maschinenbau und Elektrotechnik
12. France, P.W. and Brierley, M.C. (1991) Fluoride Fibre Lasers and Amplifiers, in P.W. France (ed), *Optical Fibre Lasers and Amplifiers*, Blackie, Glasgow and London, pp. 183-211
13. Wetenkamp, L., West, G.F. and Többen, H. (1992) Optical Properties of Rare Earth Doped ZBLAN GLasses, J. Non-Cryst. Solids **140**, 35-40
14. Shinn, M.D., Sibley, W.A., Drexhage, M.G. and Brown, R.N. (1983) Optical Transitions of Er(3+) Ions in Fluorozirconate Glass, Phys. Rev. B **27**, 6635-6648
15. Rebolledo, M.A. (1994) Analysis of Erbium-Doped Silica-Fiber Characterisation Techniques, in this book
16. Wetenkamp, L. (1991) Charakterisierung von laseraktiv dotierten Schwermetallfluorid-Gläsern und Faserlasern (Characterization of Laser-actively Doped Heavy-metal Fluoride Glasses and Fiber Lasers), Dr.-Ing. thesis, TU Braunschweig, Fakultät für Maschinenbau und Elektrotechnik
17. Schneider, J. (1994) Continuous-Wavelength Lasing at 2.7 mu-m in Er3+-doped Fluoride Fibers with Low Pr3+-codoping, Tech. Dig. CLEO'94, Anaheim, CA (U.S.A.), 133

OPTICAL FIBRES IN NUCLEAR RADIATION ENVIRONMENTS

Potential Applications - Radiation Effects - Need for Standards

F. BERGHMANS, O. DEPARIS, S. COENEN and M. DECRETON
*SCK•CEN - Research Unit Waste, Dismantling and Radiochemistry
Teleoperation Group
Boeretang 200, B-2400 Mol, Belgium*

P. JUCKER
*LETI (CEA - Technologies Avancées)
DEIN/SPE - Centre d'Etudes de Saclay
Lab for Data-Transmission in Hostile Environments
F-91191 Gif-sur-Yvette, France*

The aim of this work is to illustrate the activities related to the assessment of fibre-optic technology in hostile nuclear environment and to provide the interested reader with an introduction to this subject. Potential applications of fibre-optics in nuclear installations are summarised. Effects of nuclear radiation on optical fibres are shortly discussed. The absence of complete predictive models for the radiation induced attenuation in optical fibres is illustrated by modelling attempts and explained by the wide range of parameters which influence the induced loss. The need for standards in assessing the fibre's radiation resistance is also presented. Finally, experimental methods are illustrated with results of radiation induced attenuation in optical fibres obtained for a high-dose rate and high total dose irradiation, as not often reported in literature.

1. Introduction

The effects of nuclear radiation on optical fibres and, more generally, on amorphous silica, have been extensively studied during the last two decades. It is well-known that, though optical fibres are insensitive to electromagnetic interferences, they can be very sensitive to ionising radiation. Even at very low total doses, the optical attenuation of the fibre can significantly increase [1]. Without specific care and adaptation, this phenomenon can strongly reduce the potential for application of optical fibres in nuclear environments, especially since no satisfactory predictive model of the induced attenuation as a function of radiation type, fibre type and optical parameters is available

(∗) This paper was presented at a poster session.

yet. However, if one hopes to fully introduce the optical fibre for data-transmission and sensing purposes in nuclear installations, covering the complete nuclear fuel cycle (i.e. from uranium extraction to nuclear waste disposal sites, going through reactor dismantling operations and spent fuel reprocessing) as well as nuclear fusion equipment, military systems and space-oriented systems, it is necessary to understand the different phenomena which lead to this effect.

Radiation induced attenuation is primarily due to trapping of radiolytic electrons and holes at defect sites in the fibre's silica, i.e. the formation of the so-called "colour centres" [2]. These colour centres can absorb information-carrying photons and can be made to disappear through thermal or optical processes (thermal bleaching or photo bleaching), causing a decrease of the additionally induced attenuation [2,3]. During irradiation, both processes of new colour centre formation and disappearing of existing colour centres compete. Subsequent to irradiation, the amount of recovery depends on fibre composition, type of radiation, total dose, temperature, wavelength and optical power used to measure the attenuation [4]. Colour centre formation and the associated increase of optical attenuation will be discussed more in detail. These discussions are mainly inspired by the excellent reviews given by D.L. Griscom [2,5], E.J. Friebele [4] and R.H. West [6].

Besides radiation induced attenuation, radiation also causes other disturbing effects such as luminescence and refractive index changes to appear in optical fibres. These will only be described very briefly, since radiation induced attenuation can be considered as the most important, at least most disturbing, effect.

Two main categories of experiments and tests performed to investigate the radiation effects on optical fibres can be distinguished. The first type of experiments aims at characterising the behaviour of particular fibres under certain radiation conditions. These experiments record the radiation induced attenuation at particular wavelengths as a function of total dose and can be considered as case studies by setting the radiation hardness of a particular fibre (and sometimes the associated system) in particular conditions [7,8]. The second type of experiments has a more explanatory function, since it aims at correlating the radiation induced spectral attenuation with particular types of colour centres, in the fibre's silica. Correlations can be found through spectral attenuation measurements in conjunction with electron-spin resonance techniques to identify and characterise the various types of defects in amorphous SiO_2 [2,5,9]. Both types of experiments have established that pure silica core fibres are the most radiation resistant and therefore we will concentrate on this type of fibre. In most cases, optical fibres are doped in order to achieve a better quality of data transmission (reduced intrinsic attenuation and pulse dispersion). However, the presence of dopants in core or cladding (Ge, B, F, P) tends to increase the radiation sensitivity of the fibres by providing an increased number of traps enabling the formation of colour centres.

A remarkable disparity among the results obtained by the different research groups active in this domain led to a significant confusion and incoherence in literature, as highlighted by R.H. West [6]. The reason therefore is mainly the dependence of the radiation induced attenuation on an enormous range of experimental parameters related with the experimental conditions of both radiation and optics as well as environment (mainly temperature). This incoherence has put forward the need for defining standards for radiation testing of optical fibres in order to allow for proper comparisons between results obtained through the different fibre irradiation campaigns. This issue has been one of the main concerns of the NATO Panel IV, Research Study Group 12, Nuclear Effects Task Group (NETG) [10,11,12,13].

This text has no pretension whatsoever to act as a general review of the application of optical fibres in nuclear environments. The authors' intention is more to illustrate the difficulties of assessing the emerging fibre-optic technology in nuclear industry. Therefore, the main aspects such as the application of fibres in nuclear environments, radiation induced absorption as well as test methods and standards, together with some typical results on radiation testing of fibres will be discussed very generally in order to provide the interested reader with an introduction to a field, which definitely deserves some attention. An extended list of references can help him for further investigations.

2. Potential Applications of Optical Fibres in Nuclear Environments

Before going-over to a general treatment of radiation induced absorption, it is necessary to highlight the possible applications of optical fibres in nuclear environments. In the meantime, it allows to argue why additional research is needed in this field and to emphasise the need for the setting of specifications.

Optical fibres have proven to present many advantages as compared to conventional telecommunication or sensing means. However, their use in radioactive environments has been limited up to now mainly due to the lack of information with regard to their reliability in such a harsh environment. Generally, nuclear industry feels hostile toward delicate communication or sensing systems. The main concerns of safety and process yield strive to exclude any possible threat of personnel and environmental security as well as process shutdown. Therefore, only those means for communication and sensing which have a strong history of use and which have proven to be highly reliable in nuclear environment are considered to be suitable for application. Optical fibres and their associated systems (e.g. multiplexers) for data-transmission and sensing do not yet fully meet these requirements. Ageing conditions, for instance, are known for conventional nuclear plant instrumentation such as resistance temperature detectors (RTDs) and pressure transmitters [16,17]. However, to the authors' knowledge, this is not so for fibre-optic sensors, even in the non-nuclear process industry. No statistically significant failure analysis could be performed for fibre-optic sensors, simply because sufficient quantities have not yet been produced and used. It should be noted that this remark also applies to process industry in general [14,15].

TABLE 1. Non-exhaustive list of drawbacks and advantages of fibre-optics in nuclear environment. This table also applies to process industry in general

Drawbacks of fibre-optics	Advantages of fibre-optics
• No history of industrial use in nuclear environment • Unestablished long-term reliability • Bad, but outdated reputation with regard to radiation effects • Maintenance staff is untrained • Needs special equipment • Cost associated with implementing new technology • Lack of uniqueness • Other ...	• Immunity to EMI • High data-rate capacity • Small size, low weight • Less cabling • Low-cost potential • Intrinsically safe and passive • Functions in difficult temperature conditions • Multiplexing capabilities • Enhanced redundancy possibilities • Low amount of generated nuclear waste • Sensing with high sensitivity, wide dynamic range • High potential for remote sensing • New potential applications, unfeasible with conventional means • Other...

More generally, procedures to certify the performance of conventional nuclear equipment do exist. The same is true for fibre-optics only in relatively benign telecom environments. In the very adverse environment presented by nuclear installations, test or validation procedures are not fully defined yet. Efforts are still needed to set-up or complete acceptable test, calibration, validation and qualification procedures.

Nevertheless, the large number of potential advantages of optical fibres (cf. table 1) has rightly pushed nuclear industry to take a closer look at the opportunities offered by this technology. The fibre's small size and the high data-rate capacity together with the multiplexing capabilities would allow to significantly reduce the amount of cabling in nuclear power plants [20], in the meantime reducing the amount of possibly generated nuclear waste. Due to their flexibility and special features, fibre-optics are also expected to enable measurements which were formally unachievable with conventional instrumentation. These issues have led to the implementation of fibre-optics for different applications in nuclear industry and some of them are shortly described hereafter in view of illustrating their diversity.

Maintenance and monitoring of nuclear fusion reactors is confronted with problems such as extremely high γ-radiation dose-rates (up to 10 MGray/h), high temperatures encountered near the reactor vessel wall (typically 150 °C), very strong EMI induced by the torus magnets and the restricted number of cable penetrations allowed between control equipment and fusion reactor room [21]. Fibres may present an alternative here, due to their multiplexing capabilities and high temperature resistance. Fibres are also being studied for vision purposes with fibrescopes [22,23] and have been envisaged for use in visible spectroscopic measurements during intense neutron production in JET (Joint European Torus) [24].

Lots of optical fibre sensor types have been developed for general industrial purposes in the last 20 years [25,26] and some of these are already commercially available. However, they are not suited for direct transposition to the nuclear environment.

One of the most investigated fibre-optic sensors is probably the fibre-optic gyroscope (FOG) [27]. Related with this, nuclear survivable polarisation fibres have been studied for FOGs on spacecrafts [28].

Fibre-optic dosimeters and radiation detectors have also been intensively studied. A good review has been given by P.B. Lyons [29] and H. Bueker *et al.* [30].

Optical fibre sensors are currently investigated for monitoring of nuclear waste disposal installations [31], nuclear waste tanks [32], isotopic separation [33] and spent fuel reprocessing [34].

Robotics and telemanipulation applications in highly radioactive environments can also benefit from fibre-optic technology. Fibrescopy, fibre-based infrared sensing for proximity or collision avoidance and simply data-communication with the robots can be envisaged. Fibre-assisted high-power beam delivery can improve cutting and decontamination procedures for reactor dismantling [35,36].

The potential for application of fibre sensors in nuclear installations extends far beyond the examples cited above. Temperature measurement are the most prevalent in nuclear reactor environment. Systems similar to the York DTS [18,19] could allow for the monitoring of reactor cooling circuits. Non-contact fibre-optic infrared thermometry can be very useful for hot-spot detection. A distributed or point multiplexed security system for presence detection and fire detection (e.g. in the primary cooling circuit pump locations) can be envisaged as well. Fibre pressure sensors could enable a good monitoring of cooling circuitry and act as leak detectors.

Definitely, there is an enormous potential for application in the nuclear environment, the number of applications being only restricted to everyone's sound fantasy. In most cases, the needs are the same as in general process industry. One major problem is added to these already encountered in general process industry, i.e. nuclear radiation. Efforts are still needed to adapt existing fibre-optic systems for data-transmission and sensing purposes to the nuclear environment. These efforts can be encouraged by the fact that once a system will be suitable for a nuclear installation, it will generally also be qualified for general industrial applications considering its high level of reliability and ruggedness.

It should be noted that no attention was paid here to the components related with fibre-optic technology such as laser diodes, LEDs, optical detectors, connectors, etc. All of these elements are also sensitive to ionising radiation and they have also been subjected to radiation experiments [8,10,37]. Though it is very important to know their behaviour in radiation fields, the effects of radiation on these components may be considered as less important than on the fibre itself since where possible, sources and detectors should be isolated from radiation.

3. Effects of Ionising Radiation on Optical Fibres

Optical fibres are well-known to be very sensitive to ionising radiation. This sensitivity has been intensively studied, but, up to now, no satisfactory model has been developed which would allow to completely predict the behaviour of a particular optical fibre in a given radiation environment.

Before going over to a more detailed discussion of radiation induced attenuation, the origins of radiation damage will be very briefly mentioned. Almost each book on basic nuclear physics discusses the subject of interaction between ionising radiation and matter far more extensively than what will be presented hereunder. The reader is referred to these books for a comprehensive treatment of the subject [38,39]. Other more specialised works focus on radiation effects in solids [40] and colour centres [41], in which far more valuable information can be found than what will be discussed in this text. Radiation damage processes in optical glasses and crystals have also been summarised by D.L. Griscom [2,5] P.W. Levy [42], R.T. Williams [43], E.J. Friebele [4,44] and R.H. West [6]. The reader is referred to these surveys for a far more complete discussion than that presented below, which is merely a short summary of these works.

3.1. INTERACTION OF RADIATION WITH MATTER

The interaction of radiation with matter may lead to changes in the properties (physical, chemical, mechanical or electrical) of matter. The result of this interaction is often referred to as 'radiation damage', which, on the macroscopical level indicates changes in one or more of the material properties, and, on the microscopical level, indicates the existence of local deviations from the electronic or atomic structure present before the irradiation.

The main type of radiation that will be encountered by optical fibres in a nuclear power plant environment is gamma radiation. The primary cause of radiation damage in the case of γ-rays are radiolytic processes or ionisations. Here, chemical bonds are broken due to the energy transfer during non-radiative recombination of electron-hole pairs. Atoms may also be displaced due to momentum transfer from radiation particles (electrons released by gamma-rays and neutrons).

Other damage processes include atomic transmutations due to neutron capture (e.g. from Si to P) and electronic rearrangements due to the trapping of electron and holes by existing intrinsic defects.

The final result of the interaction of incident gamma-rays in an absorbing material can be considered as the deposition of energy by ionisations produced by secondary electrons. This deposition is described as an "absorbed dose", which measures the energy deposited per unit mass of the respective material. It is expressed in "gray" (Gy), where 1 Gy corresponds to an energy absorption of 1 J/kg. This quantity thus depends on the material itself.

3.2. SECONDARY RADIATION EFFECTS ON OPTICAL FIBRES

The main types of radiation effects on optical fibres which have been reported are luminescence, refractive index changes and dimensional modifications, changes in mechanical properties, and, of course, radiation induced attenuation. The latter effect can be considered as the most important.

3.2.1. Luminescence

When exposed to ionising radiation, an optical fibre will emit light. Such luminescence effect may be used for sensing purposes, e.g. for the fabrication of fibre-optic radiation dosimeters [29,30,45]. However, this luminescence may also compete with the light signal of interest, leading to data-transmission errors. At very high dose-rates (> 10^4 Gy/s), luminescence is mainly due to the Cerenkov effect. Cerenkov light is emitted when charged particles propagate through solids with a velocity greater than that of light in the respective material. The intensity decreases with $1/\lambda^3$. Optical receivers at 1300 nm or 1550 nm will less be influenced and filtering can be performed if necessary. As a result, luminescence may be considered as an effect of secondary importance for general purposes.

3.2.2. Refractive index changes and dimensional modifications

Nuclear radiation can give rise to microscopic structural changes, leading to dimensional modifications. As a result, the density of the material can be altered, which in turn modifies the refractive index [46,47].

3.2.3. Changes in modal properties

Due to variations in fibre composition, radiation induced absorption and refractive index modification may also vary across the fibre. Different modes may then be affected in different ways, giving rise to changes in pulse dispersion characteristics [48].

3.2.4. Changes in mechanical properties

Polymers are also sensitive to ionising radiation [49]. Radiation induced PCS fibre cladding embrittlement or core-cladding interface degradation has also been reported [35].

3.3 THE CAUSE OF RADIATION INDUCED ABSORPTION - COLOUR CENTRES IN PURE SILICA CORE FIBRES

The radiation induced absorption results from the existence of colour centres which absorb light at certain wavelengths. Colour centres are created through the trapping of radiolytic electrons and holes at defect sites in the fibres' silica. These defect sites may already exist before irradiation (intrinsic defects, e.g. induced by fibre drawing) or may be created by the radiation. They can be considered as the precursors of the colour centres. If the intrinsic defects act as precursors, one can expect that the growth of the colour centre concentration in function of the total irradiation dose will saturate at a particular dose level. This phenomenon seems to be predominant at sufficient low dose levels ($< 10^2$-10^4 Gy). At higher dose levels or when the incident radiation is sufficiently energetic (> 200 keV), new precursors are created (particularly vacancies and interstitials). In this case, colour centre concentration will monotonically increase with total dose [5]. The growth of the colour centre concentration as a function of total dose may often be analysed into a linear part and a saturating exponential part, depending on the fibre's composition.

The growth of the colour centre concentration is actually governed by the competition between the formation of new colour centres and the annealing of existing colour centres. Therefore, the magnitude of the induced attenuation depends also on the dose rate (which actually determines the production rate of the colour centres) as well as the bleaching rate of the colour centres. As a result, the induced attenuation can reach very high levels after pulsed irradiations. In the case of continuous irradiations, fibres which show good annealing characteristics will show a saturation of the induced loss when production rate of new centres equals the bleaching rate (see figures 2 and 3). This level of saturation will be higher for high dose rates and high relaxation (annealing) times. The annealing of colour centres results from thermal processes and photo bleaching. Temperature strongly influences the growth of the colour centres. Higher temperatures result in more rapid annealing and thus in a lower incremental loss. Photo bleaching can happen at optical powers as low as 100 nW.

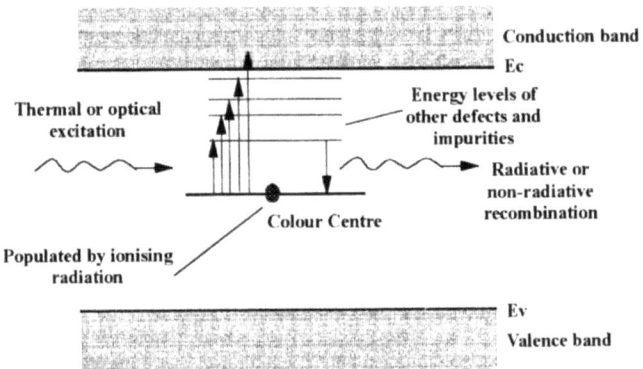

Figure 1. Schematic representation of colour centre energy level in silica bandgap [30] [This figure is used as a representation of the colour centre absorption, though one should handle this representation with particular caution when extrapolating the existence of a crystalline bandgap to vitreous (amorphous) silica]

Figure 2 describes the general shape of the growth and annealing of induced attenuation. This figure is only meant as a guide to the eye and does not represent any actual measurement result.

Besides electron or hole trapping by precursor sites, colour centres can also be created through structural relaxation of an ionised bond, radiolysis of OH radicals or diffusion of radiolytic fragments to a precursor site [5]. Great efforts were made in order to determine models for the growth and annealing kinetics of colour centres and different proposals are presented in literature. However and to the authors' knowledge, no complete explanatory model which could exclude all other possible formation mechanisms was ever set forward.

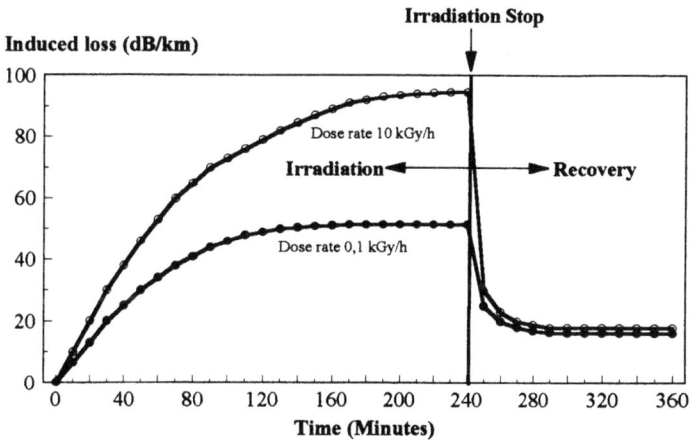

Figure 2. Typical evolution of radiation induced attenuation in the IR region, during and subsequent to irradiation. [This figure does not represent any actual measurement result but is only meant as a guide to the eye, illustrating some of the comments above]

Different types of colour centres have been identified. Identification of paramagnetic colour centres can be done using Electron Spin Resonance (ESR) techniques [5,50]. Each of these centres can be associated with a particular absorption band, through the comparison of growth and annealing behaviour of the concentration of defect centres measured by ESR with the absorption bands obtained from the gaussian resolution of the absorption spectra. These bands appear in the UV and visible part of the electromagnetic spectrum, since the absorption of the photons leads to electronic transitions. However, the band tails extend to the near IR region and thus also contribute to an increased absorption at usual fibre wavelengths. In general, the induced attenuation is lower at longer wavelengths. Except for a restricted number of colour centre types, there still exists some confusion about the correlation of certain colour centres with particular optical absorption bands. This confusion results from a number of factors related with the difficulties of establishing correct correlations between colour centres and absorption bands as well as the difficulties encountered in the identification of the colour centres. Since the fibre's glass is an amorphous material, the absorption bands are not necessarily gaussian-shaped and symmetric. This compromises the decomposition of absorption spectra into gaussian bands. Moreover, the absorption bands from different colour centres can significantly overlap. Diamagnetic colour centres can not be identified easily, since they are not detected through ESR. Finally, the kinetics of annealing and formation of colour centres is also governed by interactions between the various types of defects and by diffusion processes [5]. Initial twin defects can interact in different ways with the impurities and can be transformed into different colour centres subsequent to irradiation. Definitely, there are still a lot of outstanding questions and further research is needed in order to establish clear correlations that would help to predict the fibres' behaviour in a nuclear radiation field.

It should be noted that, due to their long path length and their high purity, optical fibres present a unique medium for the investigation of colour centres, especially those which only appear in very low concentrations and those which only absorb very little in the infrared [44].

A complete discussion on colour centre formation and annealing kinetics, as well as the correlation of colour centres with particular absorption bands is out of scope here. Therefore, the best known colour centres in pure silica fibres, together with their peak absorption wavelength will be summarised in table 2. The existence of colour centres is not only a concern in the field of radiation hardness of optical fibres, but has been extensively studied in the field of glass chemistry. Colour centers are also the starting point for the writing of photorefractive gratings with excimer (UV) laser interference patterns [63,64] and thus can be involved in photosensitivity enhancing processes used for efficient writing of Bragg-gratings into optical fibres. Colour centres and defects already appear after glass or fibre fabrication. The ionising radiation tends to increase considerably their concentration in the fibre. The summary presented in table 2 concludes a literature search conducted in November 1993 and therefore does not pretend to present the current state of the art. A far more complete but less recent table has been presented by V.B. Neustruev [65].

Irradiation experiments have been reported which led to induced losses up to 4900 dB/km in standard Ge-doped telecom fibres (wavelength 1.3 µm), at a total dose of only 10^3 Gy [66]. Pure silica core optical fibres have proven to show more interesting characteristics with regard to their radiation sensitivity as compared to doped fibres. Kyoto et al. [8] reported losses of only 6 dB/km at 1.3 µm and 2.3 dB/km at 1.5 µm for a total dose of 10^3 Gy delivered at a dose rate of 10^2 Gy/h. At a lower dose rate of 2.10^{-2} Gy/h, an induced loss of about 1 dB/km was observed for a total dose of approximately 10^3 Gy. Henschel reported an induced loss of only 7 dB/km at 850 nm after a total dose of 10^4 Gy [67]. Numerous experiments have been conducted and reported, each of which gave precious indications with regard to the radiation behaviour of particular pure silica core fibres. Distinctions have been made between low OH and high OH content fibres, as well as oxygen-surplus and oxygen-deficient silica fibres. Only the general features of these fibres will be presented here in accordance with the descriptions of Friebele et al. [4,62,68].

The fibres with pure silica cores have shown to be the least susceptible to radiation-induced attenuation in the 0.8-1.5 µm region at short times after pulsed irradiations as well as long times after continuous irradiations. The attenuation increases linearly up to a dose of about 10^2 Gy. At higher doses, a region of complete saturation appears, which makes these fibres very interesting for high dose applications. Moreover and to a certain extend, there seems to exist a radiation hardening phenomenon, i.e. induced loss is less if the fibre has first been exposed to a much higher dose and the damage has been allowed to recover. Photo bleaching has also shown to be quite effective in this type of fibres.

The induced loss at a dose of 50 Gy can decrease about 20 dB/km when the optical power is increased from 1 nW to 1 µW [10]. Figure 3 shows a typical irradiation curve of pure silica optical fibres [10].

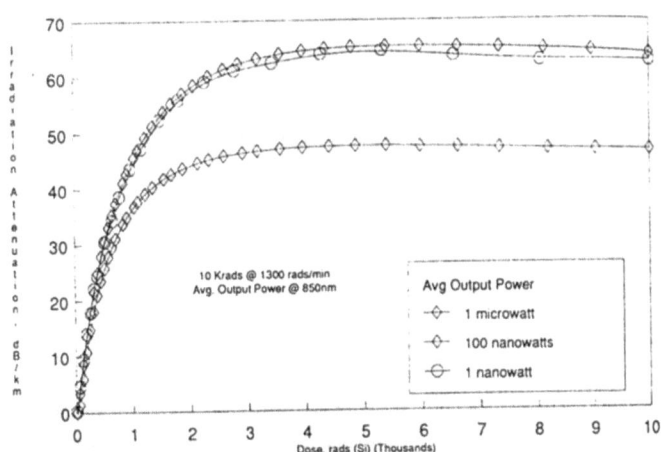

Figure 3. Induced attenuation for an irradiated pure silica core fibre at room temperature and the effect of photobleaching (figure 11 from [10])

Fibres in which Ge was the only dopant have also shown reasonable radiation resistance [71].

TABLE 2. Some well-known colour centres in pure silica and their associated absorption bands

Band E (eV)	λ (nm)	Centre name	Chemical representation	Particularity
7.6	165	E	≡Si-Si≡ + H or ≡Si-Si≡ + Cl	Band first associated with the peroxy radical ≡Si-O-O• [51], then with ≡Si-Si≡ [52] and finally with a complex ≡Si-Si≡ + H or Cl [53]
5.8	215	E_γ'	≡Si•	Correlation seems to be well established [5], numerous possible precursors
5.5	230	E_β'	≡Si•	Correlation established for quartz [5], extrapolation to fused silica is accepted
5.1	245	$B_{2\beta}$	≡Si-O-O-H	Hypothetical correlation [54], anneals only at very high temperatures [55]
5.0	250	$B_{2\alpha}$	≡Si-Si≡	Correlation seems well-established, ≡Si-Si≡ is a precursor of the E' centre [52]
4.8	260	D_0	O_3 (ozone)	First associated with the non-bridging oxygen hole centre (NBOHC) ≡Si-O•, then with ≡Si-O$^-$ [56] and finally with the "Hartley band" of ozone [57]
3.8	330	-	Cl_2	Originally correlated with ≡Si-O-O-Si≡ [58], later with Cl_2 [59]
2.0	630	-	Si_3^-	Appears in low-OH (dry) fibres after drawing [60]
2.0	630	-	≡Si-O•	Correlated with the NBOHC, the precursor seems to be different for dry and wet fibres, which induces a shift in the peak absorption wavelength from 1.975 eV to 2.1 eV [61]
1.5	850	AEC	$[Alk_nH_m]^0$	Alkali Electron Centre (AEC). This centre is an alkali-electron trap. The peak absorption wavelength seems to vary as a function of the alkali type [62]

3.4 THE ABSENCE OF PREDICTIVE MODELS

The results presented above have only a very general character. As already mentioned, exactly predicting the radiation behaviour of a certain fibre under given conditions has not been a very successful practice up to now. The reason therefore is mainly the very broad range of parameters which influence the fibre's response to radiation and which are listed in table 3. Considering all these possible influences clarifies why attempting to establish predictive models is rather challenging. It also explains why it becomes extremely difficult to compare results obtained by different research groups as well as the apparent disparity between these results [6]. As pointed out by Greenwell [10], the most important limitation is probably the available test facilities, which may dictate particular requirements on the possible radiation and temperature conditions as well as on the space available for the samples. Sometimes, it is even impossible to perform on-line measurements. Results are then only obtained some time after irradiation and thus observation of the dynamics of the processes becomes hazardous. Moreover, it is often extremely difficult and even impossible to give a complete analysis of the fibre's composition. As a result, each research group is able to give only a very limited picture of all the effects appearing in irradiated fibres.

TABLE 3. Main parameters affecting the response of optical fibres to ionising radiation

Parameters affecting radiation induced attenuation
1. Fibre fabrication (preform composition, fabrication method)
2. Fibre composition (dopants, OH-content, Cl-content, metallic impurities, cladding type)
3. Fibre dimension (core diameter)
4. History with regard to the fibre's thermal treatments
5. History with regard to previous irradiations
6. Radiation type (γ-rays, X-rays, neutrons, protons, electrons, heavy ions)
7. Dose rate
8. Total dose
9. Temperature conditions (during exposure, during measurement)
10. Stress conditions during irradiation
11. Time elapsed between exposure and measurement
12. Light launching conditions (use of a mode-stripper or mode-scrambler)
13. Optical power
14. Wavelength
15. Fibre wrap diameter
16. ...

The main result of this absence of good models is the impossibility of predicting the fibre's behaviour in radiation conditions which do not exactly match those used for the tests. Consequently, one is often limited to performing case studies of the behaviour of a certain fibre type in a particular and well-defined radiation field with well-specified environmental conditions. However, extrapolating radiation induced attenuation data obtained at high dose rates to results at low dose rates is very important. Assessing 40 year lifetimes of fibres submitted to low dose rates is only possible through accelerated testing at higher dose rates. Such extrapolations are also necessary for predicting fibre behaviour in space, for which radiation conditions cannot exactly be simulated in laboratory and where dose rates may even be time-dependent.

A model for the induced absorption growth which has been considered a couple of times in literature is a simple power-law :

$$A = \alpha D^\beta \qquad (1)$$

where A is the induced attenuation, D is the total dose and α and β are empirical constants [8,74,76]. Pure silica core fibres have shown a power-law growth of attenuation up to at least 100 Gy [76]. Equation (1) however does not take the dose rate into account neither any saturation phenomenon.

Since recovery processes were suspected to result from bimolecular processes (e.g. diffusion-limited recombination of electrons and holes or vacancies and interstitials), the recovery after irradiation has been described by n-th order kinetics, following the relation:

$$A = (A_o - A_f)(1+ct)^{-1/(n-1)} + A_f$$
$$c \equiv \frac{1}{\tau}(2^{n-1} - 1) \qquad (2)$$

where A_o and A_f are the initial and final values respectively of the induced attenuation, τ is the half-height lifetime of the recovery, i.e. the time needed for $A-A_f$ to reach $(A_o-A_f)/2$ and n is the material-dependent kinetic order [74,76,77,78]. Again this model does not take the dose rate into account. Moreover, fitting for the kinetic order parameter "n" resulted in values ranging from 3 to 10, there were n=2 was expected (cf. bimolecular processes).

Recently, Griscom et al. [74] proposed and verified a general explanation of the power-law dependencies on dose (cf. equation (1)) in case of a Ge-doped fibre. Starting from equation (1), detailed prediction of the post-irradiation recovery curves was possible, given the empirical exponent β and the irradiation time t_{irr}. The time constant of the recovery was found to be $t_{irr}/(1-\beta)$, independent of the order of the kinetics [74]. Actually, Griscom et al. related the empirical power-law equation with the kinetic theory, but the attenuation growth and recovery can still not be predicted for any arbitrary dose rate.

Based on the assumption that radiation induced attenuation growth is a linear effect, Liu and Johnston [79] derived a new growth and recovery equation. Stating that the overall absorption is a superposition of many independent events occuring sequentially in infinitesimal time intervals, they proposed the attenuation to be :

$$A(t) = \int_0^t f(t')h(t-t')dt' \qquad (3)$$

where h(t-t') is an impulse response which was assumed to be proportional to the right-hand side of equation (2) and f(t') is the time-varying dose rate function D'(t'). The resulting model is explicitly dose rate dependent and has shown reasonable agreement with experimental data when the dose rate was kept constant and when saturation was not reached yet. Saturating phenomena were indeed still not taken into account here.

4. Standards and experimental methods

In order to allow for comparison of results obtained by different research groups it is necessary to establish standard procedures for fibre testing [10,11]. When investigating what level of specification is necessary for standardising the testing of radiation effects, all the parameters present in table 3 should be fully specified. However, it would be unrealistic to expect that every laboratory capable of characterising optical fibres has the same equipment available as its peers. Moreover, commonalty with other test requirements should also be taken into account. Since optical fibres are used in conjunction with micro-electronic and opto-electronic parts, fibres should be qualified under the same conditions (or vice-versa) [10]. The principle of "balanced" radiation hardening should also be taken into account, i.e. a fibre optic data-link should only be as radiation resistant as the residual parts of a system [69]. However, when opto-electronic fibre-end components can be shielded from radiation, the fibre has to be the most rad-hardened part. In this case, full profit can be taken from the fibre's unique capacity to transport raw-data undisturbed far away from e.g. sensing-heads.

The setting of specifications for performing tests on the radiation behaviour of fibres together with assessing their reliability is complicated by the enormous variety of radiation and environmental conditions which can be encountered among the various applications. The different radiation environments encountered in military, nuclear power plant and fusion reactor applications will be summarised below. Typical average dose rates and total doses which the fibre is expected to survive will be mentioned and compared to existing standards where possible.

4.1 RADIATION ENVIRONMENTS

4.1.1. Military environment

The military radiation environment is mainly related to the explosion of nuclear weapons. When such a weapon explodes, one usually distinguishes three radiation phases : a short γ-ray pulse (high dose rate - 10^9 Gy/s, duration < 100 ns, energy 1 to 2 MeV), a neutron pulse (fluence up to 10^{13} n/cm^2, duration 10 to 100 µs, energy about 2 MeV) and finally the total dose deposition (up to 10^2 Gy in distances where equipment is not destroyed by heat and blast wave). About 80 % of the fall-out dose is deposited within 8 hours after an explosion. Military specifications also require radiation hardness guaranteed from -50 °C to +100 °C [69].

Most of the investigations on the radiation hardness of fibres were done in the case of such military scenarios and the values cited above are starting points for the definition of test procedures in this framework. The Nuclear Effects Task Group (NETG) of Research

Group 12 from NATO Panel IV has conducted a round-robin test campaign involving several well-known laboratories such as the Naval Research Laboratory (USA), the Air Force Weapons Laboratory (USA), the Commissariat à l'Energie Atomique (France), the Rome Air Development Centre (USA), the Fraunhofer-Institut für Naturwissenschaftlich-Technische Trendanalysen (Germany), the Royal Military College of Science (UK), Boeing Aerospace and Electronics (USA) and The Aerospace Corporation (USA). The aim of this test campaign was to investigate the level of specification necessary for standardised testing in order to allow for determination of the sensitivity of optical fibres with a high degree of confidence [11,12,13]. It was found that suitable control of dose, dose rate, average optical output power, wavelength, temperature and fibre material composition, good interlaboratory agreement could be obtained (e.g. measurement accuracies better than 0.35 dB/km in the case of steady-state irradiations of single-mode, multi-mode and graded-index fibres) [12]. It was also found that control of fibre wrap diameter and optical source spectra as well as a good accuracy of dosimetry would reduce measurement deviations among different experimenters.

This round-robin test campaign concurred with the evaluation of the FOTP-49A "Procedure for measuring steady-state gamma radiation-induced attenuation in optical fibres and optical cables" adopted by EIA (Electronic Industries Association) and led to the modification of this procedure [10]. This modified procedure was submitted to EIA for adoption as FOTP-64, which replaced FOTP-49A. New features include pulsed transient radiation tests. This new procedure also utilises other FOTPs as defined by EIA, e.g. FOTP-57 "Optical fiber end preparation and examination" or FOTP-78 "Spectral attenuation cutback measurement for single-mode optical fibers" [70].

For steady-state irradiations, the NETG procedure sets the use of a ^{60}Co source and the following dose / dose rate combinations [70]:

TABLE 4. Total dose / dose rate combinations for steady state tests [70]

Total dose (Gy)	Dose rate (Gy/min)
30	3
100	13
1000	13
10000	100

While these radiation levels are consistent with military specifications, they can be too low compared to those encountered in nuclear power plants and in the vicinity of fusion reactors. The use of ^{60}Co sources is indicated to avoid complications with variations in total dose as a function of depth within the fibre which appear at γ-ray energies under 500 keV. For most of these sources, 10^3 Gy/h is an upper limit. They are relatively easy to handle and allow for a good dose deposition uniformity in the samples (typically 50 to 500 m). ^{60}Co emits γ-rays with energies of: 1.17 MeV and 1.33 MeV [49]. These energy levels are compatible with military specifications and correspond also to the mean energy of γ-rays encountered in nuclear facilities. However, for particular applications such as spent fuel reprocessing, the energy-spectrum of gamma radiation can be much wider.

Neutron irradiations are not performed, since they lead to similar induced loss mechanisms [69]. Conversion factors exist between neutron fluence and deposited dose [71].

4.1.2. Nuclear power plants and fusion reactors

The radiation environments in nuclear power plants or nuclear installations can widely vary. The main difference compared to military scenarios is that systems have to be operational during the complete plant lifetime (about 40 years or possibly longer in the case of geological waste disposal sites). Temperatures as low as -50 °C have usually not to be taken into account and dose rates and total doses may exceed those encountered in military environment.

Under normal operations, fibres used in a nuclear power plant are estimated to be exposed to γ-rays at dose rates between 10^{-5} Gy/h and 1 Gy/h. They should be capable of withstanding these dose rates over 40 years, up to a total dose of 1 MGy. In accidental conditions, the dose rate may reach 10^4 Gy/h. For these conditions, numerous irradiation campaigns have been conducted by CEA on optical data-link components [80].

Typical dose rates and total doses for particular interventions such as tele-operated maintenance, dismantling, in-reactor core repair or fuel manipulations are summarised in table 5 [72]. Fibres have also to cope with high temperatures (100 °C - 200 °C), corrosive substances and will have to operate under water.

TABLE 5. Typical γ-ray dose rates and total doses encountered during teleoperated interventions in nuclear power plants [72]

Intervention	Dose rate	Total dose
Most maintenance work in low radiation environment. Light decontamination operation.	< 0.01 Gy/h	< 10 Gy
Most decontamination work. Interventions on primary loop components. Some hot-cell work.	< 10 Gy/h	< 1 kGy
Reactor vessel intervention. Dismantling work. Hot-cell tasks.	< 1 kGy/h	< 1 MGy
In-core maintenance during reactor stop. Fuel manipulation	> 1 kGy/h	> 1 MGy

Fibres are also envisaged for maintenance applications or plasma-diagnostics in fusion reactors. In this case, extreme γ-ray dose rates of 100 kGy/h and total doses of 100 MGy could be encountered.

Up to doses of 10 kGy, the standards set for military scenarios are still valid. Thus fibres used for e.g. data-transmission in normal power plant operation can be characterised using the existing FOTP-64. However, for higher dose rates and doses as encountered in particular applications and nuclear fusion, these standards should be adapted.

Assessing the reliability of fibres over lifetimes reaching 40 years is only possible through accelerated testing. This means that equal doses should be delivered in a much smaller time, i.e. at higher dose rates. Since the dose rate influences the radiation response, extrapolation of results from high dose rate to low dose rate exposures may be compromised. The NETG specifications do not readily address this issue. However, increasing dose rate has shown to increase the induced attenuation [67,73]. As a result, an increased dose rate will often set an upper limit to attenuation. Moreover, extrapolation has shown to be feasible in certain cases (see also §3.4) [73,74].

4.2 HIGH-DOSE EXPERIMENTAL SET-UP AND RESULTS

Most experiments reported in literature are usually limited to total doses about 10^4 Gy, sometimes 10^6 Gy. Irradiations are mostly performed with ^{60}Co sources. Therefore, the authors have chosen to present another irradiation facility using spent (nuclear) fuel and to discuss the results obtained from fibre irradiation campaigns using this facility. The work presented hereunder has already been published [35].

4.2.1. Description of the experiment

The experiments reported here were performed on 4 commercially available pure silica core optical fibres, which were selected mainly based on their reported radiation tolerance. They are listed in table 6. The fibres were irradiated in the so-called CMF gamma irradiation facility [75] of the BR2 reactor at SCK•CEN, which uses spent fuel elements as radiation source. The fibres under irradiation were monitored on line with a computer based data acquisition system at a dose rate of 30 kGy(Si)/h up to a total dose of 10 MGy. A schematic view of the experimental set-up is shown in figure 4. Due to the restrictions imposed by the special irradiation facility, it was not possible to closely follow FOTP-49A.

TABLE 6. List of the tested optical fibres

Ref. Label	Manufacturer	Type	ϕ Core (mm)	Numerical aperture	OH Content
M1	Mitsubishi	ST-R100C(FV)V 100	100	0.20	low
P1	Polymicro Technologies	FBP 200220240	200	0.22	low
P2	Polymicro Technologies	FHP 200220240	200	0.22	high
R1	Raychem	VSC 200/280	200	0.22	high

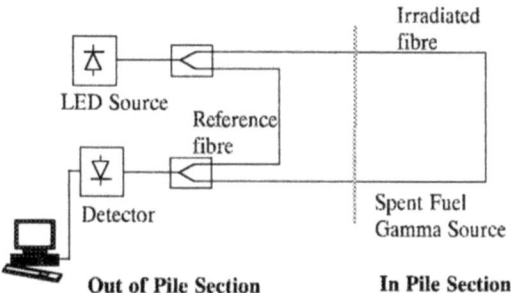

The fibres were wound on a metallic cylinder with a diameter of 70 mm and measured using two types of light sources : a LED source at 850 nm and a white light source with three bandpass filters (400 nm to 700 nm, 700 nm to 1000 nm, 1000 to 1800 nm). The light sources were used in chopped mode, with a chopper frequency of 270 Hz to limit parasitic light effects. Experimental temperature was about 60 °C.

A fibre length of 30 m was subjected to radiation. For each of them, an identical unirradiated counterpart served as a reference to measure only with high accuracy the radiation induced optical attenuation. It is important to submit the reference fibre to the same environmental conditions (bending, temperature, ...) as the fibre under test.

4.2.2. Result discussion

Figure 5 shows the radiation induced attenuation of the M1 optical fibre. The light source used is the LED at 850 nm. At the start of the irradiation the radiation induced attenuation increases very rapidly up to 0.25 dB/m. However, further increase is smaller and a fairly stable value around 0.37 dB/m is obtained at an accumulated dose of 4 MGy.

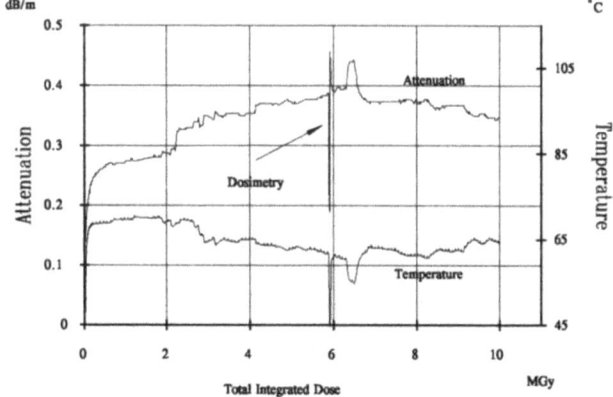

Figure 5. Radiation induced attenuation of the M1 optical fibre at 850 nm

During the dosimetry performed at 6 MGy, the fibres were removed from the gamma source for one hour, but were kept monitored on-line at room temperature. The induced attenuation decreased very fast in the absence of radiation. When the irradiation resumed, a transient increase is noticed, as was seen at irradiation start. Directly after this transient increase of attenuation, the temperature rise (from room temperature to 60° C) results in a temperature annealing, lowering the radiation induced attenuation by 0.05 dB/m. This temperature influence can be observed during the whole experiment : at 6.3 MGy, a temperature drop as small as 8 °C caused an increase of 0.05 dB/m for the induced attenuation, towards the end of the irradiation, were the temperature was slightly higher, the induced attenuation lowers.

Figure 6 presents the radiation induced attenuation for the same fibre (M1) at different wavelengths. As expected, the induced attenuation is higher for lower wavelengths, but the evolution of the induced attenuation as a function of accumulated dose remains the same over the whole measured wavelength range.

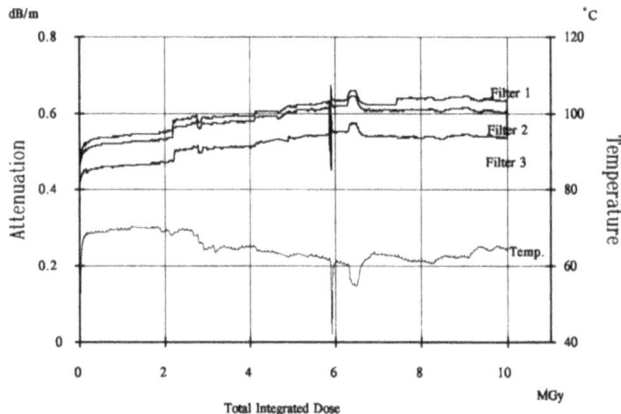

Figure 6. Evolution of the induced attenuation in the Mitsubishi fibre as a function of total dose for different wavelengths.

The results for the P1 and P2 optical fibres are presented on figure 7. The OH-content is low for P1, high for P2. P2 shows a similar behaviour to the M1 fibre (fig. 5), i.e. a sharp increase of the induced attenuation at the start of the irradiation, and a stabilisation afterwards at approximately 0.5 dB/m. The P1 fibre on the contrary shows a transitory phenomenon at irradiation start with a peak value up to 2 dB/m and a rapid stabilisation at half this value. A slight increase can be observed afterwards up to 0.9 dB/m. This is particular for low OH-content fibres where the radiation and heat induced phenomena seem to have a different relative time ratio. The temperature dependency however is very similar to the one observed for the M1 fibre. At total doss of respectively 7 MGy and 5.5 MGy, the fibres fractured.

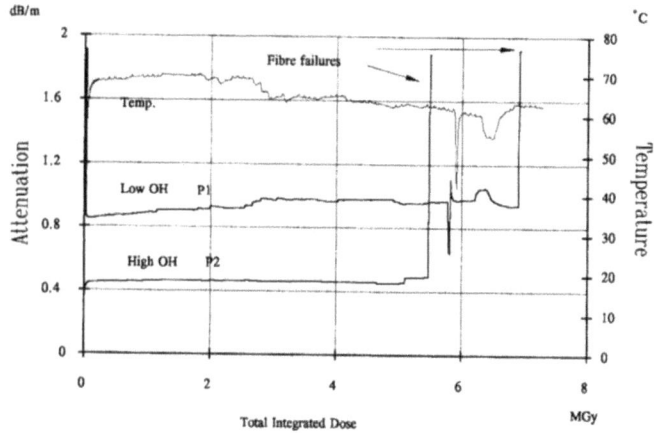

Figure 7. Radiation induced attenuation for P1 and P2 optical fibres

The R1 fibre (low OH-content) showed a similar transient behaviour, i.e. a fast increase of induced attenuation, but a mechanical fracture occurred very early at a total dose of only 20 kGy.

From these results, it is clear that radiation induced attenuation can be kept to acceptable figures with properly selected fibres. An attenuation of 0.3 to 1 dB/m allows short distance data communication and light transport for optical sensing or communication for telemanipulation purposes (fibre lengths up to 30 m). Temperature annealing or photo bleaching can even lower this attenuation to less than 0.2 dB/m. Larger core diameter enhance the transmission performance, as does a high OH content in the silica core [68]. The results present also stable values after a transient increase at irradiation start, stressing the need for pre-irradiation on all fibre components to be used in nuclear environment. The non-linear character of the radiation induced attenuation curves emphasises the validity of high total dose tests requirements, as mere extrapolation from low dose data does not necessarily predict further behaviour. Mechanical fractures occurred on uncoated fibres due mainly to thermal and mechanical stresses, and partly to the embrittlement of the polymer cladding. The only coated fibre, M1, was kept undamaged, although it was wrapped the same way as the other samples. For this fibre Kevlar-coating provides mechanical relief capabilities.

5. Conclusion

Though it was ascertained that fibre-optic communication links are suitable for direct implementation in nuclear power plants [8,80], all-optical sensing systems are however not suited for direct transposition to the nuclear environment. Moreover, there is still a lack of information with regard to these sensors' reliability once they will be exposed to ionising radiation fields. Advantages of optical fibre systems as compared to conventional communication and instrumentation have pushed nuclear industry to seriously consider their assessment in the very hostile nuclear environment and thorough research has been performed for several years on the effects of ionising radiation on the behaviour of optical fibres. Pioneers in this research were mainly governmental and military laboratories.

Due to the very broad range of parameters which influence the fibre's response to ionising radiation and the formation of colour centres, exactly predicting the radiation induced loss becomes extremely difficult. Up to now, no satisfactory predictive model was ever put forward, though some interesting attempts led to good agreements with experimental data. However, the presented models are still restricted to a low total dose or do not take dose rate effects into account. Therefore, efforts are still needed to set-up better models. Colour centre formation and annealing is a main concern in fields related to photonic technology such as glass chemistry and has major implications on the use of photorefractive materials and the creation of photorefractive gratings. Therefore, common needs among these fields should be analysed in order to catalyse the efforts necessary for gaining an even better understanding of colour centre growth and recovery phenomena.

A second result of this large variety of parameters influencing the fibres behaviour is the disparity between the results obtained by the different research groups involved in this domain. Therefore, a high level of specification of irradiation and test parameters was found to be necessary in order to allow for comparison of results obtained in different tests. The total doses and dose rates specified by existing procedures should be completed in order to meet the specifications of non-military applications also, since dose rates and total doses in nuclear power plants or fusion reactor environment can reach levels which exceed these encountered in military or space scenarios.

Tests at such very high dose rate and total dose levels (up to 10 MGy) were conducted at SCK•CEN. Even for these high doses, fibre attenuation was relatively constant and remained between acceptable limits for short distance transmission.

Clearly, lots of research and efforts were already performed in trying to assess fibre-optic technology in all kinds of nuclear environments. General results indicate that, with specific care and within certain limits, the fibres (and associated opto-electronics) might be suitable for application in these hostile environments. Nevertheless, there are still numerous standing questions and further research is definitely necessary, both on the fundamental physical aspects of radiation induced effects as on the identification of potential applications and their implementation. Future developments in fibre-optic communication and sensing will probably lead to systems with an enhanced reliability, suitable for application in very difficult conditions, including nuclear installations.

References

1. Friebele, E.J., Askins, C.G. and Gingerich, M.E. (1984) Effect of low dose rate irradiation on doped silica core optical fibers, *Applied Optics* **23**, 4202-4208.
2. Griscom, D.L. (1985) Nature of defects and defect generation in optical glasses, *Proceedings of the SPIE* **541**, 38-59.
3. West, R.H. and Lenham, A.P. (1982) Characteristics of light induced annealing in irradiated optical fibres, *Electronic Letters* **18**, 483-484.
4. Friebele, E.J., Askins, C.G., Gingerich, M.E. and Long, K.J. (1984) Optical fiber waveguides in radiation environments II, *Nuclear Instruments and Methods in Physics Research* **B1**, 355-369.
5. Griscom, D.L. and Friebele, E.J. (1982) Effects of ionizing radiation on amorphous insulators, *Radiation Effects* **65**, 63-72.
6. West, R.H. (1988) A local view of radiation effects in fiber optics, *Journal of Lightwave Technology* **6**, 155-164.
7. Chesser, J.B. (1993) Radiation testing of optical fibers for a hot-cell photometer, *IEEE Transactions on Nuclear Science* **40**, 307-309.
8. Kyoto, M., Chigusa, Y., Ohe, M., Go, H., Watanabe, M., Matsubara, T., Yamamoto, T. and Okamoto, S. (1992) Gamma-ray hardened properties of pure silica core single-mode fiber and its data link system in radioactive environments, *Journal of Lightwave Technology* **10**, 289-294.
9. Nagasawa, K., Ohki, Y. and Hama, Y. (1988) Gamma-ray induced 2 eV optical absorption band in pure-silica core fibers, in R.A.B. Devine (ed), *The Physics and Technology of Amorphous SiO_2*, pp. 165-170, Plenum Press, New york and London.
10. Greenwell, R.A. (1991) Reliable fiber-optics for the adverse nuclear environment, *Optical Engineering* **30**, 802-807.
11. Friebele, E.J., Taylor, E.W., Turquet de Beauregard, G., Wall, J.A. and Barnes, C.E. (1988) Interlaboratory comparison of radiation-induced attenuation in optical fibers. Part I : steady-state exposures, *Journal of Lightwave Technology* **6**, 165-171.
12. Taylor, E.W., Friebele, E.J., Henschel, H., West, R.H., Krinsky, J.A. and Barnes, C.E. (1990) Interlaboratory comparison of radiation-induced attenuation in optical fibers. Part II : steady-state exposures, *Journal of Lightwave Technology* **8**, 967-976.
13. Friebele, E.J., Lyons, P.B., Blackburn, J., Henschel, H., Johan, A., Krinsky, J.A., Robinson, A., Schneider, W., Smith, D., Taylor, E.W., Turquet de Beauregard, G., West, R.H. and Zagarino, P. (1990) Interlaboratory comparison of radiation-induced attenuation in optical fibers. Part III : transient exposures, *Journal of Lightwave Technology* **8**, 977-989.
14. Berthold, J.W. III (1991) Industrial applications of fiber optic sensors, in E. Udd (ed) *Fiber Optic Sensors : An Introduction for Engineers and Scientists*, John Wiley & Sons Inc., pp. 409-437.
15. Marcus, M.A. (1993) Requirements for distributed fiber-optic sensor systems in process monitoring and control applications, *Proceedings of the SPIE* **2071**, 198-208.
16. Hashemian, H.M. (1993) Aging of nuclear plant RTD's and pressure transmitters, *Proceedings of PLEX '93*, Nuclear Engineering International, pp. 85-99.
17. Druschel, R. and Schramm, W. (1993) Services for plants from other suppliers : Modernization of instrumentation and control systems in nuclear power plants, *Proceedings of PLEX '93*, Nuclear Engineering International, pp. 262-275.
18. Marcus, M.A., Hartog, A.H., Purdum, C.F. and Leach, A.P. (1989) Real-time distributed fiber-optic temperature sensing in the process environment, *Proceedings of the SPIE* **1172**, 194-205.
19. Distributed Fibre-Optic Temperature Sensor DTS-80, Product literature, York Sensors Limited, Hampshire, UK.
20. Redding, J. (1994) Advanced LWR Technology for Commercial Application, *Atomwirschaft* **Jan. 1994**, 47-52.

21. C. Ferro, M. Gasparotto and H. Knoepfel (eds), (1993) *Proceedings of the 17th Symposium on Fusion Technology* **1 and 2**, Elsevier Science Publishers, North Holland.
22. Decréton, M. (1994) Viewing systems in the fusion reactor vessel : Radiation hardened lens glasses up to 300 MGy, Accepted for presentation at the *18th Symposium on Fusion Technology*, Karlsruhe, Germany, 22-26 August 1994.
23. Hayami, H., Akutsu, T., Ishitani, T. and Suzuki, K. (1993) Radiation resistivity of pure-silica-core image guides, *Journal of Nuclear Science and Technology* **30**, 95-106.
24. Morgan, P.D. (1993) Irradiation of optical fibres at JET through 14 MeV neutron production, in C. Ferro, M. Gasparotto and H. Knoepfel (eds) *Proceedings of the 17th Symposium on Fusion Technology* **1**, Elsevier Science Publishers, North Holland, pp. 722-726.
25. E. Udd (ed), (1991) Fiber Optic Sensors : An Introduction for Engineers and Scientists, John Wiley & Sons Inc.
26. J. Dakin and B. Culshaw (eds), (1988) Optical Fiber Sensors : Systems and Applications, Artech House Inc, Boston and London.
27. W.K. Burns (ed), (1993) Optical Fiber Rotation Sensing, Academic Press Ltd., London.
28. Greenwell, R.A., Scott, D.M. and McAlarney, J.J. (1992) Nuclear survivable polarization fibers for fiber gyroscopes on spacecraft, Presented at *SPIE OE/Fibers Symposium*, Boston, USA, 8-11 September 1992.
29. Lyons, P.B. (1984) Review of high bandwith fiber optics radiation sensors, *Proceedings of the SPIE* **566**, 166-171.
30. Bueker, H. and Haesing, F.W. (1992) Physical properties and concepts for applications of attenuation-based fiber optic dosimeters for medical instrumentation, *Proceedings of the SPIE* **1648**, 63-70.
31. Roman, G.W. and Shultz, J.R (1994) Development of a Long-Term, Post-Closure Radiation Monitoring System, Product Literature, Babcock & Wilcox, USA.
32. Greenwell, R.A., Addleman, R.S., Crawford, B.A., Mech, S.J. and Troyer, G.L. (1992) A survey of fiber optic sensor technology for nuclear waste tank applications, *Fiber and Integrated Optics* **11**, 141-150.
33. Chesser, J.B. (1993) Radiation testing of optical fibers for a hot-cell photometer, *IEEE Transactions on Nuclear Science* **40**, 307-309.
34. Boisdé, G., Blanc, F., Mauchien, P. and Perez, J.-J. (1991) Fiber optic chemical sensors in nuclear plants, in O.S. Wolfbeis (ed) *Fiber Optic Chemical Sensors and Biosensors* **II**, CRC Press, pp. 135-149.
35. Coenen, S. and Decréton, M. (1993) Feasibility of optical sensing for robotics in highly radioactive environments, *IEEE Transactions on Nuclear Science* **40**, 851-856.
36. Coenen, S. and Decréton, M. (1993) Use of optical fibres for remote perception in nuclear environment, in C. Eugène (ed) *Proceedings of the International Symposium on Intelligent Instrumentation for Remote and On-site Measurements : 6th TC-4 Symposium*, Brussels, pp. 371-375.
37. Breuzé, G. and Serre, J. (1992) Gamma-ray vulnerability of light-emitting diodes, injection laser-diodes and PIN photodiodes for 1.3 µm wavelength fiber optics, *Proceedings of the SPIE* **1791**, 255-264.
38. Evans, R.D. (1955) The Atomic Nucleus, McGraw-Hill Book Company.
39. Peaslee, D.C. (1955) Elements of Atomic Physics, Prentice-Hall.
40. Dienes, G.J.and Vineyard, G.H. (1957) Radiation Effects in Solids, Wiley Interscience.
41. W.B. Fowler (ed), (1968) Physics of Color Centers, Academic Press.
42. Levy, P.W. (1985) Overview of nuclear radiation damage processes : Phenomenological features of radiation damage in crystals and glasses, *Proceedings of the SPIE* **541**, 2-24.

43. Williams, R.T. and Friebele, E.J. (1986) Radiation damage in optically transmitting crystals and glasses, in M.J. Weber (ed) *CRC handbook of laser Science and Technology* **III** - *Optical Materials*, CRC Press Inc., pp. 299-499.
44. Friebele, E.J. and Griscom, D.L. (1986) Color centers in glass optical fiber waveguides, *Material Research Society Symposium Proceedings* **61**, 319-331.
45. Lyons, P.B. (1986) Radiation effects, *Proceedings of the SPIE* **648**, 128-133.
46. Bertolotti, M. (1979) Radii and refractive index changes in γ-irradiated optical fibers, *Radiation Effects Letters* **43**, 177-180.
47. Dellin T.A. (1977) Volume, Index of refraction and stress changes in electron-irradiated vitreous silica, *Journal of Applied Physics* **48**, 1131-1138.
48. Rao, R., and Mitra, S.S. (1980) Effect of neutron irradiation on the pulse dispersion in a step-index optical fibre, *Applied Physics Letters* **36**, 948.
49. Clegg, D.W. and Collyer A.A. (1991) *Irradiation effects on polymers*, Elsevier Applied Science.
50. Wertz, J.E. and Bolton, J.R. (1972) Electron-spin resonance : Elementary theory and practical applications, Mac-Graw Hill Book Company.
51. Antonini, M., Camagni, P., Gibson, P.N. and Manara, A. (1982) Comparison of heavy ion proton and electron irradiation effects in vitreous silica, Radiation Effects **65**, 41-48.
52. Tohmon, R., Mizuno, H. Ohki, Y., Sasagane, K., Nagasawa, K. and Hama, Y. (1989) Correlation of the 5.0 and 7.6 eV absorption bands in SiO_2 with oxygen vacancy, *Physical Review B* **39**, 1337-1345.
53. Trukhin, A.N., Skuja, L.N., Boganov, A.G. and Rudenko, V.S. (1992) The correlation of the 7.6 eV optical absorption band in pure fused silicon dioxide with two-fold coordinated silicon, *Journal of Non-Crystalline Solids* **149**, 96-101.
54. Nishikawa, H., Nakamura, R., Tohmon, R., Ohki, Y., Sakurai, Y., Nagasawa, K. and Hama, Y. (1990) Generation mechanism of photoinduced paramagnetic centers from preexisting precursors in high-purity silicas, *Physical Review B* **41**, 7828-7834.
55. Dohguchi, N., Munekuni, S. Nishikawa, H., Ohki, Y. and Nagasawa, K. (1991) Effect of high-temperature treatment on optical absorption bands in amorphous SiO_2, *Journal of Applied Physics* **70**, 2788-2790.
56. Tohmon, R., Shimogaichi, Y., Munekuni, S., Ohki, Y., Hama, Y. and Nagasawa, K. (1989) Relation between the 1.9 eV luminescence an 4.8 eV absorption bands in high-purity silica glass, *Applied Physics Letters* **54**, 1650-1652.
57. Awazu, K. and Kawazoe, H. (1990) O_2 molecules dissolved in synthetic silica glasses and their photochemical reactions induced by ArF excimer laser radiation, *Journal of Applied Physics* **68**, 3584-3591.
58. Nishikawa, H., Tohmon, R., Ohki, Y., Nagasawa, K. and Hama, Y. (1989) Defects and optical absorption bands induced by surplus oxygen in high-purity synthetic silica, *Journal of Applied Physics* **65**, 4672-4678.
59. Awazu, K., Harada, K., Kawazoe, H. and Muta, K. (1992) Structural imperfections in silica glasses with an optical absorption peak at 3.8 eV, *Journal of Applied Physics* **72**, 4696-4699.
60. Friebele, E.J., Griscom, D.L. and Marrone, M.J. (1985) The optical absorption and luminescence bands near 2 eV in irradiated and drawn synthetic silica, *Journal of Non-Crystalline Solids* **71**, 133-144.
61. Munekuni, S., Yamanaka, T., Shimogaichi, Y., Tohmon, R., Ohki, Y., Nagasawa, K. and Hama, Y. (1990) Various types of nonbridging oxygen hole center in high-purity silicas, *Physical Review B* **41**, 7828-7834.
62. Friebele, E.J. (1986) Radiation effects and defect centers in fiber optic materials, *Proceedings of OPTO'86*, Paris, pp. 436-439.

63. Russel, P.St.J., Poyntz-Wright, L.J. and Hand, P.D. (1990) Frequency doubling, absorption and grating formation in glass fibres : Effective defects or defective effects ?, *Proceedings of the SPIE* **1373**, 126-139.
64. Meltz, G., Morey, W.W. and Glen, W.H. (1990) Formation of Bragg gratings in optical fibers by a transverse holographic method, *Optics Letters* **14**, 823-825.
65. Neustruev, V.B. (1991) Point defects in pure and germanium-doped silica glass and radiation resistance of optical fibres, *Soviet Lightwave Communications* **1**, 177-195.
66. Friebele, E.J., Schultz, P.C. and Gingerich, M.E. (1980) Compositional effects on the radiation response of Ge-doped silica-core optical fiber waveguides, *Applied Optics* **19**, 2910-2916.
67. Henschel, H. (1990) Radiation testing of optical fibres and typical results, Presented at the *Large Hadron Collider Workshop, Physics and Instrumentation*, Aachen, Germany, 4-9 October 1990.
68. Friebele, E.J., Long, K.J., Askins, C.G. and Gingerich, M.E. (1985) Overview of radiation effects in fiber optics, *Proceedings of the SPIE* **541**, 70-88.
69. Henschel, H and Köhn, O (1989) Radiation sensitivity of Philips single mode fibres : SL 1120 B (Matched Cladding) - FL 1097 C (Dispersion Flattened), *Internal Report No. 7/89*, Fraunhofer-Institut für Naturwissenschaftlich-Technische Trendanalysen, Germany.
70. Friebele, E.J. (1993) NATO Panel 4 on Optics and Infra-Red Research Study Group 12 on Fibre & Associated Integrated Optics Technology, Procedure for measuring radiation-induced attenuation in optical fibres and optical cables, *Technical Report AC/243(Panel 4) TR/2*, NATO Unclassified - Official Information.
71. West, R.H. (1987) Interpreting radiation tests on fibre optics, *Proceedings of the SPIE* **734**, 136-142.
72. Decréton, M. (1994) State of the art in position sensing for highly radioactive environment, Presented at the *International Symposium on Intelligent Instrumentation for Remote and On-site Measurements, 6th TC-4 Symposium*, Brussels, Belgium, 12-13 May 1994.
73. Henschel, H., Köhn, O. and Schmidt, H.U. (1990) Influence of dose rate on radiation induced loss in optical fibres, *Proceedings of the SPIE* **1399**, 1990.
74. Griscom, D.L., Gingerich, M.E. and Friebele, E.J. (1993) Radiation induced defects in glasses : Origin of power-law dependence of concentration on dose, *Physical Review Letters* **71**, 1019-1022.
75. Coenen, S., Decréton, M. and Liesenborghs, R. (1992) Gamma irradiation facilities at SCK•CEN. Testing of sensors, electronics and optical components for remote handling systems, Presented at the *International Conference on Irradiation Technology*, Saclay, France, May 1992.
76. Friebele, E.J. (1992) Survivability of optical fibers in space, *Proceedings of the SPIE* **1791**,177-188.
77. Friebele, E.J., Askins, C.G., Shaw, C.M., Gingerich, M.E., Harrington, C.C., Griscom, D.L., Tsai, T.E., Paek, U.C. and Schmidt, W.H. (1991) Correlation of single-mode fiber radiation response and fabrication parameters, *Applied Optics* **30**, 1944-1957.
78. Friebele, E.J., Gingerich, M.E., Brambani, L.A., Harrington, C.C. and Hickey, S.J. (1990) Radiation effects in polarization-maintaining fibers, *Proceedings of the SPIE* **1314**, 146-154.
79. Liu, D.T.H. and Johnston A.R. (1994) Theory of radiation-induced absorption in optical fibers, *Optics Letters* **19**, 548-550.
80. Breuzé, G., Jucker, P. and Serre, J. (1993) Fibre optics compatibility with radiation environment inside PWR containment, Presented at the *IAEA Specialists' Meeting on Improvements in Nuclear and Radiation Instrumentation for Nuclear Power Plants*, Saclay, France, 18-20 October 1993.

III Optical Fibre Components

CONNECTORS AND SPLICES

J. DUBARD
Laboratoire Central des Industries Électriques
33, av. du Général Leclerc
BP 8
92266 Fontenay aux roses
FRANCE

1. Introduction

Connectors and splices are widely used in fibre optics links. They give access to control points or allow repair of broken fibre during installation or after use. Connectors and splices induce a discontinuity in the fibre link and generate therefore optical losses. These losses originate from optical properties of the fibres and mechanical defaults of both the fibre and the connector (Fresnel reflection, lateral and longitudinal displacement, ...)

This paper is dealing with the optical properties of connectors and splices, mainly the insertion loss and the return loss. After a review of the different causes of losses, we describe the principal methods for measuring the optical parameters.

2. Definition

2.1. CONNECTORS
A connector is a device which can be plugged-in and plugged-out several times (generally more than 500 times). It is composed of 2 plugs and an adaptor in case of a fibre-fibre connection or a plug and a base in the case of a fiber-active component connection. The typical insertion loss is 0.1dB to 1dB. The coupling mechanism is based on three technics: screw, bayonet or push-pull.

2.1 SPLICES
A splice is a permanent connecting device. The typical insertion loss is less than 0.2dB. Splices are based on three technics: mechanical, glue, fusion.

3. Connectors
Three types of connectors are used depending on the technics for the coupling mechanism (screw, bayonet and push-pull). The advantages and disadvantages of the different technics are summarized in table 1.

	Screw (FSMA, FO, FC)	Bayonet (ST)	Push-pull (EC, SC, FDDI)
Advantages	- Mechanical strength - Environment proof	- Plug-in/plug-out easy -Constant applied strength due to the string	- Plug-in/plug-out very easy - Compact connector - Complex shape connectors - High density connector
Disadvantages	- Insertion loss variation with screw torque - Plug-in/plug-out uneasy in tight connector assembly - Possible unscrewing due to vibration -Displacement of one plug with respect to the other while connecting (possibility of damaging the fibre end)	- Bad environment proof	- Fragile connecting mechanims in case of rough treatment

Table 1: Summary of the advantages and disadvantages of the different types of connectors

4. Optical properties of connectors

The main optical properties of connectors which are commonly measured are the insertion loss L (dB) and the return loss RL(dB).

Considering the different components shown on figure 1 which represents the fibre interface at the connector set we have:

$$P_0 = P_t + P_r + P_l \tag{1}$$

where P_0 is the incident power, P_t the transmitted power, P_r the power that is reflected back and P_1 is the loss power. The insertion loss L and the coupling ratio η are defined as:

$$L = -10 Log(\eta) \quad \text{with} \quad \eta = \frac{P_t}{P_0} \qquad (2)$$

The return loss RL is given by:

$$RL = -10 Log(\frac{P_r}{P_0}) \qquad (3)$$

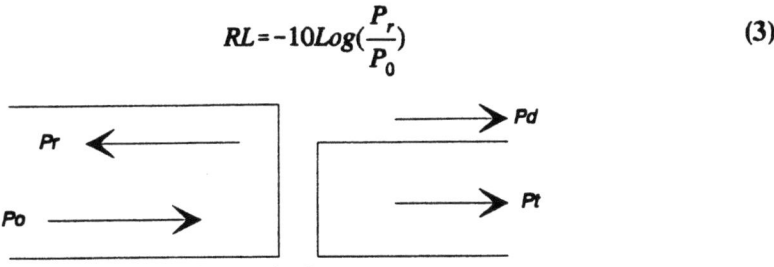

Figure 1: Power budget at a connector set

4.1 INSERTION LOSS

Insertion loss which characterizes the loss of optical power at the connector set is due to the optical phenomena and mechanical defaults of the fibres and the connector at the interface of the connector set. The losses are attributed to:
- Fresnel reflection due to the discontinuity between the two fibres
- Interferences because of the Fabry-Pérot effect between the plane surfaces of the fibre ends
- Longitudinal and transverse displacement of one fibre core with respect to the other
- Angular displacement between the two plugs
- Difference in fibre cores diameters, in numerical aperture
- Ellipticity of the fibre core

Insertion loss due to Fresnel reflection which occurs because of the change of refractive index at the fibre discontinuity is determined from the following equations:

$$L = -10 Log(1-R)^2 \quad \text{with} \quad R = \frac{(n_c - n_0)^2}{(n_c + n_0)^2} \qquad (4)$$

where n_c is the core refractive index of the fibre and n_0 is the refractive index of the medium between the fibre ends at the connector set.

The insertion loss can be reduced using index matching matter. Table 2 shows the insertion loss due to Fresnel reflection in three cases (the refractive index of the fibre core is 1.46).

Medium	Refractive index	L (dB)
air	1.0	0.310
glue	1.6	0.018
index matching (fluid, gel, film)	1.4	0.004

Table 2: Insertion loss due to Fresnel reflection as a function of the medium at the fibre interface

Interference may occur at the interface because of the Fabry-Perot effect. the coupling ratio η is given by:

$$\eta = \frac{(1-R)^2}{(1-R)^2 + 4R.\sin^2(\frac{2\pi n_0 d}{\lambda})} \quad (5)$$

where d is the distance between the two fibre ends. The variation of the insertion loss L as a function of d is a periodic function with a succession of minima and maxima. L varies from 0dB to 0.6dB. However equation 6 is valid for parallel beam. In practice the output beam of a fibre is divergent. Figure 2 shows the variation of the insertion loss for a diverging beam. For d of about 10λ the insertion loss is reduced to the one due to Fresnel reflection.

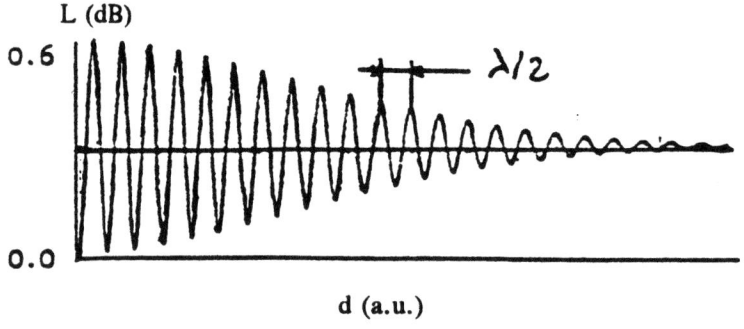

Figure 2: Insertion loss due to interferences

In multimode fibre connection insertion loss due to fibre or connector mechanical defaults is dependent on the mode power distribution at the fibre interface. Usually one encounters two types of mode power distribution:
- The fully filled mode power distribution. This is realized when using short lengths of fibres which are connected to a large aperture source such as LED.
- The equilibrium mode power distribution. This is realized when using long fibres (>500m).

The optical power carried by the fibre ([1,2] and references therein) is given by the following equation:

$$P_T = \int_0^1 P(X) \, N(X) \, dX \qquad (6)$$

where X is the normalized mode parameter ($0<X<1$), N(X) is the number of mode having the same mode parameter and P(X) is the power carried by the mode with the parameter X.

For a fully filled mode power distribution:

$$P(X) = constant \qquad (7)$$

For a equilibrium mode power distribution:

$$P(X) = A_s \, J_0(2.405\sqrt{X}) \quad \text{step index profile}$$

$$P(X) = A_p \, \frac{J_1(3.832\sqrt{X})}{\sqrt{X}} \quad \text{parabolic gradient index profile} \qquad (8)$$

where A_s and A_p are normalized constants and J_0 and J_1 are the Bessel functions of order 0 and 1 respectively.

The power carried by the input and the output fibres are calculated based on the number of mode in each fiber. The number of mode is a function of the normalized frequency V:

$$V = \frac{2\pi a (n_c^2 - n_g^2)^{\frac{1}{2}}}{\lambda} \qquad (9)$$

where n_c and n_g are the core refractive index and the cladding refractive index respectively, a is the core diameter and λ is the wavelength of the source.

Table 3 gives the value of the power loss ratio $\Gamma = 1-\eta$ for different causes of insertion loss due to either connector defaults or fibre defaults. The following parameters are defined:

Insertion loss due to connector defaults:
  - Gap between the fibre

$$s = \frac{z}{a} \frac{NA}{n_0} \tag{10}$$

  - Transverse displacement

$$u = \frac{d}{a} \tag{11}$$

  - Angular displacement

$$t = \frac{n_0 \sin\phi}{NA} \tag{12}$$

Insertion loss due to fibre defaults
  - Difference in fibre diameter

$$p = \frac{a_2}{a_1} \tag{13}$$

  - Difference in numerical aperture

$$q = \frac{NA_2}{NA_1} \tag{14}$$

  - Ellipticity

$$e = 1 - \frac{h}{g} \tag{15}$$

where z is the distance and n_0 is the refractive index of the medium between the two fibres, NA is the numerical aperture, d is the distance between the centers of the fibres, ϕ is the angle between the fibres, h and g are respectively the small axis and the large axis of an elliptical fibre core.

	case 1		case 2		case 3	
type of default	step	gradient	step	gradient	step	gradient
gap (s)	$0.42s$	$0.5s$	***	$0.6s^2$	***	$0.4s^2$
transv; displ. (u)	$0.64u$	$0.85u$	$0.64u$	$0.92u^2$	$0.64u$	$1.84u^2$
angular displ. (t)	$0.64t$	$0.85t$	$0.72t^2$	$0.92t^2$	$1.44t^2$	$1.84t^2$
differ. in diam. (p)	$1-p^2$	$1-p^2$	$1-p^2$	$2.45 \times (1-p)^2$	$1-p^2$	$1-p$
differ. in NA (q)	$1-q^2$	$1-q^2$	$1.44 \times (1-q)^2$	$2.45 \times (1-q)^2$	***	$1-q$
elliptic. (e)	$0.64e$	$0.64e$	$0.64e$	$0.6e$	$0.64e$	$0.4e$

Table 3: Power loss ratio for different connector or fibre defaults in multimode fibre connection

The calculations are performed for the following cases:
- Case 1: Insertion loss for two short cables connected together. The cables are used with a LED.
- Case 2: The input fibre is a long fibre (equilibrium mode power distribution). The output fibre is short (no equilibrium).
- Case 3: The two fibres are long. Equilibrium mode power distribution is reached in both fibres.

Figure 3 shows the insertion loss due to the transverse displacement of two $50\mu m$ diameter fibres one with respect to the other.

Figure 4 shows the insertion loss due to the difference in fibre core diameters for the three different cases discussed above.

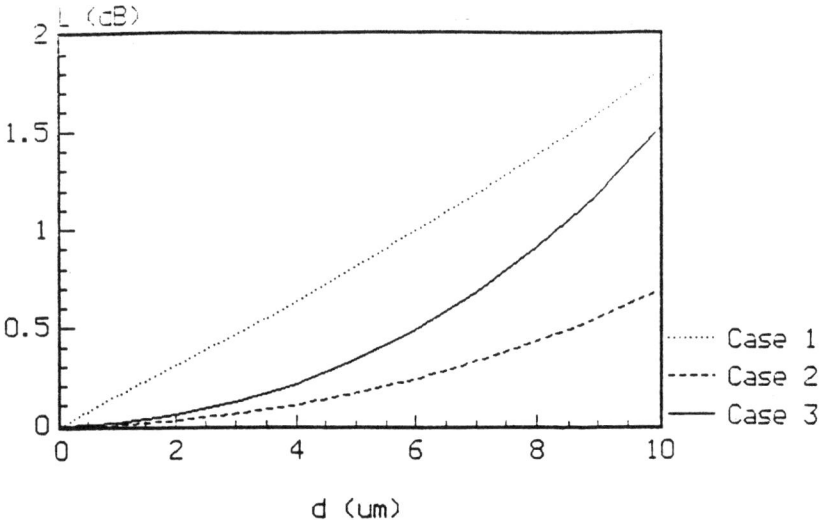

Figure 3: Insertion loss due to transverse displacement in multimode fibre connection

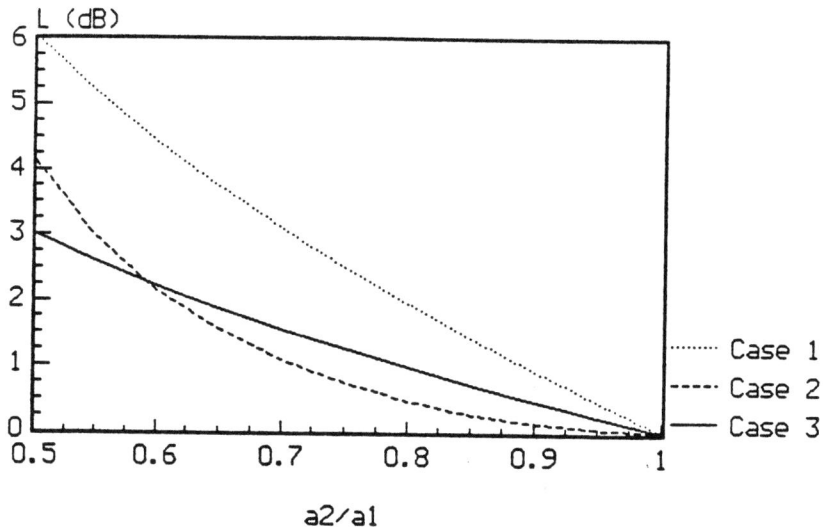

Figure 4: Insertion loss due to fibre core diameters in multimode fibres connection

Those results show that in multimode fibre insertion loss measurement the light injection conditions (mode power distribution) should be specified.

In singlemode fibre connection the power distribution, which is approximated to a gaussian beam, is independent on the length of the fibres (one propagating mode). However the wavelength at which the insertion loss measurement is performed should be greater than the cut-off wavelength to insure single mode propagation in both fibres.

Assuming a gaussian beam propagation in the fibre with a mode field radius w, the insertion loss can be calculated for the different types of connector and fibre defaults. The coupling ratio η has the following expressions:

Insertion loss due to connector defaults
 - Gap between fibres

$$\eta = \frac{1}{Z^2+1} \quad \text{with} \quad Z = \frac{z\lambda}{2\pi n_0 w^2} \qquad (16)$$

 - Transverse displacement

$$\eta = \exp(-U^2) \quad \text{with} \quad U = \frac{d}{w} \qquad (17)$$

 - Angular displacement

$$\eta = \exp(-T^2) \quad \text{with} \quad T = \frac{\sin\phi}{\lambda/n_0 \pi w} \qquad (18)$$

Insertion loss due to fibre defaults
 - Difference in mode field diameter

$$\eta = \frac{4}{(\frac{w_2}{w_1} + \frac{w_1}{w_2})^2} \qquad (19)$$

 - Difference in fibre diameter and concentricity
This is equivalent to a transverse displacement and therefore the coupling ratio η is expressed as equation 18.

Figure 5 shows the insertion loss in singlemode fibre connection as a function of transverse displacement. Because of the core diameters for singlemode fibre the insertion loss is drastically affected by any connector or fibre defaults which will induce a displacement of one fibre with respect to the other.

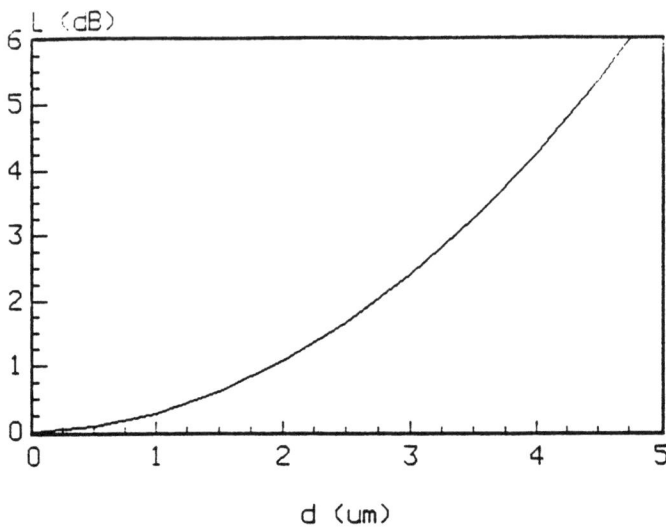

Figure 5: Insertion loss due to transverse displacement in singlemode fibre connection

According to the CCITT-G652 standard the tolerances on singlemode fibres cladding diameters is $\pm 2\mu m$. Considering these parameters in the less favourable case the insertion loss due to transverse displacement is 4.24dB.

4.2 RETURN LOSS

The return loss characterizes the amount of energy that is reflected back at the fibre interface. The energy reflected back can become a spurious signal which may disturbe any active components such as the laser or an optical amplifier. Technologies are developped to minimize the return loss. Among these technics are the use of index matching matter, the physical contact between the two fibres and the angled polishing.

Figure 6 shows the return loss as a function of the polishing angle of the fibre ends.

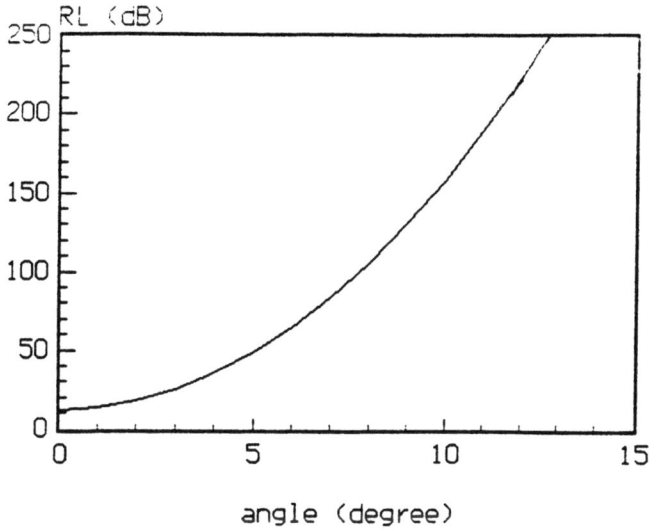

Figure 6: Return loss as a function of polishing angle of the fibre ends

5 Measurements of the optical properties of connectors

The optical properties of connectors are measured according to test methods described in the international standard IEC 874-1 [3]. This standard is dealing with the following parameters:
- Insertion loss
- Return loss
- Wavelength dependence of the insertion loss
- Susceptibility to ambient light
- Cross talk

5.1 INSERTION LOSS

Eight test methods regarding the measurement of the insertion loss are described depending of the connector set configurations. Indeed we may have to measure the insertion loss in two circumstances:

- Case 1: The two fibre ends are accessible for equipment attachment and measurement (Figure 7)

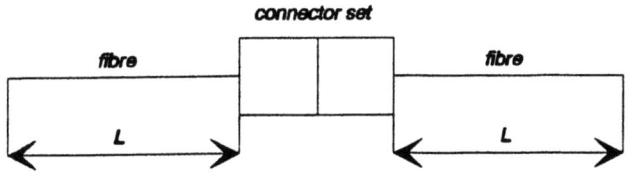

Figure 7: Connector set to be measured: case 1

- Case 2: The fibre ends are inaccessible: measurement of a cord (figure 8)

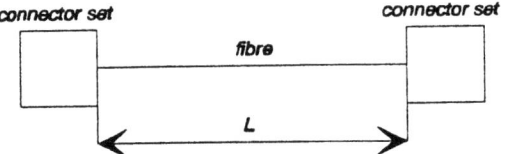

Figure 8: Connector set to be measured: case 2

Method 1 is the ideal insertion loss measurement technics. The procedure is the following these 3 steps:

1- Measurement of the output power P_0 of a fibre (figure 9). The fibre ends must be appropriately positioned and fixed with respect to the source and detector.

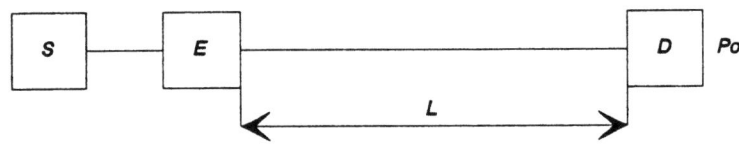

Figure 9: Method 1 (Insertion loss). Measurement of the reference power P_0.

2- The fibre is cut to allow mounting of the connector set

3- Measurement of the output power P_1 of the fibre with the connector mounted (figure 10)

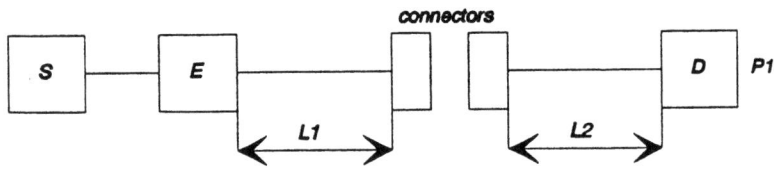

Figure 10: Method 1 (Insertion loss). Measurement of the power P_1.

The insertion loss of the connector set is given by:

$$\alpha = -10 \: Log \frac{P_1}{P_0} \qquad (20)$$

Method 1 is a destructive one. Method 2 uses a temporary joint and a reference fibre to avoid the need for new length of fibre for each measurement.

5.2 RETURN LOSS

Return loss of connector is measured using a method similar to method 1 for insertion loss measurement. Figure 11 shows the measurement set-up which includes a coupler and two detectors.

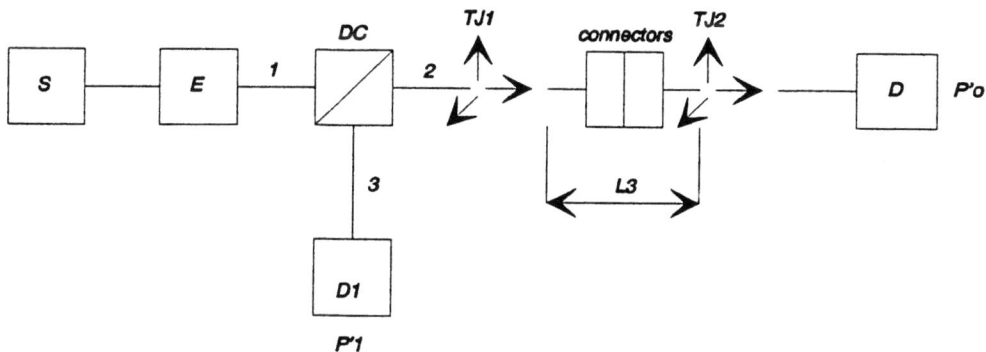

Figure 11: Return loss measurement set up.

The measurement procedure is the following:

1- Measurement of the reference powers P_0 and P_1 when the connector set is replaced by a fibre of length L_3.
2- Measurement of the powers P'_0 and P'_1 with the connector set in place.

The return loss is given by:

$$RL = -10 \: Log \frac{P'_1 - P_1}{P_0} + 10 \: Log(T_{2,3}) \qquad (21)$$

where $T_{2,3}$ is the transfer coefficient between terminals 2 and 3 of the directional coupler that is measured in accordance with Sub-clause 17.1.1 of IEC Publication 875-1.

Return loss can be measured using OTDR. Measurement procedures are actually under study in IEC working group(IEC 86B/[secretariat] 403). Figure 12 shows the return loss measurement performed on a fibre cord equipped with 3 EC connectors. The return losses (table 4) are determined using an algorithm developped at LCIE. The OTDR was calibrated using a reference reflectance set-up [4, 5] traceable to national standards by means of radiometric standards.

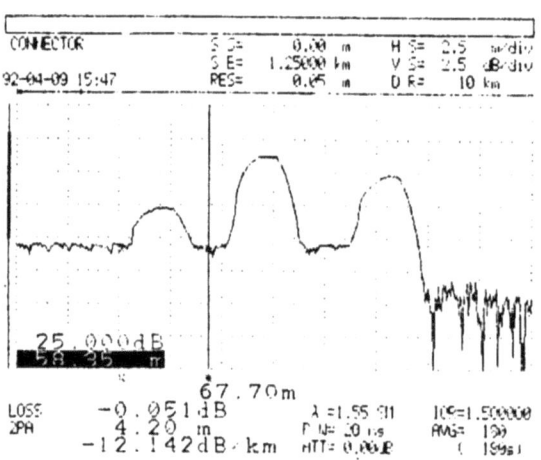

Figure 12: Return loss measurement using OTDR

	wavelength (nm)	RL (dB)
connector 1	1310	-65.4
	1550	-64.7
connector 2	1310	-51.9
	1550	-55.4
connector 3	1310	-61.7
	1550	-58.1

Table 4: Return loss measurement performed on a fibre cord equipped with 3 EC connectors and using an OTDR

5.3 OTHER OPTICAL PROPERTIES

The other optical properties for which measurement procedures are described in IEC 874-1 are:
- Spectral loss
- Susceptibility to ambient light coupling
- Cross talk

6 Splices

Splices encounter about the same optical loss mechanisms that have been described for connectors. However because of the alignment process during splicing (fusion) or the alignment mechanisms used in mechanical splices the magnitude of the different losses can be minimized. Indeed for instance in fusion splicing insertion loss can be as low as 0.03dB and there is almost no return loss.

All the optical measurement technics described in standard IEC 874-1 for connector are applicable to splices.

7 Conclusion

Optical characteristics of connectors and splices are presented. Loss mechanisms are described and measurement technics in accordance with international standard IEC 874-01 and future IEC publications are reviewed.

With tolerances on geometrical parameters of both the fibre and the connecting device measured insertion loss as low as 0.1dB for connectors and 0.03dB for splices can be reached. Return loss lower than -60dB is obtained using angled polishing technics.

8 References

[1] Gloge D. (1975), *Propagation effects in optical fibres*, IEEE Trans. Microwave Theory Tech. Vol. MTT-23, pp. 106-120

[2] Snyder Allan W., (1969), *Asymptotic expressions for eigenfunctions and eigenvalues of a dielectric or optical waveguide*, IEEE Trans. Microwave Theory Tech.,Vol. MTT-17, pp. 1130-1138

[3] International standard IEC 874-1, *Connectors for optical fibres and cables*, second edition 1987

[4] Blanchard P., Ducos L., Fournier P., Dubard J., Gauthier F., (1993), *Étalonnage en réflectance des réflectomètres optiques dans le domaine unimodal*, Opto 93, pp. 255-258

[5] Blanchard P., Zongo P. H., Facq P., (1990), *Accurate reflectance and optical fibre backscatter parameter measurements using an OTDR*, Electronics Letters, vol. 26, N° 25, pp. 2060-2062

OPTICAL ATTENUATORS AND COUPLERS CHARACTERIZATION

C. LE MEN
Laboratoire Central des Industries Électriques
33, av. du Général Leclerc
BP 8
92266 Fontenay aux roses
FRANCE

1. Introduction

During the last ten years, fibre almost reached its theoretical values regarding attenuation coefficient and dispersion. So that we can say the transmission media is now achieved. However, to build systems using fibres, one should need some other components that fall in two main categories: **Active** Components and **Passive** Components.

Active components category includes optical sources (Light Emitting Diodes, Laser Diodes,...), optical amplifiers (Semi-Conductor Amplifiers, Erbium Doped Fibre Amplifiers,...) and detectors. Passive components category includes all components that contain no optoelectronic or other transducing elements. That is to say, attenuators, connectors, branching devices (couplers).

In this article, only passive components will be treated. As the connectors characterization is a very large domain, it will be treated in an independant article. Thus, the subject we are interested in is optical attenuators and couplers characterization.

2. Optical Attenuators

2.1. CHARACTERISTICS DEFINITION

An optical attenuator is any device used in a fibre link or system to generate a constant or variable attenuation of the optical power. An attenuator could be considered as a black box with an input and an output connector. The most important parameter to be characterized is obviously the attenuation value(s) expressed in dB. Making reference to an unit calls a **calibration for attenuation** [1]. In addition with this calibration has to be measured the **polarization and wavelength dependences** of the attenuator.

2.2. ATTENUATION CALIBRATION

2.2.1. *Calibration Set-Up*

The attenuation value of a device is defined as the ratio of input and output powers. This ratio is usually expressed in dB according to the following formula:

$$A = 10 \cdot \log_{10} \frac{P_{input}}{P_{output}} \quad [dB] \qquad (1)$$

Unfortunately, this definition of attenuation includes the insertion loss of the apparatus. Now, the insertion loss is often quite difficult to measure with a good uncertainty because it depends on the type and quality of connectors and fibres which are used to connect the attenuator to your own measurement set-up. It means that a measurand proper to the attenuator, insertion loss in the present case, depends on components external to the attenuator, this is not a good way to perform measurements. Therefore, to cast off this difficulty, what is measured is not the real attenuation of the device but the variation of this attenuation compared to a reference line, this **attenuation error** ΔA_{meas} is of the form:

$$\Delta A_{meas} = A_{meas} - A_{read} \quad [dB] \qquad (2)$$

where: A_{meas} is the attenuation value measured by the calibration set-up,
A_{read} is the attenuation value read on the attenuator.

That gives us the principle of the attenuation calibration set-up [2] that shall be used. This set-up is described in figure 1.

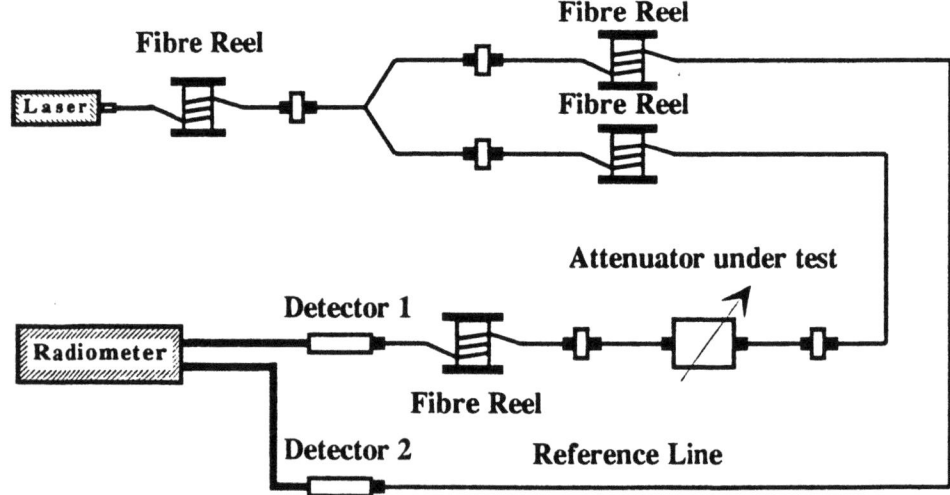

Figure 1. Attenuators Calibration Set-up

-The laser source is stabilized in optical power and regulated in temperature and polarization current. This source consists in a Fabry-Perot laser diode emitting at a wavelength that can be either 0.85 µm, 1.3 µm or 1.55 µm. Its output power level is high enough ($\approx$ 1 milliwatt) to ensure a good dynamic range of attenuation (> 50 dB).

-A 50% - 50% Y-coupler splits the source output power in two fibre lines. One constitutes the reference line, the other the measurement line in which is inserted the attenuator to be calibrated.

-Detection is performed by a two detectors radiometer. Both detectors have to be calibrated in linearity with great accuracy. The sensitive area of those detectors should be accessible in open space so that the linearity calibration could be performed by a fluxes addition method. The wavelength range allows either germanium or InGaAs detectors to be used, but very few InGaAs detectors which sensible area is accessible and large enough for radiometric measurements are available up to now. Therefore, we use germanium detectors even though their dynamic range is smaller than InGaAs detectors one.

2.2.3. *Calibration Procedure*

An attenuator is calibrated according to the following procedure:

i/ Select the "no attenuation" (or the minimum one) state corresponding to the Insertion Loss IL, then are measured the following optical power levels:
- P_{meas} (IL) the power level in the measurement line (Watt),
- P_{ref} (IL) the power level in the reference line (Watt),

the ratio of those level gives a calibration constant that cancels the Insertion Loss problem.

ii/ For each selected value of attenuation A_{read} have to be measured the following optical power levels:
- P_{meas} (A_{read}) the power level in the measurement line (Watt),
- P_{ref} (A_{read}) the power level in the reference line (Watt),

the attenuation value A_{meas} is then calculated from the following formula:

$$A_{meas} = 10 \cdot \log_{10}\left[\frac{P_{ref}(A_{read})}{P_{meas}(A_{read})} \cdot \frac{P_{meas}(IL)}{P_{ref}(IL)}\right] \quad [dB] \qquad (3)$$

Then the **attenuation error** ΔA_{meas} is calculated from the equation 2.

iii/ Step ii/ is then repeated for desired values of attenuation with the choosen interval and into the choosen dynamic range (in agreement with the attenuator owner).

2.2.3. *Calibration Results*

An attenuator calibration example [2] is given hereafter. This calibration concerns an attenuator used in single-mode fibre applications which has the following specifications:

- Attenuation range: 0.00 to 60.00 dB with a 0.01 dB step,
- Wavelength range: 1200 nm to 1650 nm with an 1 nm step.

The attenuation error ΔA_{meas} is plotted as a function of the attenuation value A_{read} selected on the attenuator for two particular wavelengths. The results of the attenuator calibration is plotted in figure 2 for a wavelength equal to 1303 nm and in figure 3 for 1544 nm.

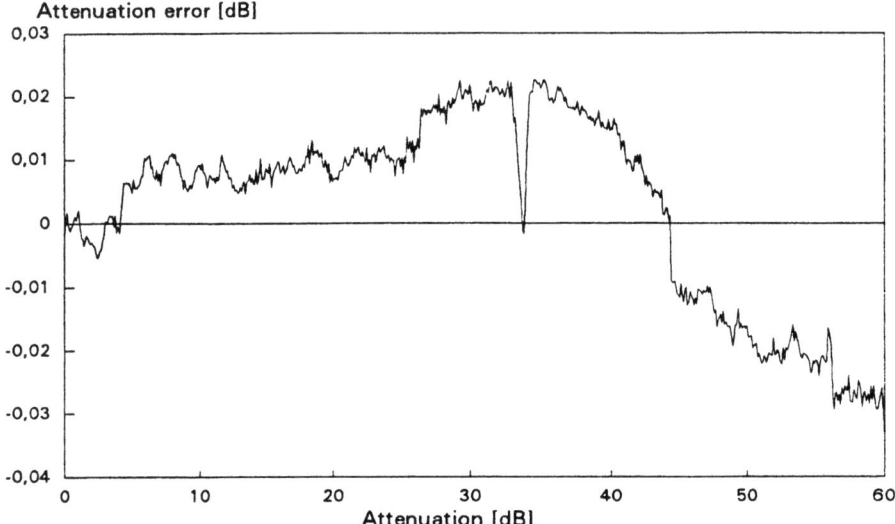

Figure 2. Attenuation error ΔA_{meas} as a function of Selected Attenuation A_{read} for a particular single-mode attenuator at a 1303 nm wavelength

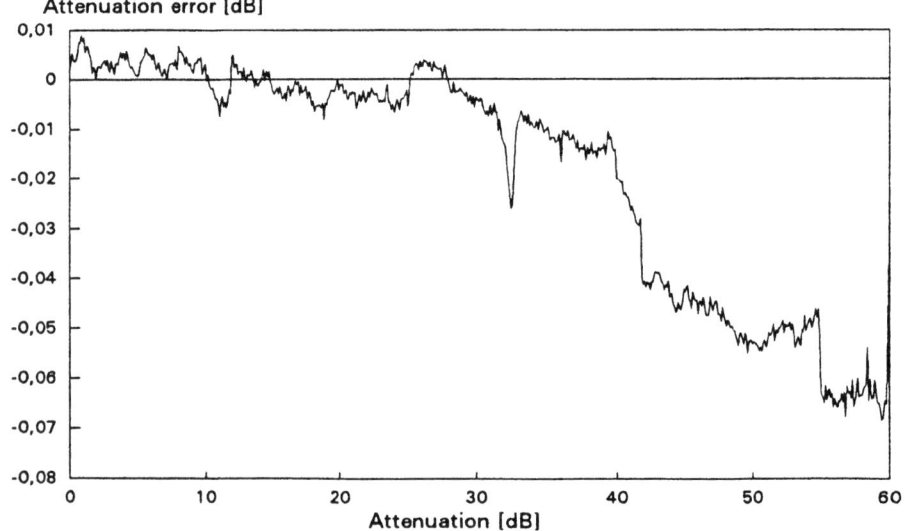

Figure 3. Attenuation error ΔA_{meas} as a function of Selected Attenuation A_{read} for a particular single-mode attenuator at a 1544 nm wavelength

Those calibration curves show a quite similar behavior. The attenuation error ΔA_{meas} is very small, let us say $\pm$ 0.01 dB, in the attenuation range 0 - 40 dB. Nevertheless, a decreasing peak of -0.02 dB does occur around 33 dB, this peak can be imputed to a change of density filters causing a "gluing" mismatch between both attenuation ranges (0 - 33 dB and 33 - 60 dB). For attenuation values greater than 40 dB, the absolute error is increasing till -0.03 dB at 1303 nm and till -0.06 dB at 1544 nm. Those maximum errors do not introduce such great penalty compared with the high attenuation values that are considered (a decrease in power of a million time !). For that particular attenuator, one can say it is "good", that is to say that read attenuation can be considered as correct. A best metrological way to conclude is to verify that no time the attenuation error is greater that the expanded uncertainty that shall be given with any measurement. For this attenuator this condition is verified. Let's see how this calibration uncertainty is calculated.

2.2.4. Calibration Uncertainties

The total uncertainty can be divided in two component types. The first component type, the so-called A component, is a statistical one, the second component type, the B component, takes into account all perturbation effects and calibration uncertainties of the set-up devices that are related to an unit. All the components are given for a confidence level of one standard deviation (1σ), considering they follow a gaussian law. Those components are detailed hereafter [2].

<u>Component of type A:</u>

A: *Uncertainty due to the calibration set-up used for attenuation measurements.* The component A is calculated as the experimental standard deviation of n measurements of attenuation errors of the **particular attenuator that is under test.**

$$A = \sqrt{\frac{\sum_{i=1}^{n}[\Delta A_{meas_i} - \Delta A_{meas}]^2}{n-1}} \quad [\%] \quad (1\sigma) \qquad (4)$$

where ΔA_{meas} is the mean value of n attenuation error measurements.

<u>Components of type B:</u>

B_1: *Stability of the calibration set-up along the calibration time.* This stability value has been found within a 1 % band for the attenuation error ΔA_{meas}. Then, considering this influence follows a rectangular law, the relative <u>gaussian</u> component B_1 is obtained by dividing this maximum deviation (1 % band) by $2 \cdot \sqrt{3}$. This yields:

$$B_1 = 0.3 \ \%.$$

B_2: *Influence of the source wavelength.* The fact that users' source wavelength could be different than the calibration wavelength must be taken into account. Then, measuring the wavelength dependence of attenuation error and applying the same method than in the B_1 component calculation, the component B_2 is:

$$B_2 = 0.12 \ \%$$

B_3: *Influence of the temperature variation ($\pm$ 1 °C)*. The effect of temperature regulation on attenuation error measurements is estimated in the same way than for B_1 and B_2 components. This yields:

$$B_3 = 0.3 \%$$

B_4: *Influence of the detectors linearity calibration uncertainty*. Both detectors have to be calibrated in linearity. This linearity calibration is given with an uncertainty. Considering that the effect of detectors linearity on attenuation error measurements obeys to a gaussian law, the component B_4 can be equalized to the linearity calibration uncertainty. So:

$$B_4 = 0.022\sqrt{N} \quad [\%] \tag{5}$$

where N is the number of octaves (3 dB) numbered in the attenuation value that is under consideration. This yields for instance **0.04 % for 10 dB** and **0.09 % for 50 dB**.

Combined standard uncertainty I:
Considering that all components are uncorrelated one to each other, their variances (the square of the components) can be add, so that the combined standard uncertainty I is given by:

$$I = \sqrt{A^2 + \sum_{i=1}^{4} B_i^2} \quad [\%] \quad (1\sigma) \tag{6}$$

Expanded Uncertainty:
The final result of a calibration is the attenuation error ΔA_{meas} and its calibration uncertainty. This calibration uncertainty, the so-called Expanded Uncertainty, is twice the Combined standard uncertainty I. This yields, for a gaussian distribution, a confidence level of two standard deviations (2σ) and therefore a 98 % probability that the "true" attenuation error is within the interval: $\Delta A_{meas} \pm$ 2.I. Two expanded uncertainties calculations are detailed in table 1 for 10 dB and 50 dB attenuations.

TABLE 1: Attenuator Calibration Uncertainties

Uncertainty sources	Attenuation 10 dB	Attenuation 50 dB
Repeatability A	0.40 % (0.022 dB)	0.70 % (0.030 dB)
Stability B_1	0.30 % (0.013 dB)	0.30 % (0.013 dB)
Wavelength B_2	0.12 % (0.005 dB)	0.12 % (0.005 dB)
Temperature B_3	0.30 % (0.013 dB)	0.30 % (0.013 dB)
Linearity Calibration B_4	0.04 % (0.002 dB)	0.09 % (0.005 dB)
Combined Uncertainty I	0.60 % (0.026 dB)	0.83 % (0.036 dB)
Expanded Uncertainty 2.I	**1.2 % (0.052 dB)**	**1.7 % (0.073 dB)**

2.3. POLARIZATION DEPENDENCE

2.3.1. *Measurement Set-Up*

The measurement Set-Up [3] that is used to measure the polarization dependence of attenuator is shown in figure 4.

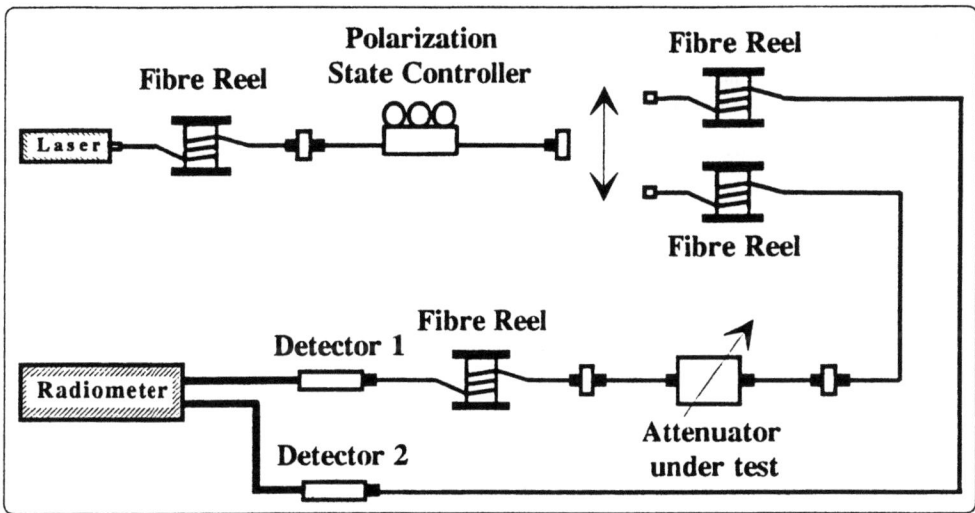

Figure 4. Polarization Dependence of Attenuators - Measurements Set-Up

-The principle of that Set-Up is the same than the attenuation calibration one. To avoid problems caused by sources fluctuations and power fluctuations due to the polarization state controller, a relative measurement is achieved by the use of both measurement and reference lines. So what is measured is a change in the attenuator Insertion Loss.

-The laser source is any of six Fabry-Perot Laser Diodes emitting at six different wavelengths. The use of many sources is justified by the need to average effects due to the type of sources and to check polarization dependence as a function of wavelength.

-The Polarization State Controller is a custom-made one, based on the Lefevre Fibre Loops principle [4]. This specific apparatus uses three successive Fibre Loops, the first and last ones respectively behave as a Polarizer and an Analyzer, the middle one constituates a quater wave plate. Those loops can be rotated, the different loops positions combinations generate all Polarization States in the output fibre. Those Polarization States are not identified (linear, circular or elliptic). But as far as the aim is not to characterize the attenuator behavior when seeing such or such identified polarization states, it is sufficient to know that they can all be swept using this device.

-The Polarization Controller output power is launched in either reference line or measurement line by means of direct connections. This manual operation is longer than when using a branching device such as a Y-coupler but is necessary to avoid an eventual polarization dependence of that coupler.

2.3.2. Measurement Results

Those measurements are performed [3] on the same attenuator than in the 2.2.3 section concerning attenuation calibration. The insertion loss variation measurements are given for different attenuation values (0, 10, 20 and 30 dB) and six different source wavelengths. Those measurements are resumed in table 2. For each wavelength and each selected attenuation is given the maximum insertion loss deviation that can be seen when scanning all possible Polarization States.

TABLE 2: Attenuator Insertion Loss Variation as a function of Polarization State for four attenuation values and six wavelength values

Wavelength	Selected Attenuation			
	0.00 dB	10.00 dB	20.00 dB	30.00 dB
	Insertion Loss Variation [dB]			
λ = 1297 nm	0.10 dB	0.10 dB	0.11 dB	0.10 dB
λ = 1303 nm	0.05 dB	0.06 dB	0.06 dB	0.06 dB
λ = 1310 nm	0.15 dB	0.16 dB	0.16 dB	0.16 dB
λ = 1540 nm	0.09 dB	0.09 dB	0.09 dB	0.09 dB
λ = 1544 nm	0.07 dB	0.08 dB	0.07 dB	0.07 dB
λ = 1554 nm	0.05 dB	0.05 dB	0.07 dB	0.05 dB
Mean Value	**0.085 dB**	**0.090 dB**	**0.093 dB**	**0.088 dB**
Standard Deviation	**0.038 dB**	**0.039 dB**	**0.037 dB**	**0.039 dB**

For each source, those results show some great variation of the insertion loss from 0.05 dB to 0.16 dB. Those variations are very repeatitive when using one particular source and changing the attenuation value. But those variations stay quite high, and in fact higher than the attenuation calibration uncertainty, for instance. If we gather all the sources measurement results, that yields an Insertion Loss Variation close to 0.09 dB (mean value) with a standard deviation close to 0.04 dB. Considered to the radiometric point of view, a 0.09 dB variation ($\approx$ 2 %) is very high and very cumbersome for some measurements. So that this attenuator can be said very sensitive to polarization. So, what about the polarization influence in uncertainty calculation for attenuation calibration ? Should this influence be considered as a relevant part of total uncertainty ? Fortunately it should not: in fact such changes of the polarization state do not occur during a attenuation calibration, for instance. And even if the polarization state changes, it is already taken into account in the repeatability (A) and stability (B_1) uncertainty components. So, one must be careful when using an attenuator and must check the surrounding of this device. Let's say that in a normal surrounding, the attenuator sensitivity to polarization does not generate great power variation. But if any device generates changes in polarization state above the attenuator then great power variations should be taken into account.

2.4. WAVELENGTH DEPENDENCE

2.4.1. *Measurement Set-Up*

The measurement Set-Up [3] that is used to measure the wavelength dependence of attenuator is shown in figure 5. This Set-Up is exactly the same than the one used for polarization dependence measurements, except for the Polarization State Controller that is obviouly useless.

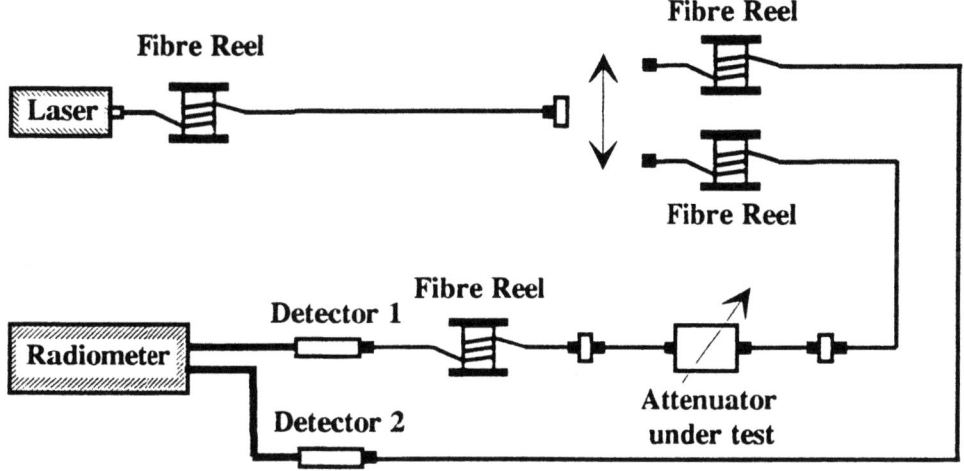

Figure 5 . Wavelength Dependence of Attenuators - Measurements Set-Up

2.4.2. *Measurement Results*

The measurement results [3] are given in table 3 under the form of Insertion Loss Variations compared with two arbitrary references. Those references are Insertion Loss Variation set equal to **0.00 at 1310 nm and 1554 nm**. The wavelength Set of the attenuator is **kept at 1310 nm** when using 1297 nm, 1303 nm and 1310 nm sources, and **kept at 1554 nm** for 1540 nm, 1544 nm and 1554 nm sources. So, this attenuator is characterized without using its internal wavelength correction. This is done that way to take into account two facts: on one hand, some attenuators do not include such a wavelength setting and on the other hand, even if they do, the source wavelength is not necessary well known by the user.

TABLE 3: Attenuator Insertion Loss Variation as a function of wavelength

Wavelength [nm]	1297	1303	1310	1540	1544	1554
Insertion Loss variation [dB]	0.04	0.02	0.00	0.13	0.08	0.00

The results show an Insertion Loss Slope close to 0.0031 dB/nm arround 1.3 μm and close to 0.0093 dB/nm arround 1.55 μm. To see the importance of such a wavelength dependence, let's consider the fabrication tolerances on laser diodes. The common tolerances for wavelength are: $1.28 \mu m \leq \lambda \leq 1.33 \mu m$ for a so-called $1.3 \mu m$ laser and $1.52 \mu m \leq \lambda \leq 1.58 \mu m$ for a $1.55 \mu m$ one. According to the previous measurements, this yields on Insertion Loss a possible variation of 0.155 dB (4%) arround $1.3 \mu m$ and 0.558 dB (14 %) arround $1.55 \mu m$! Those enormous variations are called systematic errors, those are not uncertainty sources because it is certain that the measurement is false ! So it can be seen that an attenuator is very sensitive to wavelength. If the attenuator do not include a wavelength setting that corrects those systematic errors or if the source wavelength is unknown, then those systematic errors should be included in the attenuation calibration uncertainty. In this case, let's say that to calibrate this attenuator is no use, except if you don't mind the measurement values. But even if the wavelength is supposed to be known and the attenuator is correcting those systematic errors, uncertainties on wavelength values must be smaller than **5 nm arround 1300 nm and 2 nm arround 1550 nm** to comply with the B_2 uncertainty component that was previously introduced in this article. And unfortunately this uncertainty on wavelength is the combination of the source wavelength one (at least 0.2 nm) and the wavelength correction matter one (at least 1 nm), so that matching this minimum uncertainty is not so easy to do.

2.5. SUMMARY ON ATTENUATORS

Attenuation: An **attenuation calibration set-up** is demonstrated. This set-up can perform any attenuation calibrations on attenuators designed either for **multimode or single-mode** fibres. It operates at **0.85 μm, 1.31 μm and 1.55 μm** wavelength windows with a dynamic **attenuation range greater than 60 dB**.
An attenuation calibration is instanced for a single-mode fibre designed attenuator. The **uncertainty calculation** is given for that particular attenuator. This calculation showed that the **repeatability component (A) is the relevant one**, so that actually the measurement **uncertainty figure is closely related to the apparatus** that is calibrated. For our attenuator example, uncertainties values for two standard deviations spread from **0.05 dB for a 10 dB** attenuation to **0.07 dB for a 60 dB** attenuation.

Polarization Dependence: For the same attenuator, insertion loss **variations from 0.05 dB to 0.16 dB** were measured while scanning all polarization states by mean of a Lefevre loops polarization controller. Thus is demonstrated that **polarization dependence have to be taken into account when using an attenuator in a line where polarization state is supposed to vary.**

Wavelength Dependence: For the same attenuator, spectral insertion loss **variations of 0.003 dB/nm around 1.3 μm and of 0.009 dB/nm around 1.55 μm** were measured. Without any wavelength dependence correction matter, systematic errors as great as **0.16 dB (1.3 μm) and 0.56 dB (1.55 μm)** are demonstrated. So that to comply with the uncertainties on attenuation value(s), the **wavelength has to be known and integrated in the attenuation calculation with tolerances smaller than 5 nm around 1.3 μm and smaller than 2 nm around 1.55 μm.**

3. Optical Couplers

A coupler is any branching device which has three or more ports for the input and/or output of light and that operates to share light among these ports [5]. Those ports must be optical fibres or optical fibre connectors. Branching devices such as couplers are obviously very important in most fibre applications. In networks, for instance, fibre links would have been useless if the capability to distribute the information signal wasn't possible using couplers. Using couplers made it possible to design very important and advanced active components such as Erbium Doped Fibre Amplifiers and fibre ring lasers. In those two devices, the coupler is the master way to pump the amplifying medium and in the fibre ring laser case, the coupler realizes the cavity at the same time. In instrumentation, the coupler takes a very important place. An instrument that can be instanced is the famous OTDR (Optical Time Domain Reflectometer) in which a Y-coupler is used to distinguish the launched power from the backward power.

In those multiple applications, a coupler has to be chosen for its particular characteristics. As these characteristics are function of the technology that is used to make this coupler, it is important to present first the main existing technologies. Then, some characteristics dependence to polarization and wavelength are studied.

3.1. TECHNOLOGY

The two most common technologies to realize a coupler are the Fusion-Stretching and the Planar technologies. The basis object is either a 2x2 coupler or a Y-coupler, then any NxM coupler can be made by combination of this basis.

3.1.1. *Fused Couplers*
The Fusion-Stretching technology principle is given in figure 6 for a 2x2 coupler.

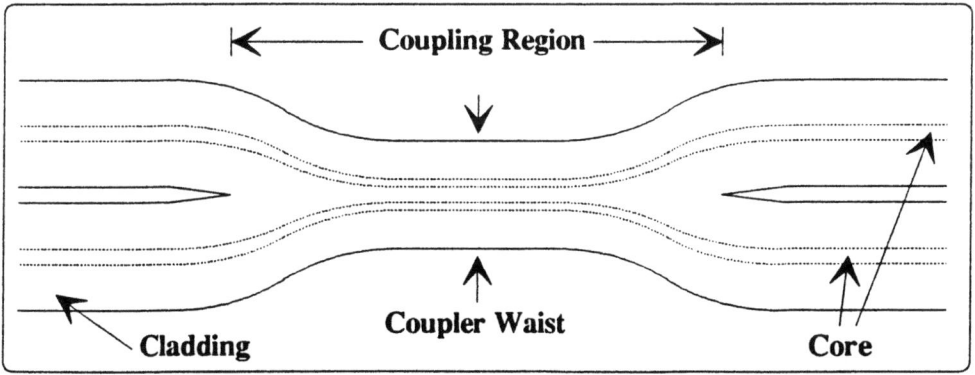

Figure 6. Principle of Fused Coupler Technology

In this technology, two fibres are placed side by side and then are heated along a given length. By stretching those fibres, the heated area collapses so that the claddings melt and the cores tend to be closer and closer. At the same time core diameters decay, so that the evanescent fields increase in the cladding. As far as cores are close one to each other, evanescent fields can be coupled from one core to the other. This evanescent fields coupling allows optical power exchanges between both fibres. The first advantage of that technology is that unbalanced couplers are feasable quite easily: 50%-50%, 90%-10%, 95%-5% and so on fused couplers are available. Another advantage is that the coupler is fully fibre-made and does not present any discontinuities. The disadvantage is that the fabrication processus is quite heavy, especially in the control. For instance, the coupling ratio of a coupler under fabrication is measured during the Fusion-Stretching process. The process is then stopped when the right coupling ratio is reached. This means that characteristics are hardly predictable. This means also that to obtain fabrication repeatability as good as possible, one should perform accurate measurement and control processus during the fabrication. This obviously yields that the coupler cost will be quite high.

3.1.2. *Planar Couplers*
The Planar technology principle is given in figure 7 for a Y-coupler.

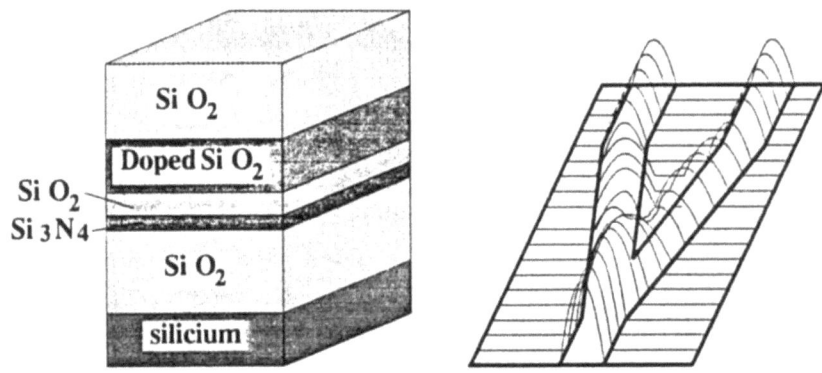

Figure 7. Principle of Planar Coupler Technology

This technology is based on integrated planar waveguides. Such waveguides are realized on silicium substrates with a ribbon shape. The core and cladding regions are made of Doped-Silica and Silica. Then the Y-coupler structure is fully designed on this substrate. The propagation modes in such ribbon waveguides are well known, so that the power repartition in both branches could be modelized as a function of some parameters such as the tilt angle beetween branches and ribbon dimensions. Once this Y-structure is achieved, some optical fibres can be placed at input and output ports. The advantage of this technology is that the coupler characteristics are "easily" predictable using mathematical modelization. So, the measurement and control processus will be easier than in the fused coupler case, thus the Planar-coupler cost could be lower than the fused-coupler one. The disadvantages are on one hand, that unbalanced couplers are quite hard to do and show high Insertion Loss, and on another hand that propagation modes parameters change at the fibre/ribbon-guide interface, so that the continuity is broken and that losses and reflections can appear.

3.2. CHARACTERISTICS DEFINITIONS

To help the coupler characteristics understanding, a scheme of a 2x2 coupler is given in figure 8. Even though this case is a particular one, all characteristics that are defined can be generalized to any coupler [5].

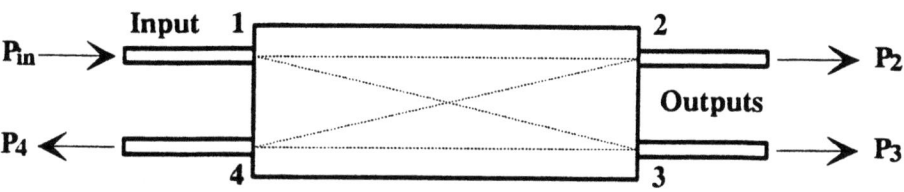

Figure 8. Coupler Characteristics Definition

3.2.1. *Excess Loss: EL*
The Excess Loss is the ratio of output and input powers expressed in dB:

$$EL = 10\log_{10}\frac{P_2+P_3}{P_{in}} \quad [dB] \tag{7}$$

3.2.2. *Coupling Ratio of one branch: CR_n*
The Coupling Ratio is the percentage of total output power in the branch:

$$CR_n = 100\frac{P_n}{P_2+P_3} \quad [\%] \quad \left(CR_n = 10\log_{10}\frac{P_n}{P_2+P_3} \quad [dB]\right) \quad n \in [2,3] \tag{8}$$

3.2.3. *Insertion Loss of one branch: IL_n*
The Insertion Loss is the ratio of the power in the branch over the input power:

$$IL_n = 10\log_{10}\frac{P_n}{P_{in}} = CR_n + EL \quad [dB] \quad n \in [2,3] \tag{9}$$

3.2.4. *Directivity: D*
The Directivity is the ratio of input power which reflected in the other input branch:

$$D = 10\log_{10}\frac{P_4}{P_{in}} \quad [dB] \tag{10}$$

3.2.5. *Uniformity: U*
The Uniformity is the maximum dispersion of all branches Insertion Losses:

$$U = IL_{max} - IL_{min} \quad [dB] \tag{11}$$

3.3. POLARIZATION DEPENDENCE

3.3.1. *Measurement Set-Up*
The measurement Set-Up [3][6] that is used to measure the polarization dependence of couplers is shown in figure 9.

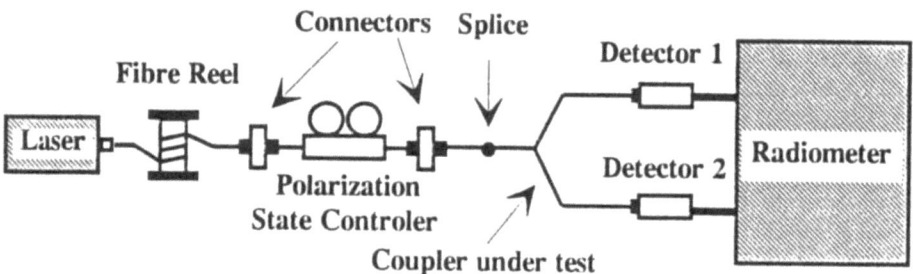

Figure 9 . Polarization Dependence of Couplers - Measurements Set-Up

-The laser source is any of six Fabry-Perot Laser Diodes emitting at six different wavelengths. The source output power is launched in Polarization State Controller by mean of a connector.

-The Polarization State Controller is based on two Lefevre fibre loops [4]. All polarization states can be observed by simply rotating those loops. The output power is then launched in an input port of the coupler under test (a Y-coupler in this example).

-The output powers of the coupler ports are measured by a two detectors radiometer.

-This set-up allows the coupling ratio measurement according to the equation 8.

3.3.2. *Measurement Results*
To avoid any connectors handling, what is measured is the coupling ratio variation as a function of the Polarization States. As well as for attenuator characterization, the polarization states are not identified but are all reviewed when scanning all rotation angles of the Lefevre loops. That way, the measurement is fully relative, so that a high accuracy can be reached with that set-up: this yields an uncertainty on coupling ratio variation smaller than 0.03 dB (0.7 %). The coupling ratio measurements were performed [3] on eight Y-couplers falling into three sets coming from differents makers as follows:

-Set 1: couplers # 1, 2, 3 and 4;
-Set 2: couplers # 5 and 6;
-Set 3: couplers # 7 and 8.

The Set 1 includes couplers from both technologies (fused and planar) and sets 2 and 3 only deal with fused-couplers.

The coupling ratio variations are gathered in table 4 hereafter. Those measurements are **absolute maximum deviations** even if they are expressed in percents. Notice that those variations aren't necessarily centered on the coupling ratio mean value.

TABLE 4: Coupling Ratio Variations of a 8 couplers Set as a function of Polarization State for six wavelength values

	Wavelength	1297 nm	1303 nm	1310 nm	1540 nm	1544 nm	1554 nm
#	Coupler Type	\multicolumn{6}{c}{Coupling Ratio Variation [%]}					
1	Planar 50% -50%	2.3	3.5	1.9	2.1	2.3	1.4
2	Planar 50% -50%	3.3	1.9	2.1	1.9	2.1	2.3
3	Fused 90% -10%	11.2	13.0	12.7	4.0	3.8	3.3
4	Fused 95% - 5%	16.4	17.2	16.1	8.4	8.6	5.4
5	Fused 95% - 5%	8.1	7.9	7.9	11.2	12.7	12.5
6	Fused 95% - 5%	8.4	8.6	8.6	11.4	12.5	11.4
7	Fused 90% - 10%	3.3	3.3	3.5	3.5	4.0	3.3
8	Fused 90% - 10%	7.2	7.6	6.9	7.6	7.6	8.4

From those results it can be seen that all couplers are sensitive to polarization states. The coupling ratio variations that are measured spread from 1.4 % to 17.2 %, and are always bigger than any radiometric measurement uncertainties. Let us look at those results more in details to understand them.

It is easy to distinguish Planar couplers which show the lowest sensitivity to polarization (arround 2 %) from Fused couplers which polarization sensitivity can reach 17 %. Even couplers coming from the same maker and having the same coupling ratio can show very different polarization sensitivities. For instance coupler # 7 has a mean sensivity close to 3.5 % and coupler # 8, which is supposed to be the same, has a mean sensivity close to twice the first coupler one (7.5 %). And this production dispersion increases while same coupling ratio couplers come from different makers. See, for instance, couplers # 3, 7 and 8 case in which sensitivities spread from 3.5 % to 12 % and also couplers # 4 and 5 case in which sensitivities spread from 8 % to 17 %.

To summarize, let's say that planar couplers seems to have a lower polarization sensivity than fused coupler. Let's say also that the dispersion of fused couplers polarization sensitivities is very high, that shows that the fabrication process of that technology is not very repeatable as it was concluded in the "technology" paragraph. So, the greatest care must be taken in using couplers in applications where polarization state is supposed to vary.

3.4. WAVELENGTH DEPENDENCE

3.4.1. *Measurement Set-Up*

The measurement Set-Up [3] that is used to measure the wavelength dependence of couplers is shown in figure 10.

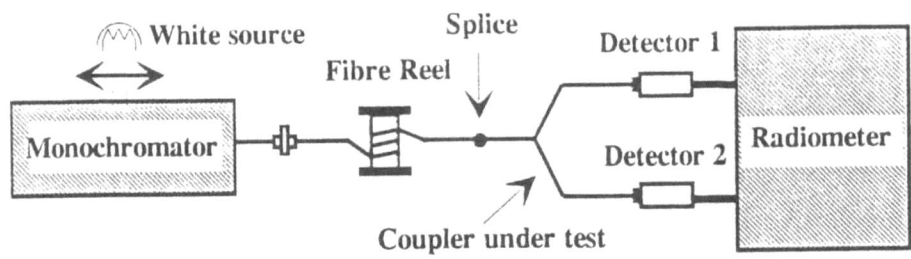

Figure 10. Wavelength Dependence of Couplers - Measurements Set-Up

The radiation emitted by a white source is launched into a grating monochromator which output is an optical fibre. This monochromator selects any wavelength values in the 1.0 - 1.6 μm range with a spectral bandwidth equal to 3 nm. The monochromator output power is launched into the coupler under test input port by splicing fibres. The output powers of the coupler ports are measured by a two detectors radiometer. Those power values allow the coupling ratio calculation according to the equation 8.

3.4.2. *Measurement Results*
The coupling ratio measurements are performed on the two couplers which showed the lowest polarization dependence in the previous paragraph. According to those measurements given in table 4, couplers # 2 and # 7 were chosen [3]. That is to say, on one hand, a 50% - 50% planar-coupler (#2) and on the other hand, a 90% - 10% fused-coupler (#7). For each coupler the coupling ratio is measured over two wavelength ranges that are: 1280 nm - 1350 nm and 1500 nm - 1580 nm. The coupling ratio as a function of wavelength is then plotted in figure 11 and 12. Three repeatability measurements are plotted using symbols, the mean value is plotted in full line.

The 50% - 50% **planar coupler** characterization (figure 11) shows some coupling ratio variations included in a 0.3 % band and that whatever is the wavelength range. The coupling ratio mean values are 50.3 % arround 1.31 μm and 50.2 % arround 1.55 μm. So, the coupling ratio on that coupler port is: 50.25 ± 0.15 % arround 1.31 μm and arround 1.55 μm. As far as the measurement uncertainty is close to 0.7 %, the only thing that can be said is that this coupler is **not sensitive** to the wavelength.

The 90% - 10% **fused coupler** characterization (figure 12) shows some coupling ratio variations included in a 0.7 % band for both wavelength ranges. But mean values increase from 8.4 % at 1.31 μm to 10.8 % at 1.55 μm. So, this coupler is not so sensitive arround a given wavelength, but can be said very sensitive if the wavelength changes from one window to the other.

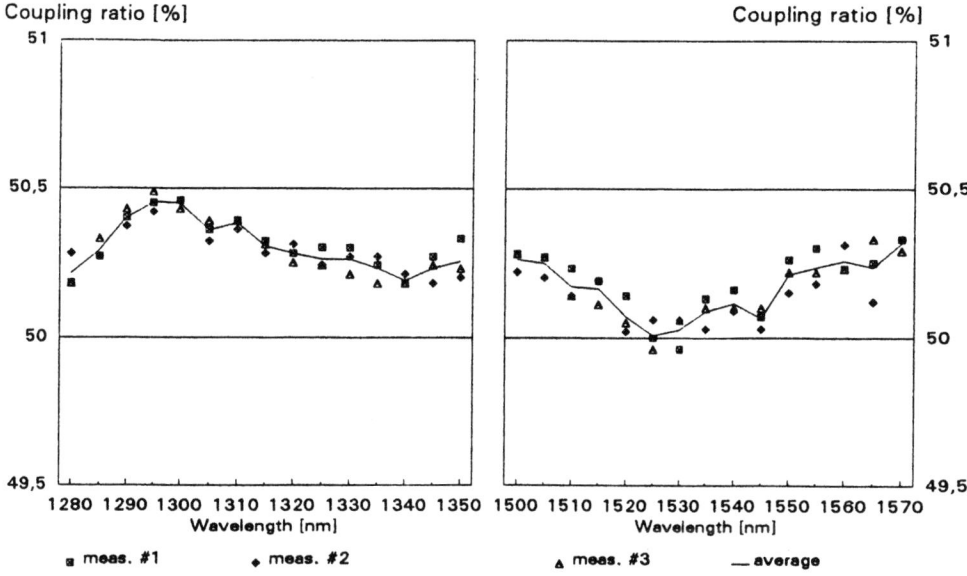

Figure 11. Coupling Ratio as a function of Wavelength for a 50% - 50% Planar coupler

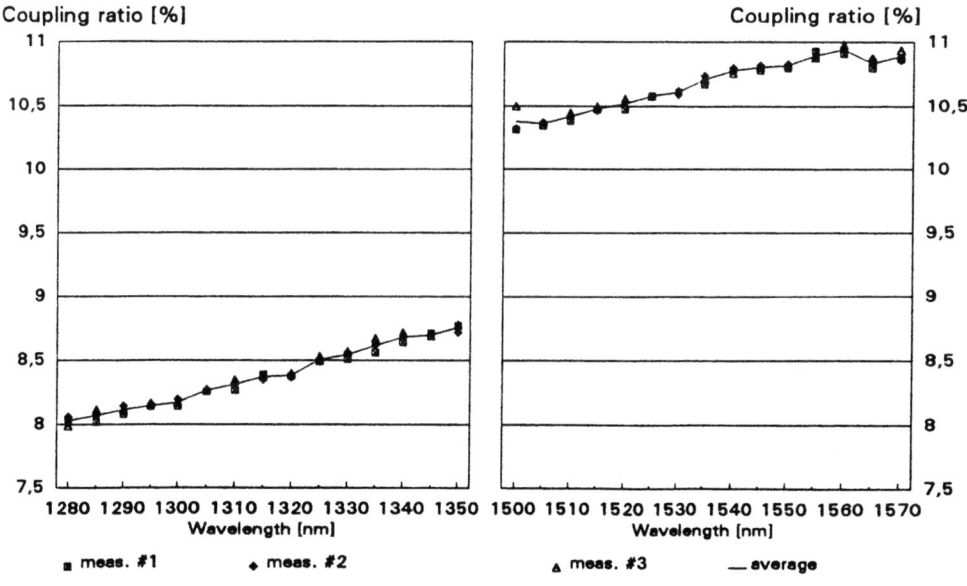

Figure 12. Coupling Ratio as a function of Wavelength for a 90% - 10% Fused coupler
(on the 10% coupling ratio port)

3.5. SUMMARY ON COUPLERS

The characteristics polarization and wavelength dependences of 8 couplers were studied. This 8 couplers set falls into 6 so-called fused couplers and two planar couplers. The measurements have shown the following results:

<u>Polarization Dependence</u>: For the 8 couplers, **coupling ratio variations from 1.4 % to 17.2 %** were measured while scanning all polarization states by mean of a Lefevre loops polarization controller. Coupling ratio variations were measured close to **2 % for both planar couplers**, but the number of measured couplers was probably too small to conclude about their polarization dependence. **Fused couplers** have shown coupling ratio variations **from 3.5 % to 17 %**. During the COST interlaboratory comparison [7] in 1990, a similar polarization dependences dispersion was found (from 1 % to 12%). Thus the dispersion in polarization dependences of couplers is demonstred to be closely related to the technology used. The fact that fused coupler fabrication process is hardly predictable and not highly repeatable seems to be evidenced.

<u>Wavelength Dependence</u>: Measurements were made on two couplers. **The first one, a 50% -50% planar coupler**, have shown no coupling ratio variations greater than **0.3 %** while scanning wavelength ranges from 1280 nm to 1350 nm and 1500 nm to 1580 nm. In both wavelength ranges the same coupling ratio was found (50.3 %). **The second one, a 90% - 10% fused coupler** have shown no coupling ratio variations greater than **0.7 %** in both wavelength ranges. However, the coupling ratio was found equal to 8.4 % around 1.31 μm and equal to 10.8 % around 1.55 μm, so that a change of wavelength window could generate **high systematic errors in many applications**, such as radiometric ones for instance.

4. Furthermore...

In this article, attenuators and couplers characterization was demonstrated. The attenuator one is fully achieved. The couplers one is limited to power splitters. It should be interesting now to go further with wavelength multiplexers and polarization splitters.

5. References

1. International Electrotechnical Commission (1993), *Generic specification for fiber optic attenuators*, Draft International Standard 86B (Central Office) 146.
2. LCIE Service Optique (1994-04), *Étalonnage des atténuateurs pour fibres optiques*, fascicule 4, Agrément n° 82-01-25 du Centre d'Etalonnage Agréé "Radiométrie-Photométrie".
3. Blanchard, P. (1994-01), *Maintien de l'étalon de réflectance*, rapport d'étude LCIE n° 1271.
4. Lefevre, H. C. (1980-09), *Single-mode fiber fractional wave devices and polarization controllers*, Electronic letters, vol. 16, no 20, pp 778-780.
5. International Electrotechnical Commission (1986), *Fiber optic branching devices. Part 1: Generic specification*, International Standard 875-1.
6. De Vries W. J., Burgmeijer J. W., Lintelo F. A. and Van Loenen E. (1991-09), *Optical characterization and measurements of single-mode branching devices*, Optical Fibre Measurement Conference Digest '91, York, pp 61-64.
7. COST 217 *Interlaboratory comparison of optical measurements on single-mode fibre couplers*, (1990-09), Technical Digest, Symposium on Optical Fiber Measurements '90, Boulder, pp 7-10.

Wavelength Division Multiplexers and the Measurement of the Channel Wavelengths

J.P. LAUDE
ISA JOBIN-YVON
16 rue du canal Longjumeau 91160 France

Abstract: *In the first part in the paper, we review different WDM integrated optics devices: in lithium niobate, on glass, on silicon or InP subtrates. It is shown that the main problem to be solved, in most of the cases, is the wavelength and transmission sensitivity to polarisation. In the second part, after an analysis of the accuracy necessary for the wavelength measurements of these devices, we review some light sources and absorption lines useful for their wavelength calibration.*

Introduction

Many multiple-channels WDM devices using integrated optics or microoptics solutions with different approaches for wavelength selection have been proposed during the last few years. In [Ref.0] is included a bibliography, with 476 references on WDM devices and technics. (The references already included in [Ref.0] for devices cited hereafter are not always given again). WDM microoptics devices are in use in optical networks for more than 10 years. Classical microoptics devices such as Stimax [Ref.1] are now available with 20 channels, 1 nm spacings, 2/3 dB losses, crosstalk lower than -30 dB, and no wavelength sensitivity to polarisation. As many papers have been devoted to such devices we will not give here more information on them.

The microoptics devices fulfil the technical and economical specifications of many optical networks, but their costs remain too high for a generalisation to all optical WDM networks. The main target of integrated optics devices is a cost reduction over microoptics devices when manufactured in very large quantities. Many different solutions remains in competition for IO WDM. The devices are numerous and need more analysis. We show hereafter different IO configurations and we compare there results to those obtained with microoptics configurations.

We analyse the accuracy requirements in the measurement of the wavelengths of the devices and give some spectral lines that are used for the calibration of our monochromators or other spectrum analysers.

Integrated optics WDM components.

Selection based, on Mach-Zehnder (MZ) interferometer elements, Fabry Perot, etched Fresnel mirrors, gratings or elliptical Bragg reflectors (EBR), or phased-array waveguides (FAW) and some other configurations give interesting results in integrated optics solutions.

Multi/demultiplexing on planar optical waveguides

Devices using lithium niobate (or lithium tantalate)

Waveguides obtained by titanium diffusion in lithium niobate may be very transparent, about 0.1 dB; this is comparable to the bulk material loss. The coupling loss between the fibre and such an optical waveguide may be relatively small. WDM, using the coherent interaction between two optical waveguides placed side by side in interaction, can be obtained.[Ref. 1]

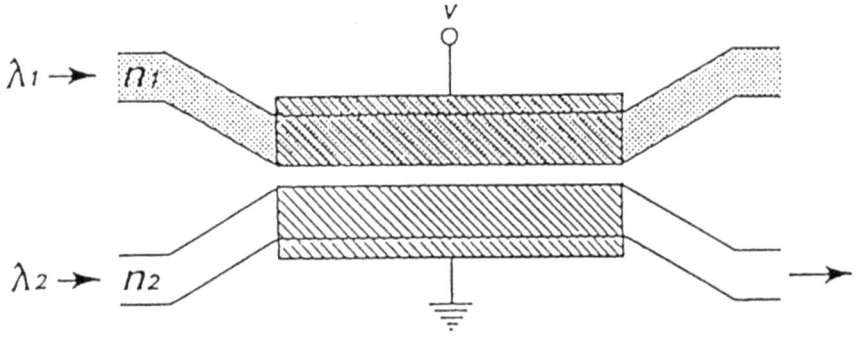

Figure 1

Tuneable integrated optic WDM structure.

An example is given in Figure 1, from Alferness and Vaselka]. Such a 1.3/1.55 μm active coupler has a 75 nm passband at half-maximum and is tuneable over 120 nm. Intrinsic multiplexing losses of 0.1 dB and a crosstalk of -17 dB were thus obtained. This crosstalk may be reduced by gradual adjustment of the distance between the guides. 4 or more channels can be obtained by cascading of elementary structures: exempla on figure 2.

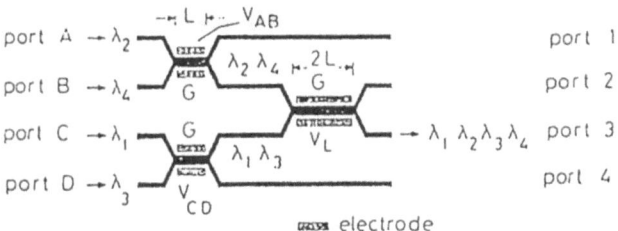

Figure 2
Four-channel WDM on Ti: Li Nb O₃
(From J.P.Lin *et al*)

As another example of results with such devices, in 1989, M. De Sario *et al* obtained a 1.33-1.55 μm multiplexer with a total insertion loss of 0.9 and 2.3 dB, respectively, and -25 dB crosstalk.

The previous devices remain polarisation dependant. However the problem can be solved to some extend. For instance, the device of GEC-Marconi developed within the RACE Programme WTDM provides both low loss and polarisation independence. A polarisation splitter couples TM field into the crossover waveguide and leaves all TE in the straight-through guide. Since the phase matched response of the polarisation converter is highly wavelength dependent, filtering is achieved and a FWHM of 1.6 nm with a wavelength extinction of 25 dB has been obtained.[Ref. 2] (Fig. 3)

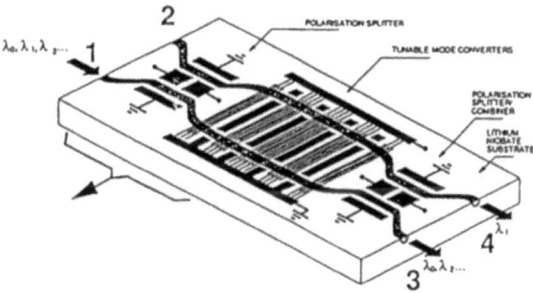

Figure 3. *Optical Filter from GEC-Marconi RACE WTDM*

Devices using amorphous dielectric or semi-conductor waveguides

Planar dielectric waveguides can be used to manufacture passive multi/demultiplexers of different types (coupling mode structure, diffraction grating structure, etc.). Different materials, such as SiO_2, TaO_5, other glasses, organic materials and semi-conductors, were proposed. These films can be deposited on glass by sputtering. Other manufacturing methods, including vacuum evaporation, plasma polymerisation, ion exchange methods, have been proposed (the last being one of the most important and often used in industrial manufacturing). [Ref. 1]

At the beginning, the proposed structures generally used 'hybrid' planar waveguides with, to some extent, prohibitive manufacturing costs.

Multi/demultiplexers on glass waveguides obtained by ion exchange

Devices using field couplings: Some examples in 1989, Goto and Yip, made a 1.3-1.55 μm multiplexer using an asymmetrical structure (Fig. 4). The loading of an Al_2O_3 strip of controlled thickness on the sodalime glass + K^+ guiding structure makes the adjustment of the wavelength range easier to separate. A 10 dB extinction ratio is theoretically obtained over 78 or 45 nm according to the Y angular structure. However, the practical results are slightly worse (5-6 dB).

In 1990, Suzuki *et al* released the results from a 1.3/1.5 μm single-mode multiplexer. The main characteristics of this device can be seen in (Fig. 5).

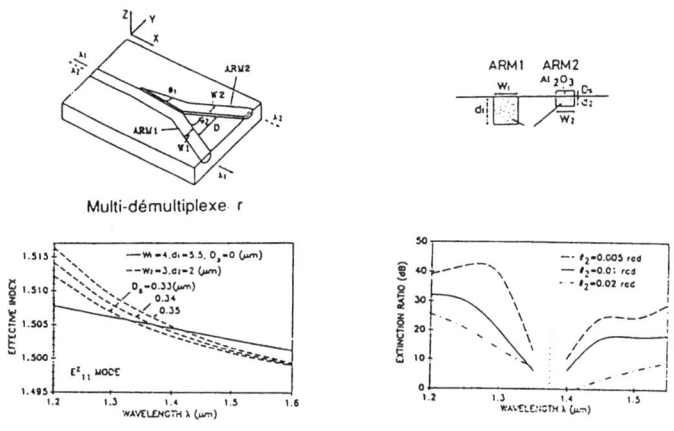

Figure 4
*Y branch wavelength multi/demultiplexer for λ = 1.30 and 1.55 μm
obtained on ion-exchanged sodalime glass.*
(After N. Goto and G. L. Yip)

Using a Mach-Zehnder interferometer structure, a 1.3/1.55 μm Y component with -30 dB crosstalk was obtained by Tervonen *et al*.

The ion exchange waveguides can also be used in diffraction grating devices. In this way, N. Kuzuta and Hasegawa manufactured a demultiplexer on a thallium ion exchange multimode waveguide. The thickness was about 70 μm. A flexible concave replica grating was bonded on the convex edge of the waveguide. This device uses 1200, 1310 and 1550 nm wavelengths with 2 to 2.4 dB losses (Fig. 6).

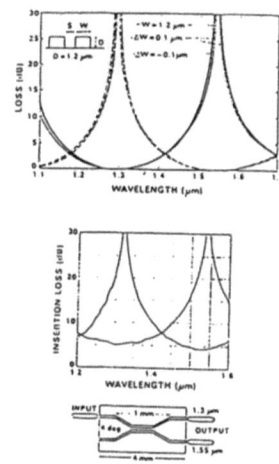

Figure 5
Single-mode multiplexer of S. Susuki et al.

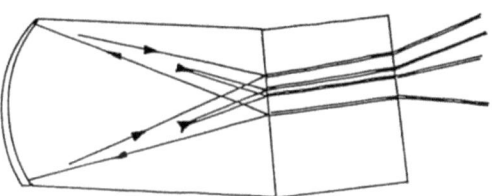

Figure 6
Multimode waveguide grating device.
Noboyuki Kuzuta and Eiichi Hasegawa.

Semi-conductor devices

Silicon substrates

Optical grating demultiplexers using a glass waveguide on a silicon substrate were proposed by Masayuki Takami almost 10 years ago.

As an example of such devices, in the multiplexer from Leti Laboratories (Fig. 7), the wavelength separation is obtained with a concave grating etched on the waveguide. The entrance and exit fibres are located in V grooves etched into the silicon.

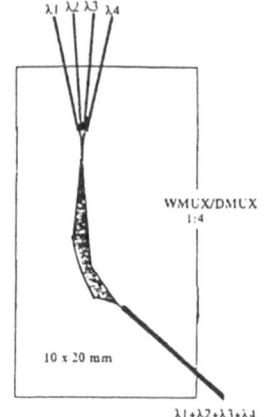

Figure 7 *Multiplexer from Leti (Leti phototech.)*
A schematic view of another silicon structure made in 1990, using a SiO₂ substrate, can be seen in (Fig.8). A concave grating is used in this device

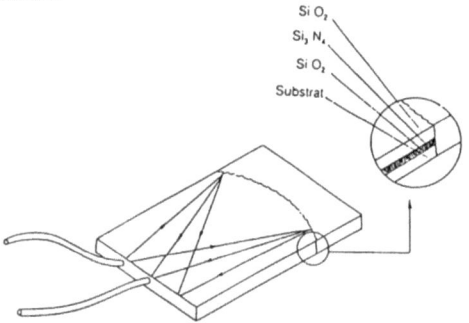

Figure 8
Spectrometer with grating integrated on a silicon substrate.
(After Nasa Tech. Briefs, April 1990.)

Another approach, based on elliptical Bragg reflectors, has been taken by Henry et al (Fig.9)[Ref.3]. This device is claimed to avoid the coupling and scattering losses of the previous devices. With a channel spacing of 5 nm on 4 channel multiplexers the rejection of crosstalk was 20 dB for single filtering. The additionnal loss was about 1 dB.

N. Takato et al designed up to 16 channel WDM using cascaded integrated optics Mach-Zehnder interferometer (Fig. 10)[Ref. 4]. The devices are claimed to have insertion losses of about 0.5 dB in the WDM region (few THz spacings) and 2 to 5 dB losses in the FDM region (few GHz spacings)

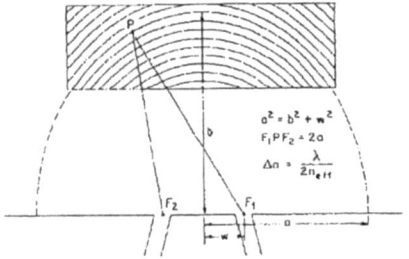

Figure 9 *EBR WDM from Henry et al [3]*

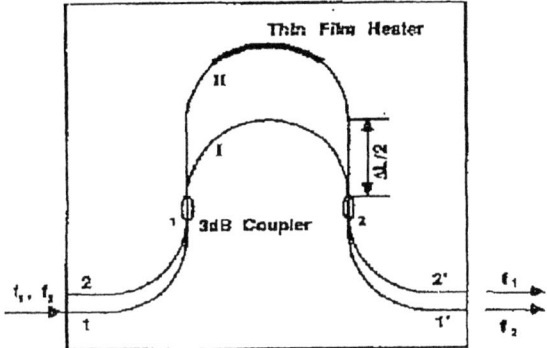

Figure 10 *Integrated optic Mach-Zehnder multiplexer of Takato et al*

InP substrate

The large losses formerly obtained with these semiconductor waveguides can be reduced by the absorption decrease of free carriers by using a high-purity substrate and weakly doped GaAs guiding layers. X couplers, made by De Bernardi, using such waveguides, demonstrated multiplexing with 35 to 100 nm spectral range between channels and an isolation better than 17 dB.

Concave grating devices made with guides of the same type were proposed. C. Cremer's device can work from 1.3 to 1.6 µm. It is compact: 1 x 3 mm^2. The grating is obtained by holographic methods. Likewise, the J. B. D. Soole device was manufactured using an etched concave grating. It can separate 50 channels with 1 nm spacing, 0.3 nm channel width and a 19 dB isolation between channels. The manufacturing of narrowband add-drop wavelength multiplexers integrated with DBR lasers has been demonstrated.

In order to increase resolving power, Takahashi *et al*, proposed to increase the optical path difference between 'diffracting' elements using the waveguide structure shown in (Fig.11) This principle is also used in the device designed on silicon by Dragone el al in 1990 (Ref. IEEE Photonics Tech. lett.) (Fig.12a), and on InP within the RACE MUNDI Programme (Fig.12b) [Ref.5]. This is a 4-channels WDM with integration of the receivers, with 2 nm wavelength spacing for 1.54 µm, with 46 arms with a path length difference of 41 µm between adjacent arms. Two Dragone waveguide arrays with an amplifier array between them were proposed by M. Zirngibl et al, as a tuneable filter with 7 channels spaced at 1.6 nm, in CLEO'94 conferences. However a wavelength shift of 1 nm between TE and TM is observed.

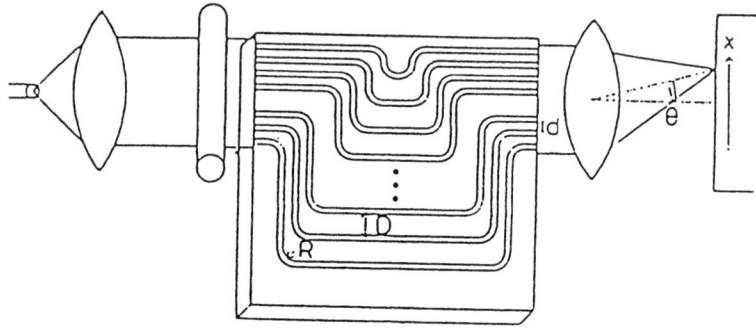

Figure 11

Arrayed waveguide grating with an increase of optical path difference between 'diffracting' elements for nanometric resolution **(Takahashi et al)**
Length difference between adjacent channels: $\Delta L = 2(D-d)$.
Transmission for wavelengths such that $n_c \Delta L + n_s d \sin\theta = m\lambda$.
Elec. Let. Jan 90

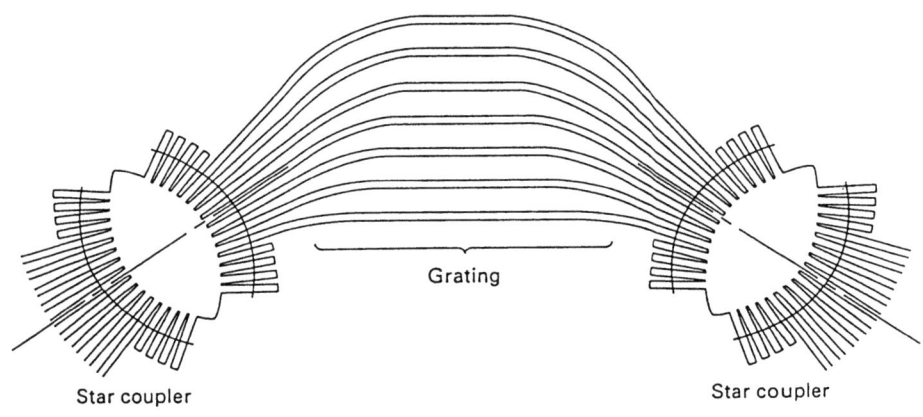

Figure 12a
Integrated N*N multiplexer in silicon from C. Dragone et al IEEE Photonics Tech. Ltrs., 3, 1, 1991.

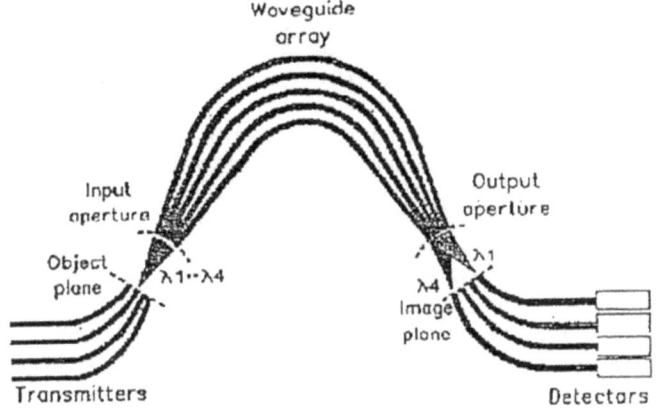

Figure 12b
WDM integrated with photodetectors in InP/InGaAsP
from M.R. Amerfoort et al (IEEE Phot. Tech. lett.6, 1, 1994)
RACE MUNDI

Comparison integrated optics versus microoptics devices.

In 3 dimension micro-optics for many channels and/or small spacings between channels (typically 20 to 0.5 nm) a grating is generally used. When the distance between channels is large and with few channels (typically 1.3/1.5 μm devices) multidielectric filters can also be used. It seems easier to get low polarisation effect and low crosstalk in the present state of the art with 3 dimension micro-optics than on integrated optic (IO) devices. On grating micro-optics configuration the same wavelengths are obtained in both polarizations, and the grating polarisation sensitivity can be cancelled at a given wavelength and reduced in a large wavelength range, typically less than 5% difference from 1500 to 1560 nm at 1 nm spacings, with the selection of the grating period and profile. On the contrary integrated optic devices have often polarisation sensitivity. For exempla, wavelength shifts from TE to TM of 4.1 nm on a 4 channels 1.8 nm spacing FAW device [Ref.5], and of 0.25 nm on a 4 channels 5 nm spacing EBR device [Ref.3]. These devices have also a loss difference between TE and TM: for exempla 3 dB difference in [Ref.3]. In [Ref. 4] on a MZ device, stess induced birefringence of 0.0004 probably leads to similar problems. Moreover the optical crosstalk between adjacent channels in IO WDM is generally -15 to -25 dB to be compared with -30 to -40 dB on bulk optic configurations. However the differences in coupling efficiency and wavelength of channels for the two polarizations in IO WDM will be probably largely eliminated in future industrial devices by appropriate choices of materials and waveguide designs.

Accuracy necessary for wavelength measurements of WDM channels

In wavelength division multiplexing it is necessary to know the absolute wavelengths of the channels with an accuracy compatible with the multi/demultiplexer channel width. This channel width will becomes smaller and smaller. However the minimum channel width and the channel spacing at a given bit rate have theoretical and practical limitations.[Ref.1-6-7-8 for instance].

If we use a CW laser, then the emission can be highly "monochromatic": theoretically the frequency width of the emission line Δv can be as small as required. But practically, in order to transmit a signal, we have to modulate and this corresponds to an enlargement of Δv. The time duration $\Delta \tau$ corresponds to the frequency spread: $\Delta v = 1/\Delta \tau$, ($v = \omega/2\pi$).

With 2.4 Gb/s bit rate near 1550 nm, channel spacings larger than 0.02 nm must be used.

The other limitations of WDM or OFDM are related to non-linear effects in the fiber such as Raman and Brillouin scattering and four wave mixing. The Raman conversion is maximum for a distance between the lower wavelength (the "pump") and the longer wavelength corresponding to a wavenumber separation of 430 cm^{-1}.

At present, the WDM solutions in use in practical systems have only 1 or few nm spacings.

Certain atomic lines as well as some molecular resonances may be regarded as primary wavelength reference standards in wavelength division multiplexed network. Exempla of secondary standards with less precision and that might drift, but less than the useful wavelength tolerance might be such as: fixed Fabry Perot cavities, Bragg reflection gratings in optical fibers, or monoblock WDM reference components. We show that nowadays, a wavelength accuracy of +- 0.05 nm is adequate in most of the WDM systems in use.

For OFDM*, the requirement will be much more severe and outside of the scope of the present paper. In the manufacturing of WDM components reference components can be manufactured first, using scanning monochromators calibrated with atomic lines with a rigorous control in temperature, pressure and humidity. As these reference components can be designed as solid silica blocks, with low thermal drifts and high long term stability, they can be utilised as practical "wavelength secondary standards" for further manufacturing with the advantage over scanning monochromators of much higher thermal and long term stability and negligible effects of pressure and humidity.

Light sources and absorption lines for calibration of wavelength-meter

Preliminary remarks:

As it is well known, at the beginning of this century the wavelength of the red light of Cadmium was measured in "mètre étalon", meter standard unity. Benoit, Fabry and Perot, using "super-imposed fringes" of several so-called Fabry Perot interferometers, were already able to measure a wavelength of 643.84696 nm at 0°C and 760 mm Hg, in dry air for Cd, an already astonishing accuracy. Much was done from that time, and the length standard is no longer a material meter standard. Now, the problem of definition of standards of wavelength call us back to the standards of time and frequency. The time reference is defined by the frequency of the transition between the hyperfine levels F=4, m_F=0 and F=3, m_F=0 of the ^{133}Cs atoms: 9192 631 770.0000 Hz [ref.9]. Nowadays, the meter is defined indirectly from this time standard and from an internationally accepted value of the speed of light c. Through a chain using different lasers and klystrons, the measurement of the frequency of a HeNe laser stabilised on CH$_4$ was reported in 1973 : 88 376 181 627 (50) THz [ref.10], see also [ref.11]. And from an interferometer comparison with the ^{86}Kr line (about 605,69 nm) that was selected in 1960 as the length standard, the wavelength of HeNe/CH$_4$ was measured as λ=3392.23140 nm. Because c=λν, c was measured to be: c = 299 792 458 m/s. A recommendation, of the CCDM (1973) [ref.12], to define the meter from this last value of c and from the time unit given by the ^{133}Cesium clock was finally internationally adopted in 1983. So we can connect all our secondary wavelengths or frequency standards for wavelength division multiplexing (WDM) or optical frequency division multiplexing (OFDM) to these primary standards.

Wavelength references in the optical telecommunication wavelength range:

Some atomic lines can be used, directly for wavelength calibration of instruments such as monochromators, spectrometers, or optical spectrum analysers, or through the frequency stabilisation of semiconductor lasers that becomes frequency transfer standard . Just as examples let us recall the stabilisation of AlGaAs laser diodes on ^{87}Rb $5S_{1/2}$ - $5P_{3/2}$ hyperfine D$_2$ levels at 780,0 nm using saturated absorption [ref 17-21], the stabilisation of InGaAsP DFB lasers on Ar $2p_{10}$ - $3d_5$ transition at 1296,0 nm or on Kr $2p_{10}$ - $3d_3$ transition at 1533,9 nm using an optogalvanic effect [ref 18 - 20]. The frequency stabilisation of the HeNe laser around 1,52 µm is also easily obtained [ref 21], and may be utilised as a secondary standard for wavelength calibration.

The stabilisation of a fiber laser on an absorption line of the rubidium was also reported [Ref 22] [Ref 23, 24]. Frequency stabilisations using the transition of Rb, $5P_{3/2}$ - $5D_{5/2}$ near 1529 nm, or of Kr, $1s_2$ - $2p_5$ were also reported [Ref 19-20][Ref 27 and 28, Kr line at 1318.102 nm)]. (Review on frequency standards [Ref 21]). It is known that absorption lines of molecules such as $H^{13}C^{15}N$, $H^{13}C^{14}N$ (Fig 13a), C^2H^2, CH_4, HF, NH_3, HI...can also be used. For exempla CH_4 is useful from 1321 to 1341 nm, and C^2H^2 from 1511 to 1541 nm.

Some frequency stabilised light sources in the 1550 to 1560 nm range with accuracy of +-10 MHz/+-0.08 pm are becoming commercially available, but they are very expensive.

However, in many applications such as the calibration of WDM component, in the most useful wavelength range 1300 to 1580 nm, relatively inexpensive laboratory lamps such as Kr, Xe, Cd and Hg can be used. Kr lines at 1523.962 nm in air (Relative intensity 160), 1533.497 (R.I. 200), 1537.204 (R.I. 100), 1547.402 (R.I.32), 1473.444 (R.I.100), 1442.679 (R.I.350), (lines from Zaidel et al Moscou 1977), can used. And the line 1547.402 nm (1547.82543 nm in vacuum or 193.68625 THz) was proposed to be used as a channel reference within the ITU-T Study Group 15. However this line is rather weak, and for a practical point of view, the cadmium, the rubidium and mercury give the best references. For the calibration of our monochromators, when the highest accuracy is necessary, the 1529,582 nm mercury line [Ref.30] (wavelength in air for natural mercury given in CRC Handbook) is very useful. It is a cheap lamp available in any optical laboratory, with well resolved lines, high power: 65 pW with a 0,1 nm resolution, compared to 7 pW given by a small tungsten halogen lamp for the same resolution on a HR 640 monochromator. Its absolute wavelength is known with a high accuracy [Ref.31-32]. Another line in the visible at 546,074 nm, can be used in 3rd order. (power : 3 pW , for the same conditions).

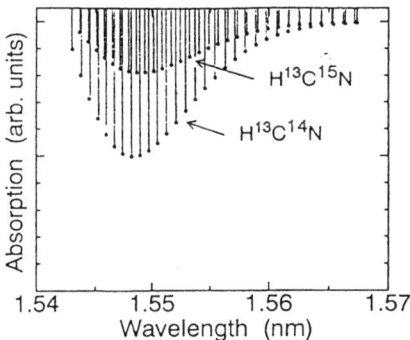

Source	λ_{air} in Å	λ_{vac} in Å	ν_{vac} in cm^{-1}
Ar	8264.5222±0.0001	8266.7939	12 096.5880
Ar	9122.9668±0.0002	9125.4707	10 958.3388
Ar	9224.4985±0.0002	9227.0299	10 837.7236
Ar	9657.7854±0.0002	9660.4341	10 351.5017
Ar	12 802.7380±0.0004	12 806.2401	7808.6932
Ar	12 956.6568±0.0007	12 960.2007	7715.9299
Ar	13 504.191 ±0.001	13 507.884	7403.084
Ar	13 622.661 ±0.002	13 626.386	7338.703
Ar	13 718.5762±0.0002	13 722.3271	7287.3937
Ar	14 093.6386±0.0004	14 097.4914	7093.4606
Ar	16 940.581 ±0.001	16 945.208	5901.373
Hg^{198}	10 139.7934±0.0002	10 142.5728	9859.4313
Hg^{198}	11 287.4068±0.0004	11 290.4974	8857.0057
Hg^{198}	15 295.973 ±0.002	15 300.153	6535.882

Fig. 13a (from YOKOGAWA)
H CN absorption lines

Fig. 13b *(From [32])*
Wavelengths of Ar and Hg^{198} lines

The isotope and hyperfine structure of the 1529.582 line of natural mercury, is fully resolved with a resolution of 0.011 cm^{-1} corresponding to 0.0026 nm or 330 MHz. [ref 31]. Moreover E. Peck et al. have proposed the Hg^{198} line :
1529.5973 -+ 0.0002 nm integrated over the profile as a secondary standard for near infrared spectroscopy [Ref.32]. (Fig. 13b).

Measurements with monochromators

In our laboratory, we use scanning monochromators for calibration of temperature stabilised solid silica block multichannel multiplexers with typically 1 or 0.5 nm spacings, 0.25 FWHM and thermal stability of 0.01 nm/°C . These components are used as reasonably good "wavelength transfer standards" in the manufacturing of other multiplexers and demultiplexers. The absolute wavelength of their channels at a given temperature, are known within 0.05 nm. Our scanning monochromators are HR1500, HR640 and Stimax instruments. HR1500 and HR640 are Czerny-Turner grating monochromators with focal length of 1500 and 640 mm respectively. The resolutions are 0.01 to 0.02 nm and 0.02 nm to 0.04 nm respectively, depending of slits width, gratings, and wavelength range. The Stimax TWS is a stigmatic fiber optic monochromator with a focal length of 80 mm. The resolution is 0,1 nm with single-mode fibers. These monochromators can be used with most optical sources and are computerised.

When the HR640 monochromator is calibrated with the 1529.582 nm line in Hg, the uncertainties in wavelength measurements, caused by defects in the scanning, can be evaluated on different known lines of Cd, Rb and Cs. We get an accuracy of ± 0.05 nm without thermal stabilisation. If a better accuracy it is necessary to control the temperature of the monochromator and CCD detectors or calibrated fibers arrays can be used. The H^{198} isotope line at 1529.5973 in air at 15 °C and 1 atm is known within +-0.002 nm.

ELEMENT	LINE in nm
Cd I	1397.9
Cd I	1433.8
Cs I	1469.491
Rb I	1475.241
Cd I	1514.40
Rb I	1528.948
Hg I	1529.582
Cd I	1571.19

Figure 14 *Some useful lines for calibration*

In order to get accurate measurements, the temperature must be stabilised since the grating spacing, mechanical adjustments and the index of air n depends on the temperature (Figures 15-16).

$$(n_0-1)\ 10^7 = A + B / (\lambda^2\ 10^{-8}) + C / (\lambda^4\ 10^{-16})$$

FOR DRY AIR

760 mm Hg	A	B	C
T = 0	2875.66	13.412	0.3777
T = 15 °C	2726.43	12.288	0.3555
T = 30 °C	2589.72	12.259	0.2576

Figure 15 *Index of air as a function of wavelength, temperature, and humidity. From CRC Physics Handbook.*

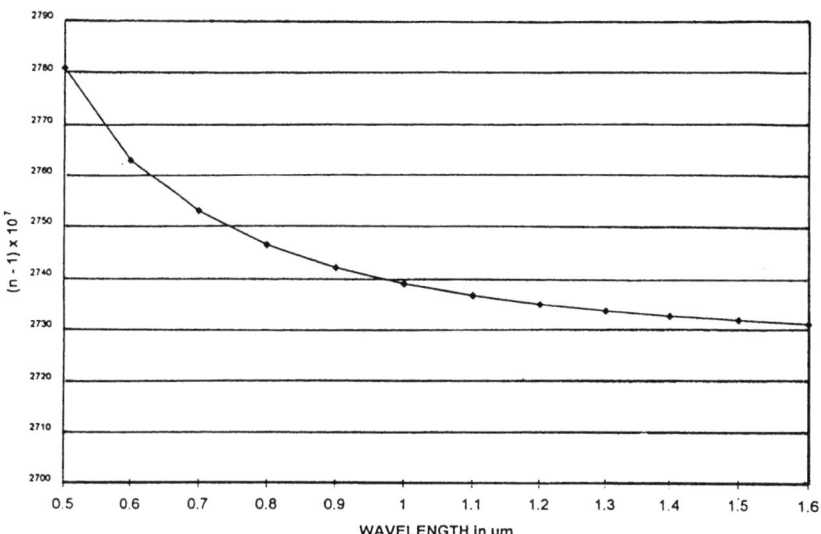

Figure 16 *Index of dry air versus wavelengths at 15 °C, 760 cm Hg.*

Moreover the index of air depends of the wavelength. So, in the 3rd order using the Hg line at $\lambda=546.074$ nm we find the position of the 3rd order component at a wavelength corresponding to $\lambda = 1638.215$ nm. In the second order using the ^{86}Kr line at 605.69 nm we have a position corresponding to 1211.38 nm and with Cd line at 643.847 nm we get 1287.691 nm

Conclusion

Microoptics WDM components with many channels and 1/0.5 nm spacings are now available. Equivalent and probably less expensive integrated optics devices may become available within few years. They require accurate wavelength measurements. This can be done with classical monochromators
calibrated with many lines of emission sources and molecular absorption lines: some are known within +- 0.2 pm or better.

Acknowledgements

This work has been supported in part by the EEC within RACE 2001 contract.

References

0: J-P. Laude: Wavelength Division Multiplexing, Prentice Hall Ed.,London and Masson Ed.Paris, (1993).

1: J-P. Laude *et al* : Stimax, a grating multiplexer for monomode or multimode fibers, ECOC'83 Proc., pp 417-420 Elsevier Sci.Pub.(North Holland),(Oct., 1983).

2: K.J.Hood, P.W.Walland, C.L. Nuttal, L.J.St Ville, T.P.Young, A. Oliphant, R.P. Marsden, J.T.Zubrzycki, G.Cannel, C.Bunney, J.P.Laude, and M.J.Anson: Optical distribution systems forn television studio applications, J. of Lightwave Tech, Vol.11, N_o 5/6, pp 680-687, (May/June 1993).

3: C.H. Henry et al : Four-Channel Wavelength Division Multiplexers and Bandpass Filters Based on Elliptical Bragg Reflectors, J. Lightwave Tech.,vol.8,n°5, pp 748-755, May 1990.

4: N. Takato et al : Silica-based Integrated optic Mach-Zehnder Multi/Demultiplexer Family with Channel Spacing of 0.01-250 nm, IEEE J. of selected areas in com.,vol.8, n°6, pp 1120-1127, Aug.1990.

5: M.R. Amersfoort et al: Low-Loss Phased-Array Based 4-Channel Wavelength Demultiplexer Integrated with Photodetectors, IEEE Photonics Tech. Let.,vol.6, n°1, pp 62-64, Jan.1994.

6 : B. Glance : Densely spaced WDM coherent optical star network,Elec. Lett., Vol 23, n° 17, pp 875-876, (Aug., 13, 1987).

7: B. Glance : Densely spaced FDM coherent optical system with random access digitally tuned receiver, Globecom 89 IEEE Dallas, (Nov., 27, 1989).

8: S. Chi, S.C. Wang : Maximum bitrate-length product in the high density WDM optical fibre communication system, Elec. Lett., Vol 26, n°18, (Aug., 30, 1990).

9: E. Beehler, R.C. Mockler, J.M. Richardson : Metrologia 1, 114, (1965).

10: Evenson, Peterson : Laser spectroscopy of atoms and molecules, Ed by H. Walter, topics in Applied Physics, Vol. 2, p 349, Springer-Verlag Ed. Berlin, (1976)

11: Evenson, Wells, Peterson, Davidson, Day, Barger, Hall : Physics Reviews Letters, 29, 1346, (1972).

12: 5th session of the consultative commitee on the definition of the meter (CCDM) BIPM Series, (1973).

13: S. Koizumi, T. Sato, M. Shimba: Frequency stabilization of semiconductor laser using absorption line under direct FSK, Elec.Rev.Lett., vol 24, n° 1-7, pp 13-14, (Jan., 1988).

14: T. Sato, M. Niikumi, S. Sato, M. Shimba : Elec. Lett., Vol 24, n°7, pp 429-431, (Mar., 31, 1988).

15: H. Furuta, M. Ohtsu : Evalations of frequency shift and stability in rubidium vapor stabilized semiconductor lasers, Appl. Opt., Vol 28, n°17, pp 3737-3743, (Sept., 1, 1989).

16: G.P. Barwood, P. Gill, W.R.C Rowley : A simple rubidium stabilized laser diode for interferometric applications, J Phys E, Vol 21, n°10, pp 966- 971, (Oct., 1988).

17: M. Suzuki, S. Yamaguchi : Frequency stabilization of a GaAs laser by use of the optical-optical double resonnance effect of the Doppler- free spectrum of the Rb D_1 lines, IEEE J Quant Elec, Vol 24, n°12, pp 2392-2399, (Dec., 1988).

18: Y.C. Chung, R.W. Tkach, A.R. Chraplyvy : Performance of a frequency locked 1.3 µm DFB laser under 50 Mbit/s FSK modulation, Elec.Lett, Vol 24, n°18, pp 1159-1160, (Sept., 1, 1988).

19: Y.C.Chung : Frequency locking of a 1.3 µm DFB laser using a miniature argon glow lamp, IEEE Photonics Tech., Vol 1, n°6, pp 135- 136,(June, 1989).

20: Y. Chung, C.B. Roxlo : Frequency locking of a 1.5 µm DFB laser to an atomic krypton line using optogalvanic effect, Elect. Let.,Vol. 24, No. 16, pp.1048-9,(Aug., 1988)

21: G. Fisher : Simple and effective frequency stabilization for He-Ne lasers at 1.52 µm, Elec. Lett., Vol 23, n°5, pp 206-208, (Feb., 26, 1987).

22: National Institute of Standards and Technology, Photonics spectra World Watch, (Dec., 1991).

23: M. Breton, P. Tremblay, N. Cyr, C. Julien, M. Têtu, B. Villeneuve, Observation and characterisation of ^{87}Rb resonances for frequency- locking purpose of a 1.53 µm DFB laser, SPIE Proc. on Frequency- Stabilized Lasers and Their Applications, Vol. 1837 p.134 (1992)

24: S.L. Gilbert, NIST research on wavelength standards for optical communications, CLEO' 94 proc. p 344, OSA Pub., (1994)

25: S.L.Gilbert : Frequency stabilisation of a fiber laser to rubidium: a high-accuracy 1.53 µm wavelength standard, Proc. SPIE, Vol. 1837, pp 146-153, (1993).

26: Y.C. Chung, Frequency-locked long-wavelength lasers for lightwave communication systems [using Kr transitions and discharche lamp], Proc.SPIE,Vol. 1837, pp 171-179, (1993).

27: Y.C. Chung, IEEE J. Lightwave Technol., Vol 8, p 869 (1990)

28: M. Guy, M. Têtu, C. Latrasse, B. Villeneuve, M. Svilans, Laser optical frequency control at 1.3 and 1.55 µm using a nonlinear multilayer semiconductor waveguide, CLEO' 94 Proc. pp 274-275, OSA Pub., (1994)

29: D.J.E.Knight, P.S.Hansel, H.C.Leeson, G.Duxbury, J.Meldau, and M.Lawrence : A review of user requirements for, and practical possibilities for, frequency standards for the optical fibre communication bands, SPIE, Vol. 1837, pp 106-114, (1993)

30: J-P.Laude : The 1529.582 nm mercury line, a cheap and accurate wavelength standard for WDM and OFDM optical fibre networks, IEE Colloquium on standards in fibre optic systems, London (February 1992). Reprint not available.

31: J. Pinard : Spectromètre de Fourier à très haute résolution, Journal de Physique Colloque C2, Supplément an n°3-4, Tome 28, p C2- 136,(Mar., Apr., 1967).

32: E. R. Peck et al : Secondary Standards of ArI and Hg^{198} in the Near Infrared, JOSA, Vol 52, n°5, pp 536-538, (May 1962).

ACTIVE FIBRE CHARACTERISATION WITH PASSIVE MEASUREMENTS*

*J. M. S. ANACLETO***
CETO, Centro de Ciências e Tecnologias Ópticas
Faculdade de Ciências, Universidade do Porto, Praça Gomes Teixeira, 4000 Porto, Portugal.

MODESTO MORAIS
IEP - Instituto Electrotécnico Português
R. S. Gens, 3717, Senhora da Hora, Portugal.

GASPAR REGO
CETO, Centro de Ciências e Tecnologias Ópticas
Faculdade de Ciências, Universidade do Porto, Praça Gomes Teixeira, 4000 Porto, Portugal.

O. D. D. SOARES
CETO, Centro de Ciências e Tecnologias Ópticas
Faculdade de Ciências, Universidade do Porto, Praça Gomes Teixeira, 4000 Porto, Portugal.

Abstract: Passive measurements of a sample of an Erbium-doped silica fibre were performed following the standard methods. A Photon Kinetics equipment was used for spectral attenuation, mode field-diameter and cut-off wavelength measurements and for geometrical characterisation.

Characterisation of Erbium-doped fibres is of great importance. The passive parameters measurements is a first step of the characterisation. The passive parameters of the fibre have to be considered to predict the active characteristics using theoretical models [3].

Passive parameters of a sample, 40m long, of an Erbium-doped silica fibre fabricated in Germany by "Bundespost Telekom" were measured. The fibre was supplied under the COST 241-WG3-SG1 programme. The measurements were performed at the Optical Fibre Laboratory of "Instituto Electrotécnico Português (IEP)".

1. Geometrical characterisation

The diameter, non-circularity of the core and cladding and the concentricity error of cladding to core were measured following IEC 793-1-A2 prescribed method. This test method determines the geometrical parameters of a fibre, analysing the near-field distribution at the cross-section at the end of the fibre [2]. A Photon Kinetics 2400 equipment was used.

The following results were obtained:

	Diameter (μm)	Non-circularity
Cladding	125.42 ± 0.1	0.914
Core	4.96 ± 0.2	7.437
Concentricity cladding/core (μm)	0.490 ± 0.01	

---

* The paper was presented at a poster session.
** Lecturer at Universidade de Trás-os-Montes e Alto Douro
   Quinta dos Prados, 5000 Vila Real - Portugal   Fax: (059) 74 480

For an Erbium-doped fibre the active properties depend, also, on the modal distribution of the propagating fields. The interaction of those fields with the Er^{3+} ions determines the active properties. Further, the modal distribution depends on the transmission properties of the fibre as well as on the geometrical parameters. The geometrical parameters are also important in view of the current standards on fibres, in particular for applications into optical fibre communications.

2. Spectral Attenuation

Spectral attenuation was measured according to the specified method on ICE 793-1-C1A. This method is based on cut-back technique [2]. First the optical power is measured at the output (or far end) of the fibre. Then without disturbing the input condition, the fibre is cut-back a small length from the source, and the output power at this new end is measured.

If P_F and P_N represent respectively the output powers of the far and shorter end of the fibre, the average loss α in decibels per kilometer is given by:

$$\alpha = \frac{10}{L} \log\left(\frac{P_N}{P_F}\right),$$

where L (in kilometers) is the separation of the two measurements points.

Figure 1 shows a plot of attenuation as a function of wavelength, obtained with PK 2500 equipment. The strong absorption peaks near 800nm, 980nm and 1480nm wavelengths corresponds to the available pumping bands. Around 1530nm an unpumped Erbium-doped fibre is a strong absorber.

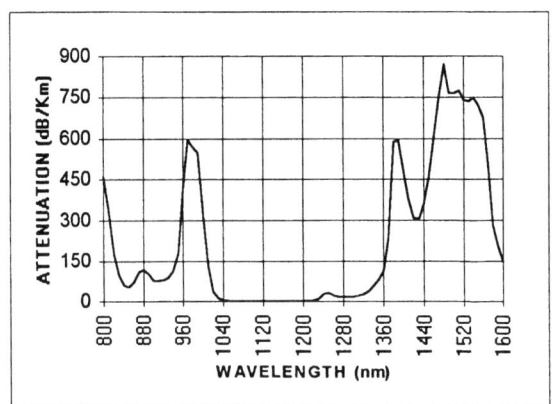

FIGURE 1 - *Graph of attenuation versus wavelength.*

3. Mode field-diameter (MFD)

The mode field diameter is a relevant parameter. For some theoretical models [3] it is assumed that the mode distribution is not modified by the presence of the Er^{3+} ions. Such assumption must be confirmed.

To measure the MFD the IEC 793-1-C9A method was followed. The mode field diameter is measured by direct far field scanning in a two step procedure. First, the far field radiation pattern of the fibre is measured. Second, a mathematical procedure based on the Petermann r.m.s. far field definition is taken to compute the mode field diameter [2].

The figures 2 and 3 show the results obtained with the PK 2500 equipment. The instrument dynamic band is 60dB so that we can detect the ringing corresponding to the diffraction pattern of the optical field emerging from the fibre.

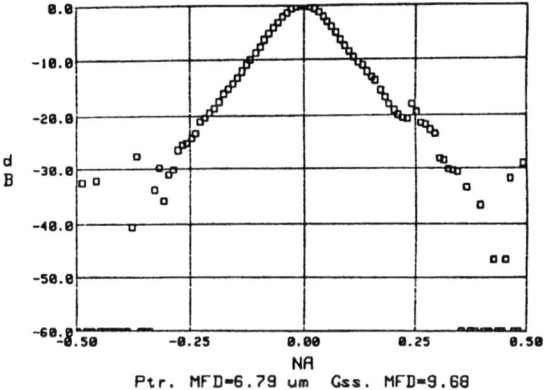

FIGURE 2 - *Far field scanning at 1555nm. The mode field diameter computed is 6.79 µm.*

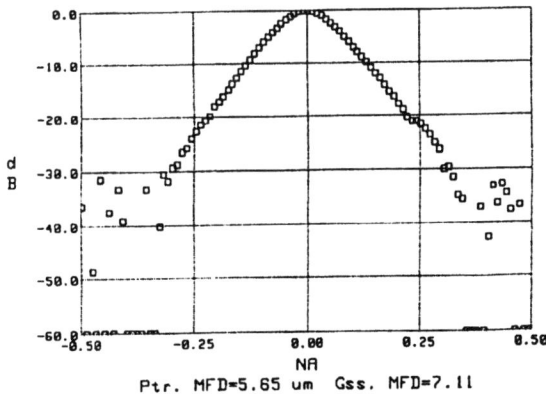

FIGURE 3 - *Far field scanning at 1306nm. The mode field diameter computed is 5.65 µm.*

4. Cut-off wavelength

In an ideal single mode fibre, cut-off wavelength λ_c of the LP_{11} mode [1] indicates that the fibre accepts a multi-mode regime for wavelengths shorter than λ_c. Theoretically the cut-off for a step-index fibre is given by:

$$\lambda_c = \frac{2\pi}{2.405} a \sqrt{n_o^2 - n_c^2},$$

where a is the core radius, n_o and n_c are the core and cladding refractive indexes.

Nevertheless, the index profile often departs significantly from a step distribution, sometimes intentionally (dispersion-optimised fibres). This implies the necessity of an experimental evaluation of the cut-off wavelength λ_c.

We use the method described in ICE 793-1-CIA. It is based on transmitted power technique [2], which measures the variation with wavelength of the transmitted power of a short length of testing fibre as compared with a reference transmitted power.

The figure 4 shows the ratio between the two transmitted powers expressed in dB as a function of wavelength, obtained with the PK 2500 equipment. The λ_c is determined as the highest value for which the ratio is equal to 0.1 dB.

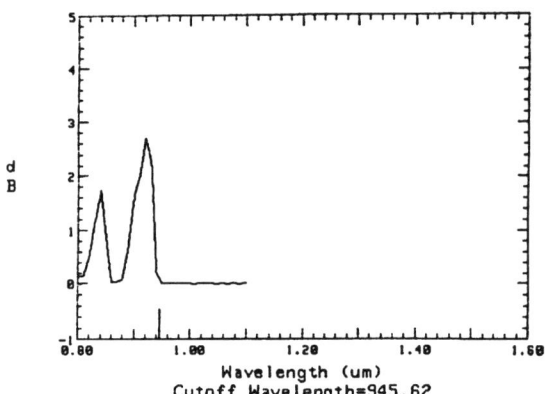

FIGURE 4 - *Ratio between the two transmitted powers as a function of wavelength to measure the cut-off wavelength. The measured value of λ_c was 945 nm.*

5. Comments and future work

These measurements represent preliminary results and are to be compared with measurements that will be done by other groups working on the COST 241 programme.

Future work envisages a complete passive and active characterisation of an Erbium-doped fibre linked with development of a theoretical models to describe the amplifier properties of devices.

6. ACKNOWLEDGEMENT

The authors are grateful to Prof. M. Marques for releasing the fibre (COST 241 programme).

7. REFERENCES

[1] - Gerd Keiser, **Optical Fibre Communications**, McGraw-Hill (1983).

[2] - Giovanni Cancellieri, Editor, **Optical Fibre Measurements: Characterization and Sensing**, Artech House (1993).

[3] - J. F. Digonnet, **Rare Earth Doped Fiber Lasers and Amplifiers**, Marcel Dekker, Inc., 1993.

IV Optical Sources

WAVELENGTH TUNABLE LASER DIODES AND THEIR APPLICATIONS

M. C. AMANN
University of Kassel, Department of Electrical Engineering
34132 Kassel, Heinrich-Plett-Str. 40
Germany

Abstract

Single mode laser diodes with an electronically tunable wavelength are among the key components needed for advanced applications in optical communications, measurement and sensing. In particular, these devices may be used as wavelength controllable transmitters in wavelength division multiplexing or as local oscillators in optical receivers using the coherent detection technique. Furthermore, distance measurements with a high accuracy and large repetition frequency as well as three-dimensional viewing are feasible using the frequency modulated continuous wave radar technique in the optical domain. Also numerous measurement techniques in optoelectronics and fiber optics can be performed by means of wavelength tunable lasers. This paper reviews the recent progress made on electronically wavelength tunable laser diodes in the InGaAsP material system at 1.5 μm wavelength. The operation principles and technological approaches of the various device concepts are presented and the laser parameters relevant for anticipated applications, such as tuning range, spectral linewidth and wavelength access, are discussed.

The continuous and discontinuous tuning modes are introduced, compared and the physical limitations with respect to the maximum tuning ranges are commented on. Particular attention is paid to device handling and the effort needed for wavelength control which are most important for practical applications.

1. Introduction

Wavelength tunable single mode laser diodes are indispensable key components of advanced photonics applications. Particularly broadband multichannel optical communications, optical switching networks, wavelength dependent measurements and several sensing techniques depend on the availability of laser diodes with an electronically tunable wavelength. With regard to the transmission properties of optical fibers, emission wavelengths around 1.55 μm are usually required in optical communications. Therefore the major development has been performed to date with the InGaAsP/InP material system. Also various optical measurement and sensing techniques become feasible or can be improved by using electronically tunable laser diodes. In most applications single mode operation of the laser diode is required, i. e. the light should be emitted in one longitudinal and transverse mode at a single frequency. Therefore, the development of the corresponding lasers has been based on the device structures of single mode lasers. This essentially comprises the well established distributed feedback (DFB) and distributed Bragg reflector (DBR) laser structures. Since the tuning range in these structures is limited to values significantly below the gain bandwidth of the semiconductor, completely different device concepts with extended tuning ranges have been applied quite recently exploiting interference effects in composite cavity structures. While continuous wavelength tuning is commonly possible in the DFB and DBR based devices, the latter device structures, however, can usually be tuned only in the discontinuous tuning mode. In many applications, e. g. coherent optical detection or optical frequency modulated continuous wave (FMCW) radar, continuous tuning is required excluding or at least limiting the applicability of widely but discontinuously tunable laser diodes. From the viewpoint of the user, finally, handling of the devices plays an important role. This means that the devices should not exhibit more control parameters than necessary; particularly, the wavelength setting should be performed by one electrical control parameter alone.

2. Applications for Wavelength Tunable Laser Diodes

Up to now the optical broadband communication has been among the most important driving forces for the development of electronic wavelength tunable laser diodes. Multichannel optical broadband communication systems have been presented both in the wavelength division multiplexing (WDM) technique [1] and by means of the coherent transmission [2, 3] scheme, as shown in Fig. 1a and b, respectively. In both cases, the signal source consists of a set of light emitters with equally spaced emission wavelengths. Owing to the large selectivity of the coherent detection, which uses the heterodyne or homodyne technique, the channel spacing can be

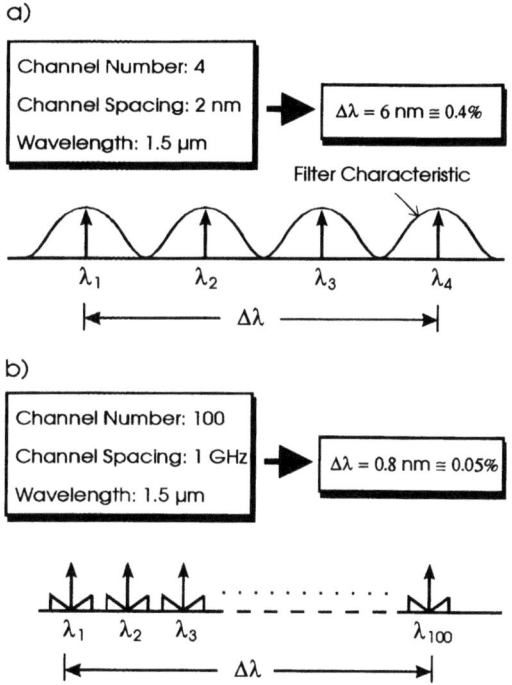

Figure 1. Schematical drawings of transmission spectra for wavelength division multiplexing (WDM) (a) and coherent optical (b) broadband communication systems. The system parameters for typical designs are given in the insets.

made as small as the bandwidth of the transmitted signal. In case of WDM systems, on the other hand, filtering on the receiver side must be performed in the optical frequency domain as indicated by the filter characteristics in Fig. 1a. So considerably larger channel spacings must be accepted as determined by the performance of the available optical filters. Accordingly channel spacings of the order $1 \cdots 4$ nm ($125 \cdots 500$ GHz at 1.5 μm wavelength) are state-of-the-art for the WDM transmission, while coherent systems may use channel spacings of only about $1 \cdots 10$ GHz ($0.01 \cdots 0.1$ nm). Typical wavelength schemes for both techniques are displayed in Fig. 1, showing that a wavelength coverage $\Delta\lambda$ of several nm commonly applies. Light emitters can be realized by a set of single mode laser diodes, for instance DFB lasers. Up to now, the exact channel spacing is difficult to achieve solely by the fabrication process, i. e. matching the DFB grating pitch or the transverse laser structure. Usually the fabrication tolerances only allow for a coarse wavelength setting, while the exact wavelength definition requires some kind of fine tuning. This is commonly performed by tuning the laser temperature in order to exploit the tem-

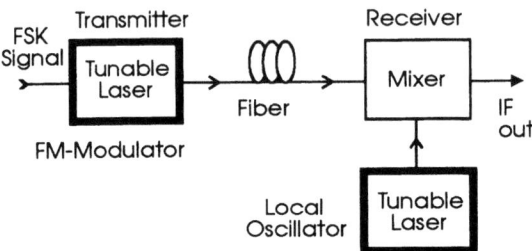

Figure 2. Simplified representation of a coherent optical communication system. The functional blocks, where wavelength tunable laser diodes may be applied are marked.

perature dependence of the emission wavelength (≈ 0.1 nm/K). A more convenient approach would be the use of wavelength tunable lasers, the emission wavelengths of which can be varied electronically over the entire emission spectrum $\Delta\lambda$. In addition, one would also benefit from the essentially smaller time constants of about several nanoseconds as compared to microseconds in thermal tuning; this significantly improves the stability and reduces the response delay of the wavelength control, which is needed in any practical system application.

In coherent transmission systems, tunable lasers are still more important for the local oscillator (LO) function in optical heterodyne receivers [4, 5] as shown schematically for a simple system in Fig. 2. As the LO must smoothly follow the transmitter wavelength, it is most essential here that the wavelength tuning is strictly continuous. This implies that no mode and frequency jumps occur within the tuning range. Otherwise the operation of an automatic frequency control (AFC) circuit would fail. Since the number of receivers in a usual transmission network is much larger than that of the transmitters, the largest numbers of lasers in a coherent transmission system are needed for the LO function. As a consequence, the development of continuously tunable laser diodes is most crucial for coherent optical communications.

The performance of these broadband communication systems as well as that of optical switching networks is ultimately limited by the wavelength coverage and the number of addressable wavelength channels. Therefore the tuning range and spectral resolution play a major role in laser development for these applications.

Continuously wavelength tunable lasers are also useful for various measurement and sensing applications [6, 7, 8]. Among these, two representative applications will be discussed briefly. A most promising application with a significant performance gain by using wavelength tunable laser diodes is the optical radar technique for distance measurements [9, 10, 11]. Due to the large frequency tuning range of several hundred GHz in the optical domain, one can expect an appreciably improved resolution as compared to microwave radar, particularly at short distances [12, 13, 14].

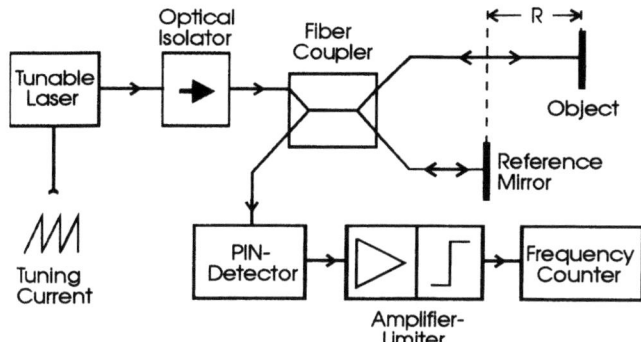

Figure 3. Optical frequency modulated continuous wave (FMCW) radar using a wavelength tunable laser diode as light source.

Accordingly, for instance, the automated testing of printed boards, where distance (height) resolutions of the order $1\cdots 10$ μm are required, appears possible. Furthermore the high modulation bandwidth of the tunable laser diodes enables more than 10^6 measurements per second so that three dimensional viewing (3D viewing) applications are feasible. A schematic outline of a simple optical radar system is displayed in Fig. 3, revealing a close relation to traditional frequency modulated continuous wave (FMCW) microwave radar systems.

A related application is the testing of high speed optoelectronic components by producing the emission spectrum of a high frequency modulated laser via interference of two frequency shifted laser beams. Referring to Fig. 4 these two laser beams can be obtained from one wavelength tunable laser diode and a Mach-Zehnder interferometer by producing a frequency ramp. As can be seen, no high frequency electronics is required and parasitic capacitances and inductances as well as the modulation bandwidth of the laser are not essential as would be the case in the high frequency current modulation of the laser diode. With frequency tuning ranges Δf

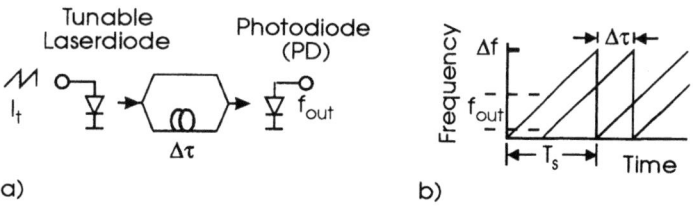

Figure 4. Bandwidth measurement of photodiodes using a tunable laser. (a) Principal test set layout and (b) illustration of beat frequency generation.

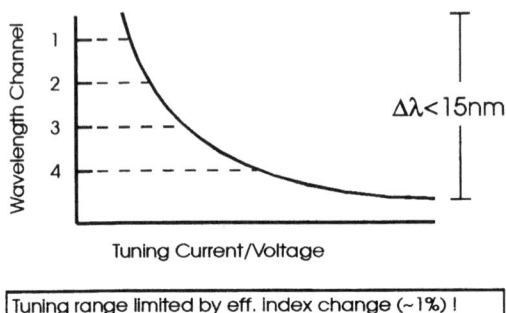

Figure 5. Schematic illustration of continuous tuning mode. The tuning range $\Delta\lambda$ should be large enough to cover the total set of wavelength channels. Note that any wavelength within the tuning range can be accessed by proper choice of tuning current or voltage. The tuning range in this preferred tuning mode is limited by the maximum refractive index change to about 15 nm at 1.5 µm wavelength.

around 500 GHz one can easily produce difference frequencies up to about 100 GHz this way, so that high speed detectors can conveniently be evaluated.

3. Continuously Tunable Laser Diodes

The continuous wavelength tuning is schematically displayed in Fig. 5. This tuning mode is the preferred one in almost every application because any wavelength within the tuning range can be accessed and because of the simplicity and unambiguity of the wavelength setting. In this wavelength tuning scheme the laser emits in the same spatial mode within the entire tuning range and mode changes or jumps do not contribute to the tuning effect. As a consequence the tuning range $\Delta\lambda$ can not be larger than that of one spatial laser mode, which is determined by the maximum change δn_e of the effective refractive index n_e achieved by the electronic control current or voltage

$$\frac{\Delta\lambda}{\lambda_0} = \frac{|\delta n_e|}{n_e}. \tag{1}$$

The idealized device function as demanded by the user is sketched schematically in Fig. 6. The continuously tunable laser should thereby exhibit two electronic controls I_a and I_t (or, equivalently, two control voltages) that allow for the independent adjustment of the output power via I_a and the wavelength via I_t, respectively, as plotted in Fig. 6b and c. Ideally, the tuning current I_t should affect only the wavelength but not the output power, while the laser current I_a should influence exclusively the output power and not the wavelength. In practice, however, such a

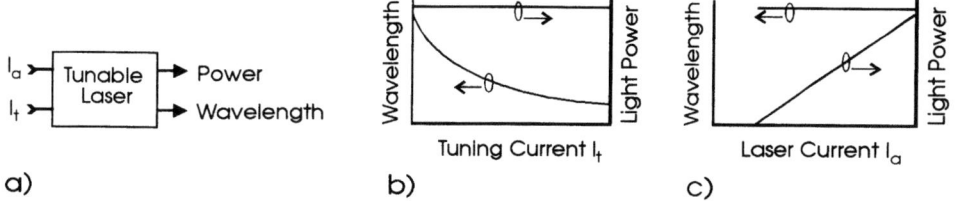

Figure 6. Idealized wavelength tunable laser diode (a) with tuning characteristics (b) and (c).

clear separation can not be achieved because on the one hand the tuning function, e. g. by carrier injection, also introduces optical losses that reduce the output power and on the other hand the changes of the laser current induce temperature changes that correspondingly affect the emission wavelength. Nevertheless, a separation of the two functions as far as possible should be aimed at in the laser development, since it essentially improves the device handling and suitability.

Since tuning is done on a single axial and transverse mode, the tuning range is proportional to the effective refractive index change induced by the tuning mechanism. So the electronic wavelength control is commonly performed by exploiting the free-carrier plasma effect [15, 16] or the quantum-confined Stark-effect (QCSE) [17], that enable the current or voltage controlled refractive index variation. Maximum refractive index changes up to about 0.04 can be achieved by exploiting the plasma effect of an electron-hole plasma injected into semiconductor lasers. Considering also that the light confinement in the tuning region is less than unity, the maximum electronic tuning range (excluding thermal heating) is therefore restricted to values less than about 15 nm at 1.5 μm wavelength [18]. With this restriction on the tuning range, the continuously tunable lasers are well suited for coherent optical communications and also for densely spaced wavelength division multiplexing (DWDM), where with channel spacings of 1...2 nm a moderate number (4...8) of channels can be covered by a single device.

Different structures have been developed so far to achieve a wide continuous tuning and to maintain a narrow spectral linewidth. Among these are the multisection DBR devices [19, 20, 21] and the TTG (Tunable Twin-Guide) laser [22, 23], in both of which tuning is performed by index change in a passive region and multisection DFB lasers, in which an axially varying bias to the active region induces the wavelength tuning [24, 25, 26].

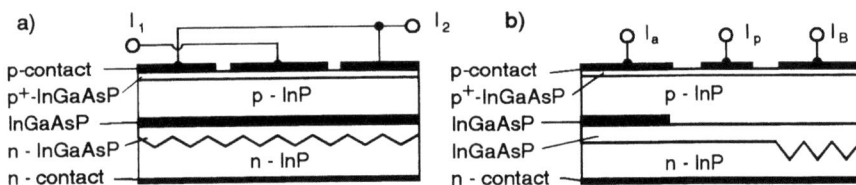

Figure 7. Schematical longitudinal sections of multisection tunable DFB Laser (a) and three-section tunable DBR Laser (b) for 1.55 μm wavelength. The active regions are marked black.

3.1. LONGITUDINALLY INTEGRATED STRUCTURES

The longitudinal sections of tunable DFB and DBR laser diode structures [18, 27] are displayed schematically in Fig. 7. The technologically most simple approach is a multielectrode DFB laser, the top contact of which is longitudinally separated into three individually biased sections (Fig. 7a). However, its wavelength tuning by nonuniform current injection is a rather complex mechanism comprising spatial hole burning effects [28, 29], thermal heating [30] and, for the case of a quantum-well active layer, also the gain-levering effect [24, 31]. In spite of the tendency to mode jumps, continuous wavelength tuning may be performed by a careful mutual adjustment of the laser currents. In this way the carrier density in the active region becomes longitudinally redistributed, which together with thermal heating changes the effective refractive index and the resonance wavelength of the laser cavity.

Owing to the random mirror phases and the inevitable waveguide perturbations the longitudinal field and carrier density distributions are different for each DFB laser (even for devices from the same wafer), so that the tuning behaviour, particularly the I_1/I_2-ratio for continuous tuning, differs from device to device. Accordingly a large amount of measurement is required to select and characterize the suited lasers. An essential disadvantage in the practical application of tunable DFB lasers is that the controls of output power and wavelength are not separated so that both parameters are affected similarly by all control currents.

The three-section distributed Bragg reflector (3S DBR) laser as shown in Fig. 7b provides a more convenient handling since an effective separation between the power control and tuning function is achieved: Current I_a mainly determines the power while both currents I_p and I_B essentially control the emission wavelength. Changing exclusively I_B or I_p, respectively, yields a discontinuous tuning by mode jumping since either the Bragg wavelength or the optical cavity length are changed solely. By a proper mutual adjustment of these two currents, however, a continuous wavelength tuning can be obtained provided that the relative changes of the Bragg wavelength exactly equal the relative changes of the optical cavity length. This handling im-

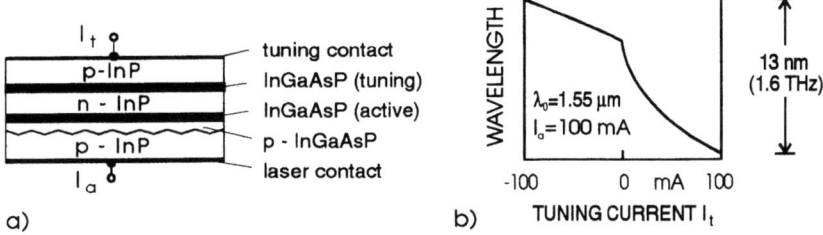

Figure 8. Schematic longitudinal section of Tunable Twin-Guide (TTG) distributed feedback laser (a) and tuning characteristic (b) including electronic tuning via the plasma effect (positive I_t) and via thermal heating of the reverse biased tuning diode (negative I_t, $U_t \approx -3$ V).

provement relative to the multisection DFB lasers is obtained by separating the laser active region from the (passive) wavelength selective phase shift and Bragg grating region. This furthermore allows the extension of the tuning range by a strong heating of the Bragg section while keeping the temperature in the gain section constant to maintain the laser action. In this way large tuning ranges up to 22 nm have recently been realized in the quasi-continuous (i. e. stepwise continuous) tuning mode [20], while the maximum continuous tuning range is around 4.4 nm [18]. Using a four-step MOVPE process and a semi-insulating InP:Fe current-blocking structure an AM modulation bandwidth of 9 GHz and a quasi-continuous tuning range of 9.1 nm were demonstrated with recent 3S DBR lasers [32]. The switching time for the transient between two successive modes can be as small as 500 ps [33].

3.2. Transversely Integrated Structures

The most convenient handling with a principally continuous tuning behaviour as well as the largest continuous tuning range have been achieved so far with the TTG laser as shown schematically in Fig. 8a [23, 34, 35]. This laser can be considered as a single mode DFB laser whose effective refractive index is tuned homogeneously along the laser axis by means of current I_t. The index changes are done by carrier injection into the passive tuning region which is collocated below or above the active region. Thereby Bragg wavelength changes are induced that exhibit a built-in synchronisation with the simultaneous optical cavity length changes. The resulting tuning behaviour is therefore inherently continuous.

Well designed TTG lasers yield continuous tuning ranges up to more than 10 nm (c. f. Fig. 8b) with spectral linewidth below about 30 MHz [36]. For ridge-waveguide TTG lasers 4 MHz linewidth with tuning ranges up to 2 nm have been obtained [34]. TTG lasers exploiting the QCSE instead of the free carrier plasma effect are well suited for broadband FM modulation yielding a flat FM response up to more

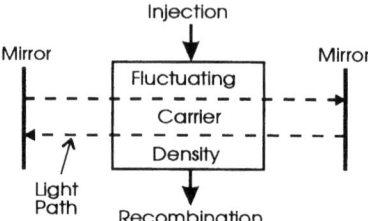

Figure 9. Simplified physical model for a laser cavity with linewidth broadening by injection-recombination shot noise (IRSN) due to the carrier injection into the tuning region.

than 2 GHz [37] with a spectral linewidth below 4 MHz. The tuning range, however, is only of the order of several Å for this device type.

Besides the large tuning range, the simple tuning scheme, which is inherently continuous and requires only one single wavelength control current, reveals as a significant advantage in practice of the TTG lasers over the multisection DBR and DFB devices with their complex tuning behaviour [38].

3.3. Spectral Linewidth

In general, electronic wavelength tuning via carrier injection causes an additional linewidth broadening. It has become clear both, experimentally [39, 40] and theoretically [41], that for most device types a trade-off between tuning range and excess linewidth broadening exists, which might limit the applicability of tunable laser diodes in demanding system concepts. The carier number fluctuations in the tuning regions due to injection-recombination shot noise (IRSN) has previously been identified as the origin for this broadening [41]. In contrast to the active region of the laser, namely, the carrier number fluctuations in the passive tuning regions are not damped out by the gain clamping in the lasing regime.

A simplified physical model illustrating this broadening mechanism is shown in Fig. 9. Even though a Fabry-Perot cavity is shown here, the model equally well applies for more complex resonator structures such as DFB or DBR laser cavities. As can be seen, the IRSN of the carriers injected into the tuning region randomly modulates the refractive index and, hence, the optical length of the laser cavity. As a consequence, the instantaneous laser frequency fluctuates yielding finally a broadening of the laser line.

The linewidth broadening $\Delta \nu_{IRSN}$ by the IRSN in the tuning regions of current tuned laser diodes (DFB, DBR and TTG lasers) was analysed theoretically [41, 42], revealing that this linewidth contribution is proportional to the square of the tuning

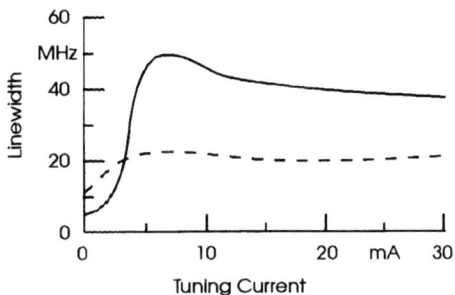

Figure 10. Total spectral linewidth of a widely (7 nm at 50 mA tuning current) tunable TTG laser versus the tuning current (solid curve). By heavily doping the tuning region n-type (n= $2 \cdot 10^{18}$ cm^{-3}), the tuning range decreases to 3 nm which yields a significant reduction of the linewidth (broken curve)

efficiency $d\lambda/dI_t$

$$\Delta\nu_{IRSN} = 4\pi e_0 \frac{c^2}{\lambda^4} \left(\frac{d\lambda}{dI_t}\right)^2 I_t. \qquad (2)$$

It is therefore difficult to achieve a large continuous tuning range simultaneously with a narrow spectral linewidth. From quantitative evaluation it turns out that, particularly for multisection DBR devices and TTG lasers with their passive tuning sections, $\Delta\nu_{IRSN}$ represents the essential contribution to the excess linewidth broadening [39, 41]. Since, for a fixed laser geometry, $\Delta\nu_{IRSN}$ is approximately proportional to the tuning range squared, it becomes most essential for the large range tunable devices. It should be noted, however, that this broadening effect does not occur by the thermal tuning, even if it is induced by the tuning current.

The total laser linewidth of a widely (7 nm) tunable TTG laser with an undoped tuning region is displayed in Fig. 10. In accordance with Eq. 2 the linewidth reveals a significant enhancement by about an order of magnitude (solid curve) at small tuning currents, where $d\lambda/dI_t$ is largest. Using the same laser structure but doping the tuning region with n= $2 \cdot 10^{18}$ cm^{-3} (broken curve) reduces the tuning range by about a factor of two but also reduces the total spectral linewidth in the tuning mode. Also in case of the three-section DBR lasers a large IRSN linewidth broadening occurs [41, 43]. The resulting linewidth broadening is typically of the order of several MHz [39], which may limit also the applicability of the three-section DBR lasers in linewidth sensitive systems. The IRSN in the tuning mode was recently measured directly in the relative intensity noise (RIN) spectrum of a tunable 3-section DBR laser [40].

A further linewidth broadening occurs in tunable DBR and TTG lasers via the cavity loss fluctuations accompanying the refractive index fluctuations in the tuning layer. These induce immediate carrier density fluctuations in the active layer

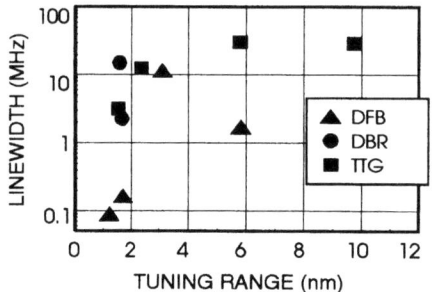

Figure 11. Spectral linewidth versus tuning range for various continuously tunable laser diode structures. In spite of the different device parameters the strong enhancement of the linewidth with increasing tuning range is obvious.

to keep the total device gain constant. Depending on the α-factor in the active region, these fluctuations are accompanied by additional index and, consequently, wavelength fluctuations. This interaction between the active and tuning region has previously been investigated theoretically for the case of the TTG DFB laser, taking into account also the longitudinal inhomogeneities of the electromagnetic field in the DFB structure. A statistical analysis was performed for as-cleaved devices, in order to treat the random grating-facet relationship. It turned out from this calculation that the IRSN linewidth broadening is enhanced in the DFB structure with large coupling-coefficient-length product κL. Compared to the Fabry-Perot limit ($\kappa L \to 0$), $\Delta \nu_{IRSN}$ is larger up to about 50 % for typical κL-values [44] at equal tuning range.

By exploiting the QCSE for tuning of the TTG laser [37, 45] no carriers are injected into the tuning region. Correspondingly no shot noise broadening occurs, so that the total spectral linewidth can be kept small. In addition the FM modulation bandwidth can be increased since the carrier lifetime limitation in the tuning region is dropped. On the other hand, however, due to the small optical confinement in the quantum wells, the tuning range is essentially smaller. Improved spectral properties and high-speed FM modulation might also be achieved in future devices using the electron-transfer within an MQW-type tuning region. With the so-called barrier reservoir and quantum-well electron-transfer structure (BRAQWETS) voltage controlled refractive index changes up to 0.02 [46] have been demonstrated and (parasitic free) switching times well below 100 ps have theoretically been predicted [47].

Using a separate confinement heterostructure quantum well (SCH QW) structure in the tuning region that localizes the holes within the wells while the electrons are distributed over the entire tuning region [48] or applying a MQW twin-active-guide [49] might enable a further simultaneous improvement of optical power and

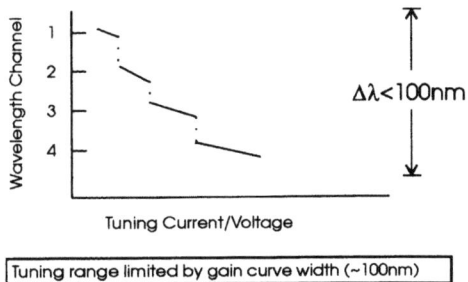

Figure 12. Schematical illustration of discontinuous tuning mode. Here the wavelength variation mainly stems from the mode jumps, so that the total wavelength coverage may be as large as the gain bandwidth ($\approx$ 100 nm). To conform with the given channel spacing, at least each one laser mode should cover one wavelength channel.

spectral linewidth of the TTG laser.

Because of the markedly smaller tuning efficiency, the IRSN linewidth broading is less important in wavelength tunable multisection DFB lasers. With properly selected devices and by exploiting also the wavelength change by thermal heating, tuning ranges up to 6 nm with spectral linewidth below 2 MHz have been reported for multisection DFB lasers[25]. Quite recently a spectral linewidth less than 100 kHz has been obtained over a (mostly thermally induced) tuning range of 1.3 nm with a corrugation-pitch-modulated multi quantum well (MQW) DFB laser [50].

Published data on the spectral linewidth for the various continuously tunable laser diodes are compiled in Fig. 11 [39, 50, 51, 36].

4. Discontinuously Tunable Laser Diodes

As mentioned above, a wavelength coverage larger than about 15 nm at 1.55 μm wavelength can not be achieved continuously, i. e. in a single transverse and axial mode. If a larger tuning range is needed, e. g. in wavelength division multiplexing (WDM) systems, one has to accept mode changes and the corresponding inconvenience during tuning. The resulting discontinuous tuning scheme is shown in Fig. 12 for the case of a multichannel application. Since the tuning range in this tuning mode benefits from the mode jumps between different axial modes, wide tunability can be achieved. The maximum of the tuning range is therefore usually only limited by the gain bandwidth of the active region, which is of the order 100 nm at 1.5 μm wavelength. Owing to the mode jumps it is extremely important in practice to ensure that the set of wavelengths that is needed for the application can be accessed by the laser (c. f. Fig. 12)

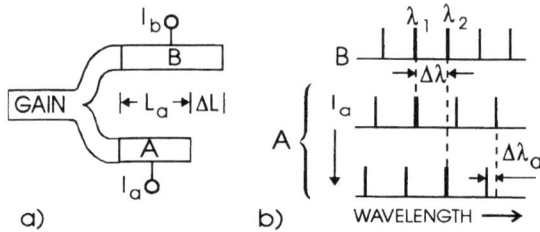

Figure 13. Schematical top view of a tunable Y-laser (a) and the comb mode spectra of arms A and B under various bias I_a (b), illustrating the tuning effect. Note that the wavelength change $\Delta\lambda$ is much larger than the current induced shift of the comb mode spectrum A, $\Delta\lambda_a$ (vernier effect).

In the discontinuous tuning mode it is of particular importance to characterize the lasers carefully over a wide range of currents and temperatures in order to know exactly under which conditions mode jumps occur and in which operation regimes stable modes exist. Major challenges resulting from the discontinuous tuning are the single mode operation and the wavelength access to as many as possible wavelengths within the tuning range. Accordingly, completely different laser structures have been developed for tuning ranges above 15 nm.

4.1. Y-Laser

Using the interferometric effect between the two arms of an (asymmetric) Y-coupled integrated laser diode, an extended discontinuous wavelength coverage can be achieved [52, 53]. The principal device structure is displayed in Fig. 13 together with an illustration of its operation. The Y-laser usually consists of an all active waveguide structure with a typical length around 1 mm. By the separation of the top p-contact into 3 to 4 sections, the two interferometer arms A and B can be biased independently. At each bias condition lasing occurs at a wavelength where the two differently spaced comb mode spectra of the two coupled laser cavities exhibit simultaneously an axial mode.

For illustration Fig. 13b shows the comb mode spectra corresponding to arm B and A, respectively, at a certain bias I_a and I_b. For small I_a lasing first occurs at λ_1. Increasing I_a slightly shifts the comb mode spectrum of arm A towards shorter wavelengths by $\Delta\lambda_a$. Lasing occurs now at wavelength λ_2, which is shifted by a much larger amount $\Delta\lambda$ towards longer wavelengths with respect to λ_1. As can be seen, the vernier effect of the two differently spaced comb mode spectra yields a significant magnification of the laser wavelength shift as compared with the induced shift of the comb mode spectrum as well the ability to cause either a blue or a red shift of the laser wavelength.

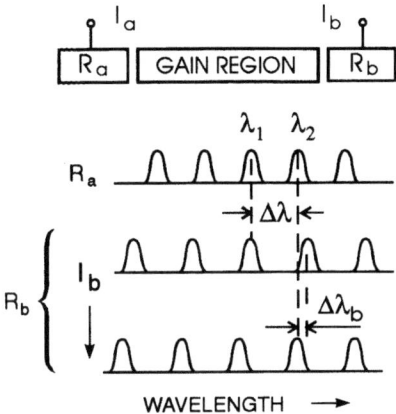

Figure 14. Schematical longitudinal sections of sampled grating (SG) and superstructure grating (SSG) tunable DBR lasers. The current tunable mirrors R_a and R_b exhibit comb mode reflection spectra each with a different mode spacing. Wavelength tuning by varying I_b is illustrated and the vernier effect is evident ($\Delta \lambda \gg \Delta \lambda_b$).

Experimentally, up to 51 nm discontinuous tuning was reported for InGaAsP/InP Y-lasers at 1.55 μm wavelength [53, 54]. Using this laser as a tunable wavelength converter, a conversion of 2.5 Gb/s data streams was demonstrated [54].

4.2. Sampled Grating (SG) and Superstructure Grating (SSG) Tunable DBR Lasers

Applying a DBR laser structure with two different Bragg reflectors at the rear and front end, that exhibit different comb reflection spectra (similar to the comb mode spectra in Fig. 13b), also yields an enhanced tuning range by the vernier effect. The corresponding super-structure grating (SSG) DBR [55, 56] and sampled grating (SG) DBR [57, 58] tunable lasers are described in Fig. 14. The absolute wavelengths of the reflection peaks of each of the two Bragg reflectors R_a and R_b can be controlled by currents I_a and I_b, respectively. In close analogy with the Y-laser, lasing always occurs at a wavelength where the two reflection spectra each exhibit a reflection peak. Consequently, small changes in any of the reflection spectra lead to large changes of the laser wavelength.

The comb reflection spectra are realized by Bragg gratings with side modes in the spatial frequency domain. This can either be achieved by a periodic spatial amplitude modulation of the grating, yielding the SG DBR lasers, or by a periodic spatial frequency modulation, which corresponds to the SSG DBR lasers. The maximum wavelength tuning achieved with the SG DBR laser so far is 57 nm [58], while record

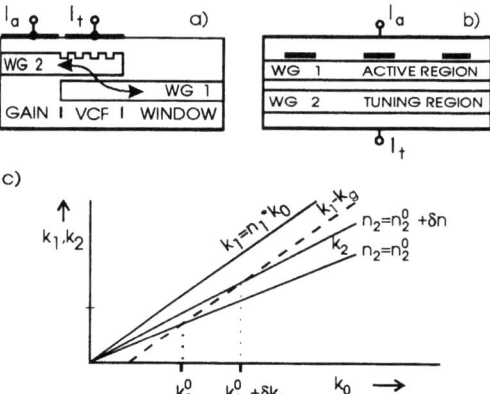

Figure 15. Schematic longitudinal view of vertical coupler filter (VCF) based lasers (a) and distributed forward coupled (DFC) laser (b). The phase-match condition $k_1 = k_2 + k_g$ is illustrated graphically (c). Tuning is performed by only one control current (I_t).

tuning ranges of 83 nm and 101 nm for single- and multimode operation, respectively, were recently presented for SSG DBR lasers [55]. Quite recently, the number of accessible wavelengths within these wavelengths ranges has been considerably increased for the SSG DBR laser to the disadvantage of the handling convenience by using both currents I_a and I_b simultaneously for tuning [59].

4.3. CODIRECTIONAL MODE COUPLING

A third principal approach for a wide tuning is the application of a codirectionally coupled two-mode twin-waveguide laser structure. The two basic structures presented so far are shown in Fig. 15. Compared with the contradirectionally coupled DFB and DBR lasers the distributed forward coupled (DFC) laser (Fig. 15b) [60] equals the longitudinally quasi-homogeneous DFB laser, while the vertical-coupler filter (VCF) type laser (Fig. 15a) [61, 62, 63] resembles more the DBR laser. Both devices rely on the interference effect of two codirectionally coupled modes as illustrated in Fig. 15c.

In the VCF type lasers, lasing can only occur at a wavelength where the phase matching condition between the two codirectionally propagating waveguide modes applies in the VCF region, since only under this condition the light couples from WG 1 to WG 2 and vice versa yielding a closed feedback loop for the light path.

In the DFC laser periodic absorbers (period ≈ 15 μm) are placed within the laser cavity. Only exactly at the phase matching wavelength the superposition of the two waveguide modes yields negligible absorption losses and a low threshold current,

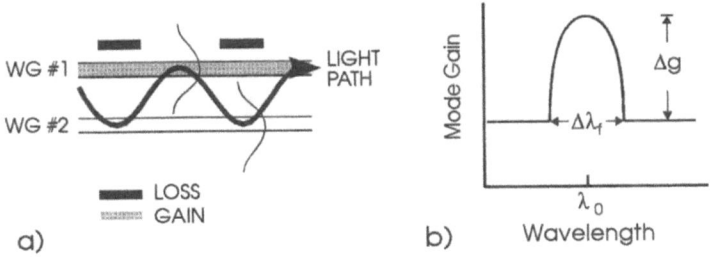

Figure 16. Light path in the DFC-Laser at the phase-match wavelength (a) and resulting filter characteristic (b).

since in this case the combined fields almost vanish at the periodic absorbers as illustrated in Fig. 16a. The calculated filter characteristic is displayed in Fig. 16b, where due to the strong filtering effect a bandwidth $\Delta\lambda_f$ below 15 nm and a gain peak Δg up to 40 cm^{-1} may be obtained [64].

Wavelength tuning is induced in these lasers by shifting the phase matching wavelength λ_0, which is defined by the phase-match condition $k_1 = k_2 + k_g$, where $k_1 = k_0 n_1$ and $k_2 = k_0 n_2$ denote the wavevectors of the two modes of the twin waveguide with effective refractive indexes n_1 and n_2, respectively. The vector $k_g = 2\pi/\Lambda$ stands for the grating wavevector with Λ being the grating pitch (typically 15-20 μm). With the effective refractive index difference $\Delta n = |n_1 - n_2|$ the phase-match condition can also be written as

$$\lambda_0 = \Delta n \Lambda. \tag{3}$$

The phase-match wavelength shift is done simply with only one control current I_t, by which in the exemplary laser structures of Fig. 15 the refractive index in WG 2 (n_2) and, as a consequence, the wavevector k_2 can be changed. Owing to the extremely strong dependence of the phase-match wavelength on the wavevectors, a large tuning effect is achieved as shown in Fig. 15c, where $\delta k_0/k_0 \gg \delta n_2/n_2$ yielding for the tuning range $\Delta\lambda$

$$\frac{\Delta\lambda}{\lambda_0} = \frac{|\delta n_2|}{\Delta n}. \tag{4}$$

Since $\Delta n \ll n_e$, where n_e equals n_1 or n_2, the tuning range is much larger than for the continuously tunable lasers (c. f. Eq. 1). Consequently, the tuning range might ultimately be limited only by the spectral width of the active region gain.

Both a red or a blue shift can be obtained by the tuning depending on whether Δn is increased or decreased by raising the tuning current (c. f. Eq. 3). This is because in contrast to the continuously tunable lasers Δn is the difference of the effective refractive indexes of the two waveguide modes. Even though the index

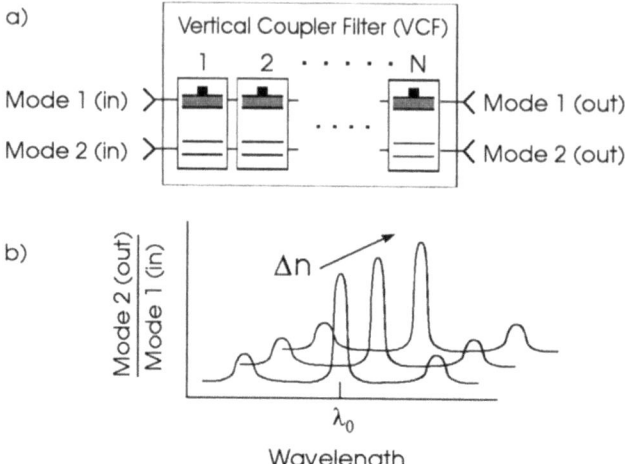

Figure 17. Schematic longitudinal section of Vertical Coupler Filter (VCF) with N periods (a). Wavelength tuning and shift of the filter characteristic of the coupling from mode 1 to mode 2 by changing the effective index difference Δn between mode 1 and 2 is illustrated (b). The change of Δn can be positive or negative and is accomplished by increasing the tuning current.

change by carrier injection is always negative, Δn can increase if the smaller one of n_1 and n_2 decreases. With the In the VCF type lasers, lasing can only occur at a wavelength where the phase matching condition between the two codirectionally propagating waveguide modes applies in the VCF region, since only under this condition the light couples from WG 1 to WG 2 and vice versa yielding a closed feedback loop for the light path. By shifting the phase-match wavelength the entire filtering curve is displaced by the same amount as shown schematically in Fig. 17 for the VCF-type devices.

With the first DFC lasers tuning ranges around 16 nm were reported (Fig. 18) with about 9 accessible wavelengths. The more mature VCF type lasers showed tuning ranges between 30 and 57 nm [62, 63] with up to about 15 accessible wavelengths (Fig. 19).

4.4. OUTLOOK

It should be stressed, that the underlying operation principles of the widely tunable laser diodes exploit the slight differences of modenumbers or reflection maxima; therefore even rather small variations of these parameters, occuring during the device fabrication, yield large relative changes of the device characteristics, particularly of the laser wavelength. The successful development of these novel components

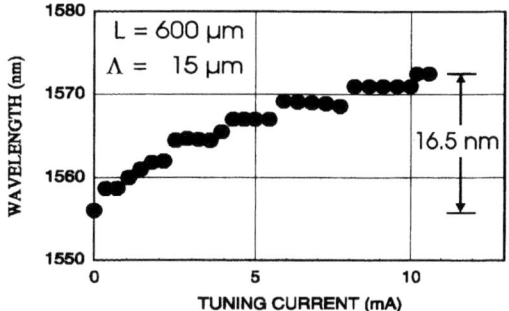

Figure 18. Wavelength versus tuning current for first DFC laser structure. The active region current is kept constant at 100 mA.

and a reasonable fabrication yield therefore basically require a highly precise and homogeneous fabrication technique; this essentially concerns the crystal growth, for which e. g. a layer thickness control on the 0.01 μm scale is demanded.

A common deficiency of all these approaches for an extended tuning range is the lack of a continuous tunability hindering the access to any wavelength within the tuning range. This still limits the practical applicability of these lasers, particularly as the number of longitudinal modes covered by the wavelength tuning typically is of the order 100. The future development is therefore challenged by improving the wavelength access, e. g. with laser structures exhibiting an enhanced wavelength selectivity or even by the practical realisation of quasi-continuously tunable devices.

5. Conclusion

The state-of-the-art of electronically tunable laser diodes and their typical applications in optical fiber communications and measurement was reviewed. For most applications the continuous tuning mode and the simple and unambiguous wavelength control revealed most essential, so that the previous development mainly focused on the achievement of large continuous tuning range and simple device handling, i. e. only one control current for the wavelength setting. For tuning ranges below about 10 nm various types of high-performance (single-mode, continuous tuning, linewidth, speed, handling) devices already exist, based on the technologically well developed DFB and DBR laser structures. For physical reasons, the maximum continuous tuning range of electronically tunable monolithic diode lasers is limited to about 15 nm, corresponding to a relative range of 1.5 % at 1.55 μm wavelength. Experimentally up to 13 nm continuous tuning range have been achieved so far with

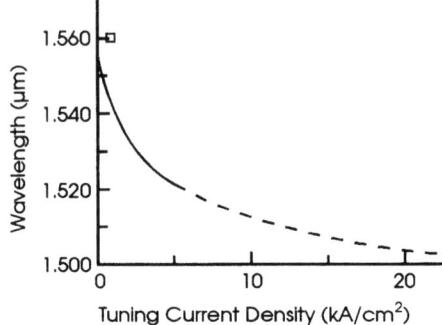

Figure 19. Wavelength versus tuning current density of VCF laser. The solid and broken curves correspond to cw and pulsed operation. The rectangle displays the wavelength for reverse bias.

the TTG laser.

In applications where one can accept the discontinuous tuning mode, novel device concepts have been presented offering discontinuous tuning up to about 100 nm, which already equals the gain bandwidth of InGaAsP at 1.55 μm. However, the wavelength access is still an issue for these new structures, since typically only about 10 to 30 channels with different spacing can individually be addressed. The essential device parameter influencing the wavelength access is the bandwidth of the filtering curve, so that the present development concentrates on the improvement of the wavelength selectivity.

References

1. Brackett, C. A. (1990) Dense wavelength division multiplexing networks: Principles and applications, *IEEE J. Selected Areas in Communications* **8**, 948–964.

2. Yamamoto, Y. and Kimura, T. (1981) Coherent optical fiber transmission systems, *IEEE J. Quantum Electron.* **QE-17**, 919–935.

3. Wagner, R. E. and Linke, R. A. (1990) Heterodyne lightwave systems: moving towards commercial use, *IEEE LCS*, 28–35.

4. Noe, R., Drögemüller, K., Rodler, H., Ebberg, A., Meißner, E., Bodlaj, V., Wittmann, J., Auracher, F., Borchert, B., Wolf, T., Amann, M.-C., Bauer, J., Albrecht, H., Pazelt, K., and I.Althaus, H. (1992) Fully engineered coherent multichannel transmitters and receivers using data-induced polarization switching (paper FC3), In *Technical Digest of Optical Fiber Conference (San Jose, USA)*.

5. Noe, R., Rodler, H., Ebberg, A., Meißner, E., Bodlaj, V., Drögemüller, K., and Wittmann, J. (1992) Fully engineered coherent multichannel transmitters and receivers with low-cost potential, *Electron. Lett.* **28**, 14–15.

6. Ebberg, A. and Noe, R. (1990) Novel high precision alignment technique for polarisation maintaining fibres using a frequency modulated tunable laser, *Electron. Lett.* **26**, 2009–2010.

7. Schell, M., Huhse, D., and Bimberg, D. (1993) Generation of short (3.5 ps) low jitter (<100 fs) light pulses with a 1.55 μm tunable twin guide laser, In *Techn. Digest of 19th European Conference on Optical Communications (Montreux, Switzerland)*, pages 229–232.

8. Slotwinski, A. R., Goodwin, F. E., and Simonson, D. L. (1989) Utilizing AlGaAs laser diodes as a source for frequency modulated continuous wave (FMCW) coherent laser radars, *SPIE Proc.-Laser Diode Technology and Applications* **1043**, 245–251.

9. Burrows, E. C. and Liou, K.-Y. (1990) High resolution laser lidar utilising two-section distributed feedback semiconductor laser as a coherent source, *Electron. Lett.* **26**, 577–579.

10. Uttam, D. and Culshaw, B. (1985) Precision time domain reflectometry in optical fiber systems using a frequency modulated continuous wave ranging technique, *IEEE J. Lightwave Technol.* **LT-3**, 971–977.

11. Strzelecki, E. M., Cohen, D. A., and Coldren, L. A. (1988) Investigation of tunable single frequency diode lasers for sensor applications, *IEEE J. Lightwave Technol.* **LT-6**, 1610–1618.

12. Economou, G., Youngquist, R. C., and Davies, D. E. N. (1986) Limitations and noise in interferometric systems using frequency ramped single-mode diode lasers, *IEEE J. Lightwave Technol.* **LT-4**, 1601–1608.

13. Dieckmann, A. (1994) FMCW-LIDAR with tunable twin-guide laser diode, *Electron. Lett.* **30**, 308–309.

14. Amann, M.-C. (1992) Phase noise limited resolution of coherent LIDAR using widely tunable laser diodes, *Electron. Lett.* **28**, 1694–1696.

15. Kobayashi, K. and Mito, I. (1988) Single frequency and tunable laser diodes, *J. Lightwave Technol.* **6**, 1623–1633.

16. Okuda, M. and Onaka, K. (1977) Tunability of distributed Bragg-reflector laser by modulating refractive index in corrugated waveguide, *Japan. J. Appl. Phys.* **16**, 1501–1502.

17. Yamamoto, H., Asada, M., and Suematsu, Y. (1985) Electric-field induced refractive index variation in quantum-well structure, *Electron. Lett.* **21**, 579–580.

18. Kotaki, Y. and Ishikawa, H. (1991) Wavelength tunable DFB and DBR lasers for coherent optical fibre communications, *IEE Proc. Pt. J.* **138**, 171–177.

19. Kotaki, Y., Matsuda, M., Ishikawa, H., and Imai, H. (1988) Tunable DBR laser with wide tuning range, *Electron. Lett.* **24**, 503–505.

20. Oeberg, M., Nilsson, S., Klinga, T., and Ojala, P. (1991) A three-electrode distributed Bragg reflector laser with 22 nm wavelength tuning range, *IEEE Photon. Technol. Lett.* **PTL-3**, 299–301.

21. Koch, T. L., Koren, U., and Miller, B. I. (1988) High performance tunable 1.5um InGaAs/InGaAsP multiple quantum well distributed feedback Bragg reflector, *Appl. Phys. Lett.* **53**, 1036–1038.

22. Amann, M.-C., Illek, S., Schanen, C., and Thulke, W. (1989) Tunable twin-guide laser: A novel laser diode with improved tuning performance, *Appl. Phys. Lett.* **54**, 2532–2533.

23. Illek, S., Thulke, W., Schanen, C., Lang, H., and Amann, M.-C. (1990) Over 7 nm (875 GHz) continuous wavelength tuning by tunable twin-guide (TTG) laser diode, *Electron. Lett.* **26**, 46–47.

24. Okai, M., Sakano, S., and Chinone, N. (1989) Wide-range continuous tunable double-sectioned distributed feedback lasers, In *Proceedings of 15th European Conference on Optical Communications (Gothenburg, Sweden)*, pages 122–125.

25. Kuindersma, P. I., Scheepers, W., Cnoops, J. M. H., Thijs, P. J. A., v. d. Hofstad, G. L. A., v. Dongen, T., and Binsma, J. J. M. (1990) Tunable Three-Section, Strained MQW, PA-DFB's with Large Single Mode Tuning Range (72 Å) and Narrow Linewidth (around 1 MHz), In *Conference Digest of 12th IEEE Semiconductor Laser Conference (Davos, Switzerland)*, pages 248–249.

26. Wu, M. C., Chen, Y. K., Tanbun-Ek, T., Logan, R. A., and Sergant, A. M. (1990) Gain-levering enhanced continuous wavelength tuning (6.1nm) in two-section strained MQW DFB lasers, In *Proceedings of Conference on Lasers and Electro-Optics (Anaheim, USA)*, page 667.

27. Koch, T. L. and Koren, U. (1990) Semiconductor lasers for coherent optical fiber communications, *J. Lightwave Technol.* **LT-8**, 274–293.

28. Kusnetzow, M. (1988) Theory of wavelength tuning in two-segment distributed feedback lasers, *IEEE J. Quantum Electron.* **24**, 1837.

29. Pan, X., Olesen, H., and Tromborg, B. (1990) Spectral linewidth of DFB lasers including the effects of spatial hole burning and nonuniform current injection, *IEEE Photon. Technol. Lett.* **2**, 312–315.

30. Okai, M., Tsuchiya, T., Takai, A., and Uomi, K. (1992) Wavelength tuning and FM response of three-section CPM-MQW DFB lasers, In *Proceedings of Conference on Lasers and Electro-Optics (Baltimore, USA)*, page 66.

31. Lau, K. Y. (1990) Broadband wavelength tunability in gain-levered quantum well semiconductor lasers, *Appl. Phys. Lett.* **57**, 2632.

32. Stoltz, B., Dasler, M., and Sahlen, O. (1993) Low threshold-current, wide tuning-range, butt-joint DBR laser grown with four MOVPE steps, *Electron. Lett.* **29**, 700–702.

33. Delorme, F., Gambini, G., Puleo, M., and Slempkes, S. (1993) Fast tunable 1.5 μm distributed Bragg reflector laser for optical switching applications, *Electron. Lett.* **29**, 41–43.

34. Wolf, T., Westermeier, H., and Amann, M.-C. (1992) Tunable Twin-Guide (TTG) Laser Diodes with Metal-Clad Ridge-Waveguide (MCRW) Structure for Coherent Optical Communications, *Europ. Trans. Telecommun. and Research. Technol.* **3**, 517–522.

35. Wolf, T., Illek, S., Rieger, J., Borchert, B., and Thulke, W. (1993) Tunable Twin-Guide Lasers with improved performance fabricated by metal-organic vapour phase epitaxy, *IEEE Photon. Technol. Lett.* **PTL-5**, 273–275.

36. Wolf, T., Illek, S., Rieger, J., Borchert, B., and Amann, M.-C. (1994) Extended continuous tuning range (over 10 nm) of tunable twin-guide lasers, In *Technical Digest of Conference on Lasers and Electro-Optics (Baltimore, USA)*, pages CWB-1.

37. Wolf, T., Drögemüller, K., Borchert, B., Westermeier, H., Veuhoff, E., and Baumeister, H. (1992) Tunable twin-guide lasers with flat frequency modulation response by quantum confined Stark effect, *Appl. Phys. Lett.* **60**, 2472–2474.

38. Kuindersma, P. I. (1989) Continuous tunability of DBR lasers, In *Technical Digest of International Conference on Integrated Optics and Optical Fiber Communication (Kobe, Japan)*, pages 19A2-1.

39. Kotaki, Y. and Ishikawa, H. (1989) Spectral characteristics of a three-section wavelength-tunable DBR laser, *IEEE J. Quantum Electron.* **QE-25**, 1340–1345.

40. Sundaresan, H. and Fletcher, N. C. (1990) Direct observation of shot noise in linewidth broadening of DBR lasers, *Electron. Lett.* **26**, 2004.

41. Amann, M.-C. and Schimpe, R. (1990) Excess linewidth broadening in wavelength-tunable laser diodes, *Electron. Lett.* **26**, 279.

42. Hamada, M., Yamamoto, E., Suda, K., Nogiwa, S., and Oki, T. (1991) Narrow linewidth characteristics in tunable twin-guide distributed feedback laser diodes, *Jpn. J. Appl. Phys. (Letters)* **31**, L 1552–L 1555.

43. dos Santos Ferreira, M. F., da Rocha, J. R. F., and de Lemos Pinto, J. (1992) Analysis of the frequency noise of tunable multisection DBR lasers, *IEEE J. Quantum Electron.* **28**, 833–840.

44. Amann, M.-C. and Borchert, B. (1992) Spectral Linewidth of Tunable Twin-Guide Laser Diodes, *AEÜ* **46**, 63–72.

45. Yamamoto, E., Hamada, M., Suda, K., Nogiwa, S., and Oki, T. (1991) Optical modulation characteristics of a twin-guide laser by an electric field, *Appl. Phys. Lett.* **59**, 2721–2723.

46. Wegener, M., Chang, T. Y., Bar-Joseph, I., Kuo, J. M., and Chemla, D. S. (1989) Electroabsorption and refraction by electron transfer in asymmetric modulation-doped multiple quantum well structures, *Appl. Phys. Lett.* **55**, 583–585.

47. Wang, J., Leburton, J. P., and Educato, J. L. (1993) Speed response analysis of an electron-transfer multiple-quantum-well waveguide modulator, *J. Appl. Phys.* **73**, 4669–4679.

48. Sakata, Y., Yamaguchi, M., Takano, S., Shim, J.-I., Sasaki, T., Kitamura, M., and Mito, I. (1993) Novel tunable twin-guide lasers with a carrier-control tuning layer, In *Proceedings of Optical Fiber Conference (San Jose, USA)*, pages 9–10.

49. Yamamoto, E., Suda, K., Hamada, M., Nogiwa, S., and Oki, T. (1991) Tunable laser diode having a complementary twin-active-guide (CTAG) structure, *Jpn. J. Appl. Phys. (Letters)* **30**, L 1884–L 1886.

50. Okai, M. and Tsuchiya, T. (1993) Tunable DFB lasers with ultra-narrow spectral linewidth, *Electron. Lett.* **29**, 349–351.

51. Illek, S., Thulke, W., Schanen, C., Drögemüller, K., and Amann, M.-C. (1990) Wavelength Tuning Efficiency and Spectral Linewidth Broadening in Tunable Twin-Guide DFB Laser Diodes, In *Conf. Digest of 12th International Semiconductor Conference (Davos, Switzerland)*, pages 60–61.

52. Schilling, M., Schweitzer, H., Dütting, K., Idler, W., Kühn, E., Nowitzki, A., and Wünstel, K. (1990) Widely tunable Y-coupled cavity integrated interferometric injection laser, *Electron. Lett.* **26**, 243–244.

53. Kuznetsow, M., Verlangieri, P., Dentai, A. G., Joyner, C. H., and Burrus, C. A. (1992) Asymmetric Y-branch tunable semiconductor laser with 1.0 THz tuning range, *IEEE Photon. Technol. Lett.* **4**, 1093–1095.

54. Schilling, M., Idler, W., Baums, D., Dütting, K., Laube, G., Wünstel, K., and Hildebrand, O. (1992) 6 THz range frequency conversion of 2.5 Gbit/s signals by a 1.55 µm MQW based widely tunable Y-laser, In *Conference Digest of 13th IEEE Semiconductor Laser Conference (Takamatsu, Japan)*, pages 272–273.

55. Tohmori, Y., Yoshikuni, Y., Tamamura, T., Yamamoto, M., Kondo, Y., and Ishii, H. (1993) over 100 nm wavelength tuning in superstructure grating (SSG) DBR lasers, *Electron. Lett.* **29**, 352–354.

56. Yoshikuni, Y., Tohmori, Y., Tamamura, T., Ishii, H., Kondo, Y., Yamamoto, M., and Kano, F. (1993) Broadly tunable distributed-Bragg-reflector lasers with a multiple-phase-shift superstructure grating (paper TuC2), In *Proceedings Optical Fiber Conference (San Jose, USA)*, pages 8–9.

57. Jayaraman, V., Cohen, D. A., and Coldren, L. A. (1992) Demonstration of broadband tunability in a semiconductor laser using sampled gratings, *Appl. Phys. Lett.* **60**, 2321–2323.

58. Jayaraman, Y., Mathur, A., Coldren, L. A., and Dapkus, P. D. (1993) Extended tuning range in sampled grating DBR lasers, *IEEE Photon. Technol. Lett.* **PTL-5**, 489–491.

59. Okamoto, K. and Yoshikuni, Y. (1994) Optical devices for wideband all-optical networks, In *Technical Digest of 5th Optoelectronics Conference (Makuhari Messe, Japan)*, pages 110–111.

60. Amann, M.-C., Borchert, B., Illek, S., and Wolf, T. (1993) Widely tunable distributed forward coupled (DFC) laser, *Electron. Lett.* **29**, 793–794.

61. Alferness, R. C., Koren, U., Buhl, L. L., Miller, B. I., Young, M. G., Koch, T. L., Raybon, G., and Burrus, C. A. (1992) Widely tunable InGaAsP/InP laser based on a vertical coupler intracavity filter (paper PD2), In *Proceedings of Optical Fiber Conference (San Jose, USA)*, pages 321–324.

62. Kim, I., Alferness, R. C., Buhl, L. L., Koren, U., Miller, B. I., Newkirk, M. A., Young, M. G., Koch, T. L., Raybon, G., and Burrus, C. A. (1993) Broadly tunable InGaAsP/InP vertical-coupler filtered laser with low tuning current, *Electron. Lett.* **29**, 664–666.

63. Illek, S., Thulke, W., and Amann, M.-C. (1991) Codirectionally coupled twin-guide laser diode for broadband electronic wavelength tuning, *Electron. Lett.* **27**, 2207–2208.

64. Amann, M.-C., Borchert, B., Illek, S., and Wolf, T. (1993) Tuning range and spectral selectivity of the distributed forward coupled (DFC) laser, *IEEE Photon. Technol. Lett.* **5**, 886–888.

ACTIVE COMPONENTS CHARACTERIZATION
Light Source Components for Optical Fibres

C. LE MEN
Laboratoire Central des Industries Électriques
33, av. du Général Leclerc
BP 8
92266 Fontenay aux roses
FRANCE

1. Introduction

Historically, the development of silica-based optical fibre started in 1970 and the greatest efforts were made to decrease the attenuation coefficient of the fibre. Now the fibre spectral attenuation curve shows a shape that depends on the impurities ratio in the fibre and especially on the hydroxyl ion (OH^-) second, third and fourth absorption harmonics. The location of those absorption lines (0.95 μm, 1.24 μm and 1.38 μm) fully determines the location of the fibre attenuation minima. Those minima were then found around 0.85 μm, 1.31 μm and 1.55 μm. If detection components were available for this wavelength range, emitting components were almost non-existent. The need to realize such emitting components then became crucial.

The first fibre applications were multimode fibres ones and at a wavelength equal to 0.85 μm. Then almost simultaneously (1973) came the first semi-conductor laser which lifetime was great enough to allow any experiment. This laser based on a GaAlAs semi-conductor compound was emitting around 0.85 μm.

At the end of the seventies the theoretical values of fibre attenuation were almost reached with values close to 2 dB/km at 0.85 μm, 0.35 dB/km at 1.31 μm and 0.2 dB/km at 1.55 μm. Obviously, the interest was to find components emitting around 1.31 and 1.55 μm to take advantage of the lowest attenuation. Then studies on new semi-conductor compounds were done, and they produced InGaAs and InGaAsP compounds. The laser structures were still simple and often limited to multimode lasers.

Then, with telecommunication systems, the single-mode fibres use increased very quickly. The attenuation of those fibres is so low that dispersion became the most limiting parameter in long distance transmissions. Therefore the new need was to realize single-mode lasers components to cancel the intermodal dispersion and also to reduce more and more their spectral width to decrease chromatic and guide dispersions. Thus, single-longitudinal mode lasers, such as Distributed Feed Back lasers, appeared.

This article deals with the technology of those multiple emitting components and finally presents their characterization in power, wavelength, bandwidth, etc.

2. Emitting components technology

2.1. LIGHT EMISSION PRINCIPLE [1] [2]

The emitting components technology is based on semi-conductor compounds. In semi-conductors, the electrons energy is quantified in energy levels that gather and fall in two bands. The lowest energy band, the so-called Valence band, gathers electrons which participate in interatomic links. The highest energy band, the so-called Conduction band, gathers electrons that can participate in the electronic conduction. Those bands are separated by an energy value called the Gap in semi-conductors. This Gap, usually expressed in eV, varies from 0.6 to 1.5 eV depending on the compound. If now we consider interactions between light (photon) and semi-conductors, then three processes can occur. Those energy exchanges between photons and electrons are described in figure 1 and are possible only if the considered photon energy is equal to the Gap. This is resumed in the well-known relation between the photon wavelength λ and the Gap [eV]: λ [μm] = 1.24 / Gap [eV].

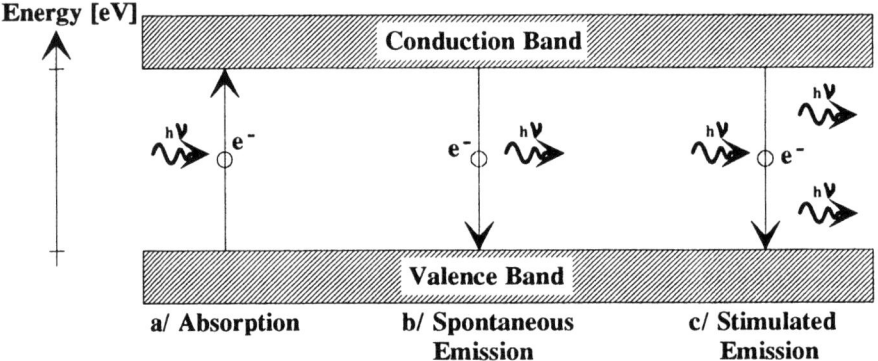

Figure 1. Transition Processes in Semi-Conductors
a/ Absorption, b/ Spontaneous Emission, c/ Stimulated Emission

The first process, the <u>Absorption</u> (fig. 1a) occurs when an electron from the Valence band is excited by an incoming photon; this electron then jumps to the Conduction band and the photon energy is absorbed. The second one, the <u>Spontaneous emission</u> (fig.1b), is a spontaneous radiative transition of an electron from the Conduction band to the Valence band; a photon which energy is equal to the gap is then emitted. Photons produced by spontaneous emission are not correlated and generate uncoherent light. The third one, the <u>Stimulated emission</u> (fig.1c), is also a radiative transition of an electron from the Conduction Band to the Valence band, **but** this transition is stimulated by an incoming photon; the emitted and incoming photons thus have the **same pulsation and phase** and constitute the so-called **coherent light**. This stimulated emission, a kind of a "negative absorption", can occur **only if** the number of electrons in the Conduction band is **greater** than the one in the Valence band, this situation is called the **population inversion** and is the basis principle of any laser. To realize a light source, we are obviously interested in radiative processes, either the spontaneous or stimulated ones. Let's see what types of semi-conductor compounds could match such radiative processes.

2.2. SEMI-CONDUCTOR COMPOUNDS [2] [3]

Semi-conductors are characterized by a periodic crystalline structure with local dislocations. This periodicity involves that each electronic energy state is associated with a wave vector in compliance with the vibrational properties of the crystal. If the last occupied energy level in the Valence band and the first free energy level in the Conduction band are associated with a zero wave vector, then the semi-conductor is said **direct**, in any other case it is said **indirect**. In indirect semi-conductors an electronic transition between the Conduction and Valence bands is coupled, because of the non zero wave vector, to a vibrational state of the crystal. In this case the energy dissipates unfortunately in heat and not in light. So a radiative transition may occur only in **direct semi-conductors**. This is the first important condition to match to select available semi-conductors. The second condition is obviously to find compounds which gap is close to the range of photon energy that interests us. That is to say gaps from 0.75 to 1.6 eV to cover the wavelength range from 0.8 μm to 1.6 μm. The very well known silicium and germanium semi-conductors have some gaps that comply with this wavelength range, unfortunately they are indirect semi-conductors, so that radiative transitions are not possible with them. Fortunately some direct semi-conductors exist in this wavelength range such as GaAs or InP, for instance. Different alloyings are also usable for emitting components: AlGaAs, InGaAs or InGaAsP, for instance. Some of those semi-conductors compounds are resumed in figure 2, that shows the fluorescence of those compounds as a function of wavelength and points to the right semi-conductor choice for the right source wavelength.

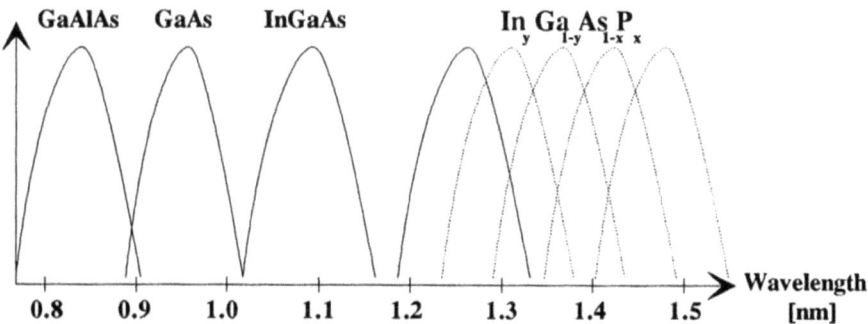

Figure 2. Available Semi-Conductor Compounds to design emitting components

As far as semi-conductors compounds are found, the next step is to design some structures that will enhance the emission processes previously demonstrated.

2.3. EMITTING COMPONENTS STRUCTURES

2.3.1. *Basis structures* [3] [4]

The common basis structure of an emitting component is a diode p-n junction. Historically, the first and simplest structure is the **homo-junction** structure in which the p-type and n-type areas are obtained by introducing appropriate impurities, the so-called dopants, in the basis crystalline structure of only one semi-conductor. A schematic of this basis structure is shown in figure 3. Under voltage condition, a depletion region is generated at the junction. The electrons and holes then drift, under electric field action, in opposite directions across the depletion region. According to the Gap energy, radiative recombinations between electrons and holes can occur in this depletion region.

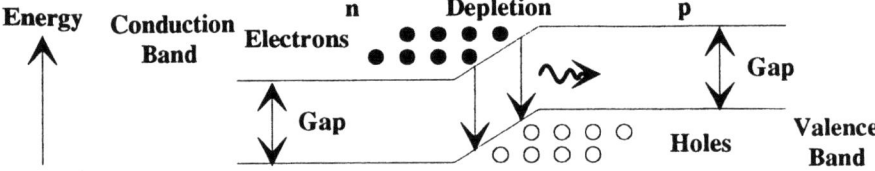

Figure 3. Homo-Junction Structure

Due to the highest mobility of carriers in the depletion region than in the diffusion region, carriers are not confined enough in the depletion region. Then the current-light conversion is not efficient enough. To increase the carriers confinement in the depletion region, the first improvement is to realize the p-n junction with p-type and n-type semi-conductors whose gaps are different, the so-called **Single Hetero-junction** is shown in figure 4a. In that structure the holes confinement is very efficient, but electrons can diffuse from the depletion to the p-type region. To increase that confinement, the second improvement is to add at that p-type region another p-type region which gap is greater. This p-p hetero-junction creates a barrier that electrons can not cross. The so-called **Double Hetero-junction**, is shown in figure 4b. Such a structure confines carriers in the middle p-type region so that the population inversion easily occurs. This active p-layer is sandwiched between two compounds whose gap **and** refractive index are higher. That involves two very important advantages: the emitted light can not be absorbed by the compounds and is confined in the active p-layer. For all those advantages, this structure is the most efficient to be used for light emission.

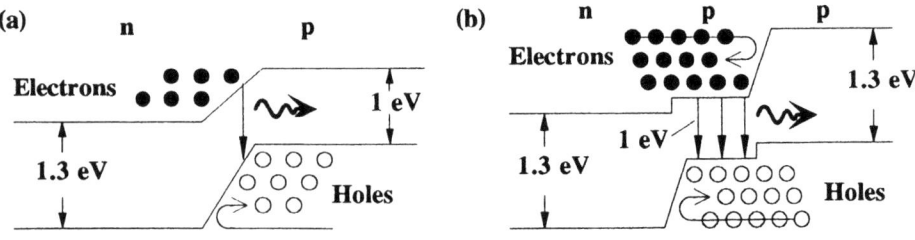

Figure 4. Hetero-Junction Structure a/ Single Hetero, b/ Double Hetero

Those structures and especially the Double Hetero-junction one are used to manufacture different components depending on the type of light you need. If the need is for uncoherent light then Light Emission Diodes (LED) have to be chosen, if the need if for coherent light then Laser Diodes are chosen. The semi-conductors associations were presented previously, let's see now how those components are realized.

2.3.2. *Light emitting diodes (LED) structure*

The LED structure depends on the practical way the current is launched in the junction. Considering the semi-conductor structure is a Double hetero-junction one for instance, the most obvious solution is to realize a sandwich of n-p-p types compounds. Then this multilayer structure is enclosed between two layers (n-type and p-type respectively) which purpose is to establish an electric contact in order to polarize the junction. The shape of those contact, and more commonly the shape of the p-contact determines the shape of the active layer emitting area. That is to say the way and the direction the light is emitted. The most common shapes for that p-contact are the **disc** and the **ribbon**. If the p-contact is a disc, the structure yields a so-called **Surface LED** (SLED). If the p-contact is a ribbon, a so-called **Edge LED** (ELED) is obtained. The typical structure of a Surface LED [5] is shown in figure 5.

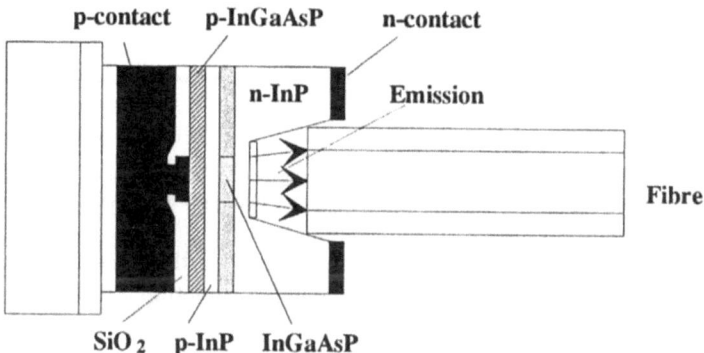

Figure 5. Structure of a Surface Light Emitting Diode (the Burrus-type Example)

This SLED is based on a double hetero-junction structure [6]. The InGaAsP layer surrounded by n-InP and p-InP layers is the light emitting layer (active layer). The n-InP and InGaAsP association constitutes the p-n hetero-junction of the diode. The InGaAsP and p-InP association stands for the hetero-isolation junction. The n and p contacts are achieved by metallisation (black layers in fig. 5). The p-InGaAsP layer is a buffer between the p-contact and the p-InP "cladding" layer. The p-contact area is isolated from the structure with a SiO_2 dielectric, except for a small disc which is directly in contact with the structure. The carriers are then launched from that disc, so that only a small part of the active layer really emits light. This light is emitted through the n-InP layer which is the surface layer of this structure. That's the reason for this component name: the "surface light-emitting diode". This SLED emits an uncoherent light in all angular directions, the so-called Lambertian source. That yields that the coupling ratio of this light in a fibre is very bad. To improve that, one can sink a well in the InP substrate to bring the fibre as closer as possible to the light-emitting area. This is called the Burrus structure as shown in figure 5.

The other solution is to shape a ribbon p-contact [7]. In that case the light-emitting area of the active layer is also a ribbon and constitutes a waveguide. Such a structure also called the gain guidance structure then emits light by its edge-face. This is shown in figure 6a and yields the so-called Edge LED (this stands for edge light-emitting diode).

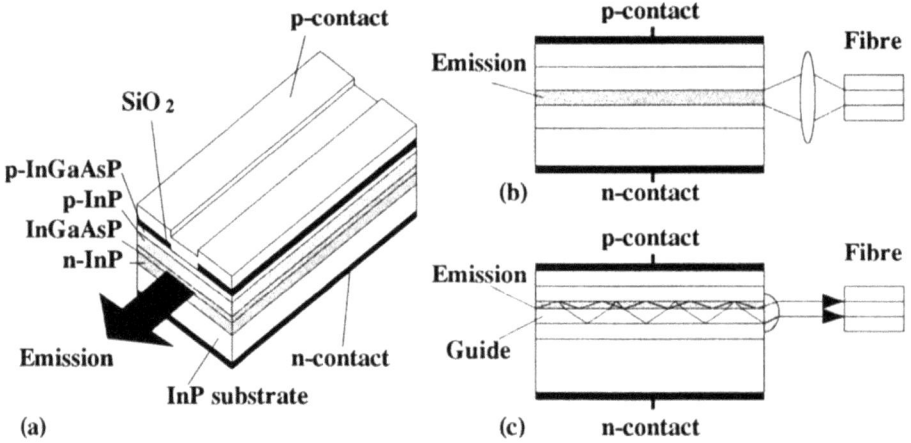

Figure 6. Structure of an Edge Light Emitting Diode: a/ the ribbon structure, b/ Fibre-light coupling with a lens, c/ Fibre-Light coupling with a hemispherical ball

In this Edge LED example, the structure is exactly the same than the SLED one, that is to say a n-InP / InGaAsP / p-InP double hetero-junction structure, except for the p-contact which non-isolated part is a ribbon. The light-emission then occurs under the form of a "pencil of light". The coupling ratio between light and the fibre is then improved. A lens (fig. 6b) or a hemispherical ball (fig. 6c) can be used in addition to increase this coupling ratio. In both examples, the active layer compound determines the component wavelength which is in the InGaAsP case around 1.31 or 1.55 μm. Due to its highest light-coupling ratio an Edge LED is used in single-mode fibre rather than a Surface LED. These are more often used for multimode fibres.

2.3.3. Laser diodes (LD) structure

The simplest structure for a laser is exactly the same than the Edge LED one, that is to say the gain guidance ribbon structure. The gain structure is placed in a Fabry-Pérot cavity [8] [9], between two plane mirrors for instance. At each return the light is amplified by stimulated emission. The population inversion is there efficient at the p-n junction. If the amplification gain is higher than the loss due to mirrors, the oscillation takes place: a semi-conductor laser is born. Such a laser obtains higher gain than a gaz laser, due to the higher density of atoms available within the cavity. Therefore such a laser can oscillate with a short cavity and low mirror reflectivity of 30 % (instead of 99 % for an He-Ne laser). Such low reflectivity mirrors can easily be obtained by simply cleaving both edge-faces of the crystal [10]. The typical structure of a Fabry-Pérot ribbon laser is described in figure 7a.

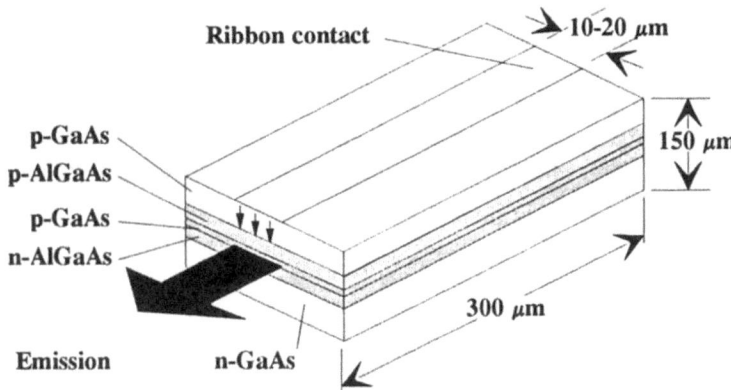

Figure 7a. Structure of a Fabry-Pérot Ribbon Laser Diode

In this instance, the laser structure is a ribbon one based on a double heterojunction diode. The p-GaAs compound, surrounded by p-type and n-type AlGaAs cladding layers, constitutes the light-emission layer. The ribbon p-contact limits the light emission in that layer to a ribbon as well. This ribbon is the part of the structure where laser oscillation takes place. Due to the gap of GaAs that is close to 1.4 eV, such a laser can oscillates at a wavelength of 0.85 µm. The laser oscillation consists in a stationnary wave in the cavity length [10]. This wave falls in a longitudinal mode and a transverse mode. The longitudinal mode expresses the quantifying condition in the cavity length direction. The transverse one expresses the quantifying condition in the axis perpendicular to the cavity length direction. This transverse mode falls in a perpendicular transverse mode which is perpendicular to the active layer and in a parallel transverse mode which is parallel to the active layer. A schematic of those ribbon laser modes is given in figure 7b.

Longitudinal modes:[10] (see fig.7b) in the cavity length axis the resonance condition involves that the stationnary wave half-wavelength $\lambda/2$ must be a dividor of the cavity length L. The number of half-wavelength included in the cavity length is the number of longitudinal modes and is equal to $2nL/\lambda$, where n is the refractive index of the active layer. For the previous GaAs laser, where λ is 0.85 µm, n is 3.5 and L is 300 µm, the cavity itself allows about 2500 longitudinal modes separated by 0.34 nm. Now this cavity function is to be multiplied by the spectral gain of the amplifying medium. Theoretically, this function is a Dirac due to the relation between the gap and the wavelength. Practically, due to thermal effects and degenerated energy levels, this spectral gain function has a given spectral spread. Therefore, within this spread a few longitudinal modes can oscillate at the same time. In Ribbon lasers, the number of longitudinal modes is about ten, depending on the wavelength.

Perpendicular transverse mode: [10] (see fig.7b) to confine carriers efficiently, the active layer thickness must be small, typically less than 0.5 μm. As far as this thickness is close to the wavelength, the light is diffracted. Therefore the laser beam divergence in the transverse axis can be as great as 60 degrees for a 0.3 μm thickness, or 20 degrees for a 0.05 μm thickness. Considering the laser beam far field pattern, its half-width angle $\theta_\perp$ typically varies from 30 to 60 degrees.

Parallel transverse mode: [10] (see fig.7b) in simple ribbon laser (gain guidance) the light is guided in the axis perpendicular to the active layer but unfortunately not in the parallel axis. The only limitation is therefore the width of the ribbon active area, which is controlled by the ribbon-shaped p-contact. The far field pattern half-width angle $\theta_\parallel$ is typically around 10 degrees.

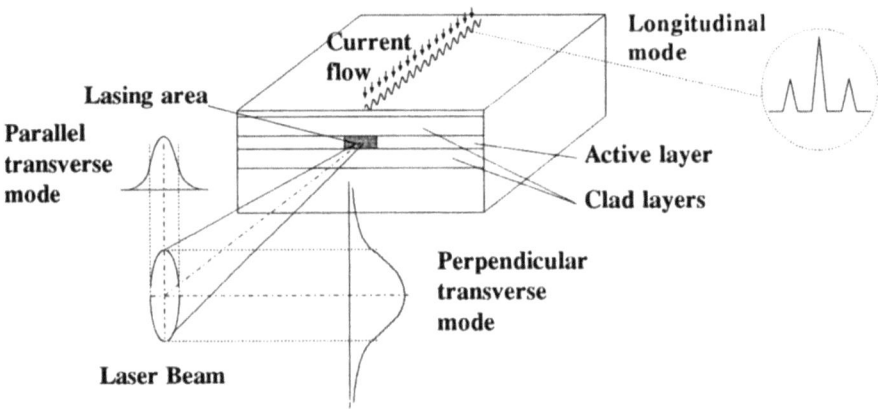

Figure 7b. Lasing Modes of a Fabry-Pérot Ribbon Laser Diode

To summarize on basis Ribbon lasers, two things can be said. First, its transverse mode divergence is wide and not very well controlled especially for the parallel transverse mode. Therefore, the fundamental transverse mode is not stabilized effectively. Second, it is not single-longitudinal mode. The final result is then a multi-transverse and multi-longitudinal mode laser.

Single-transverse mode laser: The divergence problem becomes critical while trying to launched the laser beam in a single-mode fibre. As far as the light is guided in the perpendicular axis, the idea is to guide it also in the parallel axis. Two solutions exists to realize such a guidance. The first one [11] is to build one or two guiding channels in the structure substrate. This is realized by surrounding the active area by a smallest refractive index compound. This yields the so-called Channeled Substrate Planar (CSP) lasers or the more widely used **Double Channel Planar (DCP) lasers**. If now, this cladding compound is of opposite polarity than the active area compound one, a hetero-isolation junction is realized in the parallel direction. This yields [12] the so-called **Buried Hetero-structure (BH) lasers**. Then in both the perpendicular and parallel directions, the double-heterostructure is achieved. The **light** is thus Index-Guided in both directions and the **carriers** as well are confined in the buried active area in both directions. Both effects improve the current-light conversion efficiency and therefore decrease the threshold current at which the stimulated emission starts. Such structures that realize lasers with a stable single-transverse mode are often combined to realize the famous and widely used **Double Channel Planar Buried Hetero (DCPBH)** structure [13] described in figure 8. In this structure, the first double hetero-junction is made of InGaAsP (active layer) surrounded by p-type and n-type InGaAsP. The InGaAsP active layer is also surrounded in the parallel direction by two channels of p-type InP, this association constitutes the double hetero-isolation that confines carriers and light.

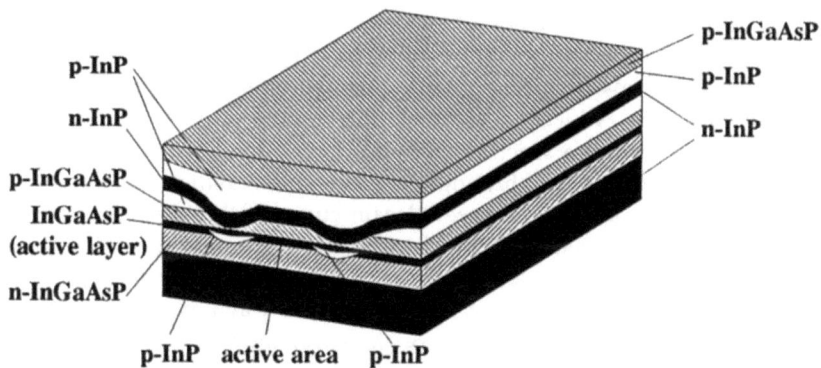

Figure 8. Schematic of a Double Channel Planar Buried-Hetero (DCPBH)-structure Laser Diode

Single longitudinal mode laser: In long distance transmissions, the optical fibre dispersion is the limiting factor. To reduce this dispersion, one must use sources whose spectral width is narrower and narrower. The theoretical solution is to select only one longitudinal mode in the cavity. Many practical structures exist that can select the right wavelength. The most widely used is the **Distributed Feed Back** structure [14] described in figure 9. The wavelength selection is achieved by a Bragg reflector built inside one cladding layer.

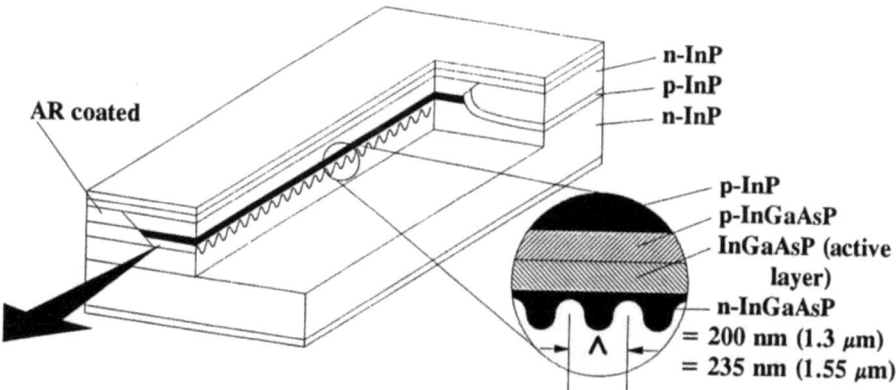

Figure 9. Structure of a Distributed Feed Back (DFB) Laser Diode

This DFB laser is based on a Double Channel Planar Buried Hetero structure in p-InGaAsP/InGaAsP/n-InGaAsP for the double heterojunction and InP/ InGaAsP/InP for the double hetero-isolation. The Bragg reflector with the adequate period is here built in the n-InGaAsP cladding layer, so in the laser structure itself.

The DFB laser is the most widely used, however some other ways to realize single-longitudinal mode lasers exist. The famous commercially available one (after DFBs) is the **Distributed Bragg Reflector (DBR)** laser in which the Bragg reflector is not inside but under the laser structure itself. That way, the Bragg reflector length can be increased without increasing the laser cavity length itself (which would act in the wrong way by increasing the number of modes). Such a longer "grating" is more selective and has a greater finesse, therefore only one longitudinal mode is selected and its spectral width is narrower than the DFB one. Another actual structure is the **External Cavity laser** in which an external mirror creates a second Fabry-Pérot longest cavity. This cavity has a spectral spacing between longitudinal modes smaller than the laser one, therefore the beating between both cavities allows only one resonance of longitudinal mode in the spectral gain curve of the semi-conductor. The advantage of this structure is that adequate changes of the external cavity length allow to tune the laser wavelength due to changes in the beating period of both cavities. Other solutions do exit, such as Grooved or Cleaved Coupled Cavities, External Grating, Injection Locked and Short Cavity for instance, however they are rarely tailored.

Advanced technology: A new type of non-vertical structure is under research nowadays. This structure is based on Multiple Quantum Wells (MQW) technique [15]. In classical three-dimensionnal semi-conductor , an **electron-hole pair** can form a bound-state in analogy with the Rydberg hydrogen atom, the **so-called exciton** with its associated Bohr radius. This exciton has a lower energy than the thermic agitation (**phonon**) one, in GaAs for instance 5 meV versus 25 meV, so excitons are very fragile and easily broken apart by phonons. As we want to create a laser, we have to achieve a good population inversion, that is to say to create as many **lasting excitons** as possible. If now some excitons are lost in phonons, the population inversion dynamically and drastically decreases. This involves an increase of the threshold current due to the lost of pumping energy and an increase of the laser risetime while generating those excitons. The need for faster and more powerfull lasers was demonstrated previously, however the limits of classical semi-conductor structures are now reached due to this excitons energy that is so low. How can this exciton energy be increased and if possible upper than the phonon one, so that excitons are captured and fully available for radiative transitions ? the Quantum Theory and the impressive Schrödinger equation give a possible answer to that question. If you realize by any mean a potential well (energy [eV] versus distance [nm]) which potential difference is much more greater than the phonons energy, and if the well thickness is smaller than the compound Bohr radius (28 nm in GaAs), then the excitons modify their structure to an elliptic one. In addition, their energy quantifies in few levels and the binding energy increases by a factor two or three. The electrons and holes are then confined (not to say compressed !) in their respective wells in both Conduction and Valence bands, and excitons are much less sensitive to phonons. A practical way to do it is for instance [16] a repetition of

$Ga_{0.7}Al_{0.3}As/$ GaAs layers. Their gap energy difference (respectively 1.81 eV and 1.42 eV) produces a 260 meV well in the Conduction band and a 130 meV well in the Valence band. For a layer thickness of 10 nm, electrons energy levels are $E_n = 50.n^2$ meV (50 meV, 200 meV, ...) and holes energy levels are $H_n = 5.n^2$ meV (5 meV, 20 meV, ...). It's easy to see that in both bands, electrons and holes are captured due to the fact their first and second energy levels keep smaller than the wells depth. Two excitons are then created (instead of one in bulk GaAs). The excitons confinement is so enhanced that it produces two "population inversion peaks" at each gap steps. Those peaks are fully resolved in the first step and are located around 1.46 eV (0.85 μm) and 1.48 eV (0.84 μm). This capture is a actual physical one, the carriers have effectively lost one degree of freedom in the direction normal to the layers. Due to that, such a MQW structure is also called a two-dimensional structure. Practical MQW structures are achieved successfully now at room temperature using a hundred periods of $Ga_{0.7}Al_{0.3}As/$ GaAs layers. Used as fast-recovery saturable absorbers, such structures make stable and reliable mode-locking laser diodes. Such diodes emitted pulse at 0.85 μm as short as 1.6 ps [17]. For the infrared range (1.3 μm and 1.55 μm), available semi-conductors compounds don't have very well-matched lattice parameters. Due to that, to grow a MQW structure is a bit more delicate. However, MQW lasers with power as high as 80 mW and linewidth as narrow as 500 MHz were demonstrated recently [18], in MQW structures of tensile strained InGaAsP-InP around 1.3 μm.

3. Emitting components characterization [19] [20]

3.1. LIGHT EMITTING DIODES

3.1.1. *Optical Power vs. Current characteristic P(i)*
The theoretical optical power vs. current characteristic of a LED is a line, but due to the LED possible heating when the current increases, this curve turns non-linear. A typical example of a LED characteristic P(i) is given in figure 10.

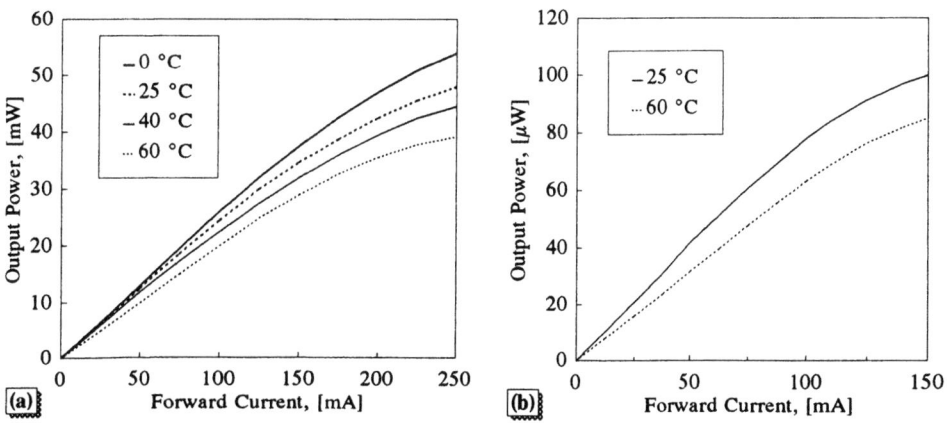

Figure 10. Optical Power vs. Current Response of LEDs: a/ without fibre, b/ with a 50/125 μm fibre

This LED is a Surface one based on a double hetero-junction structure of GaAlAs. The figure 10a shows the characteristic of the bare diode, the figure 10b shows the characteristic of the diode pigtailed with a 50/125 μm graded-index multimode fibre. This comparison between the diode characteristics with and without a fibre demonstrates the low coupling efficiency of LEDs optical power in fibre. In this instance, for a forward current of 100 mA, the coupling loss is about 25 dB. Those characteristics are also plotted as a function of the room temperature to show their high temperature dependences.

3.1.2. *Optical Spectrum*

Due to the spontaneous nature of their light-emission, LEDs show continuous and broad-band spectra. Depending on the semi-conductor compounds that is used and therefore on the emitted wavelength, the spectral half-width is ranging from 40 nm to 150 nm. LEDs emitting at a wavelength around 0.85 μm, usually based on an AlGaAs compound, have a typical spectral width of 50 nm. Those emitting around 1.3 μm are based on InGaAsP compound, and due to this compound highest gap dispersion, their spectral width is two or three time broader (typically 100 nm to 150 nm). This is illustrated in figure 11 where the relative optical spectra of two diodes are plotted. The figure 11a shows an AlGaAs - 0.85 μm one which spectral width is about 48 nm. The figure 11b shows an InGaAsP - 1.31 μm one which spectral width is about 112 nm.

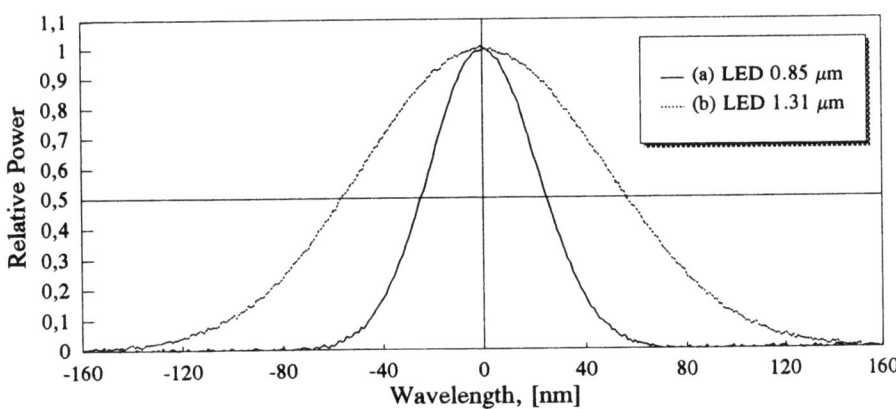

Figure 11. Optical Spectra of two LEDs: a/ an AlGaAs 0.85 μm one, b/ an InGaAsP 1.3 μm one

3.1.3. *Far Field Pattern P(θ)*

As it was demonstrated in the LED technology paragraph (§ 2.3.2.) the Far Field Pattern depends on the LED structure. Considering **bare** LEDs, that is to say without any fibre or optics, the far field pattern falls in two typical types. The surface emitting LEDs are Lambertian, therefore their optical power as a function of the angle, the so-called far field pattern P(θ), is given by the cosine of this angle. A SLED Lambertian far field pattern is shown in figure 12a, inwhich it can be seen that the half-power angle θ½ is 120 degrees. Now, the edge emitting LEDs are Lambertian only in the plane parallel to the active layer. Therefore their far field pattern falls in a parallel one and a perpendicular one. In the parallel plane, the ELED far field pattern is the same than the SLED one (fig.12a). In the perpendicular plane, the ELED far field pattern is much narrower and its half-power angle $\theta\frac{1}{2}$ is typically 30 degrees. The far field pattern of an AlGaAs Edge LED in the direction perpendicular to the layer is given in figure 12b.

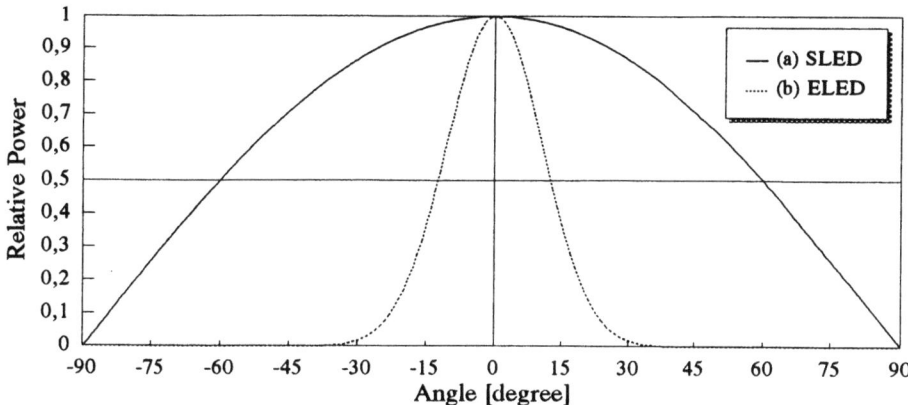

Figure 12. Far Field Patterns: a/ SLED or ELED in the // plane, b/ an ELED one in the ⊥ plane

Obviously those far field patterns are valid only for bare LEDs and must be corrected if focusing optics are added. If the LED is pigtailed with a fibre, the far field pattern is therefore the fibre one.

3.1.4. *Time Response*

The LED time response depends essentially on its structure and its dimension. The structure, whatever it is a homo-junction or single or double hetero-junction one, determines the carriers confinement efficiency and therefore the serial resistance of the p-n active junction. The more the confinement is efficient the faster is the time response. Therefore, double hetero-junction structure LEDs will be the fastest. The dimension of the active or emitting area determines the junction capacitance. The so-called RC-circuit thus constituted, limits the rise and fall times of the LEDs. Typical values of rise and fall times of double hetero-junction structure LEDs are ranging from 1.5 ns to 15 ns, more or less. InGaAsP LEDs are usually a bit faster than GaAlAs ones. Those time values yield bandwidth values about 100 Mhz. Some LEDs can be especially compensated for their time response (equalization) in order to increase their bandwidth up to values equal to 500 or 600 MHz at most.

3.2. LASER DIODES

3.2.1. *Optical Power vs. Current characteristic P(i)*

The typical optical power versus current characteristic of a Laser Diode consists in two different areas. The first area, while the current is increasing from zero to the threshold current, is the spontaneous emission characteristic. The second area, from the threshold current to upper currents, shows the lasing characteristic. An example of such characteristics is given in figure 13 for a double hetero-junction structure Fabry-Pérot Laser Diode. The lasing area of this laser consists in InGaAsP and emits at a wavelength of 1.31 μm. In this instance the laser is not pigtailed.

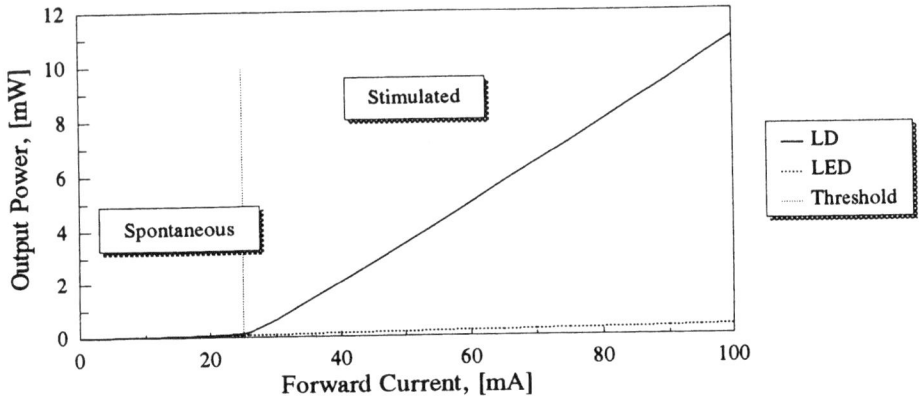

Figure 13. Typical Optical Power vs. Current Characteristic of a Laser Diode
(compared with the LED one in dotted line)

On the same figure, a LED light versus current characteristic is plotted in dotted line to compare it to the laser one. Due to small dimensions of the lasing area and to the low laser beam divergence, the light coupling in the fibre is very efficient. For instance the coupling loss is lower than 3 dB in a 50/125 μm multimode fibre and lower than 6 dB in a single-mode fibre. Therefore Fabry-Pérot LD pigtailed with a single-mode fibre easily reached power values of 2 or 3 mW, and for DFB LDs this power can reach 10 mW.

3.2.2. Optical Spectrum
A Laser Diode optical spectrum is related to the number of longitudinal modes contained in the gain curve of the LD active compound. The laser structure, whatever it is a Ribbon or a Buried or a Double Channel Planar one, determines the active compound spectral gain curve, therefore the **spectrum depends on the laser structure**. As it was demonstrated in the Laser Diode Structure paragraph (§ 2.3.3.), in a Fabry-Pérot cavity of length L and for an active compound of refractive index n, the number of modes that could oscillate is $N = 2nL/\lambda$, where λ is the wavelength in the vacuum. The spectral distance $\Delta\lambda$ between two longitudinal modes is given by: $\Delta\lambda = \lambda^2 / 2nL$, therefore, even for a given structure, the **spectrum depends also on the wavelength, on the cavity length and on the refractive index**. And finally, the number of modes **depends also on the cavity type**, whatever it is a Fabry-Pérot one or a selective one like in DFBs. It is easy to see that, due to such a great number of dependence parameters, there is one optical spectrum for each Laser Diode. In addition, due to thermal drift of the gain curve and so on, this fixed number of modes could jump to

another same number of modes which is located elsewhere in wavelength. Then, statistically, those jumping spectra are added and they produce an average spectrum that seems to contain more modes. The typical "worse" case, that is to say the spectrum that contains the greatest number of longitudinal modes, is when the simple Ribbon structure (also called Gain Guided) is used, and when the wavelength is the smallest, let's say 0.85 μm. Then , there is about 10 longitudinal modes in the spectrum and the distance between modes $\Delta\lambda$ is around 0.34 nm. Now for the same compound and structure, if you increase the wavelength the number of modes decreases, due to the increase of the distance between modes. For instance at 1.31 μm, this distance increases up to 0.82 nm and the number of modes decreases to 5. An instance of such a Ribbon GaAlAs laser emitting at 1.31 μm is shown in figure 14a. In that case the distance between modes is close to 0.75 nm and 6 or 7 longitudinal modes "seems" to oscillate. If we consider the eventual thermal drift (0. 3 nm / °C), the number of real modes is probably 4 or 5.

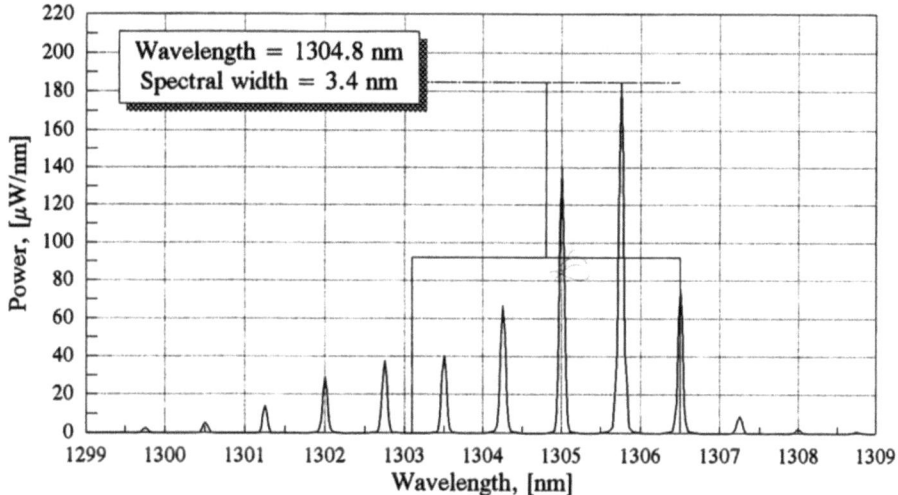

Figure 14a. Optical Spectrum of a Multiple Longitudinal Modes Laser Diode (a Gain-Guided Fabry-Pérot LD example)

The half-width $\Delta\lambda\frac{1}{2}$ of each longitudinal mode at a given wavelength, can be **theoretically** related to the spacing between modes $\Delta\lambda$ by: $\Delta\lambda\frac{1}{2} = \Delta\lambda / F$, where F is the finesse of the cavity. This finesse F only depends on the mirrors reflectivity coefficients r_1 and r_2 , and is given by:

$$F = \frac{\pi\sqrt{r_1 r_2}}{1-r_1 r_2} \quad (1)$$

In LD cavities, due to the mirrors low reflectivity coefficients, the finesse F is small (around 2), therefore the mode half-width can be as great as 0.4 nm. Fortunately, due to the stimulated emission process, this mode half-width is smaller than the cavity one, and is usually around 0.05 nm. But, the effective spectral width of the laser must take into account the multiple longitudinal modes. Due to that, the laser effective spectral width is about 4 nm at 1.31 µm and 8 nm at 1.55µm. To decrease the spectral width, one must decrease the number of modes, using for instance a Buried Heterostructure laser which usually shows 2 or even 1 longitudinal modes. However, due to the thermal drift, those modes jump and the BH laser spectrum thus contains about 2, 3 or 4 modes. If now a selective cavity is added (DFB or DBR lasers for instance), only one mode is allowed to oscillate. Therefore the laser spectral width is equal to the mode spectral width. In DFB lasers this spectral width can be as small as 0.04 nm in continuous wave operation. An instance of a DFB laser spectrum is given in figure 14b.

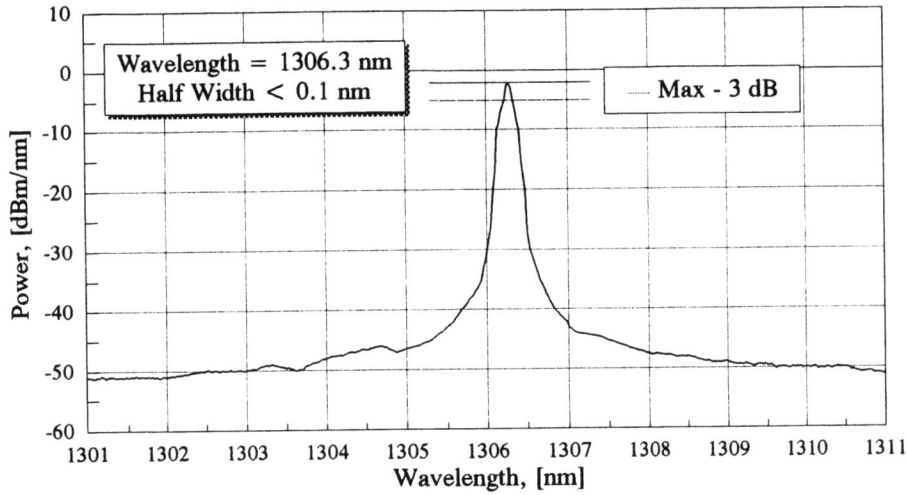

Figure 14b. Optical Spectrum of a Single Longitudinal Mode Laser Diode
(a DFB Laser Diode example)

With regard to single longitudinal mode laser, another parameter is often measured, that indicates the ratio between the highest spectral power (the single-mode peak) and the second highest spectral power (called the side-mode). The highest this parameter is, the best is the laser single-mode nature. This parameter, called the side mode suppression ratio (SMSR), is usually ranging from 30 to 40 dB.

3.2.3. *Far Field Pattern*
The bare LD typical far field pattern is the one of the perpendicular and parallel transverse mode. Now, it is already shown in figure 7b (§ 2.3.3.). Typical values for the half-power angles are $\theta_\perp$ = 35 degrees for the perpendicular tranverse mode and $\theta_{//}$ = 30 degrees for the parallel tranverse mode (in a DFB for instance). The LD lasing area dimension is usually smaller ($\approx$ 8 µm x 0.2 µm) than a single-mode fibre core one, therefore the light-coupling is easy to realize. As a consequence, available LDs are usually pigtailed with a fibre. In such a case the far field pattern is the fibre end one and to talk about the laser far field pattern doesn't make no sens.

3.2.4. Frequency Response

The frequency response, i.e. gain versus frequency curve, of a laser diode is a typical second-order one with a resonance frequency F_R. Such a frequency response is given in figure 15 for a Double Channel Planar Buried Hetero-structure DFB laser.

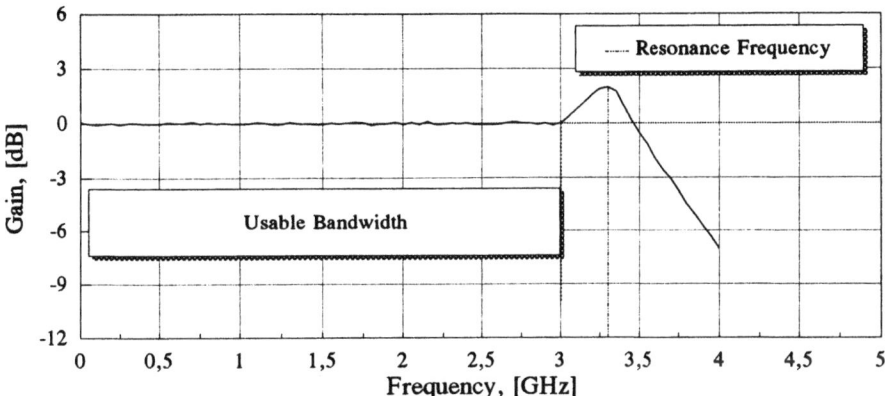

Figure 15. Frequency Response of a DCPBH-structure DFB Laser Diode
(for an output power of 5 mW)

The resonance frequency F_R is measured at the maximum gain value (in figure 15 that yields F_R = 3.3 GHz). However a LD frequency response depends on the polarization current I that is applied. Indeed, the resonance frequency F_R is related to the polarization current I by the formula:

$$F_R = \frac{1}{2\pi \tau_p \tau_s} \sqrt{\frac{I-I_{th}}{I_{th}}} \quad [Hz] \qquad (2)$$

where I_{th} is the threshold current and τ_p, τ_s are time constants proper to the laser structure itself. Therefore, it can be seen that the resonance frequency increases with the current and as a consequence that the all frequency response is shifted towards upper frequencies when the bias current increases. For the previous instance (fig.15), the frequency response is given for an output power of 5 mW, that corresponds to a bias current of 60 mA for this particular laser. Now, the threshold current is here 25 mA, close to the threshold current, let's say at 28 mA, the resonance frequency will be close to 1 GHz. If this laser has to be used at higher power, 10 mW let's say, the bias current must be about 80 mA and the corresponding resonance frequency is then about 4 GHz. Due to this dependence, an LD frequency response must be plotted for a given and mentioned bias current (or output power).

For linearity considerations, the resonance overshoot must be avoided. The laser diodes usable frequency bandwidth is then the linear one extending at lower frequencies than the overshoot. This usable bandwidth is typically ranging from 500 MHz for Ribbon lasers to 10 GHz and more for best DCPBH lasers.

In addition with the frequency response is often measured the harmonic distortion of the laser diode, that is to say the frequency spectrum of this diode modulated by a pure sine function of frequency f. If the laser response spectrum shows several harmonics of order n (frequency = n.f), ratios between the fundamental harmonic (f) and those higher order harmonics indicate the laser ability to be modulated without distortion effects. The second harmonic suppression ratio is ranging from 30 dB (bad) for ribbon lasers to 60 dB (good) for DCPBH lasers. The third harmonic suppression ratio is ranging from 30 dB to 40 dB. A good harmonic suppression becomes crucial when the laser is modulated by a signal that contains several frequencies, that is to say any signal but the sine function. Indeed, due to a bad harmonic suppression, the signal initial components and the components generated by the distortion will beat in frequency, the so-called intermodulation. Therefore new components (frequencies sums and differences) may appear and as a consequence the signal distortion will get worse and worse.

3.3. LED / LD SUMMARY

The table 1, hereafter, summarizes on LED and LD fundamental characteristics and shows the type of fibre that can be used for a given emitting components. The given values are typical ones, therefore the commercially available components values may vary around those typical values in a 10 dB range at least.

TABLE 1. A LED/ LD Characteristics Summary

Source Type	Surface LED	Edge LED	Ribbon Laser	Buried Hetero-Laser	DCPBH-DFB Laser
Fibre Type	Multi	Single / Multi	Multi	Single / Multi	Single
Coupled Power (W)	30 μW	4 μW / 60 μW	2 mW	1 mW / 2 mW	> 2 mW
Wavelength (μm)	0.85 μm / 1.3 μm	1.3 μm	0.85 μm	1.3 μm / 1.55 μm	1.3 μm / 1.55 μm
Spectral width (nm)	50 nm	100 nm	10 nm	1 - 2 nm	< 0.1 nm
Number of modes	∞	∞	10	1 - 2	1
Rise Time (ns)	10 ns	5 ns	1 ns	0.2 ns	< 0.2 ns
Bandwidth (Hz)	50 MHz	100 MHz	500 MHz	2 GHz	> 2 GHz

4. Conclusion

In this article are treated the semi-conductor compounds and the ways they could be associated in order to realize light-emitting components usable for silica-based fibre. The way that the diode junction complexity did increase is demonstrated to be closely related to the effective need of more and more efficient active components. Therefore, the homo-junction, the hetero-junction and the double hetero-junction diodes were treated in increasing order of carriers confinement. It can be seen that the most efficient confinement is obtained with the double-heterojunction diodes, and that nowadays needs involve this type of junction in almost all light-emitting components.

Both light emission processes, spontaneous and stimulated, are demonstrated in order to distinguish both types of emitting components available today: the so-called light emitting diode (LED) and the laser diode (LD). Once again, those components structures are treated in increasing order of complexity, with the intention of demonstrating how those structures solve each encountered lack of efficiency. The way the current-light conversion efficiency and the light-fibre coupling ratio were enhanced is demonstrated with the use of structures such as ribbon hetero, buried hetero (BH), channeled substrate planar (CSP), double channel planar (DCP) and finally the most widely used one: the famous double channel planar buried hetero (DCPBH) structure. The way the number of longitudinal modes was reduced is demonstrated also with the use of distributed feed back (DFB) structures or distributed Bragg reflector (DBR) ones, just to name the famoust ones.

A short introduction to a present technology is done while treating the multiple quantum wells (MQW) two-dimensional structures. This technology is shown as the most likely one to be considered to realize more powerfull and faster semi-conductor lasers. As far as semi-conductor diodes seems to reach their ultimate possibilities in power and speed, those MQW structures allow to look at the future with optimism.

Finally, the most important LED and LD characteristics are detailed and typical figures as a function of the component structure are given. An important LD characteristic pointed today, the relative intensity noise (RIN) is unfortunately not treated for the simple reason that one laser RIN becomes preponderant for high power levels (>5mW) and that the LCIE is supposed to be limited to the radiometric domain, and therefore limited to power lower than 1 or 2 mW, let's say.

5. References

1. Bertein F., (1969), *Electronique quantique*, tomes 1 & 2, Eyrolles, Paris.
2. Lasher G. & Stern F., (1964), *Spontaneous and stimulated recombination radiation in semiconductors*, Phys. Rev. , 133A.
3. Sze S.M., (1969), *Physics of semiconductor devices*, J. Willey international.
4. Kressel H. & Rutler J.K., (1977), *Semiconductor lasers and heterojunction LEDs*, Quantum electronics, Academic Press, New York
5. Burrus C. A. & Dawson R. W. , (1970), *Small-area, high-current density GaAs electroluminescent diodes...*, Appl. Phys. Letters, Vol. 17, p. 97.
6. Lee T. P. ,Burrus C. A. & Miller B. I., (1973), *A stripe-geometry double-heterostructure amplified-spontaneous-emission diode*, IEEE J. Quantum Electron., Vol. 9, p. 820.

7. Nagai H. & Noguchi Y., (1977), *InP/GaInAsP double heterostructure LEDs in the 1.5 µm wavelength region*, Integrated Optics and Optical Communication.
8. Siegman A. E., (1986), *Lasers*, University Science Book, Mill Valley, chapter 19.
9. Pérez J. P., (1988), *Optique géométrique et ondulatoire*, 2ᵉ édition, Masson, Paris, pp. 264-276.
10. Cozannet A., Fleuret J., Maître H. & Rousseau M., (1981), *Optique et télécommunications*, Eyrolles, Paris, pp. 305-312.
11. Aiki K., Nakamura M., Kuroda T. & Umeda J., (1977), *Channeld-Substrate Planar structure injection lasers*, Appl. Phys. Letters, Vol. 30, p 649.
12. Yoon S. F., (01-1994), *Observation of degradation recovery in 1.3 µm GaInAsP-InP inverted-Rib semiconductor lasers*, Journal of Lightwave Technology, Vol. 12, N° 1, pp. 55-58.
13. Valster A., Meuleman L. J., Kuindersma P. I. & Van Dongen T., (01-1986), *Improved high frequency response of InGaAsP Double Channel Buried Hetero-structure lasers*, Electronics Letters, Vol. 22, N° 2, pp. 16-18.
14. Tjassens H. & Kluitmans J. T. M., (1988), *A laser module for 4-Gbit/s optical communications*, Philips Tech. Rev., Vol. 44, N° 5.
15. ENSSAT (1990), *Cours de mécanique quantique - Propriétés des puits quantiques*.
16. Chemla D. S., (05-1985), *Quantum wells for photonics*, Physics Today.
17. Silverberg Y., Smith P. W., Eilenberg D. J., Miller D. A. B., Gossard A. C. & Wiegmann W., (1984), Optics Letters, Vol. 9, p. 507.
18. Thijs P. J. A., Van Dongen T., Tiemeijer L. F. & Binsma J. J. M., (01-1994), *High-performance $\lambda = 1.3$ µm InGaAsP-InP Strained-Layer Quantum Well Lasers*, Journal of Lightwave Technology, Vol. 12, N° 1, pp. 28-37.
19. LCIE (1993-1994), *Tests reports & certificates of calibration*.
20. Alcatel, Hitachi, Siemens, Thomson, (1994), *Optodevices Data Books & Products Specifications*.

V Optical Fibre Amplifiers

OPTICAL AMPLIFIERS

Ivan Andonovic
University Of Strathclyde
Department Of Electronic & Electrical Engineering
204 George Street
Glasgow, Scotland

ABSTRACT

The optical amplifier has made a significant impact on the design and implementation of photonic networks. This chapter provides a summary of the rapid development of a number of optical amplifier geometries; rare earth doped and non-linear fibre and semiconductor laser amplifiers. Their theory of operation and performance will be discussed; comparisons will be drawn between each type; and their application areas will be highlighted.

1. Introduction

The most significant development in the last few years in the area of lightwave systems has been the optical amplifier. These amplifiers, capable of operating in the low loss transmission regions of optical fibres, can perform a variety of simple functions without optoelectronic conversion and subsequent electronic amplification. Applied as in-line amplifiers, boosters or pre-amplifiers, capable of operating independently of data rate, and supporting, simultaneously, a number of different wavelengths, they have revolutionised network design concepts. The fundamental difference in operation compared to the conventional electronic 3R approach (reshaping, retiming, reclocking), has permitted the full exploitation of the advantages of utilising optical fibre as well as allowing a number of novel network designs to be developed with concomitant benefits with respect to network reliability and functionality. Not only have they had a significant impact in long haul submarine and terrestrial systems to achieve transmission rates at gigabits per second over thousands of kilometers, but also in many multiaccess and sensor networks in which attenuation is caused by multiple tapping or by distribution of the same signal to many points[1,2].

In general terms there are three main ways of utilising optical amplifiers (FIG 1) and all are concerned with achieving photonic systems with a large loss capability :

> power booster which amplifies the laser output power. This cannot be done indefinitely since fundamental limits to the input launch power are set by non-linear optical effects such as self-stimulated Raman and Brillouin scattering[3,4]. However the power levels which are achievable with present day high performance laser diode sources together with an increase in the use of external integrated optical modulators, has meant that this is an important

function which increases the loss capability by boosting the input launch to just below the threshold for non-linear phenomena.

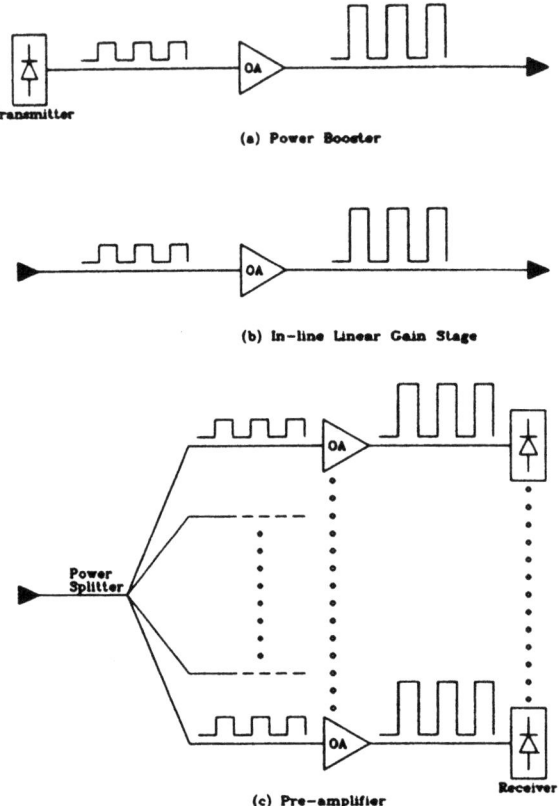

Fig.1 The three main optical amplifier application sectors

in-line linear gain stage which amplifies the signal periodically along the transmission path to compensate for fibre loss due to attenuation or splitting. Here the optical amplifier is a direct replacement for optoelectronic repeaters, resulting in a system which is effectively transparent, and future proof, since the optical amplifier functions independently of data rate or wavelength.

receiver pre-amplifier where it is used to linearly boost optical signals prior to the photodetector in an optical receiver to just below its sensitivity limit.

There are two main classes of optical amplifiers.
- semiconductor laser amplifiers
- optical fibre amplifiers which can be further subdivided into
    - rare earth doped fibre amplifiers
    - non-linear fibre amplifiers

All operate on the principle of single pass gain obtained by stimulated emission of photons from a population inversion in an optically active medium. This population inversion can be obtained either by electrical or optical pumping. The choice of amplifier is a complex question which is fundamentally driven by the application, the technological trade-offs and system design constraints. A number of key amplifier parameters must be considered : net gain, bandwidth, polarisation sensitivity, saturation power, noise figure. It is fair to say that over the past few years the most applied amplifier technologies have been the erbium doped fibre amplifiers (EDFA) and semiconductor amplifiers (SLA). The reason for this will become clear in later sections when a more thorough description of their operation and characteristics will be developed.

The answer can be in part given by noting that the main signal wavelengths for optical photonic systems are in the region of 1300 nm and 1550 nm, corresponding to the two lowest loss windows in silica based fibres (FIG 2). Although the 1300 nm window is slightly higher loss than its longer wavelength counterpart, standard step index fibre exhibits zero dispersion around 1300 nm, the wavelength most commonly used for installed systems to date. However the lower loss at 1550 nm and an increase in the use of dispersion shifted optical fibres - with its dispersion zero shifted to around the longer wavelength - has made this wavelength attractive for the future since, ultimately, higher data rates and longer transmission spans are feasible. Practical doped fibre amplifiers (EDFA) operate around the 1.5 µm wavelength, although alternatives which operate at 1.3 µm are being developed, and will be discussed later, whilst semiconductor amplifiers can operate easily at both 1.3 µm and 1.5 µm.

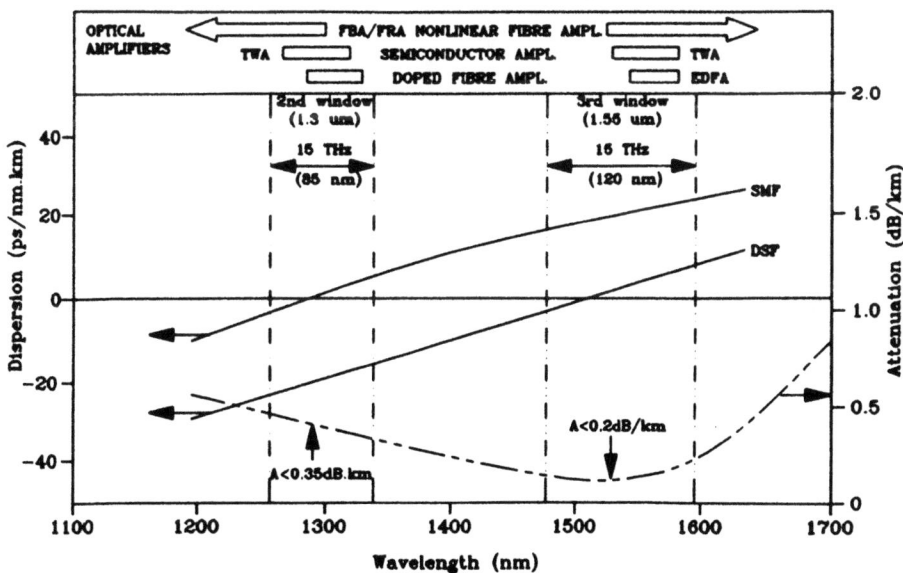

Fig.2 Dispersion and attenuation characteristics of conventional single mode fibre and dispersion shifted fibre.

The aim of this Chapter is to provide a summary of the current state of optical amplifier technology. Semiconductor amplifiers will be discussed firstly, presenting the basic theory of operation and performance. A similar description of rare earth doped fibre amplifiers will follow. Being the two most important amplifier geometries, this will form the bulk of the Chapter. Other alternatives viz non-linear fibre, will also be discussed and comparisons will be drawn between each alternative. The applications of optical amplifiers will be highlighted throughout the Chapter. A small section will be included which will describe the latest developments in the rare earth doped area.

2. Semiconductor Laser Amplifiers

2.1 Introduction

Semiconductor laser amplifiers (FIG 3)[5,6] are basically standard semiconductor laser structures. As such there are numerous advantages which result and are well documented; small power consumption and size - which in turn allows monolithic integration with other optical circuits to realise more complex functions on one chip, so called optoelectronic integrated circuits, - permits a number of optical processing tasks, in addition to amplification, to be implemented. Although their single mode waveguide structure is compatible with single mode optical fibre geometries, there is still the problem of low loss interfacing to optical fibre.

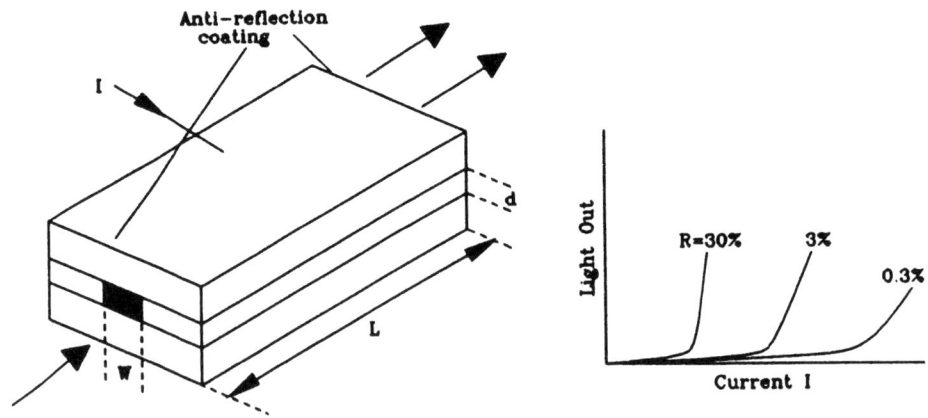

Fig.3   Schematic of a semiconductor laser amplifier (SLA)

The fundamental differences in operation between a laser and an amplifier is that (a) the normal facet reflectivities (~ 30%) have been reduced by the use of anti reflection coatings and (b) the device is operated just below the lasing threshold which creates a significant population inversion of light incident in one facet, stimulating the emission of photons (amplification) at the other with the addition of noise. There are two types of semiconductor amplifiers, the Fabry Perot (FP) amplifier and the travelling wave (TW) amplifier. The Fabry Perot is a classical resonant amplifier with facet reflectivities between 0.01-0.3, giving resonant internal gains of between[7] 25-30 dB. The highly resonant characteristics give rise to narrowband transmission characteristics (bandwidth ~ 6 GHz) and high internal fields. Hence these structures are mostly applicable in non-linear applications such as bistable operations. The TWA structure is identical to the FPA but has anti reflection coatings applied to reduce the reflectivity near to zero, thereby reducing internal feedback (the near travelling wave amplifier NTWA). The light that is amplified travels through the device only once. Consequently NTW amplifiers exhibit wide bandwidths (~75 GHz) and are capable of yielding near linear gain characteristics[8]. Their disadvantage is the added level of spontaneous noise. The NTW amplifier structure is more relevant to the remainder of the Chapter and the following discussions refer to this device, since it is the more attractive candidate for network applications.

2.2 Gain/Bandwidth Characteristics

Semiconductor materials can be designed to cover the wavelength range from 1.1 μm - 1.6 μm by proper tuning of the band gap through alteration of the percentages of alloys present in the material. Since amplification occurs in the active region, a knowledge of the factors governing the material gain is necessary to understand device performance. FIG 4 is a simple schematic of the factors influencing gain within the structure.

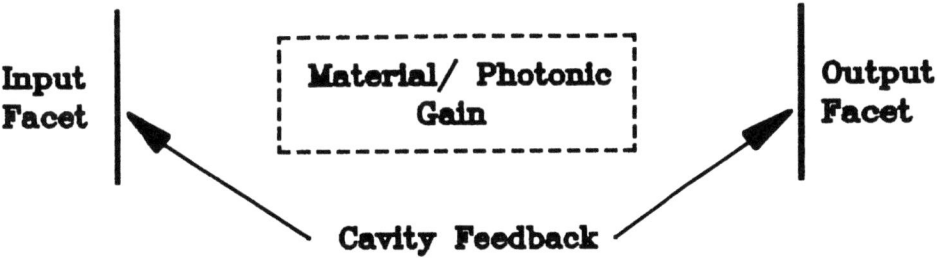

Fig.4   Factors influencing gain within the structure

The material gain coefficient per unit length g_m is given by[9]:

$$g_m = a_1 n \left[1 - \left(\frac{\lambda - \lambda_p}{\Delta} \right)^2 \right] - a_2 n_o \qquad (1)$$

where a = gain constants
n = carrier density
n_o = transparency density
λ_p = peak gain wavelength = $\lambda_o + a_3(n - n_o)$
Δ = gain curve bandwidth

Thus an increase in bias current (and hence carrier density) increases the peak material gain and decreases the peak wavelength, (FIG 5).

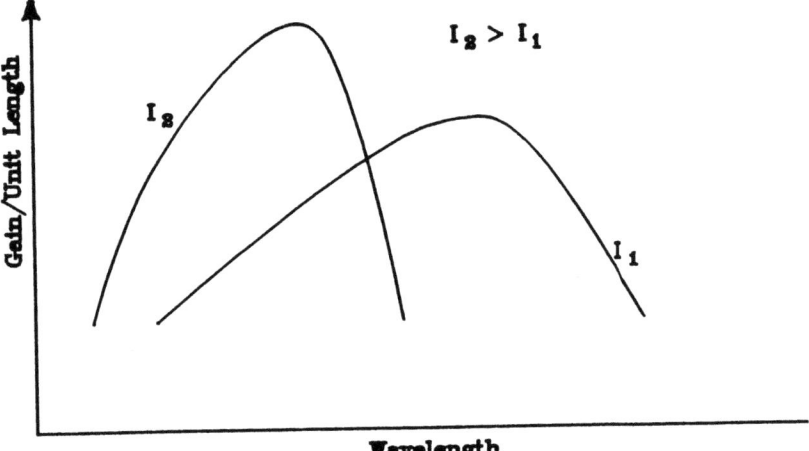

Fig.5   Gain/unit length as a function of wavelength for SLA's

The net gain per unit length is defined in terms of the material gain coefficient, the optical confinement factor Γ and the effective loss coefficient per unit length, α

$$g = \Gamma g_m - \alpha \qquad (2)$$

In the steady state, the carrier density is related to the optical intensity by[10]:

$$\frac{J}{ed} = \frac{n}{\tau} + \frac{\Gamma g_m}{E} I \qquad (3)$$

where   I = intensity = $\dfrac{dI(z)}{dz} = g(z)I(z)$

E = photon energy
τ = carrier lifetime
z = distance along amplifier length
J = injected current density
n = carrier density
g = net gain/unit length = $\Gamma g_m - \alpha$
L = amplifier length
e = electron charge

The material gain coefficient can be rewritten in terms of the averaged signal intensity and the expression for the single pass gain becomes[7,11,12,13]:

$$G_s = \exp[gL] = \exp[\Gamma g_m - \alpha]$$

$$= \exp\left[\left(\frac{\Gamma g_o}{1 + I/I_s} - \alpha\right)L\right] \quad (4)$$

where I_s = saturation intensity = $\frac{E}{\Gamma a \tau}$

g_o = unsaturated material gain coefficient in the absence of input signal, n_o being the corresponding carrier density

Therefore from the above equation, the single pass gain decreases with increasing intensity, ie once the saturation level is exceeded, the gain will decrease rapidly from its maximum (small signal) value. Since the signal power is too large, there is insufficient excited carriers to provide amplification.

The phase shift suffered on a single pass of the amplifier is[12]:

$$\phi_s = \phi_0 + \phi_b \quad (5)$$

where $\phi_0 = \frac{2\pi LN}{\lambda}$ = normal phase shift on transmission in a material of refractive index N

$\phi_b = \frac{g_o L b}{2}\left[\frac{I}{I+I_s}\right]$ = additional phase due to change in carrier density

b = line broadening factor

Although TWA should ideally exhibit zero reflectivity, in practice some residual reflectivity will occur giving rise to an optical cavity resulting in amplifier transmission characteristic with resonant peaks. Thus effectively, the optical cavity exhibits classical Fabry Perot modes whose wavelength and spacing depends on cavity length and reflectivity. FIG 6 shows typical gain curves for a Gaussian shaped single pass gain G_S, Fabry Perot amplifier (R~0.3) and near travelling wave amplifier (R ~ 0.02)

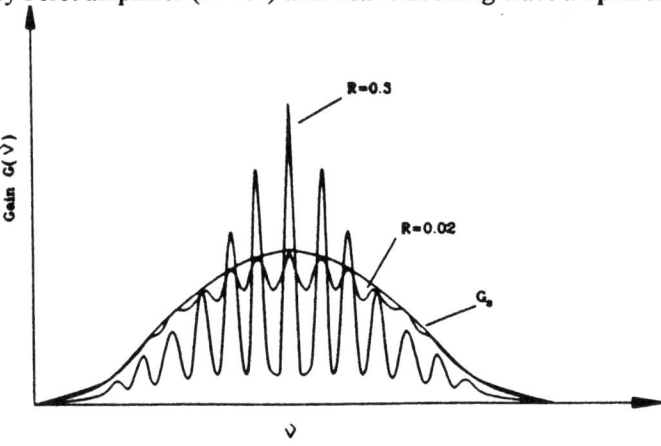

Fig.6 **Gain curves for (a) Gaussian shaped single pass gain G_s (b) Fabry Perot amplifier (R~0.3) (c) near travelling wave amplifier (NTWA) (R~0.03)**

The gain G is given by[7]:

$$G = \frac{(1-R_1)(1-R_2)G_S}{(1-\sqrt{R_1R_2}G_S)^2 + 4\sqrt{R_1R_2}G_S\sin^2\phi_S} \quad (6)$$

the peak to through ratio V by

$$V = \left[\frac{1+\sqrt{R_1R_2}G_S}{1-(\sqrt{R_1R_2}G_S)^{0.5}}\right]^2 \quad (7)$$

and the mode bandwidth

$$\Delta = \frac{c}{\pi NL}\sin^{-1}\left[\frac{1-\sqrt{R_1R_2}G_S}{2(\sqrt{R_1R_2}G_S)^{0.5}}\right] \quad (8)$$

c = speed of light

From the above equations the gain, the bandwidth and ripple are determined by G_s (effectively the cavity length L) and facet reflectivities (R). Reducing the reflectivity by applying suitable coatings, which, with present day technologies can be as low as 10^{-4} or 10^{-5} [14], allows amplification curves very close to the gain curve of the amplifier medium. The trade off between L and R can be utilised for engineering either large bandwidths with small gain or vice versa e.g. increasing L, while keeping small residual reflectivities results in an increase in the gain.

Although single pass gains in excess of 25 dB have been obtained, the large interfacing losses when coupling to optical fibre (~ 5 dB per facet) has resulted in fiber-to-fiber gains of not more than 20 dB[15]. The bandwidth of these devices is in the order of 40 nm. The maximum available single gain is determined by gain saturation. The saturation limit is set by the level of population inversion and a typical gain profile is shown in[16] FIG 7. For ordinary device structures the saturated output power is in the order of a few mW.

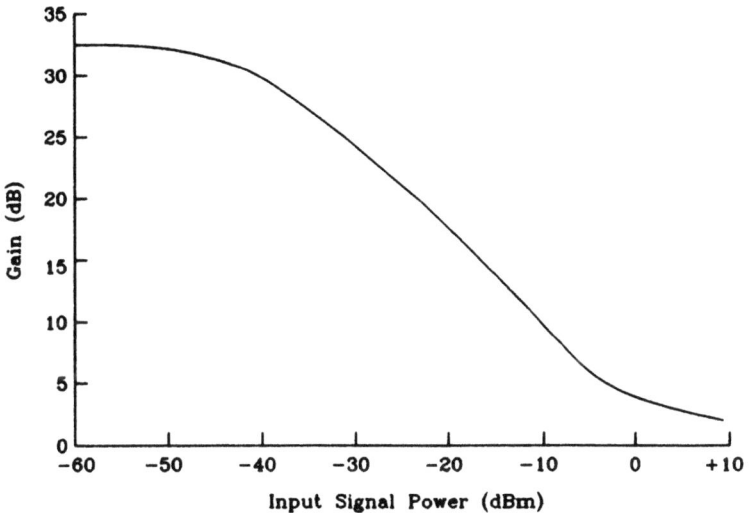

Fig.7   Gain saturation effect of a NTWA. [16]

The gain of NTWA is ordinarily polarisation sensitive. Since the optically active region is rectangular in cross section, due to the different confinement factors for the two orthogonal polarisations, the single pass gain is different (FIG 8). New device structures are beginning to yield better polarisation characteristics but at the expense of other parameters[17]. Attention must also be paid to temperature, the devices exhibiting similar temperature dependences to semiconductor lasers. As little as 5°C change in temperature causes the gain to change by as much as 3 dB; the temperature must be controlled to within 0.1°C.

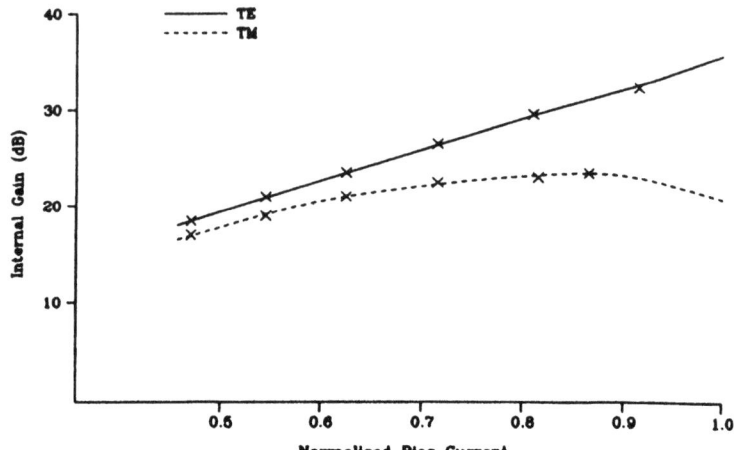

Fig.8 **Internal gain as a function of normalised bias current for TE and TM polarisations [17]**

Crosstalk also occurs when separate optical channels are superimposed in time e.g. multi-wavelength systems, and amplified simultaneously[16]. Under these conditions the gain experienced by a channel is affected by the intensity levels of other channels. The speed at which this saturation induced crosstalk occurs is governed by the carrier recombination times which are in nsec timescales. Thus crosstalk becomes a major problem for high speed multi-channel systems ranging from Mbits/sec up to several Gbits/sec.

2.3 Noise Characteristics

The analysis presented thus far has ignored the effects of spontaneous noise. Since these amplifiers operate below threshold, the levels of spontaneous noise, especially at low signal intensities can be significant which results in an over optimistic estimate of system performance.

Early results from a 200 km system[8] with a single amplifier in the centre, give an indication of the various components contributing to the noise figure (FIG 9); receiver thermal noise, amplified signal shot noise, spontaneous emission shot noise and signal spontaneous shot noise. The dominant noise generated is amplified spontaneous noise where the broadband spontaneous emission noise due to the recombination of electrons and holes in the amplifier region, is subject to amplification. This random process is given by[18,19,20] the following power spectral density:

$$P_{ase} = \chi\gamma (G-1)h\upsilon \qquad (9)$$

where

γ = population inversion parameter which indicates the level of population inversion
G = gain

h = Planck's constant
υ = frequency

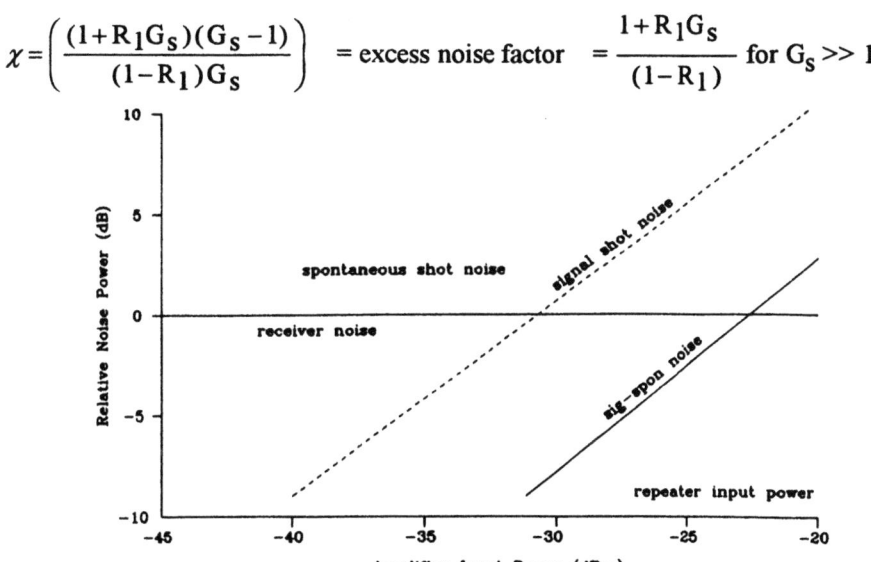

$$\chi = \left(\frac{(1+R_1 G_s)(G_s - 1)}{(1-R_1)G_s} \right) = \text{excess noise factor} = \frac{1+R_1 G_s}{(1-R_1)} \text{ for } G_s \gg 1$$

Fig.9   Summary of noise components in a NTWA system [8]

There are other noise components in addition to ASE which are created at the output of a laser amplifier stage; shot noise which is signal related and beat noise which results from signal-spontaneous and spontaneous-spontaneous beating. Collecting all these different noise components, assuming N is the number of input photons and N_0 is the number of output photons, the variance in the average number of output photons is given by:

$$\text{VAR} \langle N_0 \rangle = GN \quad \text{- signal shot noise}$$
$$+ (G-1)\gamma m \Delta f_1 \quad \text{- spontaneous shot noise} \quad (10)$$
$$+ 2G(G-1)\gamma \chi N \quad \text{- signal - spontaneous beat noise}$$
$$+ (G-1)^2 \gamma^2 m \Delta f_2 \quad \text{- spontaneous - spontaneous beat noise}$$

$\Delta f_1, \Delta f_2$ -   shot and beat noise bandwidths which are equal for NTWA
γ -   population inversion parameter
χ -   excess noise factor
m -   number of transverse modes (= 1 for fibre)

When the signal is detected by a photodiode, thermal noise has to be considered. Assuming a receiver with electrical bandwidth B, the constituent signal and noise components give a signal to noise ratio (SNR) of[21]:

$$\text{SNR} = \frac{(2eP_{sig(av)}G/h\upsilon)^2}{i^2_{sig} + i^2_{th} + i^2_{sp} + i^2_{sig\text{-}sp} + i^2_{sp\text{-}sp}} \tag{11}$$

where

i^2_{sig} = $2eGP_{sig(in)}/h\upsilon$

i^2_{sp} = $2e\gamma m(G-1)\Delta fB/h\upsilon$

$i^2_{sp\text{-}sp}$ = $2e^2(G-1)^2\gamma^2 m\Delta fB$

$i^2_{sig\text{-}sp}$ = $2e^2\gamma^2 G(G-1)P_{sig(in)}/h\upsilon$

i^2_{th} = $Fk\theta/R$

and

e = electron charge
hυ = photon energy = E
θ = temperature (Kelvin)
F = noise figure of electronics
R = equivalent resistance

For large gains beat noise powers predominate over shot noise. Since signal-spontaneous beat noise is proportional to the signal power, in contrast to spontaneous-spontaneous beat noise, the latter is the dominant noise for low input signal powers (typically <-40 dBm).

The most practical way in which to suppress both the shot and spontaneous-spontaneous beat noise components is by introducing a narrowband optical filter at the output of the amplifier stage. Correct design of the filter characteristics will limit these noise sources allowing operation in the so called "beat noise limited SNR_b" regime. In general, this detection regime can be compared to the most desired mode of operation, the shot noise limited (SNR_s) through a noise figure F defined by :

$$F = \frac{SNR_b}{SNR_s} = 2\gamma \tag{12}$$

Hence even when the population inversion parameter approaches unity, the noise figure of an ideal TWA is at a minimum, 3dB worse. In practice the best values for F are around the 5dB value[22].

2.4 Applications
2.4.1 Multilaser Amplifier Transmission

One obvious application of optical amplifiers is in the improvement of long haul transmission systems. FIG 10 is a schematic of 565 Mbits/sec transmission experiment utilising five cascaded semiconductor laser amplifiers[23]. The amplifiers exhibited fibre-fibre gains of between 7dB and 12dB giving a total of 52dB. Stable operation was achieved without isolators or filters (to suppress reflections and the build up of ASE) at each amplifier stage ensuring enhanced system bandwidth and transparency. Stability was achieved with around 15dB loss at each stage, resulting in a total loss of 85dB, which would correspond to a system span >400 km assuming standard fibre loss figures at 1.51 µm.

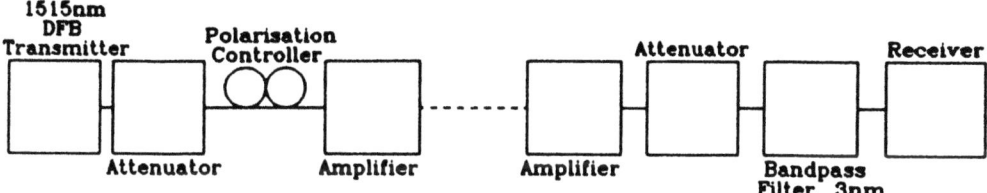

Fig.10   565 Mbits/s transmission experiment using five cascaded SLA's [23]

Since the loss between amplifiers was larger than the gain of each amplifier stage the massive build up of spontaneous emission noise was prevented. Nevertheless a 3 nm filter, which limits the beat noise terms, with 17 dB rejection was required at the receiver or an error floor rate of 10^{-3} would prevail. This basic system was upgraded to support bidirectional transmission utilising four TWA giving a system power budget of 68 dB which corresponds to ~ 350 km span assuming a fibre loss of 0.2 dB/km[24].

2.4.2 Optical Time Domain Reflectometry in Systems with Laser Amplifiers

The use of optical amplifiers to demonstrate enhanced performance in broadband communications systems has also permitted the utilisation of optical time domain reflectometry, an established technique which allows fault locations to be carried out from the terminal ends of a long span transmission system.

Since the outgoing and backscatter are both amplified - in effect a bidirectional amplification - the system does not contain in-line isolators between amplifier stages FIG 11 is an OTDR output of a system consisting of three amplifiers separated by 100 km. The gain of the last amplifier was not large enough to detect the backscatter signal from beyond 300 km, but the third amplifier reflection is clearly visible and can be relied for system fault location beyond the dynamic range[25].

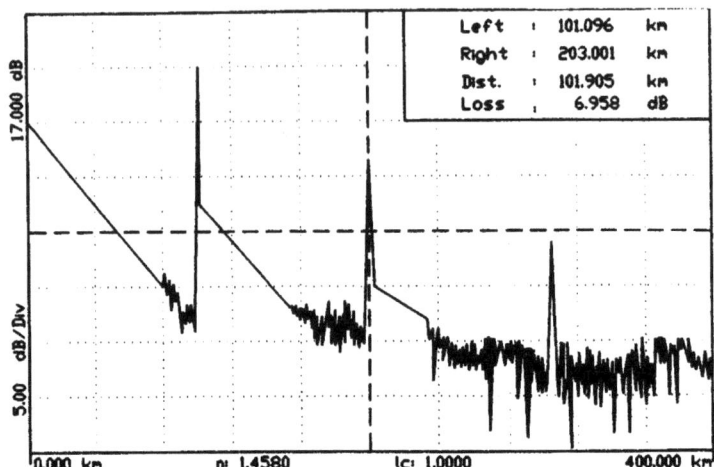

Fig.11 OTDR output of a system consisting of three amplifiers separated by 100km [25]

2.4.3 Optical Signal Processing

Most of the signal processing functions have been demonstrated in Fabry Perot type structures which in addition to providing gain can exhibit the necessary non-linearity. Since it is not the aim of this chapter to discuss this type of amplifier in any detail, they have nevertheless been used to demonstrate a variety of functions; optical switching[26], optical threshold detectors for pattern recognition and signal shaping[27], wavelength conversion for WDM systems and optical clock generation and extraction for all-optical regeneration[28]. These are extremely important functions for future optical networks but due to restrictions in space will not be covered here.

2.5 Conclusions

The semiconductor optical amplifier is based on conventional laser diode structures and is therefore highly compatible with other monomode optical technologies. The coupling losses between an optical fibre and a facet (around 3dB) has however, meant that the net available gain has been reduced to around the 20dB level. The pump requirements for NTWA are not demanding needing only a few tens of mA to achieve population inversion. The wavelength of optimum gain can be engineered by selection of the appropriate materials at the growth stage. The devices will always exhibit some resonant structure on the gain spectrum from the residual reflectivities of the anti reflection coated facets.

There are also problems with polarisation sensitivity. Although device structures can be realised which have low levels of polarisation dependence (~ 1dB), other parameters will be non-optimum. The maximum output power for a 3dB gain compression is limited to 10mW giving rise to crosstalk problems in multiwavelength operation. For NTWA carrier recombination times of ~ 1ns set a limit to their suitability for multichannel high speed systems.

Multi Quantum Wall (MQW) amplifiers offer increased performance in that large output powers (> 100 mW)[29] with very wide bandwidths (> 150 nm)[30] are attainable. In addition MQW structures lend themselves to implementing optical signal processing functions and it is in this area that semiconductor optical amplifiers will have the greatest application compared to competing amplifier technologies.

3. Optical Fibre Amplifiers
3.1 Introduction

The optical fibre amplifier has as its main advantage the fact that the basic geometry and host material is truly compatible with the transmission medium, the single mode silica fibre[31,32]. Either through the doping of standard fibres with rare-earth ions or through the use of its inherent non-linear characteristics, a linear optical amplifier can be realised. The preservation of the integrity of the transmission medium results in very low loss coupling between the transmission and amplifying (in some cases they are one and the same) fibres. As will be explained in the next sections, they also exhibit a greater tolerance to signal wavelength and the amplification mechanism is insensitive to the state of polarisation. Due to these combined characteristics, and together with its low cost potential, the fibre amplifier has been the subject of great development and deployment in the area of optical networking.

FIG 12 is a generalised schematic of an optical fibre amplifier: there are three basic components
- optical fibre (either rare-earth doped or non-linear)
- pump laser
- wavelength selective coupler which combines the pump and signal wavelengths

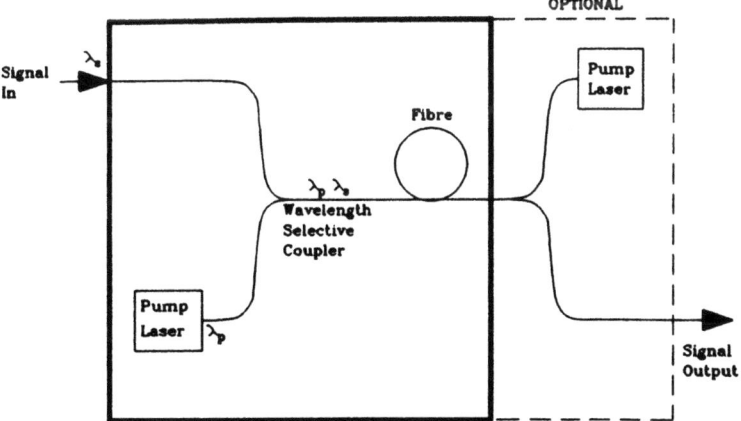

Fig.12   Generalised schematic of an optical fibre amplifier

In some cases it is necessary to reject unused pump, so a second wavelength selective coupler may be used at the opposite end of the amplifying fibre. It may also be of use as a monitor port for the transmitted pump laser, allowing the implementation of an automatic gain control mechanism. The presence of the second coupler also permits pumping of the fibre from the opposite end giving flexibility in system design, e.g. pumping simultaneously from both ends can improve both high-power and low-noise performance.

3.2 Rare-Earth-Doped Fibre Amplifiers

The most mature fibre amplifier technology thus far is the erbium doped fibre system (EDFA) suitable for the 1.5µm transmission window[33]. Fibres have also been doped with other rare-earth ions e.g. neodymium, praseodynium, but for the purposes of this Chapter, the description will be confined to EDFA's. Other doped systems, as yet not being used in practical network demonstrations extensively, will be covered later under future developments.

The net effect of doping optical fibres with rare-earth ions is to drastically affect their absorption characteristics. Removal of any optical feedback from the ends of the doped fibre results in a single pass gain from the population inversion created by optical pumping of the dopant[34,35]. Thus due to the strong absorption process, photons at the pump wavelength λ_p are absorbed by the erbium ions and are promoted from the ground state to some higher lying state. The relaxation from the high energy levels back to the ground state can occur in different ways, either radiatively or non-radiatively. A closer look at the energy level diagram of erbium (FIG 13) indicates that fast non-radiative decay occurs from the high energy levels to an intermediate level[36]. Further de-excitation from this level can be stimulated by signal photons, provided the bandgap corresponds to the energy of the photon (the 1536 nm spectral line arises from the $^4I_{13/2}$ to $^4I_{15/2}$ transition). In this case amplification of the signal wavelength (λ_s) will occur. Reference to FIG 13 also indicates the pumping wavelengths viz 807nm, 980 nm and 1480 nm required for efficient operation. Fortunately these wavelengths are readily provided by semiconductor devices giving reliable, practical optical pumping sources. Comparison between the three pumping wavelengths show that both the 980 and 1480 nm are very efficient pumping wavelengths (with the former more efficient due to a higher state absorption cross-section), much greater than the 807 nm pump where excited state absorption limits performance[37].

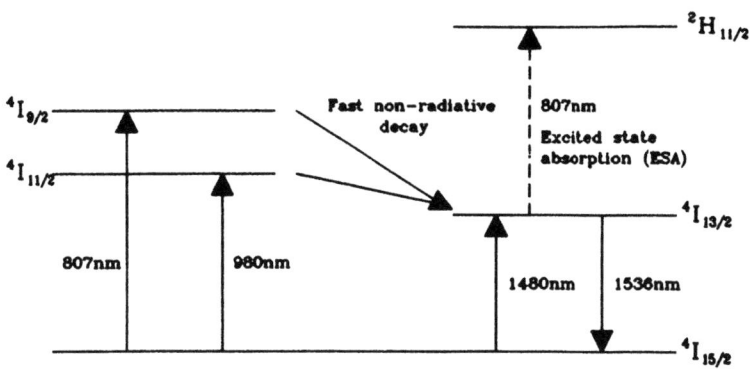

Fig.13 Energy level diagram for erbium

When 980 nm is used the population inversion occurs in two stages, the $^4I_{11/2}$ level is excited first and after a time (~ 1 μsec) decays to $^4I_{13/2}$ energy level which then relaxes to $^4I_{15/2}$ (with a fluorescence decay time of 14 ms). For the case of 1480 nm the population inversion occurs in one step.

At present 1480 nm pumps are most widely used because they are more readily available and hence are more reliable. However practical 980 nm pumps have been developed recently and are becoming the pump choice because they yield improved amplifier noise figures. It is also a far easier task to design wavelength selective couplers with pump-signal wavelength relatively far apart; although 1480 nm is still an attractive pump where pump transmission losses are to be minimised, such as remote optical pumping.

Typical pump efficiencies obtained from the 980 nm pump are around 5 dB/mW for a 5 mW pump power (net gain ~ 24 dB) requiring only 60mA bias current[38,39]. The variation of gain with pump wavelength is relatively stringent with a ± 1 nm change from the most efficient wavelength (981 nm) degrading the gain/pump power efficiency by 10%. For 1480 nm, the restrictions are relaxed and the ions may be pumped over a 50 nm range, the tolerance being ~20 nm for the 10% variation. Gains of > 40 dB for a launched optical pump power of > 130 mW[40] have been obtained.

3.2.1 Gain/Bandwidth Characteristics

It is fortuitous that the spectral width of erbium ions in a glass host is much broader than that of bulk erbium. In germania co-doped fibres, bandwidths of up to 10 nm have been obtained[41] in a 3 metre length of fibre doped with 400 ppm of erbium [FIG 14] (narrowband amplifier). With alumina co-doping, further broadening results.

FIG 15 is the gain spectrum of an alumina co-doped EDFA pumped at 1480 nm as a function of pump power[42]. The useable bandwidth extends to around 40 nm, with small signal gains in excess of 20 dB (for ~ 50 mW pump) obtained flat over this spectrum. Further engineering of the erbium gain spectrum can be achieved but will not be covered here.

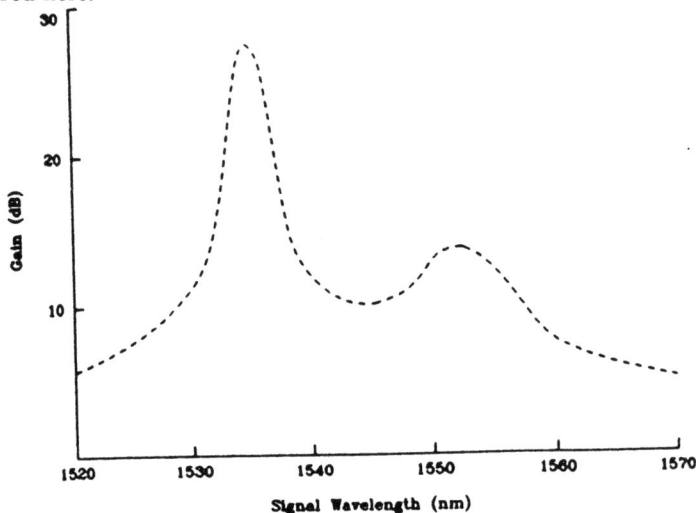

Fig.14   Gain as a function of wavelength for germania co-doped erbium doped silica fibres. (3m fibre, 400ppm erbium) [41]

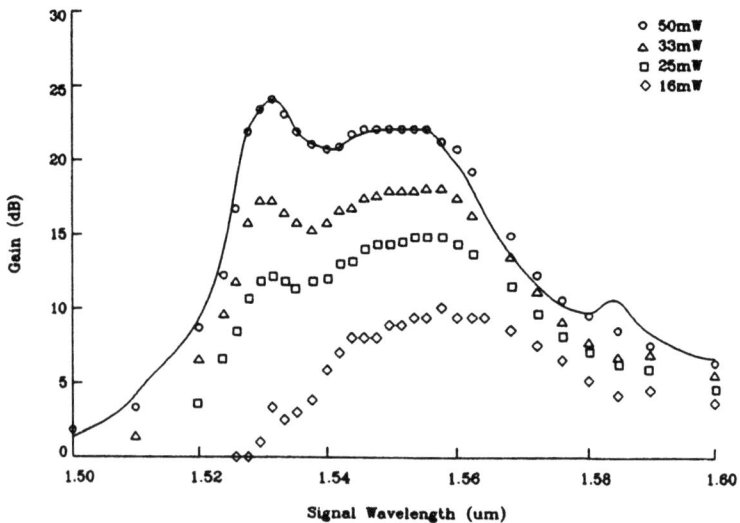

Fig.15   Gain spectrum of an alumina co-doped EDFA (10m fibre, 11000ppm erbium [42])

In addition to spectral broadening, alumina co-doping allows higher erbium doping concentrations, permitting the design of a number of different fibre amplifier geometries e.g. long length/low dopant or short compact devices with high dopant levels. Doping levels around 9000 ppm can be achieved, giving an amplifier with > 20 dB gain for a 1m length[43]. Since the erbium dopant is randomly orientated in the circular symmetric fibre core, the EDFA gain is not a function of polarisation state[44].

The gain of EDFA also saturates whenever the amplified signal power is of comparable level to the pump power, due to the substantial depletion of the population inversion. However as a consequence of the three level system, it is possible to obtain relatively high saturated output powers. FIG 16 illustrates the saturation behaviour of an EDFA, highlighting that the output saturation power is directly proportional to the pump power (3dB compression of the small signal gain increases with pump power) and is thus limited by the availability of powerful pump sources. Saturated output powers in excess of 500 mW have been achieved[45]. Even although there is this linear relationship between saturated output power and pump power, the gain nevertheless saturates for high pump powers. This may be advantageous in that high gains can be achieved with a high tolerance to pump power fluctuations. Practically, the best route to realising high efficient EDFA's is to use a modest pump power for high gain and one route to reaching this goal is to reduce the diameter of the fibre core (typically ~ 2.5µm) at the expense of an added fusion splice loss of between 0.1 and 0.5dB.

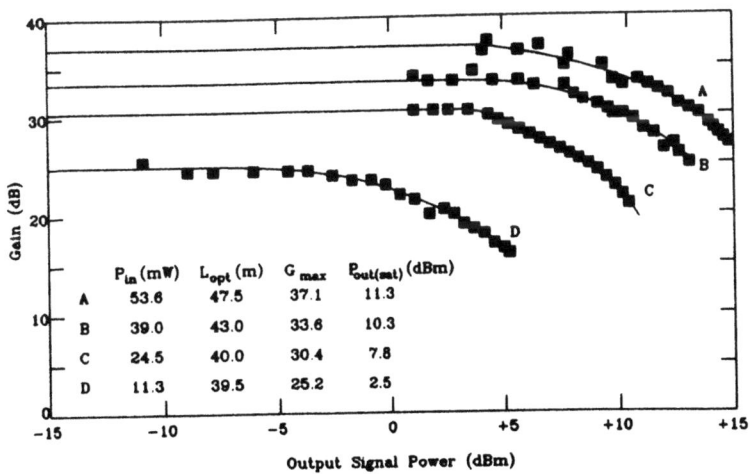

Fig.16 Signal gains as a function of amplified output power for various pump power. As input power increases, the gain remains constant (for a particular pump power) until saturation occurs.

The dynamics in erbium fibre amplifiers have important implications. The lifetime of upper states is in ms timescales which has the following ramifications:

- immunity to high frequency (>100 kHz) noise since electrical noise on the bias to the semiconductor optical pump devices introduces a degree of modulation in the amplifier gain
- can only use erbium amplifiers as slow optical switches
- saturation induced crosstalk is negligible for high speed systems. Measurement of crosstalk using two signal lasers 3.5 nm apart (one CW, other 100% square wave modulation)[46,47](FIG 17) shows that crosstalk is negligible for frequencies > 100 kHz.

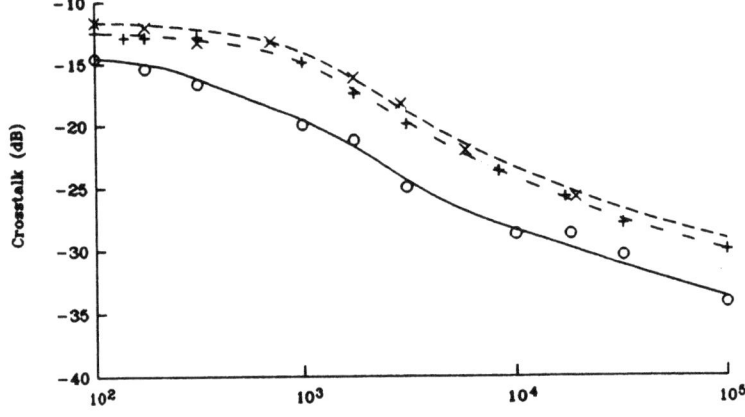

Fig.17   Crosstalk levels as a function of modulation frequency

3.2.2 Noise Characteristics

As in the case of semiconductor laser amplifiers, noise accompanies optical amplification in EDFA's, the limiting component again being amplified spontaneous emission. Equation 9 describes the effect of ASE, with $\chi=1$ since EDFA's are essentially travelling wave devices. Investigation of the resulting SNR, indicates once again that at high signal input power, signal-spontaneous beat noise predominates whilst at low input powers shot and spontaneous-spontaneous beat noise are the dominant noise sources.

In EDFA's which have been designed for maximum gain, for a co-propagating pump configuration where the population inversion at the input is low, a noise figure of 3.2dB has been achieved[48] which only represents a 0.2 dB penalty over the value expected for an amplifier with a complete population result ($\gamma=1$). FIG 18 shows typical curves of gain and noise figure as a function of doped fibre length highlighting the trade-off between the two. With a 1480 mn pump, where the pump and signal wavelengths are close together, the emission stimulated by the pump yields worse noise figures (5.2 dB) than those for a 980 nm pump[49,50].

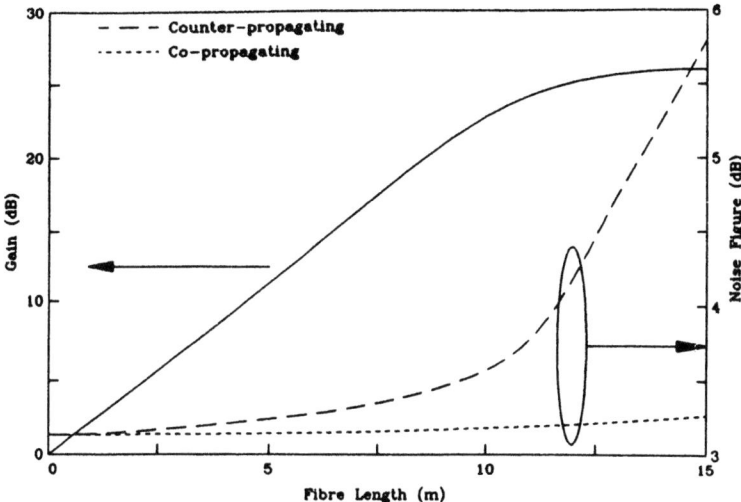

Fig.18 Gain and noise figure as a function of doped fibre length

In order to illustrate the build up of ASE in a chain of repeater amplifiers, note that the accumulated noise is proportional to the gain of each amplifier and to the number of amplifiers. Assuming a total transmission length L with amplifier spacing L_A, the number of amplifiers is therefore $M = L/L_A$ and the amplifier gain G is $(G_{sys})^{\frac{1}{M}}$ where G_{sys} represents the system loss margin. Inspection of Equation 9 indicates that the ASE noise is smallest for $L_A \to 0$ and increases slowly with L_A up to a distance about the effective attenuation length of the fibre (reciprocal of the fibre loss coefficient) FIG 19 is an illustration of the power required as a function of L_A for a system of L = 1000 km operating at 2.5 Gbits/sec for a receiver SNR = 200. The required power scales with L, data rate and SNR[51].

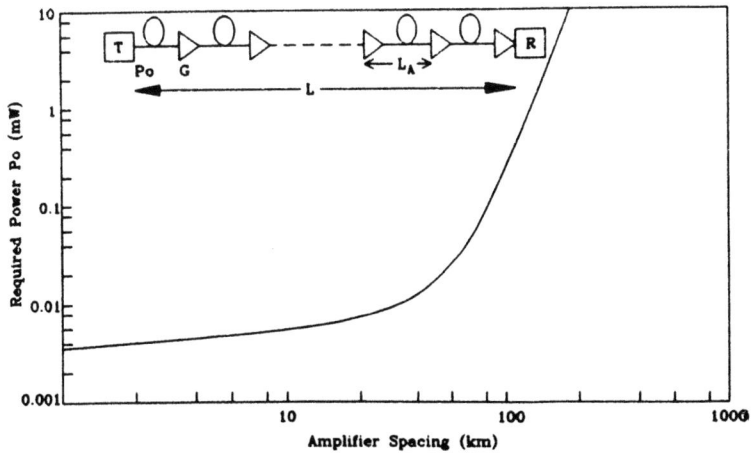

Fig.19 Required optical power in a multi optical repeater-amplifier system; amplifier output power P_0 as a function of amplifier spacing[51]

3.3 Non-linear fibre amplifiers

There are a number of non-linear effects that occur within a silica optical fibre which cause degradation of the transmitted signals and fundamentally limit the power levels that can be injected into the fibre. Conversely some of these same processes can be used to advantage as amplification mechanisms. Two non-linear scattering mechanisms, Raman and Brillouin[52] can be harnessed to amplify optical signals within the fibre. These stimulated effects arise from parametric interactions between light and acoustic photons due to lattice or molecular vibrations within the core of the fibre.

3.3.1 Raman Amplification

FIG 20 depicts the Raman scattering energy level diagram, presenting the vibrational energy levels of the core material in a fibre. If a photon of energy not equal to that of a vibrational transition is carried by the fibre, it can create a 'virtual' excited state (not absorption) which in psec timescales will re-radiate to the ground state with the emission of a photon of the same wavelength. Although this is classical Rayleigh scattering, there is a small probability that the virtual excited state will emit a photon of lower energy, leaving the molecule in a different final state - spontaneous Raman scattering. The difference in frequency is known as the Raman shift. If sufficient power is coupled into the fibre core, the degree of population inversion can be high enough for amplification of the spontaneous emission through stimulated Raman scattering (SRS). Since the spectrum of the Raman gain coefficient is wide (~ 15 THz FIG 21), for example, channels in multi wavelength systems will be coupled by SRS. Signals in the lower frequency channels can be 'amplified' at the expense of channels in the higher frequency and this effect becomes a limiting factor in optical fibre transmission.

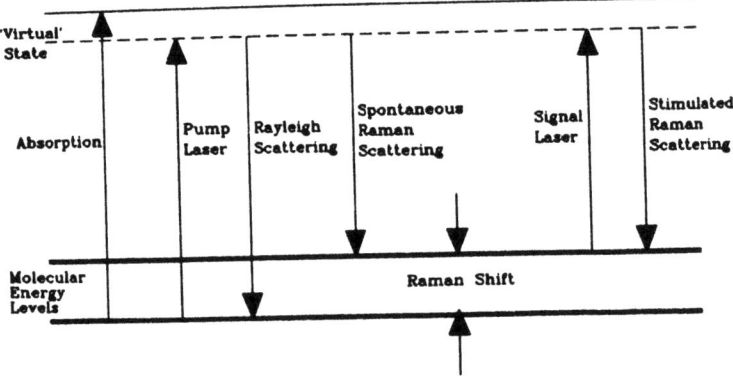

Fig.20 **Energy level diagram for Raman scattering**

However Raman amplification of a signal laser beam can be engineered via SRS if the fibre is pumped at the appropriate wavelength. Referring to the spectrum of FIG 21 indicates that if the wavelength of the signal is known then the optimum pump wavelength can be calculated. The Raman spectrum can be modified by incorporation of different molecules into the fibre core[53]. For example the introduction of germania causes enhancement of the scattering cross section. Standard telecommunications fibres contain modest levels of germania giving a maximum gain for shifts of 420-450 cm^{-1} equivalent to wavelength shifts of 80 nm and 100 nm for the 1300 nm and 1500 nm transmission windows.

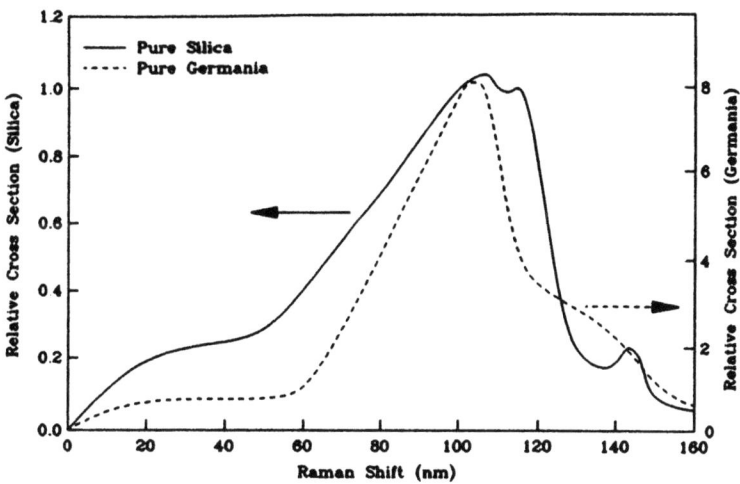

Fig.21 Raman scattering cross sections

The gain of a Raman amplifier R_G is given by

$$R_G = 10 \log_{10} e \left[\frac{gPL_e K}{A} \right] (dB) \qquad (13)$$

where

        g = Raman gain coefficient
        P = mean optical pump power
        A = fibre spot size (implying amplification is proportional to pump intensity)
        K = parameter that caters for polarisation dependence of the fibre,
                (K=1 for polarisation maintaining fibre, K = 1/2 otherwise)
        L_e = effective length which accounts for the attenuation of the pump power
                through normal loss mechanisms

$$= \frac{[1-\exp(-\alpha_p L)]}{\alpha_p}$$

and α_p = attenuation at pump wavelength

For highest gains the fibre should ideally have a small core with a large gain coefficient. Both of these designs call for a core with high germania doping but at the expense of increasing the loss of the fibre.

For standard optical fibre, gains at 1.5μm for a 100 mW pump power and 100 km fibre are around 2.5 dB for step-index fibre and 5 dB for dispersion shifted fibre[54], the difference caused by the higher germania doping levels in the latter case. FIG 22 shows a typical pump power versus gain characteristics for Raman amplifiers. Gain saturation also occurs in Raman amplification due to the depletion of the pump by the intense amplified signal. The results for a small core power amplifier with the signal at 1580 nm and pump at 1480 nm is shown in FIG 23. 105 mW of signal power was amplified to 145 mW output power which represents a 60% conversion of pump to amplified signal. A typical gain bandwidth of approximately 25 nm can be obtained.

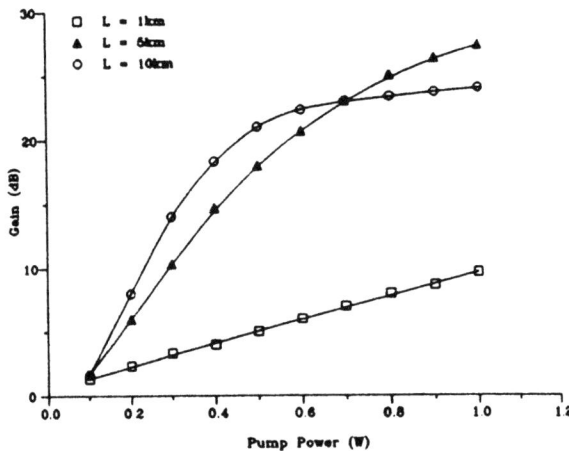

Fig.22 Gain as a function of pump power for Raman amplifiers of different length

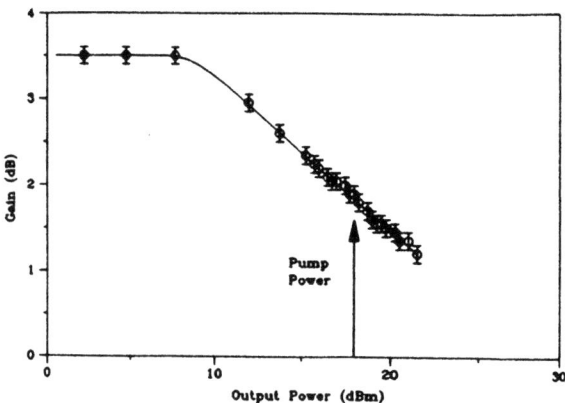

Fig.23 Gain saturation in Raman amplification (for small core power amplifier)

The implementation of a non-linear fibre amplifier is identical to that of the EDFA, the major difference being that the amplifying fibre can be standard single mode optical fibre although special fibre designs have been used (high germania doping).

There are two options available to system designers :

(a) use of standard low loss fibre in which case the Raman gain directly improves the system budget
(b) use of highly doped fibres with small spot sizes which exhibit much more loss than standard fibre. For the same pump power, the Raman gain obtained is much greater.

In general therefore, there is an optimum fibre length for which the net gain (Raman gain/linear loss trade off at signal wavelength) is a maximum. It may be that a combination of optimum gain (and length) of a short highly doped fibre for an available pump power and a long length of standard transmission fibre is the optimum design approach.

3.3.2 Brillouin Amplification

Due to the small Raman gain coefficient of optical fibre relatively high pump powers are required to achieve useful gains. An alternative approach centres on the utilisation of stimulated Brillouin scattering whose gain coefficient is four orders of magnitude larger than the value for Raman scattering.

The description of SBS is identical to SRS (see Equation 13) and again three processes are relevant : spontaneous Brillouin scattering; self-stimulated Brillouin scattering and stimulated Brillouin scattering, the amplification mechanism. However in SBS the pump photons couple to acoustic waves (phonons); a mixing of a pump photon with a phonon to create a photon of lower energy (FIG 24). In addition energy and momentum conservation conditions must be satisfied for spontaneous scattering, only valid in fibre when the resultant scattering is in opposite direction to the pump[55]. The Brillouin shift is equal to the wavelength of sound waves in the core material, approximately 0.1 nm at 1.5 μm (≈ 11 GHz) with a bandwidth of around 0.001 nm (≈ 100 MHz), due to the thermal distribution of the photons in the core. These are the major differences of SBS when compared to SRS; the former is undirectional and limited in intrinsic bandwidth.

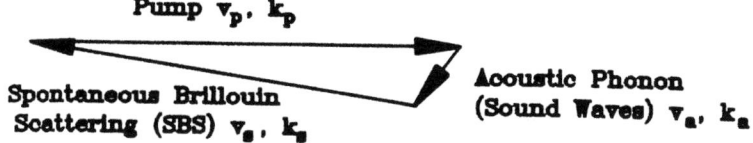

Fig.24 Brillouin scattering condition

The Brillouin gain spectra for two fibres of similar properties - pure silica core with fluorine doped depressed cladding and germania doped core with fused silica cladding - were investigated with DFB lasers for the pump and signal (30 MHz linewidths)[56]. FIG 25 illustrates the narrowband nature of the gain bandwidth, the pure silica fibre showing only one peak and the germania exhibiting several peaks, the fibre core effectively being of a multimode waveguide for sound waves in the latter case. In general it is very difficult to realise a fibre where the SBS spectrum can be predicted accurately.

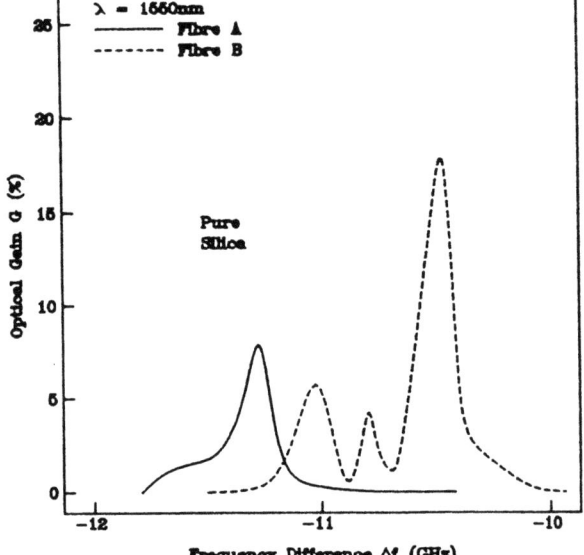

Fig.25 Brillouin gain spectra for (a) pure silica core and fluoride doped cladding (b) germania doped core, fused silica cladding [56]

Gains of the order of 25 dB have been obtained for 5mW pump powers[57] by pumping the fibre below the SBS threshold and injecting a signal laser in the opposite end of the fibre. Despite these high efficiencies the use of SBS is limited by the small gain bandwidth. Modification of the spectrum by incorporation of different materials in the core or by use of a pump laser of finite linewidth greater than the intrinsic Brillouin gain spectrum does not improve the bandwidths beyond the Brillouin shift[58,59]. The need for the use of narrowband lasers separated by 11 GHz to an accuracy of only a few MHz effectively drives the applications of these amplifiers into the specialised category. One elegant example is its use in amplifying only one channel from a star network (effectively functions as a narrowband filter) where 30 dB gain of a single channel was obtained for a modest (23 mW) pump power[60].

3.4   Applications
3.4.1 Long haul high capacity transmission

There are a number of long haul transmission scenarios where the EDFA has been applied to enhance performance. In undersea cable systems the EDFA can be utilised as a transmitter power amplifier and receiver preamplifier. For embedded terrestrial long haul networks they are also used as in-line repeaters. Since most of the embedded infrastructure is at 1.3 µm, multichannel WDM operation at lower data rates (2.5 Gbits/sec) is a more economical stance to upgrading performance. For green field sites the use of dispersion shifted fibre is an attractive approach since high data rate (> 10 Gbits/sec) and multi wavelength operation can be obtained in tandem with EDFAs. Alternatively dispersion compensation on existing 1.3 µm networks can be implemented.

Early system demonstrations concentrated on direct detection, intensity modulated unrepeatered transmission, the EDFA's providing both power and pre-amplification. The NTT experiment at 1.52 µm was 310 km long at 1.8 Gbits/sec using dispersion shifted fibre, remotely pumped erbium doped fibre gain (situated 34 km from the receiver) together with some Raman gain in the system fibre (FIG 26) 180 mW pump power at 1450-1490nm, was injected at each end, 1450nm optimal for Raman gain and 1490nm for erbium gain, an example of hybrid-discrete and distributed-amplification[61].

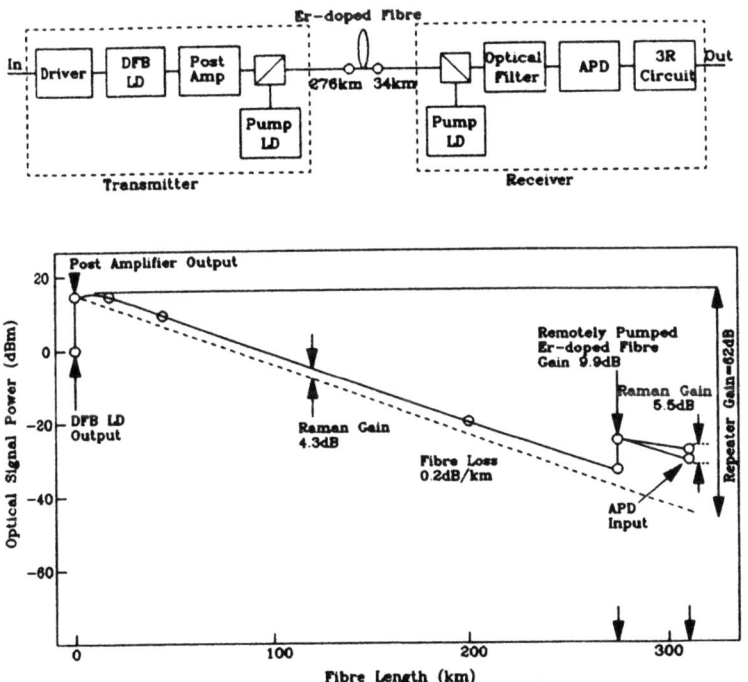

Fig.26 1.8Gbit/sec, 310km transmission in dispersion shifted fibre with remotely pumped erbium gain and Raman gain in system fibre [61]

These initial experiments quickly moved into multirepeater geometries. The first multirepeater system was demonstrated with 11 repeater EDFAs at 1.2 Gbits/sec employing 1480 pumps (FIG 27)[62]. The transmission length exceeded 900km with repeater spacings of ~ 70km. Progress has resulted in the extension of the transmission length to ~ 2200km at 2.5 Gbits/sec using 25 repeater EDFAs spaced at 80km[63]. The latter was a coherent system with a total optical gain of 440dB utilising isolators (to prevent lasing) and filters (to prevent build up of ASE). After transmission of over 2200km the excess noise from the amplifiers introduces a penalty of only 4dB. The loss compensation by repeatered optical amplification has provided substantial improvements in the bit rate-distance product[64]. 10Gbits/sec optical signals transmitted over 500km with five repeater EDFAs represents 5 Tbits/sec km[65] compared to 1 Tbits/sec km[66] for conventional systems without optical amplification.

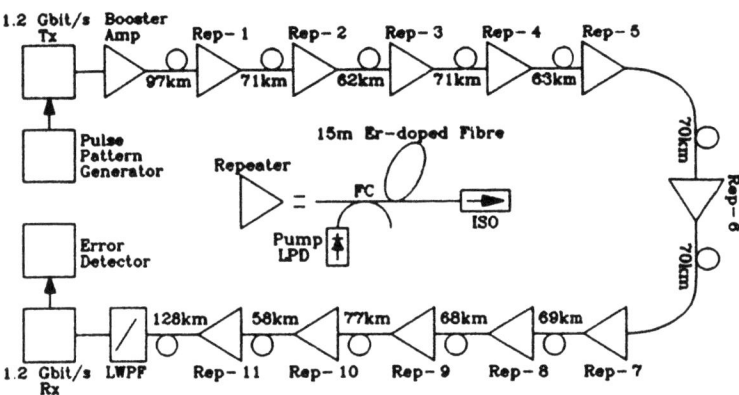

Fig.27 Schematic of a multirepeater system at 1.2Gbit/sec using 11 repeater - EDFA's with 1480 pumps [62]

Next generation undersea transmission systems will use EDFAs as repeaters and in order to minimise chromatic dispersion, dispersion shifted fibre will be used. To prevent the build up of ASE the amplifier spacings will be kept short (30-40 km). Under these conditions multigigabit per second transmission over distances of 7500 km to 9000 km[67] is achievable.

3.4.2 Multichannel Systems

Practical EDFA power amplifiers can provide >50 mW of output power and together with EDFA preamplifiers, which can enhance the sensitivity of pin photodiode receivers by ~ 15dB, will allow the realisation of nonregenerated system distances of 200 km at multigigabit rates. Capacity upgrades can then be achieved by optimising the wide gain bandwidth of EDFAs for the simultaneous amplification of WDM signals. Initially four WDM optical channels (2.4 Gbits/sec per channel) were transmitted using five repeater - EDFAs, demonstrating the feasibility of the EDFA/WDM combination[68]. The transmission distance was ~ 460km corresponding to 4.8Tbits/sec km. FM video distribution of more than 10 channels was also transmitted successfully with five repeater EDFAs over 480 km[69].

In local broadcasting a number of systems have been demonstrated; 16 optical signals at 622 Mbits/sec separated by 5 GHz have been amplified by between 15-19dB in a single amplifier. The opportunities afforded by a combination of WDM technology and optical amplifiers in a very cost effective manner, opens up new avenues of application in this sector. Since each amplifier can support many WDM channels, capacity upgrades become possible. FIG 28 is an example of this powerful combination where broadcasting of 16 2.4Gbits/sec channels to 43.8 million customers over 527 km has been demonstrated. As the transmitter signals at different wavelengths are optically multiplexed into a single fibre for amplification and transmission, optical amplifier repeaters (between 25km - 100km) can compensate for any transmission and splitting losses. At the receiver the WDM signals are optically demultiplexed into separate channels.

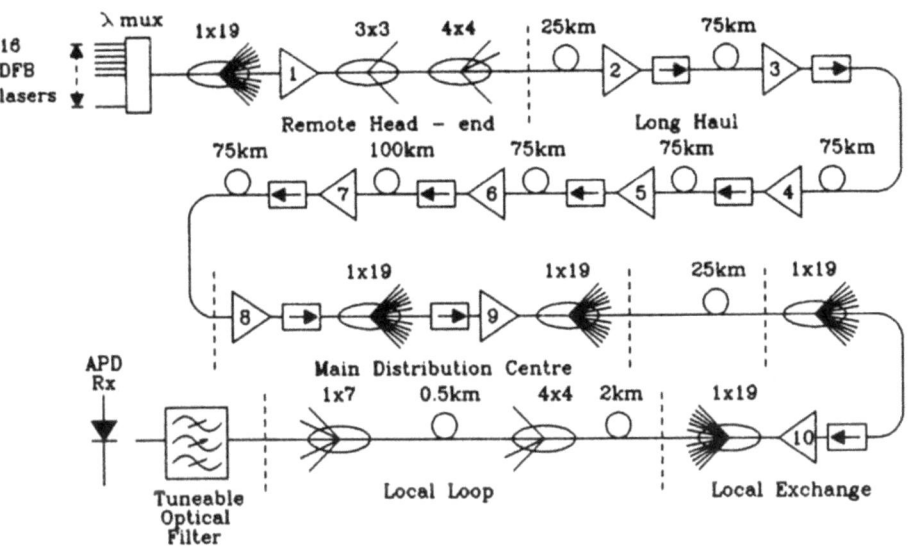

Fig.28 Demonstration of broadcasting 16 x 2.4Gbit/s channels to 43.8 million customers over 527km

3.4.3 Non-linear (Soliton) Transmission

The propagation of high intensity optical pulses in optical fibres can lead to pulse compression through the Kerr non-linearity which can, in turn, be used to offset the normal linear dispersion. This so called 'soliton' system can result in the transmission of a pulse over extremely long distances without distortion[70]. The formation of a soliton is dependent on the selection of the appropriate signal power and in a lossy fibre system periodic discrete or distributed gain must be used to maintain the signal at the critical level.

Soliton transmission systems have therefore made startling progress since the introduction of optical amplifiers. The first experiment utilised an EDFA as a power booster for a gain switched DFB at 1550 nm, sending a 2.8 Gbits/sec soliton data stream over 23km the transmission loss compensated by Raman amplification[71]. Multirepeater experiments followed, with error free transmission obtained at 5 Gbits/sec over 250km with nine repeater EDFAs. Two stage EDFAs were used for power amplifiers to generate the pulses [72]. Transmission lengths were extended to over 10000 km in fibre loop experiments[73]. The loop consisted of three 25 km long dispersion shifted fibres and three repeater EDFAs (gain of 6 dB each). The experiment suggests that 5 Gbits/sec, 10000 km systems would be possible.

3.4.4. Distributed Amplifier Systems

Investigation of the pump powers necessary to equalise the net optical gain to the transmission loss of the fibre shows that the pump power for bidirectional pumping is lower than that for unidirectional pumping FIG 29 shows that the lowest pump requirement occurs for nearly symmetric pumping from both ends, due to the uniform power distribution of the pump[74].

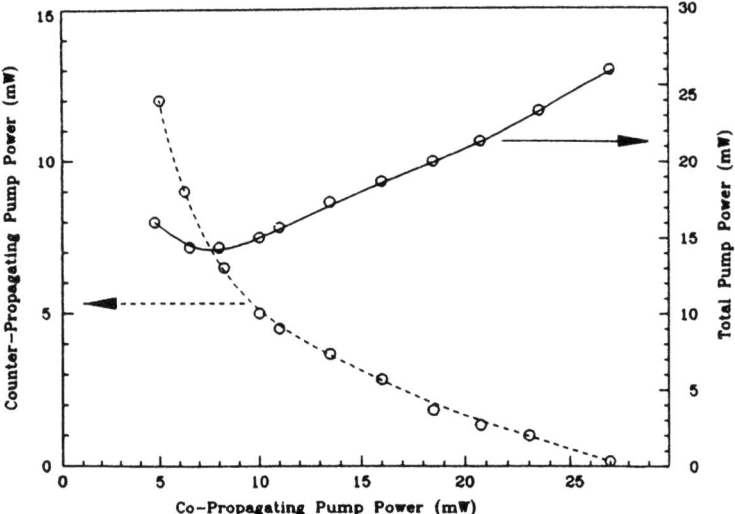

Fig.29 Bidirectional pumping characteristics for EDFA's

One application of this phenomenon has been in soliton transmission using a 9.4 km long distributed EDFA[75]. Data transmission performance has also been investigated using a 10 km long bidirectionally pumped EDFA in a 5 Gbits/sec systems[76]. A total pump power of 50 mW at 1.48 μm (30 mW forward, 20 mW backward) maintained the input signal launch power of 10 mW. The pump power requirement increases rapidly with input signal power. In optical bus networks, the utilisation of distributed EDFAs will compensate for loss due to multiple tappings to, in effect, realise a transparent bus network[77]. This approach opens up a new avenue of studies into appropriate system configurations.

3.4.5 OTDR Enhancement

The utilisation of EDFAs in OTDR measurements results in both optical post-amplification at the input, increasing input launch power and at the same time improves the receiver sensitivity by acting as a pre-amplifier for the backscattered signal. The gain of the amplifier appears as an increase in the dynamic range.

The implementation of this concept has been shown with a 1536 DFB transmitter coupled to a 1nm bandpass filter to limit the levels of ASE on the receiver[78]. An amplifier gain of 18 dB resulted in an improvement > 10dB in dynamic range when compared to the instrument alone. The promise of high resolution, high dynamic range measurement is therefore possible not only in fault location but also in distributed sensing applications.

3.5 Future Developments

The major drawback with EDFA systems is that they are restricted to operating in the 1.5 μm transmission window of optical fibres. Given that most of the embedded optical network utilises standard telecommunications fibre operating at 1.3 μm, the search for an equivalent fibre amplifier to operate at the lower wavelength has been vigorous. Inspection of the energy level diagrams of relevant rare earth ions, indicates that neodymium and praseodymium are candidates because they possess a transition energy gap which corresponds to the 1.3μm photon energy. Thus it may be expected that by doping the fibre with the above rare earths can yield a fibre amplifier which has similar performance characteristics to the EDFA but operating in the 1.3μm band.

Unfortunately neither ion can be used with silica fibre hosts since the 1.3 μm transition suffers from signal excited state absorption which limits gain in silica fibres to wavelengths above 1.36μm[79]. The excited state absorption however shifts to shorter wavelengths (FIG 30) in fluoro-zirconate fibres ($ZrF_4 - BaF_2 - LaF_3 - AlF_3 - NaF$ or "ZBLAN") allowing successful doping with these two rare earths and giving optical gain in the 1.3μm window.

In Nd^{3+} systems the existence of competitive transitions (corresponding to 1.05μm) which at high pump powers gives rise to gain saturation, necessitates suppression of the 1.05μm amplified spontaneous emission to obtain large gains. The best result to date is a 10 dB gain with 17 dBm pump power supplied by commercially available laser diodes at 0.82 μm[80]. The wavelength band (1.32-1.38 μm) over which the NDFA operates is also slightly longer than the standard 1.3 μm telecommunications band.

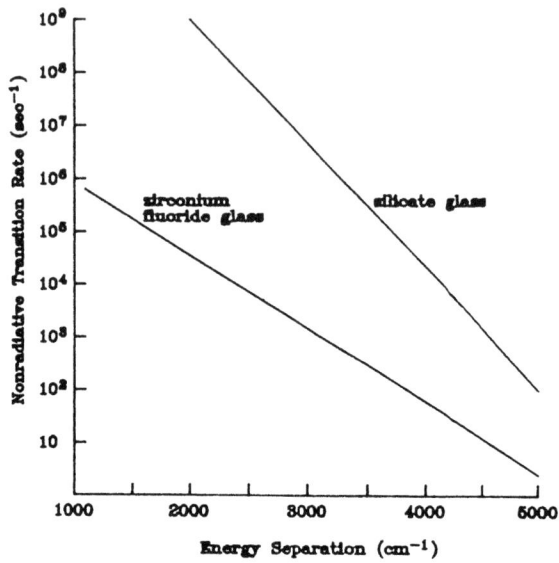

Fig.30 Non-radiate transition rate as a function of energy separation for silicate and fluoride glasses

FIG 31 is the energy level diagram for praseodymium (Pr^{3+}), the major problem being the large non-radiative decay rate due to the small emergy level difference between $^1G_4$ and $^3F_4$ transition levels which leads to complete quenching of the 1.3μm transition in silica glasses. In ZBLAN glasses the improvement in decay rate circumvents this problem albeit with small quantum efficiencies. This in turn necessitates the use of high pump powers which are not compatible with laser diode pumping. The use of small core, high NA fibres has produced a PDFA with a gain of 38dB at 1.31 μm for a launched power of ~ 25 dBm at 1.05 μm (FIG 32). In general, to increase pump efficiency, a very small core is required together with a high NA to maintain a cut-off wavelength of 1.3 μm. A 11 dB gain has been obtained with a laser diode source[81] and it is expected that improvements in pump efficiency will allow the realisation of a practical laser diode pumped amplifier.

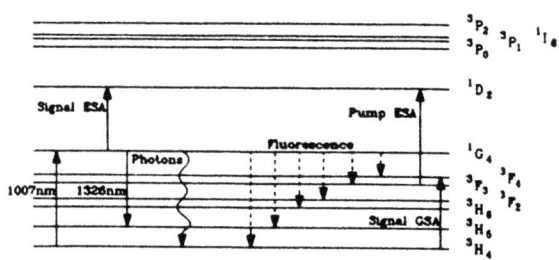

Fig.31 Energy level diagram for praseodymium

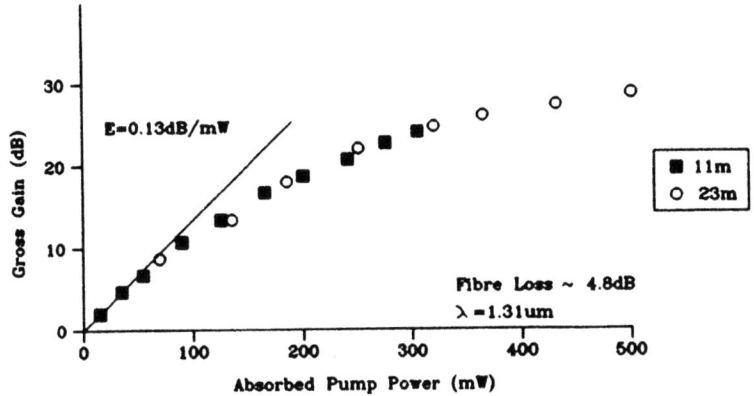

Fig.32 Gain as a function of pump power for a PDFA

FIG 33 is a representation of the saturated gain spectrum for a PDFA. Output powers > 100 mW was obtained over a 32 nm bandwidth. However notice the high level of pump power. FIG 34 shows the noise characteristics for both NDFA[82] and PDFA[83]. Although they exhibit good performances, NDFA noise figure deteriorates at wavelengths < 1.33 μm and that of the PDFA degrades > 1.32 μm. This is caused by excited state absorption (NDFA) and ground state absorption (PDFA) respectively.

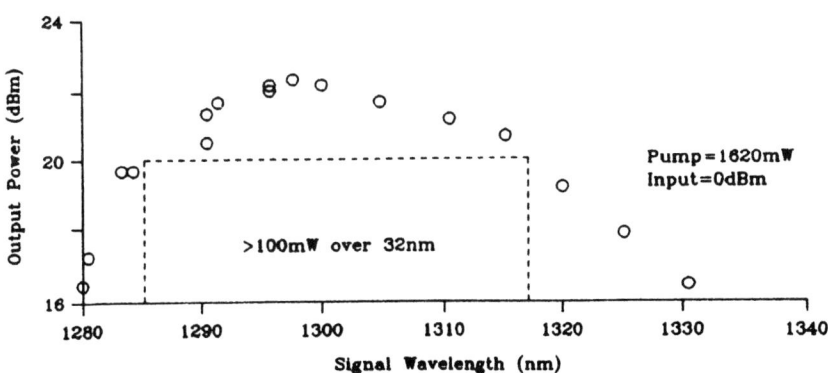

Fig.33 Saturated gain spectrum for a PDFA

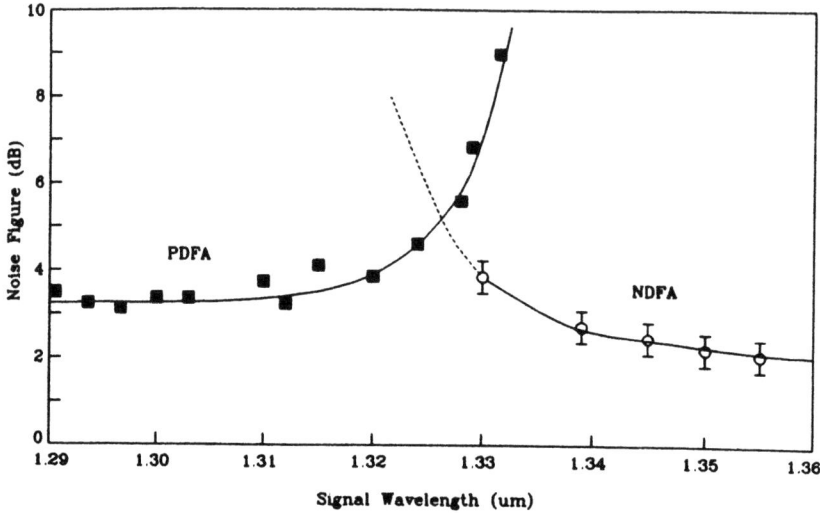

Fig.34 Noise figures for NDFA [82] and PDFA [83]

In summary, although 1.3 µm fibre amplifier systems are attractive, there are a number of observations to be made : they require more exotic glasses for useful gain; they are less efficient systems (PDFA requires > 100 mW of pump for practical gains); most suitable for power amplifier applications; they exhibit low polarisation sensitivity; mechanical properties are problematic; splicing techniques to achieve low loss interconnection to standard silica fibres need to be developed. As a result more efficient glass hosts are being investigated.

3.6 Conclusions

The optical fibre amplifier is truly compatible with the transmission medium - a stable well established glass host, - which allows easy, low loss interfacing. The rare earth doped geometry, especially erbium, is the most widely applied fibre amplifier to date.

Typical unsaturated gains greater than 20 dB are routinely available with coupling losses below 0.5 dB at each connection obtained through fusion splicing. The saturated output power is mainly set by the optical pump power, provided by semiconductor lasers of high output power at around specific wavelengths. The most efficient pump wavelength for erbium systems is 980 nm, and commercial, reliable laser diode sources are beginning to be available. The gain spectrum extends across the range 1525-1565nm range which allows amplification of several simultaneous wavelengths. Due to the dynamics of erbium systems, the crosstalk between channels is negligible for high speed transmission. In addition the fibre amplification mechanism is polarisation independent.

Non-linear fibre amplifiers have less flexibility, although there have been a number of demonstrations where this mechanism was used in tandem with doped fibre amplification to enhance performance. For transmission systems Brillouin fibre amplifiers have limited applications even though they can yield high gains for low pump powers, due to the restricted bandwidth. Raman amplifiers need large pump powers (> 100 mW) before useful gain is obtained. The pump laser wavelength needs to be ~ 100 nm with respect to that of the signal and since the availability of most high power laser diodes is restricted to wavelengths between 1470 nm - 1550 nm, the signal wavelength needs to be > 1570 nm to obtain efficient Raman gain. The gain bandwidth of Raman amplifiers is comparable to EDFAs (between 20 nm - 40 nm). Since the pump power requirements are high, the saturated output power is much higher than for EDFA. The lifetime of the virtual state is extremely short, so crosstalk due to gain saturation is negligible but if the power in any one channel in a multichannel system becomes large (> 40 mW) then it can act as a pump for Raman amplification on all other signals at longer wavelengths (Raman induced crosstalk).

For the rare earth doped fibre amplifiers designed to operate at 1.3 μm, the search for new host glasses to accommodate Nd^{3+} and Pr^{3+} ions brings the promise of increased pump efficiencies allowing higher gain for less pump power. There are also a number of other issues to be addressed before they will be considered for practical applications.

4. Summary

Optical amplifiers are the most significant photonic component to be developed in the area of optical systems for a number of years. The availability of a low cost device, compatible with the transmission medium, and performing amplification in the optical domain independently of data rate, will permit a number of network geometries to be upgraded and novel network concept hitherto not feasible, to be realised. The properties of high gain, high saturated output power, over a broad spectrum allowing many channels to be amplified simultaneously with low crosstalk and low polarisation sensitivity makes it attractive for use in a number of applications. The choice of optical amplifier is predominantly driven by the application; Table 1 is an attempt to summarise the pertinent operating parameters of the main optical amplifier geometries covered in this chapter, so that comparisons can be drawn.

For transmitter power booster applications the gain saturation characteristics would dictate; each class of amplifier could be used in this application with differing performance. Given the substantial pump power required for Raman amplification, this device gives the highest saturated output powers. However, practically, the EDFA can serve as an excellent power booster with low saturation induced crosstalk. These crosstalk effects are most severe in high speed laser power amplifiers.

All can also be used as in-line linear gain blocks. Given the compact nature of EDFAs and SLAs, they can both be used as discrete amplifying stages, with the latter again introducing greater levels of distortion in multichannel systems. The EDFA can also be used as a distributed amplifier in long distance transmission by adjustment of the erbium doping levels. Raman gain on the other hand which requires many kilometres of fibre is best suited to provide distributed gain in the transmission fibre. Since the fibre is already performing its function as a transmission medium Raman amplification can be obtained without significant perturbation and provide another option for system designers.

Preamplification can be provided by both SLAs and EDFAs with Raman fibre amplifiers unsuitable due to the low gains available. Inspection of the noise figures indicate that EDFAs can nearly provide the theoretical quantum limited 3 dB figure since they can be coupled very efficiently and almost full population inversion can be achieved. SLAs have also a slight disadvantage in that the residual FP resonances can give small changes in sensitivity for small variations in the signal wavelength.

In summary it is clear that the EDFA can provide excellent performance in the three functions important in transmission systems; power, repeater and pre-amplifier (FIG.35). However it should not be forgotten that SLAs can be used in a variety of signal processing functions where the nature of the technology offers the prospect of multi function OEICs. These new chips will increase the functionality of future optical networks.

It is not only in the long haul transmission scenarios that the optical amplifier will have a significant impact. Any photonic based network e.g. multisensor highway will be amenable to upgrade through the use of optical amplification. The fundamental barriers that the system loss budget places on designs will be more or less removed, allowing their utilization in shorter distance, higher volume applications. The result will be increased system design freedom and flexibility bringing closer to reality the concept of broadband optical networking.

	Semiconductor	Fibre Amplifiers		
		Erbium	Raman	Brillouin
Small Signal Gain (dB)	>20dB	>20dB	5-10dB	20dB
Signal Wavelength λs (μm)	N/A	1.53	any	any
Pump Wavelength λp (nm)	N/A	807, 980 1480	Stokes shift below signal 100	0.1
Gain Bandwidth (nm)	20-50	10-40	20-40	0.001
Optical Pump Power (mW)	N/A	20-50	100-200	~10
Electrical Bias Current(mA)	50	>100	>500	>50
Saturated Output Point (mW)	10	15	Limited by pump power >100	
Coupling Loss (dB)	5-6	<1	<1	<1
Polarisation Sensitivity (dB)	<1	0	0	0
Noise Figure (dB)	~5	~3	~3	>20
Direction	bi	bi	bi	uni
Crosstalk/ Distortion	Significant	any below <100kHz	low	N/A

Table 1: Performance comparison of general amplifier characteristics.

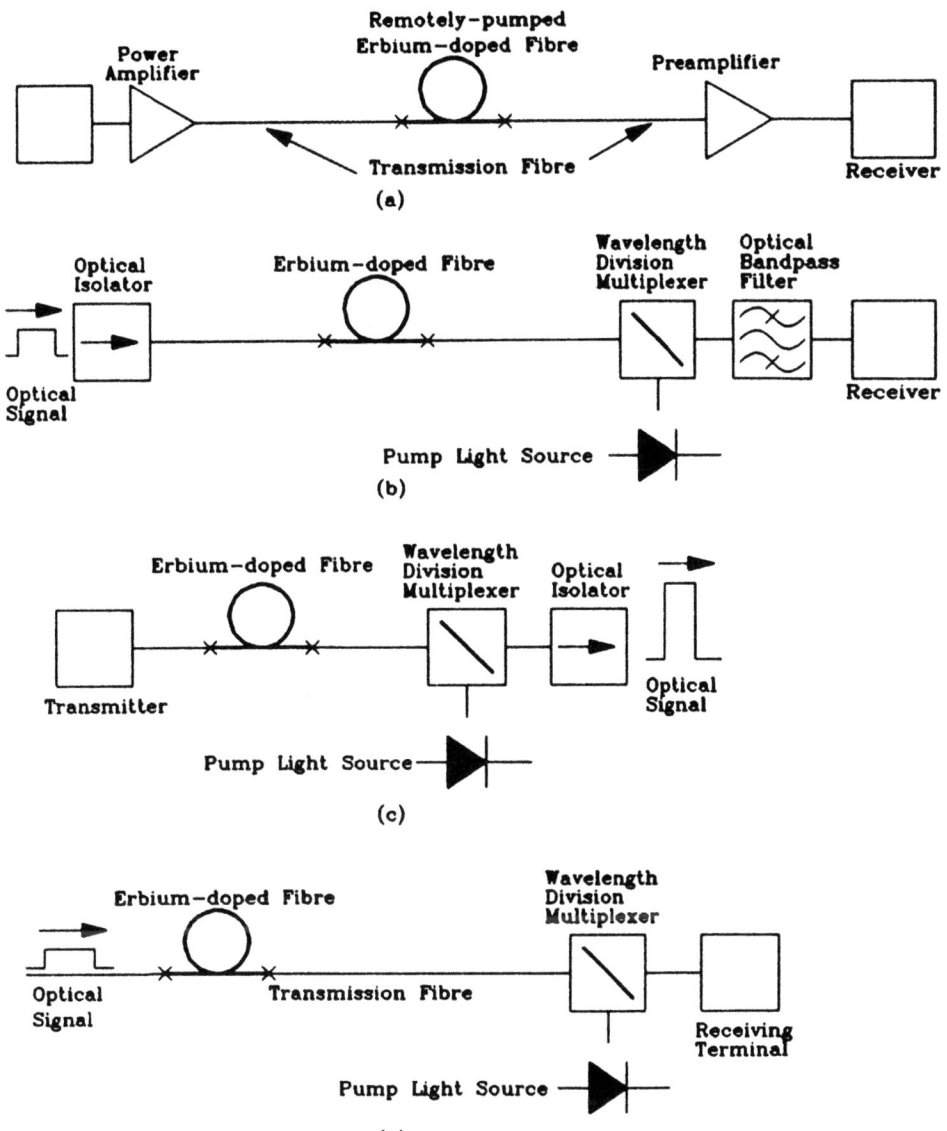

Fig.35 Schematic diagram summarising the functions that an EDFA can fulfil in transmission systems (a) repeaterless system (b) pre-amplifier EDFA (c) power EDFA (d) remotely pumped EDFA

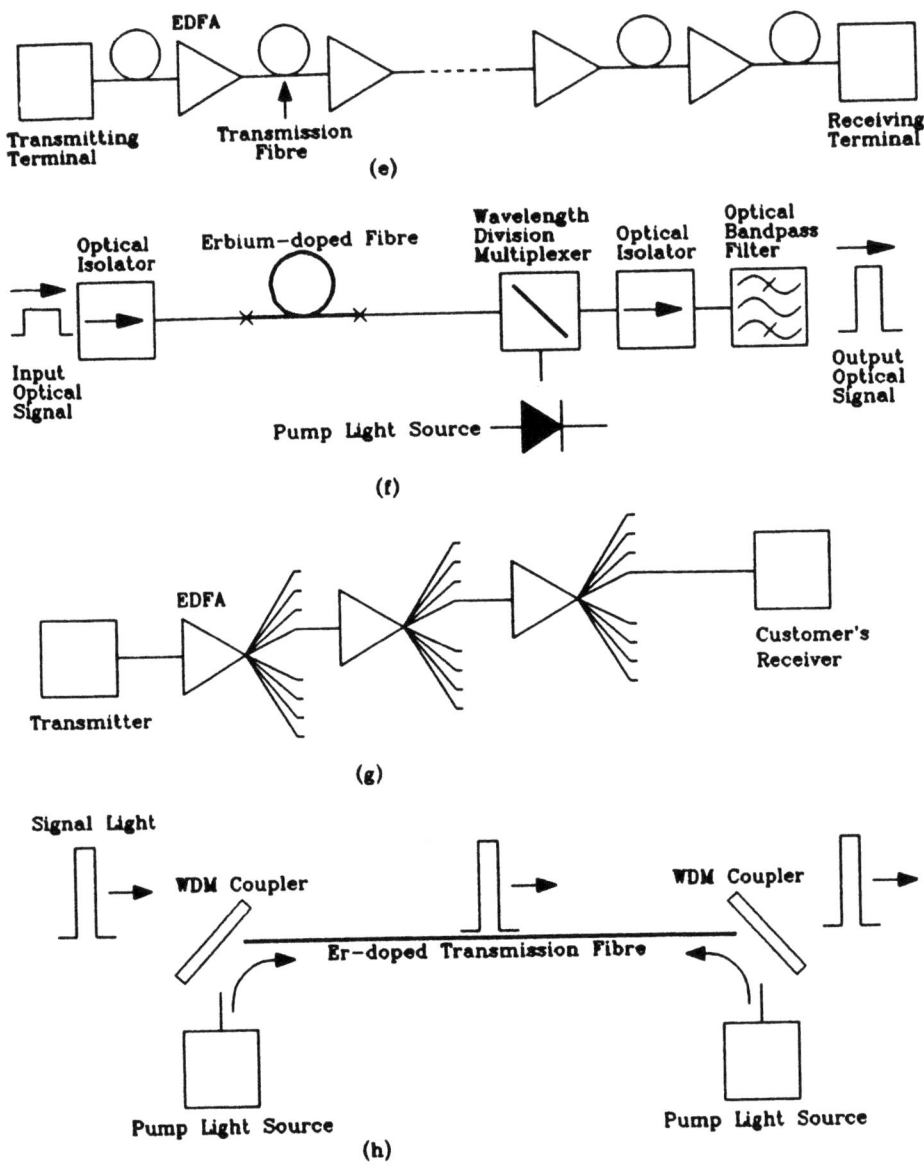

Fig.35 (Cont'd) Schematic diagram summarising the functions that an EDFA can fulfil in transmission systems (e) multirepeater (f) repeater -EDFA (g) EDFA - distribution system (h) distributed - EDFA with WDM

REFERENCES

1. Edagawa N (1993) Applications of Fiber Amplifiers to Telecommunication Systems", Chapter 12, pp536-647 in "Rare Earth Doped Fiber Lasers and Amplifiers" M F J Digonnet (Ed)
2. Li, T (1993) "The Impact of Optical Amplifiers on Long Distance Lightwave Telecommuncations" *Proc IEEE*, Vol 81, pp1568-1579
3. Smith, R G (1972) "Optical power handling capacity of low loss optical fibres as determined by stimulated Raman and Brillouin scattering" *Appl Opt*, Vol.11, pp2489-2494
4. Cotter, D (1982) "Observation of stimulated Brillouin scattering in low-loss silica fibre at 1.3μm" *Elect.Lett*, Vol.18, pp495-496
5. O'Mahony M (1988) "Semiconductor Laser Optical Amplifies for use in Future Fiber Systems", *IEEE J. Light Tech.*, Vol LT-6, pp531-544
6. Kobayashi S (1984) "Semiconductor Optical Amplifiers" *IEEE Spectrum*, pp26-33
7. Yamamoto Y (1980) "Characteristics of AlGaAs Fabry Perot Cavity type laser amplifiers" *IEEE J. Quant. Electron.*, Vol QE-16, pp1047-1052
8. O'Mahony M, Marshall, I M, Westlake, H J (1987) "Semiconductor laser amplifiers for optical communications systems", *British Telecom Tech. J.* Vol.5, No.3
9. Westbrook, L D "Measurements of dg/dN and dn/dN and their dependence on photon energy in 1.5μm InGaAsP laser diodes", *Proc IEE*, Vol 133, pp135-141
10. Adams M J, Collins, J V Henning I D (1985) "Analysis of semiconductor laser optical amplifiers" *Proc IEE*, Vol.132, pp58-63
11. Mukai, T, Yamamoto, Y, Kimura T (1983) "Optical direct amplification for fiber transmission" *Rev. Elec. Commun.Lab*, Vol.31, p340
12. Adams, M J Westlake H J, O'Mahony M, Henning I D (1985) "A comparison between active and passive bistability in semiconductors" *IEEE J Quant.Elect*. Vol QE-21
13. Mukai T (1981) "Gain, frequency bandwidth and saturation output power of AlGaAs DH laser amplifiers" *IEEE J Quant.Elect.*, Vol QE-17, pp1028-1034
14. Saitoh T, Mukai T Mikami O (1985) "Theoretical analysis and fabrication of anti reflection coatings on laser diode facets", *IEEE J Light.Tech.*, Vol LT-3, pp288-293
15. O'Mahony M (1985) "Low reflectivity semiconductor laser amplifier with 20dB fibre-to fibre gain at 1500nm" *Elect.Lett*, Vol.21, pp501-502
16. Ramaswami R (1990) "Amplifier induced crosstalk in multichannel optical networks" *IEEE J Light.Tech.*, Vol. LT-8, pp1882-1896
17. Cole S, Cooper, D M Devlin, W J, Ellis A D, Elton, D J Isaac J, Sherlock G, Spurdens P C Stallard W A (1989) *Elect.Lett.* Vol 25, pp314-316
18. Shimoda K (1957) "Fluctuation in amplification of quanta with application to maser amplifiers", *J Phys.Soc.* Japan, Vol.12, pp686-700
19. Yariv A (1991) "Optical Electronics" Holt
20. Mukai T (1982) "S/N and error rate performance in AlGaAs semiconductor laser preamplifier and linear repeater system", *IEEE J Quant. Elect.*, Vol QE-18, pp751-753
21. Olsson, N A (1989) "Lightwave systems with optical amplifiers" *IEEE J Light.Tech.*, Vol. LT-7, pp1071-1082
22. Mukai, T (1987) "5.2dB noise figure in a 1.5μm InGaAsP travelling wave laser amplifier", *Elect. Lett*, Vol.23, pp216-217.
23. Malyon, D J, Elton, D J, Regnault, J C, McDonald, S J, Devlin, W J, Cameron, K H, Bird, D M, Stallard, W A (1989) "Direct detection transmission experiment at 565 Mbits/s using cascaded in line optical amplifiers", *Elect.Lett.*, Vol.25, pp236-237
24. Malyon, D J, Stallard, W A (1989) "Bidirectional multilaser amplifier systems experiment" Elect. Lett., Vol.25, pp1366-1368
25. Blank, L C, Cox, J D (1989) "Optical time domain reflectometry on optical amplifier systems" *IEEE J. Light.Tech.*, Vol LT-7, pp1549-1555
26. Kalman, R F, Kazovsky, L G, Goodman J W, "Space division switches based on semiconductor optical amplifiers", *IEEE Phot.Tech.Lett.*, Vol, 4, pp1048-1051

27 Westlake H J, Adams, M J, O'Mahony M "Measurement of optical bistability in an InGaAsP laser amplifier at 1.5μm" *Elect.Lett.*, Vol.21, pp992-993

28 Weich, K, Hover, J, As, D J, Eggemann, R, Mohrle, M, Patzak, E "2.5 Gbits/s all-optical clocked decision and retiming circuit using bistable semiconductor lasers" (1994) *Elect.Lett.*, Vol.30, pp784-785

29 Sherlock G (1991) "1.3μm MQW semiconductor optical amplifiers with high gain and output power", *Elect.Lett.*, Vol.27., pp165-166

30 Tabuchi M (1990) "External grating tunable MQW laser with wide tuning of 240nm", *Elect.Lett.*, Vo.26, pp742-744

31 Digonnet, M J F (Ed) (1993) "Rare earth doped fiber lasers and ampliifers", Dekker

32 France P (Ed) (1992) "Optical fiber lasers and amplifiers", Blackie

33 Miniscalco W J (1991) "Erbium-doped glasses for fibre amplifiers at 1500nm" *IEEE J Light.Tech*, Vol. LT-9, pp234-250

34 Armitage, J R (1988) "Three level fiber laser amplifier: A theoretical model" *Appl.Opt.*, Vol.27, pp4831-4836

35 Desurvire E, Simpson, J R (1989) "Amplification of spontaneous emission in erbium doped single mode fibres", *IEEE J. Light.Tech.*, Vol LT-7, pp835-845

36 Dieke, G H, Crosswhite, W H (1963) "The doubly and triply ionised rare earths" *App. Opt.*, Vol.2, pp675-680

37 Laming, R I, Poole, S B, Tarbox, E J (1988) "Pump excited state absorpion in erbium doped fibers", *Opt. Lett.*, Vol13, pp1084-1086

38 Shimuzu, M, Horiguchi M, Yamada, M, Nishi, I, Uehara G, Noda, J, Sugita, E (1990) *Proc Int. Conf. Optical Fiber Communciations*, Paper PD17

39 Shimuzu, M, Yamada, M, Horiguchi, M Takeshita, T, Okayazu, M (1990) "Erbium doped fiber amplifiers with an extremely high gain coefficient of 11dB/mW" *Elect. Lett.*, Vol.26, pp1641-1643

40 Desurvire, E, Giles, C R, Simpson, J R, Zyskind, J L (1989) "Efficient erbium doped fibre amplifier at a 1.53μm wavelength with a high output saturation power", *Opt. Lett.*, Vol14, pp1266-1268

41 Mears, R J, Reekie, L, Jauncey, I M, Payne, D N (1987) "Low noise erbium doped fibre ampifier operating at 1.54μm". *Elect.Lett.*, Vol.23, pp1026-1028

42 Atkins, C H, Massicott, J F, Armitage, J R, Wyatt, R, Ainslie, B J, Craig-Ryan, S T (1989) "High gain, broad spectral bandwidth erbium doped fiber amplifier pumped near 1.5μm" *Elect.Lett.*, Vol.15, pp910-912

43 Kimura, Y (1992) "Gain characteristics of erbium doped fiber amplifiers with high erbium concentration" *Elect.Lett.*, Vol.28, pp1420-1422

44 Giles, C R, Desurvire, E, Talman, J R, Simpson, J R, Becker, P C (1989) "2 Gbit/sec signal amplification at λ=1.53μm in an erbium doped single mode fiber amplifier", *IEEE J Light.Tech*, Vol LT-7, pp651-656

45 Massicott, J R, Wyatt, R, Ainslie, B J, Craig-Ryan, S P (1990) "Efficient, high power, high gain Er^{3+} doped silica fiber amplifier". *Elect.Lett.*, Vol16, pp1038-1039

46 Pettitt, M J (1989) "Crosstalk in erbium doped fiber amplifiers", *Elect.Lett*, Vol.25, pp416-417

47 Meli, F (1992) "Gain crosstalk in saturated EDFA for WDM applications", *Elect.Lett.*, Vol.28, pp1896-1897

48 Zyskind, J L (1992) "Erbium doped fiber amplifiers and the next generation of lightwave systems", *A T & T Tech. J.* Jan-Feb, pp53-62

49 Desurvire, E (1990) "Analysis of noise figure spectral distribution in erbium doped fiber amplifiers pumped near 980nm and 1480nm", *Appl.Opt.* Vol.29, pp3118-3123

50 Desurvire, E (1990) "Spectral noise figure of Er^{3+} doped fiber amplifiers", *IEEE Photonics Tech.Lett.*, Vol.2, pp1071-1073

51 Tkach, R W (1992) "System implications of optical fiber non-linearities", Tech Dig. OSA Annual Meet, paper WX1

52 Stolen, R H, (1979) "Nonlinear properties of optical fibers" in *Optical Fiber Telecommuncations*, S E Miller, A Chynoweth (Eds) pp125-150, Academic
53 Davey, S T, Williams, D L, Spirit, D M, Ainslie, B J (1989) "The fabrication of low loss high NA silica fibres for Raman amplification". *Proc. SPIE* Vol.1191, paper 19
54 Spirit, D M, Blank, L C (1989) "Raman asssisted long distance optical time domain reflectometry" *Elect.Lett*, Vol.25, pp1687-1688
55. Ippen, E P, Stolen, R H (1972) "Stimulated Brillouin scattering in optical fibers", *Appl. Phys. Lett.*, Vol.21, pp539-542
56. Shibata N, (1988) *Opt.Lett*, Vol.13, pp595-596
57 Atkins, C G, Cotter, D, Smith D W, Wyatt, R (1986) *Elect.Lett.*, Vol.22, pp556-557
58 Shibata, N, Waarts, R G, Braun, R P, (1988) *Opt.Lett.*, Vol.13, pp269-270
59 Olsson, N A, Van der Ziel, J P, (1987) *IEEE J Light Tech.*, Vol.LT-5, pp147-152
60 Tkach, R W, Chraplyvy, A R (1989) *Proc Int. Conf. Optical Fibre Communcations*, Paper THG2
61 Aida K, Nishi S, Sata, Y, Hagimoto, K, Nakagawa K (1989) "1.8 Gbit/s 310km fibre transmission without outdoor repeater equipment using a remotely pumped in-line Er-doped fiber amplifier in an IM/DD system", *Tec.Dig.* ECOC, paper PDA-7
62 Edagawa, N, Yoshida, Y, Taka, H, Yamamoto, S, Mochizuki, K, Wagabayashi, H (1989) "904km 1.2 Gbit/sec non-regenerative optical transmission experiment using 12 Er-doped fiber amplifiers" *Tech. Dig.* ECOC paper PDA-8
63 Saito, S Imai, T, Sugie, T, Ohkawa, N, Ichihashi, Y, Ho, T (1990) "An over 2200km coherent transmission experiment at 2.5GBit/s using erbium doped fiber ampifiers" *Proc.Tech.Dig.* OFC, paper PD2
64 Henry P S (1985) "Lighwave Primer" *IEEE J Quant.Elect.*, Vol QE-21, pp1862-187
65 Hagimoto K, Nishi, S, Nakagawa K (1990) Optical bit-rate flexible transmission system with a 5 Tbit.km capacity employing multiple in-line erbium doped fiber amplifiers" *IEEE J Light.Tech.*, Vol.LT-8, pp1387-1395
66 Fujita, S, Kitamura, M Torikai, T, Henni, N Yamada, H Suzaki, T, Takano, I, Shikada, M (1989) "10 Gbit/sec 100km optical fiber transmission exeriment using high speed MQW DFB-LD and back illuminated GaInAs APD" *Elect.Lett.*, Vol.25, pp702-703
67 Marcuse, D (1991) "Single channel operation in very non-linear fibers with optical amplifiers at zero dispersion", *IEEE J Light.Tech.* Vol. LT-9, pp356-361
68 Taga, H, Yoshida, Y, Edagawa, N, Yamamoto, S, Wakabayashi, H (1990) "459km 2.4 Gbit/s four wavelength multiplexing optical fiber transmission experiment using six Er-doped fiber amplifiers", *Elect.Lett.*, Vol.26, pp500-501
69 Eichen, E E, McCabre, J, Miniscalco, W J, Olshansky, R, Wei, T (1990) "FM microwave multiplexed broadband distribution systems using Er^{3+} fiber amplifiers and pre-amplifiers", *IEEE Photon.Tech.Lett.*, Vol.2, pp220-222
70 Hasegawa, A (1990) "Optical Solitons in Fibers" Sringer Verlag
71 Iwatsuki, K, Nishi, S Saruwatari, M (1990) "2.8 Gbit/s optical soliton transmission employing all laser diodes", *Elect.Lett.*, Vol.26, pp1-2
72 Suzuki, K, Nakazawa, M (1990) "Automatic optical soliton control using cascaded Er^{3+} doped fiber amplifiers", *Elect.Lett.*, Vol.26, pp1032-1034
73 Mollenauer, L F, Neubelt, M J, Evangelides, S G, Gordon, J P, Simpson, J R, Cohen L G (1990) "Experimental study of soliton transmission over more than 10,000 km in dispersion shifted fiber", *Opt.Lett.*, Vol.15, pp1203-1205
74 Davey, S T, Williams, D L, Spirit, D M, Ainslie, J B (1990) "Lossless transmission over 10km of low dispersion erbium doped fibre using only 15mW pump power", *Elect.Lett.*, Vol.26, pp1148-1149
75 Nakazawa, M, Kimura, Y, Suzuki, K (1990) "Soliton transmission in a distributed, dispersion shifted erbium doped fiber amplifier" *Top.Meet. Tech. Dig.* "Optical Amplifiers and their Applications", Paper TUA7.
76 Spirit, D M, Blank, L C, Williams, D L, Davey S T, Ainslie, B J (1990) "+10dBm lossless transmission in 10km distributed erbium fibre amplifier", *Elect.Lett.*, Vol.26, pp1658-1659

77 Wargner, S S (1987) "Optical amplifiers applications in fiber optic local networks" *IEEE Trans. Comm.* Vol., COM-35, pp419-426
78 Blank, L C, Spirit, D M (1989). *Elect.Lett.*, Vol.25, pp1693-1694
79 Ainslie, B J, Craig, S P, Davey, S T (1987) "The fabrication and properties of Nd^{3+} in silica based optical fibers" *Mater.Lett.*, Vol.5, pp143-146
80 Miyajima, Y (1991) "Nd^{3+} doped fluoride fiber amplifier module with 10dB gain and high pump efficiency", *IEEE Photon.Rep.*, Vol.3, pp16-19
81 Shimuzu, M (1992) "Optical amplifiers and thier applications" paper PD3
82 Pedersen, J E (1990) "Noise characteristics of a neodymium doped fluoride fiber amplifier and its performance in a 2.4 Gbit/s sytems", IEEE *Photo.Tech.Lett.*, Vol.2, pp750-752
83 Suguawa, T (1992) "Noise characteristics of Pr^{3+} doped fluoride fibre amplifier" *Elect.Lett.*, Vol.28. pp246-247.

ANALYSIS OF ERBIUM-DOPED SILICA-FIBRE CHARACTERISATION TECHNIQUES

M.A. REBOLLEDO
Department of Applied Physics. University of Zaragoza
Facultad de Ciencias. 50009 Zaragoza. Spain.

1. Introduction.

Rare-earth-doped fibre amplifiers have raised great interest in recent years because of their potential application to optical fibre communications. In particular, erbium-doped fibre amplifiers (EDFAs) have been profusely studied because of their high performance in the third transmission window.

Modelling of EDFAs can be done using rate equations (including the evolution of the modes propagating in the fibre), which must be integrated. Prior to integration, the coefficients of the variables appearing in these differential equations must be evaluated. This requires the measurement of some of the parameters of the fibre, characterising its passive and active properties. Once these parameters have been measured, the model can be used to predict the performances of the EDFA. In this way an optimised design can be obtained.

In this paper experimental methods for making a complete characterisation of an erbium-doped silica-fibre are introduced and analysed in detail. Particular attention is paid to determining the best experimental conditions. It is shown that large errors appear when the experimental conditions are not carefully chosen.

2. Basic notions.

From the physical point of view, there is no essential difference between a bulk material erbium-doped amplifier and an EDFA. A scheme of at least three energy levels of the Er^{3+} ion is required, with two levels giving rise to the laser transition and a third allowing the pump transition. The analysis of the absorption spectrum gives us information on the possible pump wavelengths. Fig. 1 shows that the only possible laser transition for Er^{3+} in silica corresponds to wavelengths around 1530 nm. The main pump transitions are also shown in this figure. Fig. 2 shows the absorption spectrum of Er^{3+} in silica where the pump transitions shown in Fig. 1 can be identified. The broadening of these transitions is due to the fact that each of these levels is really a broadened Stark band. As usual, we assume that the broadening is homogeneous.

Pumping by the transition from $^4I_{15/2}$ to $^4S_{3/2}$ is usually done by way of a frequency doubled neodymium laser (530 nm) or an argon ion laser (514.5 nm). In the case of the transition from $^4I_{15/2}$ to $^4F_{9/2}$, dye lasers are used. Pumping by the transitions from $^4I_{15/2}$ to $^4I_{9/2}$, $^4I_{11/2}$ or $^4I_{13/2}$ can be done by diode lasers. Therefore these last three pump transitions are the most appropriate to obtain stable and small size EDFAs.

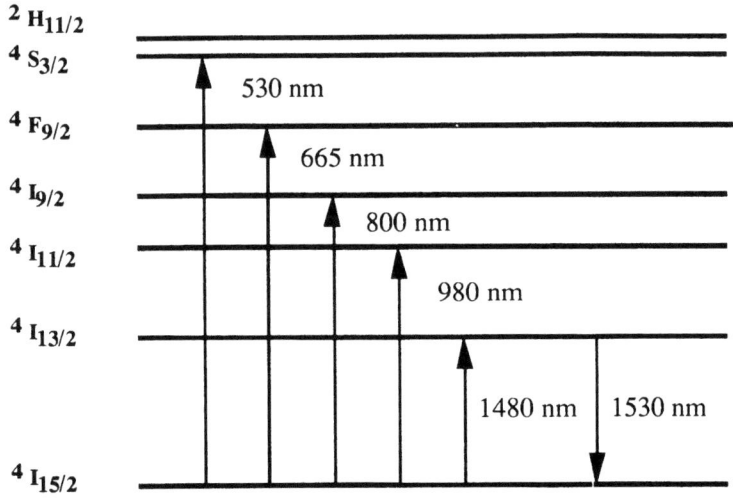

Figure 1. Laser transition and main pump transitions for Er^{3+} in silica.

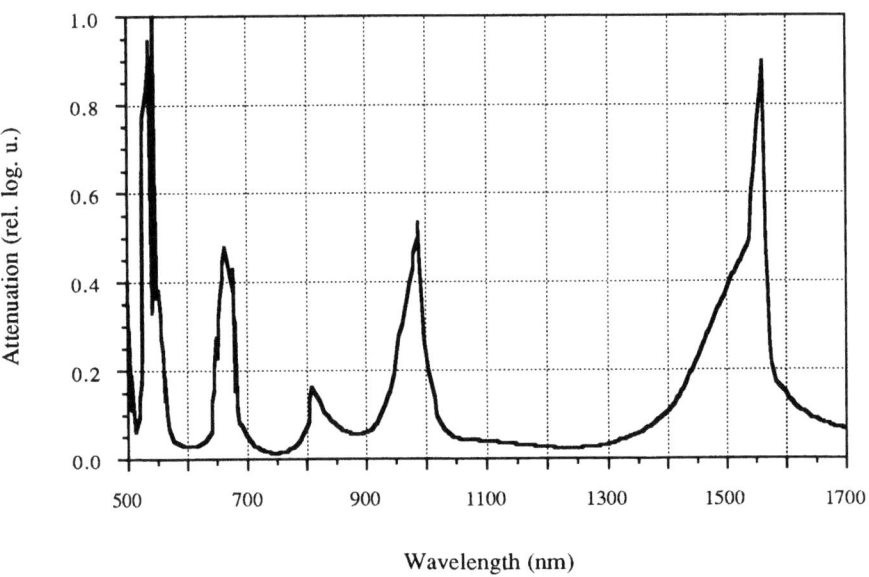

Figure 2. Absorption spectrum (in relative logarithmic units) of Er^{3+} in silica.

Pumping by the transition from $^4I_{15/2}$ to $^4I_{13/2}$ must be performed carefully because the laser transition takes place between these two bands. Therefore the absorption and stimulated emission cross sections in this spectral range must be examined to determine an efficient pump wavelength (λ_p) originating good absorption and small emission. Fig. 3 shows these cross sections. It can be observed that $\lambda_p=1480$ nm fits these requirements.

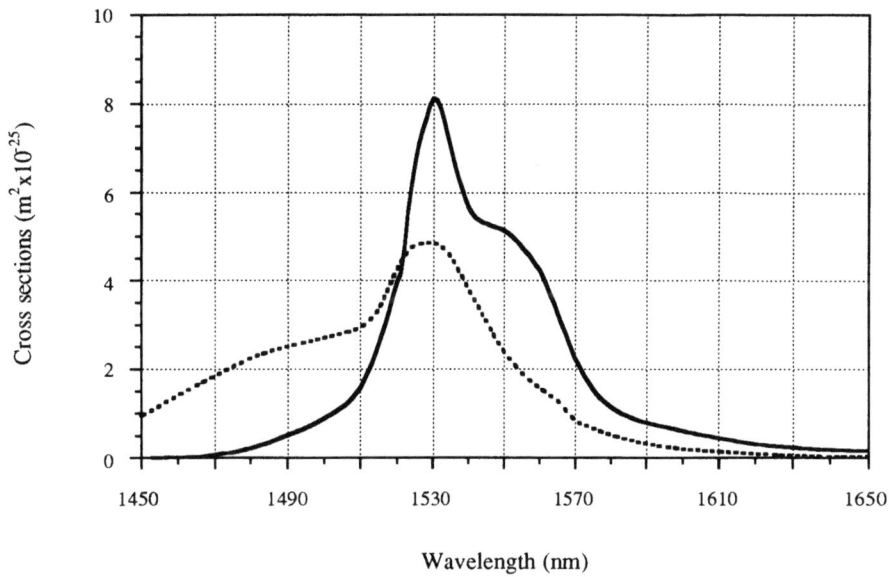

Figure 3. Absorption (dotted line) and stimulated emission (full line) cross sections of Er^{3+} in silica for the transitions between the $^4I_{15/2}$ and $^4I_{13/2}$ bands.

Once we have analysed the pump transitions, we should examine in more detail the scheme of levels (bands) taking part in the physical processes in an erbium-doped fibre (EDF). Fig. 4 shows a scheme where, besides the laser and pump bands, an additional band appears, to explain the fact that ions in the $^4I_{13/2}$ band are sometimes promoted to upper bands by absorption at the pump wavelength. This happens at pump wavelengths of about 530, 665, and 800 nm. This effect, named excited state absorption (esa), originates a diminution in the pump efficiency. As a consequence diode lasers emitting at 800 nm allow less efficient pumping than ones emitting at 980 or 1480 nm. Therefore, we shall only deal with these last two pump wavelengths.

In the particular case of pumping at 1480 nm (Fig.5), the $^4I_{13/2}$ band acts simultaneously as the upper laser band and as the pump band. Obviously, the levels inside the band acting as upper laser levels are different from those acting as pump levels. By contrast to what happens at other pump wavelengths, we must in this case consider that there is stimulated emission at pump wavelength (pe).

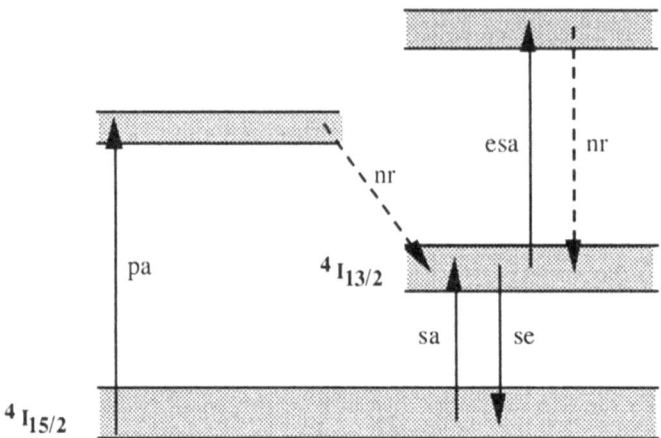

Figure 4. Scheme of bands taking part in the physical processes in an EDF; pa=pump absorption, esa=excited state absorption, sa=signal absorption, se=signal emission, nr=non-radiative transition.

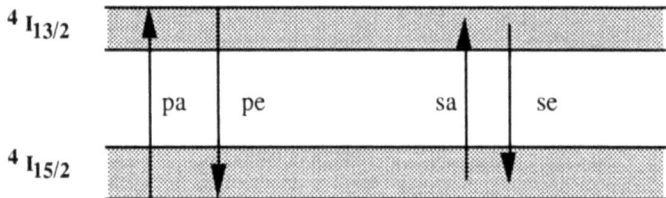

Figure 5. Scheme of bands for λ_p=1480 nm; pa=pump absorption, pe=pump emission, sa=signal absorption, se=signal emission.

Considering the schemes in Figs. 4 and 5 it is clear that pump and signal must be coupled into the EDF. Furthermore, at each point inside the doped region of the fibre a part of the spontaneous emission from the $^4I_{13/2}$ band is coupled into the fibre (Fig. 6). This originates copropagating and counterpropagating amplified spontaneous emission (ASE±). Therefore, modelling of the EDF requires the calculation of the evolution of pump, signal and ASE± powers. These powers can be represented by the functions $P^{\pm}_x(z,t,\nu)$ where z is the distance along the fibre axis, t the time and ν the optical frequency; x label is p for pumping, s for signal and f (fluorescence) for ASE; the + or - signs mean copropagating or counterpropagating waves along z axis. Pump, signal and ASE waves have a modal structure with intensity distribution $P^{\pm}_x(z,t,\nu)\,\Psi(r,\phi,\nu)$ depending on the radial, r, and azimutal, ϕ, coordinates, $\Psi(r,\phi,\nu)$ being the normalised intensity modal distribution.

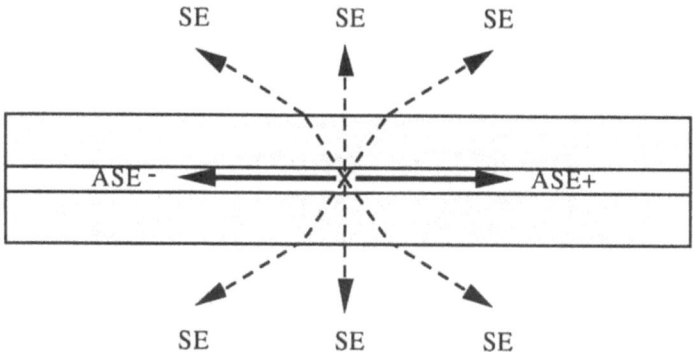

Figure 6. Generation of amplified spontaneous emission (ASE±) from the coupling of spontaneous emission (SE) at each point in the EDF.

3. Erbium-doped silica fibre modelling.

Consider a fibre which has been doped with erbium inside a circle of radius R, to obtain a radial erbium density distribution $n_T(r)$. The monochromatic pump is coupled at $z=0$ and monochromatic or polychromatic signal powers are coupled at one or both ends of the fibre. Propagation of pump $[P^+{}_p(z,t,\nu_p)]$ and signal $[P^\pm{}_s(z,t,\nu_s)]$ powers appears as a consequence of the coupled powers. Spontaneous emission originates polychromatic copropagating $[P^+{}_f(z,t,\nu_f)]$ and counterpropagating $[P^-{}_f(z,t,\nu_f)]$ ASE. To make the calculation of the ASE powers, the spectrum is divided in small slots of width $\Delta\nu$ around the frequencies ν_f, with $P^\pm{}_f(z,t,\nu_f)$ being the power in an slot.

These modes propagating along the fibre originate the transitions indicated in Fig.4 if $\lambda_p=980$ nm (no esa has to be considered) or the transitions indicated in Fig. 5 if $\lambda_p=1480$ nm. To model these transitions in a simple way they are considered as taking place between single energy levels (instead of bands) and effective cross sections are introduced which account for absorption and emission processes. In this way EDF characterisation is much more easy. By using this simple approach the transition probabilities can be calculated by the formulas

$$P_x(z,t,\nu) = P^+{}_x(z,t,\nu) + P^-{}_x(z,t,\nu) \qquad (x = p, s, f) \qquad (1)$$

$$W_p(z,r,\phi,t) = P_p(z,t,\nu_p)\, \Psi(r,\phi,\nu_p)\, \sigma_p(\nu_p) / (h\nu_p) \qquad (2)$$

$$W_a(z,r,\phi,t) = \sum_s P_s(z,t,\nu_s)\, \Psi(r,\phi,\nu_s)\, \sigma_a(\nu_s) / (h\nu_s)$$

$$+ \sum_f P_f(z,t,\nu_f)\, \Psi(r,\phi,\nu_f)\, \sigma_a(\nu_f) / (h\nu_f) \qquad (3)$$

$$W_e(z,r,\phi,t) = P_p(z,t,\nu_p)\,\Psi(r,\phi,\nu_p)\,\sigma_e(\nu_p)/(h\nu_p)$$

$$+ \sum_s P_s(z,t,\nu_s)\,\Psi(r,\phi,\nu_s)\,\sigma_e(\nu_s)/(h\nu_s)$$

$$+ \sum_f P_f(z,t,\nu_f)\,\Psi(r,\phi,\nu_f)\,\sigma_e(\nu_f)/(h\nu_f) \qquad (4)$$

In these equations W_p, σ_p account for the pump power absorption, W_a, σ_a for the signal and ASE absorption and W_e, σ_e for the stimulated emission originated by pump, signal and ASE powers; with W being transition probabilities and σ effective cross sections. In the case of λ_p=980 nm (Fig. 4), the strong non-radiative transition from the pump band to the upper laser band limitates considerably the population of the pump band. Therefore we can consider in any case that

$$n_1(z,r,\phi,t) + n_2(z,r,\phi,t) = n_T(r) \qquad (5)$$

with n_1, n_2 being the population distributions of the ground and excited laser bands ($^4I_{15/2}$ and $^4I_{13/2}$). These population distributions can be calculated from the rate equations

$$dn_1(z,r,\phi,t)/dt = -dn_2(z,r,\phi,t)/dt = [\,W_e(z,r,\phi,t) + \tau^{-1}\,]\,n_2(z,r,\phi,t)$$
$$- [\,W_p(z,r,\phi,t) + W_a(z,r,\phi,t)\,]\,n_1(z,r,\phi,t) \qquad (6)$$

where τ is the lifetime of the upper laser band. In the case where pump and signal powers do not depend on time, the last equation gives the steady-state solution

$$n_2(z,r,\phi) = [\,W_p(z,r,\phi) + W_a(z,r,\phi)\,]\,n_T(r)$$
$$\times [\,W_p(z,r,\phi) + W_a(z,r,\phi) + W_e(z,r,\phi) + \tau^{-1}\,]^{-1} \qquad (7)$$

If we now introduce the parameters

$$N_i(z,t,\nu) = \int_0^R r\,dr \int_0^{2\pi} d\phi\; n_i(z,r,\phi,t)\,\Psi(r,\phi,\nu) \qquad (i=1,2) \qquad (8)$$

(R being the doping radius of the EDF) the rate equations governing the evolution of $P^{\pm}_x(z,t,\nu)$ along the doped fibre can be expressed as

$$dP^+_p(z,t,\nu_p)/dz = P^+_p(z,t,\nu_p)\{\sigma_e(\nu_p)\,N_2(z,t,\nu_p) - \sigma_p(\nu_p)\,N_1(z,t,\nu_p)\} \qquad (9)$$

$$dP^{\pm}_s(z,t,\nu_s)/dz = \pm P^{\pm}_s(z,t,\nu_s)\{\sigma_e(\nu_s)\,N_2(z,t,\nu_s) - \sigma_a(\nu_s)\,N_1(z,t,\nu_s)\} \qquad (10)$$

$$dP^{\pm}_f(z,t,\nu_f)/dz = \pm 2h\nu_f(\Delta\nu)\sigma_e(\nu_f)N_2(z,t,\nu_f)$$

$$\pm P^{\pm}_f(z,t,\nu_f)\{\sigma_e(\nu_f)N_2(z,t,\nu_f) - \sigma_a(\nu_f)N_1(z,t,\nu_f)\} \quad (11)$$

in this way we have a set of coupled equations [(1) to (5) , (6) or (7) and (8) to (11)] that can be numerically integrated.

Figs. 7 and 8 give an example of the results obtained from the explained model. In our example we have assumed a fibre (f_{ex}) which is a step-index silica-fibre with uniform erbium density in the core (2×10^{24} m^{-3}), core and doped area radius R=4.5 µm, numerical aperture NA=0.1, τ=10 ms and σ_a, σ_e taken from Fig. 3. Copropagating pump (16 dBm at 1480 nm) and signal (-40 dBm at 1530 nm) modes are coupled into a 20 m long fibre. Fig. 7 shows the values of the pump, signal and total ASE± powers as functions of the distance z along the fibre axis. Fig. 8 shows the spectra of ASE±.

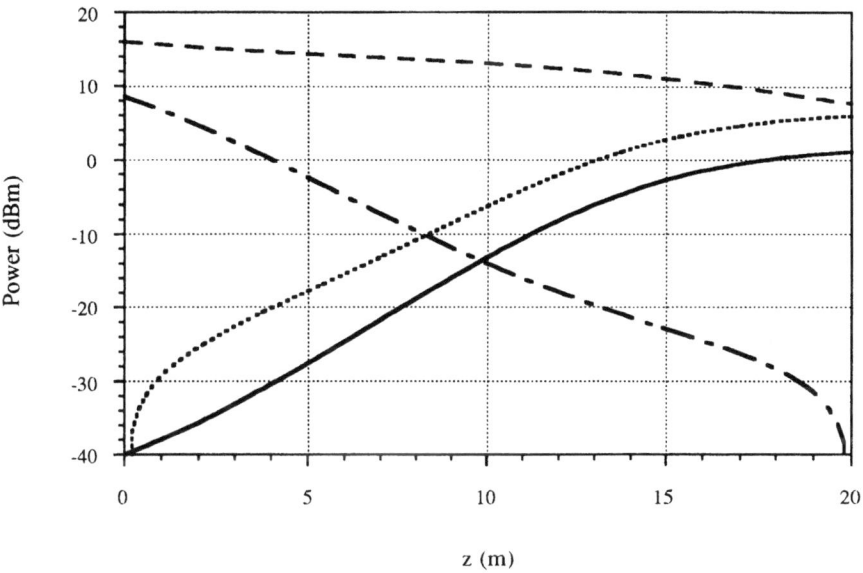

Figure 7. Evolution of pump (- - - -), signal (———),total ASE+ (········) and total ASE- (—·—·—) powers along the fibre axis for f_{ex}.

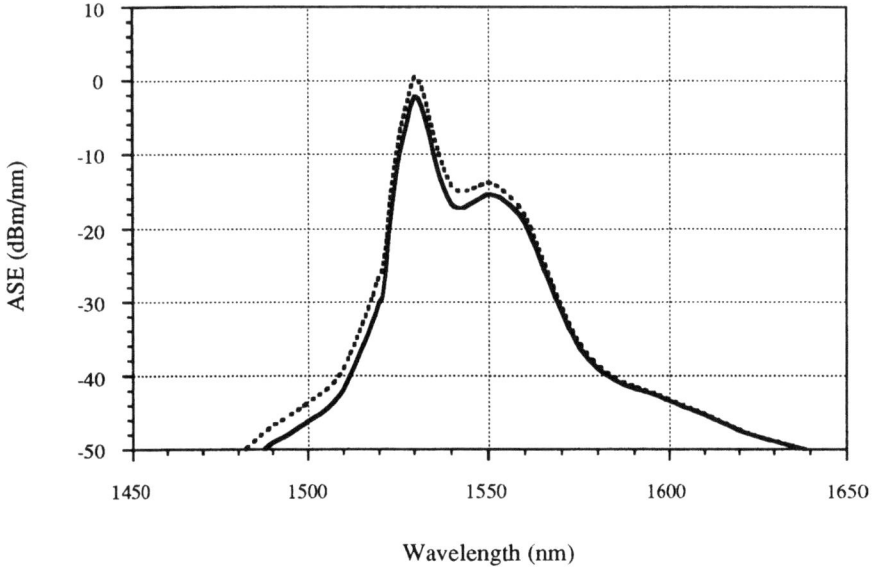

Figure 8. Spectrum of ASE+ (full line) and of ASE- (dotted line) for f_{ex}.

4. Fibre characterisation.

The equations detailed in the last section allow the calculation of the pump, signal and ASE± powers propagating in an EDF. However, prior to this calculation it is necessary to measure the unknown parameters appearing in the equations. This means measuring σ_p, σ_a, σ_e, the lifetime τ, the erbium density distribution $n_T(r)$ and the refractive index profile (the last to obtain the normalised intensity modal distribution $\Psi(r,\phi,\nu)$). Since the values of these parameters are not the same in the preform and in the fibre, it is very important to find techniques allowing for accurate measurements from a fibre sample.

The measurement of the refractive index profile in fibres can be done with a good accuracy from near-field or far-field modal analysis. The measurement of the erbium density distribution is mostly made on the fibre preform using techniques such as micro-Raman spectroscopy, secondary ion mass spectroscopy, electron probe microanalysis and X-ray fluorescence. However, the erbium density distribution changes when the fibre is drawn and direct measurements from the fibre should be expected to improve the accuracy. Recently some techniques have been studied such as near-field microscopy, differential mode launching and confocal optical microscopy operating in the fluorescence detection mode, but spatial resolution must be improved in order to get the required accuracy. In any event, these are well-known techniques and we shall not deal with them here.

As far as the measurement of cross sections and lifetimes in the EDF are concerned, special measurements techniques must be developed to determine these from fibre samples. We shall deal with these techniques in the following sections.

5. Absorption cross section measurement.

5.1. CHARACTERISATION TECHNIQUE.

Consider a small copropagating monochromatic constant power coupled into the EDF at the pump or signal wavelength, being able to meet the conditions

$$n_1(z,r,\phi) = n_T(r) \quad , \quad n_2(z,r,\phi) = 0 \qquad (12)$$

If we introduce the parameter

$$\eta_0(\nu) = \int_0^R r\,dr \int_0^{2\pi} d\phi \; n_T(r) \, \Psi(r,\phi,\nu) / N_T \qquad (13)$$

where N_T is the mean erbium density, we can easily obtain from Eqs. (9) and (10)

$$P^+_y(z+L,\nu) = P^+_y(z,\nu) \exp[-q_y(\nu)] \qquad (14)$$

where

$$q_y(\nu) = \sigma_y(\nu)\eta_0(\nu) N_T L \qquad (15)$$

and y can be p, s when used as subscrit of P^+ or p, a when used as subscrit of σ.

This means that $\sigma_y(\nu)$ can be determined if the attenuation coefficient

$$\alpha(\nu) = 10 \log[P^+_y(z,\nu) / P^+_y(z+L,\nu)] / L \qquad (\text{dB/m}) \qquad (16)$$

is measured by the cut-back technique and the following formula is used

$$\sigma_y(\nu) = \alpha(\nu) / [\, 10 \log(e)\, \eta_0(\nu)\, N_T \,] \qquad (17)$$

It can be observed that the refractive index profile (to determine $\Psi(r,\phi,\nu)$) and the erbium density distribution $n_T(r)$ have to be previously measured.

5.2. SYSTEMATIC ERRORS.

The formulas derived in the last paragraph are true if the conditions (12) are fulfilled. However, it is clear that those conditions can only be approximately accomplished because the power coupled into the fibre originates pump or signal absorption processes which populate the upper laser level. It is therefore interesting to study the solutions of equations (9) and (10) in the real conditions, in order to determine the systematic errors involved in the above technique.

For a small coupled power, the absorption and emission probabilities can be considered negligible when compared to the spontaneous emission rate τ^{-1}. In these conditions Eq. (7) can be written with a good accuracy as

$$n_2(z,r,\phi) = W_y(z,r,\phi)\, \tau\, n_T(r) \qquad (18)$$

If we introduce the parameter

$$\beta_0(\nu) = \int_0^R rdr \int_0^{2\pi} d\phi \; n_T(r) \; [\Psi(r,\phi,\nu)]^2 / N_T \qquad (19)$$

and consider Eqs. (1) , (2) , (3) and (8) we find that

$$N_2(z,\nu) = [\; P^+_y(z,\nu) / P^*(\nu)\;] N_T \qquad (20)$$

where $P^*(\nu)$ is an effective power defined as

$$P^*(\nu) = h\nu / [\; \sigma_y(\nu)\; \beta_0(\nu)\; \tau\;] \qquad (21)$$

If we replace now $n_1(z,r,\phi)$ by $n_T(r) - n_2(z,r,\phi)$ in Eqs. (9) and (10) we easily find that

$$dP^+_y(z,\nu) / P^+_y(z,\nu) = -\sigma_y(\nu)\; \eta_0(\nu)\; N_T\, dz$$

$$+ [\; \sigma_y(\nu) + \sigma_e(\nu)\;] [\; P^+_y(z,\nu) / P^*(\nu)\;]\; N_T dz \qquad (22)$$

A good approximation of Eq. (22) can be obtained if $P^+_y(z,\nu)$ is replaced in the right hand member by the expression given by Eq. (14). Once the resulting equation is integrated we get

$$P^+_y(z+L,\nu) = P^+_y(z,\nu)\; \exp[\; -\sigma'_y(\nu)\; \eta_0(\nu)\; N_T\; L\;] \qquad (23)$$

where

$$\sigma'_y(\nu) = \sigma_y(\nu) - [\;\sigma_y(\nu) + \sigma_e(\nu)\;]\{\; P^+_y(z,\nu) / [\eta_0(\nu)\; P^*(\nu)]\;\}$$

$$\times \{\; 1 - \exp[-\sigma_y(\nu)\; \eta_0(\nu)\; N_T\, L]\;\} / [\sigma_y(\nu)\; \eta_0(\nu)\; N_T\, L] \qquad (24)$$

Therefore the relative systematic error involved in the determination of $\sigma_y(\nu)$ through the technique explained in 5.1 can be calculated by

$$e_S[\sigma_y(\nu)] = [1+\sigma_e(\nu)/\sigma_y(\nu)]\; P^+_y(z,\nu)\{1-\exp[-q_y(\nu)]\}/[\eta_0(\nu)\; P^*(\nu) q_y(\nu)] \qquad (25)$$

To get an idea of the magnitude of this error, we can make some calculations with the fibre described in section 3 (f_{ex}). If a small signal at 1530 nm (y=s for power and a for cross section) is coupled into the EDF, it follows that $[1+\sigma_e(\nu)/\sigma_a(\nu)] = 2.68$, $\eta_0(\nu) = 0.693$ and $P^*(\nu) = 3.15$ mW. Since the maximum value of $\{1-\exp[-q_a(\nu)]\}/q_a(\nu)$ is one for very small fibre lengths, the maximum systematic error at 1530 nm is

$$e_S[\sigma_a(\lambda=1530\text{ nm})]_{max} = 1228\; P^+_s(z,\nu) \qquad (26)$$

where $P^+_s(z,\nu)$ is expressed in W. This means that about 1% error is obtained for $P^+_s(z,\nu) = 8$ µW. If this calculation is repeated for different wavelengths from 1450 to 1650 nm, it is found that systematic errors smaller than 1% are obtained if powers not larger than 1 µW are coupled into the fibre to measure $\sigma_a(\nu)$. Since powers of this magnitude can be easily detected, systematic errors can be neglected if the coupled power is less than the maximum allowed limit.

5.3. RANDOM ERRORS.

We shall now take into account the random errors involved in the measurement of $P^+_y(z+L,\nu)$ and $P^+_y(z,\nu)$, which are due to faults at the end surface of the fibre and inaccuracies of the power meter due to misalignment of the fibre, polarisation and temperature fluctuations, non-uniformities of the detector surface,..... We can define the random error as the relative standard deviation (sd) of the values of the corresponding parameter

$$e_R[\sigma_y(\nu)] = sd[\sigma_y(\nu)] / \sigma_y(\nu) \qquad (27)$$

Following the theory of propagation of errors we have

$$e_R[\sigma_y(\nu)] = [\sigma_y(\nu)]^{-1} \{ ([\partial \sigma_y(\nu) / \partial P^+_y(z,\nu)] \, sd[P^+_y(z,\nu)])^2$$

$$+ ([\partial \sigma_y(\nu) / \partial P^+_y(z+L,\nu)] \, sd[P^+_y(z+L,\nu)])^2 \}^{1/2} \qquad (28)$$

From Eqs. (14) and (15) it follows that

$$\sigma_y(\nu) = [\eta_0(\nu) \, N_T \, L]^{-1} \ln[P^+_y(z,\nu) / P^+_y(z+L,\nu)] \qquad (29)$$

which can be used in Eq. (28) to evaluate $e_R[\sigma_y(\nu)]$. If the relative standard deviations of $P^+_y(z,\nu)$ and $P^+_y(z+L,\nu)$ are assumed to be equal, it is found that

$$e_R[\sigma_y(\nu)] = \{ \sqrt{2} / \ln[P^+_y(z,\nu) / P^+_y(z+L,\nu)] \} \, e_R[P^+_y(\nu)] \qquad (30)$$

Finally if Eqs. (14) and (15) are taken into account in the last equation we get

$$e_R[\sigma_y(\nu)] = \{ \sqrt{2} / q_y(\nu) \} \, e_R[P^+_y(\nu)] \qquad (31)$$

Eq. (31) means that $e_R[\sigma_y(\nu)]$ decreases as L increases. However the attenuation increases as L does and large values of L can lead to excessively small values of the output power (consider that in our example the input power must be about 1 µW to prevent us from large systematic errors). These effects are shown in Figs. 9, 10 and 11, wher it can be observed that a good choice to measure the absorption cross section at wavelengths near 1530 nm would be L=10 m. In this way an output signal of about 1 nW would be obtained and the cross section would be determined with about 1% error (consider that the errors involved in the measurement of P^+_y are usually a few percents).

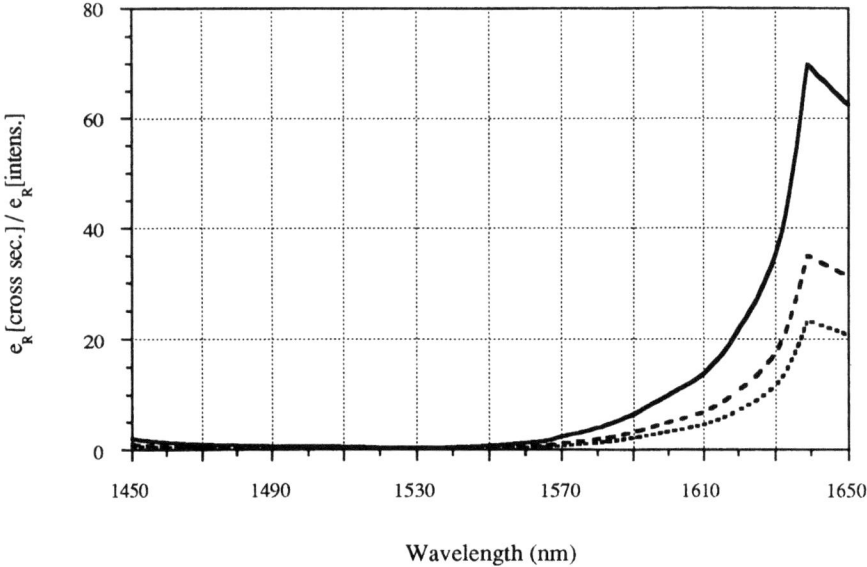

Figure 9. Random error in $\sigma_a(\lambda)$ for f_{ex} when L=5m (———) , L=10 m (- - -) and L=15 m (······).

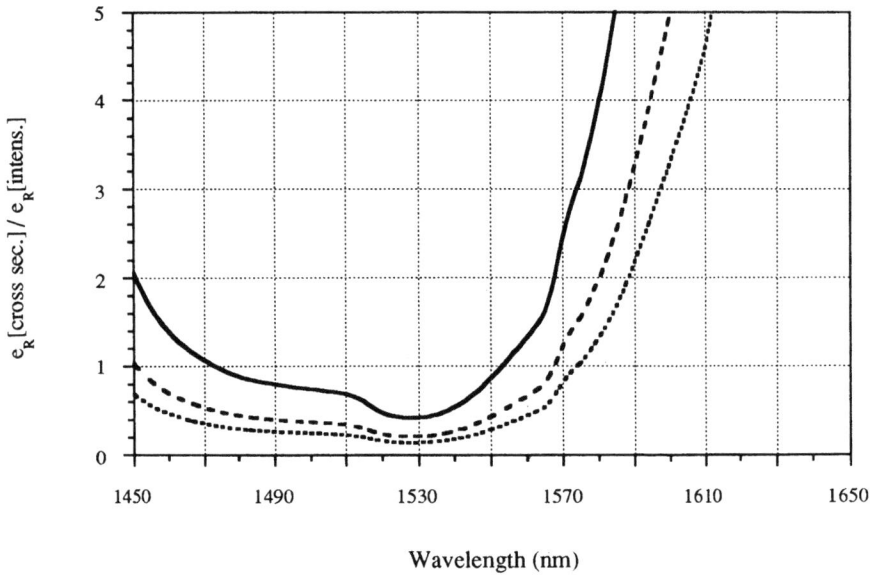

Figure 10. Detail of the minimum in Fig. 9.

It can be observed that for L=15 the improvement in the error is very small and however the output signal decreases strongly. It can also be observed that very large errors are involved in the measurement of $\sigma_y(v)$ at some optical frequencies. For those cases larger values of L should be appropriate to improve the results. As a consequence several values of L should be used in order to obtain small errors over the whole range of the absorption spectrum.

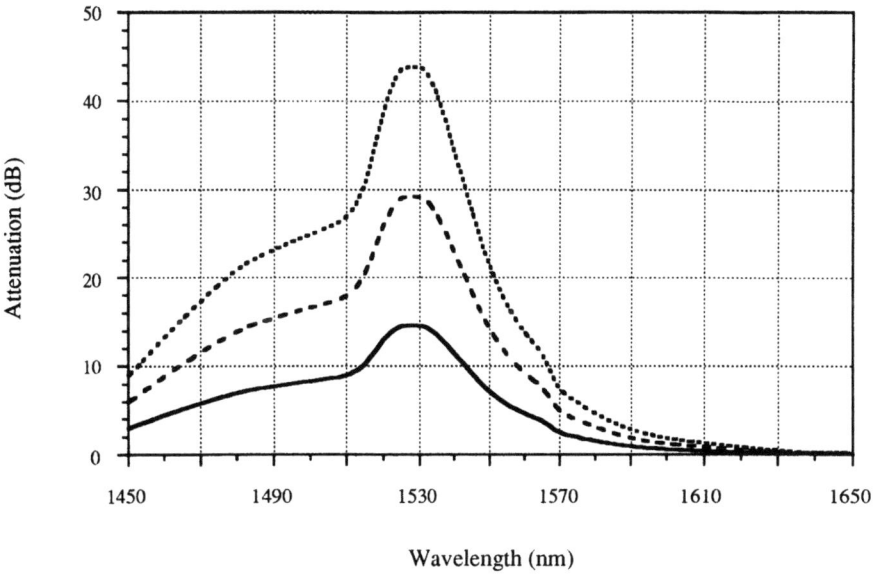

Fig 11. Signal attenuation for f_{ex} when L=5m (———), L=10 m (- - -) and L=15 m (······).

6. Stimulated emission cross section measurement.

6.1. FLUORESCENCE MEASUREMENT.

Since the stimulated emission cross section σ_e appears in Eq. (11), we should be able to obtain it from ASE measurements. To get an easy solution of Eq. (11) we can now consider an EDF (where only pump power is coupled) of length L, sufficiently short, to get values of $n_1(z,r,\phi)$ and $n_2(z,r,\phi)$ which remain unchanged along the fibre axis. In this case $N_1(z,v)$ and $N_2(z,v)$ (Eq. (8)) also remain constant along the fibre axis and we easily find from Eqs. (9) and (11) that

$$P^+_p(L,v_p) = P^+_p(0,v_p) \exp[-\gamma_p(v_p) L] \qquad (32)$$

with

$$\gamma_p(v_p) = \sigma_p(v_p) N_1(0,v_p) - \sigma_e(v_p) N_2(0,v_p) \qquad (33)$$

and

$$P^+_f(L, \nu_f) = [\zeta(\nu_f) / \gamma_f(\nu_f)] \{\exp[\gamma_f(\nu_f) L] - 1\} \quad (34)$$

with

$$\zeta(\nu_f) = 2h\nu_f (\Delta\nu) \sigma_e(\nu_f) N_2(0, \nu_f) \quad (35)$$

$$\gamma_f(\nu_f) = \sigma_e(\nu_f) N_2(0, \nu_f) - \sigma_a(\nu_f) N_1(0, \nu_f) \quad (36)$$

Since $\sigma_a(\nu_f)$ appears in $\gamma_f(\nu_f)$, it means that this quantiy must be previously measured and introduced in Eq. (36) to be able to obtain $\sigma_e(\nu_f)$. As this is a source of error we should try to simplify the technique, trying to avoid previous measurement of $\sigma_a(\nu_f)$. This can be done for very small fibre lengths because we can approximate Eq. (34) by

$$P^+_f(L, \nu_f) = \zeta(\nu_f) L \quad [\text{if} \quad \exp[\gamma_f(\nu_f) L] - 1 \cong \gamma_f(\nu_f) L] \quad (37)$$

which allows the determination of $\sigma_e(\nu_f)$ more easily. For fibre lengths fulfilling the above condition it is clear that $P^+_p(L, \nu_p)$ can be approximated by $P^+_p(0, \nu_p)$ (Eq. (32)). If the fact that small ASE powers are obtained in these conditions (as we will see in the next paragraph) is also considered, it follows that the variation of the populations n_1, n_2 with z is negligible.

To find out if the condition in Eq. (37) leads to experimental situations which can be acceptable, we shall now study as an example the case where the population of the upper laser level is saturated. This obliges a large pump power for which only the pump terms are non-negligible in Eq. (7) and therefore we obtain

$$n_{1sat} = \sigma_e(\nu_p) n_T(r) / [\sigma_p(\nu_p) + \sigma_e(\nu_p)]$$
$$n_{2sat} = \sigma_p(\nu_p) n_T(r) / [\sigma_p(\nu_p) + \sigma_e(\nu_p)] \quad (38)$$

and from Eq. (8)

$$N_{1sat}(0,\nu) = \sigma_e(\nu_p) N_T \eta_0(\nu) / [\sigma_p(\nu_p) + \sigma_e(\nu_p)]$$
$$N_{2sat}(0,\nu) = \sigma_p(\nu_p) N_T \eta_0(\nu) / [\sigma_p(\nu_p) + \sigma_e(\nu_p)] \quad (39)$$

If we consider a wavelength of 1530 nm for the fibre f_{ex}, we obtain $\gamma_f(\nu_f)=0.9642$ m^{-1} (about its maximum value). This means that, in order to get an applicability of Eq. (37) with less than 1% error for 1530 nm, we must take L=2 cm. For this length we get from Eqs. (35) and (37) that $P^+_f(1 \text{ cm}, 1530 \text{ nm})=0.68$ nW. At other wavelengths $\gamma_f(\nu_f)$ takes smaller values and therefore the approximation works better than at 1530 nm, but the values of P^+_f are smaller. In any event there are light detectors capable of detecting these small powers without any problem.

From the above comments it follows that the measurement of the ASE+ spectrum seems to be useful in the measurement of $\sigma_e(\nu)$. In fact Eqs. (1), (2), (4), (7), (8), (35) and (37) give

$$\sigma_e(\nu_f) = P^+{}_f(L,\nu_f) / [\, 2h\nu_f (\Delta\nu)\, L\, N_2(0,\nu_f)\,] \qquad (40)$$

with

$$N_2(0,\nu_f) = \int_0^R r\,dr \int_0^{2\pi} d\phi\, n_T(r)\, \Psi(r,\phi,\nu_f)$$

$$\times \{\, 1 + \sigma_e(\nu_p)/\sigma_p(\nu_p) + h\,\nu_p / [\tau\, P^+{}_p(0,\nu_p)\, \Psi(r,\phi,\nu_p)\, \sigma_p(\nu_p)]\,\}^{-1} \qquad (41)$$

It can be seen that the refractive index profile and the erbium density distribution have to be previously measured, as we saw in section 5.1 where the technique to determine $\sigma_y(\nu)$ was studied. Furthermore, τ, $\sigma_p(\nu_p)$ and $\sigma_e(\nu_p)$ must be previously determined. In the particular case of $\sigma_e(\nu_p)$ (which is non-null only for $\lambda_p = 1480$ nm), the determination could be done from $P^+{}_f(L,\nu_p)$ by using (40) and (41). But considering that if $\lambda_p = 1480$ nm the pump power at the end of the fibre is much larger than the ASE+ power at the same wavelength, it follows that $P^-{}_f(L,\nu_p)$, instead of $P^+{}_f(L,\nu_f)$, should be measured and used in Eq. (40). Although Eqs. (40) and (41) seem to show that absolute values of $\sigma_e(\nu_f)$ can be obtained from fluorescence measurements, we must make some remarks on the factor $2h\nu_f$ (the power of two photons per unit frequency) which appears in Eq. (40). This is the same factor included in Eq. (11) allowing the calculation of spontaneous emissions taking place in the direction of the stimulated emission, that give us the number of spontaneously emitted photons which are coupled into the fibre in the positive or negative direction of the fibre axis. This factor is obtained considering that the radiation field is well represented by n-photon states where the number of photons are not allowed to fluctuate. However, photon statistics clearly show that light beams cannot, in general, be described by such states, which means that we should introduce some correction in this factor. Since this fact is not included in the usual models on EDF, this prevents us from making absolute measurements of $\sigma_e(\nu_f)$. We can, of course, make uncalibrated measurements, which means that this technique allows for the determination of the shape of $\sigma_e(\nu_f)$.

As far as random errors are concerned, the same reasons and methods given in section 5.3 are valid here, and it is evident from Eq. (40) that

$$e_R[\sigma_e(\nu_f)] = e_R[P^{\pm}{}_f(L,\nu_f)] \qquad (42)$$

which means that $\sigma_e(\nu_f)$ can be obtained by this technique with a few percents error. However, as we have seen earlier, this technique only allows the determination of the shape of $\sigma_e(\nu_f)$. Therefore, we have to study other possibilities which allow for the absolute measurement of σ_e, at least at a given frequency.

6.2. GAIN MEASUREMENT.

Since the stimulated emission cross section σ_e also appears in Eq. (10), it can be determined from gain measurement. The most easy solution of this equation corresponds to the case of a short EDF, where copropagating monochromatic strong pump and small signal powers have been coupled. In this case $n_1(z,r,\phi)$, $n_2(z,r,\phi)$ take the saturation values given by Eq. (38) and $N_1(z,\nu_s)$, $N_2(z,\nu_s)$ take the values given by Eq. (39). The solution of Eq. (10) is in this case

$$P^+_s(z+L,\nu_s) = P^+_s(z,\nu_s) \exp[\,\sigma(\nu_s)\,\eta_0(\nu_s)\,N_T\,L] \qquad (43)$$

Where

$$\sigma(\nu_s) = [\,\sigma_e(\nu_s)\,\sigma_p(\nu_p) - \sigma_a(\nu_s)\sigma_e(\nu_p)\,]\,/\,[\sigma_p(\nu_p)+\sigma_e(\nu_p)] \qquad (44)$$

Therefore, if the gain coefficient defined as

$$g(\nu_s) = 10\,\log[\,P^+_s(z+L,\nu_s)\,/\,P^+_s(z,\nu_s)\,]\,/\,L \qquad (\text{dB/m}) \qquad (45)$$

is measured by the cut-back technique, we have

$$\sigma(\nu_s) = g(\nu_s)\,/\,[\,10\,\log(e)\,\eta_0(\nu_s)\,N_T\,] \qquad (46)$$

Since for $\lambda_p=980$ nm is $\sigma_e(\nu_p)=0$ then $\sigma(\nu_s)=\sigma_e(\nu_s)$. For $\lambda_p=1480$ nm is $\sigma_e(\nu_p)\neq 0$ and $\sigma_p(\nu_p)$, $\sigma_a(\nu_s)$, $\sigma_e(\nu_p)\,/\,\sigma_e(\nu_s)$ have to be measured previously to determine $\sigma_e(\nu_s)$ from $\sigma(\nu_s)$. In the particular case of $\sigma_e(\nu_p)\,/\,\sigma_e(\nu_s)$, its value can be obtained from fluorescence measurements as explained earlier. We have to note that, as in the case of the earlier discussed techniques, the refractive index profile and the erbium density distribution have to be previously measured, because they are necessary in order to calculate $\eta_0(\nu_s)$.

Since the conditions assumed at the beginning of this section require high pump powers which cannot in general be reached, we are going to extend the solution to more realistic conditions. Consider a pump power which is able to obtain values of $n_1(z,r,\phi)$, $n_2(z,r,\phi)$ near the saturation. We can assume that n_1 and n_2 continue being proportional to $n_T(r)$ as in Eq. (38), but allowing a dependence on z

$$n_i(z,r,\phi) = C_i(z)\,n_T(r) \quad (i=1,2) \qquad (47)$$

In these conditions Eq. (8) gives

$$N_i(z,\nu_s) = \eta_0(\nu_s)\,C_i(z)\,N_T = \eta_0(\nu_s)\,n_i(z) \qquad (48)$$

where $n_i(z)$ is the mean value of $n_i(z,r,\phi)$ in the doped area. A simple relation between $n_1(z)$ and $n_2(z)$ can immediately be obtained from Eq. (5)

$$n_1(z) + n_2(z) = N_T \qquad (49)$$

To find an expression for $n_2(z)$, we must integrate Eq. (7) over the doped area. Since the signal power is small, we can neglect its effect on $n_i(z)$. Furthermore, considering the small length of the fibre, we can assume a negligible effect of the ASE powers on $n_i(z)$. With these assumptions we find

$$n_2(z) = n_{2sat} / [1 + a/P^+_p(z,v_p)] \qquad (50)$$

where

$$a = [\pi R^2 h v_p] / \{\tau [\sigma_p(v_p) + \sigma_e(v_p)] n_0(v_p)\} \qquad (51)$$

If we now replace Eq. (32) in Eq. (50) and use the result to integrate Eq. (10), we find

$$P^+_s(z+L,v_s) = P^+_s(z,v_s) \exp[\sigma'(v_s) n_0(v_s) N_T L] \qquad (52)$$

where

$$\sigma'(v_s) = \sigma(v_s) + [\sigma_a(v_s) + \sigma_e(v_s)] \sigma_p(v_p) / \{[\sigma_p(v_p) + \sigma_e(v_p)] \gamma_p(v_p) L\}$$

$$\times \ln[(b+1) / \{b \exp[\gamma_p(v_p) L] + 1\}] \qquad (53)$$

with

$$b = a / P^+_p(z,v_p) \qquad (54)$$

Finally from Eqs. (33), (48) and (50) we find

$$\gamma_p(v_p) = \sigma_p(v_p) n_0(v_p) N_T \{b / [b+1]\} \qquad (55)$$

From the above formulas we can evaluate the systematic error involved in the approximation (43). It can easily be found that

$$e_s[\sigma(v_s)] = \{[\sigma_a(v_s) + \sigma_e(v_s)] / [\sigma_e(v_s) \sigma_p(v_p) - \sigma_a(v_s) \sigma_e(v_p)]\}$$

$$\times \{\sigma_p(v_p) / [\gamma_p(v_p) L]\} \ln[(b+1) / \{b \exp[\gamma_p(v_p) L] + 1\}] \qquad (56)$$

Fig.12 illustrates the values of $e_s[\sigma(v_s)]$ for the fibre f_{ex} when pump (at 1480 nm) and signal (-40 dBm at 1530 nm) are coupled into a 2 m long fibre and L is equal to 1.9 m. It is evident that the errors are excessive. Therefore the approximation (43) does not work well for realistic values of the pump power. It can also be observed that there is a good agreement between the values given by Eq. (56) and the systematic errors obtained from numerical integration of Eq. (10), which means that Eqs. (52) to (55) give an accurate description of the EDF for realistic values of the pump power. As a consequence these formulas can be used to obtain a measurement technique for $\sigma_e(v_s)$. This can be done by measuring the gain coefficient [$g(v_s,P_p)$] with a small value of L, small signal and several values of the pump power [P_p] at the input of the length L, by the cut-back technique

$$g(v_s,P_p) = 10 \log[P^+_s(z+L,v_s) / P^+_s(z,v_s)] / L \quad (dB/m) \qquad (57)$$

For each value of P_p the parameter $\sigma'(v_s,P_p)$ is obtained by

$$\sigma'(v_s,P_p) = g(v_s,P_p) / [10 \log(e) n_0(v_s) N_T] \qquad (58)$$

and then the function

$$\sigma'(v_s,P_p) = \sigma(v_s) + [A(v_s)/D] \ln\{(C+1) / [C \exp(D) + 1]\} \qquad (59)$$

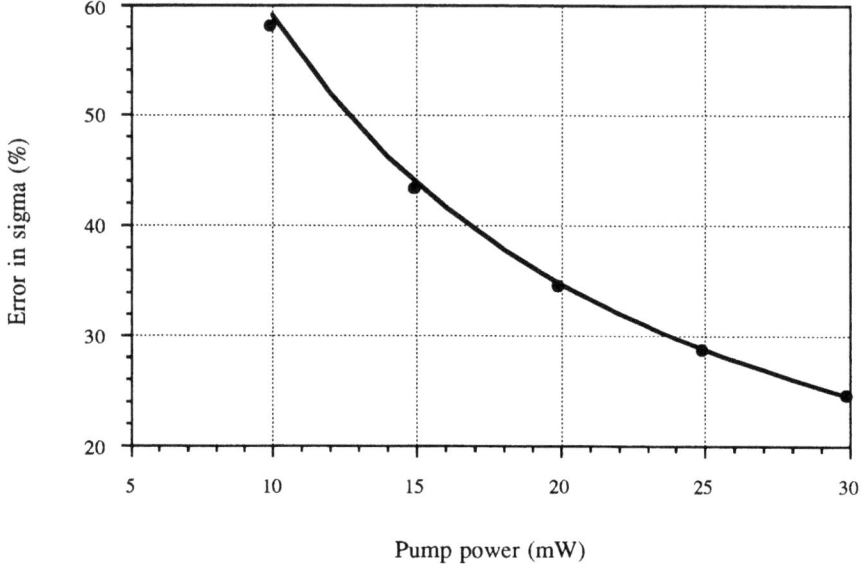

Figure 12. Systematic error in the determination of $\sigma(\nu_s)$ from Eqs. (44), (45) and (46) for f_{ex} when L=1.9 : values from Eq. (56) (full line) or from numerical integration of Eq. (10) (dots).

with
$$C = B / P_p \quad , \quad D = \sigma_p(\nu_p)\, \eta_0(\nu_p)\, N_T L\, [\, C / (C+1)\,] \qquad (60)$$

is fitted to the experimental values of $\sigma'(\nu_s, P_p)$. The parameters which must be varied to obtain the best fitting are $\sigma(\nu_s)$, $A(\nu_s)$ and B. The value of $\sigma_e(\nu_s)$ is obtained from the one of $\sigma(\nu_s)$, as explained earlier. Computer simulation shows that, in the case of f_{ex} with the conditions of Fig. 12, the value of $\sigma(\nu_s)$ can be obtained with a systematic error of about 1%. Therefore, the technique as described allows the absolute determination of $\sigma_e(\nu_s)$ with a small systematic error.

It is now the appropriate moment to study the random errors involved in this technique. Considering Eq. (52) and proceeding as in section 5.3, it is easily found that

$$e_R[\sigma'(\nu_s)] = \{\, \sqrt{2} / q_e(\nu_s)\, \}\, e_R[P^+_s(\nu_s)] \qquad (61)$$

where
$$q_e(\nu_s) = \sigma'(\nu_s)\, \eta_0(\nu_s)\, N_T L \qquad (62)$$

Eq. (61) is equivalent to Eq. (31) but the consequences are not the same in the cases of absorption and emission. In the case of attenuation measurement, the systematic error in the model used to obtain the absorption cross section decreases as L increases. However, when gain is measured to determine the stimulated emission cross section, the systematic

323

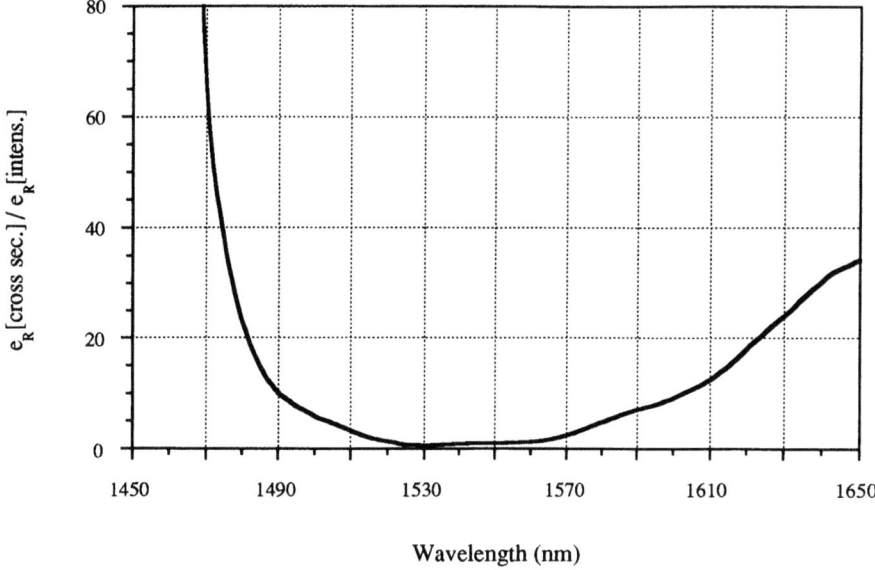

Figure 13. Random error in $\sigma'(\lambda)$ for the fibre f_{ex} when $L=1.9$ m.

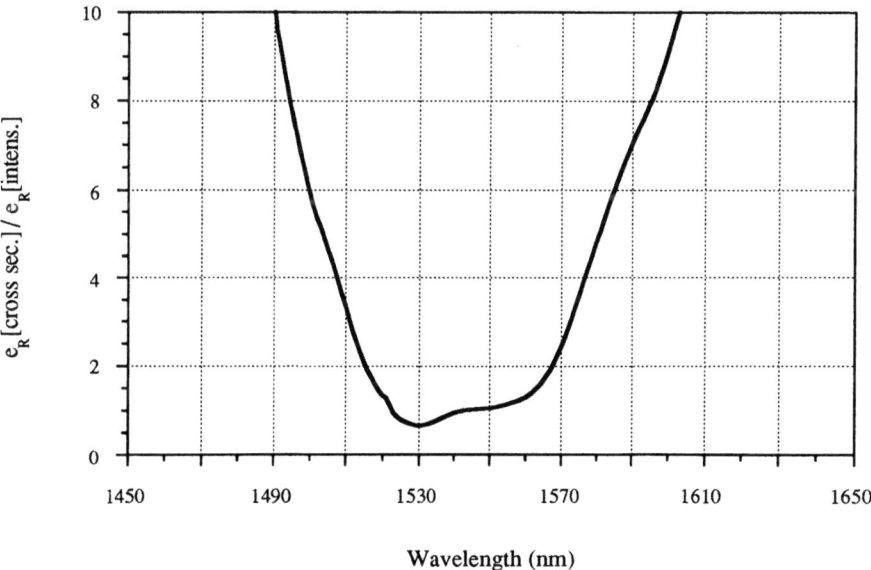

Figure 14. Detail of the minimum in Fig. 13.

error in the model increases as L does. This is the reason why we cannot make an uncontrolled increase of L. The fibre f_{ex} meets the conditions assumed in the model for L=1.9 m, as it can be seen in Fig. 12. Figs. 13 and 14 illustrate the values of $e_R[\sigma'(\lambda)]$ for this fibre length. Fig. 15 shows the gain spectrum when n_2 is saturated. It can be observed that the measurement of the signal powers required to determine $\sigma'(\lambda)$ can be easily done, because a dynamic range of less than 10 dB is enough to take this measurements. It can also be seen that σ' can be measured with small error at wavelengths around 1530 nm. However, large errors are involved at other wavelengths.

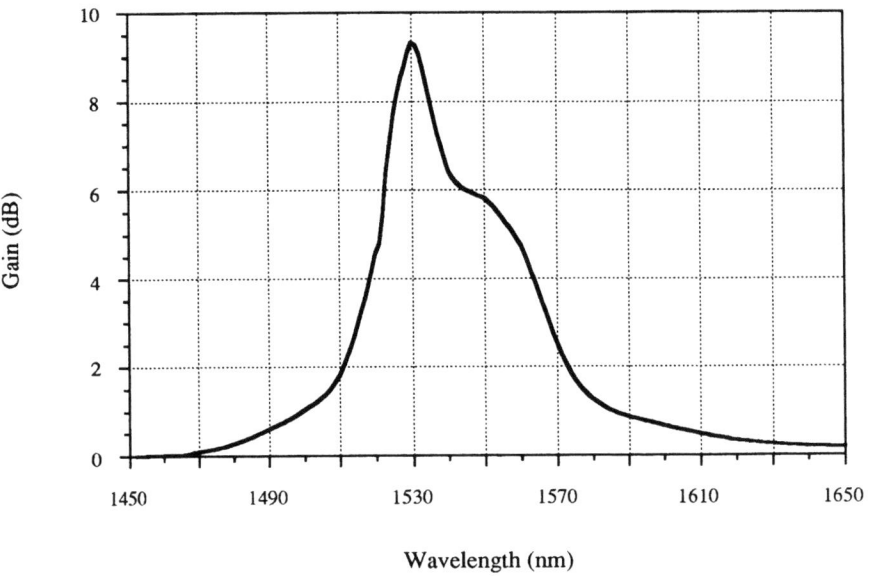

Fig 15. Gain for f_{ex} when L=1.9 m and n_2 is saturated.

From all the results set out in this section, it follows that the measurement of σ_e in the most accurate way should be done by first using the fluorescence technique in order to get the shape of σ_e as a function of wavelength and then measuring the exact value of σ_e, at a wavelength near to 1530 nm, from gain measurement.

7. Lifetime measurement.

Lifetime measurement of metastable levels in a bulk material is easily done by measuring the fluorescence decay when the pump beam is suddenly switched off. In the case of an EDF, the sample can be much longer than in the case of a bulk material and care must be taken in order to avoid large systematic errors in the results.

A small analysis of what happens inside the EDF gives us a clear idea of the errors which can appear if the lifetime is not properly measured. If the fluorescence decay in a direction perpendicular to the fibre is measured, we must realise that this fluorescence is proportional to the population of the upper laser level. This leads us to the analysis of the dynamic properties of n_2. Eq. (6) shows clearly that an exponential decay from which τ is easily measured can only be obtained if the probabilities W_a and W_e are negligible (W_p is null because the pump beam is switched off to measure τ). This obliges to use small pump power or short fibre length in order to avoid non negligible values of W_a and W_e due to ASE.

However, even in the case where we can assume W_a and W_e negligible, the solution of Eq. (11) does not depend linearly with n_2, which means, in general, that the measurement of the decay of the ASE do not give correct values for the lifetime. Only under the conditions explained in section 6.1 (to get a negligible gain on the fluorescence) could the lifetime be properly measured from the fluorescence at the end of the fibre.

8. Other models and techniques.

In the preceding sections we have discussed some techniques to measure the pump, absorption or stimulated emission cross sections. Optimal experimental conditions have been determined to avoid systematic errors and minimise random errors. However, it should be recalled that in every case the refractive index profile and the erbium density distribution should be previously measured. As discussed in section 4, the measurement of the erbium density distribution presents a problem and significant errors can be introduced in the determination of cross sections.

It is for this reason that an important effort has been made in the last few years to find simplified models where detailed information such as refractive index profile or erbium density distribution is replaced by less detailed information such as numerical aperture or mean erbium concentration. Models have been found to reduce considerably the systematic errors and to allow accurate characterisation of the EDF without the above mentioned problems.

9. Acknowledgements.

The author would like to express his gratitude to J.M. Alvarez, J. Vallés, S. Jarabo, J.C. Martín, J.A. Lázaro and M.P. Bernal who provide the "active fibres" laboratory with a stimulating atmosphere.

10. References.

1. M.J. Digonnet, ed., *Fiber laser sources and amplifiers*, (SPIE   Vol. 1171, 1989).
2. M.J. Digonnet, ed, *Fiber laser sources and amplifiers II*, (SPIE   Vol. 1373, 1990).
3. M.J. Digonnet, ed, *Fiber laser sources and amplifiers III*, (SPIE   Vol. 1581, 1991).
4. M.J. Digonnet, ed, *Fiber laser sources and amplifiers IV*, (SPIE   Vol. 1789, 1992).

5. *Optical amplifiers and their applications* (summaries of papers presented at the Optical Amplifiers and their Applications Topical Meeting, Monterey, California, 1990), (OSA, Technical Digest Series Vol. 13 ,.1990).
6. *Optical amplifiers and their applications* (summaries of papers presented at the Optical Amplifiers and their Applications Topical Meeting, Snowmass Village, Colorado, 1991), (OSA, Technical Digest Series Vol. 13 ,.1991).
7. *Optical amplifiers and their applications* (summaries of papers presented at the Optical Amplifiers and their Applications Topical Meeting, Santa Fe, New Mexico, 1992), (OSA, Technical Digest Series Vol. 17 ,.1992).
8. *Optical amplifiers and their applications* (summaries of papers presented at the Optical Amplifiers and their Applications Topical Meeting, Santa Fe, Yokohama, Japan, 1993), (OSA, Technical Digest Series,.1993).
9. P.W. France, ed., *Optical fibre lasers and amplifiers*, (Blackie, 1991).
10. M.J.F. Digonnet, ed., *Rare earth doped fiber lasers and amplifiers*, (Marcel Dekker, 1993).
11. D.W Oblas, F. Pink, M.P. Singh, J. Connolly, D. Dugger and T. Wei (1992) *Digest of Materials Research Society Annual Meeting* (MRS Pittsburg, Pa), pp J6.3.
12. J.K. Trautman, E. Betzig, J.S. Weiner, D.J. DiGiovanni, T.D. Harris, F. Hellman and E.M. Gyorgy (1992) Image contrast in near field optics, *J. Applied Physics* **71**, 4659-4663.
13. A.M. Vengsarkar, D.J. DiGiovanni, W.A. Reed, K.W. Quoi and K.L. Walker (1992) *Optics Letters* **17**, 1277-1279.
14. D. Uttamchandani, A. Othonos, A.T. Alavie and M. Hubert (1994) Determination of erbium distribution in optical fibers using confocal optical microscopy, *IEEE Photonics Technology Letters* **6**, 437-439.
15. M.A. Rebolledo and S. Jarabo (1994) Erbium-doped silica fiber modeling with overlapping factors, *Applied Optics* **33** (in press).

OPTICAL FIBRE AMPLIFIERS STANDARDISATION

P. DI VITA
CSELT
Via Reiss Romoli 274 - Torino - Italy

Abstract

The present status of standardisation in the field of optical fibre amplifiers is reviewed, together with the perspectives of development. The role played by the various International Standardisation Bodies is underlined, summarising the content of the documents either produced or under preparation in this field. Standardisation of the characterising parameters of the optical amplifier, its specifiable characteristics and its introduction in the various networks, is discussed.

1. Introduction

The tremendous increase in the performances of optical transmission networks observed in the last 5 years has been possible thanks to the introduction of a new revolutionary function: the optical amplification. Optical amplifiers, particularly Optical Fibre Amplifiers (OFAs) based on active silica fibres doped with Erbium ions, have disclosed to optical communication not only an enormous improvement of performances, but also the challenging possibilities of new network functions and architectures.

Therefore OFAs have experienced a very rapid development from basic resarch to commercially available devices in less then 4 years, and now more than a dozen of manufacturers are regularly producing OFAs, while their applications are progressively multiplying exceeding the original field of optical transmission.

This rapid development posed since the beginning the problem of standardisation both of OFAs and their applications. Consequently standardisation activities were started in the various National and International Standardisation Bodies more than three years ago and many documents have been already prepared while many others are well in progress, even if considerable activity is still needed.

It was soon recognised that a more meaningful OFA Standardisation activity could have been developed only in close connection with that relative to their applications, with particular regard to transmission systems. In fact it was well understood that the consistent development of both kinds of standardisation, represent a key condition for a rational evolution of modern telecommunication networks.

These considerations underline the twofold motivation for a standardisation activity in this case. Firstly, there is the need to establish uniform requirements in order to guarantee the compatibility among different devices ensuring the functional characteristics in the different assemblages. In second place, there is a need for a proper characterisation of OFAs in view of their introduction in a telecommunication network.

A delicate aspect of OFA standardisation is related to the rapid development of both OFA products and applications. In fact, from one side, this rapid pace should require a similarly rapid development of the corresponding standardisation, in order to avoid uncompatibilities among devices already commercially available. From the other side, it is just this rapid development to indicate a certain prudence in issuing norms which, in a field not yet well consolidated, could refrain or distort further technological developments. Therefore the various OFA standardisation committees should have a special sensibility in dealing with these two opposite requirements, preparing norms sufficiently effective to meet the market needs, but not so restrictive to prevent further developments in the field.

2. OFA Standardisation Committees

Three International Standardisation Bodies are at present active on OFAs: two, ITU-T and IEC, at world level, one, ETSI, at European level. ITU-T and

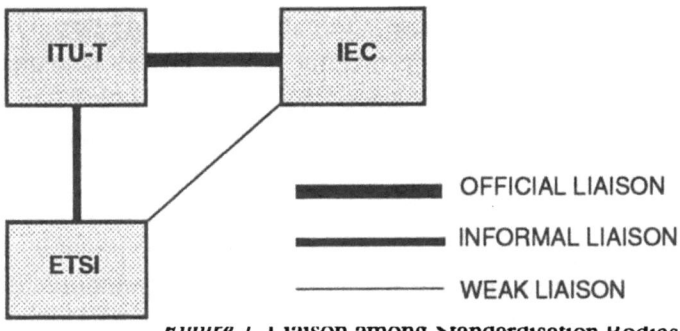

Figure 1. Liaison among Standardisation Bodies

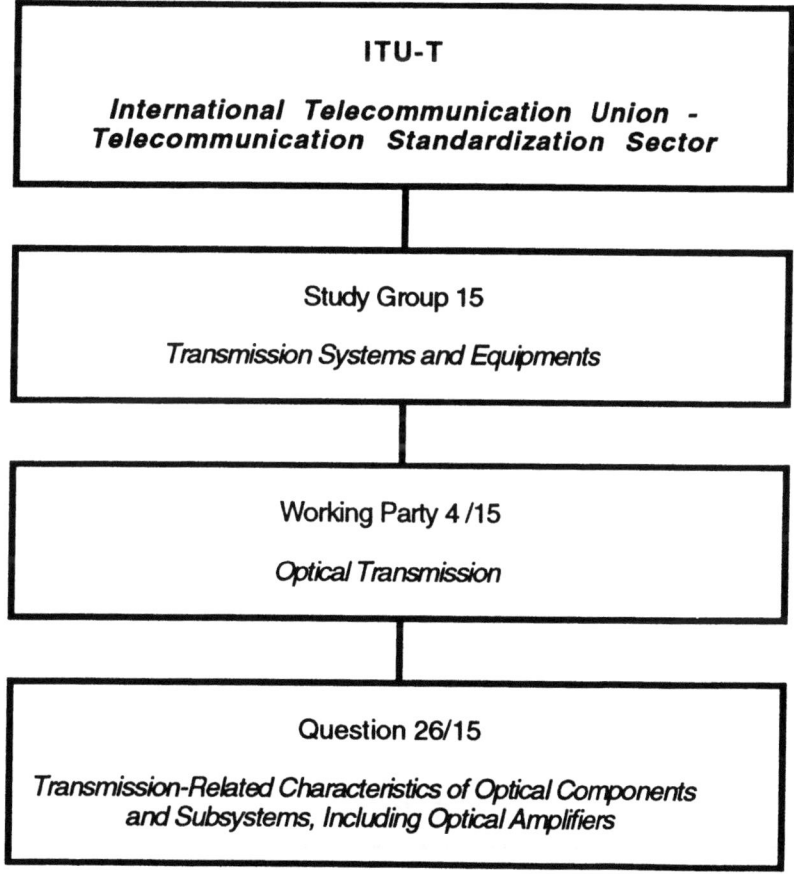

Figure 2. Structure of ITU-T Groups for OFAs

ETSI are more oriented to the applications in the field of telecommunications, while IEC is prevalently oriented towards the product. There are effective liaisons among these three bodies (see Figure 1). In particular ITU-T and IEC have set-up an official liaison aimed at avoiding inconsistencies between the documents issued by the two Bodies. On the contrary, the liaison between ITU-T and ETSI is informal, and many norms issued by ETSI specialise the corresponding ITU-T documents to the European peculiar needs.

ITU-T (International Telecommunication Union - Telecommunication Standardization Sector) generally issues Recommendations which, although not having the strength of a norm, are taken worldwide in the maximum consideration. ITU-T sometimes prepare also informative documents (such as Guidelines or Handbooks) on particular subjects. ITU-T studies OFAs (see Figure 2) in its Study Group 15 (Transmission Systems and Equipments), Working Party 4/15 (Optical Transmission), under Question 26/15 (Transmission-Related Aspects of Optical Components and Sub-Systems, Including Optical Amplifiers).

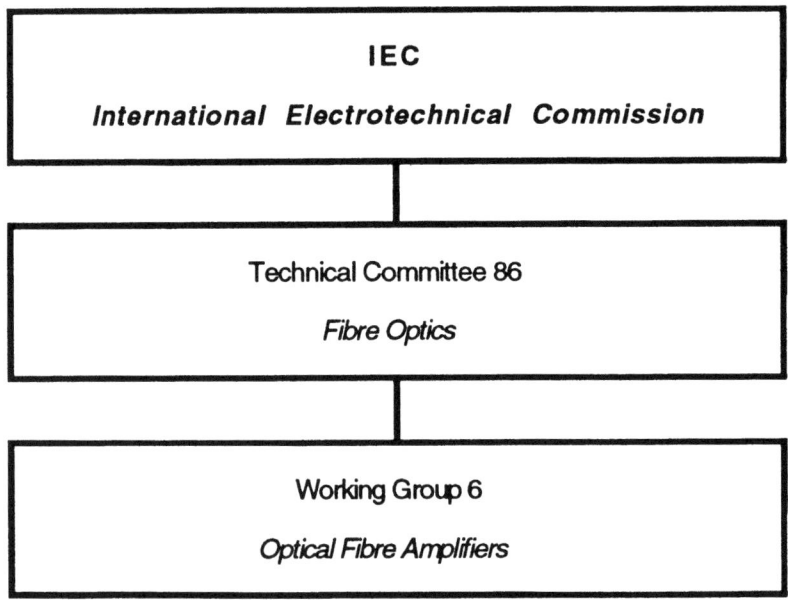

Figure 3. Structure of IEC Groups for OFAs

IEC (International Electrotechnical Commission) generally issues Standards with the strength of norms for its Country members. Also IEC emits informative (not normative) documents (Technical Reports) on some particular aspects. IEC considers OFAs (see Figure 3) in its Technical Committee 86 (Fibre optics), Working Group 6 (Optical fibre amplifiers).

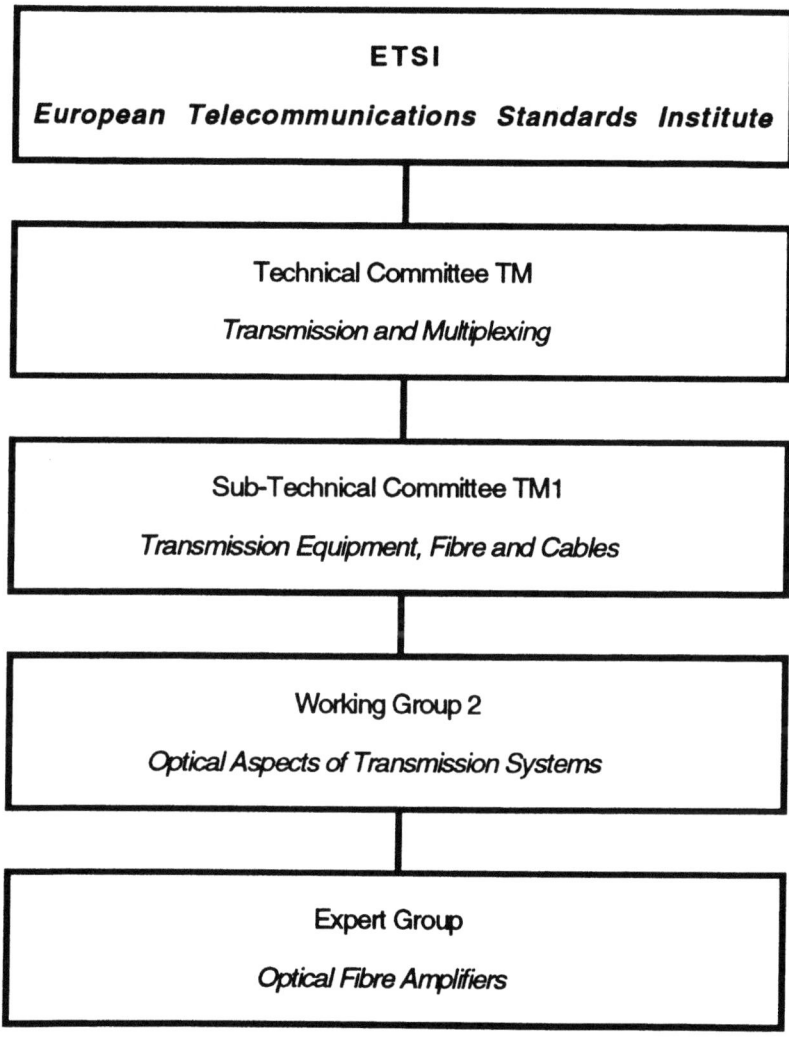

Figure 4. Structure of ETSI Groups for OFAs

ETSI (European Telecommunications Standard Institute) issues Standards recognised by the Country members, as well as informative Technical Reports. ETSI studies OFAs (see Figure 4) in its Technical committee TM (Transmission and multiplexing), Sub-Technical Committee TM1 (Transmission Equipment, fibre and cable), Working Group 2 (Optical aspects of transmission systems), Expert Group on OFAs.

These Standardisation Groups generally operate in three directions. The first consists in identifying the correct parameters relevant for the OFA characterisation, providing proper and unambiguous definitions for these parameters together with consistent test procedures. The second consists in providing those specifiable characteristics necessary in the various applications of OFA devices and sub-systems. The third direction is directly related to the introduction of the OFAs in the various kinds of optical networks and consists in providing the permitted value ranges for transmissive, reliability and safety parameters. The activities along these three directions in the various Standardisation Bodies will be examined in detail in the following sections.

It is to be stressed that all the standardisation activities developed so far concern the OFAs utilising Erbium-doped silica-based fibres as active medium (EDFAs). These OFAs represent a quite consolidated product largely diffused in the market and the OFA system applications are at present mainly referring to them. Other kinds of optical amplifiers are under development in many laboratories (e.g. OFAs utilising Praseodymium-doped fluoride-based active fibres, or semiconductor optical amplifiers) and the first products are becoming commercially available. However, the standardisation documents provided so far are based on the experience gained with EDFAs, which represent a mature technology. Nevertheless, exclusion of other kinds of optical amplifiersit is not intended. In fact generally each document was prepared so as to permit an easy introduction of these new amplifiers, once they become sufficiently mature.

3. Standardisation of OFA parameters

The activity finalised at the identification, definition of the relevant parameters and preparation of their test methods has lead so far to the issue of ITU-T Recommendations G.661 and G.662 [1, 2], the draft IEC Generic Specification 1291-1 [3] and 8 draft IEC Basic Specification of the 1290 series for the test methods [4-11].

The ITU-T Recommendation G.661 provides the definitions of the following families of relevant transmission parameters:

- Gain parameters;
- Input and output signal power parameters;
- Noise parameters;
- Reflectance parameters;
- Pump leakage parameters;
- Out-of-band insertion loss parameters.

The ITU-T Recommendation G.662 provides, among the other things, the subdivision of the OFAs in OFA devices (intended as stand-alone OFAs) and OFA sub-systems (intended as integrated OFAs). Furthermore, according to the OFA applications the following 5 categories are identified:

- The *Power (Booster) Amplifier* (BA) is a high saturation-power OFA device to be used directly after the optical transmitter to increase its signal power level.

- The *Pre-Amplifier* (PA) is a very low noise OFA device to be used directly before an optical receiver to improve its sensitivity.

- The *Line Amplifier* (LA) is a low noise OFA device to be used between passive fibre sections to increase the regeneration lengths or in correspondence of a point-multipoint connection to compensate for branching losses in the optical access network.

- The *Optically Amplified Transmitter* (OAT) is an OFA sub-system in which a power amplifier is integrated with the laser transmitter, resulting in a high power transmitter.

- The *Optically Amplified Reveiver* (OAR) is an OFA sub-system in which a pre-amplifier is integrated with the optical receiver, resulting in a high sensitivity receiver.

The IEC Generic Specification 1291-1 [3] contains the definitions of the transmission parameters of Rec. G.661, of analogous transmission parameters specific of OFA sub-systems and of some additional parameters concerning the the following OFA non-trasmissive characteristics:

- Mechanical properties;
- Environmental properties;
- Reliability properties;
- Safety properties.

Concerning the Test Methods (TMs), thanks to an agreement between the ITU-T and IEC OFA Groups, they are studied by IEC only and ITU-T Recommendations will fully refer to the relative IEC Basic Specification produced on OFAs. 8 TMs have been developed by IEC so far, as listed in Table 1 [4-11], while some others are under study (as also reported in Table 1).

TABLE 1. Test methods for OFAs

Group of parameters	IEC Basic Specification number: Test Method (TM)
Gain parameters	1290-1-1: Optical spectrum analyzer TM 1290-1-2: Electrical spectrum analyzer TM 1290-1-3: Optical power meter TM
Optical power parameters	1290-2-1: Optical spectrum analyzer TM 1290-2-2: Electrical spectrum analyzer TM 1290-2-3: Optical power meter TM (under study)
Noise parameters	1290-3-1: Optical spectrum analyzer TM (under study) 1290-3-2: Electrical spectrum analyzer TM (under study)
Reflectance parameters	1290-5: (under study)
Pump leakage parameters	1290-6-1: Optical demultiplexer TM
Out-of-band insertion loss	1290-7-1: Filtered power meter TM

From an inspection of Table 1 it is also evident that more than one TM is (or are going to be) provided for a single parameter. This is the case of Gain and Optical power parameters. This, according to the rules of both ITU-T and IEC, poses the problem to choose one, among the various test methods for each single parameter, as Reference Test Method (RTM), the others being considered Alternative Test Methods (ATMs). The RTM should be directly related to the definition, should have the maximum intrinsic accuracy and it is considered of absolute validity. An ATM should be sufficiently accurate, should be relatable to the RTM and could be more practical than the RTM. At present there is not sufficient experience to decide about the RTM for the various OFA parameters, and the IEC Group on OFAs has started an interlaboratory measurement campaign to evaluate the performances of each TM. The RTM for each OFA parameter will be chosen according to the results of this measurement campaign,

So far, ETSI has not produced any document on OFA parameters, fully endorsing the corresponding ITU-T Recommendations. However very recently, also in consideration of the European Directives, it has been decided to prepare a standard specialising the content of ITU-T Recommendations G.661 and G.662 to the European requirements.

The general approach in OFA standardisation is based on the "black box" concept, according to which all the parameters are specified at the input and/or the output OFA optical port without any reference to internal components. This approach is the most convenient from a functional point of view, nevertheless, also in consideration of the very recent development of OFAs, the IEC has considered suitable to issue an informative Technical Report [12] (without normative value) about the relevant parameters of the following internal optical components the OFA:

- Active fibre;
- Pump laser;
- Wavelength Division Multiplexing (WDM)coupler;
- Optical isolator;
- Amplified Spontaeous Emission (ASE) filter;
- Pump filter;
- Optical connectors.

4. Specification of OFA characteristics

The main results of the standardisation activity aimed at providing those characteristics specifiable for the different OFA applications, are at present the already mentioned ITU-T Recommendation G.662 [2] and an informative ETSI Technical Report [13].

ITU-T Recommendation G.662 firstly establishes the criteria of specification of OFA devices and sub-systems in a way to ensure, as far as possible, the maximum compatibility with the existing Recommendations G.955 and G.957 [14, 15], for Plesiochronous Digital Hierarchy (PDH) and Synchronous Digital Hierarchy (SDH) line systems, respectively.

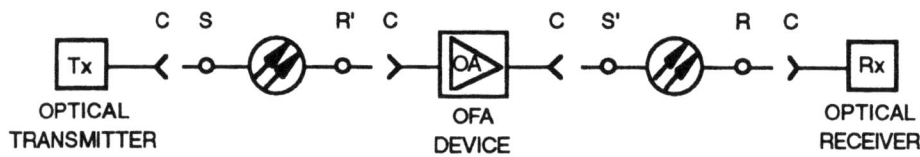

S - reference point of the fibre just after the connection (C) of the transmitter
R - reference point of the fibre just before the connection (C) of the receiver
S' - reference point of the fibre just after the connection (C) of the OFA device
R' - reference point of the fibre just before the connection (C) of the OFA device

Figure 5. Scheme of insertion of an OFA device

According to this criterion, such a compatibility is in principle possible for OFA devices (with a restriction on the operating wavelength), provided that the OFA (BA, PA or LA), inserted along an optical path, shall be considered a separate element placed between the reference points S and R defined in ITU-T Recommendations G.955 and G.957 for line terminals and regenerators, as shown in the scheme of Figure 5. (Note that the input and output characteristics of the OFA device shall be specified at reference points R' and S', before and after the OFA device, respectively).

On the contrary, in the case of OFA sub-systems the integration imply that the connection between the transmitter or the receiver and the OFA is proprietary and shall not be specified. Consequently a reference point S only can be defined for the specification of the OAT output characteristics after the OFA, as shown in Figure 6, and a reference point R only can be defined for the for the specification of the OAR input characteristics before the OFA, as shown in Figure 7. The compatibility with existing ITU-T Recommendations on line systems is therefore more difficult, and this has suggested the preparation, just started, of new Recommendations on line systems using OFA devices and sub-systems.

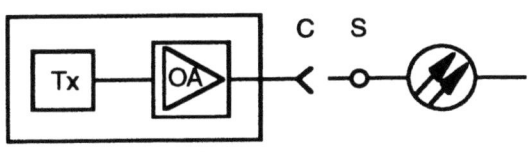

S - reference point of the fibre just after the connection (C) of the OAT

Figure 6. Scheme of insertion of an OAT

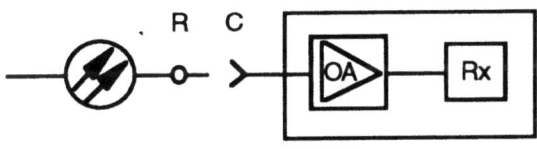

R - reference point of the fibre just before the connection (C) of the OAR

Figure 7. Scheme of insertion of an OAR

In second place ITU-T Recommendation G.662 provides a minimum list of relevant parameters specifiable for BA, PA, LA, OAT and OAR, respectively.

IEC is developing a similar activity finalised to the issue of a Sectional (family) and 5 Blank Detail Specifications, for OFA to be used in digital applications. Preparation of Sectional and Blank Detail Specifications on OFAs to be used in analogue applications, is also planned.

The ETSI Technical Report [13] is not normative and concentrates on various kinds of OFA applications in long distance links and in optical access networks. It includes aspects concerning operation, supervision, maintenance and safety, as well as considerations about the OFA integration in the existing standards on line systems.

5. Specification of OFA in optical networks

The standardisation activity aimed at the specification of those aspects directly related to the introduction of the OFA devices and sub-systems in the various kinds of optical networks was started recently, particularly in ITU-T and in ETSI. ITU-T decided to issue by 1996 a Recommendation (provisionally denominated G.OA3) on application-related aspects of OFA devices and sub-systems which should address the following topics, common to the generality of OFA applications:

- transmisssion-related aspects;
- Operation, Administration and Maintenance;
- environmental conditions (e.g. climatic and EMC/ESD);
- optical safety aspects
- parameter values and ranges for OFA devices (e.g. power and wavelength value ranges).

Recommendation G.OA3 will also incorporate a "Guideline in the use of OFAs", indicating the possible "secondary" effects induced by OFAs on optical fibre systems, including optical non-linearities, polarisation and cromatic dispersion, noise accumulation, etc. For each of these effects this Guideline will indicate the onset conditions, the limitations imposed to the transmission and possible techniques to avoid these problems.

Those aspects more specific to each kind of system applications will be developed in some dedicated ITU-T Recommendations, whose provisional denominations are given in the following:

- G.SCS   for optically amplified single-channel long distance line systems;

- G.MCS   for optically amplified WDM long distance line systems;
- G.LON   for the functional characteristics of optically amplified interoffice line systems;
- G.OASS   for optically amplified submarine line systems.

Different approaches will be considered for systems which include or do not include line OFAs, due the substantial differences in the OFA supervision systems. The preparation of these Recommendation was recently started and they should be ready by 1996 as well.

ETSI started recently the preparation of a Standard about the Optical Access Network in which the specifications of OFA introduction in this kind of network is considered. In particular it was established that each OFA sub-systems will be considered as a part of the corresponding line terminal, while each OFA device (BA, PA or LA) will not be considered part of the Optical Distribution Network (ODN), but will sub-divide it in partial ODNs (ODN.0, ODN.1, ODN.2, etc.), as shown in Figure 8.

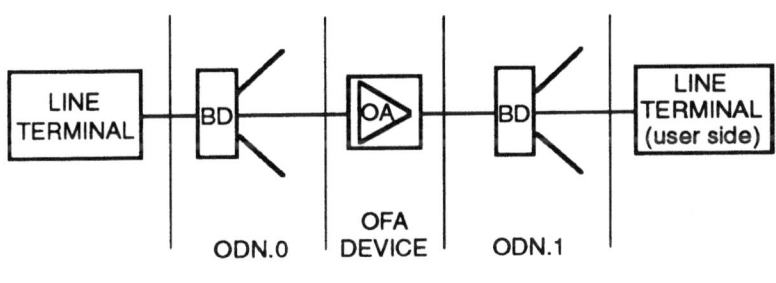

BD   - Branching Device
ODN.0 - Optical Distribution Network level 0
ODN.1 - Optical Distribution Network level 1

Figure 8. Example of insertion of an OFA device in an Optical Access Network

6. Conclusions

In the current globalisation of the economy, telecom operators and manufacturers need to deploy common solutions. In this scenario the role of standardisation is fundamental. It definitely influences the regulation process in progress and is crucial in the development of an open telecommunication market. This is particularly true for Optical Fibre Amplifiers (OFAs), a new component which has introduced an authentic revolution in the already recent technology of optical communications. Standardisation will have a key relevance in the development of the OFAs in the networks and in the innovations that they make possible.

In this paper a summary of the present status of OFA standardisation has been presented, trying to emphasise the role played by the international Standardisation Bodies in this field. It is evident that a considerable amount of work has been done and this has lead to a certain number of consolidated Recommendations, Standards and Technical Reports. The aspect of identification of the key characterising OFA parameters has been abundantly developed, while the one concerning the OFA characteristics is going to be fulfilled. However additional work is still needed, particularly for the development of norms concerning the introduction of the OFAs in the various networks; activities in this direction have been just started.

Due to the rapid development of OFAs and their applications their standardisation in this field is requiring a judicious progressive setting of norms, balancing the market needs and the exigency of avoiding possible distortions of further progress. This task is not easy, but in consideration of the results obtained so far, it is felt that these efforts are allowing an harmonised evolution of the whole technology, which is one of the primary objectives of OFA standardisation.

7. References

1. ITU-T Recommendation G.661 (1993) Definition and test methods of the relevant generic parameters of optical fibre amplifiers.

2. ITU-T Recommendation G.662 (Draft 1994) Generic characteristics of optical fibre amplifier devices and sub-systems.

3. IEC Document 1291-1 (Draft 1994) Generic specification - Optical fibre amplifiers.

4. IEC Document 1290-1-1 (Draft 1994) Basic specification for Optical fibre amplifier test methods - Test methods for gain parameters - Optical spectrum analyser test method.

5. IEC Document 1290-1-2 (Draft 1994) Basic specification for Optical fibre amplifier test methods - Test methods for gain parameters - Electrical spectrum analyser test method.

6. IEC Document 1290-1-3 (Draft 1994) Basic specification for Optical fibre amplifier test methods - Test methods for gain parameters - Optical power meter test method.

7. IEC Document 1290-2-1 (Draft 1994) Basic specification for Optical fibre amplifier test methods - Test methods for optical power parameters - Optical spectrum analyser test method.

8. IEC Document 1290-2-2 (Draft 1994) Basic specification for Optical fibre amplifier test methods - Test methods for optical power parameters - Electrical spectrum analyser test method.

9. IEC Document 1290-3 (Draft 1994) Basic specification for Optical fibre amplifier test methods - Test methods for noise parameters.

10. IEC Document 1290-6-1 (Draft 1994) Basic specification for Optical fibre amplifier test methods - Test methods for pump leakage parameters - Optical demultiplexer test method.

11. IEC Document 1290-7-1 (Draft 1994) Basic specification for Optical fibre amplifier test methods - Test methods for out-of-band insertion losses - Filtered power meter test method.

12. IEC Document 1292-1 (Draft 1993) IEC technical report type 3 - Parameters of optical fibre amplifier components.

13. ETSI Technical Report DTR/TM1019 (1993) Applications of optical fibre amplifiers to long distance and optical access networks.

14. ITU-T Recommendation G.955 (1993) Digital line systems based on the 1 544 kbit/s and the 2 048 kbit/s hierarchy on optical fibre cables.

15. ITU-T Recommendation G.957 (Draft 1994) Optical interfaces for equipments and systems relating to the synchronous digital hierarchy.

OPTICAL AMPLIFICATION MODELLING FOR Er-DOPED FIBRE*

*JOAQUIM M. S. ANACLETO**, O. D. D. SOARES*
CETO, Centro de Ciências e Tecnologias Ópticas
Laboratório de Física, Faculdade de Ciências - Universidade do Porto
Praça Gomes Teixeira - 4000 Porto - Portugal

Abstract: A model for optical amplification in an Er^{3+} doped fibre was treated, including saturation, fibre mode theory and ASE profile. The propagation equations were integrated numerically using a 4th order Runge-Kutta method and a computer C programme. The evolution of the pump, signal, and ASE optical powers along the fibre length was obtained, considering different fibre lengths and input conditions for pump and signal.

The present paper presents a modelling for an Er^{3+} fibre amplifier in the 1500-1600nm spectral region pumped at 1480nm.

The known models vary on the type of approximation [1-3]: small signal regime, no amplified spontaneous emission (ASE), gaussian mode approximation. Analytical solutions were derived for some of the models [4].

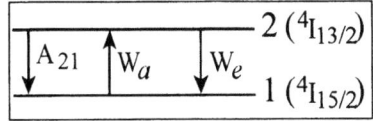

FIGURE 1 - *Diagram of laser levels and corresponding transitions rates.*

The saturation is included in this model. The model is based in modal analysis with ASE but excluding excited state absorption (ESA), inhomogeneous broadening and transient phenomena. Once considering homogeneous broadening the amplifier behaves as a quasi-two level laser system, Fig. 1.

The transition rates corresponding to stimulated phenomena are denoted by W_β, with the subscript $\beta = a$ for absorption and $\beta = e$ for emission. Spontaneous transitions from level 2 to level 1 is denoted by A_{21}.

1. Rate equations

An optical fibre is taken as a cylindrical guiding structure, in which a cylindrical coordinate system $\{z, r, \phi\}$ can be defined with the z axis along the axis of the fibre.

Denoting $n_i(z,r,\phi)$ ($i = 1,2$) for the population densities, the rate equations for levels 1 and 2 come:

$$\frac{dn_2(z,r,\phi)}{dt} = W_a(z,r,\phi)\, n_1(z,r,\phi) - [W_e(z,r,\phi) + A_{21}]n_2(z,r,\phi)\ ,$$

$$\frac{dn_1(z,r,\phi)}{dt} = -\frac{dn_2(z,r,\phi)}{dt}\ ;\quad n_1(z,r,\phi) + n_2(z,r,\phi) = n(r)\ ,$$

where $n(r)$ is the total ionic population.

---

* The paper was presented at a poster session.
** Lecturer at Universidade de Trás-os-Montes e Alto Douro
  Quinta dos Prados, 5000 Vila Real - Portugal   Fax: (059) 74 480

Considering a steady state regime operation, one has:

$$\frac{dn_1(z,r,\phi)}{dt} = \frac{dn_2(z,r,\phi)}{dt} = 0,$$

and the solutions of the rate equations come:

$$n_1(z,r,\phi) = \frac{W_e(z,r,\phi) + A_{21}}{W_e(z,r,\phi) + W_a(z,r,\phi) + A_{21}} n(r)$$

$$n_2(z,r,\phi) = \frac{W_a(z,r,\phi)}{W_e(z,r,\phi) + W_a(z,r,\phi) + A_{21}} n(r)$$

The dependence of the transition rates W_β on the intensity of the optical fields that interact with ions, cross sections associated to the transitions, and photon energy, is given by:

$$W_\beta(z,r,\phi) = \frac{\sum_\alpha P_\alpha^\pm(z,\nu_\alpha)\,\sigma_\beta(\nu_\alpha)\Psi(r,\phi,\nu_\alpha)}{h\nu_\alpha}$$

where $P_\alpha^\pm(z,\nu_\alpha)$ represents the optical field powers, and the subscript α will be p for pump, s for signal and f (fluorescence) for ASE. The signs + and − relate to copropagating and counter-propagating waves, respectively. The $\Psi(r,\phi,\nu)$ is the normalised intensity distribution calculated using passive fibre methods [6]. In monomode regime $\Psi(r,\phi,\nu)$ is circularly symmetric, therefore independent of ϕ. The emission and absorption cross sections, σ_β, are defined as:

$$\frac{dI}{dz} = \pm n\,\sigma_\beta(\nu)I$$

where I is the intensity of the wave field at frequency ν, z is the direction of propagation and n is the ionic population density. The + sign is taken for an emission process and the − sign for an absorption process.

Pumping and signal waves are considered monochromatic and unidirectional, but the ASE term is polychromatic and propagates in both directions. Spectral bandwidth of ASE is taken in small intervals of width of $\Delta\nu$ (1nm). Therefore, $P_f^\pm(z,\nu_f)$ is the power in the interval $\Delta\nu$ centred in ν_f.

2. Propagation equations

The evolution of $P_\alpha^\pm(z,\nu_\alpha)$ along the length of the fibre is then given by:

$$\frac{dP_p^\pm}{dz}(z,\nu_p) = \pm P_p^\pm(z,\nu_p)\left[\sigma_e(\nu_p)N_2(z,\nu_p) - \sigma_a(\nu_p)N_1(z,\nu_p)\right]$$

$$\frac{dP_s^\pm}{dz}(z,\nu_s) = \pm P_s^\pm(z,\nu_s)\left[\sigma_e(\nu_s)N_2(z,\nu_s) - \sigma_a(\nu_s)N_1(z,\nu_s)\right]$$

$$\frac{dP_f^\pm}{dz}(z,\nu_f) = \pm 2h\nu_f\,\Delta\nu\,\sigma_e(\nu_f)N_2(z,\nu_f)$$

$$\pm P_f^\pm(z,\nu_f)\left[\sigma_e(\nu_f)N_2(z,\nu_f) - \sigma_a(\nu_f)N_1(z,\nu_f)\right]$$

The cross sections σ_β ($\beta = a$ for absorption and $\beta = e$ for emission) take non-zero values within the band 1400nm to 1650nm (see figure 2 extracted from ref. [1]). The parameters $N_i(z, \nu)$ are given by:

$$N_i(z, \nu) = \int_0^R r\,dr \int_0^{2\pi} n_i(z,r,\phi)\,\Psi(r,\phi,\nu)\,d\phi \quad (i = 1,2)$$

The above propagation equations are a set of coupled ordinary non-linear differential equations with 504 dependent variables, $P_\alpha^\pm(z, \nu_\alpha)$, and one independent variable, z.

The boundary conditions we have to impose are $P_f^+(0, \nu_f) = 0$, $P_f^-(L, \nu_f) = 0$ (L is the length of the fibre), $P_s^+(0, \nu_s)$ = signal power coupled to the fibre, and $P_p^+(0, \nu_p)$ = pump power coupled to the fibre. These conditions represent an additional difficulty because all power values at the input end of the fibre are unknown.

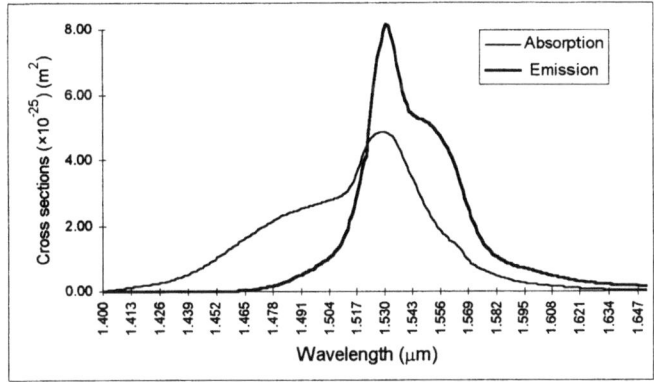

FIGURE 2 - *Emission and absorption cross sections for Er^{3+} in a host glass* [1].

3. Results

In order to integrate numerically the propagation equations, a computer program (in C language) was developed. The method used to integrate the equations was a 4th order Runge--Kutta method [5]. A step index fibre was considered in the simulations with the following characteristics:

- ρ (core radius) = 2.2 µm;
- NA (numerical aperture) = 0.19;
- n_{co} (core index) = 1.45;
- R (doped area radius) = 2.2 µm;
- τ (life time of level 2 - $1/A_{21}$) = 11 ms;
- n (Erbium concentration) = 8×10^{24} m^{-3};

The aim in device modelling is to understand the performance while the different parameters vary, in view of the optimisation of the required properties of an oscillator or amplifier. The following figures present results obtained from simulations.

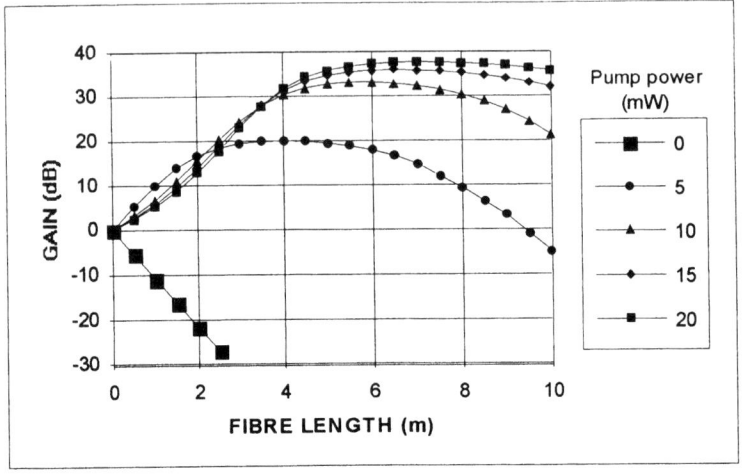

FIGURE 3 - *Fibre gain as a function of fibre length for different pump powers obtained with a input signal power of* 0.1 µW.

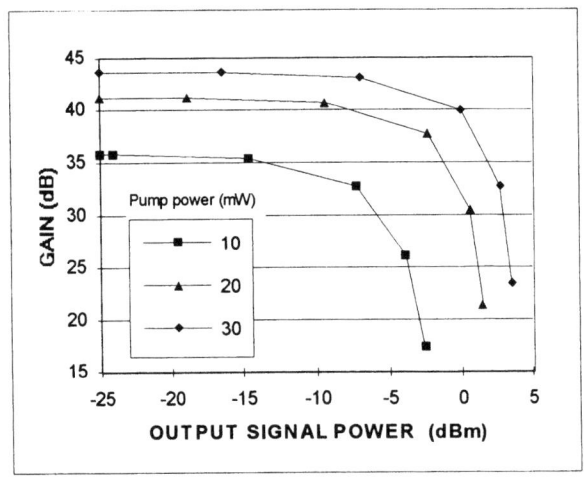

FIGURE 4 - *Fibre gain as a function of output signal power for different pump powers obtained with a fibre length of* 5m.

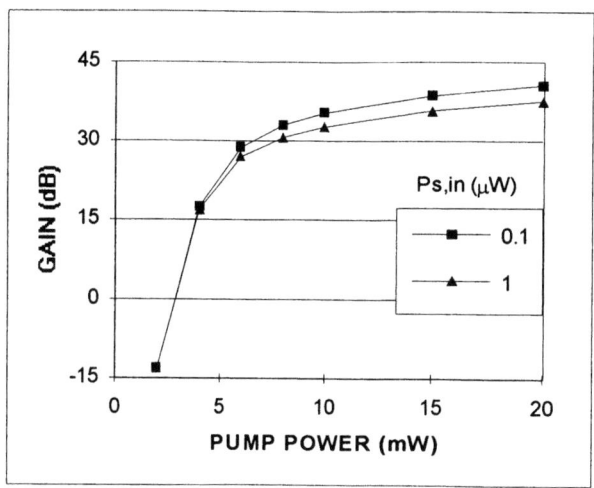

FIGURE 5 - *Fibre gain as a function of pump power for different input signal powers obtained with fibre length of 5m.*

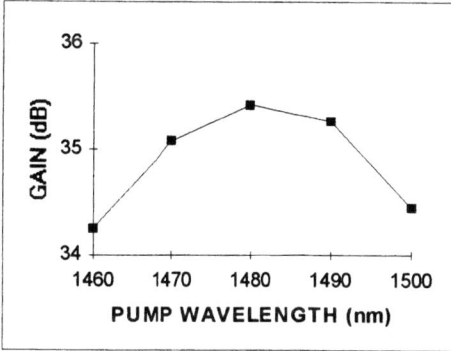
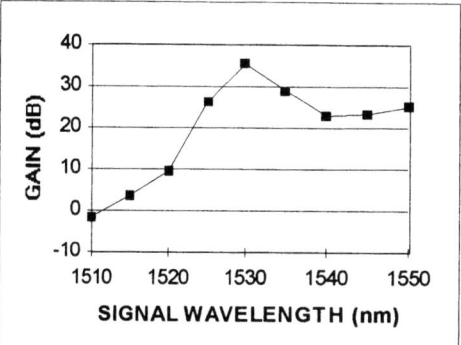

FIGURE 6 - *Fibre gain as a function of pump and signal wavelengths obtained with a fibre length of 5m and a pump and a signal input powers of 10mW and 0.1µW, respectively.*

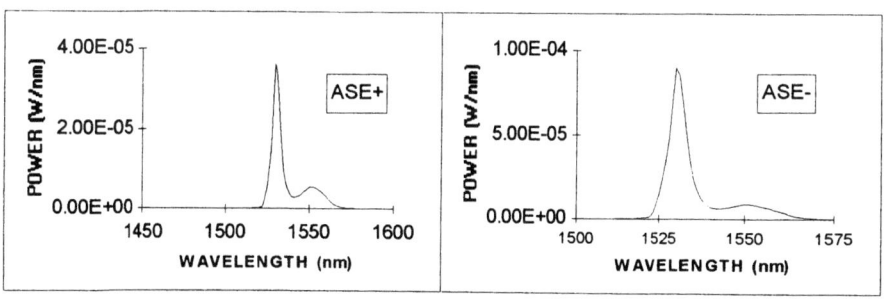

FIGURE 7 - ASE *spectral power profile at the input face of the fibre* (ASE-) *and at the output end of the fibre* (ASE+). *The fibre length was 5m, the input signal power 0.1µW, and the pump power of 6 mW.*

4. Discussion of the results

An unpumped fibre is a strong absorber as expected, Fig. 3. During pumping, the gain increases. It reaches a optimum value (pump power dependent) and then decreases.

The pumping efficiency of the amplifier is determined by the gain coefficient which is the maximum ratio of gain to launched pump power. This ratio is represented by the slope of the tangent to the gain curve (in figure 5) that intersects the origin. There is a pump power value for which the fibre is transparent (gain = 0 dB), approximately 3mW in our simulation, Fig. 5. As the pump increases, the population inversion tends toward a maximum that depends on the absorption and emission cross sections at pump wavelength.

According to figure 4, the amplified signal causes saturation, imposing then a limit on the gain. We can define a saturated output power P_{OUT}^{SAT} as the output power for which the gain decreases 3dB from the unsaturated value. This parameter is important as it determines the transition from linear to non-linear regime for the gain. The P_{OUT}^{SAT} depends on pump power (see figure 4).

Variation of the gain with wavelength of the signal and pump, is shown in figure 6. The dependence is weak for the pumping around 1480nm and very significant for a signal around 1530nm. This may be expected from the relations between emission and absorption cross sections at these wavelengths as shown in figure 2.

Finally, the spectral profile of ASE is shown in figure 7. The ASE degrades the amplifier performance as a noise component and then limits the gain.

5. Conclusions

A model for an Er^{3+} fibre amplifier pumped into $^4I_{13/2}$ was developed. It included saturation, fibre mode theory and ASE, but excluded inhomogenous broadening, and excited state absorption. The method used can be extended to other fibre amplifier types and it is believed to form a basis for developing a model that covers dynamical regimes.

6. ACKNOWLEDGEMENTS

I am grateful to Prof. M. A. Rebolledo, S. Jarabo, J. Martin, and J. António, for valuable suggestions during my stay at Department of Applied Physics of University of Zaragoza, Spain.

7. REFERENCES

[1] E. Desurvire, J. Simpson, **Amplification of Spontaneous Emission in Erbium-Doped Single-Mode Fibers**, Jornal of Lightwave Technology, 5 (1989), 835-845.

[2] E. Desurvire, C. Giles, J. Simpson, **Gain Saturation Effects in High-Speed, Multichannel Erbium-Doped Fiber Amplifiers at $\lambda = 1.53 \mu m$**, Jornal of Lightwave Technology, 12 (1989), 2095-2104.

[3] J. F. Digonnet, **Rare Earth Doped Fiber Lasers and Amplifiers**, Marcel Dekker, Inc., 1993.

[4] Pierluigi Franco, Michele Midrio, **Quasi-analytic solution of erbium-doped fiber-amplifier power equations**, Applied Optics, 36 (1993), 7442-7445.

[5] William Press, Brian Flannery, Saul Teukolsky, William Vetterling, **Numerical Recipes, The Art of Scientific Computing**, Cambridge University Press.

[6] Gerd Keiser, **Optical Fiber Communications**, McGraw-Hill (1983).

VI Optical Fibre Characterisation, Calibration Standards

FIBRE CHARACTERIZATION AND MEASUREMENTS

C. LE MEN
Laboratoire Central des Industries Électriques
Service optique
33, av. du Général Leclerc BP 8
92266 Fontenay aux roses - FRANCE

1. Introduction

Propagation of light in free-space can not be reliably used due to the significant wave diffraction and to the dependence of the attenuation on atmospheric conditions. Therefore, it is to use the guided propagation of light by using a convenient medium: the optical fibre. The story of optical fibres started in 1963 with the first realization of a semiconductor laser diode emitting at 0.8 μm. In 1970, these laser diodes became stable, powerfull and reliable enough to start the first studies on optical fibres. From 100 to 1000 dB/km in glasses, the fibre attenuation at 0.85 μm then decreased to 20 dB/km in 1970 and down to 0.5 dB/km in 1973. At the beginning of the eighties the theoretical attenuation limit of 0.15 dB/km at 1.55 μm was almost reached. In the middle of the eighties, optical fibres could easily be tailored with common attenuation values of 0.35 dB/km at 1.3 μm and 0.2 dB/km at 1.55 μm. As a consequence, in 1988, the first transatlantic link was realized with 6000 km of single-mode fibre using a signal rate of 280 MBit/s at 1.3 μm. To take advantage of the lowest attenuation, the interest was to work at 1.55 μm. However, the fibre dispersion is greater at 1.55 μm than at 1.3 μm: 20 ps/(nm.km) compared with a maximum of 5 ps/(nm.km). So the dispersion became the limiting factor in high speed transmission. Therefore, new kinds of fibre were designed with a dispersion lower than 5 ps/(nm.km) at 1.55 μm. An instance for their use is the transatlantic link planned in 1996 that will use dispersion shifted fibres supporting a rate of 5 GBit/s at 1.55 μm. Such high performance evidences the need for accurate attenuation and dispersion measurements. The characterization of those parameters will be described in two different sections of this article. If we consider the practical use of fibre, the fibre to fibre coupling losses become the actual limiting factor. That points out the need for the fibre geometrical characteristics measurements that are presented first in this article. Finally, single-mode fibre measurements, cut-off wavelength and mode field diameter, will be presented. Due to their specific status and role, calibration laboratories acting in the optical fibre domain, must use test methods which comply with the international standards given in the references [1 to 6], to issue certificates of calibration for their customers. Therefore, this article presents results obtained by these standard methods.

2. Opto-geometrical measurements

2.1. OPTO-GEOMETRICAL CHARACTERISTICS DEFINITIONS

2.1.1. Optical fibre structure

The optical fibre structure consists of two concentric glass cylinders of different refractive indexes. A scheme of this structure is shown in figure 1. The inside cylinder, the so-called **core** region, has a maximum refractive **index** n_1 and a **radius a**. The external cylinder, the so-called **cladding** region, has usually a constant refractive **index** n_2 and a **radius b**.

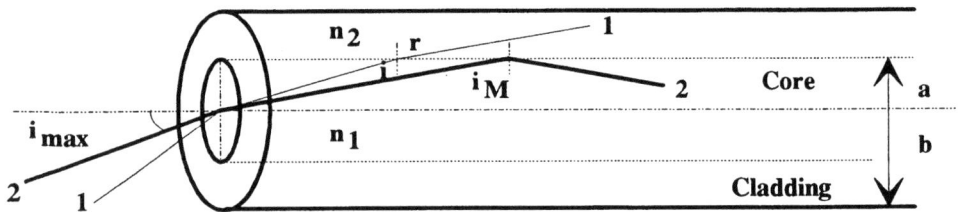

Figure 1. Structure of an optical fibre.

Applying the ray theory at the core/cladding interface, one can use the Snell-Descartes Law to relate angles between rays and the perpendicular to the interface in the core and cladding regions. The limit between refraction and reflection is obtained when the angle r is equal to 90°. The substitution in the Snell-Descartes law yields the limit angle value i_M, also called the critical angle, which exists only if the core index is greater than the cladding index.

$$\sin i_M = n_2 / n_1 \quad \text{if } n_1 > n_2 \tag{1}$$

2.1.2. Numerical Aperture NA

The propagation -reflection- is thus obtained if the ray angle i inside the core is greater than the critical angle i_M (figure 1 - ray #2), in the other case (i < i_M) the ray is refracted in the cladding (figure 1 - ray #1). Applying this condition at the input face of the fibre, one obtains the maximum acceptance angle i_{max} that can be launched in the fibre. This angle is more often described by the **Numerical Aperture NA** that can be related to the **relative index difference** Δ between core and cladding:

$$NA = \sin i_{max} = \sqrt{n_1^2 - n_2^2} = n_1\sqrt{2\Delta}$$

$$\text{with } \Delta = \frac{n_1^2 - n_2^2}{2n_1^2} \approx \frac{n_1 - n_2}{n_1} \tag{2}$$

2.1.3. Normalized frequency parameter V

The normalized frequency parameter V is defined by:

$$V = ka\,NA \quad \text{with } k = 2\pi/\lambda_0 \tag{3}$$

in which k is the propagation constant of a plane wave in vacuum and λ_0 the wavelength. This normalized parameter V is of great importance because the modal properties of the fibre are dependent primarily on this parameter.

2.1.4. Refractive-Index Profile n(r)

The refractive-index profile of a fibre is the curve that plots the index value n with respect to the radial position in the cross section of the fibre. The most common types of refractive index profiles are the Step-Index one shown in figure 2 and the Graded-Index one shown in figure 3.

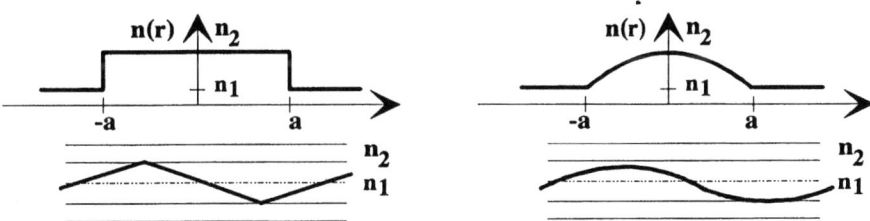

Figure 2. Step-Index Profile   Figure 3. Graded-Index Profile

For multimode fibre, the Step-Index profile results in a very high intermodal dispersion so that it is more often used for single-mode fibre. The Graded-Index profile reduces the differential propagation time between axial rays and paraxial rays and thus decreases the intermodal dispersion by a factor of up to 200 compared to the equivalent Step-Index fibre. To compare these different index profiles, Gloge and Marcatili [7] introduced a general formula describing several profiles by a change in a single parameter, the so-called profile parameter α. Index-profiles as a function of this α parameter are plotted in figure 4. The Step-Index profile is obtained for $\alpha = \infty$, the case $\alpha = 2$ yields a Parabolic-Index profile.

$$n(r) = n_1[1-2\Delta(r/a)^\alpha]^{1/2} \quad \text{for } |r| \leq a \tag{4}$$
$$n(r) = n_1[1-2\Delta]^{1/2} = n_2 \quad \text{for } r \leq -a \vee r \geq a$$

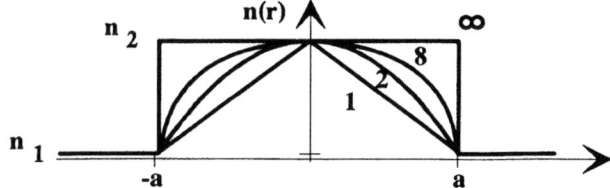

Figure 4. Index-Profiles for different values of profile parameter α

2.1.5. *Geometrical characteristics*

The geometrical characteristics of the fibre are those related to the physical dimensions of the fibre. That is to say, core, cladding and coating diameters, core and cladding non-circularities and finally core/cladding and cladding/coating concentricity errors. The way those characteristics are defined [1] [3] is shown in figure 5.

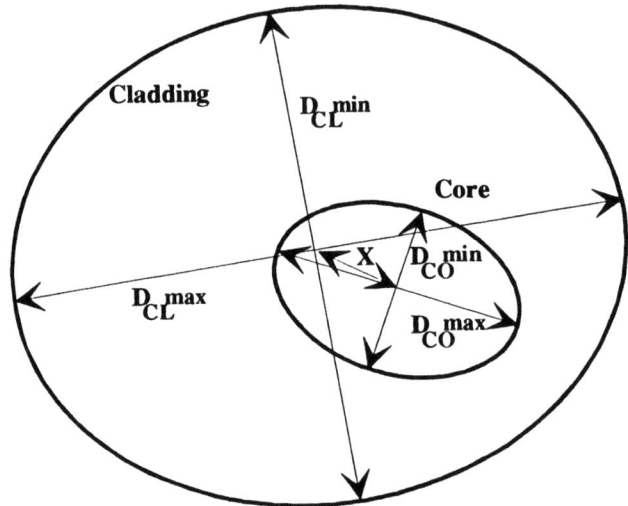

Figure 5. Geometrical characteristics definitions

- Core, cladding and coating diameters D_{CO}, D_{CL} & D_{PC} are defined as the average values of the maximum and minimum associated diameters:

$$D_{XX} = \frac{D_{XXmax} + D_{XXmin}}{2} \quad [\mu m] \quad \text{where } XX = CO, CL, PC \quad (5)$$

- Core & cladding non-circularities N_{CO} & N_{CL} are defined as the differential values of the maximum and minimum associated diameters normalized by the corresponding diameter:

$$N_{XX} = 100 \cdot \frac{(D_{XXmax} - D_{XXmin})}{D_{XX}} \quad [\%] \quad \text{where } XX = CO, CL \quad (6)$$

- Core/cladding & cladding/coating concentricity errors $C_{CO/CL}$ & $C_{CL/PC}$ are defined as the distance separating the corresponding centers normalized by the diameter:

$$C_{CO/CL} = 100 \cdot \frac{X}{D_{CO}} \quad [\%] \quad , \quad C_{CL/PC} = 100 \cdot \frac{Y}{D_{CL}} \quad [\%] \quad (7)$$

Fibres used in current systems are standardized for nominal values of diameters and have to be fabricated with specified tolerances for the geometrical characteristics that were described in figure 5. Fibre manufacturers previously followed recommendations of different international committees. The most well-known committees are the *International Electrotechnical Commission* (**IEC**) which publishes International Standards and the *Union Internationale des Télécommunications* (**UIT**) which issues Recommendations dealing with the Telecommunication domain. Not to be limited to telecommunication applications, the fibre geometrical characteristics are here given after the IEC International Standard [2]. Multimode fibre characteristics are shown in table 1, those of single-mode fibres are shown in table 2.

TABLE 1. Multimode fibre dimensions, nominal values & tolerances after IEC 793-2

FIBRE TYPE	A1a	A1b	A1c	A1d
Core Diameter D_{co} [μm]	50 ± 3	62.5 ± 3	85 ± 3	100 ± 5
Cladding Diameter D_{CL} [μm]	125 ± 3	125 ± 3	125 ± 3	140 ± 4
Coating Diameter D_{PC} [μm]	250 ± 15	250 ± 15	250 ± 15	250 ± 15
Core non-circularity N_{co}	≤ 6 %	≤ 6 %	≤ 6 %	≤ 6 %
Cladding non-circularity N_{CL}	≤ 2 %	≤ 2 %	≤ 2 %	≤ 4 %
Core/Cladding concentricity error $C_{CO/CL}$	≤ 6 %	≤ 6 %	≤ 6 %	≤ 6 %

In table 2 are gathered nominal values and tolerances for step-index single-mode fibres after IEC 793-2 [2]. Those after UIT G652 [5] are given in bold when they are different. Specifications for dispersion shifted and flat dispersion fibres are different for the mode field diameter nominal values, but are the same for tolerances.

TABLE 2. Single-mode fibre dimensions, nominal values & tolerances after IEC 793-2 (after UIT G 652)

FIBRE TYPE	B1 (Step Index)
Mode Field Diameter (MFD)	9-10 μm ± 10 %
MFD/Cladding concentricity error	**1 μm**
Cladding Diameter D_{CL} [μm]	125 ± 3 (**± 2**)
Cladding non-circularity N_{CL}	≤ 2 %
Coating Diameter D_{PC} [μm]	250 ± 15

2.2. TEST METHODS

The tolerances that are specified on geometrical characteristics are of great importance for fibre users. In fact, the fibre geometry is, today, the most critical problem in standardisation and the greatest source of conflict between fibres manufacturers and users such as cable or connector manufacturers. A simple way to approach this problem is to do the Passed / Failed Test consisting in trying to put a given fibre in a given connector ferrule. The tolerances on connector ferrules are very severe so that you can consider those ferrules as standard pin-holes. If the fibre doesn't go into the ferrule, it indicates that the tolerance on the cladding diameter (125 ± 3 μm) is not met. Unfortunately, anyone who has ever tried to mount a connector at a fibre end knows that sometimes this test fails. This test is simple to apply to cladding diameter, but what about the other characteristics which can not be externally measured ?

Another problem is not to misunderstand what these tolerances mean. These tolerances are not measurement uncertainties: to ensure such tolerances, the geometrical measurements have to be done with **negligible uncertainties with respect to those tolerances**. That is to say, uncertainties must be about ten times less the tolerance. That gives us, for instance, a **0.2-0.3 μm uncertainty** on diameter and cladding non-circularity measurements. For the corresponding test method, that means a measurement accuracy and also a calibration uncertainty better than this uncertainty.

The test methods that are recommended in the IEC International Standard 793-1 and in the UIT recommendations G651-G652 are gathered in table 3, following.

TABLE 3. Opto-geometrical characteristics Test Methods

TEST METHOD	TESTS	CHARACTERISTICS
IEC 793-1 A1	**Refracted Near Field**	- refractive index profile - theoretical numerical aperture - core & cladding diameters - non-circularities - concentricity errors
IEC 793-1 A2	**Near Field Light Distribution**	- core, cladding & coating diameters - non-circularities - concentricity errors
IEC 793-1 A3	**Four Concentric Circles**	- compliance with tolerances
IEC 793-1 A4	**Mechanical Diameter Measurement**	- cladding & coating diameters - non-circularities

2.2.1. Refracted Near Field

Measurement Principle and Set-Up: this method is based on the principle that, in the fibre, the refracted (non-guided) light is closely related to the refractive index distribution n (r). So by focusing incident light on a fibre section with a numerical aperture greater than the fibre one, guided **and** refracted "modes" are excited. Then the refracted light that is detected is proportional to the refractive index of the fibre cross section where the focused spot is launched. This principle is illustrated in figure 6a.

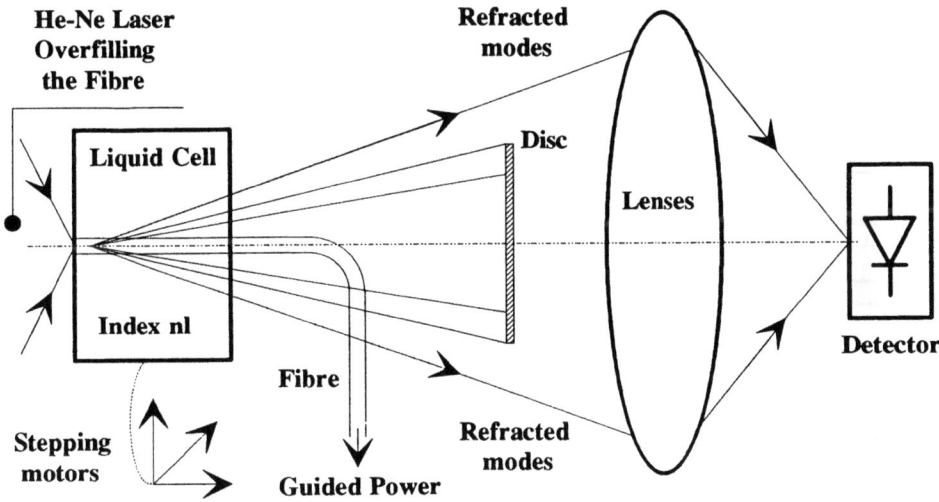

Figure 6a. Principle of the Refracted Near Field Method

-*Launching*: the beam of a He-Ne laser is focused on the fibre end face with a large numerical aperture (typ. > 0.6) in order to overfill the fibre core and cladding. The focusing **spot size is achieved as small as possible** (≈ 1.5 μm).

-*Cell*: the fibre is placed into a closed cell filled with a **known refractive index liquid**. This refractive index n_l is greater than the fibre cladding index, so that rays refracted in the fibre are then refracted in the liquid. The cell is mounted on stepper motor translation stages (x, y, z) in which displacement resolution is close to 0.1 μm.

-*Detection*: an opaque disc is placed on the spot axis in order to filter the evanescent modes. This way, only refracted modes can be collected by a lens combination and focused on a detector. The power on the detector is the so-called Refracted Near Field. This power is proportional to the refractive index which is "seen" by the focused spot.

-*Scanning*: by the mean of the translation stages, the fibre cross section is scanned in front of the incident spot, this scan yields the Refracted Near Field Distribution which is proportional to the **refractive index profile n (r)** of the fibre.

The greatest sources of uncertainty in this set-up are: on one hand, the focusing spot size, location and alignment with respect to the fibre end face, and on the other hand, the liquid index fluctuations with temperature. To limit those uncertainty sources as much as possible, some devices are added to the previous set-up. The achieved refracted near field measurement set-up is then shown in figure 6b.

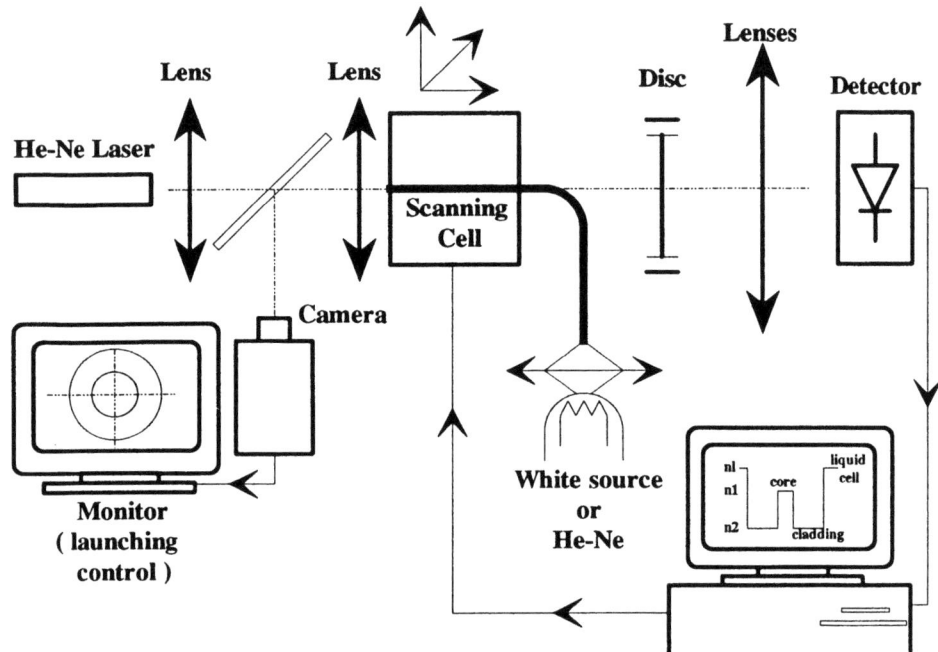

Figure 6b. Refracted Near Field Measurement Set-Up

-*Video Monitoring*: to ensure that the launching conditions are good, the fibre input face is monitored by a video camera after being magnified. This fibre is illuminated by launching the light from a white light source at the free end face. The output optical path is then distinguished from the input He-Ne light path using a separating slide. This device allows the control, before starting the manipulation, the focusing spot size, the alignment and cleanness of the fibre end face and also locates the focusing spot on a fibre cross diameter. During the sweep of this device, the displacement of the fibre with respect to the input spot, is shown.

-*Temperature Monitoring*: to obtain the refractive index profile of the fibre, it's obviously very important to calibrate the refractive index measurements. This is assumed by using the liquid refractive index n_l and the silica cladding index n_2. This liquid refractive index is well kwown but sensitive to temperature fluctuations (-4.10^{-4} °C^{-1}). So the liquid cell temperature is monitored by a temperature probe and regulated.

Measurement Results: after being calibrated in refractive index and in displacement (μm) the direct result of that test method is the fibre **refractive index profile n (r)**. An example of a multimode fibre index profile measured by the Refracted Near Field method is given in figure 7.

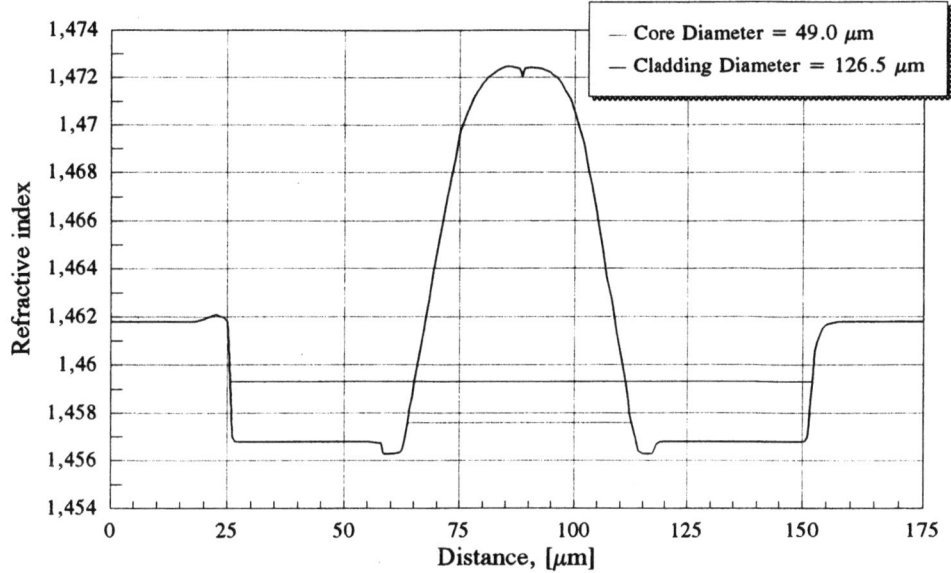

Figure 7. Refractive Index Profile of a multimode fibre measured by the Refracted Near Field Method

From the absolute refractive index profile, the following parameters can be determined:
- the **core maximum index**: $n_1 = 1.4725 \pm 0.0002$,
- the **theoretical numerical aperture**: $NA = 0.213 \pm 0.002$ (see equation 2),
- one **core diameter** which is obtained at the intersection of the index profile with the straight line $n = 0.05 (n_1 - n_2)$: $D_{co} = 49.0 \ \mu m$,
- one **cladding diameter** which is obtained at the intersection of the index profile with the straight line $n = 0.5 (n_1 - n_2)$: $D_{cl} = 126.5 \ \mu m$.

Unfortunately those diameters measurements do not comply with the diameter definition given in equation 5. What is measured is **one particular diameter** of the fibre cross section. Now, what has to be measured is the <u>**mean diameter**</u>. This measurement can be achieved by scanning all the chords of the fibre cross section, then the mean diameters, the non-circularities and the concentricity errors of the core and cladding can be calculated according to the equations 5, 6 and 7. However, to cover all the fibre cross section, with a 0.1 μm step for instance, the number of scans must be very high and the entire measurement will last a very long time. Actually, this Refracted Near Field Method is almost always limited to the first measurements, that is to say particular core and cladding diameters and **not "true" diameters**.

Uncertainties: according to the previous conclusion, no uncertainty calculations can be made on "true" diameters. However, if the set-up is calibrated in length with an uncertainty better than 0.1 μm then the uncertainty on "particular" diameter is the repeatability one and is close to 0.3 μm. The expanded uncertainty is thus **0.6 μm (2 σ)**.

2.2.2. Near Field Light Distribution

Measurement Principle and Set-Up: It was demonstrated that a relation exists between the refractive index profile and the transmitted power distribution in the cross section of the fibre. So, this power distribution, the so-called near field light distribution, gives an idea of the relative index profile and above all a means to measure the core and cladding geometrical characteristics. The principle is thus to image the fibre end face cross section. The measurement set-up is shown in figure 8.

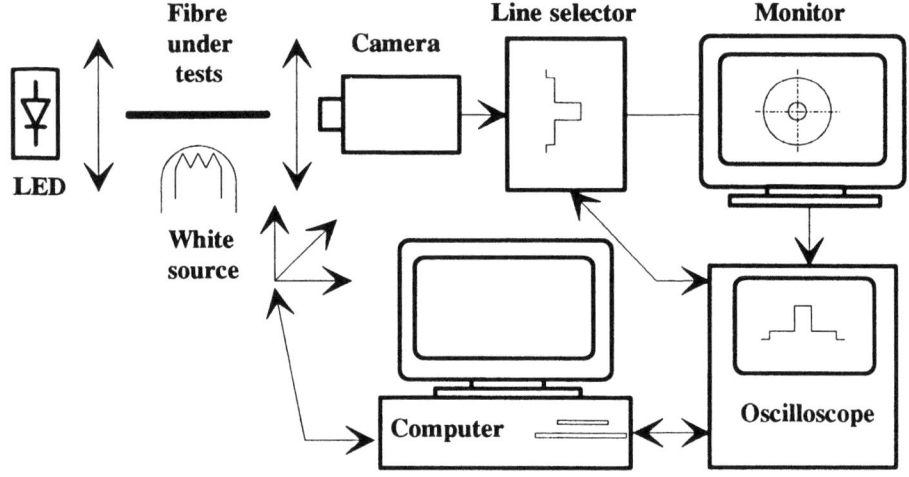

Figure 8. Far Field Light Distribution Measurement Set-Up

-*Launching*: a first light source is launched axially into the fibre core, a second source illuminates the fibre cladding transversely. Both sources should be incoherent so that the near field distribution is related to the index profile.

-*Microscope*: the cross section of the fibre output end is imaged on a video camera with a large magnification (up to 600 x) and large numerical aperture objective. The combination of the magnifying optic and the video camera realizes a video microscope whose resolution is as close as possible to the diffraction limit (Airy spot).

-*Monitoring*: to improve the accuracy and not to be dependent on contrast and luminosity, the video signal is sent line by line to a digital oscilloscope and to a monitor.

-*Calibration*: this video microscope is calibrated in length using any artifact whose physical dimensions are known and **related to the near field light distribution**.

Measurement Results: Each line of the video signal represents a chord of the fibre cross section. As well as in the refracted Near Fied method, the diameter line can be analysed separately to determine **one** core diameter and **one** cladding diameter. But it should be better if the mean diameters could be measured. Now if those lines are gathered and analyzed, one obtains the near field light distribution in a three-dimensional way, that is to say the power at "each" point of the fibre cross section. In that power distribution, the maximum power in the core P_{core} and the average power in the cladding P_{clad} can be measured easily. Then, to obtain the mean diameters, a powerful method to be used is to analyse the intersection of this near field distribution with two cross section planes describing the core and cladding areas. The "cladding plane" is defined as a constant power equal to half the average power in the cladding: $P = 0.5\ P_{clad}$. The "core plane" is defined as a constant power equal to 5 % of the power difference between the core and the cladding: $P = 0.05\ (\ P_{core} - P_{clad}\ )$. The intersection points between those planes and the video lines gives a representation of the **fibre core and cladding perimeters**. An example of this representation is given in figure 9 for a multimode fibre.

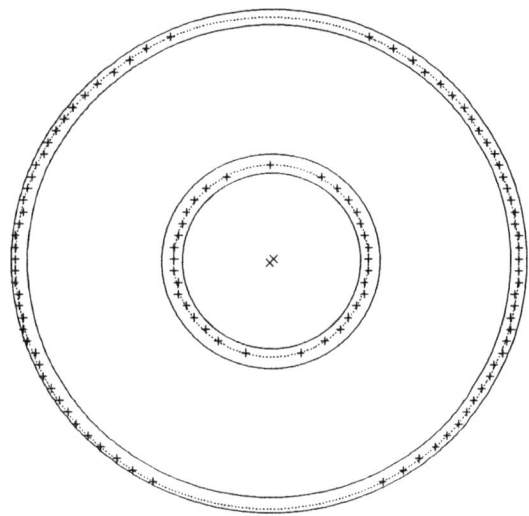

Figure 9. Geometrical measurements by the Near Field Light Distribution Method:
A multimode fibre example

From the cross section analysis one **can fit** the most likely circles that represent the core and cladding perimeters. From those fitted circles and their centers can be determined all core and cladding geometrical characteristics. This yields for the previous instance:
- the core diameter: $D_{co} = 49.2\ \mu m$, and its non-circularity: $N_{co} = 4.3\ \%$,
- the cladding diameter: $D_{CL} = 125.8\ \mu m$, and its non-circularity: $N_{CL} = 1.7\ \%$,
- the core/cladding concentricity error: $C_{CO/CL} = 0.8\ \%$.

Uncertainties: to calibrate the set-up in length an artifact whose <u>dimensions</u> are known <u>and related to the near field light distribution</u>, must be used. Such an artifact was developped by the NPL in collaboration with York Instruments. It allows a transfer calibration uncertainty close to 0.05 μm. The uncertainty is thus the repeatability which is essentially due to the diffraction limitation of the microscope (1-1.5 μm). Fortunately the mean component, obtained in a statistical way, is lower than the diffraction limitation, and is often close to 0.4 μm (1σ). Thus, even if the calibration in distance can be neglected, the **expanded uncertainty is close to 0.8 μm (2σ)**.

2.2.3. *Four Concentric Circles*

This method is not really a measurement one, and it can not be used to measure the values of geometrical characteristics individually. But it can be used to verify the compliance of those geometrical characteristics with the mean values and their given tolerances specified in international standards. For instance, the IEC 793-2 specifications on category A1 multimode fibres are:

- core diameter: $D_{co} = (50 \pm 3)$ μm and its non-circularity: $N_{co} \leq 6\%$,
- cladding diameter: $D_{CL} = (125 \pm 3)$ μm and its non-circularity: $N_{CL} \leq 2\%$,
- core/cladding concentricity error: $C_{co/CL} \leq 6\%$.

Those values can be resumed in only two greater values on diameter tolerances, those new values are then:

- core diameter: $D_{co} = (50 \pm \Delta D_{co})$ μm with $\Delta D_{co} = 4$ μm,
- cladding diameter: $D_{CL} = (125 \pm \Delta D_{CL})$ μm with $\Delta D_{CL} = 5$ μm.

Those values are sufficient to comply with all the tolerances if the four circles so-defined are concentric. Then, those four concentric circles form two rings as shown in figure 10.

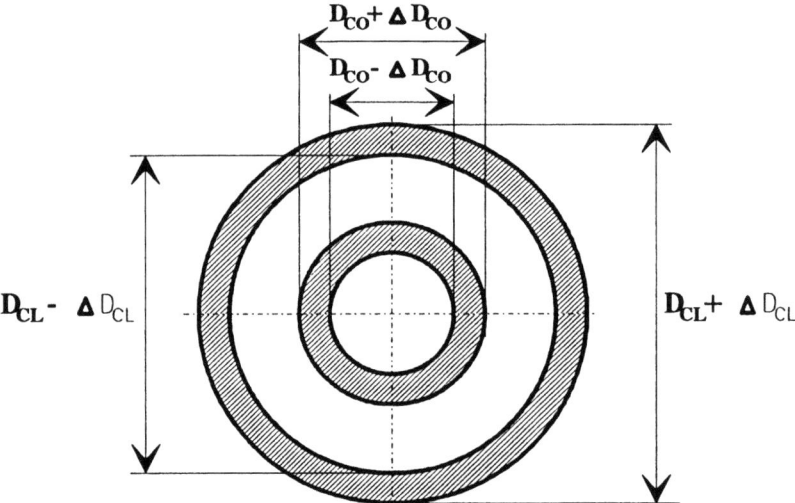

Figure 10. Four Concentric Circles Method of Compliance

To pass this method of compliance, all core and cladding measurements values must be inside those rings. This method is typically applied to the core and cladding perimeter measurements of the near field light distribution method (see figure 9).

2.2.4 *Mechanical Diameter Measurement*

Measurement Principle and Set-Up: this method can be used to measure external diameters of the fibre, that is to say either cladding or coating diameters. The principle of this method is to contact the fibre under test with two surfaces and to measure the separation between these surfaces with a great accuracy. A schematic of one possible set-up is shown in figure 11. In that set-up, the fibre is contacted with a stationnary steel cylinder, the so-called anvil, and a moveable flat surface, the so-called spindle, that is pressed against the fibre with a known force. At the opposite end of the spindle is fixed a mirror that constitutes the measurement arm of a Michelson interferometer. The difference between the spindle position when the fibre is in place, and its position when it is pressed against the anvil is measured by the interferometer and gives one fibre diameter. Practically this measurement has to be corrected for the compression, especially at the fibre-anvil contact point. This could be achieved by measuring the fibre diameter at several contact forces, and then by extrapolating to the zero contact force.

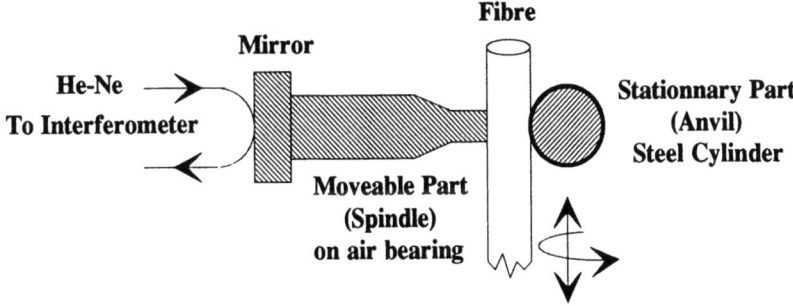

Figure 11. Mechanical Diameter Measurement Set-Up: the Interferometric Micrometer

Uncertainties: on one diameter measurement, a combined uncertainty of **0.02 μm (1σ)** was obtained [8]. However, this is only one diameter and the diameter definition requires measurement of several diameters by rotating the fibre and then computing the mean diameter. Therefore the uncertainty on the fibre mean diameter is necessarily greater because it must take into account the rotation step of these several measurements.

2.3. OPTO-GEOMETRICAL MEASUREMENT SUMMARY

These three methods show uncertainties that are small enough to control the fibre geometrical specifications at one given cross section. However, the compliance with these specifications must of course be verified along the total fibre length. And that's the problem because none of these techniques is appropriate [9] [10], or easily usable for that. Unfortunately, this inadequate approach was demonstrated during an international fibre geometry round robin [11], that showed some dispersions on the cladding diameter greater than the tolerance. And to worsen this fact, nowadays tolerances are smaller (2 μm against 3 μm). This mismatch is understandable due to the fact that, with these three test-methods, the diameter control is not assured during, but after the fabrication process. Due to that, the relation between diameter measurements during and after the fabrication process is quite hard to do. Therefore, the solution is not to enhance those post-fabrication test methods, but to improve the in-line techniques that are used to control the fibre diameter during the fabrication. A few proposals that follow this approach are being studied presently, but are not published yet.

3. Attenuation measurements

3.1. ATTENUATION IN FIBRES
3.1.1. *Definition*

The power loss between two cross sections of the fibre is defined as the ratio, expressed in dB, of the energy fluxes $P(z_1)$ and $P(z_2)$ crossing those sections (see figure 12). Then the attenuation coefficient α is the ratio of this power loss divided by the distance L separating these sections (equation 8).

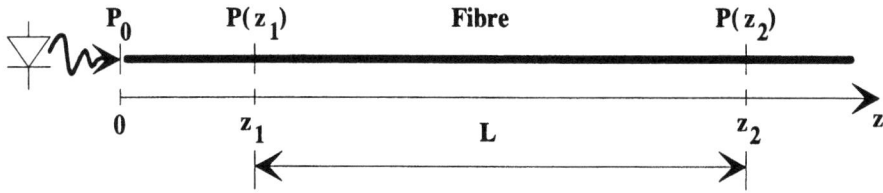

Figure 12. Attenuation Coefficient definition

$$\alpha = \frac{10}{L} \log_{10} \frac{P(z_1)}{P(z_2)} \quad [dB/km] \tag{8}$$

3.1.2. *Power Loss Components*

The fibre attenuation coefficient is due to several losses components [12]. These components are **absorption, scattering, microbending and bending losses**, that are described in the following and summarized in figure 13 for a multimode fibre.

Absorption losses: [13] Due to the compound (silica), impurities and dopants used in the fibre, absorption bands and local absorption lines can be distinguished:

- Absorption "queues" are those of ultraviolet and infrared absorption in the silica. The UV absorption, due to electronic transitions, decreases exponentially with the frequency ($\propto e^{-A\nu}$, where ν is the frequency and A a constant) and results in typical attenuation coefficients of 0.1 dB/km at 0.8 μm and 0.01 dB/km at 1.5 μm. The IR absorption, due to vibrational and rotational transitions (multiphonon absorption), increases exponentially with the frequency ($\propto e^{B\nu}$, where ν is the frequency and B a constant) and results in attenuation coefficients of about 0.01 dB/km at 1.5 μm and as high as 1 dB/km at 1.8 μm. Both UV and IR absorption queues are given in figure 13a and 13b respectively where the attenuation coefficient is plotted in dB/km on a logarithmic scale as a function of wavelength.

- Absorption "lines" are localised in wavelength and correspond to resonance frequencies of impurities present in the silica. These impurities are, for instance, metallic ions and especially the iron (Fe) that generates a strong absorption of 130 dB/km/ppm of Fe at 0.85 μm. Fortunately, today's processes allow elimination of this Fe ion (<< ppb of Fe) and reduce its absorption line to a negligible value. However, another impurity, the hydroxyl ion (OH⁻), presents three strong absorption lines in the useful wavelength range (0.8 μm - 1.6 μm): line intensities are about 60 dB/km/ppm of OH⁻ at 1.38 μm, 3 dB/km/ppm of OH⁻ at 1.24 μm and 1 dB/km/ppm of OH⁻ at 0.95 μm. The wavelength location of these lines historically determined the choice of the so-called fibre "windows". In figure 13, an instance of those lines is plotted for a very poorly dehydrated fibre. The first window, 0.85 μm, is just before the 0.95 μm OH⁻ line; the second, i.e. 1.31 μm between 1.24 μm and 1.38 μm OH⁻ lines and the third, 1.55 μm, after the 1.38 μm line. Today, a good dehydration is achieved during the fibre manufacturing processes and the concentration of OH⁻ is reduced to less than ten ppb. Therefore, lines are reduced at most to 0.6 dB/km at 1.38 μm, to 0.03 dB/km at 1.24 μm and 0.01 dB/km at 0.95 μm. Obviously, these lines can not be neglected in intensity, but their spectral widths are reduced enough to separate them and to cancel their influence on the "windows", especially for the 1.31 μm and the 1.55 μm windows.

Scattering losses:[13] Due to thermal variations during the fibre cooling process (from 2000 °C to 20 °C), the fibre presents index, density and dopants concentration fluctuations that scatters the light. The size of those inhomogeneities with respect to the wavelength determines the type of scattering that takes place. If this size is greater than the wavelength, the scattered-light ratio do not depend on the wavelength and the light scatters in the propagation direction. If the wavelength λ is about the inhomogeneities size, the light scatters in the propagation direction and the amount of scattered-light varies in λ^{-2}, the so-called Mye scattering. Now, and that's the silica fibre case, if the wavelength is much more greater than the inhomogeneities size, the light scatters in all directions (isotropic light) and its intensity varies in λ^{-4}. This scattering is the well-known **Rayleigh** scattering. In optical fibres, the Rayleigh scattering losses α_S are modelled by the relation:

$$\alpha_S = \frac{8\pi^3}{3\lambda^4}(n^2-1)k_B T \beta_c \quad (9)$$

where n is the refractive index, β_c is the thermal compressibility, k_B is the Boltzmann constant and T is the absolute vitrification temperature. For pure silica, this Rayleigh scattering loss α_S is 0.75 dB/km at 1.0 μm. For germanium doped-fibres, due to dopants addition, this coefficient α_S increases with the refractive index difference Δn between the core and the cladding, according to the relation:

$$\alpha_S = \frac{0.75 + 66.\Delta n}{\lambda^4} \quad \text{[dB/km]} \quad (\text{for } \lambda \text{ in } \mu m) \quad (10)$$

In a step-index single-mode fibre, for $\Delta n = 4.10^{-3}$, typical values for Rayleigh scattering losses are 1.94 dB/km at 0.85 μm, 0.345 dB/km at 1.31 μm, 0.175 dB/km at 1.55 μm. The entire curve (eq.10) is plotted in figure 13c for a multimode fibre. Those values added with absorption losses yield the theoretical smallest attenuation coefficients of a fibre and represent the ultimate limit of attenuation. Nowaday, a step-index single-mode fibre commonly tailored shows attenuation coefficients very close to those theoretical minima, let's say **0.35 dB/km at 1.31 μm and 0.2 dB/km at 1.55 μm**.

Non-linear scattering processes, such as Raman and Brillouin ones, must be considered for high power density (>1 MW/cm²). However, the non-linearity limit corresponds to a power of 640 mW in a common single-mode fibre (9 μm-core). Such a high power density is not available in telecommunication applications, except for very narrow linewidth sources. Therefore, these non-linear scattering processes are not easily observed in germanium-doped fibre.

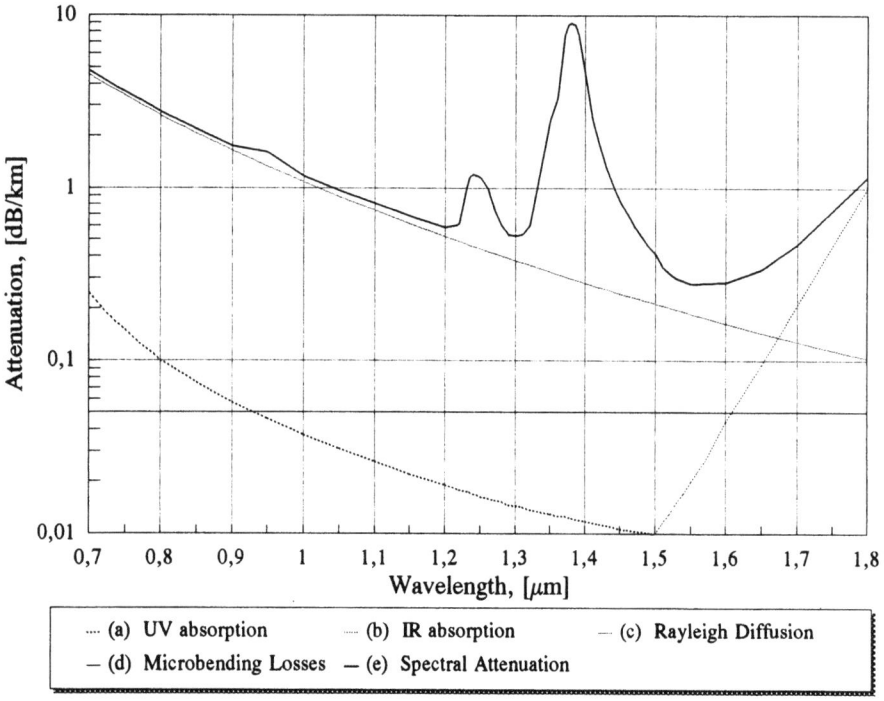

Figure 13. Attenuation Coefficient of a multimode fibre: a/ UV absorption queue, b/ IR absorption queue, c/ Rayleigh scattering, d/ Microbending Losses, e/ Spectral Attenuation Curve

Microbending losses $\alpha_{\mu c}$: [13] [14] In cables or on reels, the fibre is subjected to local stresses that generate random microbendings. At a transition between a straight and curved part of the fibre line, the matching error between fields generate losses. Those microbending losses are very hard to separate from other loss components. However, a typical instance of multimode fibre microbending losses is given in figure 13d, that shows microbending losses of 0.05 dB/km. With regard to single-mode fibres, microbending losses are usually proportional to $w_0^{10} + b^8$, where w_0 is the mode field radius and b the cladding radius. It's commonly admitted that those losses can be neglected at the wavelength λ if the mode field radius is smaller than 4λ. In single-mode fibres, the mode field radius is smaller than 5 μm, therefore microbending losses can be neglected at any wavelength greater than 1.25 μm.

The addition of absorption losses, scattering losses, microbending losses with respect to wavelength, yields the fibre spectral attenuation curve (see figure 13e).

Bending losses: Due to small curvature radii of the fibre, evanescent modes appear in the cladding. Those evanescent modes obviously generate losses that increase very quickly while the curvature radius decreases. An instance of bending losses as a function of the curvature radius is given in table 4, in the case of a "classical" single-mode fibre. From table 4, it can be seen that bending losses can be neglected if one takes care in using fibres without any curvature radius smaller than about 4 cm. Since such a small curvature radius of the fibre can be considered as an operating error, those bending losses are not included in the attenuation coefficient.

TABLE 4. Bending losses of a single-mode fibre ($\lambda=1.0$ μm V=2.4 Step-index 2a=9 μm)

Curvature Radius, [cm]	1.0	2.0	3.0	3.5	4.0
Attenuation, [dB/km]	10^6	10^3	1	10^{-2}	10^{-4}

3.2. ATTENUATION MEASUREMENTS

The test methods that are recommended in the IEC International Standard 793-1 [1] and in the UIT recommendations G650 [3] are the cut-back technique, the insertion loss technique and the backscattering technique. The cut-back technique is the reference test method and allows measurement of the spectral attenuation curve. The insertion loss technique can be used to measure the attenuation coefficient at a given wavelength. Finally, the backscattering technique allows measurement of the attenuation coefficient of a fibre and checking its homogeneity when only one fibre end is accessible.

3.2.1. Cut-Back technique
Measurement Principle and Set-up:

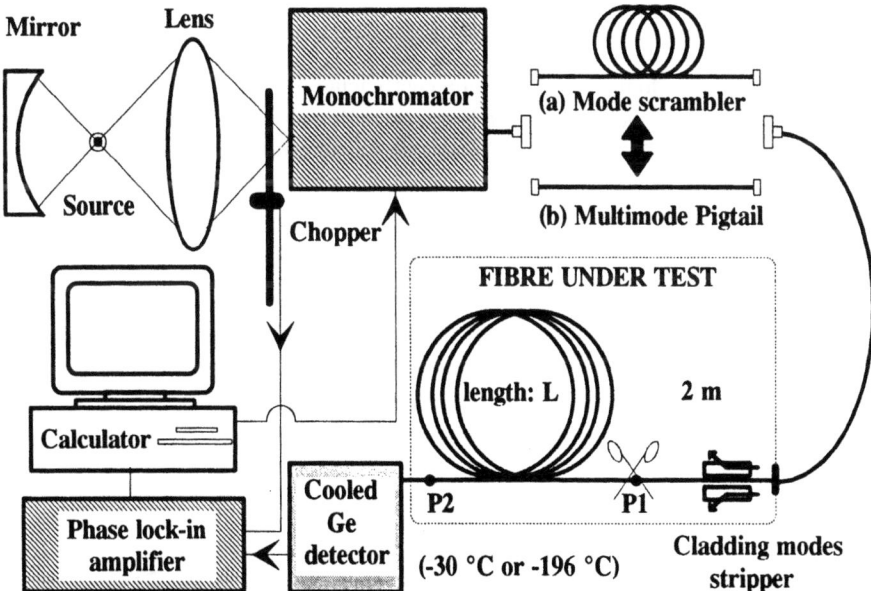

Figure 14. Cut-back technique experimental Set-Up a/ Multimode Fibre, b/ Single-mode Fibre

The principle of the cut-back technique follows the attenuation coefficient definition; namely to measure the power ratio through two fibre cross sections separated by a known distance. To apply this definition, the power is first measured at the end of a long length of fibre, yielding the power P_2. Then the fibre under test is cut to a short length (typically 2 meters) and the power P_1 is measured without making any change in the launching conditions. A schematic of a cut-back technique set-up that allows determinination of the total spectral attenuation curve, is given in figure 14. In this set-up a monochromator allows selection of any wavelength from 0.8 μm to 1.6 μm with a spectral width of 3 nm. The monochromator exit slit is pigtailled with a multimode fibre used as a launch fibre. The optical power detection is achieved by a germanium detector. In such a set-up, the power levels are very low, typically lower than 1 nW. To improve the signal/noise ratio, the light source is modulated by a chopper and synchronous detection is achieved by a lock-in amplifier connected to the detector. In addition, the germanium detector is cooled to decrease its noise density. Usually, the detector is cooled down to -30 °C by Peltier stages. However, to measure very long single-mode fibres for instance, it has to be cooled down to -196 °C by means of a dewar filled with liquid N_2. This set-up can be used whether the fibre is multimode or single-mode. However, with regard to the launching conditions, the greatest care must be taken for multimode fibres. It's of great importance that the mode equilibrium is reached in the fibre whatever its length is, and especially when the power is measured at the short length (2 m) after the cut. To achieve this mode equilibrium, a mode scrambler is inserted in the line just before the fibre under test. This mode scrambler can consist in a great length of multimode fibre (fig.14a) or in a combination of step-gradient-step index profiles fibres. For a single-mode fibre, the launching fibre simply consists in a multimode pigtail (fig.14b).

Measurement Results: the spectral attenuation curve of a multimode fibre of 1600 meters length is given in figure 15a. Typical values of 2.2 dB/km at 0.85 μm and 0.54 dB/km at 1.31 μm are found for attenuation coefficients. In this instance, it can be said that the fibre is not very well dehydrated because the three hydroxyl absorption lines can be seen at 1.38 μm, 1.24 μm and 0.95 μm, which is unusual nowadays.

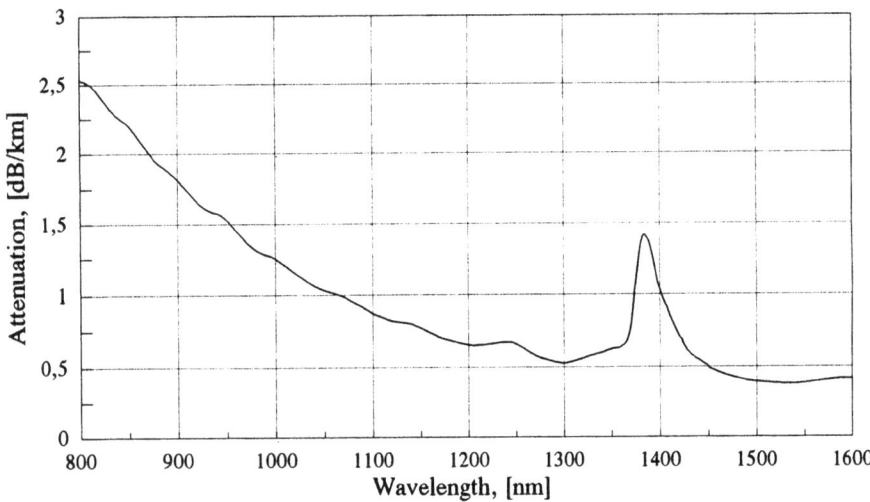

Figure 15a. Spectral Attenuation Curve of a 50/125 μm multimode fibre by the Cut-back technique

The spectral attenuation curve of a single-mode fibre of 2600-meters length is given in figure 15b. Typical values of 0.338 dB/km at 1.31 μm and 0.204 dB/km at 1.55 μm are found for attenuation coefficients. The loss band from 1.1 μm to 1.27 μm shows the limit of the single-mode propagation in that fibre. In this instance, the fibre cut-off wavelength is close to 1.27 μm. For smaller wavelength the fibre becomes multimode. Therefore below 1.27 μm, the attenuation coefficient is not significant because no care is taken of the mode equilibrium.

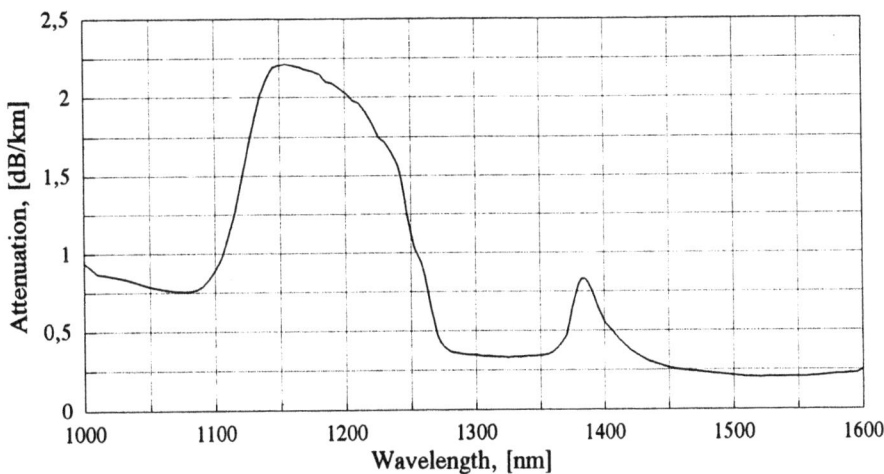

Figure 15b. Spectral Attenuation Curve of a Step Index single-mode fibre by the Cut-back technique

Uncertainties: the uncertainty components of this set-up are summarized as follows. Typical figures are given in table 5 for the usual "windows": 0.85 μm and 1.31 μm for multimode fibres, 1.31 μm and 1.55 μm for single-mode fibres. Those uncertainties are greater around the water line and around the cut-off wavelength.
A: repeatability, B_1: influence of the radiation wavelength uncertainty, B_2: influence of the source spectral width, B_3: influence of the launching conditions in a multimode fibre, B_3': influence of the polarization in a single-mode fibre, B_4: calibration of the detector linearity, B_5: calibration of the fibre length, I: combined standard uncertainty. All these components are expressed in percents for one standard deviation (1σ).

TABLE 5. Uncertainties on attenuation measurement by the cut-back technique

Fibre type	A (%)	B_1 (%)	B_2 (%)	B_3 (%)	B_4 (%)	B_5 (%)	I (%)
Multimode	0.15	0.08	0.01	0.10	0.01	0.05	**0.20**
Single-mode	0.025	0.008	-	0.020	0.010	0.025	**0.045**

For **multimode** fibres, the combined uncertainty I is essentially due to the repeatability component A. The expanded uncertainty is thus about 0.4 % or **0.02 dB/km (2σ)**. For **single-mode** fibres, the expanded uncertainty (2.I) is about 0.09 %, that is to say **0.004 dB/km (2σ)**.

3.2.2. *Insertion Loss technique*

Measurement Principle and Set-up: the principle of this technique is given in figure 16. The launching conditions are the same as for the cut-back technique. Here, the cross sections where powers are measured are the fibre ends. The power P_1 is first measured without the fibre or on a short length of the fibre under test, as shown in figure 16a. The fibre under test is then inserted to measure the power P_2 as shown in figure 16b. The ratio of those powers yields the fibre loss. Unfortunately, this ratio also includes the changes in the coupling losses of the fibre to fibre coupling devices (splices, connectors,...). For that reason, this method usually yields an uncertainty greater than the cut-back technique one.

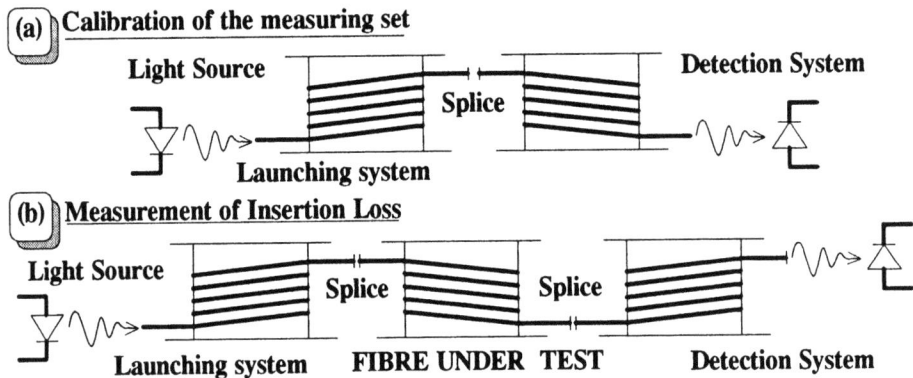

Figure 16. Insertion Loss technique Principle & Set-Up

3.2.3. *Backscattering technique*

Measurement Principle and Set-up: this method uses the Rayleigh scattering demonstrated in § 3.1.2, and illustrated by equation 9. The principle is to launch an optical pulse in the fibre and then to measure the amount of backscattered power. This is illustrated in figure 17. The backscattered power can be related to the fibre loss and its flight time can be related to the distance along the fibre length. Using two backscattered powers measured at differents times, and thus separated by a known distance, one can compute the attenuation coefficient.

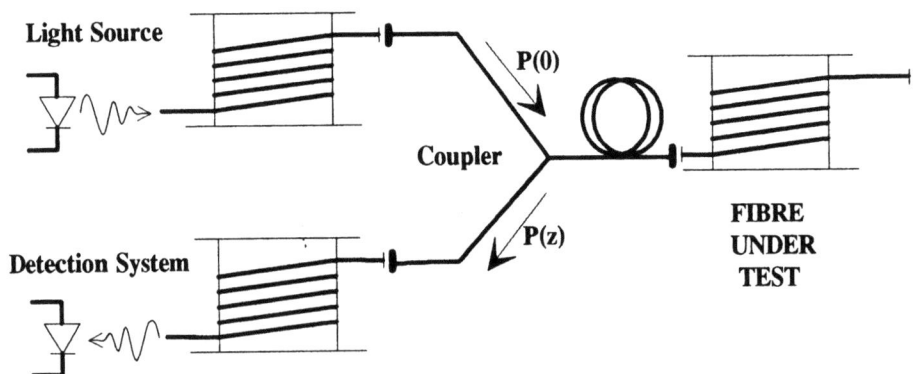

Figure 17. Backscattering technique Principle & Set-Up

This principle is used in the so-called optical time-domain reflectometer (OTDR). This apparatus converts the backscattered power to decibels and the pulse flight-time to distance. The backscattered power can be plotted as a function of distance, yielding a so-called backscattering signature of the fibre.

Measurement Results: the backscattering signature of a 2600-meters length single-mode fibre is given in figure 18. This signature is obtained at 1.31 µm. The slope of the curve shows the attenuation coefficient of this fibre. In this case an attenuation coefficient of 0.33 dB/km is measured from 300 m to 2410 m of fibre. The first peak corresponds to the Fresnel reflection at the fibre input (fibre to fibre coupling) and is the so-called dead zone where no measurement is possible. The second peak corresponds to the Fresnel reflection at the fibre output and is much stronger due to the poor coupling at the fibre to air interface.

Uncertainties: with regard to attenuation coefficient, the combined uncertainty is usually greater than 0.01-0.02 dB (1σ) at least. This is due to the great number of unknown parameters. First, the wavelength and the spectral width of the OTDR laser are not always well known. Second, the apparatus must be calibrated in distance which means that the fibre index must be known. And third, the apparatus must be calibrated in attenuation and in linearity of the attenuation scale. Therefore, even if the repeatability of the measurements is currently smaller than 0.01 dB/km, great systematic errors can occur. As a consequence, the extended uncertainty is, for a very well calibrated OTDR, at least about 0.02 dB/km (2σ) for the single-mode fibre attenuation coefficient.

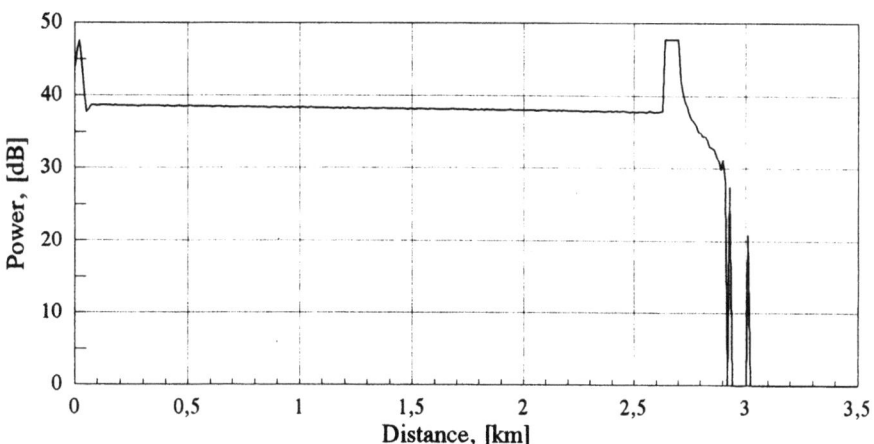

Figure 18. Backscattering Signature of a single-mode fibre

3.3. ATTENUATION MEASUREMENT SUMMARY

None of these methods can be considered better than any other one. Each one presents its advantages that depend on the need for a given level of uncertainty. The cut-back technique yields the best results, but is destructive. The insertion loss method is not destructive. And the backscattering method not only provides the attenuation coefficient, but also is a very powerful way to locate any attenuating or reflecting event.

4. Dispersion measurements

4.1. DISPERSION IN FIBRES
4.1.1. *Definitions* [7] [12]

The propagation modes in the fibre are characterized by their propagation constant β, that is usually expressed as the normalized propagation constant B:

$$\beta = kn_1\left[1 - 2\Delta\left(\frac{u}{V}\right)^2\right]^{\frac{1}{2}} \qquad B = 1 - \left(\frac{u}{V}\right)^2 \qquad (11)$$

where V is the normalized frequency and u and w are reduced parameters introduced during the solving of the dispersion equation. The parameters u and w are given by:

$$u = \sqrt{a^2(k^2n_1^2 - \beta^2)} \qquad w = \sqrt{a^2(\beta^2 - k^2n_2^2)} \qquad u^2 + w^2 = V^2 \qquad (12)$$

Each mode propagates with its own group velocity V_g, which is defined as:

$$v_g = \frac{d\omega}{d\beta} \quad \text{with} \quad \omega = 2\pi f \qquad (13)$$

where ω is the radian frequency of the transmitted signal. Therefore, for a distance L in the fibre, the transit time τ is:

$$\tau = \frac{L}{v_g} = L\frac{d\beta}{d\omega} = L\frac{d\beta}{d\lambda}\frac{d\lambda}{d\omega} \quad [s] \qquad (14)$$

From equation 14, it can be seen that a time pulse will expand along the fibre as a function of the wavelength and the propagation constant. This effect is the so-called dispersion. The dispersion may be divided into several components: the material dispersion, the intramodal dispersion and the intermodal dispersion.

4.1.2. *Material dispersion M_2*

This component is due to the non linear wavelength dependence of the core refractive index n_1. For that reason, this component is commonly called the chromatic dispersion. This chromatic dispersion M_2 and the resulting pulse delay $\Delta\tau_m$ are then given by:

$$M_2 = -\frac{\lambda}{c}\frac{d^2n_1}{d\lambda^2} \qquad [\text{ps}/(\text{nm.km})]$$

$$\Delta\tau_m = L\lambda M_2\frac{\Delta\lambda}{\lambda} = -\frac{L}{c}Y_m\frac{\Delta\lambda}{\lambda} \quad [s] \qquad (15)$$

where $\Delta\lambda$ is the source spectral width.

In equation 15, the chromatic dispersion coefficient in [ps/ (nm.km)] expresses the resulting pulse delay in [ps] per source spectral width unit [nm] and per fibre-length unit [km]. To normalize this coefficient, one also uses the dimensionless Y_m coefficient expressed in equation 15. Some material dispersion coefficients M_2 [ps/(nm.km)] of fused-silica, germanium doped-silica and boron doped-silica [13] are plotted as a function of wavelength in figure 19.

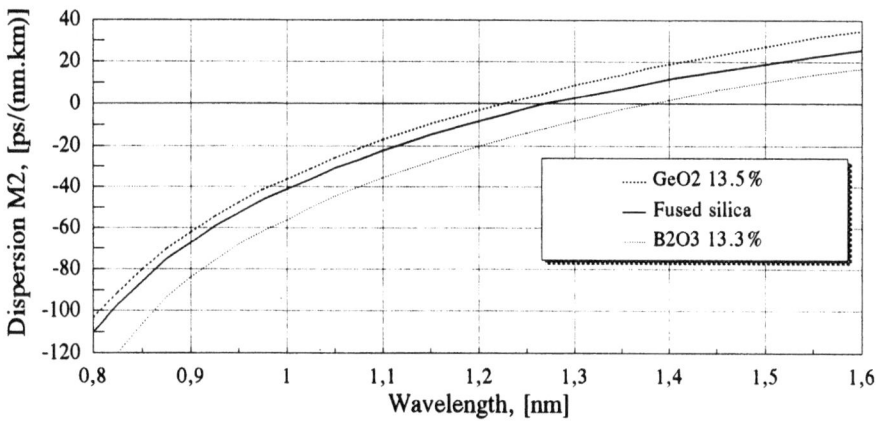

Figure 19. Material dispersion coefficient M_2 of silica-fibre with respect to the wavelength.

On this figure, the index non linearity is evidence. Typical values for chromatic dispersion coefficient are 85 ps/(nm.km) around 0.85 μm, 5 ps/(nm.km) around 1.3 μm and 20 ps/(nm.km) around 1.55 μm. In addition with the small attenuation coefficients, these values indicate the choice of the 1.31 μm and 1.55 μm windows for telecommunications. The zero dispersion is obtained at 1.27 μm for fused-silica, but is shifted when the fibre is doped.

4.1.3. *Intramodal dispersion*

For a given mode, the propagation constant β depends on the normalized frequency V (see eq.11). Due to the non linearity of this dependence, a dispersion occurs that is the so-called intramodal dispersion. The normalized frequency V is a parameter related to the shape of the fibre itself. For that reason, this dispersion is commonly called the waveguide dispersion. This waveguide dispersion M_g and the resulting pulse delay $\Delta\tau_g$ are then given by:

$$M_g = -\frac{\Delta.n}{\lambda c}\frac{Vd^2(VB)}{dV^2} \quad [ps/(nm.km)]$$

$$\Delta\tau_g = L\lambda M_g \frac{\Delta\lambda}{\lambda} = -\frac{L}{c}Y_w\frac{\Delta\lambda}{\lambda} \quad [s]$$

(16)

As well as for the chromatic dispersion case, the coefficient M_g is expressed in terms of the dimensionless guide dispersion coefficient Y_w in equation 16. This guide dispersion is a function of the normalized frequency V and therefore a function of the fibre core radius a. A few guide dispersion coefficients Y_w are plotted in figure 20 as a function of the wavelength and for several values of this core radius a [13]. The material dispersion coefficient Y_m is also plotted to be compared to the guide dispersion.

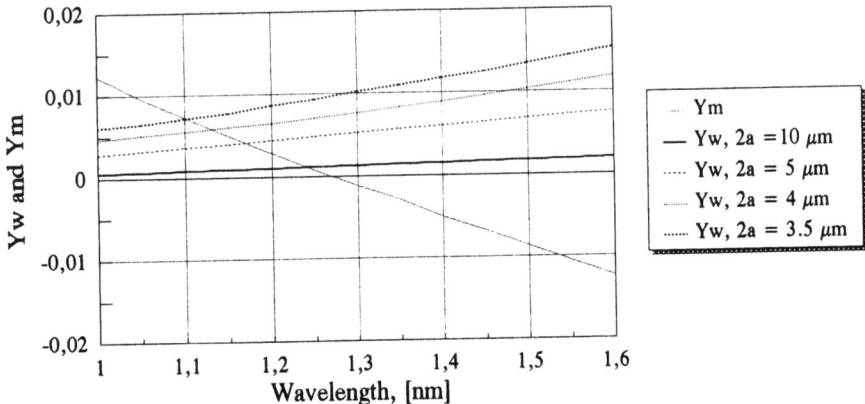

Figure 20. Guide dispersion coefficient Y_w of fused-silica with respect to the wavelength

On this figure, it's important to see that for any wavelength greater than 1.27 μm, the chromatic and the guide dispersion act in opposite ways. Therefore, they will compensate to reduce the total dispersion. That shows a possible way to solve the dispersion problem, by using a 3.5 core diameter fibre for instance. However, such a small core diameter would involve other problems, especially on curvature losses and coupling losses. Due to that, typical values for the core diameter are about 9 or 10 μm.

4.1.4. *Intermodal dispersion*

The intermodal dispersion [12] is a consequence of the mode number dependence of the group velocity V_g. The fastest mode will be the fundamental one, that is to say the axial one associated to the velocity $V_g(0)$. The slowest one is the one that suffers from a great number of reflections, that is to say the most paraxial one. This mode propagates with the velocity $V_g(N)$. For an index profile n(r) described by the Gloge and Marcatili relation (eq.4), the resulting pulse delay $\Delta\tau_i$, after a fibre-length L, is given by:

$$\Delta\tau_i = L\left(\frac{1}{v_g(0)} - \frac{1}{v_g(N)}\right) \quad [s] \tag{17}$$

The pulse delay as a function of the index-profile parameter α is plotted in figure 21a, for a 50/125 μm multimode fibre at 0.85 μm with $n_1 = 1.5$ and $\Delta = 1\,\%$. This curve [13] shows a minimum for a index-profile parameter α close to 2. This corresponds to a parabolic profile and points the right choice for such a profile in typical multimode fibres. From a step to a parabolic index multimode fibre, the pulse delay is divided by a factor of about 1000. Actually the index profile depends on the wavelength also. This wavelength dependence increases the pulse delay and yields the new delay curve that is plotted in figure 21b. This curve is then shifted and the minimum dispersion is obtained for a profile parameter $\alpha = 2.3$. Now, from a step to a parabolic index profile fibre, the dispersion is only divided by 25. This factor increases to 200 with a profile parameter of 2.3. The intermodal pulse delay can be expressed easily for profile parameters $\alpha = \infty$ (step-index) and $\alpha = 2.3$:

$$\Delta \tau_i \big|_{\alpha = \infty} = L \frac{n_1 \Delta}{c} \quad \text{(step-index)}$$

$$\Delta \tau_i \big|_{\alpha = 2.3} = L \frac{n_1 \Delta^2}{2c} \quad \text{(graded-index)} \tag{18}$$

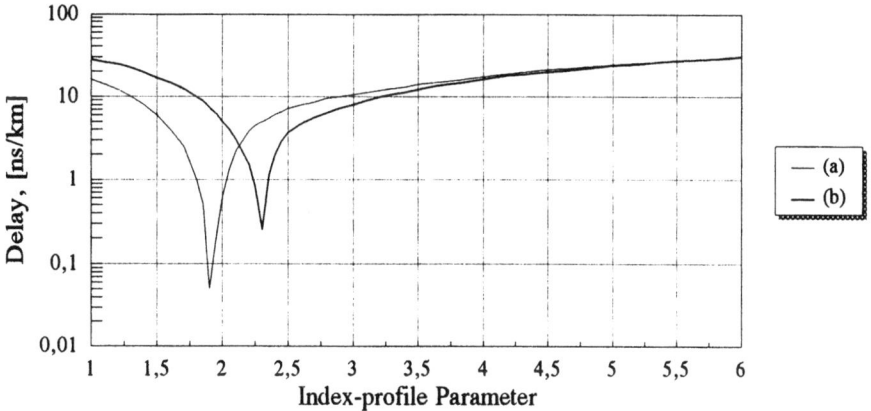

Figure 21. Intermodal dispersion of multimode fibres as a function of the index-profile parameter α
a/ Intermodal delay curve, b/ Wavelength dependence of the intermodal delay curve
($\lambda = 0.85$ μm, $n_1 = 1.5$, $\Delta = 0.01$)

4.2. MULTIMODE FIBRE BANDWIDTH MEASUREMENTS

4.2.1. *Bandwidth Definition*

In multimode fibres, the main dispersion components are the intermodal dispersion and the material dispersion. For a gaussian pulse, the pulse delays are added in quadrature:

$$\Delta \tau = \sqrt{\Delta \tau_i^2 + \Delta \tau_m^2} = \sqrt{\Delta \tau_i^2 + (LM_2 \Delta \lambda)^2} \quad [s] \tag{19}$$

Assuming **gaussian shaped pulses**, the multimode fibre bandwidth B_f is defined by:

$$B_f = \frac{0.44}{\Delta \tau} \quad [Hz] \tag{20}$$

where 0.44 is defining bandwidth to be at the -3 dB point of the frequency response.

Therefore, for sources such as light emitting diodes or Fabry-Pérot lasers, the product of bandwidth and distance in the fibre is a constant and is given by:

$$B_f \cdot L = \frac{0.44}{\sqrt{\left(\frac{\Delta \tau_i}{L}\right)^2 + (M_2 \Delta \lambda)^2}} \quad [\text{Hz.km}] \quad (21)$$

It is sometimes said that chromatic dispersion can be neglected in multimode fibres. To check that point, the relative weight of intermodal and material dispersions is given in table 6 for the three classical wavelength "windows" and for the spectral width 40 nm and 2 nm associated respectively with LEDs and Fabry-Pérot laser diodes. From this table, it's obvious that the chromatic delay can not be neglected with respect to the intermodal one when a light emitting diode is used. At 0.85 μm and 1.55 μm, the chromatic dispersion is even greater than the intermodal one. In fact the chromatic dispersion can be neglected only at 1.31 μm and 1.55 μm if the source is a laser diode.

TABLE 6. Multimode fibre Bandwidth: influence of the source spectral width

Source wavelength [nm]	850		1310		1550	
Material dispersion M_2 [ps/(nm.km)]	85		5		20	
Source spectral width [nm]	40	2	40	2	40	2
Material delay $M_2 \cdot \Delta\lambda$ [ps /km]	3400	170	200	10	800	40
Intermodal delay $\Delta\tau_i/L$ [ps /km]	250		250		250	
Total delay [ps /km]	3410	302	320	250	840	253
Bandwidth x Distance [MHz.km]	130	1460	1375	1760	524	1740

To measure the fibre bandwidth [1], the approach is to measure either the impulse response g (t) or the frequency response G (ω) of the fibre. The way these responses are defined is shown in figure 22.

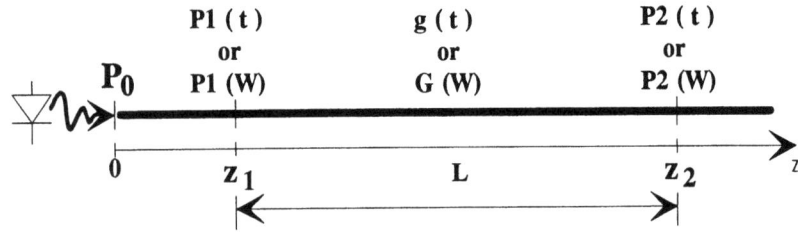

Figure 22. Impulse & frequency responses definition

In the time domain, the output power $P_2(t)$ is the convolution product of the input power $P_1(t)$ by the impulse response $g(t)$. In the frequency domain, the output power $P_2(\omega)$ is the product of the input power $P_1(\omega)$ by the frequency response $G(\omega)$:

$$g(t) \otimes P_1(t) = P_2(t) \quad \text{and} \quad G(\omega) \cdot P_1(\omega) = P_2(\omega) \tag{22}$$

The impulse and frequency responses are related to each other by the Fourier transform:

$$G(\omega) = \int_{-\infty}^{+\infty} g(t) \exp(-i\omega t) dt = FT[g(t)] = \frac{FT[P_2(t)]}{FT[P_1(t)]} \tag{23}$$

4.2.2. Impulse Response Method

Principle and set-up: the principle of this method is to measure the fibre response in the time domain. The impulse technique experimental set-up is shown in figure 23. To measure the fibre response, a short optical time pulse is launched in the fibre through a mode scrambler. The input time pulse $P_1(t)$ is supposed to be gaussian and its half width Δt_1 is measured first by a sampling oscilloscope. Then the fibre output time pulse $P_2(t)$ is measured by the oscilloscope to find its half width Δt_2. The input pulse could be measured after the mode scrambler or after a short length of fibre under test (fig.23).

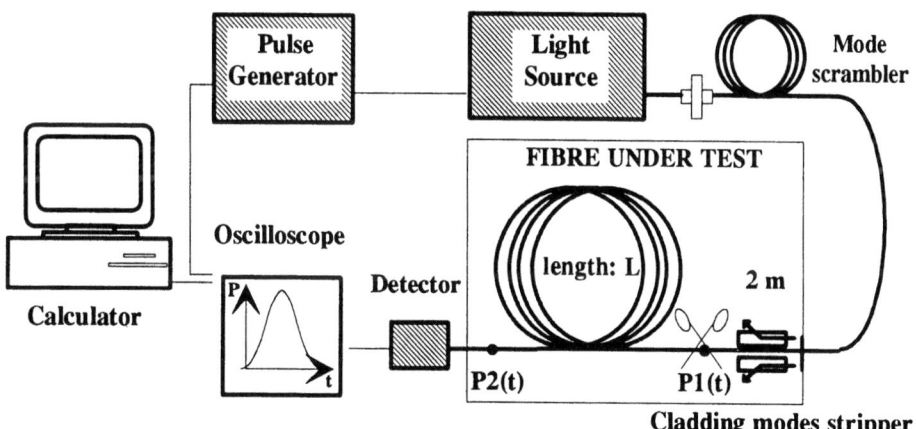

Figure 23. Impulse response technique experimental set-up

Measurement results: from the respective widths of the input and output time pulses (Δt_1, Δt_2), the bandwidth-distance product is simply given by (for a gaussian pulse):

$$B_T \cdot L = \frac{0.44}{\sqrt{\Delta t_2^2 - \Delta t_1^2}} \cdot L \quad [\text{Hz.km}] \tag{24}$$

Another way to find the bandwidth is to compute the ratio of the output and input pulses Fourier transforms (eq.23) that yields the frequency response G(ω). An instance of such a frequency response computed from the Fourier transforms of time pulses is given in figure 24. The frequency response is computed on a 1150 meters-length 50/125 μm multimode fibre at 0.85 μm and 1.3 μm. The bandwidth values are then measured at -3 dB (for optical detector) and are in this case, 270 MHz (310 MHz.km) at 0.85 μm and 1150 MHz (1320 MHz.km) at 1.31 μm.

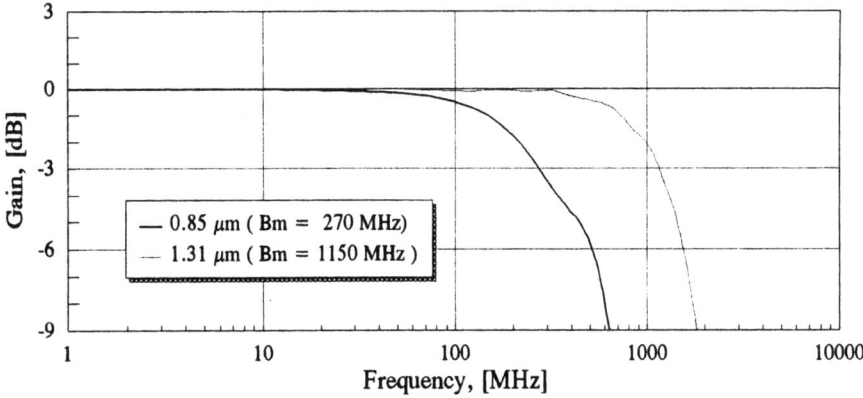

Figure 24. Modal baseband computation from the impulse response Fourier transform at 850 nm & 1310 nm (50/125 μm multimode fibre of 1150 meters-length)

4.2.3. Frequency Response Method

Principle and set-up: the fibre response can be measured in the frequency domain to obtain without any computation the frequency response G(ω). The optical source is now modulated by a sine function generator. The ratio of the output and input powers $P_2(\omega)$ and $P_1(\omega)$ gives the fibre frequency response. The simplest way to realize this is to use a sweep generator that scans the appropriate frequency domain, and to measure directly the fibre response with a frequency spectrum analyser. This is described in figure 25.

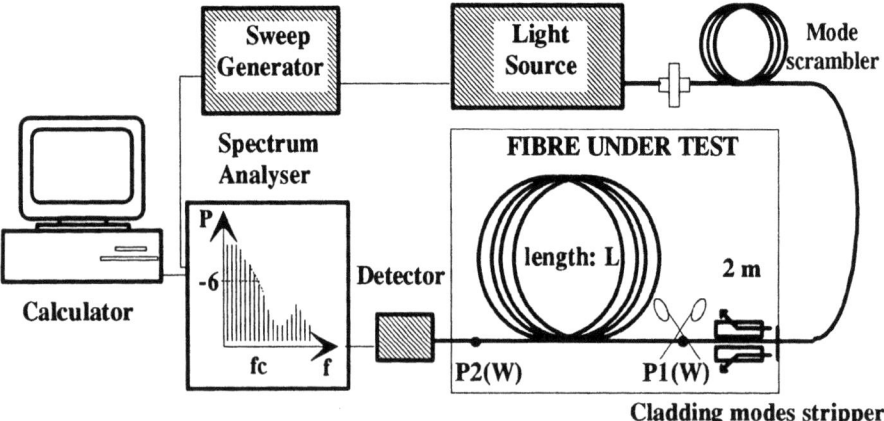

Figure 25. Frequency response technique experimental set-up

Measurement results: the frequency response of a 2540 meters-length 50/125 μm multimode fibre is given in figure 26, at 0.85 μm and 1.3 μm. The bandwidth values are here measured at -6 dB due to the apparatus computation in electrical decibels. In this instance, the bandwidth values are 193 MHz (490 MHz.km) at 0.85 μm and 365 MHz (927 MHz.km) at 1.31 μm.

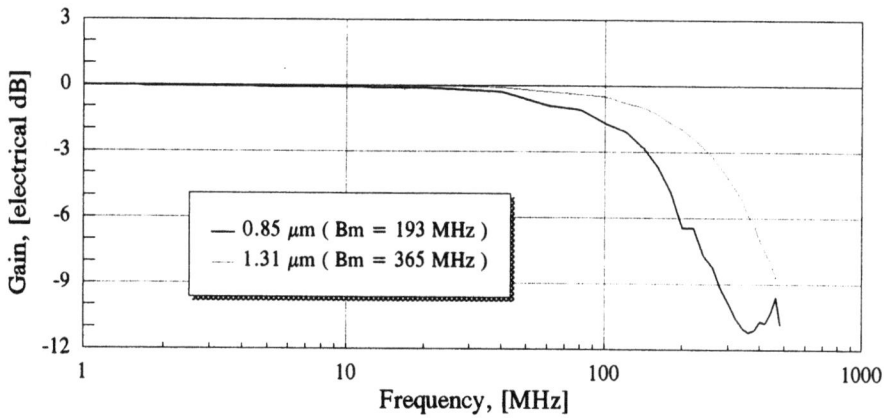

Figure 26. Modal baseband response from the frequency response at 850 nm and 1300 nm (50/125 μm 2540 meters-length multimode fibre)

4.2.4. Summary

To be consistent in bandwidth measurements, what is presented as a final result is the modal bandwidth of the fibre, that is to say the bandwidth only due to the intermodal dispersion (see § 4.1.4.). As was demonstrated in table 6, the fibre bandwidth is only the modal bandwidth if the chromatic dispersion can be neglected. For that reason, either the optical sources used must have a small spectral width ($\ll$ 1 nm), or the fibre chromatic dispersion must be known and substracted from the total dispersion. This chromatic dispersion influence is often forgotten and is a great source of systematic errors in the bandwidth measurement. Another problem is to make sure that the modal equilibrium is reached in the fibre. Obviously, as we talk about modal dispersion, this modal equilibrium is of great importance. A bad launching condition in the fibre could yield a relative error greater than a hundred percent on the final bandwidth. In addition, both methods are based on a discrete computation. Either the Fast Fourier Transform is used on digital measurements made by the sampling oscilloscope, or it's the spectrum analyser that gives digital measurements. Therefore, the result is closely related to the number of measuring points either in the time pulse, or in the frequency spectrum. This leads to an uncertainty on the -3 dB bandwidth as great as 5 %. Also, the calibration problem is not very easy. The measurement apparatus that we need must have bandwidths as great as 40 GHz. And to calibrate them with a good enough uncertainty requires great care. Finally, if great care is taken to avoid systematic errors due to the chromatic dispersion and to the launching, the combined uncertainty on bandwidth measurement is usually as great as 5 % (1σ). However, nowadays specifications on multimode fibres bandwidth are not very severe and often consist in a simple verification that the modal bandwidth is greater than a given value.

4.3. SINGLE-MODE FIBRE BANDWIDTH MEASUREMENTS
4.3.1. *Bandwidth Definition*

In single-mode fibres, the total dispersion depends on the material dispersion $\Delta\tau_m$ (eq.15) and the guide dispersion $\Delta\tau_g$ (eq.16). The pulse delay is then given by:

$$\Delta\tau = \Delta\tau_m + \Delta\tau_g = \frac{L}{c}\left|-Y_m-Y_w\right|\frac{\Delta\lambda}{\lambda} \quad [s] \tag{25}$$

where $\Delta\lambda/\lambda$ is the <u>effective</u> spectral bandwidth that depends on the source spectral width $\Delta\lambda_{source}$ <u>and</u> on the signal-modulation bandwidth B_f. The effective spectral width is then given by the equation 26, where f is the light frequency:

$$\Delta\lambda/\lambda = \sqrt{(\Delta\lambda_{source}/\lambda)^2 + (B_f/f)^2} \tag{26}$$

The maximum signal-modulation bandwidth is obviously the fibre bandwidth itself, therefore it is <u>also</u> related to the pulse delay $\Delta\tau$ by:

$$B_f = 0.44/\Delta\tau \quad [Hz] \tag{27}$$

By combining these equations, the bandwidth B_f can be easily expressed in two cases. The relation depends on the relative weigths of the ratio B_f/f (about 10^{-4}) and of the source relative spectral width $\Delta\lambda_{source}/\lambda$, that is to say on the type of sources used.

1 - If the source spectral width is greater than 1 nm, which is the case for light emitting diodes and Fabry-Pérot laser diodes, then the effective spectral width does not depend on the bandwidth B_f. The product (bandwidth x distance) is then a constant given by:

$$\text{if} \left(\frac{\Delta\lambda_{source}}{\lambda} > \frac{B_f}{f}\right) \text{ then } B_f.L = \frac{0.44}{\frac{1}{c}\left|-Y_m-Y_w\right|\frac{\Delta\lambda_{source}}{\lambda}} \quad [Hz.km] \tag{28}$$

2 - Now, if the source spectral width is smaller than 0.05 nm, which is the case for single-longitudinal mode laser diodes such as distributed feed back (DFB) or distributed bragg reflector (DBR) lasers, then the effective spectral width is the fibre bandwidth itself. Therefore, it is the product (Bandwidth² x distance) that is a constant:

$$\text{if} \left(\frac{\Delta\lambda_{source}}{\lambda} < \frac{B_f}{f}\right) \text{ then } B_f^2.L = \frac{0.44}{\frac{1}{cf}\left|-Y_m-Y_w\right|} \quad [Hz^2.km] \tag{29}$$

To summarize, the single-mode fibre dispersion is illustrated in table 7 that gives for several wavelengths the chromatic dispersion coefficient Y_m, the guide dispersion coefficient Y_w and the total dispersion for a step-index single-mode fibre whose core diameter is about 9 µm. Then the fibre bandwidth is given for both the cases discussed previously. The first calculation is done with a line width of 1 nm (a Fabry-Pérot LD). The second calculation is for a line width of 0.01 nm (DBR lasers).

TABLE 7. Source dependence of single-mode fibres bandwidth

λ [nm]	Y_m	Y_w	Y_m+Y_w	$\Delta\lambda_{Source}$: 1 nm $B_f \cdot L$ [GHz . km]	$\Delta\lambda_{Source}$: 0.01nm $B_f \cdot \sqrt{L}$ [GHz.$\sqrt{km}$]
1270	0.00015	0.0037	0.0039	20	90
1310	- 0.0015	0.0037	0.0022	40	120
1350	- 0.0028	0.0037	0.0009	100	180
1550	- 0.0100	0.0037	0.0063	15	65

The table shows the limitation for the fibre bandwidth. If the bandwidth-distance product was a constant for the second type of source, it would have been a hundred times greater than for the first source, that is to say 10 THz.km at 1.35 µm for instance. Of course such a great bandwidth is not possible with regard to the light frequency itself (200 THz). The fibre bandwidth is then only limited by this fact. However, this bandwidth decreases as a function of the distance square-root now, which is an improvement with regard to the linear relation (eq.28). For instance at 1.35 µm, the 180 GHz.$\sqrt{km}$ result yields a 18 GHz bandwidth over 100 km of fibre (instead of 10 km for the linear relation). It can be seen that the best bandwidth is obtained around 1.31 µm. To take advantages of the best attenuation coefficient, it is interesting to work at 1.55 µm. Therefore, special kind of fibres, the so-called dispersion shifted (DS) fibres with double or triple cladding are designed to shift the zero-dispersion wavelength around 1.55 µm. And to work at both wavelengths (1.31 and 1.55 µm), flat dispersion (FT) fibres can be used, in which dispersion is less than 5 ps/(nm.km) from 1285 nm to 1575 nm. The dispersion curves of these fibres [12] are plotted in figure 27.

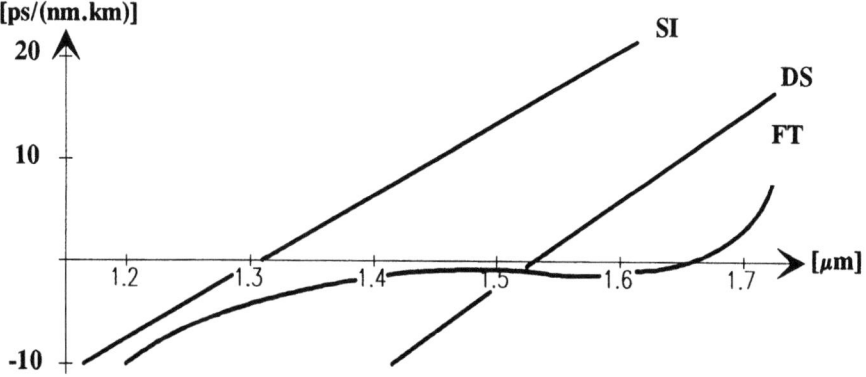

Figure 27. Dispersion of step-index (SI), dispersion shifted (DS) & flat dispersion (FD) single-mode fibres

It is seen that the bandwidth definition depends on the source spectral width. Consequently, it is recommended to give the dispersion as a final result. To obtain the dispersion, the natural mesurand [1] [3] is the delay curve $\tau(\lambda)$ expressed in [ps/km]:

$$\tau(\lambda) = |M_2 + M_g|\Delta\lambda = \frac{1}{c}|-Y_m-Y_w|\frac{\Delta\lambda}{\lambda} \quad [\text{ps/km}] \quad (30)$$

From the measurement values the delay curve is fitted by a polynomial of the form:

$$\tau(\lambda) = A + B\lambda^2 + C\lambda^{-2} + D\lambda^4 \quad [\text{ps/km}] \quad (31)$$

Then, the dispersion curve $D(\lambda)$ in [ps/(nm.km)] is the derivative of the delay curve with respect to the wavelength:

$$D(\lambda) = \frac{d\tau(\lambda)}{d\lambda} = M_2 + M_g \quad [\text{ps/(nm.km)}] \quad (32)$$

The delay curve $\tau(\lambda)$ can be determined by three main methods described in the following.

4.3.2. *Phase Shift Technique*

Principle and Set-up: the principle of this technique is to modulate the optical source sinusoidally and then to measure the signal phase shift introduced by the fibre under test with respect to the wavelength. These relative phase shifts can then easily be related to the relative time delays. The optical sources are typically laser diodes, but also light emitting diodes filtered to reduce their spectral width [16] or tunable lasers. An instance of a phase shift set-up using multiple laser diodes is given in figure 28.

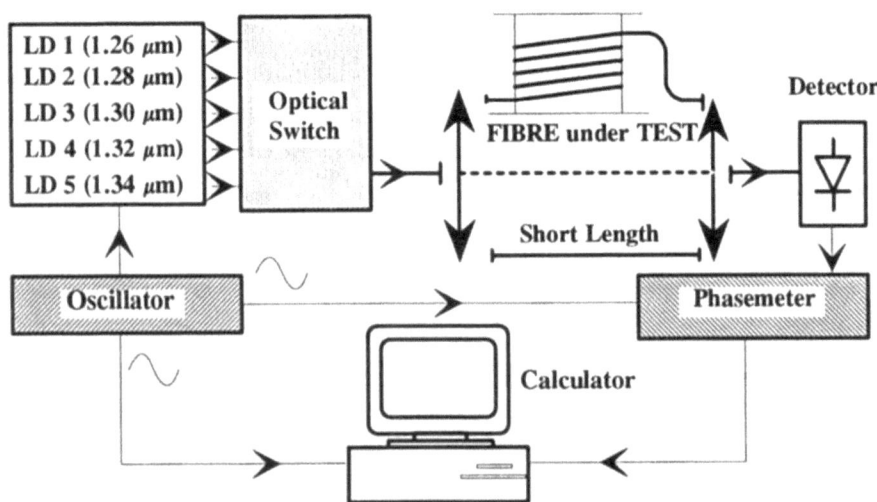

Figure 28. Phase shift technique experimental set-up (Laser Diodes)

In this set-up five laser diodes of different wavelengths around 1.31 μm are used to measure the dispersion of classical step-index fibres. The set-up is first calibrated on a short length of fibre to measure the absolute phase shifts when the fibre under test is inserted. These five measured points are then used to fit the delay curve by using a polynomial function (eq.31).

Measurement results: The so-fitted delay curve of a 1.03 km-length step-index fibre is given in figure 29. The delay is given in [ps/km] on the left axis. The dispersion curve, computed from the delay curve (eq.32) is given on the same figure but on the right axis expressed in [ps/(nm.km)]. For this fibre, the zero dispersion wavelength is found to be 1309 nm. Typical dispersion values are found, that is to say less than 4 ps/(nm.km) in the 1.3 μm window.

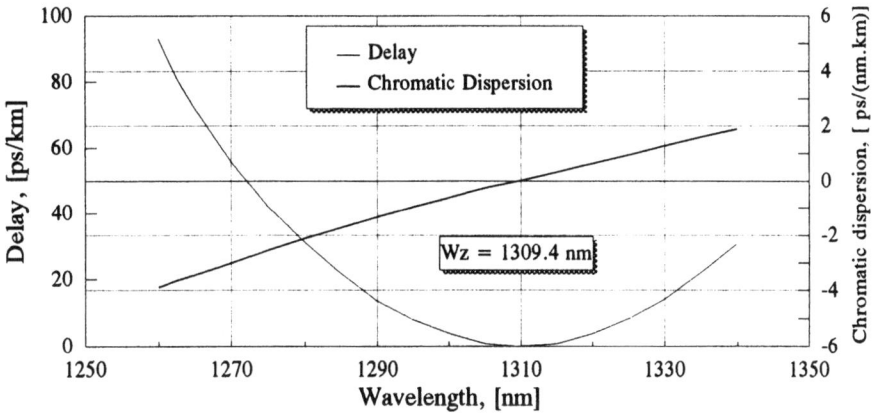

Figure 29. Delay & dispersion curves of a 1.03 km-length single-mode fibre

The limitation of this technique is in the small number of wavelength values that can be launched. The repeatability of this few points measurement then has a great influence on the delay curve fitting, and as a consequence on the dispersion curve. This could be solved by using a tunable laser source, that allows a wavelength scanning with a small step. However, in this technique the source has to be very stable in wavelength, optical power and especially in phase. This was not the case of available tunable sources until two years ago. Nowadays, a few tunable lasers that are stable and repeatable enough to replace the multiple laser diodes or the filtered LED are now available.

4.3.3. *Pulse Delay Technique*

Principle and Set-up: in this technique the fibre delay curve is obtained directly by measuring the propagation delay of an optical time pulse as a function of the wavelength. Depending on the wavelength range to be characterized, the optical sources can be either pulsed laser diodes, pulsed tunable lasers, or a fibre Raman laser. These sources must have a time pulse duration short enough to be smaller than the average time delay to be measured. To use pulsed laser diodes or tunable lasers only allows the fibre dispersion to be characterized over a small wavelength range of 60 nm in the wavelength bands around 1.31 and 1.55 μm. The fibre Raman laser technique, described in figure 30, uses the non-linear Raman effect in a fibre pumped by a Nd:YAG Q-switched laser. Under this high pump energy, the fibre emits a fluorescence spectra, the so-called Raman spectra, that extends from 1.06 μm to 1.7 μm. Then combined with a monochromator, this Raman fibre pumped by a Nd:YAG laser constitutes an optical source, tunable in wavelength from 1.06 μm to 1.7 μm with a chosen wavelength step and a chosen spectral width. In addition, a mode-locked Q-switched Nd:YAG laser can emit short optical pulses whose width can be easily smaller than 400 ps.

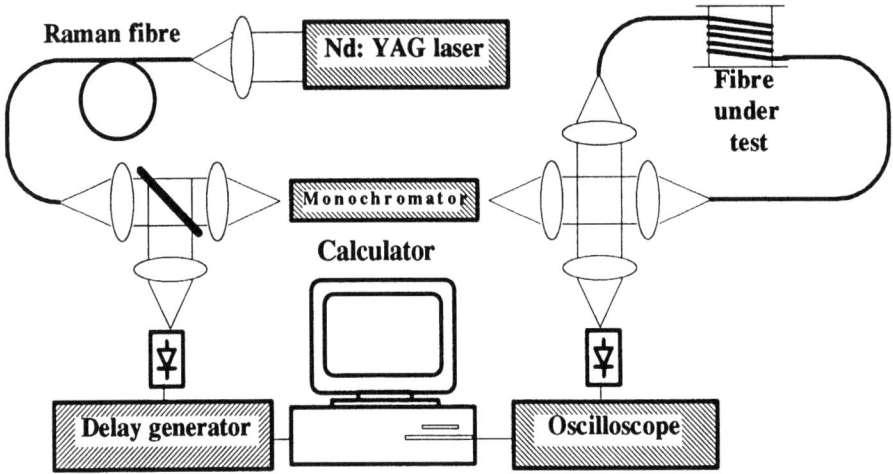

Figure 30. Pulse delay technique experimental set-up (fibre Raman laser)

Measurement results: [15] in figure 31 are plotted the measured delay curve and the computed dispersion curve obtained by the fibre Raman laser technique, for a 1.03km-length step-index fibre. The time delay is expressed in [ns/km] on the left axis and the dispersion in [ps/(nm.km)] on the right axis. This fibre is the same as in the phase shift technique instance and its zero dispersion wavelength is once again 1309 nm.

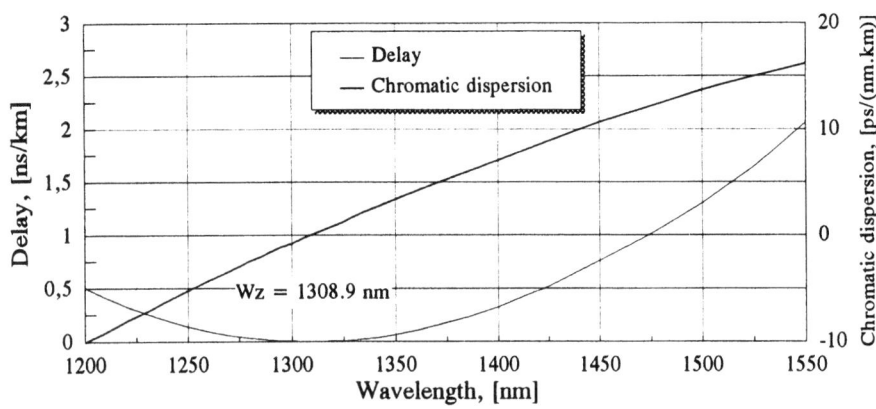

Figure 31. Delay & dispersion curves of a 1.03 km-length fibre (fibre Raman laser)

Both techniques presented previously must be used with sufficient fibre lengths to ensure a differential time delay great enough to be measured with a good uncertainty. And these time delays of a few nanoseconds are to be compared with the propagation times of a few microseconds. This difference in both time scales involves difficulties, such as the calibration of the sampling oscilloscope, for instance. A solution is to directly measure the dispersion curve by the differential phase shift method [17]. Another solution is to realize homodyne detection to reduce the calibration to the time delay scale only. In the optical domain, homodyne detection is simply realized by an interferometric method. This is described in the next section.

4.3.4. *Interferometry Technique*
Principle and Set-up: the principle of this technique is to split the source power into two arms. The fibre under test is inserted in one arm, the so-called measurement arm. The other arm, that constitutes the reference arm, is made of any component whose dispersion is well-known, air for instance. The two signals are then recombined in order to make them interfere. Such a set-up constitutes the well-known Mach-Zehnder interferometer. Now, to obtain the maximum visibility of the interferogram at a given wavelength, it's enough to equalize the optical paths in both arms. This could be easily obtained by changing the reference arm length by any means. Now if the wavelength is changed, the pulse delay in the fibre $\tau(\lambda)$ is related to the reference arm length variation Δl that is necessary to recover the interferogram maximum visibility. Since the air dispersion can be neglected with respect to the fibre dispersion, this simple relation is $\tau(\lambda) = \Delta l / c$. This realizes a convenient change of scale, that is to say that small time delays are transformed into convenient length variations: for instance a delay of 1 ps is measured by a 300 μm-length variation. Since a 300 μm-length displacement can "easily" be realized and calibrated with an accuracy of 0.1 μm, this technique can measure a time delay with an accuracy as small as a femtosecond. A schematic of a set-up using the optical interferometry technique is given in figure 32.

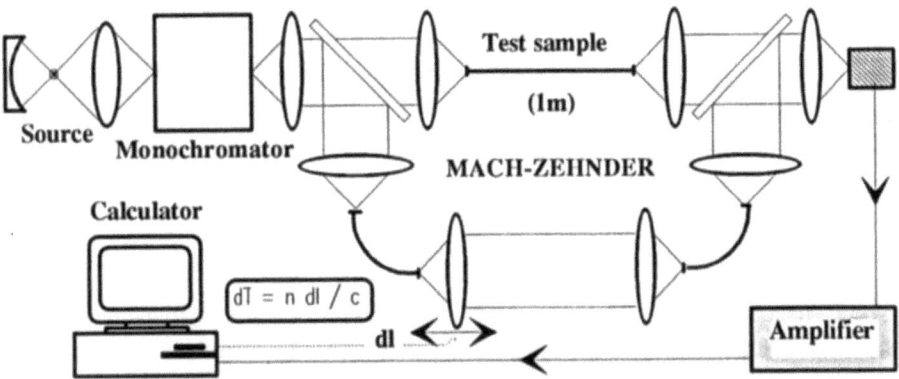

Figure 32. Optical interferometry technique experimental set-up

An important point for this method is to realize a good compromise between the source spectral width and the fibre length. Indeed, to obtain any interference, the optical source coherence-time (or length) must be greater than the maximum delay (or length-variation) that is to be measured. The coherence-length L_c of a source is related to its wavelength λ and its spectral width $\Delta\lambda$ by:

$$L_c = \lambda^2/\Delta\lambda \quad [m] \tag{33}$$

For a wavelength of 1.55 μm for instance, this coherence length ranges from 240 μm to 240 mm when the source spectral width ranges from 10 nm to 0.01 nm. Now, typical delays that are to be measured at 1.55 μm are about 2 ns/km, that is to say a 0.6 meter-length variation per kilometer of fibre. The comparison between coherence lengths and length-variations shows that the fibre length must be less than 0.4 m for a spectral width of 10 nm and less than 400 m for a spectral width of 0.01 nm. In the set-up described in figure 32, a white light source combined with a monochromator is used in order to be able to cover the spectral range 1.2 μm-1.6 μm. Over this spectral range, the spectral width of 3 nm gives a coherence-length that allows a fibre length of 1.4 m at most. For that reason a fibre test sample of one meter is chosen in this set-up.

Measurement results: to show the basis principle of the measurement, an instance of interferograms obtained at 1.24 μm, 1.22 μm and 1.20 μm is drawn in figure 33. The upper scale gives the delay in [ps/m] and the lower scale the length-variation of the reference arm in [μm/m]. The relative delays measured are -0.21 ps/m from 1.24 μm to 1.22 μm and -0.29 ps/m from 1.22 μm to 1.20 μm.

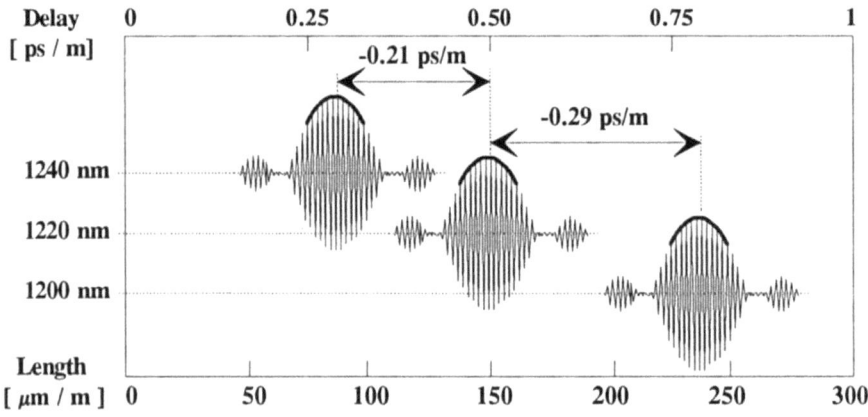

Figure 33. Dispersion measurement using optical interferometry

4.3.5. *Summary*

On the time delay, the absolute uncertainty is hardly smaller than 1 ps. Around the zero-dispersion wavelength which is the range that interests us, the relative delays are, of course, the smallest (from 0 to 50 ps/km around 1.31 ± 0.03 μm). So, on the dispersion curve, to obtain a relative uncertainty smaller than 1 %, the measured delays must be always greater than 100 ps, that means a fibre length greater than 10 km for a spectral width of 1 nm. The interferometric method yields the same uncertainty if the result on the short fibre length is extrapolated to one kilometer.

5. Single-mode fibre specific measurements

5.1. THE FUNDAMENTAL MODE

5.1.1. *Dispersion Equation*

To solve the propagation problem, one must find the **dispersion equation** that gives the allowable longitudinal propagation constants β as a function of the wave vector k. In step-index fibres, the propagation equation is solved in the weakly guided mode approximation, that is to say for a step-index Δ smaller than 1 %. This approximation, also called the scalar approximation, consists in looking for a transverse solution for the electric field E and magnetic field H. This yields a dispersion equation that can be simplified into the linearly polarized (LP) modes dispersion equation [18]. This LP m,μ mode dispersion equation is given below. Each LP m,μ mode combines several modes which propagation constants are equal, the so-called degenerate modes.

$$u \frac{J_{m-1}(u)}{J_m(u)} = -w \frac{K_{m-1}(w)}{K_m(w)}$$

$$m=1 \rightarrow LP_{1,\mu} = TE_{0,\mu} + TM_{0,\mu}$$

$$m \neq 1 \rightarrow LP_{m,\mu} = EH_{m-1,\mu} + HE_{m+1,\mu}$$

(34)

where u and w are reduced transverse propagation constants described in equation 12, and J_m, K_m are respectively Bessel and Hankel (Mc Donald) functions of order m. This equation 34 also shows how each LP mode falls in exact modes that are either transverse electric (**TE**), transverse magnetic (**TM**) or hybrides such as quasi-transverse magnetic (**EH**) or quasi-transverse electric (**HE**).

The **parameter** μ is the radial mode number and (μ-1) is physically the number of zeros of the Bessel function that represents the electric field in the fibre radius.

The **parameter m** is the azimuthal mode number and represents the angular dependence of the field. Physically the electric field has (2m) zeros in 2π and is described either by a cosine or a sine function. These two orthogonal functions represent the two degenerate states of polarization of each LP mode.

TABLE 8. LP modes as a function of the normalized frequency V

Range of V (V= k.a.NA)	LP m,μ mode	Degenerated modes	Number of modes	Total number of modes
0 - 2.4048	LP 01	HE 11 x 2	2	2
2.4048 - 3.8317	LP 11	TE 01, TM 01, HE 21 x2	4	6
3.8317 - 5.1356	LP 21 LP 02	HE 31 x 2, EH 11 x 2 HE 12 x 2	6	12

Table 8 gives, for the four first LP modes, their degenerate modes (exact solutions of the dispersion equation) as a function of the normalized frequency V. The LP 21 and

LP 02 modes are almost degenerate. It is important to notice that the total number of modes (TE, TM, HE and EH) increases very quickly with the normalized frequency V. For instance in a 50/125 μm Step-Index fibre, the normalized frequency V is about 24 at 1.3 μm and yields about 290 modes ($\approx V^2 / 2$ modes). This total number of modes is divided by two ($\approx V^2 / 4$ modes) in a 50/125 μm Parabolic-Index fibre. In figure 34 are plotted the normalized propagation constants B of the same first four LP modes as a function of the normalized frequency V. These **B (V) curves are fundamental** because they are the normalized graphical representation of the fibre **dispersion equation β (k)**.

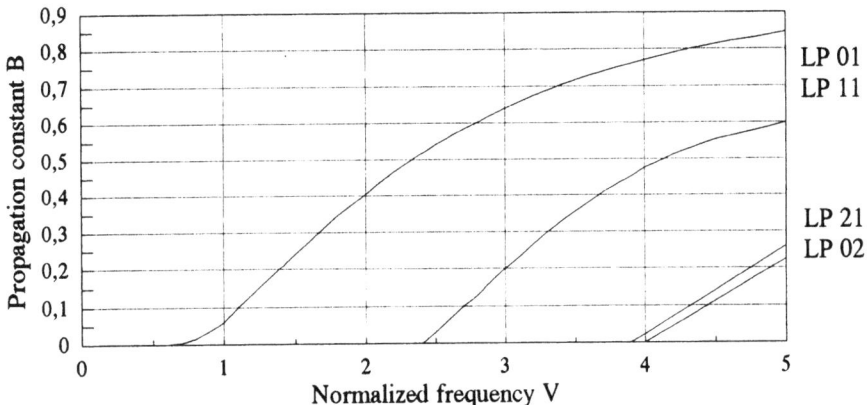

Figure 34. LP modes normalized propagation constants B as a function of the normalized frequency V

From figure 34, it can be seen than the the cut-off frequency of the second mode (LP11) is $V_c = 2.4048$. Therefore, to realize a single-mode propagation fibre, it's enough to satisfy the condition: $V < V_c$. The LP01 mode is then the only one to propagate. This mode is the **fundamental one** [12] [19] which is more often described by it's exact solution: the **HE11** mode that is twice degenerate.

5.1.2. *Cut-off Wavelength*

For a given fibre, the cut-off frequency V_c corresponds to an equivalent wavelength value, the so-called **cut-off wavelength λ_c**. The single mode propagation is then obtained if the wavelength is always greater than this cut-off wavelength. By using the normalized frequency V expression, the cut-off wavelength can be expressed as:

$$\lambda_c = \frac{2\pi a \, NA}{2.4048} \quad \text{and} \quad \lambda > \lambda_c \tag{35}$$

For a 50/125 μm fibre (2a=50 μm, NA=0.2), this cut-off wavelength is about 13 μm. Now, to realize a single-mode fibre that can be used at 1.31 μm, the fibre cut-off wavelength must be smaller than 1.31 μm. This is only possible if the fibre core radius and/or the numerical aperture are reduced. That explains the choice for the single-mode fibres parameters (2a $\approx$ 9μm, NA $\approx$ 0.1) that yield a cut-off wavelength of about 1.2 μm.

5.1.3. *Mode Field Diameter*

The condition on the cut-off wavelength requires that single-mode fibres have a small core radius. As a consequence, the fundamental mode HE11 is not totaly confined in the core. The transverse electric field (Ex or Ey) is resonant in the core and is described by the Bessel function $J_0[ur/a]$ and evanescent in the cladding under the form of a Hankel function $K_0[wr/a]$. So, knowing the core diameter is not very useful in predicting coupling losses or microbending losses for instance. Then, the idea of a mode diameter was introduced to allow any comparisons between single-mode fibres. The first definition that matches the real field with a gaussian was proposed by Marcuse [20]. The maximum recovery integration is then obtained for a field of the form:

$$E_g = \sqrt{\frac{E_x}{H_y}} \cdot \frac{2}{\sqrt{\pi}} \cdot \frac{1}{w_0} \exp\left(-\frac{r^2}{w_0^2}\right) \tag{36}$$

where **$2w_0$ is the mode field diameter (MFD)** at the $1/e^2$ power.

Other mode field diameter definitions were proposed by Petermann. Those definitions express the mode field diameter from the far field power [21] or/and from the near field intensity average quadratic diameters (see § 5.3).

5.2. CUT-OFF WAVELENGTH MEASUREMENTS

The cut-off wavelength measurement principle [1] [3] is to measure the power wavelength dependence of a short length of test fibre. A measurement set-up that applies this principle is given in figure 35.

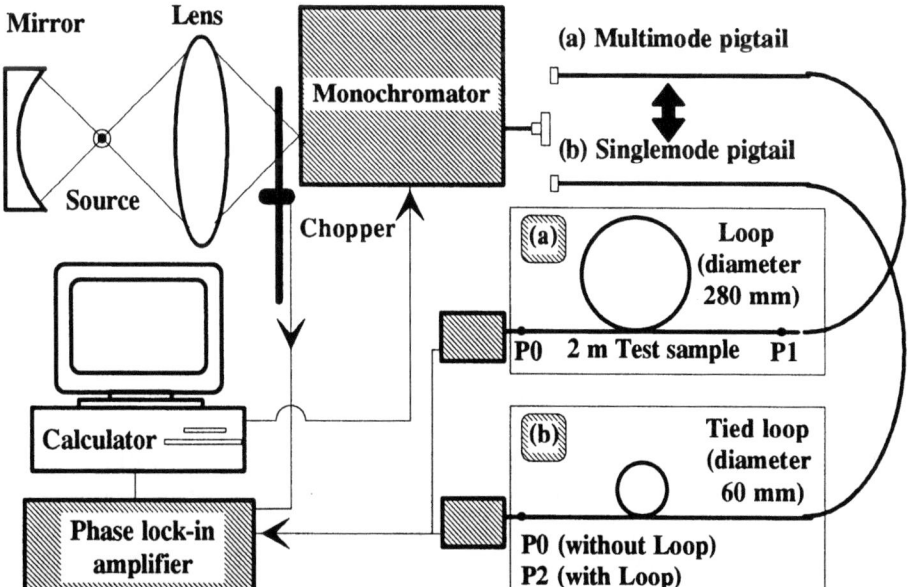

Figure 35. Cut-off wavelength measurement set-up: a/ LP 01 & LP11 stimulation b/ LP11 filtering

The basic set-up is the one used for the attenuation coefficient measurement. This set-up allows the scanning of a wavelength range from 1 μm to 1.6 μm, in order to consequently stimulate the multimode and the single-mode states in the fibre. Two methods are then possible, depending on the launching condition in the single-mode fibre under test. As a final result, each method gives a curve that represents a power ratio [dB] as a function of the wavelength. Due to the abrupt change of state when the fibre becomes single-mode, the power ratio curve will present an abrupt variation. This power variation locates the cut-off wavelength.

5.2.1. *Transmitted Power Technique # 1*
The first method consists in stimulating both LP01 and LP11 modes by using a multimode launching fibre (fig.35a). This stimulation is assumed if, below the cut-off wavelength, the fibre under test presents no small curvature radius. To verify this condition, a big diameter loop (280 mm) is made with the fibre. Now, by measuring the input power $P_1(\lambda)$ and the fibre under test output power $P_0(\lambda)$ as a function of the wavelength, one can compute the power ratio $R_1(\lambda)$ as follows:

$$R_1(\lambda) = 10 \log_{10} \frac{P_1(\lambda)}{P_0(\lambda)} \quad \text{[dB]} \tag{37}$$

The determination of the cut-off wavelength from this ratio is shown in figure 36. This ratio $R_1(\lambda)$ is measured on a 2600-meters length step-index single-mode fibre and the cut-off wavelength is found to be 1272 nm ± 5 nm (2σ).

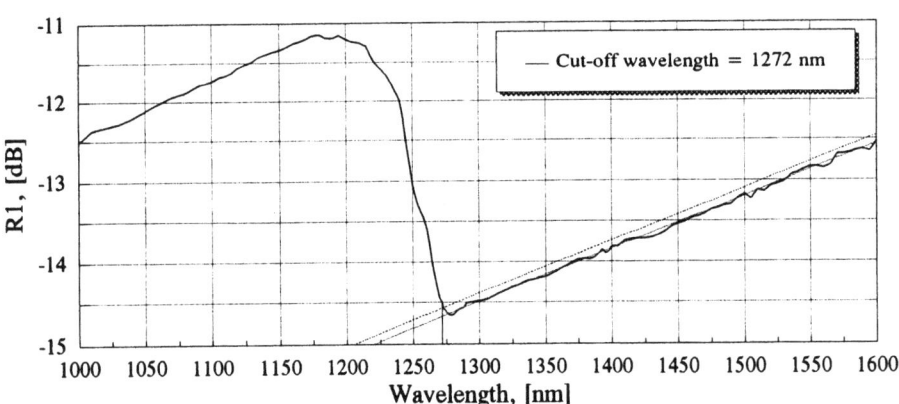

Figure 36. Cut off wavelength determination by the multimode launching fibre technique

The minimum of this curve shows the wavelength at which the LP11 mode disappears, that is to say the cut-off wavelength. However, to obtain repeatable measurements, the cut-off wavelength is determined by the following method. The first step is to compute the equation of the linear part of this curve (from 1.3 μm to 1.6 μm in this case). The intersection of the $R_1(\lambda)$ curve with the line parallel to the previous one, but shifted up by 0.1 dB, gives the location of the cut-off wavelength.

5.2.2. Transmitted Power Technique # 2

The second method consists in filtering the LP11 mode to only allow the LP01 mode propagation. This is shown in figure 35b. The LP11 mode filtering is realized by simply making a tied loop of 60 mm-diameter in the test-fibre. The output powers $P_2(\lambda)$ and $P_0(\lambda)$ are then measured respectively with the tied loop and without any loop. Then, the power ratio $R_2(\lambda)$ is computed as follows:

$$R_2(\lambda) = 10 \log_{10} \frac{P_2(\lambda)}{P_0(\lambda)} \quad [dB] \tag{38}$$

This power ratio $R_2(\lambda)$ is plotted in figure 37, for the same fibre as in the previous case. The intersection of this curve with the line R = 0.1 dB yields two points. The point for which wavelength is the greatest gives the cut-off wavelength. In this instance, this cut-off wavelength is found at 1272 nm again.

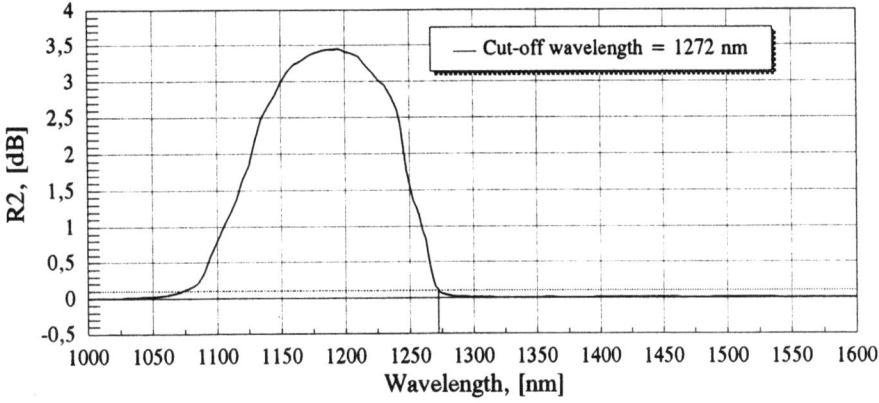

Figure 37. Cut off wavelength determination of a single-mode fibre by the LP11 filtering technique

Due to the high slope (dB/nm) of the curve (whatever it is R_1 or R_2) around the cut-off wavelength, the measurement uncertainties are closely related to the step of the spectral scanning. In fact the measurement repeatability is almost equal to this step. And this step is at least the source spectral width. The measurement set-up uses a white light source filtered by a monochromator, so that the power launched in a single-mode fibre is very small (less than 1 nW). Therefore, the monochromator spectral width is commonly a few nanometers (10 nm is many cases). In the previous instance, the spectral width is reduced to 3 nm and a repeatability of 2 nm was found. The expanded uncertainty on cut-off wavelength measurements is, for this spectral width of 3 nm, about 5 nm (2σ). The same measurements were realized with a spectral width of 6 nm, that demonstrated an expanded uncertainty of 9nm. So the greatest care must be taken on this spectral width choice [22], which in this test technique case, contributes to almost half the expanded uncertainty (2σ).

5.3. MODE FIELD DIAMETER MEASUREMENTS

The principle of the mode field diameter measurement is based on the Petermann definitions [21]. These definitions use either the far field power distribution, or the near field intensity distribution that can be measured at the fibre output. A third method, the Variable Aperture Launch, is also mentionned in the IEC 793 international standard but is now canceled in the UIT G650 recommendation. Therefore, we will focus our interest only on the far field and near field scanning techniques.

5.3.1. *Far field scanning technique*
Principle and Set-up: the far field power distribution $P_m(\theta)$ is the angle θ dependence of the fibre output power P_m. A simple way to measure this far field power is described in figure 38a. The detector is placed at a distance far enough from the fibre end and is actuated by a rotating stage with an angle step smaller than 0.5°. This distance should be great enough compared to the detector diameter to substend an angle smaller than the rotating step. The measurement is performed around 1.31 μm or/and 1.55 μm from -25° to +25° which involves a dynamic range better than 50 dB. To ensure such a dynamic range, synchronous detection is used.

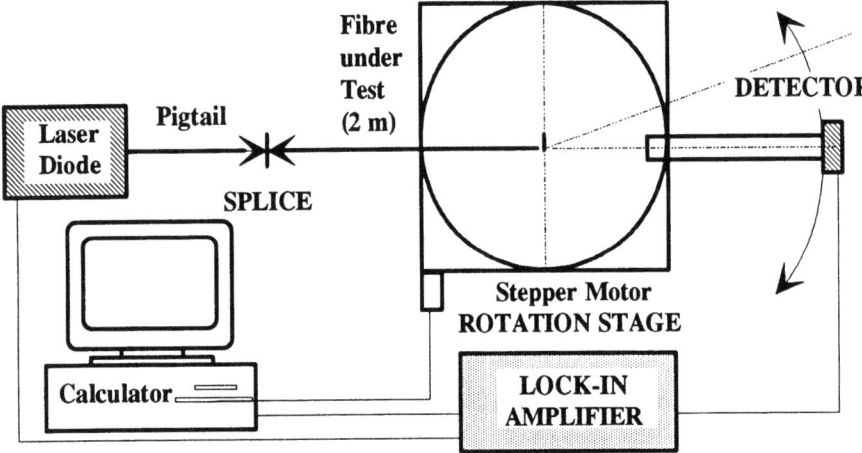

Figure 38a. Far field scanning measurement set-up

Following the Petermann definition, the Mode-Field Diameter $2W_0$ is then computed from the far field power distribution $P_m(\theta)$, by the following formula [1]:

$$2W_0 = \frac{\sqrt{2}\lambda}{\pi} \left(\frac{\int_0^\infty P_m(\theta) \sin\theta \cos\theta \, d\theta}{\int_0^\infty P_m(\theta) \sin^3\theta \cos\theta \, d\theta} \right)^{\frac{1}{2}} \quad [\mu m] \quad (39)$$

Measurement results: an instance of far field power distribution is plotted in figure 38b.

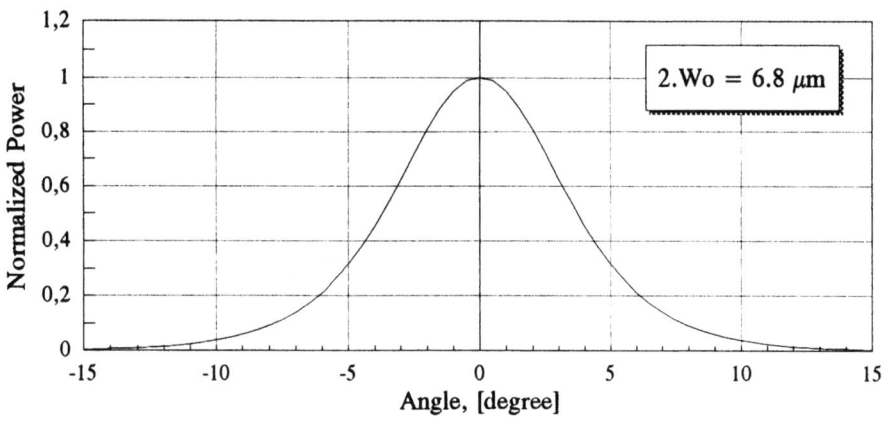

Figure 38b. Far field power distribution of a single-mode fibre

After computation (eq.39), this far field power distribution yields a mode field diameter $2W_0$ of 6.8 μm ± 0.1 μm (2σ) at 1.55 μm.

5.3.2. Near field scan technique

Principle and Set-up: the near field intensity pattern $f^2(r)$ is the radius r dependence of the intensity expressed at the fibre output face. This pattern is measured by magnifying the fibre output face in a detector plane. A schematic of two means to realize the detection is given in figure 39a. The fibre image plane can be scanned by a fibre pigtail, or a vidicon or CCD camera. The near field intensity pattern in this plane is then measured and corrected for the magnification to obtain the near field intensity pattern $f^2(r)$ in the fibre face plane. As a consequence, great care must be taken on the fibre and detector positioning with respect to the magnifying optic.

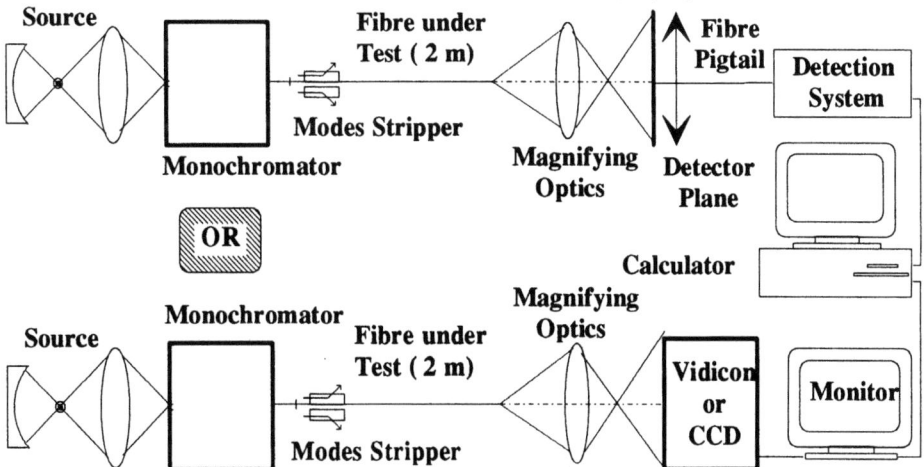

Figure 39a. Near field scan measurement set-up

In addition to that focusing problem, the magnification must be known with good accuracy to ensure the right correction. Then, from the near field intensity distribution $f^2(r)$, the mode field diameter $2W_0$ is computed by the following formula [1] that follows the second Petermann definition:

$$2W_0 = 2 \left(2 \frac{\int_0^\infty r f^2(r) \, dr}{\int_0^\infty r \left[\frac{df(r)}{dr} \right]^2 dr} \right)^{\frac{1}{2}} \quad [\mu m] \qquad (40)$$

Measurement results: an instance of a near field intensity pattern is plotted in figure 39b for a dispersion unshifted single-mode fibre.

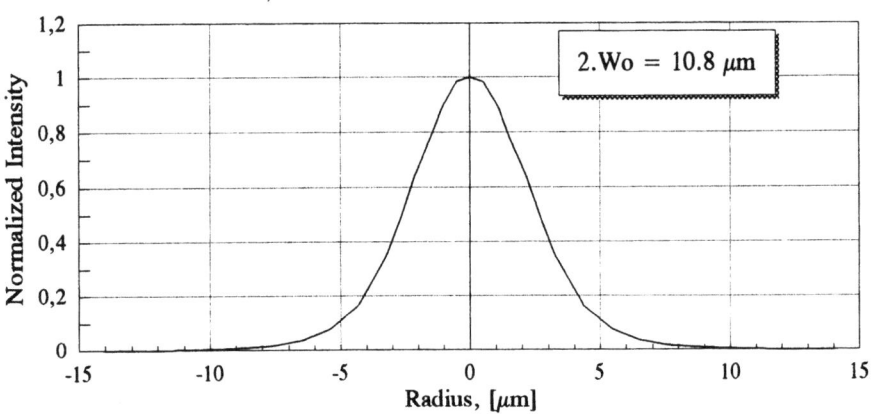

Figure 39b. Near Field Intensity Pattern of a single-mode fibre

After computation (eq.40), this near field intensity distribution yields a mode field diameter $2W_0$ of 10.8 μm at 1.3 μm. Due to the magnification and its limitation (Airy spot), the expanded uncertainty is hardly better than 0.4 μm (2σ).

6. Conclusion

In this article the relevant characteristics of optical fibres are given. Those characteristics are the opto-geometrical parameters, the fibre attenuation, the fibre dispersion and finally single-mode fibres parameters such as the cut-off wavelength and the mode field diameter. Each parameter is theoretically explained in order to demonstrate the real need for its characterization. Therefore, several test techniques are described in their principle and some measurement results are given in order to appreciate the measurement uncertainties. Those uncertainties are detailed in each chapter summary. An important measurement is not treated in this article, the so-called Polarization Mode Dispersion. Indeed, this dispersion between both degenerate polarization states of the HE11 fundamental mode is about the chromatic dispersion value in single-mode fibres applications where the source spectral width is about 0.01 nm. This measurement can be achieved by using the interferometric technique described in § 4.3.4. if the spectral source is consequently replaced by these two polarization modes. Treatment of this matter could lead to another article.

7. References

1. International standard IEC 793-1, (11 / 92), *Optical Fibres - Part 1: Generic Specification*, International Electrotechnical Commission (IEC).
2. International standard IEC 793-1, (11 / 92), *Optical Fibres - Part 2: Products Specifications*, International Electrotechnical Commission (IEC).
3. Recommandation UIT-T G.650, (03 /93), *Définition et méthodes d'essai des paramètres pour les fibres monomodes*, Union Internationale des Télécommunications (UIT).
4. Recommandation UIT-T G.651, (03 / 93), *Caractéristiques d'un câble à fibres optiques multimodes à gradient d'indice 50/125 µm*, Union Internationale des Télécommunications (UIT).
5. Recommandation UIT-T G.652, (03 / 93), *Caractéristiques des câbles à fibres optiques monomodes*, Union Internationale des Télécommunications (UIT).
6. Recommandation UIT-T G.653, (03 / 93), *Caractéristiques des câbles à fibres optiques monomodes à dispersion décalée*, Union Internationale des Télécommunications (UIT).
7. D. Marcuse, (1981), *Principles of Optical Fibre Measurements*, Academic Press.
8. M. Young, (1991), *Fibre cladding diameter measurement by contact micrometry*, Conference Digest, OFMC' 91, York (UK), pp. 123-126.
9. COST 217 Group, (1989), *Interlaboratory measurement campaign on single-mode fibres*, IEEE Proceedings, pp. 307-314.
10. M. Young, (1990), *Standards for optical fibre geometry measurements*, Technical Digest, Symposium of Optical Fibre Measurements, Boulder, Colorado (USA).
11. W. T. Kane, (1991), *The First international fibre geometry Round Robin*, Conference Digest, OFMC' 91, York (UK), pp. 119-122.
12. M. Monerie, (1990), *Optique unimodale et mesure*, CNET, cours ENSSAT, Lannion.
13. P. A. Perrier, (1994), *Systèmes de transmission sur fibres optiques*, tech. pub. Alcatel CIT.
14. A. Cozannet & al., (1981), *Optiques et Télécommunications: Transmission et traitement optique de l'information*, CNET-ENST, Eyrolles, Paris.
15. J. C. Bizeul & al., (1984), *Mesures de transmission et de géométrie*, Note technique, Centre National d'Etude des Télécommunications (CNET).
16. F. M. E. Sladen & al., (1986), *Chromatic dispersion measurements on long fibre lengths using LEDs*, Electronic Letters, Vol.22, pp. 841-842.
17. A. J. Barlow and I. Mackenzie, (1987), *Direct measurements of chromatic dispersion by the differential phase shift technique*, Technical Digest OFC/IOOC'87, Reno (USA).
18. D. Gloge, (1976), *Weakly Guiding Fibres*, Optical Fibre Technology, IEEE Press, New York (USA), pp. 178-184.
19. E. Miller, E. A. J. Marcatili and T. Li, (1976), *The transmission medium*, Optical Fibre Technology, IEEE Press, New York (USA), pp. 5-26.
20. D. Marcuse, (1978), *Gaussian approximation of the fundamental mode of fibres*, J.O.S.A., Vol.68, pp. 103-109.
21. K. Petermann, (1983), *Constraints for fundamental mode spot size for broadband dispersion-compensated single-mode fibres*, Electronic Letters, Vol.19, pp. 712-714.
22. G. Kuyt, COST 217 Group, (1991), *COST 217 Interlaboratory comparison of uncabled and cabled fibre cut-off wavelength measurements*, OFMC' 91 Conference Digest, York (UK), pp. 12-15.

MECHANICAL AND ENVIRONMENTAL TESTING

J. DUBARD, P. FOURNIER, P. BLANCHARD, F. GAUTHIER
Laboratoire Central des Industries Électriques
33, av. du Général Leclerc
BP 8
92266 Fontenay aux roses
FRANCE

1. Introduction

Active and passive fibre optics components are used in various mechanical and environmental conditions. Their characteristics and behavior may vary when they are subject to conditions that are far from the standard or laboratory conditions. In order to evaluate the effects of such conditions, mechanical and environmental testing are required. They are intended to demonstrate the ability of a component to withstand different kind of stresses simulating either transport and storage conditions or operating conditions.

The tests are performed under a well established monitoring technique depending on the type of component under investigation and using well characterized monitoring components in order to assess the performances of the component under test.

The tests are currently performed on fibres, cables and connectors because they are the components in a fibre link which are the most subject to external stresses during their life time. This paper in particalar reviews the mechanical and the environmental tests set to investigate the characteristics of optical fibre connectors. The different tests are described and illustrated by experimental results.

2. Mechanical and environmental testing of connectors

2.1. MONITORING TECHNIQUE

Mechanical or environmental testing of connectors are performed under a monitoring technique depicted on figure 1 [1]. The test set-up included the following elements:
  - a Multi-channel excitation/source unit. This unit is composed of multiple stable LED's or lasers each connected to a fibre, or a source coupled to a multichannel coupler.
  - one or more independent monitoring channels.

- a reference channel in order to take account for changes in power source for instance.
- a Multi-channel detector unit. This unit may include an oscilloscope to monitor the dynamic variation of the insertion loss of connector.

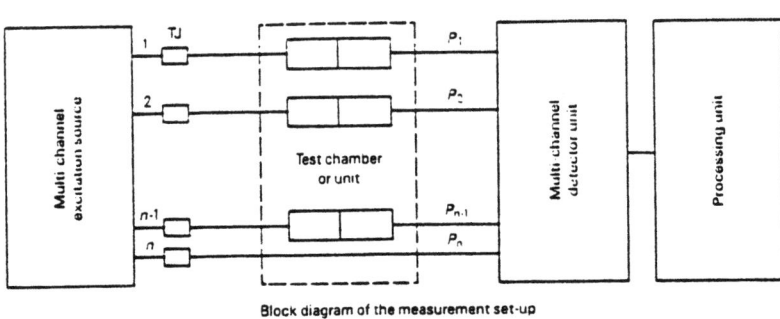

Block diagram of the measurement set-up

Figure 1: Monitoring set-up for mechanical and environmental testing

The principle of the measurement is the following:
- Measurement of the initial power P (t1) of each monitoring channel (i=1, n-1) and P(t1) of the reference channel
- Measurement of the power P(t2) of each monitoring channel and P(t2) of the reference channel during and/or after the test.

The insertion loss variation (ILV) is given by:

$$ILV = 10 \, Log \frac{P_i(t2)}{P_i(t1)} - 10 \, Log \frac{P_n(t2)}{P_n(t1)} \qquad (1)$$

The first term of equation 1 is the change in insertion loss due to the combined effect of the tested connector, the fibres or cables used to connect the optical elements, the source, the temporary joints and the detector unit.

The second term of equation 1 is the change in insertion loss due to all components but the connector under test.

The variation of the insertion loss can be attributed to changes of the optical properties of the fibre ends or to mechanical changes (longitudinal and transverse displacement, angular displacement,...) [see chapter "connectors and splices", same book].

2.2 MECHANICAL TESTING OF CONNECTORS

Mechanical testing procedure and parameters for connectors are those described in standard IEC 874-1 [1].

2.2.1 *Vibration*

The testing conditions are the following:
- The connector shall be attached to a suitable length of fibre (multimode or single-mode)
- Both the connector and the cable are clamped or the connector only is freely suspended
- Vibration shall be performed in at least two perpendicular directions.

The parameters to be specified are:
- The frequency
- The constant acceleration
- The endurance duration

Figure 2 shows the spectrum of the vibration test performed with the following parameters:
- Frequency range: 10-3000 Hz
- Acceleration: 10g (98 m/s^2)
- Endurance duration: 2 hours per direction
- Directions: 2 perpendicular direction (one being paralell to the connector axis)

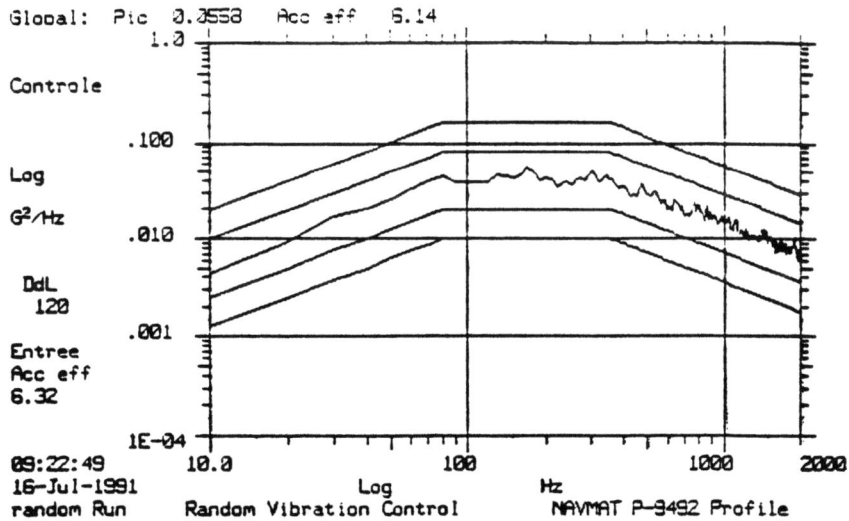

Figure 2: Spectrum of the vibration test

ILV measurement during a vibration test shows a negligable change in the insertion loss on the actual connector sets. Those tests are performed with CW sources. It will be interesting to performed these tests while doing bit error rate measurement.

2.2.2 Effectiveness of fibre or ferrule retention

The force should be applied to the free end of the cable along the common axis of the cable and the connector (figure 3).

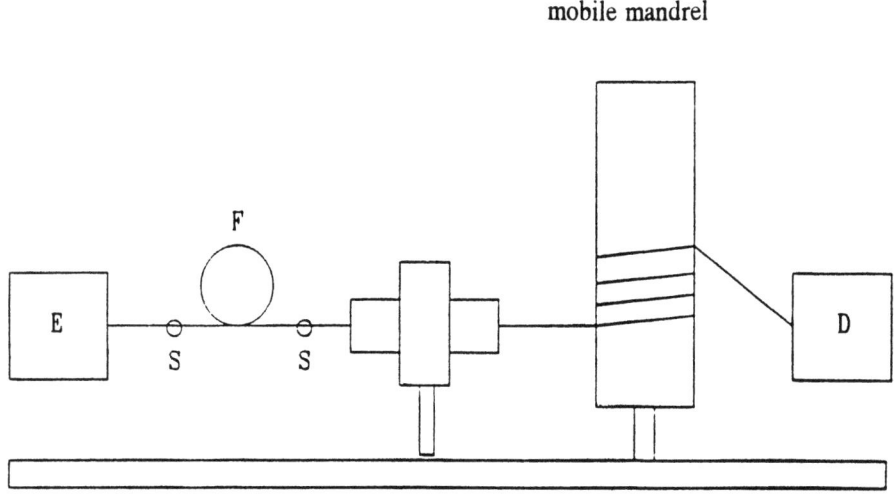

Figure 3: Effectiveness and fibre or ferrule retention test set-up

On figure 3, E is the source, F a fibre, S the splices and D a detector. The connector set is fixed and the force is realized by rotation of a mobile mandrel.

The parameters to be specified are the direction, the duration, the rate and the magnitude of the applied force. The final measurements include the visual inspection and the ILV.

Experimental test parameters are:
- Direction: along the axis of the connector
- Duration of the exposure to the maximum stress: 5s
- Rate: 1 N/s
- Magnitude: 10N

The test was performed on 5 connector sets. For each of them the ILV was less than 0.1dB during the test. The ILV was measured for each 1 N of force increment and after relaxing the stress. The test was carried out up to the breakage of the fibre. The retention force at breakage was 17 N. However in one case the ILV was less than 0.1dB for retention forces below 6 N. For a retention force of 6 N the fibre was extracted from the ferrule.

2.2.3 Bump

This test is intended to determine the ability of the connector to withstand a series of bumps. The bump severity shall preferably be 4000 ± 10 bumps at 390 m/s^2 with a pulse duration of 6 ms.

Experimental data are obtained while the bump test is performed on a cable head equipped with 6 VFO type connectors. The characteristics of the pulse is shown on figure 4.

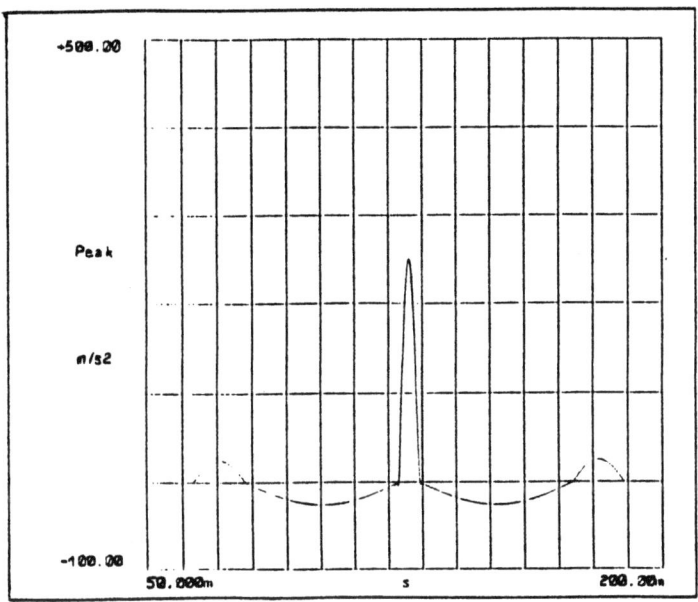

Figure 4: Characteristics of the pulse for the bump test

The experimental bump test parameters are:
- Acceleration: 250 m/s
- Pulse duration: 6 ms
- Number of bump: 1000 (over 15 minutes)
- Directions x, y, z directions

The results of the ILV measurements are shown on figure 5

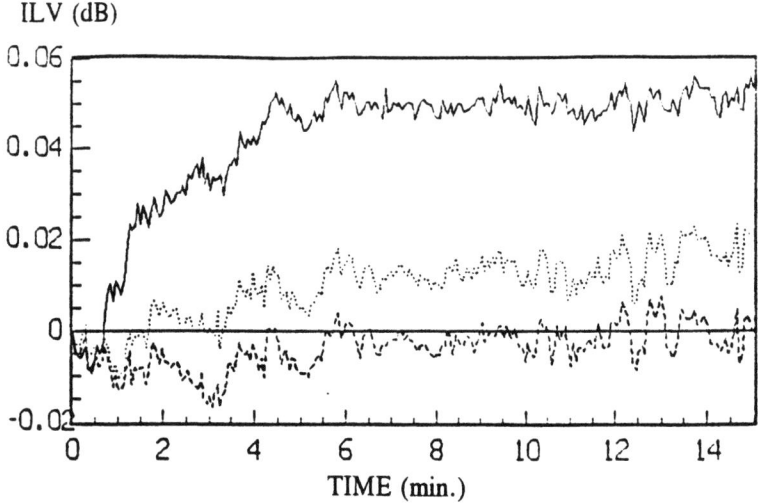

Figure 5: ILV measurement of a connector sets during a bump test

2.2.4 *Impact*
This test is intended to determine the ability of a connector to withstand a localized impact or a series of impacts with hard objects. The test apparatus is depicted on figure 6. The parameters to specified are:
- The radius of the semi-circular cylindrical hammer face
- The mass of the hammer
- The height of the hammer
- The form of the test specimen
- The number of impacts
- The point of impact on the connector.

The results of the ILV measurements on a connector set and for a impact energy of 0.5 J is shown on figure 7.

Impact 1 is done on the adaptor. Impacts 2 to 5 are done on one connector. Between each impact the connector is rotated by 90°. Impacts 6 to 9 are done on the other connector, following the same procedure.

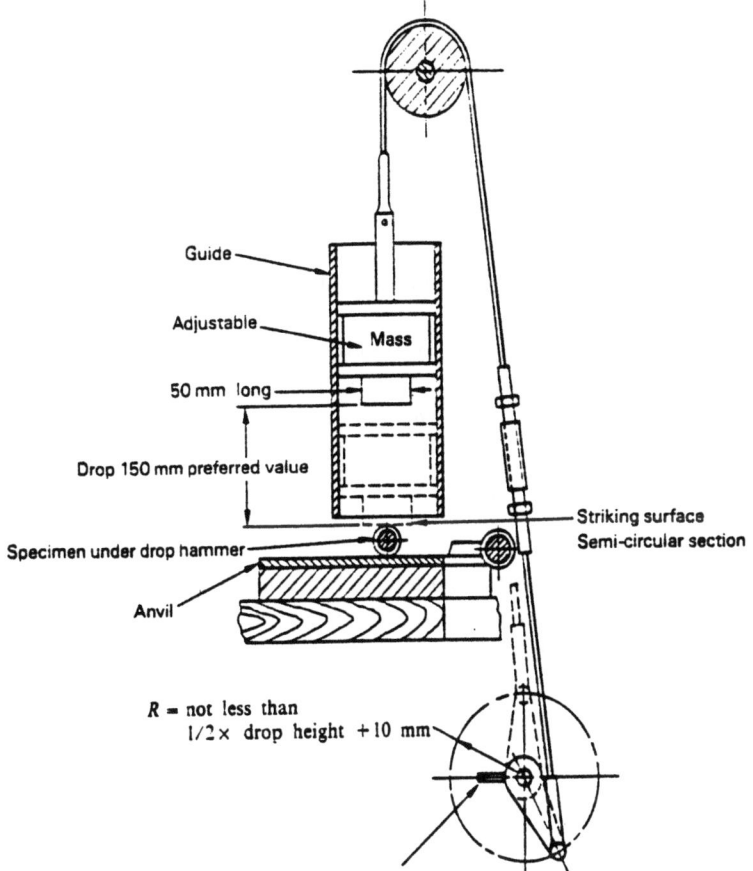

Figure 6: Apparatus for the impact test

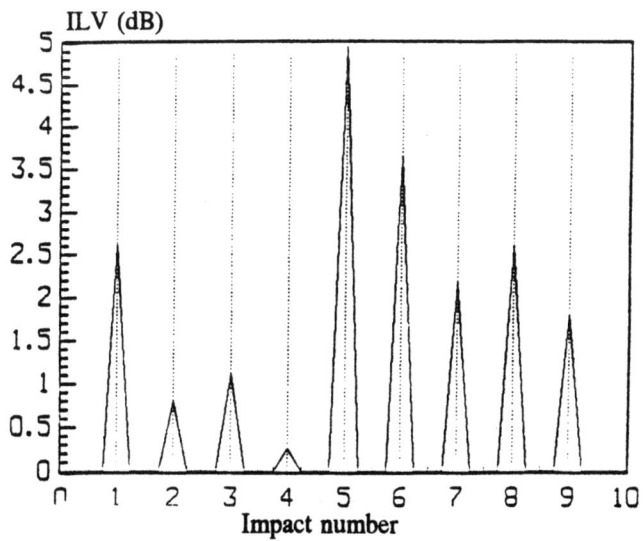

Figure 7: ILV of a connector set during the impact test

2.2.5 *Other mechanical tests*

The other mechanical tests described in the standard IEC 874-1 are:

- Static load test: intended to determine the ability of a fixed connector used in position where it may be subject to shearing force.

- Engagement and separation force test: intended to determine the force or torque necessary to overcome friction or compression of spring and other resilient parts in order to engage and disengage the coupling mechanism.

- Strength of cable retention and cable entry test: intended to determine wether the device for fixing and clamping the cable is effective when tensile force and/or torque are applied to the attached cable. This test includes a cable pulling test (a tensile force is applied to the free end of the cable along the common axis of the cable and the connector) and a cable torsion test (a torque is applied to the cable while the connector is held fixed).

- Strength of coupling mechanism test: intended to determine the ability of the connector set to withstand a tensile force applied to one part of the connector set while the other part is securely fixed

- Bending moment test: intended to determine the ability of a connector set to withstand a bending moment in such a way that the coupling mechanism is stressed.

- Shocks test: intended to determine the ability of a connector set to withstand a series of shocks.

- Crush resistance test: intended to simulate the effect of loads applied to a free connector set such as might occur when the connectors are in a vulnerable position such as on the ground or on the floor surface.

- Axial compression test: in order to determine the optical effects of an applied force tending to thrust the optical fibre into a connector shell or housing. It is intended to simulate the effect of rough handling of a free or panel-mounted connector set.

- Drop test: intended to assess the ability of an optical fibre connector to withstand the impacts it could receive when dropped onto a hard surface.

2.2 ENVIRONMENTAL TESTING OF CONNECTORS

2.2.1 *Cold*

The cold test procedure for connector set is the following:
- The specimen is placed in a chamber at room temperature
- The temperature is decrease down to the test temperature at a rate of 1° C per hour
- The specimen is tested at the test temperature for a specified test duration
- The specimen is kept in the chamber while the chamber temperature is raised to ambient

The severity of the test as far as the test temperature and duration are concerned, is determined by the parameters indicated in table 1.

Temperature (°C)	Duration (hours)
-65	2
-55	16
-40	72
-25	96
-10	-
-5	-
+5	-

Table 1: Cold test severity parameters

Test duration for temperatures of -10 °C, -5 °C and +5 °C is not specified.

The results of the ILV measurement on three connector sets during a cold test at a temperature of -25 C and for a test duration of 96 hours are shown on figure 8.

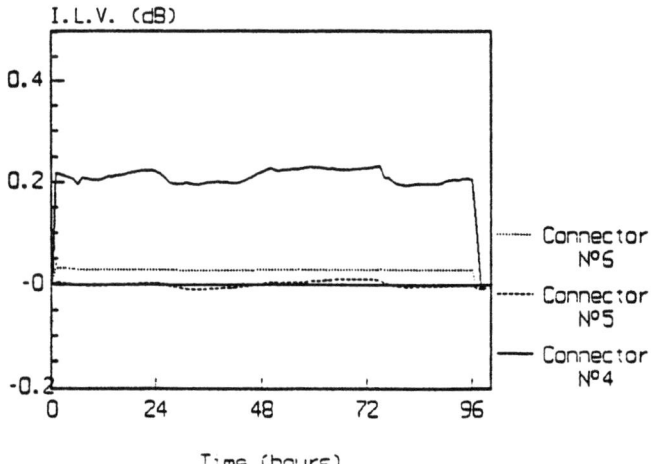

Figure 8: ILV measurement of three connector set during a cold test.

2.2.2 *Dry heat*
The dry heat test procedure for connector sets is the following:
- The specimen is placed in a chamber at room temperature
- The temperature is increase up to the test temperature and kept in the chamber for the specified duration
- At the end of the specified test duration the specimen is kept in the chamber while the temperature is lowered to ambient.

The dry heat test parameters are indicated in table 2.

Temperature (C)	Duration (hours)
30 | 155	2
40 | 175	16
55 | 200	72
70 | 250	96
85 | 315	-
100 | 400	-
125 |	-

Table 2: Dry heat test parameters

The results of the ILV measurement on three connector sets during a dry heat test at a temperature of 85 °C and for a test duration of 1000 hours are shown on figure 9.

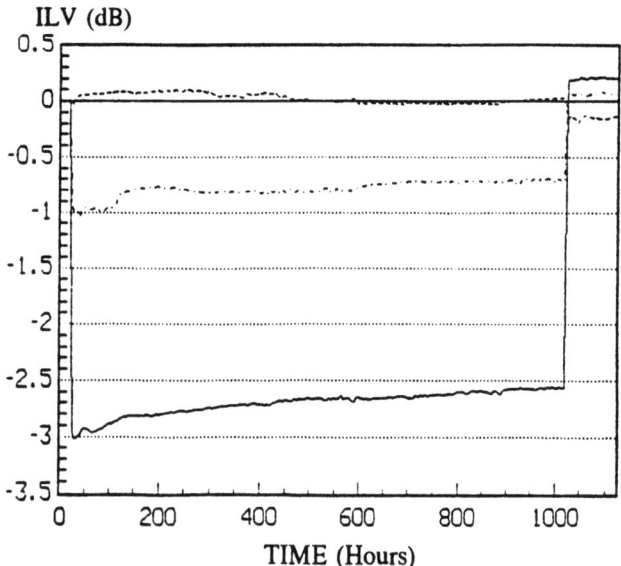

Figure 9: ILV measurement of three connector sets during a dry heat test

2.2.3 Climatic sequence

The purpose of this test is to provide standard climatic sequences consisting of the sequential application dry heat, damp heat, cold and low air pressure. The test procedure is based on two methods.

Method 1:
- Step 1: the specimen is exposed to a high temperature test
- Step 2: the specimen is exposed to a cycle of damp heat test at 55 °C
- Step 3: the specimen is exposed to a cold test

A low air pressure test completes the cycle by checking the sealing of the specimen.

Method 2:

Method two interposes a cold test between each of the damp heat cycles. This method is more severe than method 1.

An exemple of the temperature change in a method 1 climatic sequence test is shown on figure 10. The cold test is performed after the first damp heat test cycle. The results of the ILV measurement is depicted on figure 11.

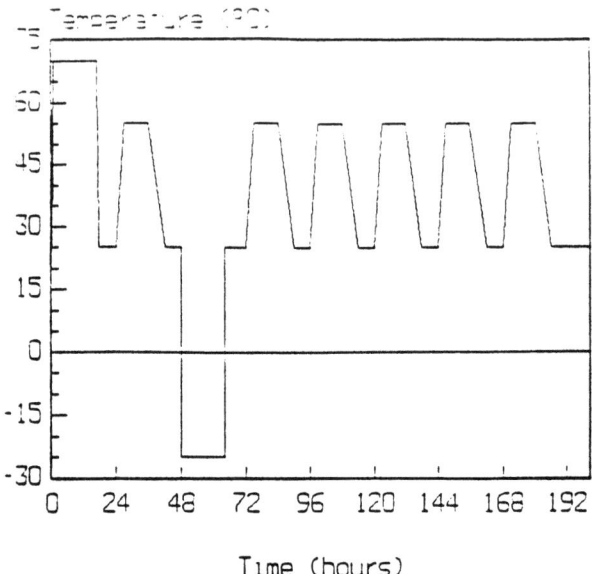

Figure 10: Temperature change in a method 1 climatic sequence test

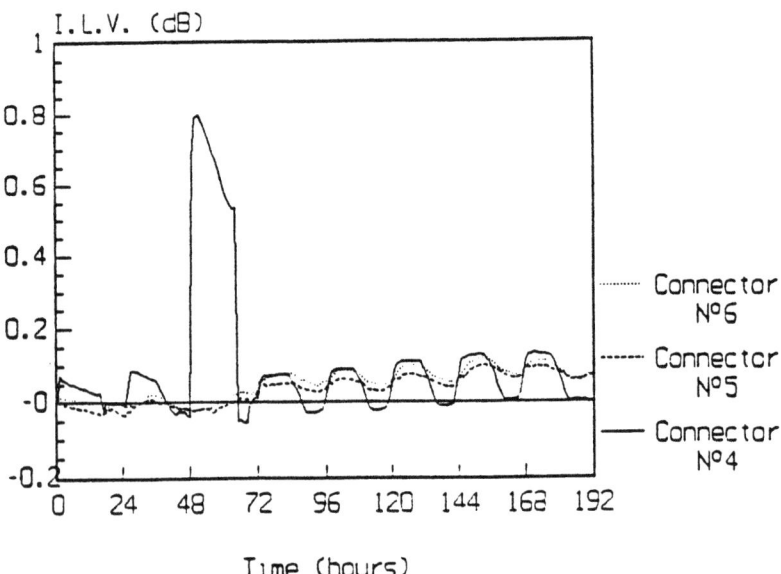

Figure 11: ILV of four connector sets during a method 1 climatic sequence test

2.2.5 Change of temperature

This test is intended to determine the suitability of a specimen to withstand the effects of change of temperature or a sucession of changes of temperature. It is conducted according to the standard IEC 68-2-14 whic defines two types of tests: the Na type test and the Nb type test.

Na test:
This test subjects the specimen to extreme temperatures with a very short change over time then inducing thermal shocks. A cycle of the test sequence is depicted on figure 12.

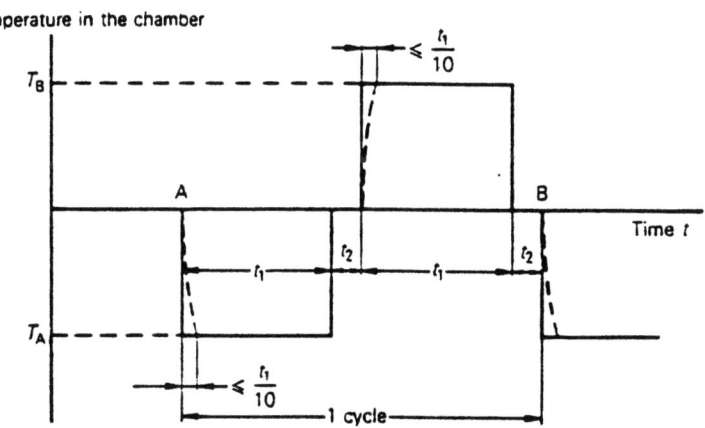

Figure 12: Cycle sequence in a type Na change of temperature test

Test Nb:
This test subjects the specimen to a gradual variation of the temperature which stresses the specimen but does not induce thermal shocks. A cycle of the test sequence is depicted on figure 13.

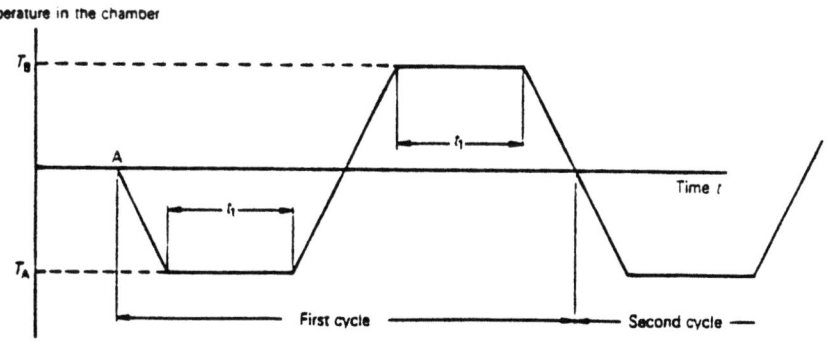

Figure 13: Cycle sequence in a type Nb change of temperature test

The test severity parameters are given in table 3.

High temperature (°C)	Low temperature (°C)	Duration (min.)
55 | 155	10 | -25	10
70 | 175	5 | -40	30
85 | 200	0 | -55	60
100 | 250	-5 | -65	120
125 | 315	-10 |	180

Table 3: severity parameters for a change of temperature test

Figure 14 and 15 show respectively the change of temperature sequence and the ILV measurement on three connector sets during a Na type test.

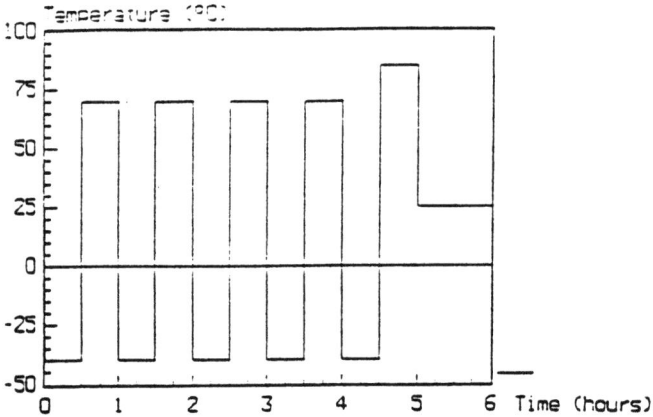

Figure 14: Sequence of a Na type change of temperature test

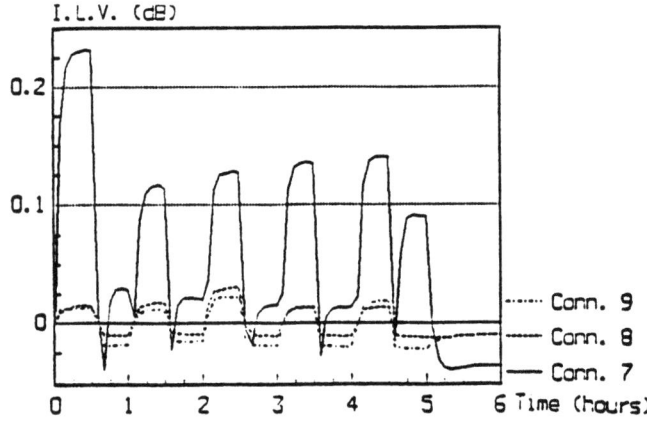

Figure 15: ILV measurement of three connector sets during a Na type change of temperature test

2.2.5 Other environmental tests

The other environmental tests described in the standard IEC 874-1 are:

- The damp heat test: the specimen is placed in a chamber at a temperature of 40 °C, a relative humidity of 93% and for a period of time of 4, 10, 21 or 56 days.

- The salt mist test: the specimen is exposed to a salt mist environment within a test chamber maintained at a temperature of 35 °C.

- The condensation test: it is intended to provide a composite test procedure to determine in accelerated manner the resistance of the specimen to the deteriorative effects of the high temperature/humidity and cold conditions. The specimen is placed in a humidity chamber and subjected to 10 temperature/humidity cycles of 24 hour duration each. During any five of the first nine cycles and after exposure to the high temperature/humidity sub-cycle the specimen shall be subjected to a cold test.

- The dust test: the specimen is exposed to a specified dust concentration within a conditioning chamber in which the air is circulated over a period of time.

3 Conclusion

Mechanical and environmental testing of connectors show that the insertion loss variation can be important (as high as 5dB for the impact test and 3dB for the dry heat test). However recovery of the initial insertion loss is usually observed when the stress is relieved. This is not always the case. For instance bending test performed on optical fibre or cable can lead to an increase of the attenuation as a function of the number of bendings and the rate of the variation of the attenuation is dependent upon the radius of curvature of the fibre. Tests on fibres and cables are described in standard IEC 793-1.

The IEC standards test parameters may not be suitable for all type of applications (spatial or nuclear environments for instance). Therefore new parameters or test apparatus should be defined.

4 References

[1] International standard IEC 874-1, *"Connectors for optical fibres and cables"*, 1993.

[2] International standard IEC 793-1, *"Optical fibres"*, 1992.

FIBRE RIBBON MEASUREMENTS

LAURI OKSANEN
Nokia Cables, Fiber Optics
P.O. Box 77, FIN-01511 Vantaa, FINLAND

1. Ribbon Construction and Advantages

Optical fibre ribbons are linear arrays of fibres, held together by a common acrylate coating, often called the matrix material. A ribbon would typically contain 2, 4, 6, 8, 10 or 12 fibres. Even 16 fibre ribbons are being contemplated. Figure 1 is a photograph of a 4 fibre ribbon end.

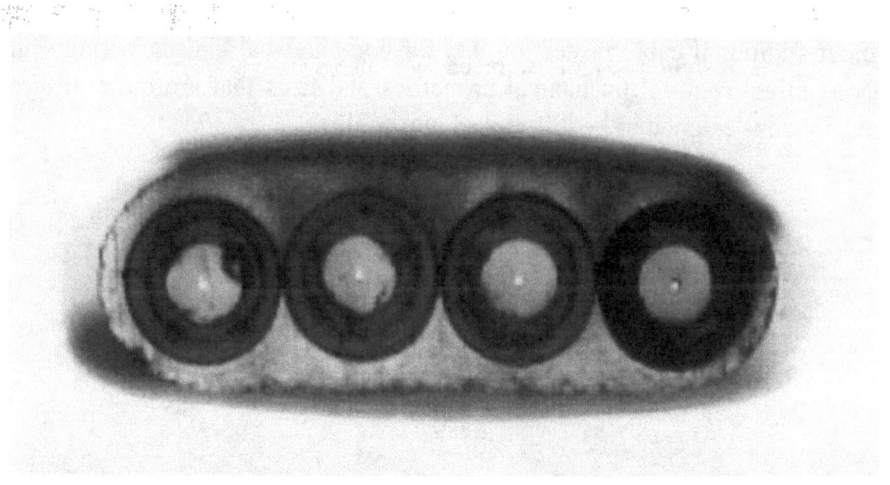

Figure 1. Photograph of the end face of a 4 fibre encapsulated ribbon.

Ribbons are commonly manufactured in two stages which can be tandemised. First each fibre is coloured with a thin layer of UV curable material for identification. In the second stage the fibres are pulled through a die where the common acrylate coating is applied and then through UV lamp curing units [1]. Thus in a finished ribbon the glass fibre is surrounded by 4 layers of different materials: the first and second layers of the primary coating, the colouring layer and the ribbon matrix material.

Fibres for ribbons require suitable primary coating and colouring materials to fulfil all the requirements for a ribbon, such as easy strippability and separability. There is also a shift in standards to lower the nominal value of the fibre coating diameter to 245 +/-10 µm [2] to have a coloured fibre diameter close to 250 µm.

Fibre ribbons were first used in cables already in 1977 [3] in the AT&T Chicago Lightwave Project. In recent years there has been renewed interest in ribbons as the fibre count in cables has grown. In Europe the use of ribbons is relatively new with some countries now having a few years experience. The main benefits of ribbons are high packing density in cables and mass splicing [1], [4]. All the fibres in a ribbon can be stripped at the same time and cleaved in one operation. Then they can be fusion spliced in one operation. It is also possible to terminate a ribbon with an MT type connector.

There are two basic types of ribbons, the encapsulated structure, Figures 1 and 2a, and the edge bonded type, Figure 2b. Obviously, the encapsulated ribbon is a more robust structure. There are even plans to use ribbons as such without additional cable protection e.g. for computer back plane wiring. Such ribbons might require mechanical properties and tests that are quite different from ribbons designed to be introduced into cables.

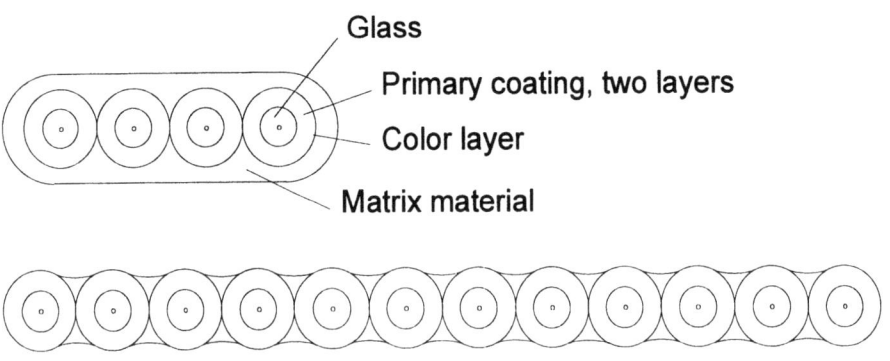

Figure 2. Ribbon types: a) encapsulated, b) edge bonded. In-between types can be useful in some applications.

Even though there exist multimode fibre ribbons, the main emphasis of this text is on single-mode fibres as they account for the majority of ribbon applications.

2. Fibre Ribbon Measurements

The main emphasis on fibre ribbon test development has been to characterise the ribbon with regard to jointing capability, including stripping, splicing, and behaviour of ribbon in splice enclosures or trays. For splicing to succeed, the alignment of the fibres within the ribbon must be sufficiently regular. An example of requirements for ribbon dimensional tolerances is given in Table 1. For definitions see Figure 3. The values of Table 1 are those that are at present being considered in international standardisation in IEC. As they are maximum allowable values they cover both encapsulated and edge bonded ribbon types. For a specific application the required tolerances could be tighter and consist of both minimum and maximum limits.

TABLE 1. An example of ribbon dimension specifications. All values are maximum limits.

Number of fibres	Width w (µm)	Height h (µm)	Fibre alignment		Planarity p (µm)
			Horizontal separation		
			Adjacent d (µm)	Extreme b (µm)	
2	700	480	280	280	-
4	1220	480	280	835	50
6	1770	480	280	1385	50
8	2300	480	300	1920	50
10	2850	480	300	2450	50
12	3400	480	300	2980	50

Most optical measurements on ribbons are basically similar to single fibre measurements. Thus mode field diameter, cut-off wavelength, glass geometry, and dispersion measurements are performed much as with individual single-mode fibres. These parameters are usually measured by the fibre manufacturer before the ribbon process. In some ribbons the fibre cut-off wavelength may be lower than that of the fibre before the ribbon process. Otherwise, the results of these measurements should be the same as for the fibres prior to being included in the ribbon.

Polarisation mode dispersion (PMD) is known to be sensitive to the environment of the fibre. Indeed there is evidence that PMD depends on the position of the fibre in a ribbon, i.e. whether the fibre is in the centre or on the edge of the ribbon, and the position of the ribbon in the cable [5]. Therefore

PMD measurements are best performed in the finished cable. The test procedure is the same as for single fibres.

The ribbon structure makes increased measurement efficiency possible, due to reduced need for handling. Care should be taken in laying of the ribbon since it readily bends in only one plane. Stressing the ribbon with forced bends might affect attenuation and cut-off results.

The definitions and test methods explained here are in accordance with the current standardisation proposals within IEC, where such exist.

3. Ribbon Production Measurements

3.1. ON-LINE DIMENSION MEASUREMENT

During production of the ribbon it is possible to measure the outside dimensions, that is the height and width of the ribbon with optical methods. The ribbon must be well aligned to the measurement axes.

3.2. ATTENUATION

The ribbon is usually checked for attenuation and freedom of point defects with an OTDR. This can also be done later in the finished cable. The measurement is performed as for single fibres. Again, it is possible to take advantage of the handling efficiency of the ribbon. There are equipment on the market which will, after the ribbons have been loaded, automatically align an input fibre to a given fibre in a given ribbon one at a time. Thus it is possible to automate the OTDR measurement of an entire cable. In good quality ribbons the fibre attenuation is not affected by the ribbon process.

3.3. RIBBON GEOMETRY

The geometry of a manufactured ribbon needs to be measured. The definition of ribbon dimensions is illustrated in Figure 3. The width and height of the ribbon are the dimensions of the minimum rectangular area enclosing the ribbon cross section. The basis line goes through the centres of the first and last fibre. The horizontal centre to centre separations are measured along the basis line and the planarity deviations orthogonal to it. Some proposed dimensional tolerances are in Table 1.

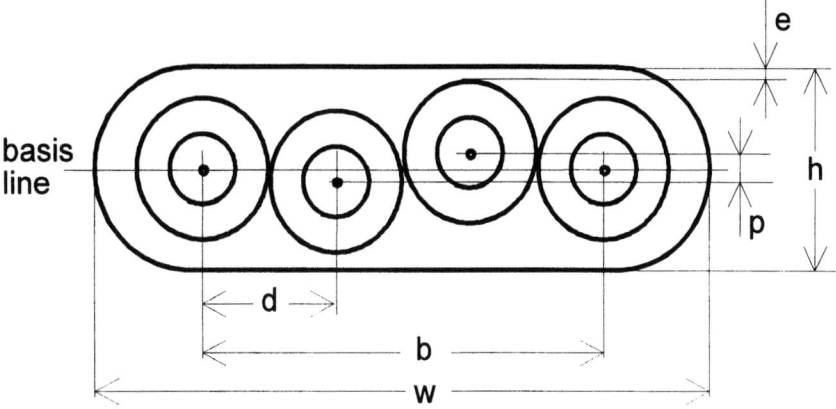

Figure 3. Ribbon dimension definitions. h - height, w - width, p - planarity, d - adjacent fibre centre-centre distance, b - centre-centre distance between first and last fibre, e - common coating minimum thickness.

A quick method for a pass/fail test on the dimensions is to use a well defined aperture. If the ribbon passes through this aperture it is considered to pass the test. This test is usually sufficient for edge bonded ribbons.

The outside diameter of encapsulated ribbons can be measured with a dial gauge. This does not give much information on the alignment of the fibres within the ribbon, especially if the common coating is very thick, but may be sufficient for production control purposes if the process has been ascertained to be stable.

All the dimensions in Figure 3 can be measured by microscope analysis of the ribbon end face. This requires that the end face is polished to a smooth, flat surface, perpendicular to the ribbon axis. One must take care in grinding to ensure that the end face is not distorted. To facilitate grinding the ribbon is usually placed upright in a mold with resin which is then cured. It is also possible to use a special ribbon holder for grinding.

4. Non Routine Tests

4.1. STRIPPABILITY

To realise the handling advantages it must be easy to strip the ribbons so that all fibres come clean. The special purpose stripping tool usually includes a heating element which heats the ribbon to over 80 °C to decrease the adhesion between the glass and primary coating. The heating time is typically between

15 and 30 seconds. Then the common coating and the fibre primary coatings are stripped in one operation. The stripping length is usually restricted by the tool to less than 30 mm.

In stripping tests the stripping tool is mounted in a tensile test equipment and the strip force recorded as a function of distance. Typical results for a 4 fibre ribbon are shown in Figure 4. Figure 5 shows the stripping force as function of heating time. The quality of the stripping result is judged by visual inspection. It is often possible to get a better result by manual stripping, even when the stripping speed is higher than with the tensile test equipment. Therefore a test program might also include manual stripping with evaluation of the stripping quality.

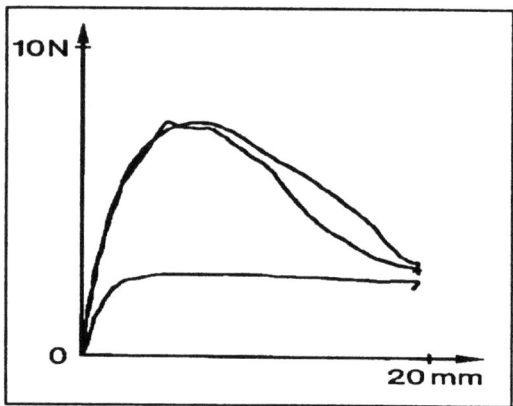
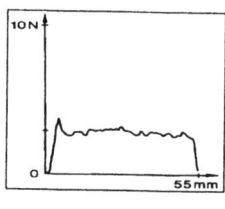

Figure 4. Two stripping force plots as function of distance for a 4 fibre ribbon. Stripping tool Sumitomo JR-4, temperature 90 °C, speed 500 mm/min, preheating time 15 s. The stripping result was good, no debris left on the fibres. The lowest curve is the tool without any ribbon. The small picture on the right shows a single fibre stripping force plot for comparison.

As seen in Figure 4 the strip force versus distance curves look quite different from curves obtained for single fibres with conventional stripping tools without heating. In heated stripping the primary coating typically comes off as a tube, whereas in "cold" stripping the coating is peeled off in bits and slices. The definition of strip force used for cold stripping of single fibres is "the mean stripping force, excluding the first peak" [6]. This does not suit ribbons very well. For ribbons the more appropriate measure could be the maximum force from the force versus distance plot.

"Bumps" in the ribbon strip curve have been found to correlate with poor strip quality, i.e. coating debris left on the fibres. This correlation is, however, not good enough to be used as criteria for strip quality. A visual examination of the stripped ribbon is therefore necessary in stripping tests. The quality could then be determined by grading the result based on how many fibres are completely clean and what is the amount of residue possibly left on some fibres.

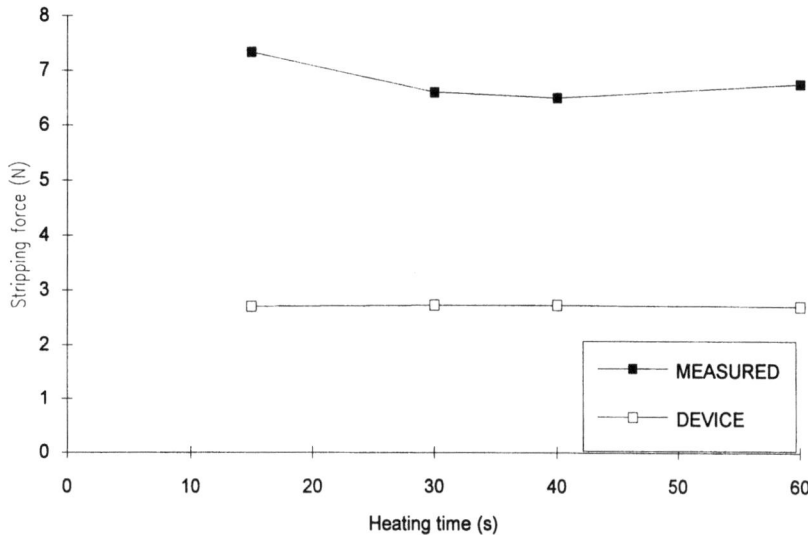

Figure 5. 4 fibre ribbon peak stripping force as function of heating time. All results with good stripping quality. Speed 500 mm/min.

4.2. SEPARABILITY

In some applications it is required that individual fibres or fibre groups can be separated from the ribbon. In this case the fibre primary coating with colour layer should remain undamaged and not be delaminated from the glass. In any case the ribbon should withstand some handling without losing integrity.

There is yet no good quantitative test for separability. Typically, an experienced craftsperson experiments manually with the fibre and gives a subjective evaluation based on the ease of separating fibres and on the damage criteria above.

A possible test for comparing separation forces between ribbons is as follows. One or more fibres are separated from the ribbon and fixed to one clamp of a tensile testing equipment. The ribbon end from which they were separated is fixed to the other clamp. The separation is continued with the tensile tester and the required force recorded.

4.3. TORSION

For ribbons torsion for example in splice boxes or in cable manufacturing can be much more of a problem than for single fibres. Torsion tests are made as follows. A 0.1 to 1 m ribbon is fixed at one end and rotated at the other end while being held under a minimum tension of 1 N. After each rotation the ribbon is examined for delamination. It is also possible to monitor the attenuation of the fibres during the test. An alternative test configuration, which facilitates attenuation measurements as both ends stay fixed is shown in Figure 6. Here the rotation is applied at the middle point of the ribbon.

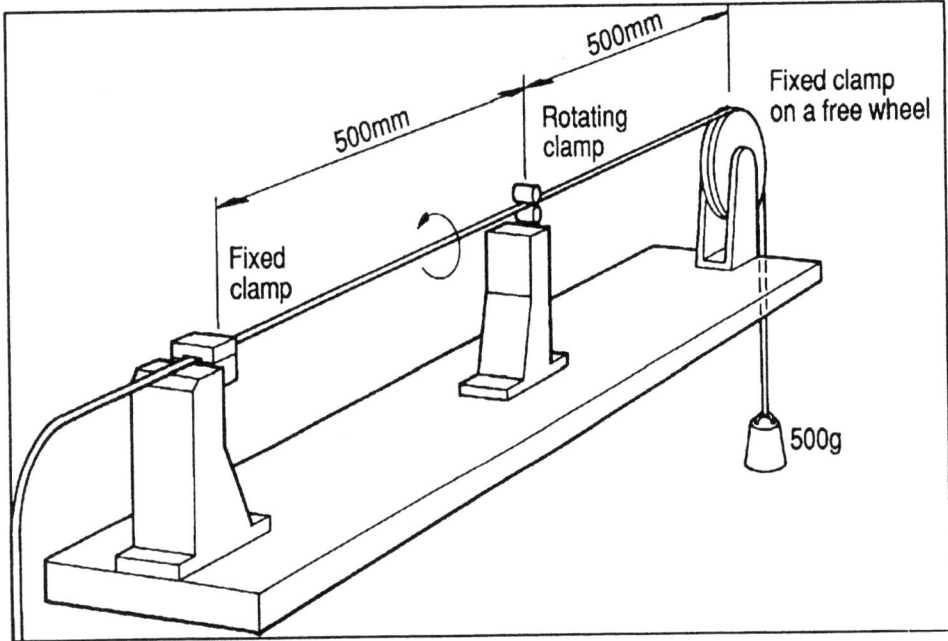

Figure 6. Set-up for ribbon torsion test.

Typically ribbons can tolerate many complete turns per meter without any damage or increase in attenuation. Lower fibre count ribbons tolerate more turns than wide ribbons. Figure 7 shows a typical result of attenuation versus number of turns. In this case no structural damage to the ribbon was observed.

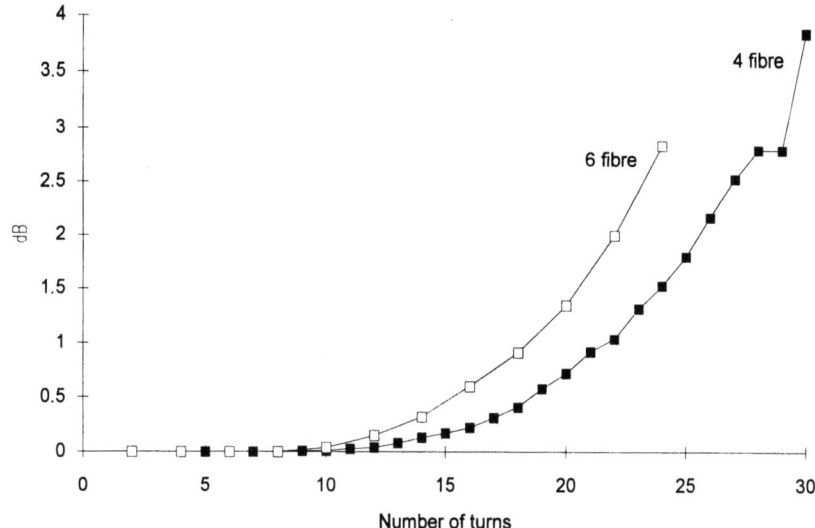

Figure 7. Attenuation change of edge fibres in a 4 and 6 fibre ribbon at 1550 nm as function of number of turns.

4.4. MACROBENDING

A macrobending test for ribbons is made as for individual fibres by winding the ribbon around a mandrel of specified diameter and monitoring transmitted power changes. Care must be taken to avoid torsion on the ribbon. When doing the test on a short piece of ribbon, a small loop (e.g. 60 mm diameter) should be placed on the ribbon to eliminate possible disturbance from higher order modes in the straight condition. The same applies for the torsion and crush tests.

4.5. CRUSH

For some applications the crush sensitivity of the ribbon may need to be measured. For normal cable jointing purposes this test is usually not required. In the crush test the ribbon is placed flat between two parallel plates. The load on the plates is increased while monitoring the attenuation of the fibres. A typical 4 fibre encapsulated ribbon can tolerate several hundreds of Newtons of load (for 50 mm plate length) before the attenuation starts to increase.

4.6. ENVIRONMENTAL TESTS

If the ribbon is considered an integral part of the cable then its tolerance to environment can be measured in the finished cable. For some uses the ribbon might be tested by itself. In this case the test would follow the guidelines for similar testing of an individual fibre. Of course the cable manufacturer should ensure that the ribbon is compatible with the cable materials such as cable jellies.

Environmental tests for ribbons could include some of the following:
- Water immersion
- Temperature cycling
- Accelerated ageing

Parameters to test are fibre attenuation during the test and stripping and separability after each test phase.

5. Implications of Ribbon for Fibre Measurements

In mass fusion splicing the corresponding fibres in the opposing ribbons are butted against each other, guided by V-grooves, and the surface tension of the glass aligns the claddings of the fibres. It is not presently possible to optimise individual fibre alignment. For the cladding alignment to work well the fibre diameter should be close to nominal and the core concentricity error and cladding non-circularity small.

Single fibre glass geometry measurement should be able to support a cladding diameter tolerance of +/-1 µm. With modern grey scale measuring equipment this is achievable as far as repeatability is concerned but calibration of these test sets is yet to be verified on an international level. Another issue is fibre curl which may be a problem in mass splicing.

5.1. FIBRE CURL

Optical fibres exhibit some degree of curvature when stripped from their coating. If this curvature or curl is excessive, it may become a problem in fusion splicing of single-mode fibres with equipment which relies on passive V-groove alignment of the fibres.

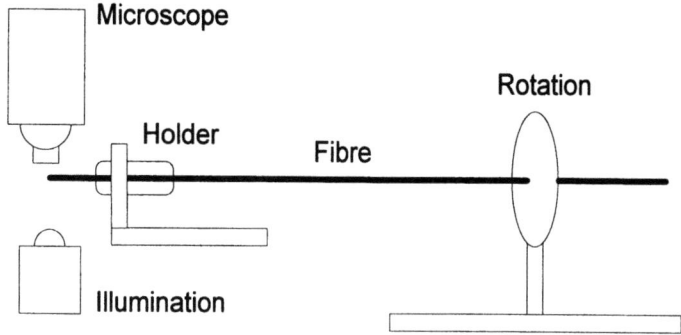

Figure 8. Microscope based set-up for measurement of fibre curl.

Several methods have been proposed for fibre curl measurement, see e.g. [7] and [8]. Perhaps the simplest to implement with normal laboratory hardware is shown in Figure 8. The fibre is stripped and put into a holder which can be e.g. a V-groove or connector ferrule. The fibre is rotated about its axis in the holder and the deviation of the position of the fibre end is observed with a microscope. The curl can be obtained simply from the maximum deviation. Another method is to fit the deviation as function of rotation angle to a sine function and calculate the curl from the amplitude of the fitted curve. Both methods yield sufficient resolution.

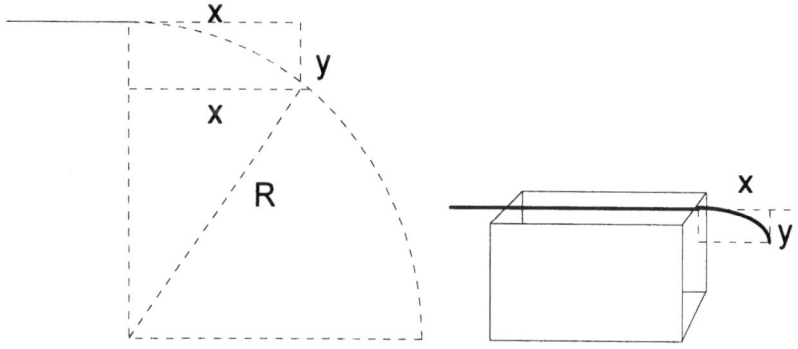

Figure 9. Circular model for fibre curvature.

$$R^2 = (R-y)^2 + x^2 = R^2 - 2Ry + y^2 + x^2$$
$$\Leftrightarrow R = \frac{x^2 + y^2}{2y} \approx \frac{x^2}{2y} \quad , \text{when} \quad y \ll x \tag{1}$$

x = free length of fibre , y = deviation from straight line

To magnify the deviation the free length or overhang of the fibre is typically chosen to be from 10 to 20 mm. In splicing equipment the overhang is on the order of 3 mm. The measurement result can be reduced to the splicing conditions by assuming that the curvature is circular, Figure 9. To make measurements with different set-ups directly comparable it has been proposed that the curl should be specified as the radius of curvature R, assuming the circular model. Some values suggested for the acceptable radius of curvature are around 2 to 3 m [9], [10].

6. Standards

International standardisation work is going on in IEC TC 86 and also in CECC WG 28 on the European level. Generic ribbon standards are now in the first phases of voting in standards organisations. Emphasis of the work has been on those ribbon properties which are important for jointing of ribbon cables. The ribbon specific test methods in the current IEC proposal are shown in Table 2.

TABLE 2. Ribbon tests being standardised in IEC TC 86.

Ribbon geometry definition and measurement methods
– End face measurement
– Aperture gauge
– Dial gauge
Separability of individual fibres from a ribbon
Torsion test

7. Summary

Basic measurement methods for fibre ribbons are fairly well developed. Most optical properties such as attenuation, mode field diameter, cut-off, fibre geometry, and dispersion are measured much as for single fibres. Usually it is enough to measure these at the fibre stage and the attenuation then in the finished cable. A more efficient method for full ribbon geometry measurement would be useful. A better understanding of user requirements is needed to develop detailed test procedures for many handling oriented properties of ribbons. These include ribbon stripping and separability. For some applications ribbons might require environmental characterisation which is still under study.

8. References

1. Greco, F. and Ragni, A. (1992) High-Performance optical fiber ribbon by UV-curing method, *Wire Journal International*, Dec. 1992, 63-72
2. *Interim European Telecommunication Standard prI-ETS 300 227*, ETSI, Jan. 1993, article 7.2.1
3. Schwartz, M.I., Reenstra, W.A., and Mullins, J.J. (1978) The Chicago Lightwave Project, *Bell System Technical Journal*, Vol. 57, No. 6, 1181-1188
4. Tomita, S., Matsumoto, M., Nagasawa, S., Tanifuji, T., and Uenoya, T. (1992) Preliminary Research into High-Count Pre-Connectorized Optical Fiber Cable,*Proc. International Wire and Cable Symposium*, Reno, 5-12
5. Caltarossa, A., Schiano, M. (1993) Polarization mode dispersion in high-density ribbon Cables, *Technical Digest, 2nd Optical Fibre Measurement Conference*, OFMC'93, Torino, 181-184
6. Method IEC 793-1-B6 - Strippability, *International Standard IEC 793-1, Optical fibres, Part 1: Generic specification*, IEC, Fourth edition, 1992-11
7. Ingles, A.L. (1992) An Automated Method for Measuring the Radius of Curvature of an Optical Fiber, *Technical Digest, Symposium on Optical Fiber Measurements*, Boulder, NIST Special Publication 839, 127-130
8. Emig, K.A. (1993) A comparison of manual and automated fiber curl measurement systems, *Technical Digest, 2nd Optical Fibre Measurement Conference*, OFMC'93, Torino, 85-88
9. Robbins, S.D. (1993) Fiber curl - a concern for mass-fusion splicing,*Telephony*, Jan. 1993, 52
10. Jonsson, K. and Björk, A. (1993) Comparative measurements of curl and splice loss for an optical 4-fibre ribbon, *Technical Digest, 2nd Optical Fibre Measurement Conference*, OFMC'93, Torino, 73-75

CALIBRATION ARTEFACTS, TRACEABILITY AND ACCURACY

The use of calibration artefacts for fiber test set calibration

A. J. BARLOW
EG&G Fiber Optics
Sorbus House
Mulberry Business Park
Wokingham, RG11 2GY, UK.

1. Introduction

In general, calibration is defined as the process in which the relationship between the values indicated by an 'infant' test set and a known standard one is established [1]. Calibration also includes the estimation (quantification) and correction of all uncertainties where possible. Calibration of the 'infant' set is updated by *transferring* the calibration of the parent set to the infant. This process is repeated all the way down from standards lab to the working set in the factory or quality control laboratory, by a network or chain of reference standards, standard test sets and transfer standards or artefacts. An example of such a chain is shown in Figure 1.

Obviously, the use of these artefacts is a central part of the process. The use of calibration artefacts is essential to the accurate calibration of many types of fiber optic measurement instrumentation, often in their installed (i.e. factory/QA lab environment) states. Recently, of course, the upsurge of the adoption of ISO9000 and similar quality systems has generated a need for regular, on-site traceable calibration of test sets of all types. Considerable activity in IEC and other standards bodies to provide standard calibration procedures (as opposed to standard fiber measurement procedures) has occurred in the last three years as a result [1],[2],[3]. Similarly, a lot of work in generating hardware calibration artefacts has been done, and some of this work is reported here. In view of other papers in this volume, we will focus only on CD, Length, strain, group index and Polarisation Mode Dispersion (PMD) calibration in detail.

2. General Requirements

The requirements for artefacts intended for fiber test set calibration are not only that the calibration be transferred to the working test set on-site in a very short time (i.e. less than half a day) with minimal disruption to the test set and its measurement schedules, but also to provide all the usual needs such as traceability, known (preferably high) accuracy etc, to satisfy ISO9000 requirements. This places severe demands on the artefact, some of which are summarised in Table 1.

Figure 1. Example of a calibration chain for a Chromatic Dispersion (CD) Test Set

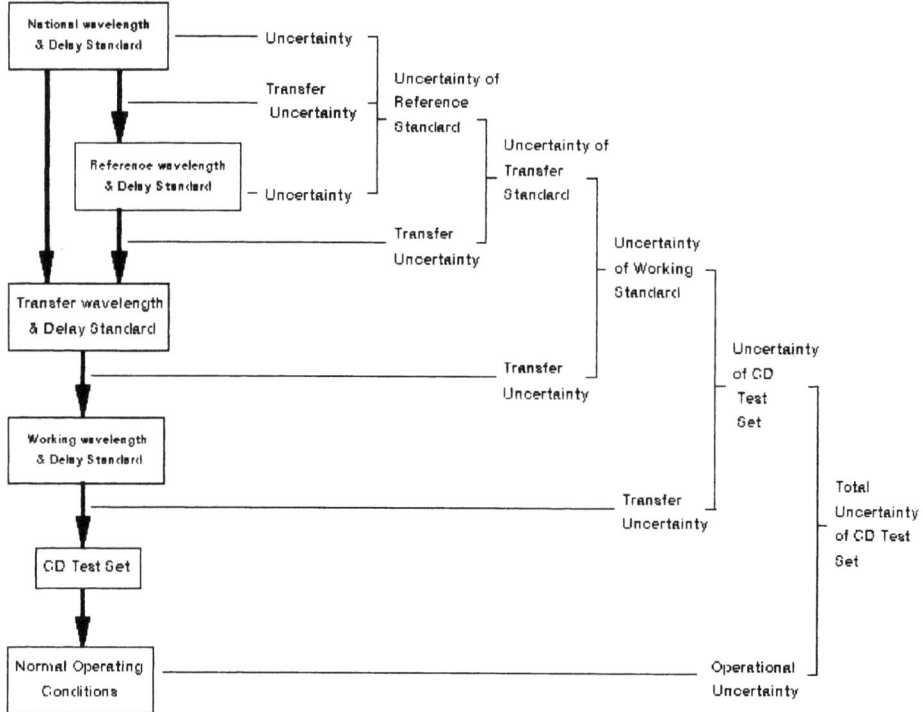

3. Traceability and Accuracy

Of course, the accuracy that the artefact provides is paramount to the performance of the test set to which it is applied, but this can only be achieved alongside traceability to the best possible standards. In general, traceability *per se* is relatively simple to achieve, using an unbroken chain of standards and transfers from the Standards Laboratory to the artefact, such as in Figure 1. Each transfer must have the associated uncertainties accounted for. However, as shown in Figure 1, each time a transfer is made, uncertainty grows, and the impetus is to try to reduce the number of links in the chain. However, by using calibration artefacts as a means of on-site calibration may in fact add a transfer stage to the normal process (eg return-to-factory re-calibrations) and therefore impact accuracy. This problem can be overcome if the normal processes of test set manufacture and calibration include the artefact, with minimal overall number of links. Often, however, this may result in artefact calibrations and other suitable standardisation work being required at the Standards laboratory.

TABLE 1. General Requirements for Calibration Artefacts

* *high accuracy*   - to allow best possible field accuracy of test set

* *traceabilty to given standards*   - required for ISO9000

* *easy interface to the test set*   - fiber connections to the test set

* *high stability*   - re-calibration periods greater than one year

* *portability* (e.g. hand-carry on aircraft)   - preferably solid-state

* *high stability*   - over time, temperature, humidity etc

* *rapid warm up time* (if applicable)

* *high turn-on repeatability*

* *ease of use by normally trained service staff* - automatic software

Overall there has been a trend in recent years to develop artefacts that are part of the internal manufacturing processes for test sets, and simultaneously to obtain direct transfer to the artefact(s) at the Standards laboratory, so that both factory-shipped new sets and re-calibrated existing sets receive exactly the same calibration process. By this means, the number of steps in the calibration chain are reduced, and yet maximum standardisation of test sets world-wide is obtained.

Many of the other requirements listed in Table 1, can be met by suitable optical design, artefact (phenomenon) selection, or engineering, of the actual artefact unit, usually referred to as a "calibrator". What is unique to this application is that the calibrator must be easy to interface to the test set in question. In most cases, this means that the calibrator has to replace the normal fiber presented to the test set in the normal way. This has the advantages that not only is the calibration process speeded up, the instrument remains undisturbed, and moreover is used in its entirety during the calibration procedure, at normal temperature, so reducing the chances for errors. The calibrator unit is therefore often some kind of optical transmission device, which fortunately usually lends itself to easily being made solid-state.

4. Examples of Artefacts

A large variety of artefacts have been proposed for fiber test sets. Some of these are summarised in Table 2. Many of these artefacts are now commercially available and are either being used in 'round robin' work to establish world-wide consensus, or are in fact providing on-site calibration services. It is beyond the scope of this paper to deal with all these in detail, and this paper will cover chromatic dispersion, group index, length and strain since these are very similar measurement instruments, but also PMD for which calibrators are only just emerging.

TABLE 2 : Artefacts proposed for fiber optic test set calibration

TEST SET	ARTEFACT
Power meters	Calibrated light sources
OTDR's	Recirculating delay lines Golden fibers Splice simulators External Sources Fiber attenuators etc
Optical Spectrum Analysers	Light sources Optical filters
Fiber Geometry	chrome on glass artefacts
Chromatic Dispersion	wavelength sources, filters dispersion calibrators delay lines
Length	frequency counters, delay artefacts
Group Index	Standard fiber sample
Strain	frequency counters, delay artefacts, attenuators
PMD	birefringent fibers or wave plates

5. Chromatic Dispersion, Strain, Length and Group Index Artefacts

In general, all these measurements use the propagation delay in the fiber, and in particular the commercial offerings all work using the phase shift method in various forms. A generic diagram of the method is shown in Figure 2. In this method, a broad band source such as an LED is intensity modulated by an oscillator at high frequency. By detecting the arrival phase of the modulation on the light after passing through the fiber delay, the total delay of the fiber can be detected. Some variants of chromatic dispersion set use a differential phase shift method [4] where the difference in delays between two wavelengths is detected by modulating also the wavelength and then using a lock-in amplifier to detect the phase meter ac component, or the differential phase signal {D}. This yields a direct dispersion measurement and considerable immunity to temperature drifts as the wavelength switching occurs at some 200 Hz.

Figure 2: Schematic Diagram of Phase Shift Test Sets for Chromatic Dispersion, Strain, Length and Group Index

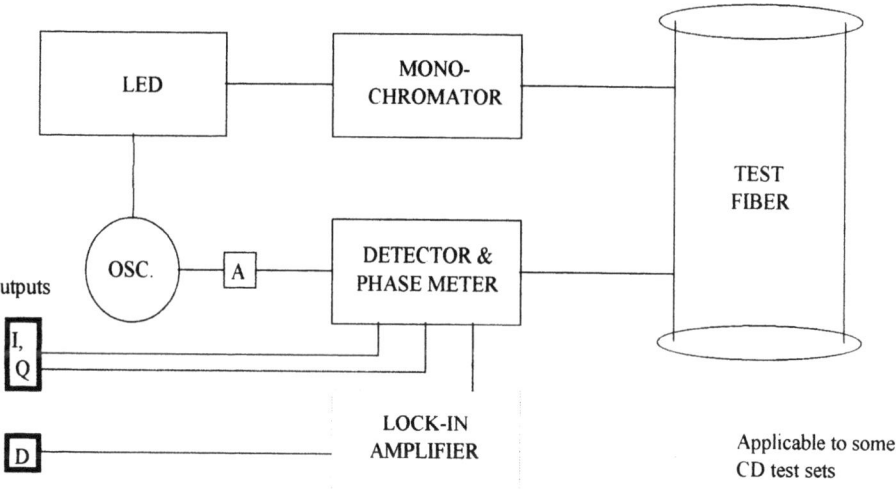

For strain the phase meter only is used, and the monochromator may be omitted. Now total fiber delay is monitored as a function of applied cable strain using a tensile test jig. By relating delay to physical length change, the fiber mechanical strain and therefore its likely fatigue or fracture characteristics may be determined, under given cable loading conditions. The Quadrature Phase Shift principle [5], uses both quadrature channels {I,Q} of the phase meter to simultaneously detect the delay and power change caused by a strain. The delay change is related to the axial strain (elongation) of the fiber, and also the group index and the photo-elastic effect within the fiber activated by axial tension.

Group index is determined by measuring the total delay in a short known length of fiber cut back from a longer spool, and then relating time to length to give group index. Careful selection of the cut back length to be one cycle of the phase meter allows very high accuracy and resolution to be achieved [6].

Fiber length on a spool is measured by relating time delay and the assumed group index value from the spool. In order to measure with high resolution, the oscillator frequency must be made high, giving many cycles of phase shift at this frequency. To determine the integer number of cycles without ambiguity, the oscillator frequency is stepped down to a very low value progressively, each time computing the cycle ambiguities.

From the common elements of these systems, it is easy to see that there are several common parameters that affect calibration accuracy. These are listed in Table 3.

TABLE 3: Common Elements of CD ,Strain, Group Index and Length Calibration

Chromatic Dispersion	Wavelength
	phase meter alignment
	dispersion/ differential phase response
	(oscillator frequency-small error)
Strain	phase meter alignment
	oscillator frequency
	power change
	delay change slope and linearity
	elongation value & stress optic effect [7]
Group Index	phase meter alignment
	oscillator frequency
	delay change linearity
	Group index verification
Length	phase meter alignment
	oscillator frequency
	delay /phase linearity
	(group index is needed)

6. Phase Meter Alignment

An important pre-requisite of the calibration process is that the two phase meter channels (if used) have exact quadrature and matched sensitivities to the input signal. This is necessary for the vector processing to perform correctly. Therefore a periodic adjustment of these two parameters is necessary, achieved using a special accessory consisting of a variable phase shifter is inserted in the phase detection system reference path at point A in Figure 2 to artificially move the relative phase of the signal and reference modulation waveforms. Varying the phase shift enables minute errors in gain and orthogonality to be detected and adjusted. The phase meter device is a simple manually controlled electrically powered unit, which is used at the beginning of any system calibration operation. Access to point A may require instrument covers to be removed, and these must be replaced and the instrument left to stabilise before proceeding with other calibration operations.

7. Oscillator Frequency

The master oscillator frequency defines the scale factor (accuracy) of all the methods, but since dispersion slope must be calibrated separately (see below), it is not necessary to calibrate oscillator frequency alone; it can be included in the overall slope calibration. For all but dispersion, the frequency is calibrated using a portable frequency counter. The accuracy of the frequency counter must be commensurate with the required test set accuracy, but this is typically easily achieved. To carry out the calibration the frequency signal is taken from point A in Figure 2.

8. Wavelength

The IEC document entitled "Procedure for calibration of Single Mode Fiber Chromatic Dispersion Test Sets" [1] currently under development by the TC86 WG4 Dispersion Group details calibration of the wavelength axis using a calibration artifact that is itself traceable to international standards (eg NIST, NPL etc).
In one commercially available artefact, a set of four calibrated interference filters as wavelengths to be used to build up a calibration curve for the monochromator used within the CD test set. Each interference filter is carefully calibrated by the supplier to 0.1nm accuracy and we take great care to ensure that they are mounted so that the optical beam is normal to each filter when in use. These filters are specified to maintain a specified accuracy for a period of five years and have a low temperature coefficient so that the specified accuracy can be achieved at normal ambient temperature without temperature control. A five year recalibration period is typical of these wavelength calibrators.

9. Dispersion/Delay

The calibration of the delay or dispersion response of a CD test set is achieved [1] using a second artifact which contains a variable optical delay line formed by a moving mirror/ lens system as the time delay standard. The delay line delay can be adjusted by the movement of the mirror. Accurately quantified movement of the mirror is achieved using a calibrated micrometer system. By carefully selecting a suitable precision micrometer movement, it is possible to obtain an movement accuracy of a few microns over a travel range of 50 -500mm, with micron resolution. This translates to an overall delay change of several hundred picoseconds, with sub-picosecond resolution. This far exceeds the normal requirements for CD test sets so therefore, due to the high inherent accuracy and quality of such a micrometer, the calibration period of the dispersion calibrator device can easily be of the order of five years. Even though lead screw wear over this period will increase uncertainty, the overall levels of uncertainty will remain well below that needed for CD test set calibration.

10. Total Delay Change

Note that since a strain/ Group index system always measures relative to an initial starting delay value of zero, only relative delay change must be considered. This change must be traced to an NPL or NIST (US) measurement.

The linearity and overall accuracy of the strain measurement can be established using a delay artefact consisting of a series of optical delay lines formed by a series of loops of fiber of calibrated optical lengths (delays). The loops are all contained in a metal box and are connectorised. The loops have nominal lengths of 1.0, 1.2, 1.4, 1.6, 1.8 and 2.0 m in order to cover the full delay range required. The exact time delay of each loop, between connector ferrule end faces is given by an optical measurement traceable to a standards laboratory. Each fiber is presented in turn to the strain measurement system and the strain reading recorded. The next loop of fiber is presented and the process repeated for all the remaining lengths. A linear regression of the recorded delay data points yields a slope of the delay versus loop delay change line which can be compared to unity (the exact value for a calibrated system). This test provides a traceable means of checking the delay change calibration, relying only on the fundamental parameters of fixed fiber lengths, speed of light in fiber, and the modulation frequency, f, whose accuracy has been established using the frequency counter (see above).

11. Elongation Calibration

In order to convert the delay change to an elongation, two new factors are brought into play, namely the group index, $n(\lambda)$, and the stress optic correction factor K. Any uncertainty in these two parameters will cause a corresponding uncertainty in the elongation readings.

The formula used to convert delay change $\tau(\lambda)$, at a wavelength λ, to elongation ΔL and hence to strain ϵ is given by [7]:

$$\epsilon = \Delta L/L = \Delta\tau(\lambda).c \ / \ (n(\lambda).L.K) \qquad \qquad --(1)$$

The determination of the stress optic correction factor K, is described in ref. 7. based on measuring the time delay change produced by a known physical elongation in a sample of the same fiber type used for the strain test. The overall uncertainty achieved using the method is 1% [7]. Group Index is calibrated as described below.

12. Power Change

As mentioned above only a consideration is needed of the relative power change. This can be calibrated by means of a fiber optic attenuator which has been calibrated using a light source and some form of calibrated (traceable) optical power meter. The attenuator calibration curve can then be used as a transfer for calibration of the test set, over the normal range of power change experienced in normal use of the instrument.

13. Group Index Verification

Having calibrated the phase meter and delay calibration, the test set will give accurate group index results. However, for proof of traceability, it is necessary to use a group index standard artifact. This contains a short sample of fiber of known group index, derived in a traceable way from a calibrated test set in the factory.

14. PMD calibration Artefacts

There are several methods for PMD measurement, ranging from interferometric to wavelength scanning techniques. In the case of wavelength scanning methods, the calibration of wavelength can provide the basis for traceable PMD calibration, but there is now considerable activity in traceable PMD artefacts that are applicable to all the methods currently in use. One of the difficulties is that PMD in fibers is a statistical property with a distribution of delays and not a single valued one, due to the mode coupling that occurs with bends and twists etc. Thus both the scale factor of each axis of the distribution display must be calibrated, over their entire range. Many of the proposals for artefacts are based around either simple birefringent wave plates which provide excellent instrument calibration at one delay point, or simulators of the PMD in normal fibers, where a distribution of fixed PMD values is used to build up a more complete distribution to test the instrument display axes. The trick is to simulate the random mode coupling with a sufficient set of deterministic discrete devices, which are not environmentally sensitive. At present work is being carried out on both approaches and suitable artefacts will emerge within a few months.

15. Conclusion

Calibration artefacts can provide valuable on-going support for customers with fiber test equipment of all varieties, as well as being applicable for use in the instrument factory. The artefact is used to transfer calibration to the test set in its normal operating environment, and by being traceable back to standards laboratories, means that the test set calibration is also traceable. Many artefacts have been developed and several are commercially available, which can travel world-wide, providing traceability, accuracy, and long term stability to installed test sets. Some of these artefacts have been reviewed here. By careful attention to the artefact design, extremely high reliability of the devices and re-calibration periods in excess of five years are now commonplace.

Calibration artefacts permit easy conformance to the requirements of ISO9000 calibration procedures in respect of traceability, and as this standard is widely adopted world-wide, the need for a full range of calibration services can be envisaged. Calibration of fiber test sets is an on-going requirement with probable significant market growth likely over the next few years accompanying the adoption of ISO9000. Many improvements to calibrators, the adoption of new phenomena, improved calibration services etc will become available over the next three years.

16. References

1. "Procedure for calibration of Single Mode Fiber Chromatic Dispersion Test Sets", IEC TC86, in preparation.
2. "Calibration of Fiber Optic Power meters" IEC TC86, in draft.
3. "Calibration of Optical Time Domain Reflectometers" IEC TC86, in draft
4. Barlow, A, J. (1987) "Techniques for the Absolute Calibration of Chromatic Dispersion Measuring Instruments", *Proc. SPIE* 841.
5. Barlow, A. J. (1989) "Quadrature Phase |Shift Technique for measurement of Strain, Optical Power transmission and Length in Optical Fibers, *Jnl. Lightwave Technology*, 7, 1264-1269.
6. Barlow, A. J., Ives, D. & Dodd, S. (1991) "Measurement of Group Index using Fiber Cut back", *Proc. OFMC*, York, 16-19.
7. Barlow, A.J., Ficke, W.H., & Voots, T. (1992) "Determination of Stress Optic Effect in Optical Fibers", *Proc. Symposium on Optical Fiber Measurements*, NIST Boulder, 107-109.

MIX
Papier aus verantwortungsvollen Quellen
Paper from responsible sources
FSC® C105338

If you have any concerns about our products,
you can contact us on
ProductSafety@springernature.com

In case Publisher is established outside the EU,
the EU authorized representative is:
**Springer Nature Customer Service Center GmbH
Europaplatz 3, 69115 Heidelberg, Germany**

Printed by Libri Plureos GmbH
in Hamburg, Germany

Trends in Optical Fibre Metrology and Standards

NATO ASI Series

Advanced Science Institutes Series

A Series presenting the results of activities sponsored by the NATO Science Committee, which aims at the dissemination of advanced scientific and technological knowledge, with a view to strengthening links between scientific communities.

The Series is published by an international board of publishers in conjunction with the NATO Scientific Affairs Division

A	Life Sciences	Plenum Publishing Corporation
B	Physics	London and New York
C	Mathematical and Physical Sciences	Kluwer Academic Publishers
D	Behavioural and Social Sciences	Dordrecht, Boston and London
E	Applied Sciences	
F	Computer and Systems Sciences	Springer-Verlag
G	Ecological Sciences	Berlin, Heidelberg, New York, London,
H	Cell Biology	Paris and Tokyo
I	Global Environmental Change	

PARTNERSHIP SUB-SERIES

1.	Disarmament Technologies	Kluwer Academic Publishers
2.	Environment	Springer-Verlag
3.	High Technology	Kluwer Academic Publishers
4.	Science and Technology Policy	Kluwer Academic Publishers
5.	Computer Networking	Kluwer Academic Publishers

The Partnership Sub-Series incorporates activities undertaken in collaboration with NATO's Cooperation Partners, the countries of the CIS and Central and Eastern Europe, in Priority Areas of concern to those countries.

NATO-PCO-DATA BASE

The electronic index to the NATO ASI Series provides full bibliographical references (with keywords and/or abstracts) to more than 30000 contributions from international scientists published in all sections of the NATO ASI Series.
Access to the NATO-PCO-DATA BASE is possible in two ways:

– via online FILE 128 (NATO-PCO-DATA BASE) hosted by ESRIN,
Via Galileo Galilei, I-00044 Frascati, Italy.

– via CD-ROM "NATO-PCO-DATA BASE" with user-friendly retrieval software in English, French and German (© WTV GmbH and DATAWARE Technologies Inc. 1989).

The CD-ROM can be ordered through any member of the Board of Publishers or through NATO-PCO, Overijse, Belgium.

Series E: Applied Sciences - Vol. 285

Trends in Optical Fibre Metrology and Standards

edited by

Olivério D. D. Soares
CETO - Centre de Ciências e Tecnologias Ópticas,
Universidade do Porto,
Porto,
Portugal

Springer-Science+Business Media, B.V.

Proceedings of the NATO Advanced Study Institute on
Trends in Optical Fibre Metrology and Standards
Viana do Castelo, Portugal
June 27–July 8, 1994

Library of Congress Cataloging-in-Publication Data

```
Trends in optical fibre metrology and standards / edited by O.D.D.
Soares.
      p.   cm.  --  (NATO ASI series. Series E, Applied sciences ; vol.
285)
    "Published in cooperation with NATO Scientific and Environmental
Affairs Division.
    Includes index.
    ISBN 978-94-010-4020-4      ISBN 978-94-011-0035-9 (eBook)
    DOI 10.1007/978-94-011-0035-9
    1. Optical fibers--Measurement.  2. Optical fibers--Standards.
I. Soares, O. D. D. (Olivério D. D.)  II. North Atlantic Treaty
Organization.  Scientific and Environmental Affairs Division.
III. Series: NATO ASI series.  Series E, Applied sciences ; no. 285.
TA1800.T73  1995
621.36'92'0287--dc20                                            95-5544
```

ISBN 978-94-010-4020-4

Printed on acid-free paper

All Rights Reserved
© 1995 Springer Science+Business Media Dordrecht
Originally published by Kluwer Academic Publishers in 1995
Softcover reprint of the hardcover 1st edition 1995
No part of the material protected by this copyright notice may be reproduced or utilized in any form or by any means, electronic or mechanical, including photocopying, recording or by any information storage and retrieval system, without written permission from the copyright owner.

TABLE OF CONTENTS

Preface _____ XI

Institute Programme _____ XV

Participants _____ XIX

Sponsors and Co-sponsors _____ XXVII

Committees _____ XXVIII

Opening Message
Ministro da Indústria e Energia
Eng° Luis Mira Amaral _____ XXIX

I INTRODUCTION

1.1 A Historical Survey of Optical Signals and Optical Fibers
Silvério P. Almeida and Olivério D.D. Soares _____ 3

1.2 Trends in Optical Fibre Metrology and Standards
Olivério D.D. Soares and Silvério P. Almeida _____ 19

II OPTICAL FIBRES AND MATERIALS

2.1 Polarised Light Evolution in Optical Fibres
J. Pelayo, F. Villuendas _____ 47

2.2 Nonlinear Optical Fibres
Raman Kashyap _____ 69

2.3 Exotic Fibres
F. Gauthier _____ 103

2.4 Fibers and Fiber Lasers for the Mid-Infrared
U.B. Unrau _____ 113

2.5 Optical Fibres in Nuclear Radiation Environments
F. Berghmans, O. Deparis, S. Coenen, M. Decréton and P. Jucker _____ 131

III OPTICAL FIBRE COMPONENTS

3.1 Connectors and Splices
J. Dubard _____ 159

3.2 Optical Attenuators and Couplers Characterization
C. Le Men _____ 175

3.3 Wavelength Division Multiplexers and the Measurement of the Channel Wavelengths
J.P. Laude _____ 193

3.4 Active Fibre Characterisation with Passive Measurements
J.M.S. Anacleto, Modesto Morais, Gaspar Rego, O.D.D. Soares _____ 209

IV OPTICAL SOURCES

4.1 Wavelength Tunable Laser Diodes and their Applications
M.C. Amann _____ 217

4.2 Active Components Characterization
C. Le Men _____ 241

V OPTICAL FIBRE AMPLIFIERS

5.1 Optical Amplifiers
Ivan Andonovic _____ 263

5.2 Analysis of Erbium-Doped Silica-Fibre Characterisation Techniques
M.A. Rebolledo _____ 305

5.3 Optical Fibre Amplifiers Standardisation
P. Di Vita _____ 327

5.4 Optical Amplification Modelling for Er-Doped Fibre
Joaquim M.S. Anacleto and O.D.D. Soares _____ 343

VI OPTICAL FIBRE CHARACTERISATION, CALIBRATION AND STANDARDS

6.1 Fibre Characterization and Measurements
C. Le Men _____ 353

6.2 Mechanical and Environmental Testing
J. Dubard, P. Fournier, P. Blanchard, F. Gauthier _____ 399

6.3 Fibre Ribbon Measurements
Lauri Oksanen _____ 415

6.4 Calibration Artefacts, Traceability and Accuracy
A.J. Barlow _____ 429

VII Optical Fibre Instrumentation, Measurement and Metrology

7.1 Photodetector Calibration
Antonio Corrons _____ 441

7.2 Power Meter Calibration
Pedro Corredera Guillén _____ 453

7.3 OTDR Calibration
Francis Gauthier and Lionel Ducos _____ 469

7.4 Optical Spectrum Analyzer Calibration
J. Dubard and C. Le Men _____ 489

7.5 EDFA Testing
Eric Malzahn _____ 511

7.6 Real-time Heterodyne Fibre Polarimetry by Means of Jones and Stokes Vector Detection
R. Calvani, R. Caponi, F. Cisternino _____ 527

7.7 Application of Er-Doped Fiber Amplifiers to Optical Measurement Techniques
Haruo Okamura and Katsumi Iwatsuki _____ 541

VIII Optical Communication Systems

8.1 Optical Fiber Communication Technology and System Overview
Ira Jacobs _____ 567

8.2 Modern Optical Communication Systems
Govind P. Agrawal _____ 593

8.3 An Optical Coherent Transmission System Based on Polarization Modulation and Heterodyne Detection at 155Mbit/s Bitrate and 1.55 µm Wavelength
R. Calvani, R. Caponi, F. Delpiano and G. Marone _____ 617

IX OPTICAL SENSORS

9.1 Fiber Sensor Review
D.A. Jackson _____ 629

9.2 Fibre Optic Sensors
M.R.H. Voet, Y. Verbandt, L. Boschmans, H. Thienpont and F. Berghmans _____ 647

9.3 Fiber Optic Sensors for Environmental Monitoring
A.G. Mignani, M. Brenci, A. Mencaglia _____ 691

X FUTURE TRENDS

10.1 Nonlinear Optics: Theory and Applications
U. L. Österberg _____ 711

10.2 Advanced Optical Fibre Optoelectronics
B. Culshaw _____ 727

10.3 Plastic Optical Fibres (POF)
Demetri Kalymnios _____ 747

XI FURTHER POSTERS

11.1 Refractive Index of Planar Microlenses
S. Ríos, E. Acosta, M. Oikawa and K. Iga _____ 773

11.2 Fibre Interferometer Set for Phase and Polarisation Measurement of Passive Fibre Elements
M. Szustakowski, J.L. Jaroszewicz _____ 781

11.3 FMCW-Lidar with Tunable Twin-Guide Laser Diode
A. Dieckman, M.C. Amann _____ 791

11.4 Fibre-Optic White-Light Interferometer Insensitive to Polarization Fading
L.A. Ferreira, J.L. Santos, F. Farahi _____ 803

11.5	QND Measurement of the Photon Number using Kerr Effect in Optical Fibres V. Sochor, I. Paulička	809
11.6	Laser Transitions Characterization by Spectral and Thermal Dependences of the Transient Oscillations in the Fiber Lasers Oleg G. Okhotnikov and José R. Salcedo	813
11.7	Nonlinear M-Lines Spectroscopy of DMOP-PPV Polymer Planar Optical Waveguides: Quasi-Permanent and Fast Electronic Refractive Index Changes Francesco Michelotti	825
11.8	Plastic Optical Fibre (POF) Displacement Sensor N. Ioannides, D. Kalymnios and I.W. Rogers	827
11.9	A New Reflective Plastic Optical Fibre Displacement Transducer S. Hadjiloucas, L.S. Karatzas, D.A. Keating and M.J. Usher	829
11.10	White Light Interferometer Using Linear CCD Array with Signal Processing Stephen R. Taplin, Adrian GH. Podoleanu, David J. Webb and David A. Jackson	831
11.11	Radiation-to-guided Field Coupling in Ti:LiNbO$_3$ Intensity Modulators M. Marciniak, M. Szustakowski	833
11.12	White Light Fiber Optic Interferometry a Technique for Monitoring Tissue Properties Scott Shukes, Faramarz Farahi, M. Yasin Akhtar Raja and Robert Splinter	833
11.13	Holographic Interferometry of Immersed Systems in Electrochemistry K. Habib	834
11.14	Pressure Effects of Optical Frequency in External Cavity Diode Lasers Hannu Talvitie, Johan Åman, Hanne Ludvigsen, Antti Pietiläinen, Leslie Pendrill and Erkki Ikonen	834
11.15	Methods of Measurement for Angle-Polished Optical Fiber Launching Efficients Darrin P. Clement and Ulf Österberg	835

11.16 The Application of a Fiber Optic Gyroscope for Studies of Extinction Ratio of Fiber Optic Polarizer
J.L. Jaroszewicz and M. Szustakowski _____ 835

11.17 Blocks and Architecture for a Multichannel Fibre Optic Correlator
Adrian Gh. Podoleanu, Ryan K. Harding, David A. Jackson _____ 836

11.18 Calibration of Fibre Optical Measurement Equipment at SP
R. Andersson, M. Holmsten and L. Liedquist _____ 836

11.19 Fibre Application in Absolute Ranging Interferometry with Computer Generated Holograms (CGH)
L. Wang, M. Deininger, K. Gerstner, T. Tschudi _____ 837

11.20 Monitoring Through-the-Thickness Behaviour of Composites Using Fiber Optic Sensors
W. E. Wolfe and B. C. Foos _____ 837

11.21 Fibre-Optic Fabry-Perot Sensor for Vibration and Profile Measurements
I. Paulička, V. Sochor, J. Stulpa _____ 838

11.22 Application of OTDR in the Mining Industry
V. Kumar and Dinesh Chandra _____ 838

11.23 Parametric Amplification by Four-Wave Mixing for Distributed Optical Fibre Sensors
J. Zhang, V.A. Handerek and A.J. Rogers _____ 839

11.24 Intracavity Laser Diode Pumped and Modulated Er^{3+} Doped Fiber Lasers
Oleg G. Okhotnikov and José R. Salcedo _____ 839

11.25 Mode-Locking of A 2.7 µm Erbium-Doped Fluoride Fiber Laser
Christian Frerichs _____ 840

AUTHOR INDEX _____

SUBJECT INDEX _____

PREFACE

Fibre Optics has gained prominence in: telecommunications, data transmission and distribution, cable television networks, sensing and control, light probing and instrumentation.

The 1990's shows an increased expansion of optical fibre networks which respond to the rapid growth on a world scale of long distance trunk lines combined with a family of emerging optical based services in which fibre-to-the-home will have the greatest impact. There is already evidence that optical communications are moving toward higher bit-rates, wavelength transparency and irrelevance of signal formats.

The rate of change in fibre optics and the emergence of new services will be a mere consequence of economics. The actual increasing of cost and the demand for high-date-rates or large bandwidth per transmission channels, and the lack of available space in the congested conduits in urban areas, strongly favour the technological change to fibre optics. The recognised advantages of fibre optic technologies and the unchallenged potential to respond to future needs requires the inclusion of fibre optics networking into new installations.

Concomitantly, current progress in the field of optical fibres (optical fibre amplifiers, optical fibre switching, WDM, fibre gratings, etc.) unfold major technical advances and greater flexibility in the designs and engineering of networks, optical fibre components and instrumentation. The explosion of growth in fibre sensors, fibre probes and the myriad of fibre based components shows that we are only using a fraction of optical fibre potential.

The optical fibre market expansion implies a simultaneous accretion of measurement techniques, particularly aimed at the three levels of testing, i.e. laboratory, factory and in the field, to encompass the world character of trade.

The global market and production delocalisation cannot function effectively unless specifications are described by quantities that can be universally measured by common agreed procedures. Interoperability and interconnectivity relies on the development of standards. Standardisation bodies at both national and international levels are active in setting specific standards and reference/alternative testing methods for fibre optics.

The fibre optics world-wide communications market is estimated to reach (USA $) 8,3 billion by the year 1998. The full exploitation of this projected enormous potential of fibre optics in scientific, engineering and the myriad of applications requires a precise characterisation of optical fibres and their associated systems and components, via effective and accurate measurements supported by reliable measuring standards and good practices in measurements and calibrations. In addition the assurance of universal metrological traceability requests permanent intercomparison practices both at vertical and horizontal level.

The ISO 9000 compliance for adequate quality control of products and services already specifies the requirements for the test, measurement and inspection equipment (ISO 10012-1) which stress the calibration imperatives.

Optical fibre metrology had in recent times a significant evolution to respond to the paradigm shift in telecommunications, to the explosion of applications, with the tightening of specifications and exigencies in performance: automation, portability, real-time surveillance and remote mode testing.

Finally, given that 30% of the investment in fibre optic network installations concerns metrology and related activities, there were sufficient reasons to organise the Advance Study Institute (ASI) on "Trends in Optical Fibre Metrology and Standards".

The paramount importance of providing a progressive and comprehensive presentation of current issues and trends concerning the optical fibre metrology and standards was recognised two years ago within EUROMET (European Federation of Metrological Institutes) were the design of the event was enunciated.

The programme of the ASI considered the science and methodology of optical fibre metrology and standards up to the components and instrumentation level. Equipment was also made available for demonstrations and manipulation by the ASI participants.

Despite the inevitable limitations, a broad and reasonably coherent coverage of the field was achieved, with significant contributions from the attendees some of which presented posters (30 in total) supported with written material of high interest, part of which is included in the proceedings.

The topics covered during the eleven working days include: optical fibres and materials (polarisation, non-linear effects, plastic fibres, fibre lasers); optical fibre components (connectors and splicers, WDM); optical sources (tuneable lasers); optical fibre amplifiers (Er-Dopped amplifiers, amplifier standardisation); optical fibre characterisation; calibration and standards (fibre characterisation, ribbon fibres, calibration artefacts); instrumentation (photodetectors, power meters, OTDRs, EDFAs, Er-Dopped fibres in metrology); optical communications systems (fibre communication technology, modern optical systems, coherent transmission systems); optical sensors (fibre sensors, environmental sensors); future trends (fibres evolution, solitons, non-linear effects, fibre optoelectronics).

Lack of time prevented one from dealing with other crucial aspects of optical fibre metrology such as: automation, remote metrologic systems, software validation, safety and legal aspects of metrology, as well as a family of recently developed fibre components.

The profusivity of material could not be matched by further enlargement of the ASI. The ASI was designed for a lecturers staff of around 10 to 15 lecturers assembled from academia and industry to produce cross-fertilisation and providing views from different backgrounds and orientations. Due to the interest of the topics and timeliness of the event we obtained 27 world leading experts who expressed their willingness to participate. The number of attendees was programmed for around 60. The actual number of participants was curtailed to 100 who came from four continents. These were selected, from about the 200 requests received.

The organisation of an ASI of these proportions and which was scheduled over a period of two weeks could only be met by laborious efforts of a responsive secretariat which the editor and the organising committee thanks for their dedication and the excellence of a job well done.

Only the financial assistance provided by the Scientific and Environmental Division of NATO, and of the Measurements and Testing Division of the European Commission could make it possible to engineer the ASI to the proportion it reached. Therefore the funding Institutions are thanked profusely for their support and faith in the project.

The indispensable sponsorship endorsed by the FLAD - Fundação Luso Americana para o Desenvolvimento, JNICT - Junta Nacional de Investigação Científica e Tecnológica, IPQ - Instituto Português da Qualidade, NSF - National Science Foundation (USA) is thanked with high appreciation.

The friendship and continued support given by VIEIRA de CASTRO & Filhos, Lda. is also deeply acknowledged.

The valuable cooperation and support of further co-sponsoring companies and organisations is also thanked with gratitude.

The extensive support and engagement by local Institutions, the Governo Civil of Viana do Castelo, the Camara Municipal of Viana do Castelo, Região de Turismo do Alto Minho, and in particular the enrichment brought by an ennobling and charming cultural and social programme, product of the efforts and dedication of the Local Hosting Committee is thanked with great reconnaissance.

The honouring presence of the Minister for Industry and Energy, as a strong sign of encouragement to the fostering of cooperation with Portuguese Industry and academia, is acknowledge with high appreciation.

It is hoped that researchers, production and field engineers concerned with design, manufacturing, installation and maintenance of networks and systems and instrumentation involved in the area of fibre optics whether in measurements, characterisation of materials, devices, systems, networks, standards or oriented R&D will find the proceedings a tool of direct interest to their professional activities and for the development of new directions of industrial and applied research. Furthermore, the book will serve as an excellent up-to-date source of bibliographical material.

The materialisation of the proceedings is of course, the result of considerable efforts by authors for which they deserve the credit and are cordially thanked and acknowledged for the excellence of their contributions.

The editor is also indebted to Mrs. Nel de Boer of KLUWER Academic Publishers for the devoted work to the publication of the proceedings.

The success of the ASI was the direct result of the excellence of the work of all the active members, lecturers and participants but it is the time ahead that shall demonstrate the relevance of the impulse, the benefits of full endeavour and the strength of the outcome toward the traced challenges.

Olivério D.D. Soares
CETO - Centro de Ciências e Tecnologias Opticas
Universidade do Porto, Portugal
August 1994

INSTITUTE PROGRAMME

TRENDS in OPTICAL FIBRE METROLOGY and STANDARDS
VIANA do CASTELO - Hotel do Parque**** - PORTUGAL
Cultural and Congress Center SANTIAGO da BARRA CASTLE

1st Week	1st Day 27th June-Monday	2nd Day 28th June-Tuesday	3rd Day 29th June-Wednesday	4th Day 30th June-Thursday	5th Day 1st July-Friday	6th Day 2nd July-Saturday	7th Day 3rd July-Sunday
9-10	REGISTRATION	Fibre Optical Metrology & Standards S.P.Almeida / O.Soares	Fibre Characterisation and Measurement C. le Men	Passive Comp. Characterisation C. le Men	Active Comp. Characterisation C. le Men	PAUSE for INFORMAL GROUP DISCUSSIONS	EXCURSION
10-11		Optical Fibre Communication Technology Ira Jacobs	Fibre Characterisation and Measurement C. le Men	Tunable Lasers and Applications M.C. Amann	WDM Technologies and Applications J.P. Laude		
Coffee Break 11.30 - 12.30		Optical Fibre Communication Technology Ira Jacobs	Ribon Fibers Measurements L. Oksanen	Tunable Lasers and Applications M.C. Amann	WDM Technologies and Applications J.P. Laude		
Lunch							
Further Activities	16. OPENING CEREMONY	16. Fibre Characterisation and Measurement C. le Men	15-17 Heterodyne Polarimetric Measurements and Communications R. Calvani		16h30 Optical Amplifiers Ivan Andonovic		
17-18	Modern Optical Communications System G.P. Agrawal	Measurements and Testing Framework Programme 4 P. Salieri (BCR)	Mechanical and Environmental Testing F. Gauthier	Optical Fibre Connectors & Splices F. Gauthier			
Coffee Break							
18½ / 19½	Trends in Optical Fibres S. Mustafa	Fibre Characterisation and Measurement C. le Men	Mechanical and Environmental Testing F. Gauthier	Optical Fibre Connectors & Splices F. Gauthier	19-19h30 Effects Laser RIM on Optical Amplifier Noise Figure I. Jacobs		
19½ / 20½	Welcome	Posters presentation and discussion	Posters presentation and discussion	Posters presentation and discussion	Posters presentation and discussion		
Dinner	Dinner						

TRENDS in OPTICAL FIBRE METROLOGY and STANDARDS
VIANA do CASTELO - Hotel do Parque**** - PORTUGAL
Cultural and Congress Center SANTIAGO da BARRA CASTLE

2nd Week	8th Day 4th July - Monday	9th Day 5th July - Tuesday	10th Day 6th July - Wednesday	11th Day 7th July - Thursday	12th Day 8th July - Friday	13th Day 9th July - Saturday
9-10	Optical Fibre Amplifiers Standardisation P. di Vita	Fibre Sensors Metrology Marc Voet	Photodetector Calibration A. Corrons	Er Doped Fibres Characterisation M. Rebolledo	Polarisation in Fibres J. Pelayo	
10-11	Plastic Fibers K. Kalymnios	OTDR Calibration F. Gauthier	Calibration Artifacts, Traceability and Accuracy A. Barlow	Er Doped Fibres Characterisation M. Rebolledo	Application Non-Linear Effects U. Österberg	
Coffee Break						
11.30 - 12.30	Plastic Fibers K. Kalymnios	OTDR Calibration F. Gauthier	Glass Fibres and Fibre Lasers Mid-IR U.B. Unrau	Polarisation in Fibres J. Pelayo	Application Non-Linear Effects U. Österberg	
Lunch						
Further Activities	16 HP EDFA - Testing Eric Malzahn		16h30 Analog Communications on Fibres I. Jacobs	16 Non-linear Fiber Optics R. Kashyap		
17-18	Fibre Sensors Review D. Jackson	Power Meter Calibration P. Corredera	AFA - Instrumentation H. Okamura		Exotic Fibers F. Gauthier	
Coffee Break						
18½ / 19½	Fibre Sensors Metrology Marc Voet	Optical Spectrum Analyser C. le Men	AFA - Instrumentation H. Okamura	Solitons in Optical Fibres G.P. Agrawal	Advanced Fibre Optoelectronics B. Culshaw	
19½ / 20½	Posters presentation and discussion	Posters presentation and discussion	Posters presentation and discussion	Posters presentation and discussion	CLOSING of SCHOOL	
Dinner				Farwell Gala Dinner		

Participants

Lecturers

AGRAWAL, PROF. GOVIND P.
The Institute of Optics
University of Rochester
206 WalmoBuilding
Rochester NY 14627
USA
Telf: 1-716-275-48 46
Fax: 1-716-244 49 36
E-mail: gpa@optics.rochester.edu

ALMEIDA, PROF. SILVÉRIO P.
University of North Carolina
at Charlotte
Charlotte, N.C. 28223
USA
Telf: 704-547 20 40
Fax: 704-547 31 60

AMANN, PROF. MARKUS-CHRISTIAN
Institute of Technical Electronics
University of Kassel
Heinrich-Plett-Str. 40
34109 Kassel
Germany
Telf: 00-49-561-804 44 85
Fax: 00-49-561-804 41 36

ANDONOVIC, PROF. IVAN
Strathclyde University
Dept. Electronic & Electrical
Royal College Building
204 George Street
Glasgow G1 1XW
Scotland
Telf: 44-41-552 44 00
Fax: 44-41-552 24 87
Telex: 7742 UNSLIB G

BARLOW, DR. ARTHUR
EG&G Fiber Optics
Sorbus House
Mulberry Business Park
Wokingham
Berkshire RG11 2GY
United Kingdom
Telf: 44-734-77 30 03
Fax: 44-734-77 34 93
Telex: 847164

CALVANI, DR. RICARDO
CSELT
Via G. Reins Romoldi, 274
10148 Torino
Italy
Telf: 39-11-22 85 111
Fax: 39-11-22 85 085
Telex: 220539 cselt

CORREDERA, DR. PEDRO G.
Instituto de Optica "Daza de Valdés"
Serrano, 121
28006 Madrid
Spain
Telf: 34-1-561 68 00
Fax: 34-1-564 55 57
Telex: 42182 CSICE E

CORRONS, PROF. ANTONIO
Instituto de Optica (CSIC)
c/ Serrano, 121
28006 Madrid
Spain
Telf: 34-1-561 68 00
Fax: 34-1-564 55 57
Telex: 42182 CSICE E

CULSHAW, PROF. BRIAN
Strathclyde University
Optoelectronics Division
Dept. Electronic & Electrical
Royal College Building
204 George Street
Glasgow G1 1XW
Scotland
Telf: 44-41-552 44 00
Fax: 44-41-552 24 87
Telex: 77472 UNSLIB G

DUBARD, DR. J.
LCIE
33, Av. General Leclerc
F 92266 Fotenay-aux-Roses
France
Telf: 33-1-40-95 60 60
Fax: 33-1-40 95 60 50
Telex: labelec 634 147 F

GAUTHIER, DR. F.
LCIE
33, Av. General Leclerc
F 92266 Fotenay-aux-Roses
France
Telf: 33-1-40-95 60 60
Fax: 33-1-40 95 60 50
Telex: labelec 634 147 F

JACKSON, PROF. DAVID A.
Applied Optics Group
University of Kent
Canterbury
Kent CT2 7NR
United Kingdom
Telf: 44-227-475 423
Fax: 44-227-764 000
E-mail: daj1@ukc.ac.uk

JACOBS, PROF. IRA
Virginia Polyt. Inst Bradlex
Elect. Eng. Dept.
Fiber & Elect. Opt. Div.
Blacksburg, VA 24061
USA
Telf: 1-703-231-56 20
Fax: 1-703-231 33 62
E-mail: ijacobs@vtm1.cc.vt.edu

KALYMNIOS, PROF. DEMETRI
School of Electrom
Univ. North London
166 - 220 Holloway Road
London N7 8DB
UK
Telf: 44-71-753 51 26
Fax: 44-71-753 7002

KASHYAP, DR. RAMAN
BT Laboratories
DTN12, B55/131
Marthesham Heath
Ipsswich, IP5 7RE
United Kingdom
Telf: 44-473-64 53 63
Fax: 44-473-64 68 85

LAUDE, DR. JEAN PIERRE
ISA / JOBIN - YVON
16-18, Rue du Canal
91163 Longjumeau Cedex
France
Telf: 33-1-64 54 13 00
Fax: 33-1-69 09 07 21
Telex: 602882 F

MALZAHN, DR. ERIC
Hewlett Packard
Herrenberger Strasse 130
71034 Böblingen
Germany
Telf: 49-7031-14 4662
Fax: 49-7031-14 7023
Telex: labelec 634 147 F
E-mail: Eric_Malzahn@hpgrmy desk.hp.com

MEN, DR. C. LE
LCIE
33, Av. General Leclerc
F 92266 Fotenay-aux-Roses
France
Telf: 33-1-40-95 62 50
Fax: 33-1-40 95 60 50
Telex: labelec 634 147 F
E-mail: flyer@email.teaser.com

MUSTAFA, DR. SYED AHMED
CORNING
Telecommunication Products
Corning NY 14831
USA
Telf: 1-607-974 48 02
Fax: 1-607-974 70 41
E-mail:

OKAMURA, DR. HARUO
NTT
Transmission Systems Labs
1-2356 Take, Yokosuka-shi
Kanagawa 238-03
Japan
Telf: 81-468 59 32 19
Fax: 81-468 59 33 96
E-mail: okamura@nttsd.ntt.jp

OKSANEN, DR. LAURI
Nokia Cables
PO Box 77, Virkatie 8
01511 Vantaa
Finland
Telf: 358-0-68 251
Fax: 358-0-870 13 59

ÖSTERBERG, PROF. ULF
Thayer School of Engineering
Dartmouth College
8000 Cummings Hall
Hanover
NH 03755-8000
USA
Telf: 1-603-646-34 86
Fax: 1-603-646 38 56
E-mail:
Ulf.Österberg@dartmouth.edu

PELAYO, PROF. J.
University of Zaragoza
Ciudad Universitaria
50009 Zaragoza
Spain
Telf: 34-76-56 07 35
Fax: 34-76-56 52 00

REBOLLEDO, PROF. J.
University of Zaragoza
Ciudad Universitaria
50009 Zaragoza
Spain
Telf: 34-76-56 07 35
Fax: 34-76-56 52 00
E-mail: rebolledo@msf.unizar.es

SOARES, PROF. OLIVÉRIO D.D.
CETO - Centro de Ciências e Tecnologias Opticas
Fac. Ciências - Univ. do Porto
Praça Gomes Teixeira
4000 Porto
Portugal
Telf: 351-2-310 290
Fax: 351-2-319 267
E-mail: ceto@fis1.fc.up.pt

UNRAU, PROF. ING. U.
Institut für Hochfrequenztechnik
Technische Universität
Braunschweig
PO Box 3329
D-38023 Braunschweig
Germany
Telf: 49-531-391 24 58
Fax: 49-531-391 58 41
Telex: 952526 tubsw-d

VITA, DR. PIETRO DI
CSELT
Via G. Reins Romoldi, 274
10148 Torino
Italy
Telf: 39-11-22 85 278
Fax: 39-11-22 85 520
Telex: 220539 cselt i

VOET, DR. MARC R.H.
Identity E.E.I.G.
Inspralaan 75
B-2400 Mol
Belgium
Telf: 32-14-58 11 91
Fax: 32-14-59 15 14

Attendees

Anacleto, Dr. Joaquim
CETO
Faculdade de Ciências
Universidade do Porto
Praça Gomes Teixeira
4000 Porto - Portugal
Telf: 351-2-310290
Fax: 351-2-319267
E-mail: ceto@fis1.fc.up.pt

Araújo, Dr. Francisco M.C.
INESC
Rua José Falcão, 110
4000 Porto - Portugal
Telf: 351-2-2094301
Fax: 351-2-2008487
Telex: 23023 INESC P

Bao Varela, Dr. Maria del Carmen
Lab. de Óptica - Dept. Física Aplicada
Facultade de Fisica
Campus Universitario
15706 Santiago de Compostela
Spain
Telf: 34-81-521984
Fax: 34-81-21984

Barcelos, Dr. Sérgio
University of Southampton
Optoelectronics Research Group
Southampton SO17 1BJ
UK
Telf: 44-703-59 31 72
Fax: 44-703-59 31 49
E-mail: s.barcelos@ieee.org

Beguin, Dr. Claude
Swiss Telecom
PTTResearch & Development
Fibre Optic Metrology FE 144
Ostermundigenstrasse 93
CH - 3000 Bern 29
Switzerland
Telf: 41-31-338 31 86
Fax: 41-31-338 57 47
Telex: 911 031 vpttch

Berghmans, Dr. Francis
SCK.CEN, DEPT BR3
Boeretang 200 - B-2400 MOL
Belgium
Telf: 32-14 33 26 37
Fax: 32-14 31 19 93
E-mail: fberghma@vnet3.vub.ac.be

Boccardi, Dr. Pasquale
Via LVII Strada a denominarsi, 6
70059 Trani
Italy
Telf: 39-883 44 969

Brinkmann, Dr. Sven
Hindenburgstr. 14
D-91054 Erlangen
Germany

Cervasio, Dr. Alberto
S.V. Giagumona, 90
07040 (Ottava) - Sassari
Italy
Telf: 39-79-390709

Clement, Dr. Darrin P.
Thayer School of Engineering
Dartmouth College
8000 Cummings Hall
Hanover, New Hampshire 03755
USA
Telf: 1-603-646 1466
Fax: 1-603-646 3856
E-mail:darrin.clement@dartmouth.edu

Cloninger, Dr. Todd L.
1725 Ratchford Drive
Dallas, N.C. 28034-9555
USA
Telf: 1-704-547 25 23
Fax: 1-704-547 31 60

Cortes, Dr. Santiago David Armando Reyes
Universidade da Beira Interior
Departamento de Física
R. Marquês d'Ávila e Bolama
6200 Covilhã - Portugal
Telf: 351-75-25141/4
Fax: 351-75-26198
Telex: 53733 UBI P

Costa, Dr. Manuel Filipe P.C.
Universidade do Minho
Dept. de Fisica
Largo do Paço
4719 Braga Codex - Portugal
Telf: 351-53-60 43 27
Fax: 351-53-60 43 39

Cui, Dr. Guoqi
AILUN
Via della Resistenza, 39
08-100 Nuoro - Italy
Telf: 39-784-203409
Fax: 39-784-203158

Dias, Dr. Ireneu Manuel Silva
INESC
Rua José Falcão, 110
4000 Porto
Portugal
Telf: 351-2-2094300
Fax: 351-2-2008487
Telex: 23023 INESC P

Dieckmann, Dr. Andreas
SIEMENS AG
ZFE ST KM 42
Oho-Hahn-Ring 6
81739 Munich - Germany
Telf: 49-89-63644486
Fax: 49-89-6363832

Dopazo, Dr. Juan Félix Roman
Begonias 31 - 1° D
La Coruna
Spain
Telf: 34-81-293311 / 256410

Ducos, Dr. Lionel
LCIE
33, Av. General Leclerc
F 92266 Fotenay-aux-Roses
France
Telf: 33-1-40 95 62 50
Fax: 33-1-40 95 60 50

Ferreira, Dr. Ana Cristina Bento
TELECOM
Av. Fontes Pereira de Melo, 40
1089 Lisboa Codex - Portugal
Telf: 351-1-540020
Fax: 351-1-523614

Ferreira, Dr. João M.C.
INESC
Rua José Falcão, 110
4000 Porto - Portugal
Telf: 351-2-209 4000
Fax: 351-2-318 692
Telex: 23023 INESC P

Ferreira, Dr. Luis Alberto
INESC
Rua José Falcão, 110
4000 Porto - Portugal
Telf: 351-2-2094300
Fax: 351-2-2008487
Telex: 230023 INESC P

Figueira, Dr. Ana Rita L. C.
CETO
Faculdade de Ciências
Universidade do Porto
Praça Gomes Teixeira
4000 Porto - Portugal
Telf: 351-2-310290
Fax: 351-2-319267
E-mail: ceto@fis1.fc.up.pt

Fiore, Dr. Marina
C.so L. Kossuth, 71
10132 Torino
Italy

Fonseca, Dr. António Fernando Figueiredo
TELECOM
Av. Fontes Pereira de Melo, 40
1089 Lisboa Codex - Portugal
Telf: 351-1-3504094
Fax: 351-1-523614

Foos, Dr. Bryan C.
Research Scientist
WJ/FIVEC Bldg 45
Wright-Patterson Air Force Base
334 Aberdeen Avenue
Oakwood - OHIO
USA
Telf: 1-513-2553021
Fax: 1-513-2551633
E-mail: foosbc@fivmailgw.flight.wpafb-af.mil

Frerichs, Dr. Christian
Institut für Hochfrequenztechnik
Technishe Universität Braunschweig
PO Box 3329
38023 Braunschweig - Germany
Telf: 49-531-391 24 23
Fax: 49-531-391 58 41
E-mail: C.Frerichs@tu.bs.de

Fuchs, Dr. Richard
European Patent Office
Gitschinerstrasse 103
Berlin D-10958 - Germany
Telf: 49-30-25901-201
Fax: 49-30-25901-845

Gray, Dr. George R.
University of Utah
Dept. of EE
3280 Merril Eng. Bldg.
Salt Lake City, UT 84112
USA
Telf: 1-801-585 6157
Fax: 1-801-581 5281
E-mail: gray@ee.utah.edu

Guerrero, Dr. Hèctor
Dept. Optica
Facultad de Ciencias Fisicas
Universidade Complutense
Ciudad Universitaria s/n
28040 Madrid - Spain
Telf: 34-1-394 4402/3
Fax: 34-1-394 4683
Telex:47273 FFUC
E-mail: w653@emducm11

Habib, Dr. Khalid
Materials Application Department
KISR
PO BOX 24885 SAFAT
13109 Kuwait
Kuwait
Telf: 965-4830-432
Fax: 965-57237-19

Ioannides, Dr. Nicos
109 Uxbridge Road
Hanwell, London W7 3ST
United Kingdom
Telf: 44-71-607 27 89 / 607 25 96
Fax: 44-71-753 70 02

Iodice, Dr. Mario
IRECE - CNR
Via Diocleziano, 328
80124 Napoli
Italy
Telf: 39-81-5707999
Fax: 39-81-5705734
E-mail: irece::iodice
E-mail: iodice@dimes.tudelft.nl

Karafolas, Dr. Nikos
University of Strathclyde
Optoelectronics Group
204 George Street
G1 1XW Glasgow
Scotland
Telf: 44-41-5524400
Fax: 44-41-5522487

Karoutis, Dr. Athanase D.
Univ. Creete, School of Health Sciences
Dept. Medicine, Sector Neurology
and Sense Organs
PO BOX 1352
711 10 Heraklion Crete - Greece
Telf: 30-81-54 20 70
Fax: 30-81-54 21 16

Konstantaki, Dr. Maria
Strathclyde University
Dept. E.E. Engineering
Royal College Building
204 George Street
Glasgow G1 1XW
Scotland
Telf: 44-41-552400
Fax: 44-41-5522487
E-mail: m.konstantaki@uk.ac.strath

Kumar, Dr. Virendra
Dept. of Electronics
and Instrumentation
Indian School of Mines
Dhanbad 826 004
India
Telf: 326-822273
Fax: 326-832 040
Telex: 0629-214

Lago, Dr. Maria Elena López
E.U. Optica
Univ. Santiago de Compostela
Campus Sur
15706 Santiago de Compostela
Galicia - Spain
Telf: 34-81-52 19 84 / 56 31 00
Fax: 34-81-52 19 84

Liedquist, Dr. Leif
Physical Measurements
Physics & Electronics
Sveriges Provnings-och
Forskningsinstitut
Statens Provningsanstalt
BOX 857 S-501 15 BORAS
Sweden
Telf: 46-33 16 50 00
Fax: 46 33 13 83 81

Mangia, Dr. Maria
Via Iglesias, n.9
08-100 Nuoro
Italy
Telf: 39-784-204187 / 202024

Marciniak, Dr. Marian
Military Academy of
Telecommunications
05-131 Zegrze 300/11
Poland
Telf: 48-2-6883536
Fax: 48-2-6883413
E-mail:wsowl.frodo.nask.org.pl

Martins, Dr. Maria Raquel V.
CETO
Faculdade de Ciências
Universidade do Porto
Praça Gomes Teixeira
4000 Porto - Portugal
Telf: 351-2-310290
Fax: 351-2-319267
E-mail: ceto@fis1.fc.up.pt

Marttila, Dr. Pekka
Nokia Cables
PO BOX 77
01511 - VANTAA
Finland
Telf: 358-0-68251
Fax: 358-0-8701359
Telex: 123475

Michelotti, Dr. Francesco
Univ. di Roma "La Sapienza"
Dept. di Energetica
Via A. Scarpa, 16
I-00161 Roma - Italy
Telf: 39-6-49 76 65 62
Fax: 39-6-44 24 01 83
E-mail: bertol88@itcaspur.caspur.it

Mignani, Dr. Anna Grazia
IROE - CNR
Via Panciatichi 64
50127 Firenze
Italy
Telf: 39-55-423 1 / 4235262
Fax: 39-55-4379569

Morais, Dr. Modesto Cerqueira
IEP
Rua de S. Gens, 3717
Senhora da Hora
4450 Matosinhos - Portugal
Telf: 351-2-9529675
Fax: 351-2-9530594

Navarrete, Dr. Mª Cruz
Universidade Complutense de Madrid
Dept. Optica - Fac. Ciencias Fisicas
Ciudad Universitaria s/n
28040 Madrid - Spain
Telf: 34-1-39 44403
Fax: 34-13944683

Okhotnikov, Dr. Oleg G.
INESC
Rua José Falcão, 110
4000 Porto - Portugal
Telf: 351-2-2094300
Fax: 351-2-2008487
Telex: 230023 INESC P

Olszak, Dr. Artur
Warsaw University of Technology
Optical Engineering Division
Dept. Precision Mechanics
8 Chodkiewicza St.
02-525 Warsaw
Poland
Telf: 48-22-499871
Fax: 48-22-490 392
E-mail: olszak@mp.pw.edu.pl

Otero, Dr. José Lazaro Gato
E.U. Optica
Dept. Fisica Aplicada
Facultad de Fisica
Univ. Santiago de Compostela
Campus Universitario
15706 Santiago de Compostela
Galicia - Spain
Telf: 34-81-52 19 84
Fax: 34-81-52 19 84

Pervan, Dr. Ogus
Turkish Atomic Energy
Authority
Ankara Nuclear Research and
Training Center
Saray, 06105 - Ankara
Turkey
Telf: 312-8154300/8154390
Fax: 312-8154307

Pieraccini, Dr. Massimiliano
IROE - CNR
Via Panciatichi 64
50127 Firenze - Italy
Telf: 39-55-4235214
Fax: 39-55-4223889
E-mail: brenci@iroe.iroe.fi.cnr.it

Podoleanu, Dr. Adrian
University of Kent
at Canterbury
Physics Laboratory
Kent CT2 7NR
United Kingdom
Telf: 44-227-764000
Fax: 44-227-475423
E-mail: physics-lab@ukc.ac.uk

Pousa, Dr. José Marcelino
CET
Rua Engº José Ferreira Pinto Basto
3800 Aveiro - Portugal
Telf: 351-381831
Fax: 351-24723
Telex: 37371

Ravet, Dr. Fabien
Faculte Polytechnique de Mons
Electromagnétisme et
Télécommunications
Boulevard Dolez, 31
7000 Mons
Belgium
Telf: 32-65-374191
Fax: 32-65-37 41 99
E-mail: ravet@fpms.fpms.ac.be

Rego, Dr. Gaspar Mendes do
Fiopos - Barroselas
4905 Barroselas - Portugal
Telf: 058-972215

Ribeiro, Dr. A.B. Lobo
INESC
Rua José Falcão, 110
4000 Porto - Portugal
Telf: 351-2-2094300
Fax: 351-2-2008487
Telex: 230023 INESC P

Rodriguez, Dr. Jorge
U.P.C. ETSI
Campus Nord UPC
Edifici D-3
C/ Sor Eulàlia Anzizu s/n
E - 08034 Barcelona - Spain
Telf: 34-3-401 6795
Fax: 34-3-401 7232

Rodriguez, Dr. Susana Rios
E.U. Optica e Optometria
Dept. Fisica Aplicada
Facultad de Fisica
Univ. Santiago de Compostela
Campus Sur
15706 Santiago de Compostela
Galicia - Spain
Telf: 34-81-52 19 84
Fax: 34-81-52 19 84

Romolini, Dr. Andrea
IROE - CNR
Via Panciatichi 64
50127 Firenze - Italy
Telf: 39-55-4235233
Fax: 39-55-4223889
E-mail: falciai@iroe.iroe.fi.cnr.it

Santos, Dr. Fernando M. Ferreira
INEGI
Fac. Engenharia do Porto
Rua dos Bragas
4099 Porto Codex - Portugal
Telf: 351-2-2007505

Santos, Dr. José Luis C.
INESC
Rua José Falcão, 110
4000 Porto - Portugal
Telf: 351-2-2094300
Fax: 351-2-2008487
Telex: 230023 INESC P

Sartori, Dr. Giovanni
AILUN
Via della Resistenza, 39
08-100 Nuoro - Italy
Telf: 39-784-20 34 09
Fax: 39-784-20 31 58

Seker, Dr. Selim
Bogaziçi University
Dept. of Electrical-Electronic
P.K.2 80815 Bebek
Istanbul - Turkey
Telf: 212-2631540
Fax: 212-2575030

Serra, Dr. Giovanni
Via F.M.Brundu, 1
07040 Codrongianus (SS)
Italy
Telf: 39-79-435203

Sharer, Dr. Deborah
103 South Fork Road
Indian Trail - NC 28079
USA
Telf: 1-704-547 23 02

Shukes, Dr. Scott
1409 Richmond Place
Charlotte, N.C. 28209
USA
Telf: 1-704-527 0821

Sillas, Dr. Hadjiloucas
University of Reading
Dept. Cybernetics - Instrum.
Measurement Research Group
Whiteknights,
PO BOX 225
Reading - Berks RG6 2AY
UK
Telf: 44-734-318219
Fax: 44-734-318220
E-mail: shrhadji@uk.ac.reading

Sochor, Dr. Vaclav
Fac. of Nucl. and Physics Eng.
Csech Techn. Univ.
V Holesovickach 2
180 00 Prague 8
Czech Republic
Telf: 422-85762285
Fax: 422-66414818
E-mail: sochor@troja.fjfi.cvut.cz

Sousa, Dr. Fernando J. Pelicano
Portugal TELECOM
DCRS / RGT
Rua Tomás Ribeiro, 2, 3° DCRS
1000 Lisboa - Portugal
Telf: 351-1-540020
Fax: 351-1-526110
Telex: 60256 DIRS P

Sporea, Dr. Dan G.
Bucharest
P.O. BOX 31-53
Romania
Fax: 40-1-312 11 54
E-mail: sporea@roifa - Bitnet
sporea@ifa.ro - Internet

Szustakowski, Dr. Mieezislaw
Institut of Technical Physics
2 Kahskiego St.
01-489 Warsaw
Poland
Telf: 48-22-36 93 53
Fax: 48-22-36 22 54
Telex:812 535 WAT pl

Talvitie, Dr. Hannu
Helsinki Univ. of Technology
Fac. Electrical Engineering
Otakaari 5 A
SF - 02 150 Espoo
Finland
Telf: 358-0-451 23 36
Fax: 358-0-460 224
E-mail: hannu.talvitie@hut.fi

Taplin, Dr. Stephen R.
University of Kent
Room 116 - Physics Dept.
Applied Optics Group
Canterbury - Kent CT2 7NR
United Kingdom
Telf: 44-227-764000
Fax: 44-227-475423
E-mail: srt1@ukc.ac.uk

Teixeira, Dr. José Manuel Feliz
CETO
Faculdade de Ciências
Universidade do Porto
Praça Gomes Teixeira
4000 Porto - Portugal
Telf: 351-2-310290
Fax: 351-2-319267
E-mail: ceto@fis1.fc.up.pt

Traian, Dr. Dumitrica
Rudului 129
Ploiesti 2000
Romania
Telf: 40-44-140371

Varga, Dr. Jozsef
PKI Telecommunications Inst.
Budapest IX
Zombori u. 1
H-1456 P.O.B. 2 - Hungary
Telf: 36-1-147 1560
Fax: 36-1-127 5075

Velasco, Dr. Mª Lourdes Pedraza
Escuela Universitária de
Enfermeria Fisioterapia Y Podologia
3es Piso Fac. De Medicina
Univ. Complutense de Madrid
Ciudad Universitaria
Madrid 28040 - Spain
Telf: 34-1-3941525
Fax: 34-1-3941539

Vinhais, Dr. Carlos Alberto A.
CETO
Faculdade de Ciências
Universidade do Porto
Praça Gomes Teixeira
4000 Porto - Portugal
Telf: 351-2-310290
Fax: 351-2-319267
E-mail: ceto@fis1.fc.up.pt

Vorropoulos, Dr. G.
European Patent Office
Gitschinerstrasse 103
Berlin
D-10958 - Germany
Telf: 49-30-25901614
Fax: 49-30-25901845

Wang, Dr. Lingli
Institute of Applied Physics
Technishe Hochschule
Darmstadt, Hochschulstr. 6
D-6100 Darmstadt - Germany
Telf: 49-6151-16 30 17
Fax: 49-6151-16 41 23
E-mail: lingli@trudel.iap.physik.th-darmst

Wolfe, Dr. William
Ohio State University
470 Hitchcock Hall
2070 Neil Avenue
Columbus, OH 43210-1275
USA
Telf: 1-614-292 7338
Fax: 1-614-292 3780
E-mail: wolfe.10@osn.edu

Zhang, Dr. Jian
Dept. Electronic Electrical
Engineering
Kings College of London
Strand
London WC2R 2LS
UK
Telf: 44-71-873 2371
Fax: 44-71-836 4781
E-mail: zdee486@bay.cc.kcl.ac.uk

SPONSORS

Scientific and Environmental Affairs Division - NATO
Commission of the European Union (M &T Division)
CETO - Centro de Ciências e Tecnologias Opticas
The European Institute for Advanced Studies in Optics
FLAD - Fundação Luso-Americana para o Desenvolvimento
JNICT - Junta Nacional de Investigação Científica e Tecnológica
IPQ - Instituto Português da Qualidade
NSF - National Science Foundation
Governo Civil de Viana do Castelo
Câmara Municipal de Viana do Castelo
RTAM - Região de Turismo do Alto Minho
IEP - Instituto Electrotécnico Português
MARCONI - Companhia Portuguesa Rádio Marconi, SA
IVP - Instituto do Vinho do Porto
Vieira de Castro & Filhos, Lda.
CLUB TOUR - Agência de Viagens

CO-SPONSORS

EOS - European Optical Society
HP - Hewlett Packard Portugal
DECADA - Equipamentos de Electrónica e Científicos, S.A.
BA - Fábrica de Vidros Barbosa & Almeida, SA
Comissão de Viticultura da Região dos Vinhos Verdes
CANON - Copicanola
Hotel do Parque - Viana do Castelo
M. T. Brandão, Lda.
Câmara Municipal de Ponte da Barca
PERCON-Computadores
Adega Cooperativa de Monção
BCI - Banco de Comércio e Indústria
SOPETE - Casino da Póvoa de Varzim

COMMITTEES

Organising Committee

PROF. OLIVÉRIO D.D. SOARES (Chairman)
CETO - CENTRO DE CIÊNCIAS E TECNOLOGIAS OPTICAS
Lab. Fisica - Fac. Ciências - Univ. Porto
Praça Gomes Teixeira
4000 Porto - Portugal

PROF. S. P. ALMEIDA
Department of Physics
University of North Carolina at Charlotte
Charlotte, N.C. 28223
USA

DR. RICARDO CALVANI
CSELT
Via G. Reins Romoldi, 274
10148 Torino
Italy

Local Hosting Committee

Mr. Roleira Meirinho
Governo Civil de Viana do Castelo

Dr. Defensor Moura
Câmara Municipal de Viana do Castelo

Dr. Francisco Sampaio
RTAM - Região de Turismo do Alto Minho

Secretariat and Technical Committee

José Sousa Fernandes
Fernanda Campos
Luis Vilaça

O
MINISTRO da INDÚSTRIA
e
ENERGIA

Excelentíssimo Senhor Governador Civil de Viana do Castelo

Excelentíssimo Senhor Presidente da Câmara Municipal de Viana do Castelo

Excelentíssimo Senhor Presidente da Região de Turismo do Alto Minho

Excelentíssimo Senhor Presidente do CETO - Centro de Ciências e Tecnologias Opticas

Ladies and Gentlemen

The time is now upon us to leave the past and join the challenge of the future before it becomes too late. As individuals, as enterprises and as a country, we must continuously upgrade our educational and vocational training, replace obsolete equipment, and develop effective management methods. Portugal is now on the brink of great opportunities to modernise itself and to prepare for the 21st Century. It must not after in its pursuit of the challenges which lie ahead.

Recent advances in telecommunications, information processing and electronic industries represent the driving forces behind the present day industrial revolution. In order to keep abreast of these rapidly developing areas, Portugal must develop new and stronger partnerships between its industries and universities. We must take advantage of the position universities are in to educate and train the much needed students and industrial personnel in areas of research and applications which are vital to the future of the country's development. The Portuguese Ministry for Industry and Energy recognises these needs and has taken a deep interest in the Advanced Institute on "Trends in Optical Fibre Metrology and Standards".

The Advanced Institute on "Trends in Optical Fibre Metrology and Standards" provides a thorough coverage of fibre technology and an insight into opportunities offered to various industries. The theme of the Advanced Institute "Optical Fibre" is central to industries associated with communication, data transfer, sensors, instrumentation; and is a key element in many optoelectronic components.

The telecommunication network represents one of the largest human systems ever developed in the world. It is also one of the most complex in that it handles in excess of 1,000 billion calls each year. This, in addition to, an immense amount of transmitted data. Its very existence represents an important element of personal freedom, progress and economic globalization. Every nations' defence, trade, commerce, industry and social practices are inextricably dependent on the extent and quality of their communication and data transfer networking.

Optical networks, namely, optical fibre based links, offer the potential of gigabit switched bandwidth. Recently developed commercial optical amplifiers provide yet another enormous step forward in transmission and data processing thereby lowering the cost of both terrestrial and submarine transmission. Other important industries which make use of fibre optics include: aerospace, computing, broadcasting, consumer electronics and instrumentation. Also, a diverse range of activities broadly defined as the information technology sector make use of fibre optics.

European market experts project for Portugal, an expanded use of about 100,00 km of fibre optics cabling by the year 1995. The estimated cost for this expansion is of the order of 21 billion dollars. A fraction of 30% of the installation costs goes towards characterisation and metrological uses.

The Portuguese Ministry for Industry and Energy has recognised the need to modernise and support the electronics and information technology industries. In particular, this has recently led to the implementation of an integrated program for information and electronics technologies - PITIE, whose main goal is the setting up of a motivating environment for the development of that industry. Also the strategic program for the dynamic development of the Portuguese industry - PEDIP I was set up to help integrate industrial and entrepeneurship linkages.

The Ministry for Industry and Energy, through PEDIP I, has supported technological innovations in three portuguese optical fibre cable manufacturers. It has provided adequate funding for the instalment of optical fibre metrological facilities both in the manufacturing industries and at related technological centers and laboratories. In addition it has supported, under the supervision of the Portuguese Institute for Quality - IPQ, the Laboratório Primário de Optica - LPO, as part of the Central Laboratory for Metrology - LCM. The LPO which is currently under construction at Vila da Feira, is part of the basic infrastructure for optical metrology. It will provide calibrations, the realisation and dissemination of fibre optical standards. At international level, it will participate in metrological inter-comparisons. In particular, within the Federation of the European National Metrological Institutes - EUROMET and the Bureau Communitaire de Reference - BCR, for adequate traceability. The LCM and CETO - Centro de Ciências e Tecnologias Opticas will participate in the scientific and technological programmes of LPO. They will also provide the necessary interface to the industrial users.

Recently, a new strategical program, PEDIP II was formed. PEDIP-II was developed to insure a continued modernisation and strengthening of the portuguese industry. It will lend support to the entrepreneurs' goal to guarantee quality control and sustain its competitiveness within the global economy. In addition, it will support efforts, as those of CETO, to motivate portuguese companies to develop innovative technologies and manufacture of high-tech products and to promote the establishment of foreign companies in Portugal in the field of optoelectronics and related areas.

In conclusion, the Ministry for Industry and Energy welcomes the Advanced Institute on "Trends in Optical Fibre Metrology and Standards" for its foreseeable contribution in promoting contacts and ultimately the internationalisation of related industrial activities, and for stimulating innovating ideas in the latest technologies.

Personally, I would like to welcome all the participants and wish them a very pleasant stay and fruitful outcome.

Engº Luis Mira Amaral
VIANA do CASTELO
27 de Junho 1994

VII Optical Fibre Instrumentation, Measurement and Metrology

PHOTODETECTOR CALIBRATION

ANTONIO CORRONS
Instituto de Optica (C.S.I.C.)
Serrano 121, 28006 MADRID. SPAIN

1. Introduction

Detection is any process in which incident optical radiation produces a measurable physical effect. Detectors are instruments that convert the radiant energy in a measurable signal, generally an electrical signal.

In most cases it is not enough to detect the radiation, as in the case of photoelectric cells used in security systems or in automatization, but it is necessary to know the quantity of incident photons or any other characteristic of the incident radiation. In this case the detector is a measurement instrument. To use it properly, it is necessary to know the accuracy and precision of the measurements, and consequently to study the detector characteristics and to calibrate it in the adequate units.

In this paper there is a short description of the detectors fundamental characteristics and the most common calibration techniques.

2. Photodetectors Fundamental Characteristics

Detectors fundamental characteristics are:
- Linearity
- Stability
- Spectral Responsivity
- Spatial Responsivity
- Noise
- Detectivity
- Dark current

- Time constant
- Temperature dependence

2.1. LINEARITY

A detector is linear when its response I, is proportional to the signal E it receives into a magnitud range in which the detector is usefull.
$I = K.E + I_0$. I_0 is the dark signal.

The linearity range of all detectors has two limits. It should be $I = 0$ when $E = 0$, but this is not the normal case, because there is a dark current, I_0, or signal with no irradiation. In the other side all the detectors reach a saturation current, I_s, from which if the irradiance E is increased, the signal I does not increase.

The more usual methods to measure the linearity of a detector are:

1) Method based on the inverse Square Law

This method is based on $E = I/r^2$, which states that the irradiance E caused by a point source is inversely proportional to the square of the distance r. To test the linearity, the irradiances E_1 and E_2 are measured for two different distances r_1 and r_2. If the detector is linear, it must be
$$E_1/E_2 = r^2_2/r^2_1.$$
Inherent assumptions for the method to work according to this simple equation are the following:
(a) All distances must be sufficiently large or the source accordingly small so that the source is virtually a point source. The inverse square law is applicable only for a point source.
(b) Distances must be known, with sufficient accuracy.
(c) The source must emit radiation uniformly within the cone delimited by the detector at its smallest distance.

2) The superposition or addition method

The basic principle of this method may be stated as follows: first a flux ϕ_1 irradiates the detector and a signal S_1 is recorded; then a flux ϕ_2 irradiates the detector and a signal S_2 is recorded; then both fluxes $\phi_1 + \phi_2$ irradiate the detector and the signal $S_{1,2}$ is recorded. If it is found that

$S_{1,2} = S_1 + S_2$ the detector is linear between the minimum value of these signals (S_1 or S_2) and the maximum $S_{1,2}$. If $S_{1,2} \neq S_1 + S_2$ a correction factor is obtained from $S_{1,2}/(S_1 + S_2)$. Since one pair of fluxes ϕ_1 and ϕ_2 cover only a short range, stepwise attenuation of both fluxes is often used to expand the range.

Experimental verifications of this principle have been used with multiple sources or with a single source with multiple apertures.

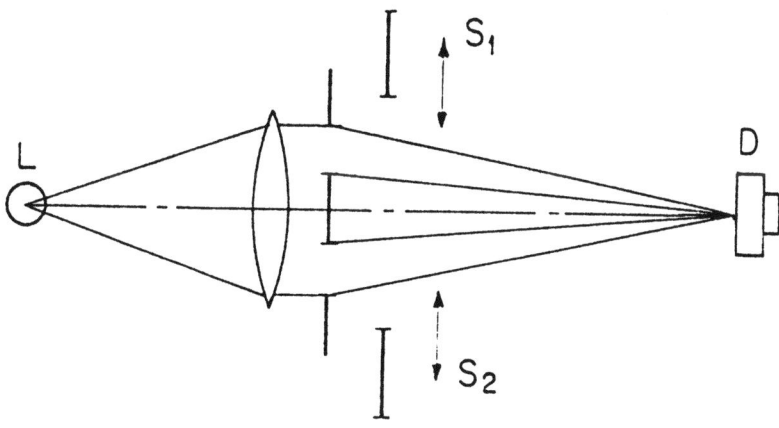

Figure 1.- Double-aperture linearity test. L=source; S_1 and S_2=shutters; D=detector.

2.2 STABILITY

A detector is stable when being invariable the irradiance on the sensible surface, the response is constant in the time. When the output decreases progressively, the detector presents fatigue. The Stability is a characteristic of the detector together with the associated electronic of the measurement.

2.3 SPECTRAL RESPONSIVITY

The spectral responsivity $S(\lambda)$ is the ratio of the wavelength-dependent detector output quantity $dY(\lambda)$ to the spectral detector input quantity

$dX_e(\lambda)$. In more practical terms the equation
$$S(\lambda) = I_{ph}(\lambda)\phi_\lambda(\lambda)$$
describes the spectral responsivity $S(\lambda)$ as the ratio of the photocurrent $I_{ph}(\lambda)$ to the incident monochromatic flux $\phi_\lambda(\lambda)$. In this case the units of input and output are ampere and watt, respectively, and the unit of the responsivity is ampere per watt. It is very important to realize that the dimension or units of responsivity depend on the particular application.

The spectral responsivity function of a detector describes then its spectral responsivity as a function of wavelength. In the case of so-called black or nonselective detectors the spectral responsivity is constant, at least over a certain wavelength range. However, all other detectors have a spectral responsivity function for a limited wavelength range with one or more maxima.

The measure of the spectral responsivity is very important as if it is known in energetic absolute values, the detector is calibrated as a radiometer.

To measure $S(\lambda)$ it is necessary to use a known spectral reference function. Two methods are the most used:

a) The comparison of the unknown or sample detector with a reference detector of known spectral responsivity.

b) Filters method that requires a standard source of a known spectral power distribution and a set of filters of a known spectral transmittance function.

Method **a** has been widely used.

a) Comparison method

This system require a radiation source, an instrument for selecting various wavelengths, a device that switches the radiation from the reference detector to the sample detector, and a reference detector.

The source is usually an incandescent lamp, such as a quartz halogen lamp driven by a highly regulated or controllable power supply. Current regulation rather than voltage regulation is preferred because contact resistances in the lamp base may change as a function of temperature, but it is unlikely that considerable leakage currents arise. A current regulation overcomes such changes of contact resistance.

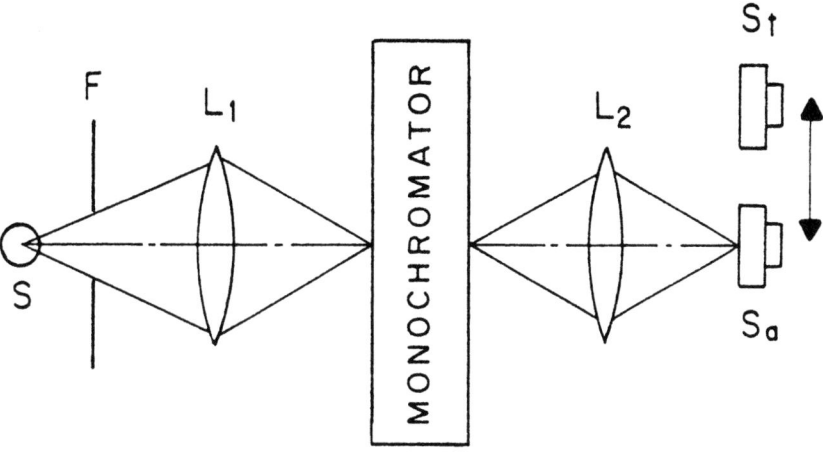

Figure 2.
Schematic diagram of system for spectral responsivity measurements. S = lamp; F = aperture; L_1 and L_2 = lens; S_a = sample detector; S_t = standard detector.

Wavelength selection is usually implemented by means of grating monochromators. To select a monochromator for this purposes, spectral purity is of greater importance than resolution because most of spectral responsivity functions are rather smooth curves. Other rather important consideration in the selection of the monochromator is the efficiency or throughoutput. If a thermopile is used as reference detector, the low detectivity requires a high flux.

The most commonly used reference detectors are the nonselective or flat detectors: thermopiles or pyroelectric detectors, which are assumed to be nonselectives. Depending on the wavelength range, self-calibrated silicon detectors or devices of 100% quantum efficiency can be used.

The spectral responsivity of the sample detector will be
$$S(\lambda) = (I(\lambda)/I_{st}(\lambda)) \cdot S_{st}(\lambda),$$
where $I(\lambda)$ and $I_{st}(\lambda)$ are the signals of the sample and standard detector, and $S_{st}(\lambda)$ is the spectral responsivity of the standard detector. If S_{st} is calibrated in absolute values, $S(\lambda)$ will be an absolute spectral responsivity.

2.4 SPATIAL RESPONSIVITY

The spatial responsivity is the variation of the response of a detector with respect to the incidence angle of the radiation. To measure it a photogoniometer is necessary. This measurement can be of great importance, as in the case of luxmeters.

2.5 NOISE AND DETECTIVITY

Detectivity is a term quantifying the minimum amount of radiation a detector can measure with certainty. The detectivity of a detector strongly depends on the noise inherent in the detection process, and therefore it is necessary to discuss the various types and sources of noise first:

a) Shot noise occurs whenever a phenomenon can be considered as a series of independent events occurring at random. In photoemissive detectors the emission of electrons from a photocathode is such a process.

b) Thermal noise (Johnson noise) results from the random motion of carriers in any conductor or resistor.

c) Flicker noise has various causes and is characterized by a noise intensity spectrum that has a const/f form, where f is the frequency.

d) Temperature fluctuation noise results from the irregular heat exchange between the detector and its environment. It is more pronounced for smaller detectors.

e) Generation-recombination noise occurs whenever free charge carriers are generated or recombined in a semiconductor. This process can be considered as a shot noise process. However, it is useful to consider the fluctuation dn of the carrier density n as causing a resistance fluctuation dR, which can be detected by applying a bias voltage and measuring the resulting current fluctuation.

Noise in semiconductors

The semiconductors are the base material in most of the present detectors. The noise is generally proportional to the areas of sensitive surface in them.

With no external polarization, there is an intern minimum absolute noise, the Johnson noise. This type of noise occurs in resistors and is caused by the random thermally excited vibration of discrete charge carriers, mainly electrons, in a conductor. The thermal noise current is given by

$$I_N = (4KT\Delta f/R)^{1/2},$$

where K is the Boltzmann constant, T the absolute temperature, Δf the bandwidth of measurement and R the resistance.

The second type of excess noise in resistors is the Flicker or 1/f noise, characterized for a spectrum in which the noise power depends on an inverse function of the frequency. The general expression for the intensity of the current is:

$$I = ((KI_B^\alpha B)/f^\beta)^{1/2},$$

where K is a proportionality factor,
I_B is the polarization current,
f is the frequency,
$\left. \begin{array}{l} \alpha - 2 \\ \beta - 1 \end{array} \right\}$ constants.

The other form of excess noise is the G–R (Generation–Recombination) noise. There are many forms of the G–R noise expression, depending on the semiconductor intern properties. One of the more usable is

$$I_N = 2I_B[\tau B/N_0(1+\omega^2\tau^2)]^{1/2},$$

where R is the resistance,
τ is the average life of the free charge carriers,

ω is the angular frecuency, and
B is the bandwidth.

At low frecuency dominates the flicker noise, the G-R noise dominates at median frecuencies and at high frecuencies the Johnson noise is the more important, depending on the semiconductor and the fabrication process too.

Signal-to-Noise Ratio and Dark Current

Two aspects of the noise need clarification and definition: signal noise and dark current noise. Fundamentally there is no difference between them.

Signal noise is the noise inherent in a signal, and the important application of this term is not by itself but rather to quantify the signal-to-noise ratio (S/N ratio). This ratio describes the maximum precision in the measurement of radiation by means of a photoelectric detector.

Noise Equivalent Power (NEP)

The dark current noise or dark noise is the noise that appears superimposed on the dark current of the detector, that is, the signal appearing without any incident radiation. In semiconductor junction diodes used in photovoltaic mode this dark current is often in the order of 10^{-10} or 10^{-12} A, and the shot noise of these currents is considerable. In addition, thermal noise appears. The noise of the dark current is, of course, a limiting factor for the detection of very weak signals, as oftenly occurs in optical fibre detection, because it is difficult to detect with certainty a signal that is smaller than the noise of the dark current. The weakest signal detectable is that for which the signal-to-dark-noise ratio is 1.0. This minimum power is called the noise equivalent power (NEP).

Detectivity

Unfortunately, NEP is undesirable as figure of merit because a detector

with a larger NEP is less suitable for the detection of small inputs. Consecuently the detectivity D has been defined as the inverse of NEP
$$D = 1/NEP$$
However, even this figure of merit is unsuitable for comparing different types of detectors, because in many instances detectivity is linearly proportional to $(A\Delta f)^{1/2}$. The **normalized detectivity D*** was defined by
$$D^* = (A\Delta f)^{1/2}/NEP$$
where A is de active area of detector and Δf the bandwidth of electronic equipment.

Measurement of Noise

The measurement of noise from a photodetector can hardly be made directly with an rms voltmeter because the noise voltages or currents are too small. The combination of the detector-amplifier noise is measured.

Sine-wave method. Three steps are necessary for the implementation of this method.

1) The transfer voltage gain of the amplifier is measured by means of a sine-wave voltage/generator that is conected in series with the signal source. This measurement is done at a signal level considerably higher than the noise signal level. The transfer voltage gain is different from the voltage gain because it is measured with the given input impedances.

2) The output noise of the detector-amplifier combination is measured by means of either a wave analyzer if the noise spectrum is to be determined, or a combination of a band pass filter and an rms meter with a bandwidth wider than the filter.

3) The source noise is calculated dividing the total noise output by the transfer voltage gain. Separate measurements may be required to separate the equivalent input noise of the amplifier from the source (detector) noise.

2.6 TIME CONSTANT

Assuming that the irradiance on a detector at a time t_0 instantaneously jumps from a level E=0 to a level E≠0, the irradiance is said to follow a step function. Usually the response of the detector is not a similar step function, but the signal increases first rapidly then with decreasing speed and approaches the final value asymptotically. The signal generally follows an exponential function

$$I(t) = I(1 - e^{-t/\tau}),$$

where I(t) is the instantaneous current at the time t, t the time, beginning at t_0, and I the final value of I(t). The constant τ is the so-called "time constant" or "response time". This is a constant typical for each detector type and its unit is the second. Usually τ is defined as the time when the signal reaches (1–1/e) of its final value.

2.7 TEMPERATURE DEPENDENCE

In the photodiodes the number of couples electron–hole thermally excited increases exponentially with the temperature, with a decrease of current as consequence. With the increase of the temperature the noise also inverse, and in some detectors it is necessary to decrease the temperature to that of the liquid He to measure very weak signals.

The change of responsivity with temperature is quantitatively described by the temperature coefficient C_t, which is defined as the relative change of responsivity per degree centigrade:

$$C_T = (I(T) - I(T_0))/I(T_0) \cdot (T-T_0),$$

where I(T) is the response at temperature T and $I(T_0)$ the response at temperature T_0.

3. Detectors Calibration

The calibration of a detector consists in knowing the equivalence between the output signal, in electric units, and the input amount of incident photons or irradiance on the sensitive area of the detector. To calibrate a photodetector is to know with accuracy the spectral responsivity, $S(\lambda)$, in absolute values. In that case the calibrated detector becomes a Radiometer.

To calibrate a detector first we must know its linearity range, stability, detectivity, noise, time constant and temperature dependence. If all of them are appropriated, the detector can be calibrated in absolute values.

4. References

1. Budde, W. and Dodd, C.X. (1971) Measurements of relative spectral distributions of photoelectric receivers, Appl. Opt., **10**, 2607.
2. Commission Internationale de l'Eclairage (CIE) (1982) "Methods of characterizing the Performance of Radiometers" Publ. No. 53, CIE, Paris.
3. J. Campos and A. Corróns (1986) "Radiometría absoluta basada en detectores de silicio". Publicación del Instituto de Optica nº 45. Madrid.
4. Zalewski, E.F. and Geist, J. (1980) "Silicon Photodiode Absolute Spectral Response Self-Calibration". Appl. Opt. **19**, 3795-3799.
5. Geist, J. and Baltes, H. (1989) "High Accuracy Modeling a Photodiode Quantum Efficiency". Appl. Opt. **28**, 3929-3939.
6. Campos, J. Corredera, P. Pons, A. and Corróns, A. (1991) "Spectral responsivity calibration of Ge photodiodes with respect to an electrically calibrated pyroelectric radiometer and to black body source". Metrologia, **28**, 141-144.

POWER METER CALIBRATION

Pedro Corredera Guillén.
Instituto de Optica "Daza de Valdés". C.S.I.C.
c/ Serrano 121. 28006 Madrid (Spain)

Abstract

One of most important fibre optic test instrument used in the characterization and analyses of fibres is the power meter. The background on the accuracy and precision of the optical power meter measurements in fibres is described and the possible error sources are discussed. Some techniques for the absolute calibration of power meters are described.

1. INTRODUCTION

The rapid development of the optical fibres and optoelectronic industries with these multiple applications in communications, sensors, non linear propagation, etc., has been accompanied with the development of systems and instrumentation projected "ad hoc" to measuring the interest properties of the optical fibres.

The knowledge of the optical fibres properties and their characterization is carried out from three different measurement environments or levels: laboratory measurements, mainly addressed to the research and development, and which accuracy and sensitivity are required; factory measurements, addressed to quality control of products, which mainly require is the automation and the repeatability, and field measurements, guided to installations testing and its maintenance, which need portable instrumentation of practical use.

The measurement of the optical power transmitted or reflected on the fibre is basic in all these levels, and the power meters are essential for measurement, control and evaluation of lasers and LED characteristics, also they are used for the verification of power budget of optical links, and the most important of all, the power meter is the key instrument in the characterization and performance test of optical fibre components (fibres, connectors, attenuators, switches, etc).

The power meters operate in the near infrared using the silica transmission windows at 850, 1300 and 1550 nm, over the power range from mill-watts (0 dBm) to pico-watts (-90 dBm), and are manufactured with photodiodes of germanium or InGaAs (exceptionally with Si photodiodes in the first spectral transmission window). The application which have been designed these instruments (the optical power measurement in fibres) often makes necessary to incorporate an appropriate connector adaptor to the optical cable or fibre used. These connectors are of different types and simplifies the field measurements.

In this context, the maintenance of optical power meter accuracy has became very important. The power meters require calibration against National Standard Responsivity Scales and because is difficult to couple optical power to the meters from traditional responsivity calibration facilities and transfer standards have been development.

The propose of this lecture is to discuss some factors that limits the accuracy and reproducibility of optical power measurements in fibres. From the knowledge of this factors we will conclude the necessity of power meter calibration, object to the second part of the lecture, in which some techniques designed for the power meters calibration are described.

2. FACTORS THAT LIMITING THE ACCURACY AND PRECISION OF OPTICAL POWER MEASUREMENTS IN FIBRES

The optical power measurement in fibres seems very simple, however, some limitations and inaccuracies doing inconsistencies this measurement. These problems may be caused by the meter, but there is an equally high probability of problems outside of the meter, such as reflections, interferences and laser sensitivity to backreflections.

2.1. SPECTRAL RESPONSIVITY

The responsivity of a Ge and InGaAs photodiodes is different whit the wavelength. The responsivity rises until cut-off wavelength (next to 1530 nm in Ge detector and 1650 nm in InGaAs detector, depending of work temperature and manufacture), where the responsivity drops rapidly, because the photon's energy is lower than the bandgap energy. The actual responsivity curve of a detector is determined by the reflectance, the absorbance and recombination process in the detector material. Large variations between different photodiodes of the same type can be observed.

2.2. TEMPERATURE DEPENDENCE OF RESPONSIVITY

The response of a photodiode under constant irradiance varies with the temperature of work because the electronic levels changing with the temperature. These variations do not affect for equal to every wavelengths (they are more intense near the cut-off wavelength), and are described by the temperature coefficient (C_t) defined as the relative change of responsivity for grade centigrade. The temperature coefficient could be positive or negative, this is the responsivity can increase or decrease with the temperature. Several detectors have temperature coefficients positive in a spectral range and negative for others.

The figure 1 show the temperature coefficient of Ge and InGaAs detectors. This effect very important in the germanium photodiodes at wavelengths next to 1550 nm (a spectral window of silica fibres) and can be minimized with temperature stabilized systems.

2.3. DETECTOR SIZE AND NON-UNIFORMITY OF RESPONSE

To use small detectors (<2 mm diameter) is not interesting in power meters because the fibres have different numerical apertures and the small detector could cut-off tails of the power distribution.

Large size photodiodes present non-uniformity response in the active area. The superficial response of Ge diode shown differences typically of 1% across the active area and 2 % or 3% can be observed in InGaAs photodiodes.

2.4. NON-LINEAR RESPONSE AT HIGH POWERS

A detector is linear when its response is proportional to the incident optical power. At low optical powers the Si, Ge and InGaAs photodiodes shown excellent figures of linearity, however, does not happen the same at high powers (between 100 µW (-10 dBm) and 10 mW (10 dBm)), where the material saturation can occur, and fall down the output signal. A similar result could be observed if is used a small light spot with excessive density of energy. The figure 2 show a typical figure of linearity for an InGaAs photodiode at room temperature. The linearity ranges are differences with the temperature [1].

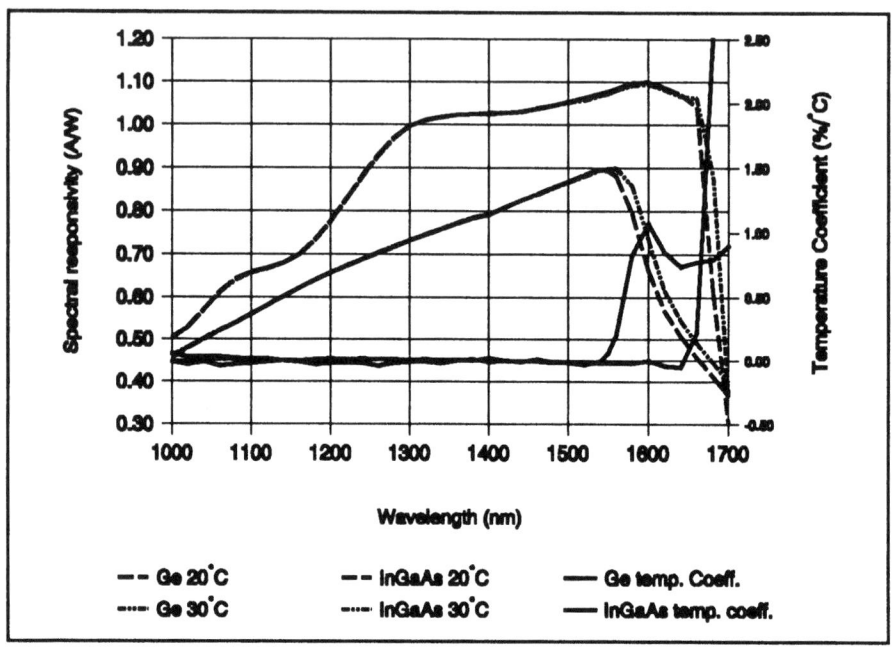

Figure 1.- Spectral responsivity of Ge and InGaAs photodiodes at two temperatures and temperature coefficient

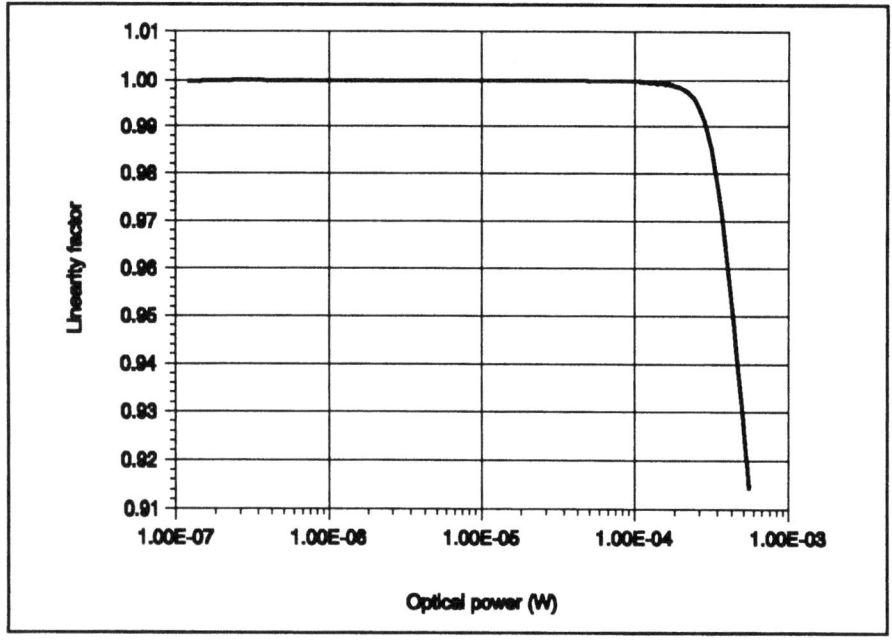

Figure 2.- Linearity test results of an InGaAs detector at 1300 nm and room temperature.

2.5. CHANGE OF DETECTORS RESPONSIVITY WITH TIME: AGING

The photodiodes used in optical power measurement in fibres have an aging due to the continua degradation that the detector material suffers. Quantitative measurements of this degradation are not now very well-known for the photodiodes of Ge and InGaAs photodiodes. The manufacturers claim a 10% change of the responsivity over 5-years period, although at short wavelengths the curve of responsivity is affected more strongly. Until best information and technological advances are made, regular calibration of the power meters is the only solution of this problem.

2.6. DEPENDENCE OF THE DETECTOR RESPONSIVITY WITH THE ANGLE OF INCIDENT OF LIGHT

Another measurement error is due to angle dependence of the responsivity. The response of Ge and InGaAs photodiodes depend on angle of incidence of light. Thus, the numerical aperture of the fibre or connector end may affect the calibration and its response. This dependence may be due to interference effects in the passivation layer of the detector surface. Moreover, the angular response change will depend on the polarization state of the radiation light as is shown in figure 3 [2,3]. The different angular response is inversely related to de reflectance change of the detector surface, so if the angular reflectance change was know it would be possible to predict the response change. However, internal structure of the detector surface is not known and it is not possible to calculate the reflectance angular change and therefore the response changes. Use of parallel beam with the help of collimating lens eliminate this problem.

2.7. INTER-REFLECTIONS BETWEEN THE DETECTOR AND THE FIBRE CONNECTOR CAP

Solid-state photodiode detectors can have high reflectivities even when anti-reflected coated are used. Typical reflectance are between 20% to 30% for Ge detectors. Care must be taken, therefore, where connector caps are fitted (Figure 4), because they may significantly increase the response of power meter by inter-reflections. Increase as large as 16% (0.64 dB) can be observed. An additional problem is caused because the different fibres (monomode and multimode) creates different spots on the detector surface and consequently different patterns of reflection.

These multi-reflections are very difficult to eliminate. The intents are based to use anti-reflection windows and coatings in the detector surface (but anti-reflection coating have wavelength dependence) and to paint back connector adapter black to

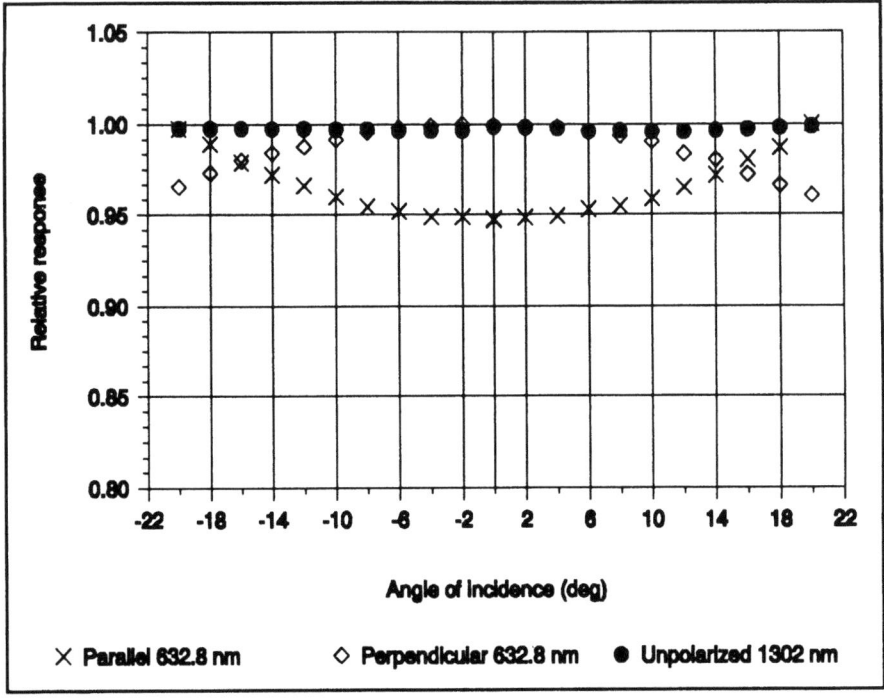

Figure 3.- Angular response of Ge photodiode for parallel and perpendicular polarised ligth at 632.8 nm and unpolarised ligth at 1302 nm.

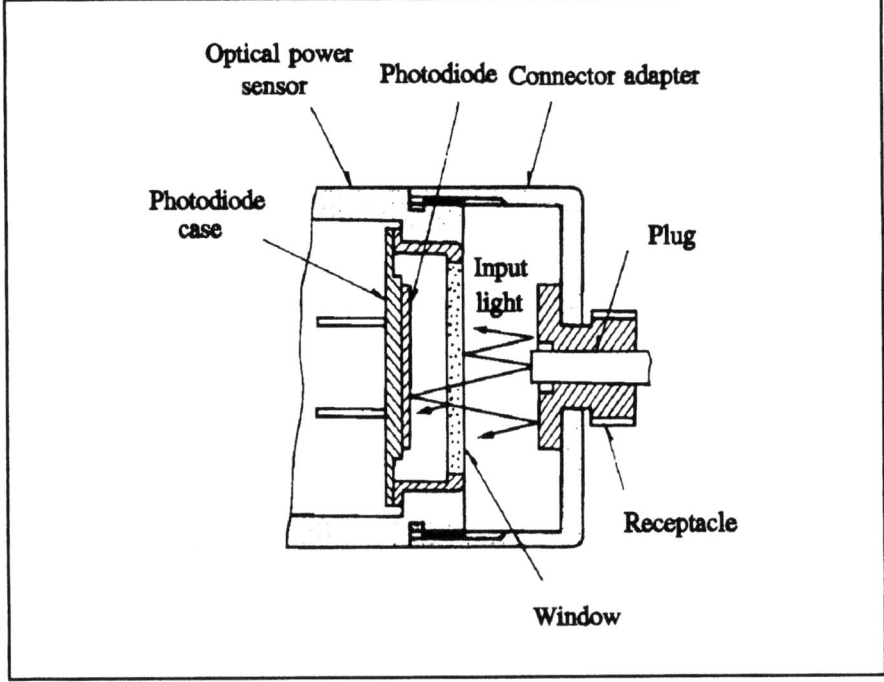

Figure 4.- Inter-reflections on connector cap and detector surface in power meter.

prevent reflections. The reflectance reduction also reduce the measurement errors produced by the different numerical apertures of fibres.

In the table I is shown as summary mode the different factors that could limit the accuracy of the optical power measurement in fibres and its possible solution.

Table I.- Factors that limiting the accuracy of optical power measurements, and possible solutions.

CAUSE	EFFECT	SOLUTION
Different spectral responsivity	Different measurements with different detectors	Spectral calibration
Temperature dependence of spectral responsivity	Measurements depend on temperature	Temperature stabilized detector
Non-linearity of response at high powers	Inaccuracy at high power levels	Avoid focused beams, Verify linearity
Non-uniformity of response	Power measured is different rotating the detector	Verification of uniformity
Aging	Slow degradation of accuracy	Regular calibration
Angle dependence of response	Measurements depend on numerical aperture	Parallel beans Integrating spheres
Reflections and inter-reflections in the optical heat	Non-reproducible measurements	Coating surfaces Integrating spheres
Modes unknown	Measurements depend on length and curvature of fibre	Mode filters Long fibres

A good example that quantify the influence of these possible error sources is the intercomparison carried out in 1986 by R. L. Gallawa and S. Yang in the NIST (formerly NBS) [4], which compared the fibre optical power measurements from different laboratories and industries of United States at 850 and 1300 nm wavelengths. In this intercomparison and excluded detectors of area minor 1 mm^2 show differences between laboratories of 0.6 dBm (15%) at 850 nm and 0.8 dBm (20%) at 1300 nm.

These results are unacceptable compared with the current state of uncertainties in the measurement of the optical power and show the necessity of calibrating the power meters to the National Responsivity Scales (uncertainties lower than ±1%) [5].

3. POWER METERS CALIBRATION

From a radiometric point of view an optical power meter is a calibrated detector that is used at some specify wavelength range (850, 1300 1550 nm) and in a very concrete conditions, as are with the proprieties of light in fibres and connectors ends (numerical aperture, modal distribution).

The radiometric calibration of a power meter, as any other detector, consists of four different calibrations:

A) Absolute responsivity calibration.
B) Linearity measurement.
C) Uniformity measurement.
D) Spectral relative responsivity calibration.

The calibration strategy must be selected to minimize the requirement for absolute responsivity calibration (most difficult calibration with most severe conditions) to a few measurements conditions. The other measurements required to fully characterization meter can be made relative to these absolute measurement points, relaxing some measurement conditions. The absolute measurements are selecting to do in high power level at each of the silica fibre spectral windows: 850, 1300 and 1550 nm.

Spectral relative calibration is necessary for extend the calibration to any wavelength and to calculate the optical power of LED using the following mathematical correction for spectral LED width [6]:

$$P = P_m * \frac{\int_{\Delta\lambda} p(\lambda)d\lambda}{\int_{\Delta\lambda} p(\lambda)S_r(\lambda)d\lambda} = P_m * \frac{\int_{\Delta\lambda} f(\lambda)d\lambda}{\int_{\Delta\lambda} f(\lambda)S_r(\lambda)d\lambda}$$

where P and P_m are the power of LED and power measured in the power meter, $P(\lambda) = P_0 f(\lambda)$ is the spectral emission power of LED, P_0 is the pick power and $f(\lambda)$ and dimensionless spectral emission curve of LED and $S_r(\lambda)$ the spectral responsivity of the power meter.

3.1. PRIMARY STANDARD: CRYOGENIC RADIOMETER

Absolute response calibration is based on comparison with the absolute responsivity scale. The absolute responsivity scales are established by primary optical standards. The primary standards could be sources (Black body and synchrotron radiation) o absolute radiometers (electrical substitution radiometers or predictable response detectors). From the appearance of cryogenic radiometers most of important National Laboratories had selected the cryogenic radiometers as primary standard.

A cryogenic radiometer is a electrical substitution radiometer operating at the temperature of liquid helium (4.2 K). The most accuracy of these radiometers was designed by Martin and Fox in the NPL (UK) [7], that could measure the optical power of intensity stabilized lasers with an uncertainty of less than ±0.0005%, and has been tested measuring the Stefan-Botzman constant with only 0.02% difference from the theoretical value. Other designs commercially available could obtain uncertainties better than ±0.02% [8].

In a measurement with an electrical substitution radiometer (ESR) the unknown radiant flux is absorbed by the detector (generally a thermal detector). The resulting temperature rise cause an electrical output signal, which is there reproduced by heating the detector with a know amount of reference electrical power. In this ideal process the incident radiant flux can be derived from measuring of electrical power.

The conventional ESR operating at room temperature have radiative and convective energy flow between the receiver and its enclosure, and joule losses in the heater connectors that product "non-equivalence" errors.

The cryogenic radiometers operate in high vacuum at liquid helium temperature, and the radiative and convective coupling between the receiver and its enclosure can be neglected at the typical laser irradiance levels of the calibrations.

In a cryogenic radiometer the receiver is a pure copper tube fitted internally with an inclined plane to receive the laser radiation. This cavity is blackened with a specular paint whose reflectance has been accurately measured over 0.2 to 40 µm. This cavity is at liquid helium temperature. Several important proprieties of materials are affected by liquid helium cooling, enabling the cryogenic radiometers to achieve important performance improvements over convectional ESR's. The high thermal conductivity and extremely low heat capacity (1000 time lower than at room temperature) of pure copper at 2-4 K dramatically increase the thermal diffusivity of the receiver reducing the non-equivalence errors and improve the sensibility of radiometers.

The heaters are provided in the same received, opposite to the blackened surface that absorbed the laser radiation. This heaters are superconductive wires at liquid helium temperature eliminates Joule heating losses.

The rise of temperature is detected by a germanium resistance thermometer of high resolution and sensibility. The temperature resistance are measured with a AC bridge which all electrical connection are made with superconducting NbTi wire to avoid lead resistance losses.

The possible radiometers budges errors and uncertainties are: window transmission, scattering light, cavity absortance and heater power measurements. The window transmission loss is minimized using a fused silica disc mounted to Brewster angle for the wavelength used, to allow maximum transmission for the high perpendicular polarized intensity-stabilized laser beam. Scatter light is controlled using high collimated beams and aperture and baffles.

3.2. POWER METERS CALIBRATION AT "INSTITUTO DE OPTICA (CSIC)"

3.2.1 *Absolute responsivity response*

The calibration of the absolute responsivity of the fibres power meters is carried out by comparison with the Absolute Scale Responsivity of the "Instituto de Optica", established in a cryogenic radiometer with a minor uncertainty of 0.02%, when are used stabilized and high polarised beam lasers.

As transfer standards detectors, to calibrated the power meters against the absolute responsivity scale, integrating sphere radiometers are used (Figure 5) [9]. These transfer standards are constructed with 5 mm diameter germanium photodiode mounted on the exit port of a 50 mm diameter sphere. An Fc-type connector is mounted on the input port. In this design the sensibility of response of photodiode to the size and position of image, to numerical aperture, to inter-reflections between the detector and the connector caps, are eliminated. This eliminates also uncertainties caused by non-uniformities in the response across the surface of germanium photodiode, and sensitivity to the polarization state of light. The only disadvantage of these transfer standards is the reduction of the signal by three orders of magnitude. However, its calibration must be carried out at a power level of 100 µW (-10 dBm) to obtained small uncertainties.

The calibration of these integrating spheres radiometers is carried out according to the following steps:

1.- Using stabilized lasers (by means of external stabilization of the optical power using electro-optical modulators) is calibrated a pyroelectric radiometer at the wavelength of the He-Ne (or Krypton laser). This pyroelectric radiometer have a contrasted plane response between 0.2 to 10 µm with differences of response inferior to 0.01.%. Against this radiometer are calibrated the integrating sphere radiometers in an experimental system that is shown in figure 6, using stabilized laser diodes at wavelengths close to 850, 1300 1550 nm and whit known spectral emission.

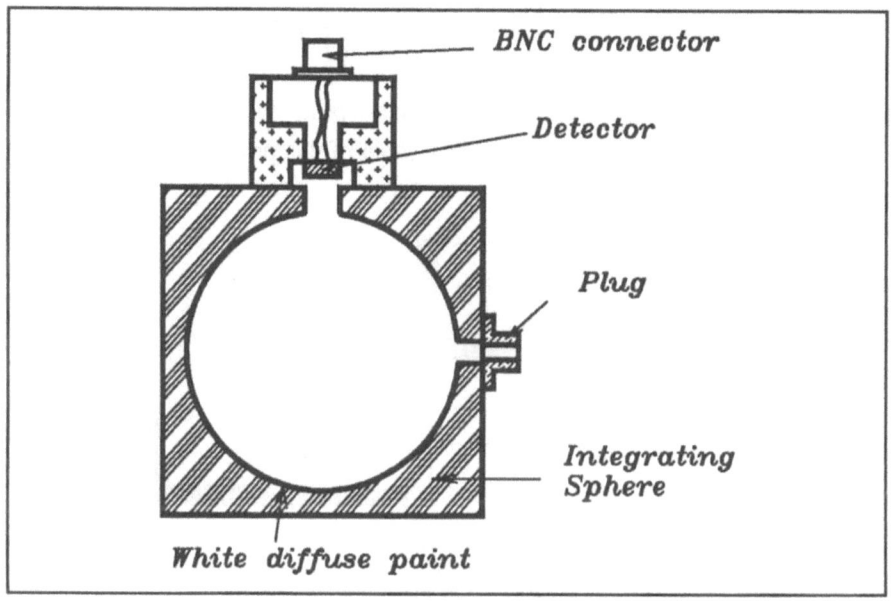

Figure 5.- Diagram of the integrating sphere radiometer used as transfer standard

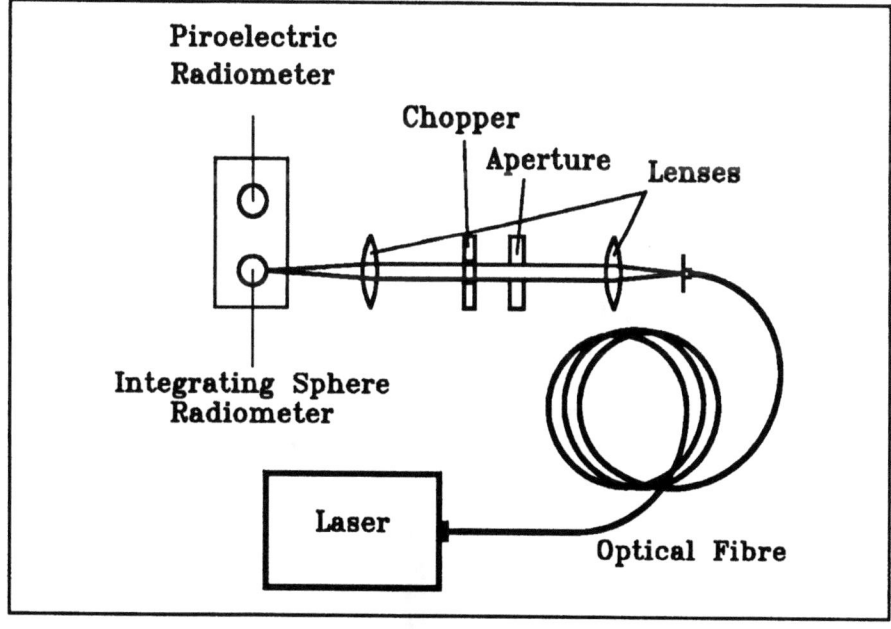

Figure 6.- Experimental setup used to calibrate the transfer standards

2.- The calibration of commercial optical power meters is carried out using the same standards stabilised laser diodes and selected monomode pigtail, by direct comparison with the integrating sphere radiometers. The maximal uncertainty, in the strategy calibration is produced in this comparison due to non-repeativity of the connections to the power meter. Figure 7 shows the traceability chain of this process.

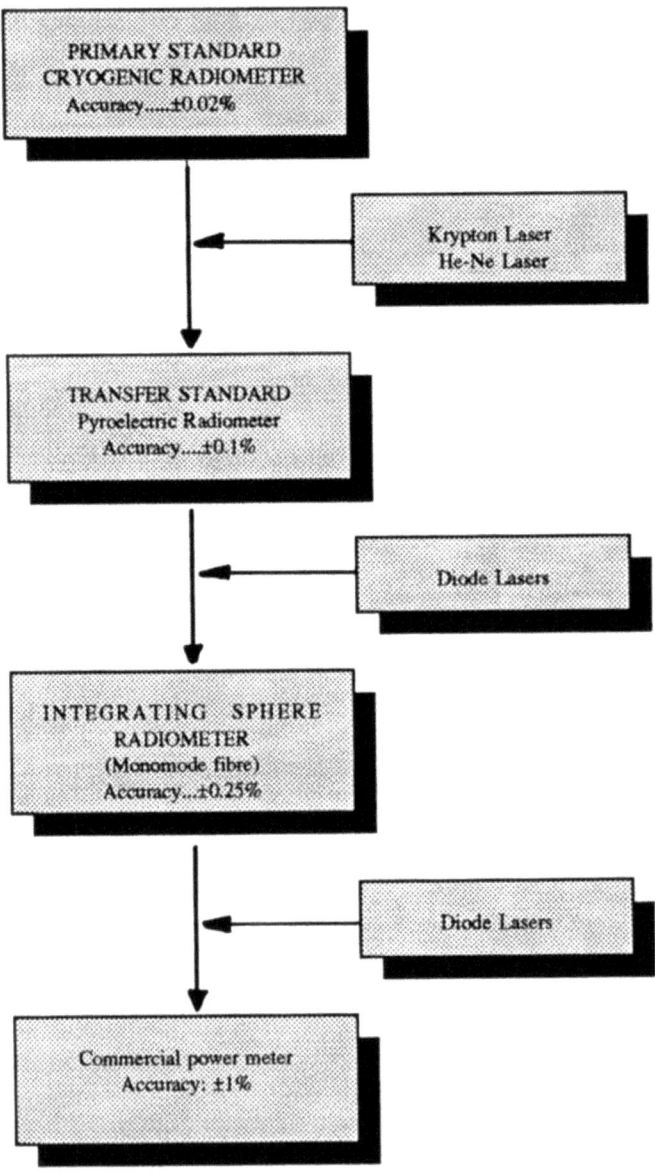

Figure 7.- Traceability chain process.

3.2.1. Measurement of linearity of power meter

The calibration described in the previous section is only possible to carry out at high power levels (>50 µW), to kept very low uncertainties. For using power meters at different levels of power is necessary to determine the linearity range of power meter. The measurement of the linearity of the power meters is carries out by comparison of the response of the meter with the response of a Germanium photodiode of known linearity (determined by a method of addition of stimulus). The comparison is carry out using stabilised laser diodes and programmable attenuators. Starting at the most high power level the response of the power meter and the linear detector is recorder changing the power level with the programmable attenuator. After the linearity curve is related to the absolute calibration power level.

3.2.3. Uniformity of response of power meter

The verification of the uniformity of response of the power meters is necessary for the realization of precise measurements. The measurement procedure consist in recording the response of various part of the detector area to the same amount of optical power. The experimental system is shown in the figure 8. The output of a stabilized lased diode connected to a fibre is filtered, collimated and imaged to 0.5 mm of diameter spot size onto the detector. This type of illumination is spatially more uniform than produced by directly imaging the laser output. The detector is mounted on a translation stage that allowed to move the detector surface in a plane perpendicular to de optical axis. A monitor detector control the power variations in the laser.

3.2.4. Measurement of spectral relative responsivity

The extension of the calibrated absolute of the power meter to any other wavelength could be made determining the spectral relative responsivity of the power meter. The spectral relative responsivity of a photodiode is the variation of it responsivity at each wavelength related to a reference wavelength. The determination of the spectral relative responsivity is made using a pyroelectric detector, who has a plan response over the wavelengths from 0.2 to 10 µm, in a comparison system, composed by a monochromator, a stable halogen lamp and the suitable imaging optic. From spectral scan of the pyroelectric response and the spectral scan of the power meter response the spectral relative responsivity is obtained.

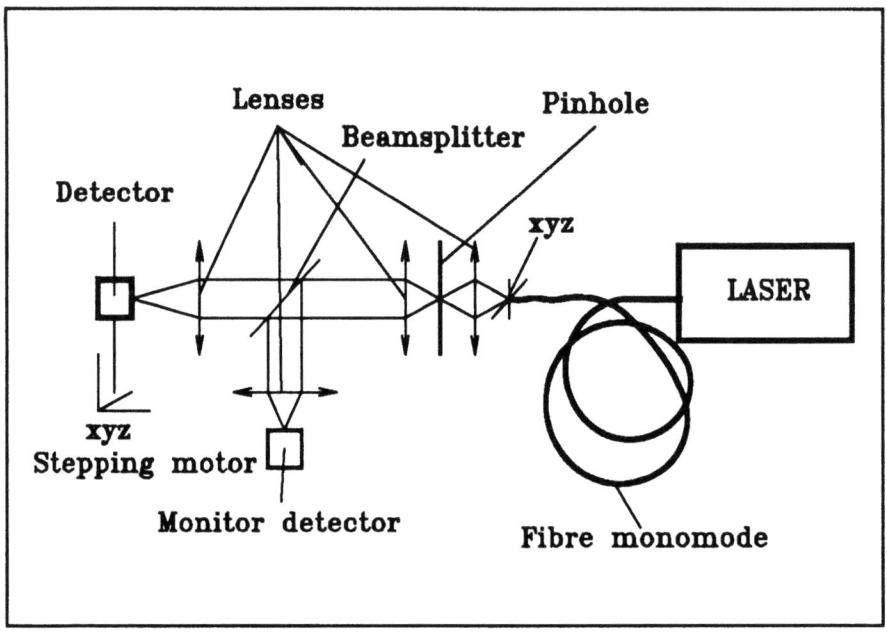

Figure 8.- Experimental setup used to evaluate the non-uniformity of response of power meters

4. Future developments

Recognising the limited accuracy that can be achieved with the integrating sphere radiometer, because this detector need regular recalibration as the stability of response depends critically on the constancy of the reflectance of the sphere paint, and changes in the responsivity of Ge detector can be produced, the new developments are addressed to obtained InGaAs trap detectors. The trap detectors consist of an arrangement of three InGaAs photodiodes of large area [10]. The arrangement of the trap detector is such that the radiation incident into the detector is either absolved by the photodiode or reflected onto another. The result is that the radiation reflected by the system after five reflections is negligible.

Whit this arrangement if each photon product an electron the diodes can be non-uniform in response, and if the quantum efficiency is unity its spectral responsivity ($S(\lambda)$) can be calculate from the following equation by knowing only the wavelength λ, of the incident radiation and fundamental constants (e, the electronics charge, c speed of light and h Planck constant):

$$S(\lambda) = \frac{\varepsilon(\lambda)\lambda e}{hc}$$

The first result has been obtained at NPL whit special designed InGaAs photodiodes that arranged in the trap present quantum efficiency of 99% from 940 to 1640 nm.

5. References

[1] Stock K. D. (1989) "Calibration of Fibre Optical Power Meters at PTB". *New Developments and Applications in Optical Radiometry.* 157-165. London.
[2] Campos J., Corredera P., Pons A. and Córrons A. (1991). "Germanium Photodiodes as Standards of Optical Fibre System Power Measurements". In *Fibre Optic Metrology and Standards.* The Hague. 66-74.
[3] Gallawa R. and Li X. (1987). "Calibration of Optical Fibre Power Meters: The Effects of Connectors". *Appl. Opt.*, vol. 26, nº 7, . 1170-1174.
[4] Gallawa R. and Yang S. (1986). "Optical Fibre Power Meters: Round Robin Test of Uncertainty". *Appl. Opt.*, vol. 25, nº 7, . 1066-1068.
[5] Gardner J.L., Gallawa R.L., Stock K.D. and Nettleton D.H. (1992) "International intercomparison of detector responsivity at 1300 and 1550 nm". *Appl. Opt.* 31 nº34, 7226-7231.
[6] Hewlett-Packart (1987). Application note 1031.
[7] Martin J.E., Fox N.P. and Key P.J. (1986). "A cryogenic radiometer for absolute radiometric measurements". *Metrologia* 21. 147-155.
[8] Foukal P.V., Hoyt C., Kocchling H. and Miller P. (1990). "Cryogenic absolute radiometers as laboratory irradiance standards, remote sensing detectors, and pyroheliometers". *Appl. Opt.* 29 nº7 983-993.
[9] Nettleton D. (1989). "Application of Absolute Radiometry to the Measurement of Optical Power in Fibre Optic System." *New Developments and Applications in Optical Radiometry.* 93-97. London.
[10] Fox N. P. (1993). "Improved near infrared detectors". *Metrologia* 30, 321-326.

OTDR CALIBRATION

Francis GAUTHIER, Lionel DUCOS
Laboratoire Central des Industries Electriques
33 avenue du Général Leclerc
92260 FONTENAY-AUX-ROSES
FRANCE

1. Introduction

Optical time domain reflectrometry is the primary measurement technique for the characterization of single-ended optical fibre. Easy to use, it allows to determine magnitudes and locations of faults and reflections as well as fibre length and lineic attenuation of a fibre network. Nowadays, the OTDR (Optical time domain reflectometer) is in current use in laboratory as well as in site.

The first section of this document describes the conventional OTDR through its block diagram, its measurement technique and main characteristics. The second section describes the OTDRs calibration through the methods LCIE has developped to calibrate distance, attenuation and reflectance parameters.

2. OTDR fundamentals

2.1. PHYSICAL BASIS

Reflectometry is based on the measurement of the reflected part of the light travelling along a fibre link. Light decay results from absorption loss (αa) and diffusion loss (αd)

$$\alpha = \alpha a + \alpha d \qquad (1)$$

Rayleigh scattering is the preponderant loss cause in a single-mode fibre and is identified with α in this document.

Rayleigh scattering occurs because of inhomogeneities in the local refractive index of the fibre. A small fraction of the incident light is guided back in the fibre in relation with its numerical aperture.

This phenomenon, called backscattering, allows to determine the attenuation of the fibre under test. The dependance of Rayleigh scattering on wavelength is $1/\lambda^4$ and has motivated fibre system designers to use longer transmission wavelengths : typically 1.31 μm (the wavelength of minimum dispersion) and 1.55 μm (the wavelength of minimum loss) [1].

2.2. OTDR BLOCK DIAGRAM

Figure 1 shows the block diagram of a generic OTDR [2]

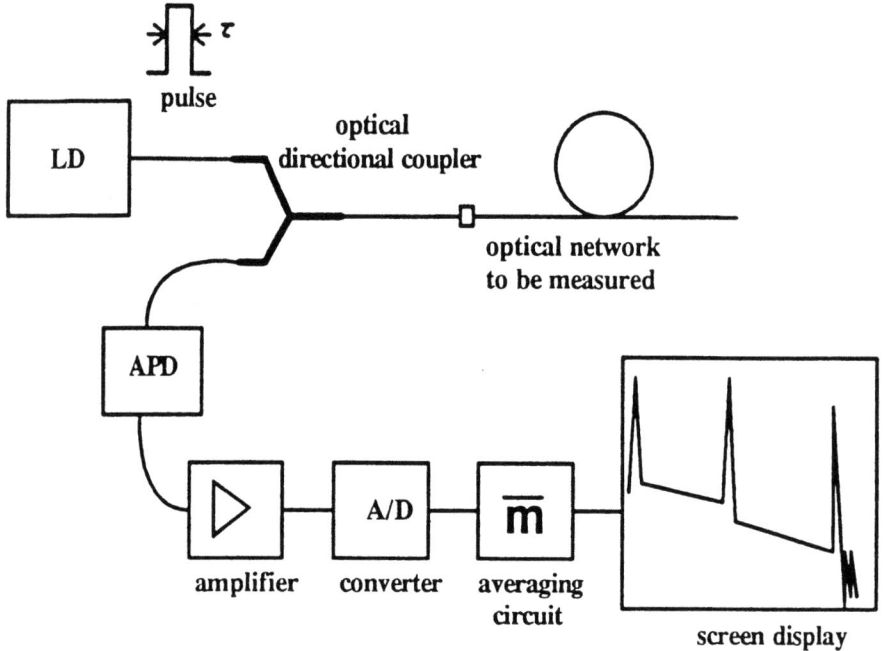

figure 1 : block diagram of an OTDR

The optical pulse from the laser diode (LD) is launched into the optical network to be measured through an optical directional coupler. The backscattered and reflected light generated in the network returns to the emission end, and is fed via the optical directional coupler into the avalanche photodiode (APD) where it is converted into an electrical signal. This electrical signal is amplified and then converted into a digital signal by an A/D converter. This signal is further processed by an averaging circuit to improve the S/N ratio before being displayed on a screen [2].

2.3. BACKSCATTERED POWER EXPRESSION

Hartog et al [3] have reported the mathematical expression of the backscattered power at the z abscissa of an optical link :

$$P(t) = Vg \cdot S(z) \cdot \alpha d(z) \cdot Po \cdot \tau \cdot e^{-\int_0^z (\alpha'(u) + \alpha''(u)) du} \qquad (2)$$

where Vg is the light velocity in the fibre, S(z) is the local recapture ratio of the Rayleigh scattered power back into the fibre, αd (z) is the local power attenuation coefficient due to Rayleigh scattering, Po is the peak power of the optical pulse of width τ, α' and α'' are the go and back attenuation in the fibre respectively.

It can be shown that the derivation of the Neperian logarithm of P(t) is directly proportional to the attenuation of the fibre under test :

$$\frac{d}{dt}\left\{ Ln\ [P(t)] \right\} = -2\alpha\ [z(t)] \cdot \frac{dz(t)}{dt} = -2Vg\ \alpha\ [z(t)] \qquad (3)$$

Consequently, a reflectometric signature displayed on a logarithmic-scale present a straight decreasing line relative to the tested optical fibre attenuation.

2.4. OTDR MEASUREMENT CURVE

Figure 2 shows a general measurement curve displayed on the screen when the OTDR is measuring a fibre link. The horizontal axis indicates distance, that is fibre length or event location. The vertical axis indicates the strength of reflected light and is expressed in dB.

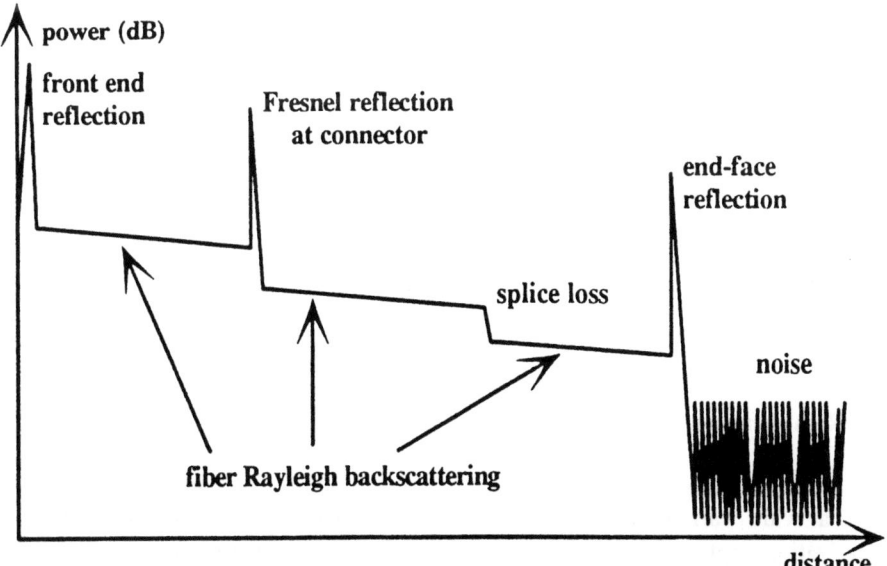

This measurement curve results from both Rayleigh scattering and Fresnel reflection.

As seen before, Rayleigh backscattering is represented by a straight decreasing line, indicative of the exponential decay due to propagation loss, which allows to determine fibre characteristics (attenuation, length, homogeneity...). When a fusion-connected fibre is observed, a step wise difference results at the location of the fusion-connected point.

The magnitude of this step difference indicates the magnitude of the splice loss.

If the optical fibre has a break (fibre end, connector, mechanical splice...), a Fresnel reflection is generated, due to mismatch in the optical refractive indices of two different materials. The energy measurement of a Fresnel reflection peak (as displayed on an OTDR screen) allows to determine the reflectance (in dB) of the tested event.

2.5. DYNAMIC RANGE - SPATIAL RESOLUTION COMPROMISE

The dynamic range is defined by international experts [4] as "the amount of fibre attenuation that causes the backscatter signal to equal the noise level". It can be represented by the difference between the extrapolated point of the backscattered trace (taken at the intercept with the power axis) and the noise level expressed in dBs. Then, the dynamic range is proportional to the energy $Fo = Po \cdot \tau$ of the OTDR emitted pulse, where Po is the power of the emitted pulse, and τ is its duration.

The spatial resolution ΔL is a measure of the minimum distance between two close defaults so that they can be distinguished from one another. ΔL is directly related to the pulse duration τ : $\Delta L = C\tau/2N$ where C is the velocity of light in vacuum and N is the group index of the tested fibre. The smaller ΔL is, the better the spatial resolution is.

Consequently, when τ is increasing, the dynamic range increases as well even though the spatial resolution decreases.

It shows it is necessary to make a compromise between the dynamic range and the spatial resolution of an OTDR to keep good measurements' performances on both parameters.

3. OTDR calibration

OTDRs are able to make measurement from one access point. Consequently they are used in a wide variety of applications like fibre length determination, fibre characterization, optical network maintenance, splice and connector maintenance, remote data acquisition from optical sensors, etc...

An OTDR measurement relies upon the interpretation of scattered and reflected signals depending upon the optical network and the OTDR performances (signal generator, detector, signal processing etc...).

Consequently, it is necessary to calibrate OTDRs. OTDR calibration is in progress since 1988 on the bosom of working group 4 of the technical committee 86 of the International Electrotechnical Commission (IEC/TC86/WG4).

The aim of the prepared document [4] is to provide procedures for calibrating single-mode OTDRs. It only covers OTDR measurement deviations and uncertainties. It doesn't cover correction of the OTDR response.

Presently, three fundamental OTDRs' measured parameters are defined : distance, loss and reflectance. A selection of methods developped or set at LCIE to calibrate OTDRs relatively to these parameters is presented.

3.1. DISTANCE CALIBRATION

3.1.1. *General principle*

The objective of distance calibration is to determine deviations (errors) between the measured and actual distances between points on a fibre, and to characterize the uncertainties of these deviations.

An OTDR measures the location L of a feature from the point where a fibre is connected to the instrument, by measuring the round-trip transit time T for a light pulse to reach the feature and return. L is calculated from T using the speed of light in vacuum, c (2.99792458 x 10^8 m/s), and the group index N of the fibre :

$$L = \frac{c\,T}{2\,N} \qquad (4)$$

Errors in measuring L will result from scale errors and offsets in the timebase of the OTDR and from errors in locating a feature relative to the timebase. Placing a marker in order to measure the location may be done manually or automatically by the instrument. The error will generally depend on both the marker placement method and the type of feature.

In order to characterize location error, a specific model will be assumed that describes the behavior of most OTDRs. Let L_{ref} be the reference location of a feature from the front panel connector of the OTDR, and let L_{otdr} be the displayed location. It is assumed that the displayed location, L_{otdr}, using OTDR averaging to eliminate noise, depend functionally on the reference location L_{ref} in the following way :

$$L_{otdr} = S_L\,L_{ref} + \Delta L_0 + f(L_{ref}) \qquad (5)$$

Here S_L is the scale factor, which ideally should be 1. ΔL_O is the location offset, which ideally should be zero.

Finally, $f(L_{ref})$ represents the distance sampling error, which is also ideally zero. The distance sampling error is a periodic function with a mean of zero and a period equal to the distance interval between sampled points on the OTDR.

Equation (5) is meant to characterize known errors in location measurements, but there may still be an additive random uncertainty. This will affect both the distance measurements and the accuracy with which parameters describing the errors can be determined by the procedures below.

S_L and ΔL_O may be determined by measuring L_{otdr} for different values of L_{ref}, then fitting a straight line to the data by the least squares method. S_L and ΔL_O are the slope and intercept, respectively.

Equivalently, a line may be fitted to the difference function $L_{otdr} - L_{ref}$, in which case ΔS_L is the slope, and ΔL_O is still the intercept ; this is illustrated in figure 3. After finding the linear approximation, the distance sampling error, $f(L_{ref})$, respectively its half amplitude, ΔL_{sampl}, may be determined by measuring departures from the line for different values of L_{ref}.

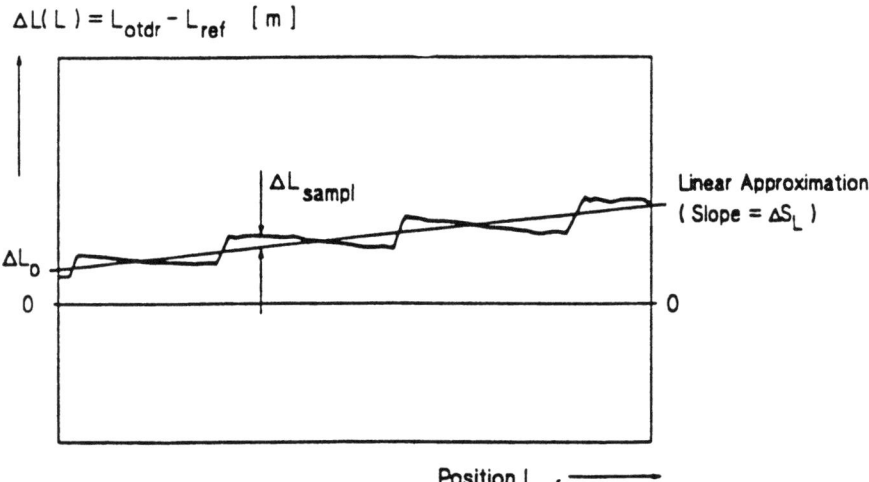

figure 3 : representation of the location error, $\Delta L(L)$

In this document, the distance sampling error amplitude, ΔL_{sampl}, is treated as part of the (random) location readout uncertainty.

Therefore, the result of the distance calibration shall be stated by the following parameters :

$\Delta S_L, \sigma_{\Delta SL}$: the distance scale deviation and its uncertainty ;
$\Delta L_O, \sigma_{\Delta LO}$: the location offset and its uncertainty ;
$\sigma_{Lreadout}$: the location readout uncertainty.

Note that the uncertainty will depend on the distance, response displayed power level, and on the instrument settings.

3.1.2. Recirculating delay line method

A fibre-type recirculating delay line to be used as calibration artifact for OTDR distance calibration is described here below.

As illustrated in figure 4, the device is constructed from the following :

a) a four-port coupler with a long fibre (typically several km) fusion spliced between a pair of input and output ports ;
b) a lead-in fibre (typically 1 km) connectorized to the OTDR, which is spliced to the second input port ;
c) an output fibre which is spliced to the second output port of the coupler and which is terminated with a low back-reflection connector. The length of this fibre piece is kept short, e.g. < 1 metre ;
d) a gold-plated connector or front surface mirror, preferably with a reflectivity of nearly 100 %, which can be used to create a reflection at the end of the output fibre.

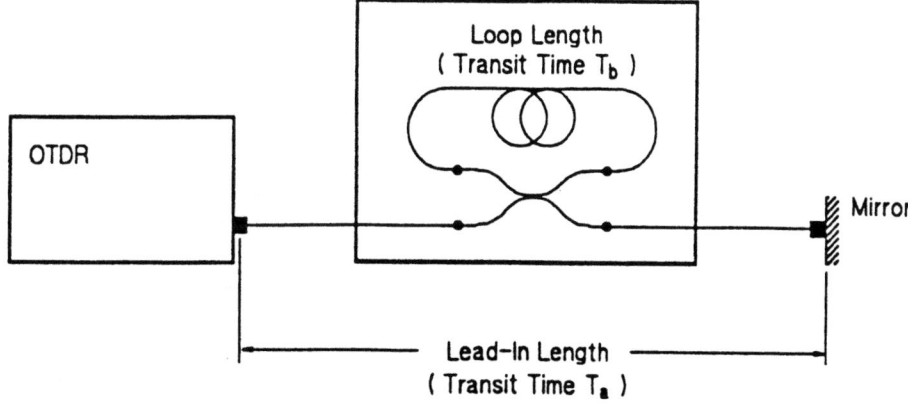

It is necessary to calibrate the recirculating delay line for two parameters, the transit time of the fibre in the loop T_b and the transit time of the lead-in length T_a. The latter is the sum of the transit times of the lead-in fibre, the coupler pigtails and output fibre. Transit times T_a ant T_b are measured in accordance with IEC793-1 standard [5]. For a group index N set to 1.46, it is possible to calculate lengths L_a and L_b as :

$$L_a = \frac{C \cdot T_a}{N} \quad \text{and} \quad L_b = \frac{C \cdot T_b}{N}$$

The recirculating delay line places a number of reflective features on the OTDR display. The first feature is the one obtained from the optical pulse directly travelling to the miror, then directly back to the OTDR. The second feature is generated by the optical pulse travelling once through the loop, then to the mirror, then directly back to the OTDR (this pulse coincides with the pulse travelling directly to the mirror, then back through the loop, then back to the OTDR). The third pulse travels through the loop twice, etc...

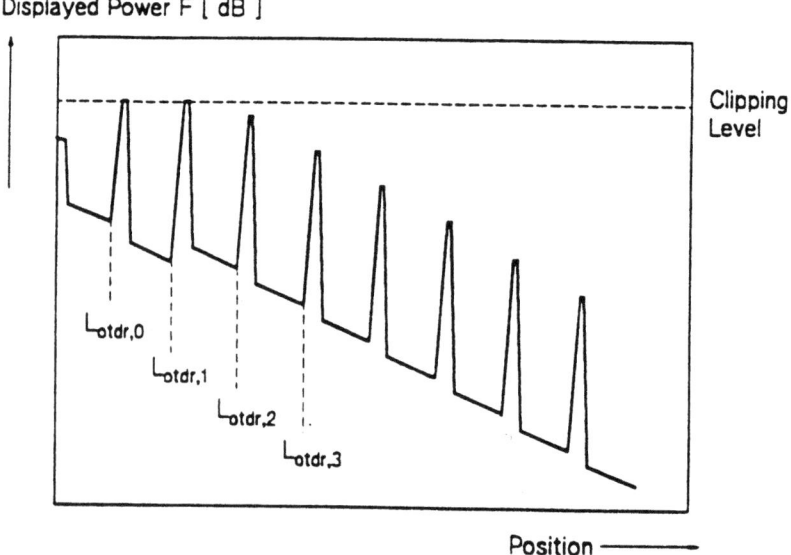

figure 5 : OTDR trace produced by the recirculating delay line

Accordingly, the ideal displayed locations would be:

$L_{otdr,0} = L_a$,
$L_{otdr,1} = L_a + L_b/2$,
$L_{otdr,2} = L_a + L_b$ etc.

where L_a is the length of the the lead-in fibre, and L_b is the length of the fibre loop.

The measurement procedure assumes that no incremental fibres are used and is described hereafter:

- Measure the locations of successive reflections from the recirculating delay line with the OTDR. Record these as $L_{otrd,i}$, where the index i goes fom O to k and represents the number of pulse flights through the loop.

- Using the calibration data of the recirculating delay line, T_a and T_b, the series of reference locations is:

$$L_{ref,i} = \frac{c \left(T_a + i \cdot \frac{T_b}{2} \right)}{N} \qquad (6)$$

where N is the group index set to 1,46.

- Use the displayed locations, $L_{odtr,i}$, and the reference locations to calculate the series location errors, ΔL_i:

$$\Delta L_i = L_{otdr,i} - L_{ref,i} \qquad (7)$$

- Use a least-squares procedure to fit a line to the data. Specifically, choose ΔS_L and ΔL_O so that the summation

$$\sum_i (\Delta L_i - \Delta S_L L_{ref,i} - \Delta L_O)^2 \qquad (8)$$

is minimized.

- Record the location offset, ΔL_O, and the distance scale deviation, ΔS_L, obtained from the approximation.

Regarding the uncertainty calculation, experiment and theory show that the nature of the processed parameters allows to apply the "combined standard uncertainty", used to collect a number of individual uncertainties into a single number. The combined standard uncertainty is based on statistical independence of the individual uncertainties; this leads to a root-sum-square of their standard deviations. The

$$\sigma = \left[\sum_{i=1}^{k} \sigma i^2 \right]^{1/2}$$

where σi is the well known experimental standard deviation and i the number of individual uncertainties to take into account.

Applying the standard formula for the propagation of errors yields the distance scale uncertainty, $\sigma_{\Delta SL}$:

$$\sigma_{\Delta SL} = \left[\left(\frac{\sigma_{Dotdr}}{D_{ref}} \right)^2 + \left(\frac{\sigma_{Dref}}{D_{ref}} \right)^2 \right]^{1/2} \quad e.g. \ [m/km] \tag{9}$$

where :

σ_{Dotdr} = the uncertainty of the displayed OTDR distances.
σ_{Dref} = the uncertainty of the reference distances.

The location offset, ΔL_O, is equal to the intersect of the least-squares approximation with the vertical axis. Using $i = 1$ as the representative sample, equation (8) yields a location offset which can be used as the basis of the uncertainty calculation :

$$\Delta L_O \approx L_{otdr,1} - \frac{c \, T_a}{N} \tag{10}$$

The location offset uncertainty, $\sigma_{\Delta Lo}$, can be calculated from (10) by using the standard formula for the propagation of errors :

$$\sigma_{\Delta LO} = \left[\sigma_{Lotdr,1}^2 + \left(\frac{c}{N} \right)^2 \sigma_{Ta}^2 \right]^{1/2} \tag{11}$$

$\sigma_{Lotdr,1}$ = the uncertainty of the first displayed distance.
σ_{Ta} = the documented uncertainty of the lead-in time of recirculating delay line.

Two other methods can be used to calibrate the distance range of an OTDR and are not further discussed here. They are called the external source method and the concatenated fibre method.

3.2. LOSS CALIBRATION

3.2.1. *General principle*

The objective of the loss calibration is to determine the loss deviation, $\Delta S_A(F)$ for power levels within the OTDR's backscatter regime and to evaluate the measurement uncertainties. $\Delta S_A(F)$ is a function of the displayed power level, F; it includes both the inaccuracy of the displayed loss and the nonlinearity of the OTDR's power scale.

For each measured loss, the displayed power level or an equivalent parameter that can be used to reproduce the vertical position of a measurement sample is to be determined. This level is termed F.

Unless otherwise specified, the OTDR's clipping level shall be used as the (default) reference point for determining F : $F_{ref} = 0$ dB. All values of F shall be stated in relation to this reference point. The clipping level was chosen because it represents the most reproducible level on most OTDRs' power scales.

The principle of the loss calibration is to apply a device of known (reference) loss, A_{ref}, to the OTDR and to measure the displayed loss, $A_{otdr}(F)$, as a function of the displayed power level F, as shown in figure 6.

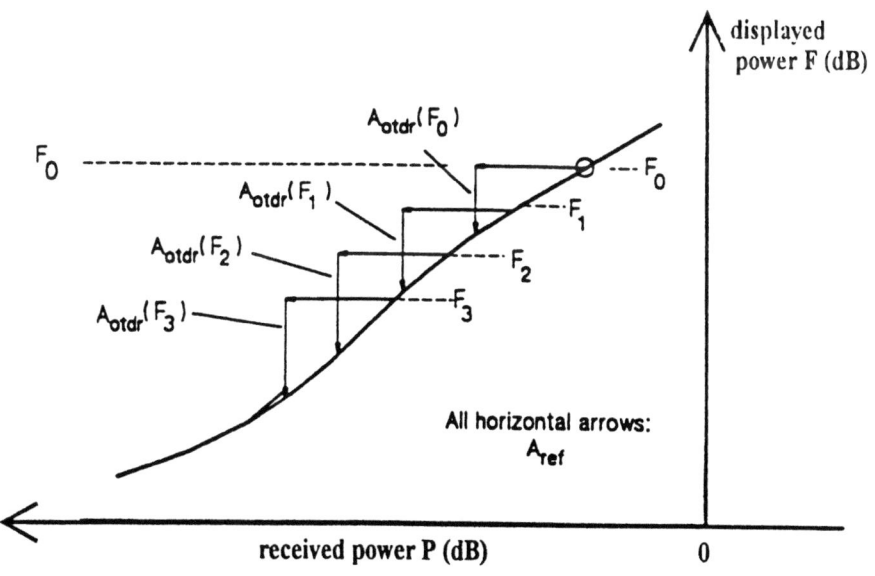

figure 6 : measurement of the ODTR's loss function

Notice that the F values denote the high power ends of $A_{otdr}(F)$. The displayed power level is to be varied over the whole backscatter range of the OTDR, with approximately equal vertical spacings ; a spacing of 0.5 - 1 dB, and not more than the reference loss, A_{ref}, is recommended. The loss function may be distance, as well as power level, dependant.

From a theoretical point of view, an infinitesimally small value of A_{ref} would be desirable to perform the OTDR loss calibration. In practice, values of A_{ref} being too small may result in additional measurement uncertainty due to OTDR noise, whereas large values of A_{ref} tend to obscure fine details. Therefore, the specific value of A_{ref} used in the calibration has to be documented. The recommended range for A_{ref} is 0.5 to 2 dB.

The reference loss can be an actual or simulated fibre optic component. Note that fibre optic components usually exhibit wavelength-dependant losses. Therefore, it is necessary to know the OTDR's center wavelength.

The detail of the calculation of the calibration results is described hereafter.

From the measured values $A_{otdr}(F_i)$ calculate the loss scale factors, $S_A(F_i)$:

$$S_A(F_i) = \frac{A_{otdr}(F_i)}{A_{ref}} \quad [dB/dB] \tag{12}$$

and the loss deviations, $\Delta S_A(F_i)$:

$$\Delta S_A(F_i) = \frac{A_{otdr}(F_i) - A_{ref}}{A_{ref}} = S_A(F_i) - 1 \quad [dB/dB] \tag{13}$$

The loss uncertainties, $\sigma_{\Delta SA}(F_i)$, are discussed in conjunction with the calibration method and are described hereafter.

The error of a loss value measured with the OTDR, $\Delta A(F) = A_{otdr}(F) - A_{ref}$, and its uncertainty can be calculated from the calibration results by the following formula, in which the recommended confidence level of 95 % is used :

$$\Delta A(F) = [\Delta S_A(F) \pm 2\,\sigma_{\Delta SA}(F)]\, A_{ref} \quad [dB] \tag{14}$$

In this formula, A_{ref} can be replaced by the displayed loss, $A_{otdr}(F)$, without serious consequence. Notice that this error applies, in a rigorous sense, only to measured losses at the displayed power level and displayed location for which $\Delta A(F)$ was recorded.

3.2.2. *Loss calibration with the shuttle pulse system method*

This method uses a shuttle pulse system which generates a train of descending and regularly spaced pulses from a pulse emited by the tested OTDR. The top of each pulse displayed on the OTDR's screen consitutes a measurement sample. The combination of the shuttle pulse system with a variable attenuator allows complete characterization of the OTDR power scale over the entire dynamic range. The method is well suited to fully automated laboratory testing under computer control. No electronic equipment is needed for the calibration.

In addition to the tested OTDR, the measurement equipment includes :
a. a variable optical attenuator (attenuator # 1)
b. an optical attenuator (attenuator # 2) with good repeatability to toggle out a specific quantity of attenuation (the recommended value is in the range 0.5 - 2.0 dB)
c. a shuttle pulse system as described hereafter.

The shuttle pulse system (see figure 7) is constituted by an optical fibre (length 500 - 4000 m) set between two mirrors M1 and M2. The first mirror reflective coefficient is such that pulses' amplitude is of the same order than the backscatter level usually detected by the OTDR. Mirror M1 is made of a 98 % reflective coating on the cavity fibre face of a half pitch selfoc lens S1. The other face of S1 is coupled with a lead-in fibre, launching the emited OTDR's pulse into the cavity. This coupling selfoc is equipped with light traps to reduce clutter echoes as low as possible (diaphragm D, index liquid, surface processing). Mirror M2 is made of a 100 % reflective coating on the cavity fibre face of a quarter pitch selfoc lens S2. The attenuation between two successive pulses depends on the transmission and reflection coefficients at the cavity ends, on the losses due to the fibre and on the optical couplings quality. The cavity fibre length has to be chosen so that the attenuation between two successive measurement samples doesn't exceed 2.0 dB.

The test setup is shown in figure 7.

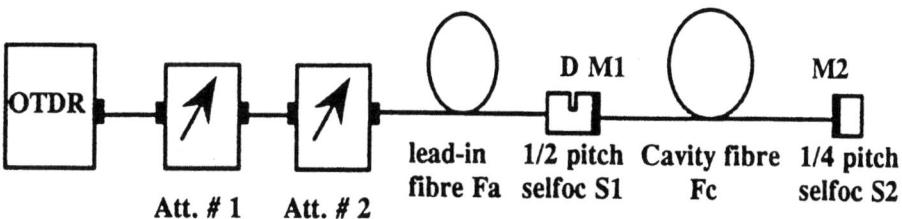

figure 7 : setup for loss calibration with the shuttle pulse system

The purpose of attenuator # 1 is to set measurement samples in region A defined as an approximation to the region where the user normally takes measurements (see figure 8). The purpose of attenuator # 2 is to generate the reference attenuation in the range 0.5-2.0 dB, according to calibration requirement. Note that attenuator # 1 is optional, considering that attenuator # 2 can perform both actions. The shuttle pulse system signature on the OTDR display is shown in figure 8.

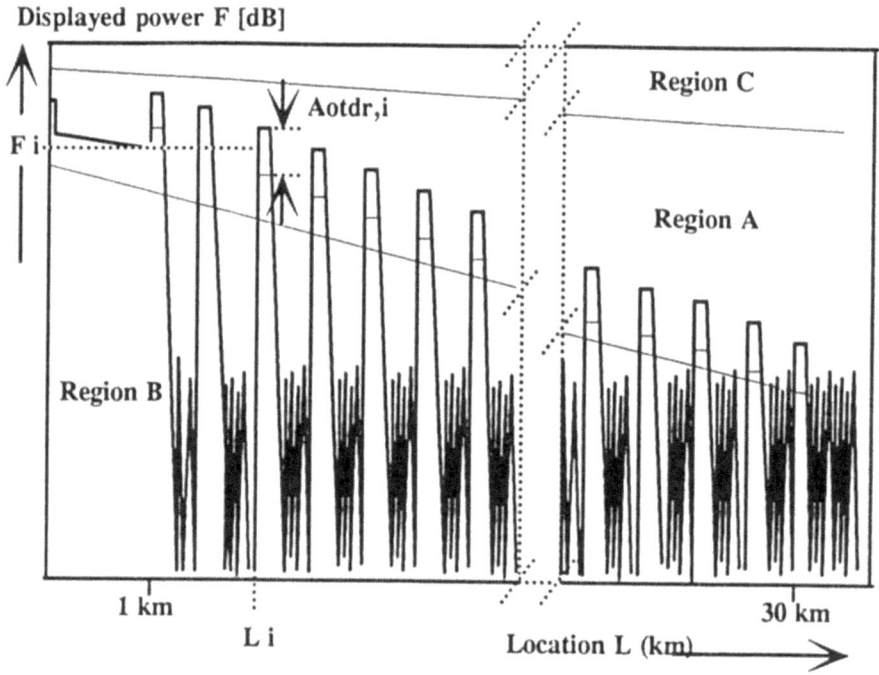

figure 8 : OTDR display with the shuttle pulse system

Calibration of the reference loss is performed with the help of an optical source which wavelength corresponds to the desired wavelength calibration, and a power meter.

A typical radiometric measurement allows to calibrate attenuator # 2, and then the reference loss.

This calibration can be completed with measurements concerning the attenuator dependence to polarization.

The measurement procedure to determine the loss scale deviation $\Delta S_A(F)$ is described hereafter.

Connect the shuttle pulse system to the OTDR output via attenuators # 1 and # 2. Select an appropriate attenuation of attenuator # 1 to set measurement samples within region A of the OTDR display. Set an attenuation of 0 dB on attenuator # 2 and record the power level $F_i(0)$ for each i order pulse. Set an attenuation of X dB which gets a calibrated value and record the power level $F_i(X)$ for each i order pulse.

$A_{otdr,i}$ is calculated with the following formula :

$$A_{otdr,i} = F_i(0) - F_i(X) \qquad [dB] \tag{15}$$

where $A_{otdr,i}$ referes to the power level F_i calculated with the following formula :

$$F_i = \frac{F_i(0) + F_i(X)}{2} \qquad [dB] \tag{16}$$

For each F_i level measurement, calculate the loss deviation $\Delta S_A(F)$ with the formula :

$$\Delta S_A(F) = \frac{A_{otdr}(F) - A_{ref}}{A_{ref}} \qquad [dB/dB] \tag{17}$$

Record the displayed power level, F_i, and the location, L_i, associated with each loss deviation value. It may also be advisable to plot the loss deviation values as a function of the displayed power level F_i.

The standard deviation characterizing the loss uncertainty can be calculated from (17), using the standard formula for the propagation of errors :

$$\sigma_{\Delta SA}(F) \approx \frac{1}{A_{ref}} \left(\sigma^2_{A_{ref}} + \sigma^2_{A_{otdr}} \right)^{1/2} \qquad [dB/dB] \tag{18}$$

The uncertainty σ_{Aref} should be accumulated, by root-sum-squaring, from the following contributions :

$\sigma_{A,pm}$ = the (systematic) uncertainty in dB due to calibrating the reference loss with the power meter, e.g. due to the power meter's nonlinearity, polarization dependence, inhomogeneity and noise. This uncertainty should itself be accumulated by root-sum-squaring ;

$\sigma_{A,step}$ = the (random) loss uncertainty in dB introduced by the instability of the attenuation step ;

$\sigma_{A,\lambda}$ = The (systematic) uncertainty in dB caused by the difference between the optical source center wavelength used to measure A_{ref} and the OTDR's

The uncertainty $\sigma_{A,otdr}$ gets only one contribution :

$\sigma_{A,otdr}$ = the random uncertainty in dB introduced by the variability of the attenuation step measurements, e.g. caused by limited readout resolution and displayed power levels when they approach the noise limit.

Three other methods can be used to perform the loss calibration of an OTDR and are not further discussed here. They are called the external source method, the fibre standard method, and the splice simulator method.

3.3. REFLECTANCE CALIBRATION

3.3.1. *General principle*

The objective of the reflectance calibration is to determine the reflectance deviation, $\Delta R(R_{ref})$, and to evaluate the measurement uncertainties. $\Delta R(R_{ref})$ is a function of the measured reference reflectance R_{ref}, it includes both the inaccuracy of the displayed reflectance and the non linearity of the OTDR's power scale.

The principle of the reflectance calibration is to apply a standard device generating a known (reference) reflectance, R_{ref}, to the OTDR and to measure the displayed reflectance, $R_{otdr}(R_{ref})$, as a function of the reference reflectance value.

The power level F (see loss calibration) in relation with the reflectance measurement has to be mentionned to clearly define the reflectance measurement conditions. The level F can be measured just before the beginning of the leading edge of the reflected pulse to be measured.

A large number ($\simeq 10$) of reference reflectances R_{ref} would be desirable to perform the OTDR reflectance calibration. The reference reflectance can be an actual or simulated fibre optic component. Note that fibre optic components usually exhibit wavelength dependant losses. Therefore, it is necessary to know the OTDR's center wavelength.

To compute the calibration results, proceed as follows : from the discrete measured values $R_{otdr}(R_{ref})$ calculate the discrete reflectance deviations $\Delta R(R_{ref})$:

$$\Delta R(R_{ref}) = R_{otdr}(R_{ref}) - R_{ref} \quad [dB] \tag{19}$$

The reflectance uncertainties, $\sigma \Delta R(R_{ref})$, are discussed in conjunction with the calibration method and are described in the next paragraph.

3.3.2. Reflectance calibration with the reflectance simulator method

The OTDR is calibrated using a reflectance simulator which generates reflectances in the range - 10 dB to - 74 dB with a step of 0.1 dB.

The reflectance simulator (see figure 9) is constituted by a lead-in fibre of length L_o launching the OTDR's emitted pulse in the input port of a Y type coupler. Output ports 1 and 2 of the coupler are connected together (making a loop) via a variable attenuator and two similar fibres of equal length L fusion-welded at point M.

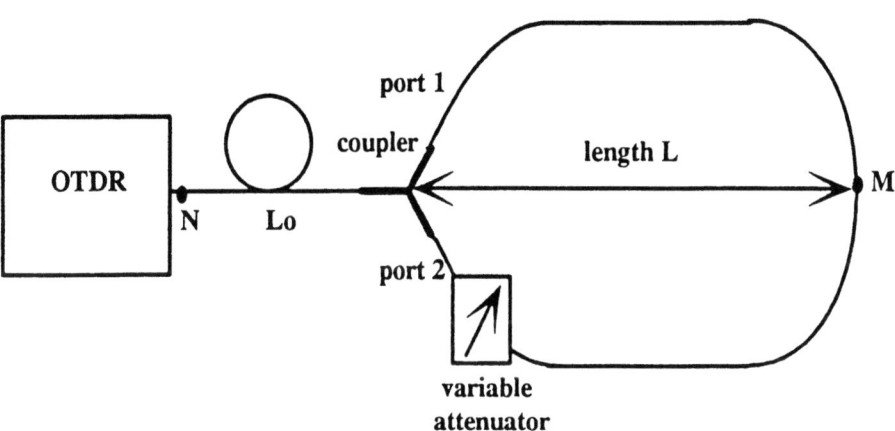

figure 9 : setup for reflectance calibration with the reflectance simulator

Such a device creates a simulation of a reflection peak at a distance $L_o + L$ (point M) from the OTDR (see figure 10). This peak is superimposed to the backscattered trace displayed on $(L_o, L_o + 2L)$ by both fibres of length L. This reflection seems to result from a fictitious event located at point M. In fact, the incident pulse arriving to the input port of the coupler is separated into two half-pulses travelling in opposite directions in the loop. Resulting pulses coming back to the OTDR after their transit in the loop are added together and are superimposed to the fibres backscattering trace.

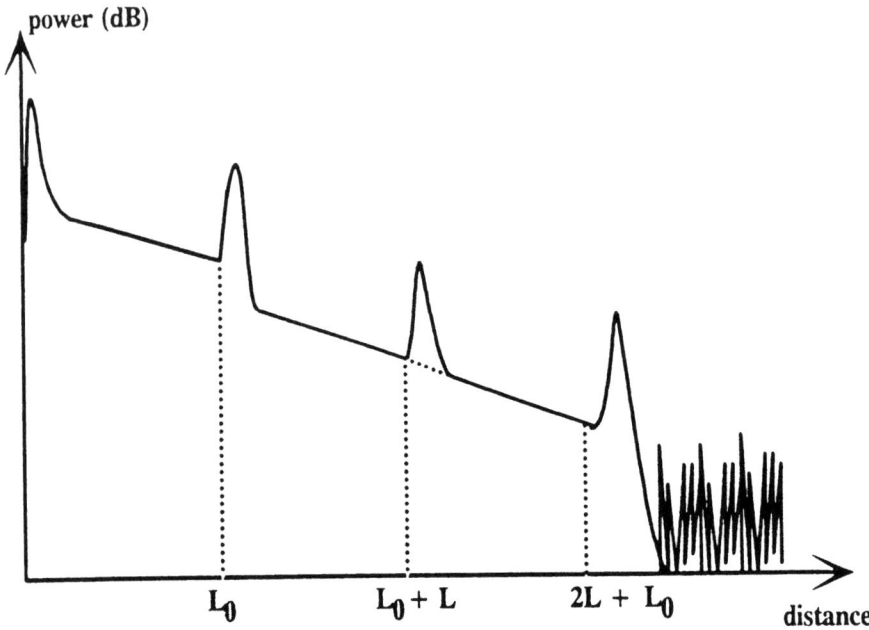

figure 10 : OTDR display with the reflectance simulator

Both half-pulses are detected by the OTDR at a time corresponding to a distance of $2(L_o + L)$. The backscattered signal is continuously detected during the time interval corresponding to the distance interval $[2L_o, 2(L_o + 2L)]$. It can be shown that the reflectance R, equivalent to the peak at point M, can be expressed as :

$$R(dB) = 10 \log_{10} \left[\frac{P_1 A_2 + P_2 A_1}{P_1 A_1 + P_2 A_2} \right] \quad (20)$$

where P_1 and P_2 are the powers received at point M from outputs 1 and 2 respectively when a continuous light source is used in place of the OTDR, and where A_1 and A_2 are the loss coefficients between M and N in the M to N direction through outputs 1 and 2 respectively. The different terms of relation (20) are determined with radiometric measurements.

Variable reflectances are generated with the help of the variable attenuator in accordance with relation (20).

The measurement procedure to determine the reflectance scale deviation $\Delta R(R_{ref})$ is described hereafter :

Connect the reflectance simulator to the OTDR.
Adjust the variable attenuator attenuation value to get the desired reference reflectance R_{ref}. With the OTDR, measure the reflectance $R_{otdr}(R_{ref})$ of each reference reflectance value R_{ref}.

For each R_{ref}, calculate the reflectance deviation $\Delta R(R_{ref})$ in accordance with relation (21) :

$$\Delta R(R_{ref}) = R_{otdr}(R_{ref}) - R_{ref} \quad [dB] \tag{21}$$

The standard deviation characterizing the reflectance uncertainty can be calculated from (21), using the standard formula for the propagation of errors :

$$\sigma \Delta R(R_{ref}) = = \sqrt{\sigma_{R_{ref}}^2 + \sigma_{R_{otdr}(R_{ref})}^2} \quad [dB] \tag{22}$$

The uncertainty σR_{ref} should be accumulated, by root-sum-squaring, from the following contributions :

$\sigma_{R,pm}$ = the (systematic) uncertainty in dB due to calibrating the reference reflectance with the power meter, e.g. due to the power meter's non linearity, polarization dependence, inhomogeneity and noise. This uncertainty should itself be accumulated by root-sum-squaring ;

$\sigma_{R,step}$ = the (random) reflectance uncertainty in dB introduced by the instability of the reflectance ;

$\sigma_{R,\lambda}$ = the (systematic) uncertainty in dB caused by the difference between the optic source center wavelength used to measure R_{ref} and the OTDR's center wavelength.

The uncertainty $\sigma_{R,otdr}$ gets only one contribution

$\sigma_{R,otdr}$ = the measurement uncertainty in dB due to the measurement repeatability.

4. Conclusion

Only a few laboratories are ready to provide calibration services for OTDRs.

Presently the parameters considered are distance, loss, reflectance and the influence of the polarization conditions. European intercomparisons are underway through EUROMET and BCR projects to validate the calibration methods for distance and loss.

The international standardization is in progress and a document has been proposed in the IEC/TC86/WG4 to define calibration of OTDRs for distance and loss. It should be completed by the reflectance calibration.

REFERENCES

[1] S.A. Newton "A new technique in optical time domain reflectometry".
RF and microwave measurement symposium and exhibition - Hewlett-Packard.

[2] Anritsu corporation. "Operation manual OTDR MW9040B". June 1992. Version IV

[3] A. HARTOG, M. GOLD. "On the theory of backscattering in simple-mode optical fibres".
Journal of Lightwave Technology, vol. LT2, n° 2, april 1984, pp. 76-82.

[4] IEC-TC86-WG4-SWG2/OTDR (Hentschel 9)
"Calibration of optical time domain reflectometers".
29 april 1994.

[5] IEC 793 "Optical fibres" ; Part one : Generic specifications (IEC 793-1) ; Part two : Product specifications (IEC 793-2)

OPTICAL SPECTRUM ANALYZER CALIBRATION

J. DUBARD, C. LE MEN
Laboratoire Central des Industries Électriques
33, av. du Général Leclerc
92260 Fontenay aux roses
FRANCE

1. Introduction

Because optical instruments responses are usually wavelength dependent, measurements performed in the fibre optics field require an accurate knowledge of the wavelength and the spectral characteristics of the optical source used. For instance, power meters have a calibration factor that is dependent on the wavelength of the source to be measured; fibre optics attenuator, and coupling ratio of couplers are strongly wavelength dependent. Therefore, there is a need to characterize the sources which are commonly fibre optics terminated.

Optical Spectrum Analyzers (OSA) are instruments developed for use in fibre optic communications. Such instruments can directly measure the optical spectrum output from an optical fibre. The fibre is connected to an input port implemented on the OSA by a fibre optic connector. In order to determine the wavelength and the spectral bandwidth of a source, the accurate measurements of the power distribution of the emitted light must be performed

Many parameters of this complex apparatus need tol be calibrated: spectral resolution, wavelength, displayed power level and linearity, mathematical function (determination of the spectral bandwidth using the envelope or the RMS techniques). For uncertainty calculation purposes, measurements of polarization and temperature dependence are investigated and evaluated. This paper is based upon a standard project. It first describes the principle of an OSA, then the set-ups to perform the calibration and investigation of the different parameters.

2. Principle of an Optical Spectrum Analyzer

2.1. CONFIGURATION

Optical Spectrum Analyzers developed for the fibre optics field are compact instruments and are composed of an optical unit and a display unit. The units can be physically

separated or integrated in the same hardware. The optical unit is made of:
- a collimating optical element
- a dispersive component (grating)
- a focusing optical element
- an entrance and an exit slit

The display unit gives access to the different functions of the OSA and displays the spectrum as well as the informations concerning the characteristics of the source measured.

2.2. TYPES OF OPTICAL SPECTRUM ANALYZERS

Two types of OSA are currently encountered on the market depending upon the monochromator configuration used: the single-pass monochromator or the double-pass monochromator.

2.2.1. *Single-pass monochromator*
A single-pass monochromator is composed of one grating (figure 1). Such configuration has the following advantages and disadvantages:

Advantages:
- High dynamic range: from +10dBm to less than -70dBm typically
- Large wavelength range: 350nm- 1700nm for the widest spectral range on the market(possibility to obtain photometric information for sources emitting in the visible)

Disadvantages:
- Low stray light rejection: limits the possibility to measure weak spectral lines close to a strong one. Typically the dynamic range away from a strong line is 40dB (50dB) within 1nm (5nm) of the peak level. Moreover, some OSA may not be equipped with sharp cutoff band selection filters which would be needed for the measurement of large spectral bandwidth sources.

Entrance slit

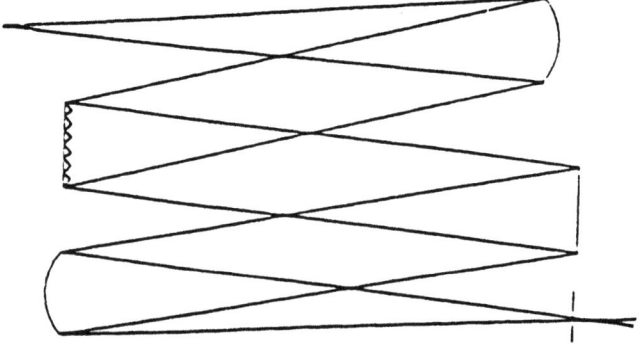

Exit slit

Figure 1: Diagram of a single-pass monochromator

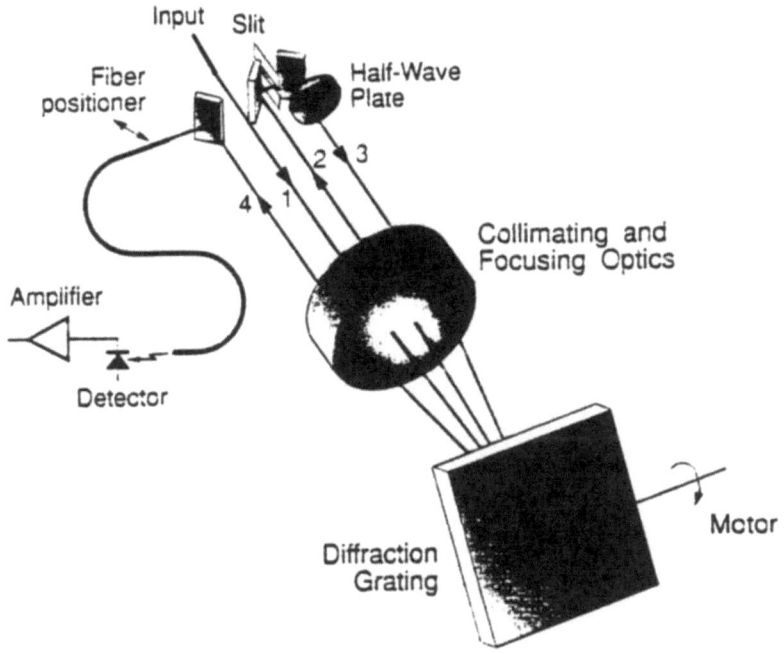

Figure 3: New double-pass monochromator design

2.3 APPARATUS FUNCTION OF A MONOCHROMATOR

The spectral resolution of a monochromator is determined by the apparatus function which is the convolution of the entrance and the exit slit widths. Three cases are encountered depending upon the combination of the width of the two slits. Figure 4 shows the three different cases and the resultant apparatus function which gives the spectral resolution of the monochromator knowing the dispersion factor in the plane of the exit slit (in nm/mm).

In any of the three cases, the width of the apparatus function is imposed by the width of the largest slit. Therefore the spectral resolution is determined by the largest slit, and the optimum throughput and spectral resolution are achieved when the widths of the entrance and exit slits are equal.

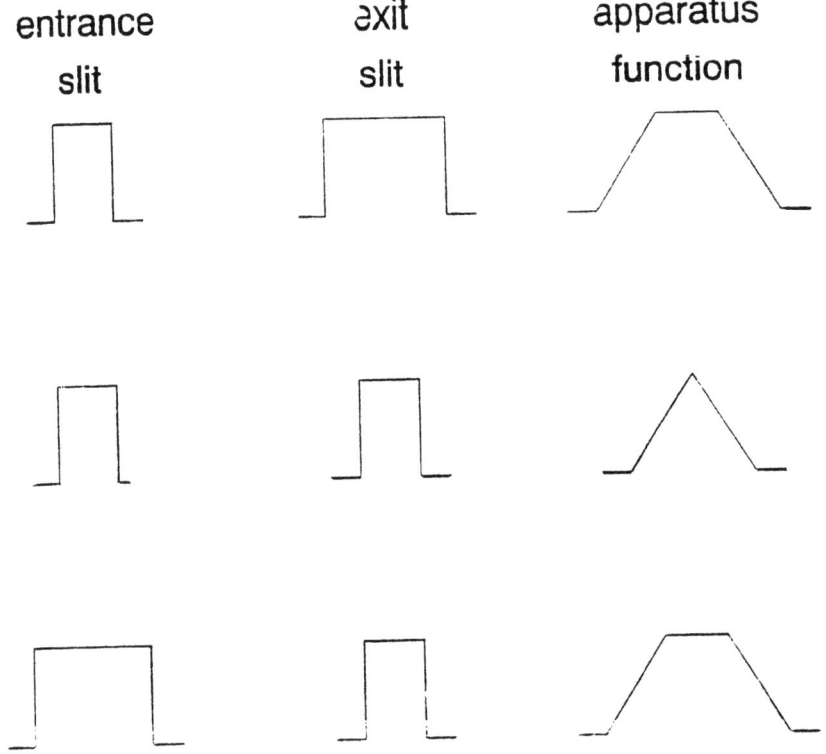

Figure 4: Apparatus function of a monochromator as a function of entrance and exit slits width combination

Figure 5 shows the spectral resolution measured for three combinations of entrance and exit slit widths. The source is a HeNe laser at 632.8nm. The largest slit width is 1mm. The dispersion of the monochromator is 1.76nm/mm.

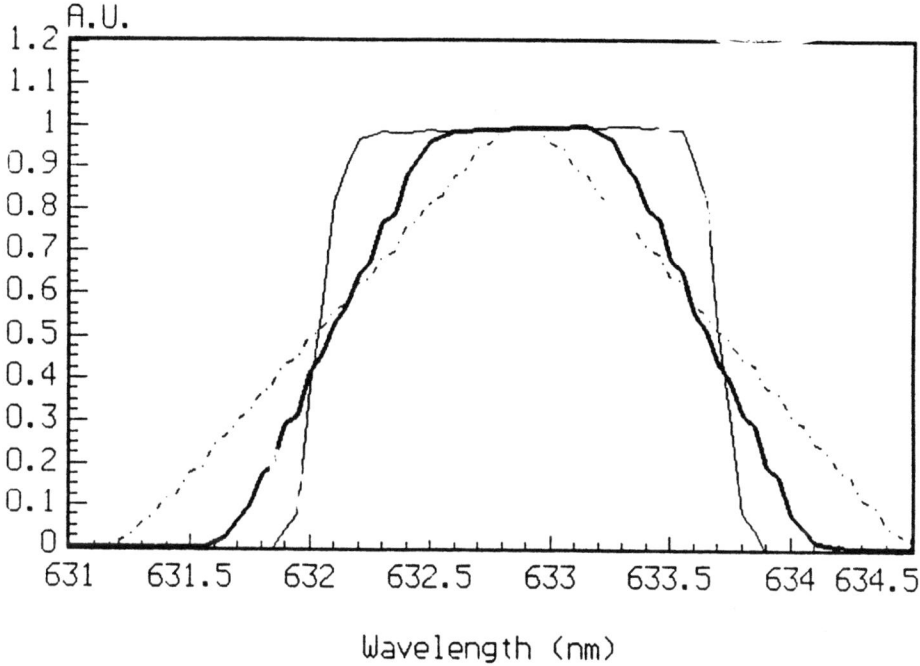

Figure 5: Spectra of a 632.8nm HeNe laser with different entrance/exit slits width combinations: dashed line 1mm/1mm, bold line 0.5mm/1mm, line 1mm/0.1mm or 0.1mm/1mm.

In the OSA, the entrance slit is the fibre itself. Therefore the spectral resolution will depend upon the fiber core diameter. This dependence can be determined using the set-up shown in figure 6.

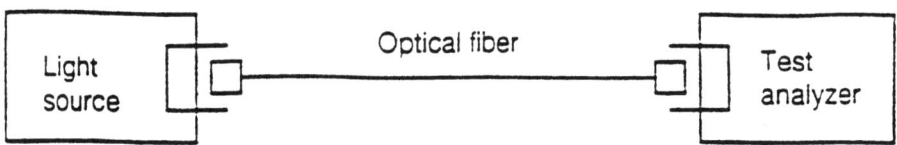

Figure 6: Set-up to emphasize the dependence of the spectral resolution of an OSA on the input fibre core diameter

The set-up uses a narrow spectral line source such as a gas laser, a DFB laser or a tunable laser diode (line width 100kHz), connected to the OSA through fibres with different core diameter. Table 1 gives the result of the spectral linewidth measured by the OSA (which is the effective spectral resolution) for four types of fibres: a singlemode fibre, 3 multimode fibres with core diameters of 50 μm, 62.5 μm and 600 μm. The source is a 632.8nm HeNe laser. The wavelength resolution setting of the OSA is 0.1nm.

Fibre	singlemode (9 μm)	multimode (50 μm)	multimode (62.5 μm)	multimode (600 μm)
Measured spectral linewidth (nm)	0.1	0.17	0.18	2.38

Table 1: Measured spectral linewidth as a function of the input fibre core diameter for a 0.1nm spectral resolution setting

The results show that the effective spectral resolution of the OSA is strongly dependent upon the fibre core diameter. In fact the spectral resolution setting of the instrument is valid for the singlemode fibre only. These results are related to the instrument under test.

Figure 7 shows the measured spectral linewidth of a DFB laser connected to the OSA through the 600 μm multimode fibre. The spectral resolution setting to get a near triangular spectral distribution is 2nm (equal entrance and exit slit widths). This setting is to be compared to the measured spectral linewidth indicated in table 1.

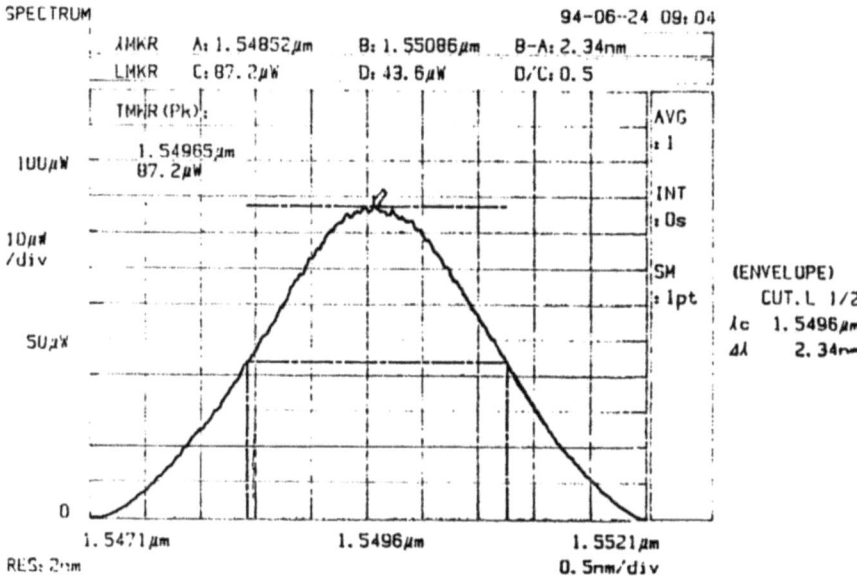

Figure 7: Spectral linewidth of a DFB laser connected to a 600 μm multimode fibre

3 Spectral resolution accuracy test

The spectral resolution accuracy test is performed using the set-up described in § 2.2 (figure 4). However the fibre used is a singlemode fibre. Table 2 gives the measured spectral linewidths of the light emitted by two HeNe lasers (632.8nm and 1523nm) as a function of the spectral resolution setting of the OSA.

Fluctuations of the measured spectral linewitdh of the two lasers (linewidth < < 0.1nm) is due to the weak signal level measured during the experiment because the power coupled into the 9 μm core diameter fibre is very low.

Spectral resolution setting (nm)	Measured spectral linewidth (nm) 632.8 nm	Measured spectral linewidth (nm) 1523 nm
0.1	0.1	0.1
0.2	0.196	0.208
0.5	0.504	0.508
1.0	1.008	1.064
2.0	2.0	2.08
5.0	5.0	5.14
10.0	9.92	-

Table 2: *Measured spectral linewidths of a 632.8nm and a 1523nm HeNe lasers as a function of the OSA spectral resolution setting.*

4 Wavelength calibration

Wavelength calibration of an OSA is important because the accuracy of the measurements will strongly depend on it. The set-up used is illustrated in figure 8.

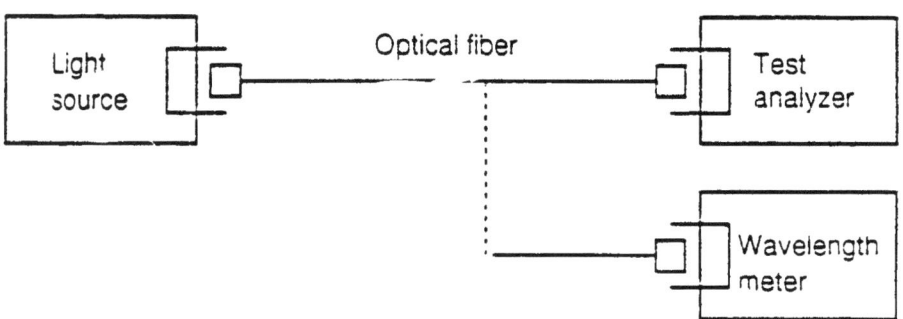

Figure 8: *OSA wavelength calibration set-up*

The source is a spectral lamp, a gas laser, or a laser diode with a narrow linewidth such as a DFB laser or a tunable laser based on an external cavity. In the latter cases the wavelength of the laser diode is measured using a wavelength-meter (10^{-6} uncertainty level). In order to achieve the best accuracy, singlemode fibre should be used (better spectral resolution).

Table 3 gives the value of the wavelength of the recommended gas laser sources. Some manufacturers commercialize a connectorized wavelength calibration source which is composed of several spectral lamps coupled to fibres. Table 4 gives the value of the wavelength of the high pressure Hg spectral lamp. The choice of wavelengths is great and can be increased using the second order dispersion when available (OSA without second order filters).

It should be noted that OSA gives the wavelength of a source as measured in the air. Consequently appropriate correction factor must be applied to wavelength calibrated sources which may be given relative to the vacuum.

Source	Wavelength [air] (nm)
Ar laser	487.9
	514.4
HeNe laser	632.8
	1152
	1523

Table 3: *Wavelengths of recommended laser sources for the wavelength calibration set-up of OSA*

The calibration procedure is the following:
- The OSA wavelength measurement range is set so that it includes the wavelength of the light source.
- The spectral resolution R is set such that:

$$R > 10 \frac{S}{N}$$

where S is the wavelength range and N is the number of displayed points.
- The center wavelengths $\lambda_{c,i}$ (i=1,n) of the source are measured.

Then the measured average value λ_{aver} of the wavelength of the source is:

$$\lambda_{aver} = \sum_{1}^{n} \frac{\lambda_{c,i}}{n}$$

The wavelength deviation can be determined using the following equation:

$$\Delta \lambda = \frac{\lambda_{aver}}{\lambda_{ref}} - 1$$

The OSA can be equiped with an external device (knob, screw) in order to perform a wavelength mechanical adjustment.

Waqvelength (nm)	Relative intensity
365.0	36
365.5	18
366.3	16
404.7	56
407.7	4
435.8	120
491.6	16
546.1	256
576.9	104
579.1	104
1014.0	340
1128.7	270
1357.0	88
1367.3	140
1395.0	32
1529.5	144
1692.1	40

Table 4: Wavelengths of a high pressure Hg arc lamp for the wavelength calibration set-up of OSA

5 Displayed power level calibration

The displayed power level calibration is performed using the set-up illustrated on figure 9.

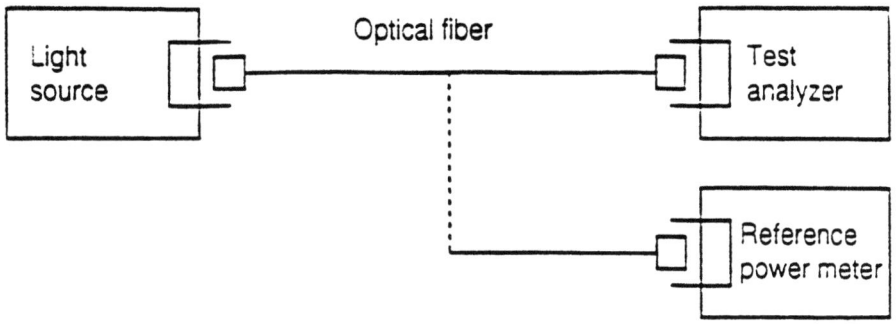

Figure 9: OSA displayed power level calibration set-up

The set-up includes:
- a light source
- a reference power meter

The displayed power level is the power of the source contained in the spectral bandwidth corresponding to the spectral resolution setting of the OSA. Therefore the spectral distribution of the source should be well known. This implies the use and the knowledge of the spectral power of a large spectral bandwidth source, or the use of a single narrow linewidth (< than the spectral resolution setting of the OSA) source such as a gas laser, a DFB laser or a tunable laser. The core diameter of the optical fibre used should be chosen according to the spectral resolution setting.

The displayed power level calibration procedure is the following:
- Measurement of the power level P_{ref} using the reference power meter
- Measurement of the displayed power level P by the OSA
- Determination of the displayed power level deviation ΔP according to the following equation:

$$\Delta P = \frac{P}{P_{ref}} - 1$$

The measurements are performed for each spectral resolution setting. The displayed power level deviation is dependent upon the input fibre core diameter because the effective spectral resolution of the OSA depends on it. Table 5 shows the displayed power level deviation as a function of the fibre core diameter for the different spectral resolution settings. The power measurements are performed using a DFB laser.

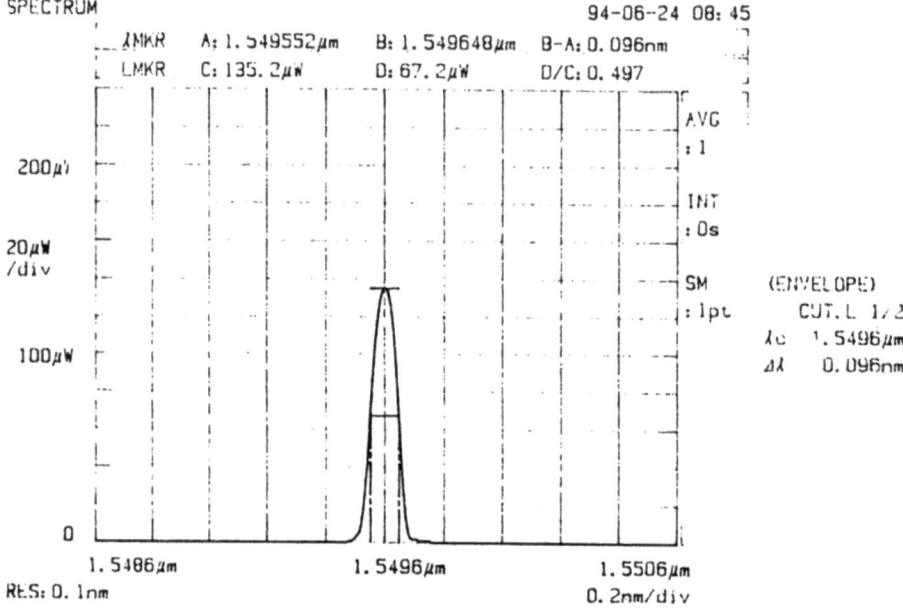

Figure 11: Spectrum of a laser diode(140 μW). Spectral resulution setting: 0.1nm

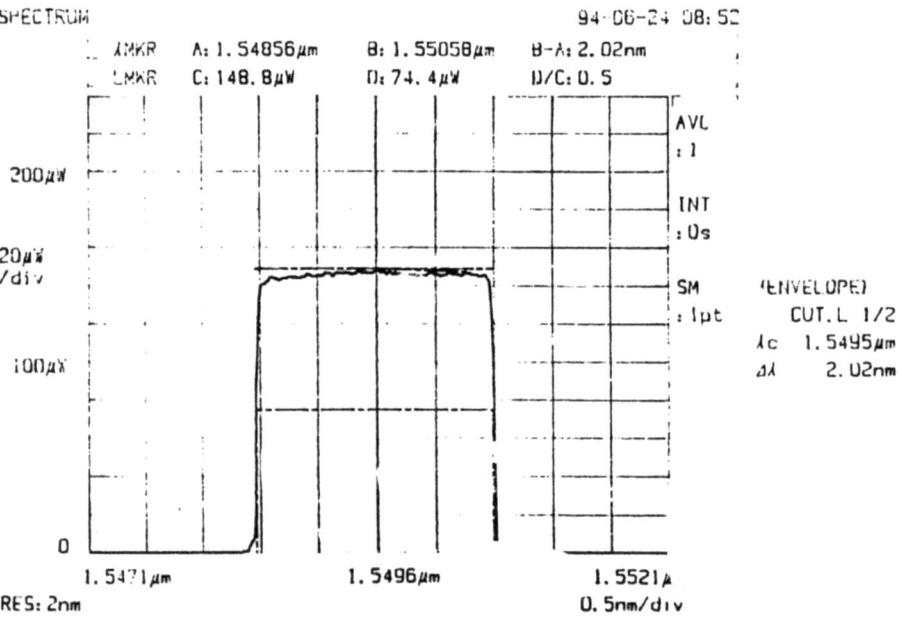

Figure 11: Spectrum ot a laser diode (140 uW). Spectral resolution setting: 2nm

	Displayed power level deviation			
Spectral resolution (nm)	singlemode 9 μm	multimode 50 μm	multimode 200 μm	multimode 600 μm
0.1	0.08	0.3	0.58	0.95
0.2	0.07	0.05	0.23	0.89
0.5	0.03	0.01	0.12	0.75
1.0	0.02	0.01	0.09	0.52
2.0	0	0	0.08	0.18
5.0	0	0	0	0

Table 5: *Displayed power level deviation as a function of the input fibre core diameter and for different spectral resolution settings.*

The displayed power level deviation of a few percent measured with a single mode fibre for the lowest spectral resolution setting is due to measurement errors because of the low signal level and reproducibility of the exit slit width. Figures 11 and 12 show the spectrum of a laser source measured at two different spectral resolution settings

6 Displayed power level linearity calibration

The displayed power level linearity calibration set-up is illustrated in figure 10.
The set-up includes:
- a light source, preferably a single narrow-line source
- a variable attenuator
- a linearity calibrated power metre

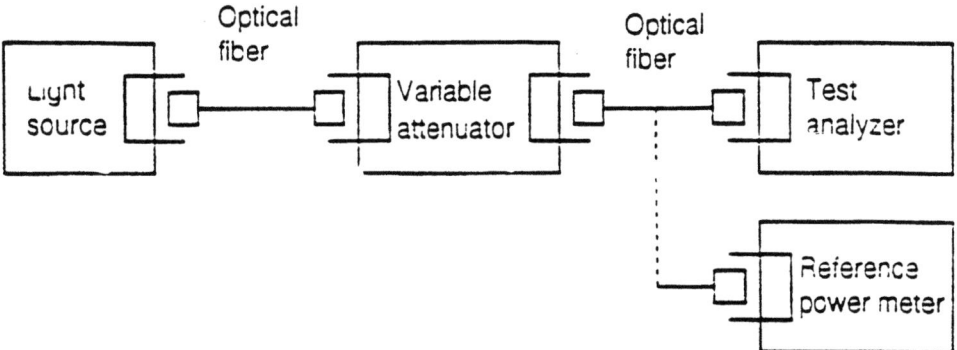

The displayed power level linearity calibration procedure is the following:
- Measurement of the power level $P_{ref,i}$ with the reference power metre for a given attenuation setting of the attenuator
- Measurement of the displayed power level $P_{d,i}$ by the OSA
- Determination of the linearity error R(i) given by:

$$\Delta R(i) = \frac{R_i - R_{ref}}{R_{ref}}$$

where

$$R_i = \frac{P_{d,i}}{P_{ref,i}}$$

and

$$R_{ref} = \frac{P_d}{P_{ref}}$$

P_d and P_{ref} are respectively the displayed power level and the reference power level measured during the displayed power level calibration.

The measurement procedure is repeated for each attenuation setting on the attenuator.

Table 6 shows the experimental data obtained when performing a displayed power level linearity calibration of an OSA using a DFB laser at a wavelength of 1306.9nm, a singlemode fibre and a spectral resolution setting of 0.1nm. These data allow the computation of R_i only. The complete analysis of the data requires the knowledge of the value of R_{ref} which is determined during the displayed power level calibration measurement procedure.

P_{ref}	Displayed power level
500 μW	745 μW
100 μW	142 μW
10.0 μW	14.0 μW
1.00 μW	1.40 μW
100 nW	140 nW
10.0 nW	14.0 nW
1.00 nW	1.15 nW

Table 6: Experimental data obtained during the displayed power linearity calibration measurement procedure

7 Spectral width measurement test

The OSA performs spectral width measurements using two methods:
- The envelope method
- The RMS (Root Mean Square) method

7.1 ENVELOPE METHOD

The spectral width is determined at half of the peak level of the envelope. The test can be done directly on the OSA using the available cursors.

7.2 RMS METHOD

This method requires the knowledge of the relative spectral power distribution P_i and the wavelength λ_i of each spectral component. The center wavelength λ_c and the spectral width B of the source are computed using the following equations:

$$\lambda_c = \frac{\sum P_i \cdot \lambda_i}{\sum P_i}$$

and:

$$B = 2.35 \sqrt{\frac{\sum P_i \cdot \lambda_i^2}{\sum P_i} - \lambda_c^2}$$

Figure 13 shows an example of the measurement of a multimode laser.

The spectral width measurement test of an OSA can be performed using such a source by comparing the width given by the OSA and the width computed from the value of the relative power distribution and the wavelengths of the different spectral components.

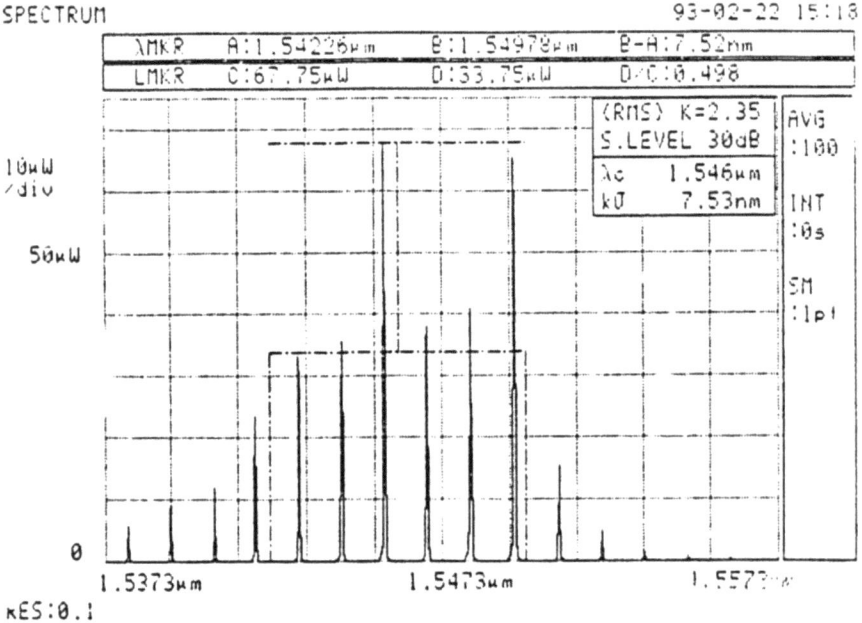

Figure 13: Spectrum analysis of a multimode source performed using an OSA.

8 Polarization dependence

The polarization dependence of the spectrum measurements performed with a particular OSA is determined using the set-up illustrated in figure 14.

The set-up includes:
        - a light source, preferably a single narrow-line light source
        - a connectorized polarization controller. The polarization state of the output of the fibre shall have an extinction ratio of 20dB. This can be measured using a polarization analyzer.

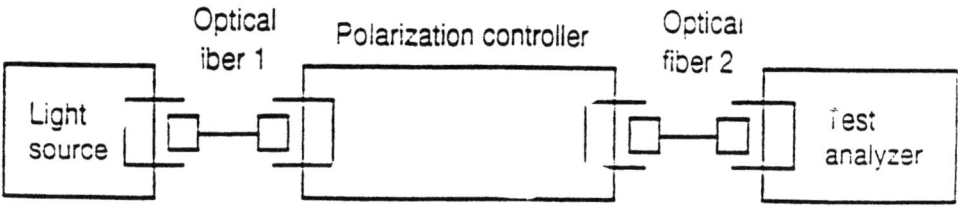

Figure 13: OSA polarization dependence measurements set-up

The procedure is the following:
- The polarization controller is set in order to get a linearly polarized state of polarization with an extinction ratio of at least 20dB
- The polarization state is rotated over a 180° range or more
- The maximum P_{max} and the minimum P_{min} readings are recorded

The polarization dependence of the OSA is determined from:

$$\Delta P_{pol} = \frac{(P_{max} - P_{min})}{P_{ave}}$$

where:

$$P_{ave} = \frac{(P_{max} + P_{min})}{2}$$

The procedure is repeated for several wavelength as polarization effects are wavelength dependent. During the measurements the fibres should be clamped to prevent them from moving.

9 Temperature dependence

Temperature dependence should be investigated when performing wavelength and displayed power level calibrations. This will provide useful informations to evaluate the uncertainty component related to temperature. Figure 15 shows an exemple of the measurement set-up used for the wavelength calibration temperature dependence test.

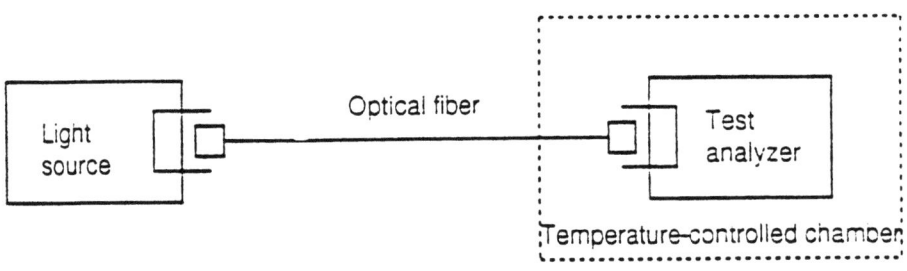

Figure 15: Wavelength calibration temperature dependence test set-up

The test procedure implies the measurement of the wavelength of a single narrow-line light source for at least five temperature settings in the operating temperature range of the OSA. For each temperature setting the deviation in wavelength is determined using the following equation:

$$\Delta \lambda_{T,i} = \frac{\lambda_{m,i}}{\lambda_{ref}} - 1$$

A similar procedure applies for the power level temperature dependence test. In this case one measures the deviation of the displayed power level for different temperature settings, with respect to the displayed power level at the reference temperature conditions.

10 Applications

Calibration of OSA is useful in order to get accurate data in many measurements. Of course OSA are designed mainly to perform spectral characterization of optical sources. They are used also in measurement set-ups such as the one developed for chromatic dispersion of fibre, Polarization Mode Dispersion (PMD) of fibre [2] or polarization dependent loss over wavelength of singlemode fibre components [3].

Figure 16 shows the experimental set-up for PMD measurement [2] using an OSA and figure 17 shows the result of the measurement performed on a singlemode fibre.

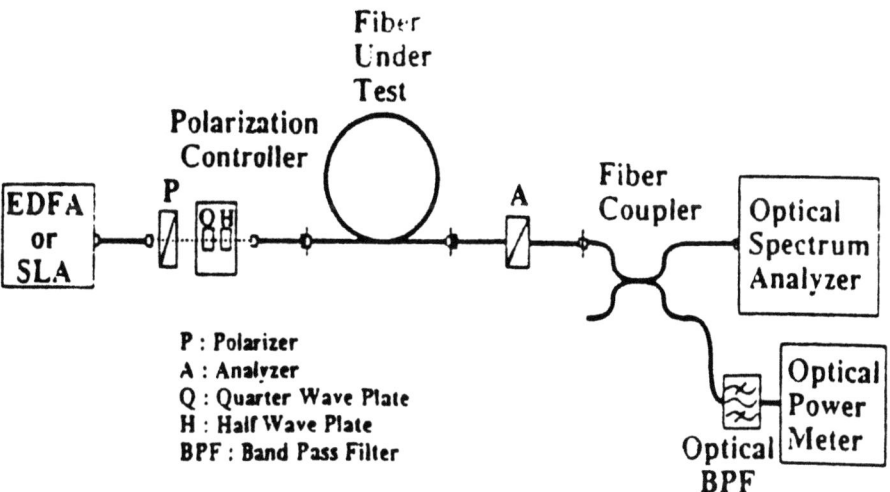

Figure 16: *Set-up for fixed analyzer PMD measurement method using an OSA*

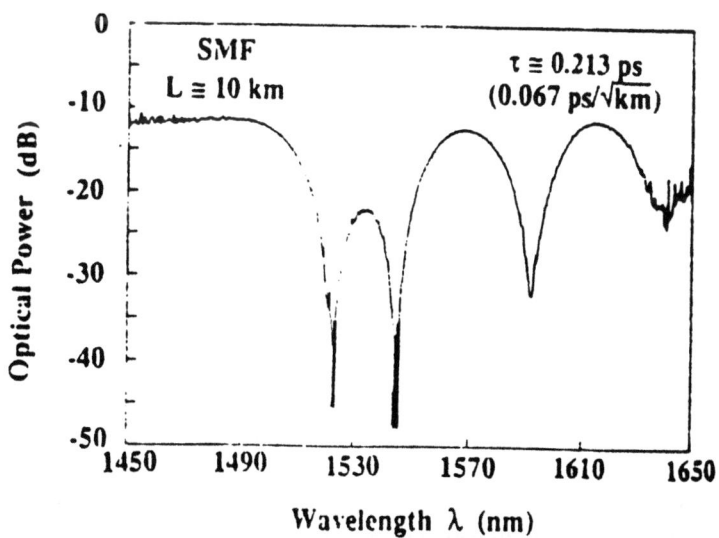

Figure 17: *Experimental result of a singlemode fibre PMD measurement using the*

Figure 18 shows the set-up used to perform the measurement of the polarization dependent loss over wavelength of singlemode fibre components. In this case the monochromatic light is available from the fibre connector at the monochromator output port.

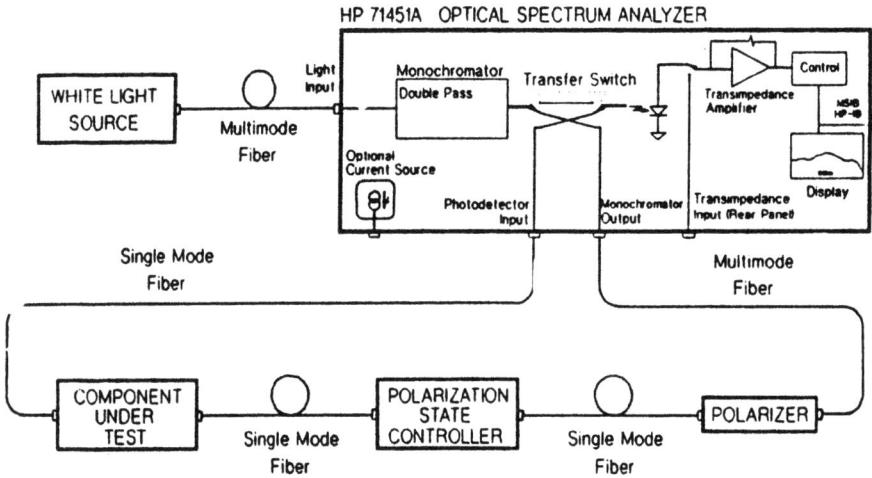

Figure 18: Set-up for measurement of polarization dependent loss over wavelength of singlemode fibre component

11 Conclusion

The Optical Spectrum Analyzer is a powerful instrument for characterizing fibre optic sources. However, accurate measurements can be performed only when such an instrument is correctly and fully calibrated. The different parameters to be calibrated are: the spectral resolution, the wavelength, the displayed power level, the linearity and the mathematical functions for the determination of the spectral bandwidth. Polarization and temperature dependence should be investigated in order to completely evaluate the uncertainty level associated to the calibration.

The different calibration procedures are presented in the performance sequence because accurate calibration of some parameters is dependent upon the previous calibration.

Calibration should be performed at ambient room temperature and using calibrated instruments which are traceable to national or international standards.

12 References

[1] Stimple Jim, (1992), *Les nouveaux analyseurs de spectre optique sont équipés d'un monochromateur original*, Opto N° 66, pp. 67-69

[2] Namihira Y., Maeda J., (1992), *Polarization mode dispersion measurements in optical fibers*, Technical digest, Symposium on optical fiber measurements, 1992, pp. 145-150

[3] Stokes L.F., (1992), *Accurate measurement of polarization dependent loss over wavelength of single mode optical fiber component*, Technical digest, Symposium on optical fiber measurements 1992, pp. 141-144

EDFA Testing

Problems and Solutions

ERIC MALZAHN
Hewlett-Packard GmbH
Böblingen Instruments Division
Herrenberger Straße 130
D-71034 Böblingen
Germany

Abstract

Erbium doped fiber amplifiers (EDFAs), purely optical amplifiers, are well on the way to replacing opto-electronic regenerators in undersea and long-distance links.
So far, their high gain and output power has enabled longer links with fewer regenerators. Their simple design has achieved higher reliability.
In future, they will allow increased system capacity because they offer wavelength division multiplex (WDM) transmission and have virtually unlimited modulation bandwidth. With their high output power, EDFAs will be used to drive upcoming fiber distribution networks in cable television (CATV) applications. This paper gives an introduction to the field of EDFA testing. Since the technology is still very new, the test methods are under ongoing investigation and are constantly being improved. Therefore this paper deals with the current problems and solutions for testing optical amplifiers.
It begins with basic measurements and definitions like output power, gain and noise figure. Well established measurement methods are explained and a turnkey solution for EDFA testing presented. Finally, the paper provides an overview of future trends.

1. Introduction

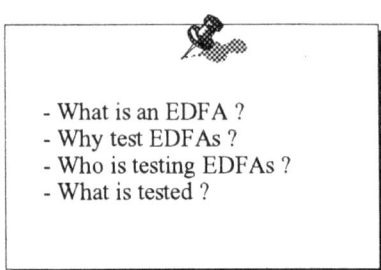

Figure 1. Introduction

Erbium Doped Fiber Amplifiers (EDFAs) are purely optical amplifiers consisting of an optical fiber which is erbium-doped, a pump-laser, which is the only active device, and some optical components like couplers, isolators and connectors. But this simple design is not the only advantage making EDFAs an attractive alternative to electric regenerators.

The power level of today's *high power* EDFAs already exceeds +26dBm (= 400mW). This high output power is needed

- to design links without repeaters, where the EDFAs are used as *booster amplifiers* at the beginning of a link.

- in cable television (CATV) and fiber to the home (FTTH) applications, where the information from one trunk line has to be delivered to many recipients.

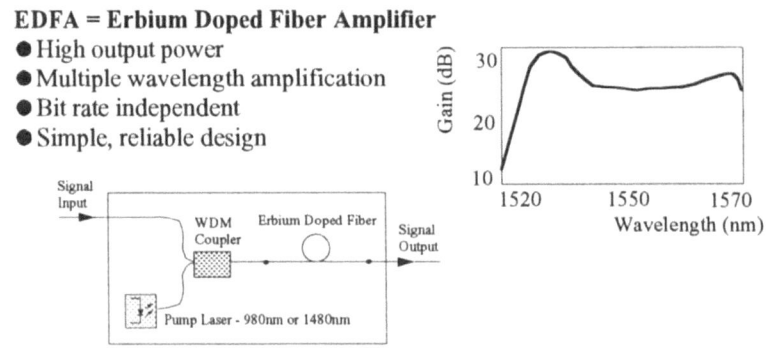

Figure 2. What is an EDFA?

As shown in figure 2, an EDFA amplifies over a broad wavelength range with almost equal gain. Today's amplifiers operate within a range of 1540nm - 1560nm. The main advantage of this multiple wavelength amplification is that today's single wavelength links will not have to be re-designed when *wavelength division multiplexing* (WDM) is applied in the future.

Finally, EDFAs are *bit rate independent*. That means that today's 2.5GBit/s links can be upgraded to 10GBit/s or even higher bit rates simply by only changing the transmitter and receiver station.

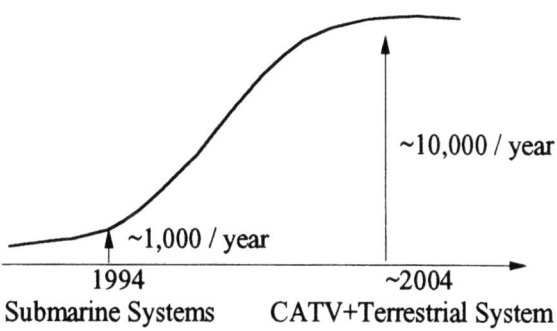

Figure 3. EDFA sales development

Figure 3 shows the expected sales development of EDFAs worldwide. Today's main market for EDFAs is *submarine systems*, and sales will grow tremendously as soon as the amplifiers are integrated into *CATV and terrestrial systems* as well.

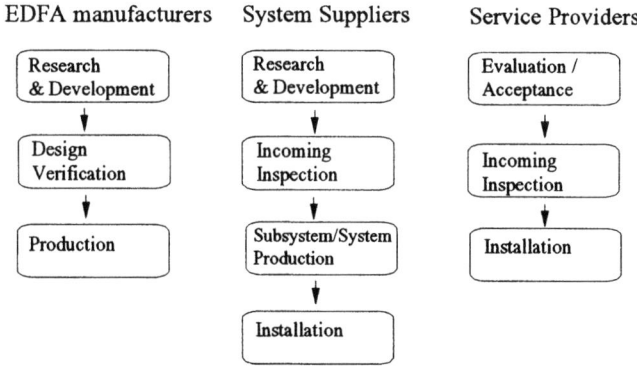

Figure 4. Who is testing EDFAs?

EDFAs need to be tested from the early stages of research, through production, up to the point where they are integrated into or maintained in an installed link. *EDFA manufacturers*, a growing number of small and large companies around the world, have to monitor the quality of their EDFAs in all phases. System suppliers, which are mostly big international companies, might develop their own amplifiers or buy them from EDFA manufacturers. Whatever they decide, they have to test the EDFA they are using, as do the end users, which are big telephone companies, who own and use links in which EDFAs are integrated.

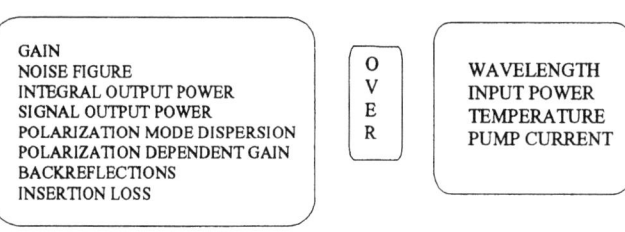

Figure 5. What is tested ?

The measurement requirements depend strongly on the *application* of the EDFAs. Since this is such a new technology the number of test parameters is increasing steadily. Figure 5 gives an overview of the current figures of interest.

People who use EDFAs as a *booster amplifier* are mostly interested in the signal output power. In between a link the EDFAs *refresh the signal*. Especially in long-haul links the noise level has to be kept as low as possible since the noise of one EDFA is amplified by all succeeding amplifiers. Before entering a receiver the signals are *preamplified*. These EDFAs have to be optimized for highest gain.

2. Basic Measurements

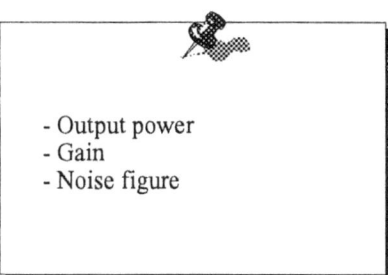

Figure 6. Basic measurements

As described in chapter 1, the three most *important test parameters* of an EDFA are output power, gain and noise figure. These figures give the manufacturer and the buyer of an EDFA a basic idea of the characteristics of the device.

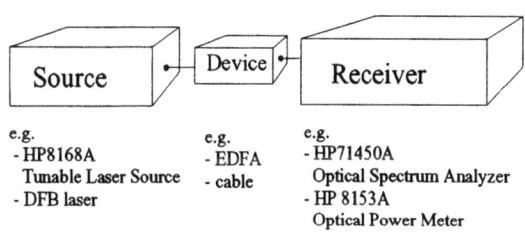

Figure 7. Basic measurement set-up

Depending on the application you can use a *distributed feedback laser* (DFB laser) or a *tunable laser source* (TLS) to measure the EDFA at one or at many wavelengths, respectively.
Figure 7 shows the set up for measuring
- the *total output power*
  (if an *optical power meter* is used as a receiver)
- the *input and output spectrum*
  (if an *optical spectrum analyzer* is used).

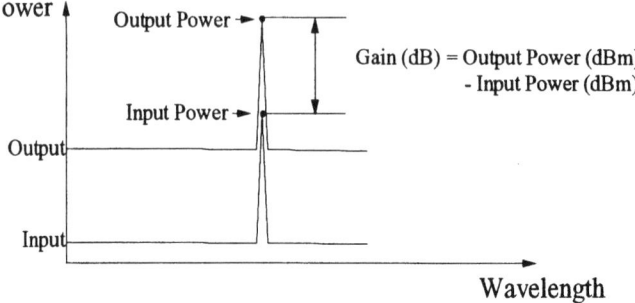

Figure 8. Gain measurement

With the basic measurement set up you can measure the optical spectrum of the EDFA's input signal (DFB or TLS) and the EDFA's output spectrum. The *gain* is defined as the difference between the signal output power and the signal input power. By IEC definition, the gain includes the loss of one connector pair as well /1/

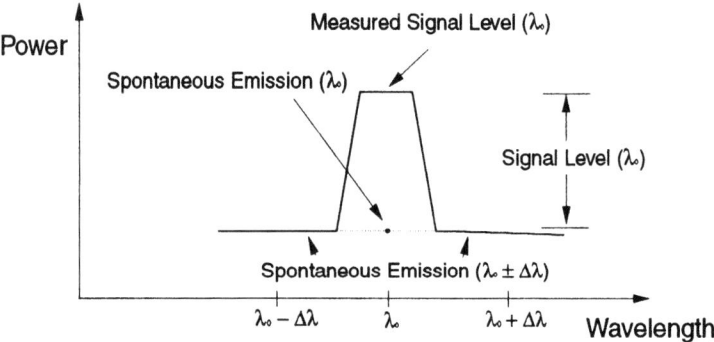

Figure 9. Signal / Noise level

Looking closer into the spectrum you can see that the *measured signal level* is not the true signal level since it also includes noise. The *true signal level* therefore is the difference between the measured signal and the noise at the signal wavelength. The noise - as shown in figure 9 - cannot be measured directly but has to be *interpolated* by measuring the *noise level* at two wavelengths, to the right and to the left of the signal wavelength.

Linear Interpolation
Spontaneous Emission (λ) = [S.E.(λ+Δλ) + S.E.(λ-Δλ)] / 2

Signal Level Correction for Spontaneous Emission
Signal Level (λ) = Measured Signal Level (λ) - Spontaneous Emission (λ)

Gain
Gain (λ) = Ouput Signal Level (λ) - Source Signal Level (λ)   (1)
 (dB)   (dBm)   (dBm)

Equation 1. Noise Correction

The *true gain* can then be calculated from the true output and input signal levels. However, for practical gain measurements this correction can be ignored since the effect of noise on the result is negligible.

N.F. = The degradation in signal-to-noise-ratio (SNR) from the input to the output of an EDFA

NF = SNR $_{INPUT}$ - SNR $_{OUTPUT}$ (dB)

Figure 10. Noise Figure Definition

The noise figure is the last and most complex of the three basic performance figures of an EDFA. The *noise figure* characterizes the degradation in the *signal-to-noise-ratio* (SNR) by comparing the input and the output of an EDFA. The assumptions for the noise figure definition are:
- shot noise limited source
- shot noise limited photodetector without
  thermal noise and a quantum efficiency of 1.

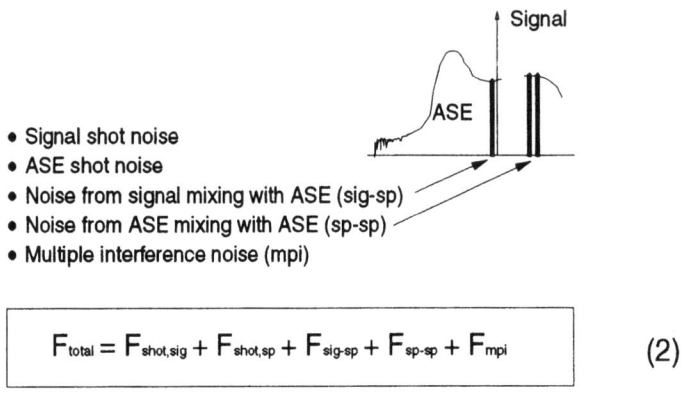

- Signal shot noise
- ASE shot noise
- Noise from signal mixing with ASE (sig-sp)
- Noise from ASE mixing with ASE (sp-sp)
- Multiple interference noise (mpi)

$$F_{total} = F_{shot,sig} + F_{shot,sp} + F_{sig-sp} + F_{sp-sp} + F_{mpi} \qquad (2)$$

Equation 2. EDFA noise figure

The *total noise figure* consists of five elements /2/. The signal and the *amplified spontaneous emission* (ASE) produce *shot noise* - a noise due to quantum effects - at the receiver. *Heterodyne mixing* of the ASE with the signal and with itself results in additional noise. Multiple interference noise occurs due to *internal reflections* between the EDFA's optical components like isolators, couplers and the pump laser.

Many different test methods can be used to measure the noise figure. The two mainstream measurements are the *optical and electrical methods*. To perform the *electrical test method* an *electrical spectrum analyzer* (ESA), a calibrated photodetector and a low noise optical source are needed. A basic optical set up was shown in figure 7.

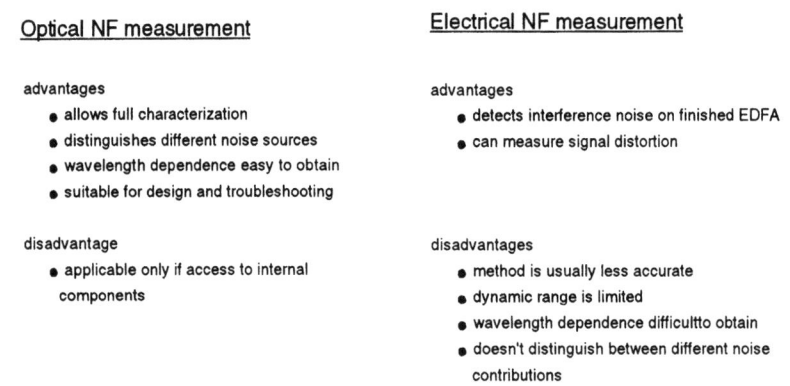

Figure 11. Optical versus electrical method

In a comparison of noise figure measurements performed by *the optical and electrical test method* it has been shown that both are capable of delivering the same results /2/. In practice, it can be very difficult to achieve the ideal conditions. Which measurement method is better depends very much on the *application*. In general, people who see the EDFA as a black box and have no need to distinguish between different sources of noise can use the *electrical test method*. These will be mostly the *service providers*. *Optical methods* are better for those people who need to know the performance data of the internal components. These will be mostly *EDFA manufacturers* and people doing *research on EDFAs*.

Since optical test methods allow distinguishing between different noise sources and deliver high accuracy, this will be the focus of this paper.

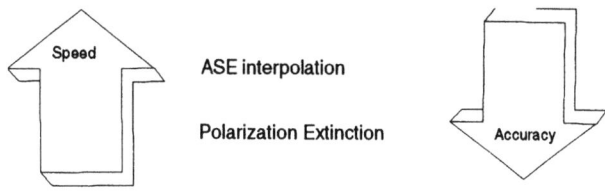

Figure 12. The right optical method

There are many optical methods, but here we will focus on the two most important: the ASE interpolation method and the polarization extinction method.

Choosing the right optical method also depends on the application. The *ASE interpolation method* allows testing at higher speed and lower costs since no polarization controller is required. However, the *Polarization Extinction method* delivers the highest accuracy.

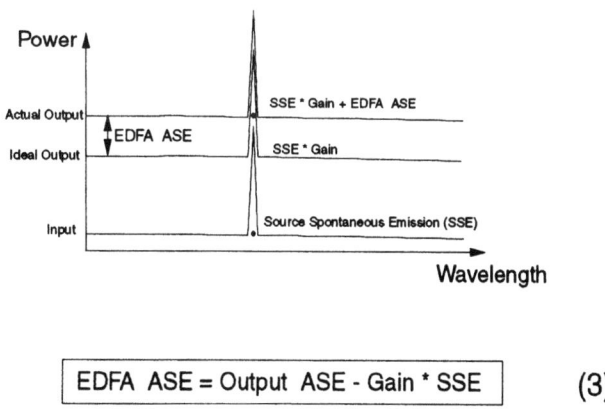

Figure 13. Equation 3. ASE Interpolation Method

You can measure the noise figure with the ASE interpolation method by using the basic measurement set up (figure 7). The source spontaneous emission (SSE) can be measured as the noise level of the input signal. The gain measurement was described above (figure 8) and the measurable output noise level is the sum of the amplified SSE and the ASE. In this way you can easily calculate the noise figure of the EDFA.

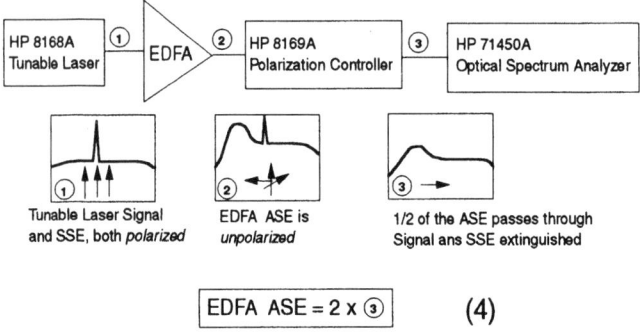

EDFA ASE = 2 × ③   (4)

Figure 14. Equation 4. Polarization Extinction Method

The second optical method presented in this paper is the polarization extinction method. This takes advantage of the fact that the tunable laser signal and the SSE are polarized whereas the ASE is unpolarized. Since the input signal is still polarized even after amplification, you can eliminate the tunable laser signal and the SSE with a polarizer. Half of the ASE still passes the polarizer since it is unpolarized. So the ASE can be calculated by doubling the measured spectrum.

$$F = \frac{1}{G} + \frac{EDFA\ ASE}{h\nu GB} \quad (5)$$

G = EDFA gain
h = Planck's constant
v = Frequency of optical signal
B = OSA bandwidth used to measure EDFA ASE
EDFA ASE = Amplified Spontaneous emission, in bandwidth B, produced by EDFA at signal wavelength

Equation 5. Noise figure determination from optical measurement

After measuring the gain and the ASE we have got all the components needed to calculate the *EDFA noise figure* (equation 5). Based on the total noise figure formula (equation 2) the first element is the shot noise. It is just the reciprocal value of the gain. The second element comes from the heterodyne mixing between the signal and the ASE. The last two elements, *ASE-ASE mixing and multiple interference* noise are *ignored* since their impact is usually very small.

3. HP 81600 EDFA test system

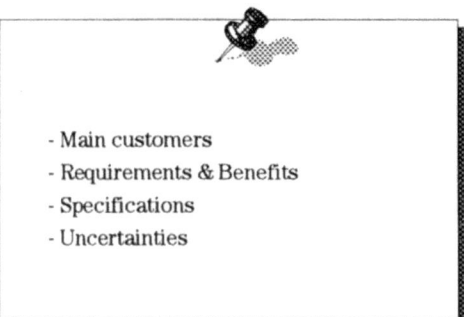

Figure 15. HP 81600 EDFA test system

Chapter 2 presented a basic measurement set up to perform EDFA testing. This way of testing is quite slow since it is a manual process. Moreover the accuracy can be improved by carefully improving the measurement methods and selecting the right equipment for EDFA testing. A *high performance solution* is presented in chapter 3.

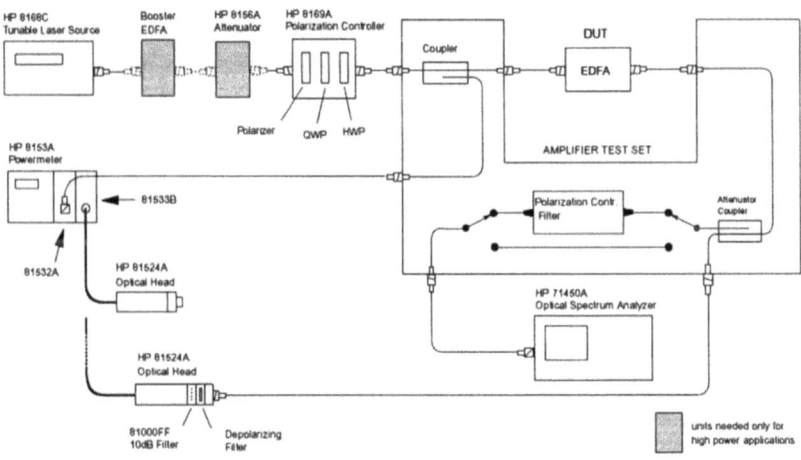

Figure 16. Block diagram of the HP 81600 EDFA test system

Building on the basic measurement set up, Hewlett-Packard developed a fully automatic EDFA test system. It also uses an HP 8168C tunable laser as source and an HP 71540A optical spectrum analyzer as receiver. The HP 8153A is used to monitor the input and output power continously with highest accuracy. To measure EDFAs at a higher input power you need to add a booster EDFA, an HP 8156A external attenuator and an HP 8169A polarization controller. This so called *"booster system"* can also measure *polarization mode dispersion (PMD)* and *polarization dependent gain (PDG)*. The system is controlled by a workstation and runs under UNIX®, which allows it to be integrated into a *local area network (LAN)* with e.g. access to a production database or performing remote measurements.

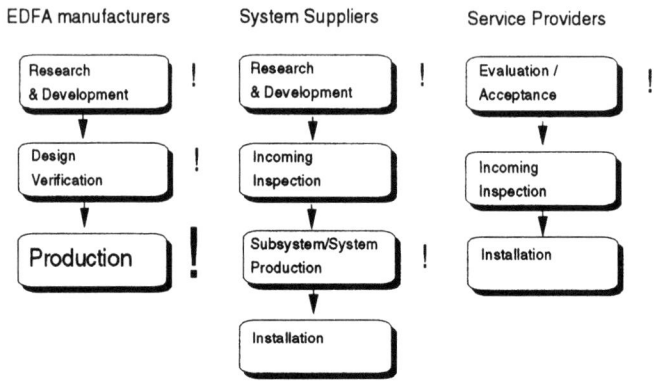

Figure 17. Main customers

The main customers for this system are *EDFA manufacturers*, people who need *high throughput with low uncertainty*. They have access to the internal components of the EDFA so there are no limitations to using this method, as described in chapter 2.

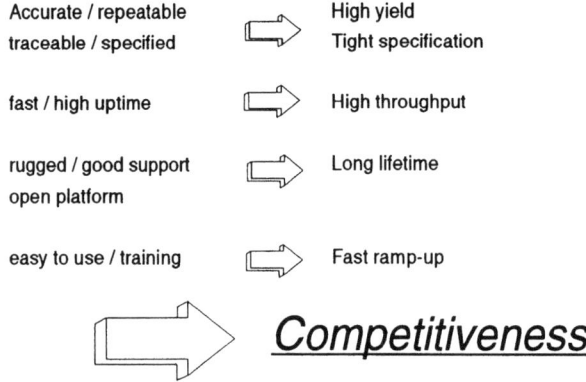

Figure 18. Requiremens and benefits

However, customers not only require high throughput; they also want a high yield to *reduce costs* and the ability to provide tight specification bandwidths to gain *competitive advantage*. To reduce the *time to the market* and *save money* during the learning curve, the ramp-up must be very fast. Finally, this investment should be *future-proof* based on good support and an upgrade path for enhancements.

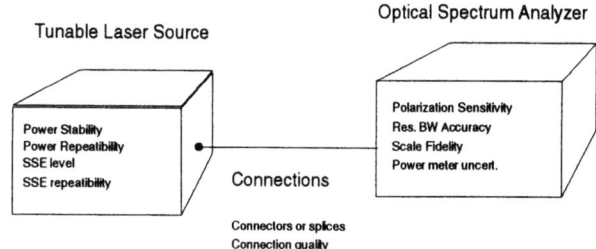

Figure 19. System uncertainties

To make EDFA test results worldwide comparable the system offers well analyzed and low uncertainties. This starts with a thorough calibration, which includes all instruments and the path losses in between them. The remaining uncertainties are caused by the light source (TLS), the device connections, the amplifier test set and the receivers (OSA and power meter).

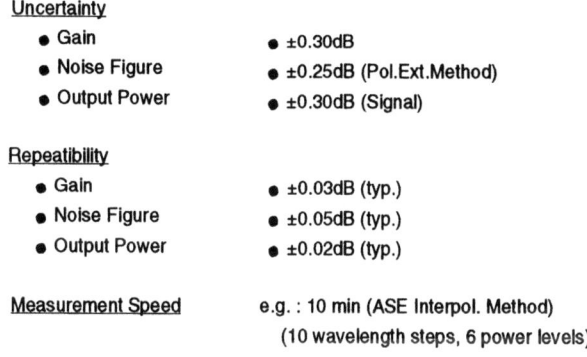

Figure 20. System specifications

As mentioned in figure 18 the requirements for the system's hardware are very high. Low uncertainty, high repeatability and fast measurement speed require *more than state of the art* technology. As well as support and training, the system provides all the benefits mentioned above.

4. Future trends

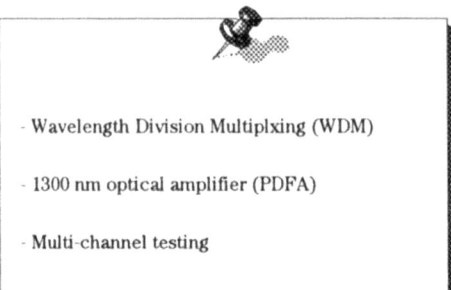

Figure 21. Future trends

Looking into the future of optical amplifier testing we can identify two main trends in the short term: *wavelength division multiplexing (WDM)* and *1300nm optical amplifiers*. Both technologies are in the transition from R&D status to production level.

Testing optical amplifiers at 1300nm requires dedicated lightwave components like the HP 8167A 1300nm tunable laser source, but the measurement methods will stay the same as for 1550nm.

Characterizing optical amplifiers for WDM communications is more difficult. The first approach would be to measure using exactly the same signals as will be used in the final link, e.g. to couple two, four or even more DFB lasers and use them as the test input signal. The certainty that the EDFA will behave in the same way int the real world is very high, but the method is expensive and inflexible. For example, when you wish to change the wavelength you would need to replace each laser.

A more sophisticated solution would be to use one laser to drive the EDFA to saturation and another one as probe laser. This set up is not only cheaper but also more flexible than the previous one. However the theory of testing EDFAs for WDM applications is not yet fully understood and presents many challenges for the future.

Acknowledgement

The author would like to thank his colleagues Dr. Christian Hentschel and Edgar Leckel for their suggestions, discussions and other contributions to this paper.

Literature

/1/ IEC Publication 1291-1
Generic Specification, Optical Amplifiers

/2/ C. Hentschel, E. Müller, E. Leckel
EDFA Noise Figure Measurements - Comparison between Optical and Electrical Technique -
HP Lightwave Symposium, May 1994

UNIX® is a registered trademark in the United States and other countries, licensed exclusively through X/Open Company Limited.

REAL-TIME HETERODYNE FIBRE POLARIMETRY BY MEANS OF JONES AND STOKES VECTOR DETECTION

R. CALVANI, R. CAPONI, F. CISTERNINO
CSELT - Centro Studi e Laboratori Telecomunicazioni S.p.A.
Via G. Reiss Romoli, 274 - 10148 Torino - Italy
Fax: +39.11.2285.085

1. Introduction

The study and measurement of the state of polarization (SOP) of the light propagating along a single-mode optical fibre, has been receiving particular attention in recent applications, such as coherent optical communications and fibre sensors [1, 2]. In both cases the polarization analysis requires a dynamical measurement of the SOP at the fiber output, which can be done on the basis of the Jones vector [3] (in terms of the amplitude ratio R and the relative phase ϕ of the two components of the electric field) or of the Stokes vector [4] (in terms of the intensity parameters usually represented on the Poincaré sphere to describe a fully polarized beam: s_0, s_1, s_2, s_3).

In this paper, the principle of the heterodyne interferometry has been reviewed, together with the fundamental polarimeter configurations suitable to detect Jones and Stokes vectors.

The inherent principle is the same as in coherent transmissions, where the optical information carrier and the local oscillator radiation give rise to an intermediate frequency (IF) beating. So, by analogy with this known technique, heterodyne interferometric polarimetry can be viewed as an optical mixing between a field, whose polarization has to be measured, and a reference beam at a slightly different frequency (linearly polarized at 45°). The superposition of these two radiations (or, in other words, their heterodyning) allows the measurement to be converted from the optical frequency into the radiowave range.

Two type of interferometers (Mach-Zehnder and Michelson) were used to demonstrate the heterodyne polarimetry principle, which is particularly suited to the dynamical exploration and real-time visualization of the SOP at the fibre output [3].

Moreover a method for studying the polarization evolution inside the fibre has been experimented. This non destructive technique, which is based on a stressing device moving along the fibre, is very accurate and can be applied to measure submillimetric beatlength or equivalently very large birefringences in highly birefringent (HB) fibres [5].

2. Theory of the SOP measurement

In all the polarimetric experiments described below, the heterodyne principle has been exploited to obtain SOP measurements in a range of frequencies allowing the use of commercial electronic instruments.

As shown in Fig. 1, the superposition of a measurand beam (characterized by a generally elliptical SOP) and a reference beam (linearly polarized at an angle of 45° in a suitable x, y frame) give rise to an optical mixing, which, after analysis with a polarization splitter, produces two beating signals at the transmission and reflection output of the Glan-Taylor prism, having x, y direction respectively.

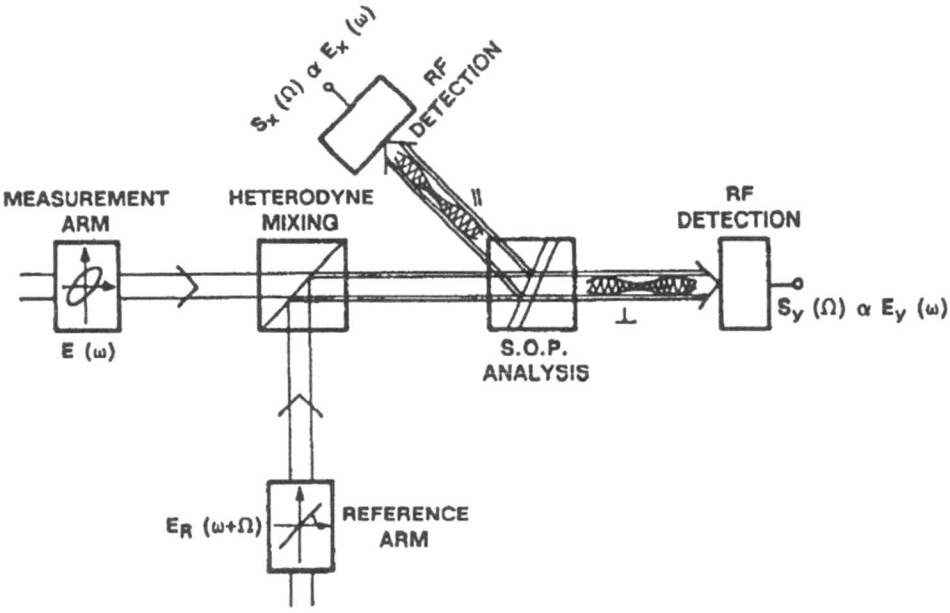

Figure 1. Heterodyne polarimetry principle.

In this way, each of the two components resulting from the SOP analysis, at the detection stage of the instrument, contain the beating between the measurand and reference beams. Furthermore, the amplitudes of these two radio frequency (RF) electrical signals are proportional to the corresponding components of the optical field, while their relative phase equals the polarization phase. So, all the relevant SOP information has been down converted from the optical wave to the IF beating signals and can be measured at RF.

To describe in detail the working principle of the proposed measurement scheme, some simple formulas become useful. In this representation the measurand $\vec{E}$ and the reference $\vec{E}_R$ fields are expressed as follows

$$E_x = E_{1x} e^{i[\omega_1 t + \phi + \phi_1(t)]}; \qquad E_{Rx} = (E_2/\sqrt{2}) e^{i[\omega_2 t + \phi_2(t)]}$$

$$E_y = E_{1x} e^{i[\omega_1 t + \phi_1(t)]}; \qquad E_{Ry} = (E_2/\sqrt{2}) e^{i[\omega_2 t + \phi_2(t)]}$$

(1)

where $\omega_1, \phi_1(t)$ are respectively the frequency and phase noise (due, in particular, to the laser source) of the measurand $\vec{E} = (E_x, E_y)$, while $\omega_2, \phi_2(t)$ are the corresponding quantities of the reference field $\vec{E}_R = (E_{Rx}, E_{Ry})$, linearly polarized at 45° (so that

$E_{Rx} = E_{Ry}$). After the superposition of $\vec{E}$ with $\vec{E}_R$ and the polarization analysis along the x, y axes on the output beam-splitter, the two detected IF signals S_x, S_y turn out to be proportional to the Jones vector components, rescaled at the beating frequency of the two interfering fields, which can be expressed in the form

$$S_x \propto E_{1x}E_2 \cos[\Omega t + \phi + \phi_{1,2}(t)]; \quad S_y \propto E_{1y}E_2 \cos[\Omega t + \phi_{1,2}(t)] \quad (2)$$

where $\Omega_2 = \omega_2 - \omega_1$ is the IF, $\phi_{1,2}(t) = \phi_2(t) - \phi_1(t)$ the relative phase noise of the two fields and ϕ the SOP phase to be measured. At this stage the relevant SOP information, carried by the IF signals S_x, S_y, can be extracted in two different ways [4].

The simplest and earliest procedure relies on the Jones polarimetry and consists in evaluating the amplitude ratio R=S_x/S_y and the relative phase ϕ of the S_x and S_y quantities with a vector voltmeter and/or in visualizing the SOP on a sampling oscilloscope operated in x-y mode [3].

A different way of determining the SOP to be measured can be obtained using the Stokes parameters in the Poincaré sphere representation. From the usual definition s_1, s_2, s_3 are related to the measurand signal components through the following equations

$$s_0 = E_{1x}^2 + E_{1y}^2; \quad s_2 = 2E_{1x}E_{1y} \cos\phi$$
$$\quad (3)$$
$$s_1 = E_{1x}^2 - E_{1y}^2; \quad s_3 = 2E_{1x}E_{1y} \sin\phi$$

where s_0 represents the optical intensity of the measurand, and corresponds (in the case of fully polarized light of the source) to the radius of the Poincaré sphere, according to the relation

$$s_0^2 = s_1^2 + s_2^2 + s_3^2 \quad (4)$$

which holds unchanged both at the input and at the output of an optical fibre, because the transmission medium does not modify significantly the polarization degree of the light during propagation.

To determine the three Stokes parameters, the electrical outputs S_x, S_y of the polarimeter are properly mixed and low-pass filtered in a suitable way, which generates (through IF subtraction) three baseband signals proportional to s_1, s_2, s_3.

It is noteworthy that the $\phi_{1,2}(t)$ phase term contained in S_x, S_y components cancels out in both Jones and Stokes versions of the heterodyne polarimeter. So the resulting SOP parameters of the totally polarized light remain unaffected by phase disturbances (especially laser noise) synchronously present in both polarization components. This allows the largest freedom in the generation of the reference beam, because its coherence with the measurand is not required. In particular it can be obtained from either an independent laser source (such as the local oscillator of a coherent transmission system) or from the same source of the measurement beam. The latter solution is the one used in the following experiments for its greater simplicity because only one laser is needed.

3. Mach-Zehnder interferometers for fibre polarimetry

3.1. NARROW-BAND SOURCES

The different ways of implementing the aforesaid heterodyne principle substantially come from the various possibilities of obtaining a reference field in the experimental schemes adopted, also taking into account the coherence properties of the real quasi-monochromatic sources available.

The simplest way to generate a reference field is to take it from the same optical source which illuminates the fiber under test. This naturally leads to a two-frequency Mach-Zehnder interferometer, as the one illustrated in Fig. 2, suitable for application to fiber characterization.

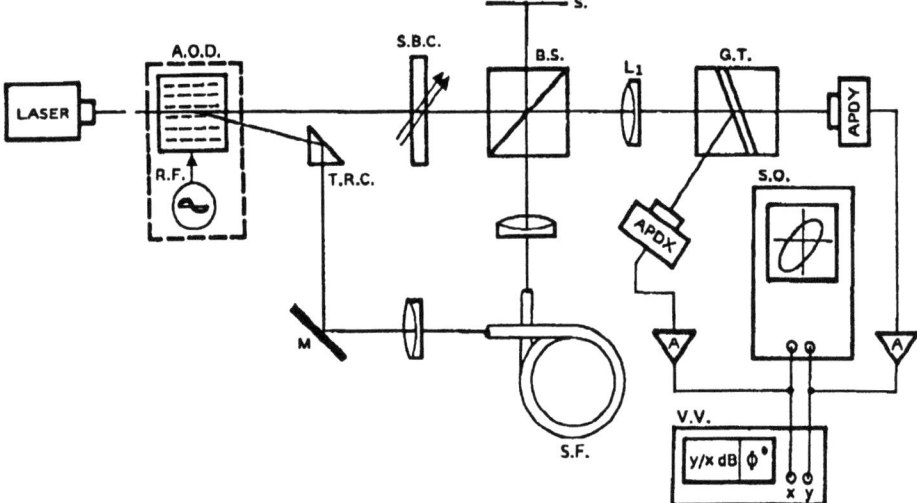

Figure 2. Two frequency Mach-Zehnder interferometer for fibre polarimetry with narrow-band sources.

The source is a linearly polarized laser with a single longitudinal mode, coupled to an acousto-optic device (AOD), which is used with reduced Bragg efficiency to provide both the main and the reference beam, with a relative frequency shift $\Omega/2\pi = 40$ MHz. The fiber, whose output SOP is to be determined, is placed in the measurement arm, while an optical compensation device is inserted in the reference path to produce the linear SOP oriented at 45°, as previously discussed. Recombination of the two beams and polarization analysis is accomplished with a beam splitter and a Glan-Taylor prism respectively. The two emerging waves are separately detected (to obtain the electric outputs S_x and S_y), amplified and conveyed to a sampling oscilloscope (operated in x-y mode), which allows the polarization ellipse to be visualized. In addition a precise measurement of Jones vector parameters, from the amplitude ratio and relative phase of the two beatings, can be made using a vector voltmeter, which assumes either of the two detection channels as reference.

The critical point of the system is that very unnoisy RF signals are required to operate the vector voltmeter. Therefore, if the optical path difference of the interferometer exceeds the source coherence length (so that the fluctuation bandwidth of $\phi_{1,2}(t)$ in eq. 2 attains its maximum, crresponding to a complete uncorrelation of ϕ_1, ϕ_2), only the qualitative observation of the SOP ellipse is possible. From this viewpoint a preliminary experiment with an He-Ne laser demonstrated the feasibility of the SOP visualization on a sampling oscilloscope operated in x, y mode [1]. The resulting trace proved to be much less noisy than the single outputs S_x, S_y (both affected by the $\phi_{1,2}$ phase), so confirming the subtractivity property of the polarimeter. However, the same characteristic could not be exploited with the vector voltmeter for the above reason.

In conclusion, to perform a precise SOP measurement from the two RF outputs (at least with commercial electronic instrumentation), the $\phi_{1,2}(t)$ term in eq. 2 should be reduced to a very slowly varying time function. This means that the fastest fluctuations of $\phi_1(t)$ must equal the corresponding oscillations of $\phi_2(t)$, or, in other words, that the laser phase noise has to be cancelled out (almost completely) even from the S_x, S_y signals. The only way to satisfy the new requirements is, of course, a strict path equalization of the interferometer within the source coherence length. Therefore, the available single-frequency semiconductor lasers limit to few meters the length of measurable fibre spans (2.5 m for a 20 MHz linewidth) and confine the possible applications to a limited range of application conditions [1].

3.2. BROAD-BAND SOURCES

The problem already discussed of path equalization becomes even heavier if the coherence degree of the source is lowered so that the associated linewidth can largely exceed the frequency shift $\Omega/2\pi$ (as in the case of multimode lasers or LEDs) and a strict path equalization between the two interferometer arms is needed. A possible solution in these conditions is to remove the fibre from the interferometer and split the two beams at the fibre output. Of course they contain the same SOP, so that the linear reference polarization must be obtained from the state to be measured through proper SOP resetting in either of the two arms.

In some sense this configuration is analogous (but more convenient from an experimental viewpoint) to the one in which path equalization is obtained inside the interferometer through the insertion of a compensating fibre.

From the viewpoint of the theoretical formulation, the above scheme is consistent with the previous general discussion, except for a slight change regarding the amplitude E_2 and phase ϕ_2 of the reference field, which are now determined (because of the polarization reset) by the equations

$$E_2 = \sqrt{(E_{1x}^2 + E_{1y}^2 + 2E_{1x}E_{1y}\cos\phi)/2}$$

(5)

$$\phi_2(t) = arctg\left(\frac{E_{1x}\sin\phi}{E_{1y} + E_{1x}\cos\phi}\right) + \phi_1(t) + \phi_0(t)$$

where $\phi_1(t)$ represents the laser phase noise, as before, while $\phi_0(t)$ accounts for the (slow) acoustic and thermal fluctuations of the relative spatial delay due to the different path of the two beams. Due to such relationships, a further dependence on the SOP to be measured is introduced in the RF detected signals S_x, S_y, but not in their amplitude ratio and relative phase. As a consequence, the experimental calibration procedure should be extended to include a new step, consisting in a $\pm 90^0$ rotation of the reset polarizer in the reference arm whenever necessary to avoid the SOP condition of power fading ($E_{1x} = E_{1y}; \phi = (2n+1)\pi$

An experimental implementation of this system version is illustrated in Fig. 3, where the broad-band source (multimode semiconductor laser with 1 nm linewidth at $\lambda = 850$ nm) is coupled to a fibre (750 m long, $\lambda_c = 800$ nm and attenuation 2.95 dB/Km at 850 nm) outside the interferometer [1].

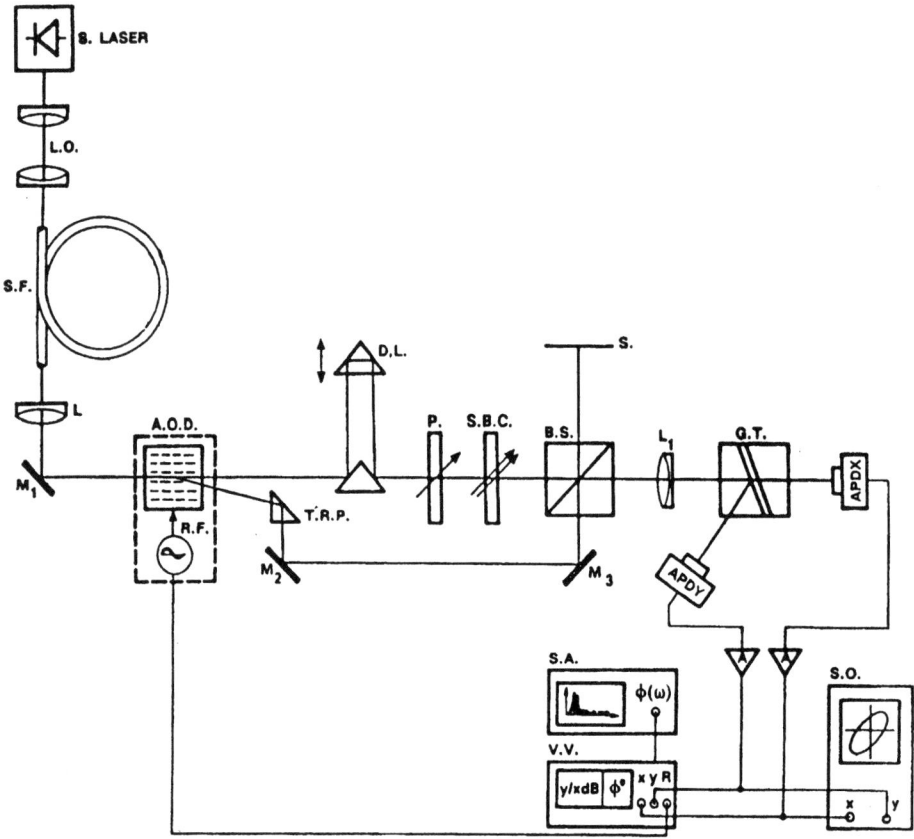

Figure 3. Mach-Zehnder interferometer for fibre polarimetry with broad-band sources.

4. Michelson interferometers for fiber polarimetry

Another polarimeter configuration for the measurement of Jones parameters on the basis of the heterodyne principle is provided by the Michelson interferometer, which allows the realization of a compact instrument with reduced alignment problems. Two version of this setup were demonstrated, where the substantial difference consists in the use of the AOD, which operates as a doubler of the frequency shift in both cases, but in one case works also as a beam splitter.

4.1. AOD USED AS A FRQUENCY SHIFT DOUBLER

The relevant arrangement shown in Fig. 4 is characterized by the same location of the fibre as in the case of broad-band sources previously described, again to avoid that the fibre length may exceed the laser coherence length.

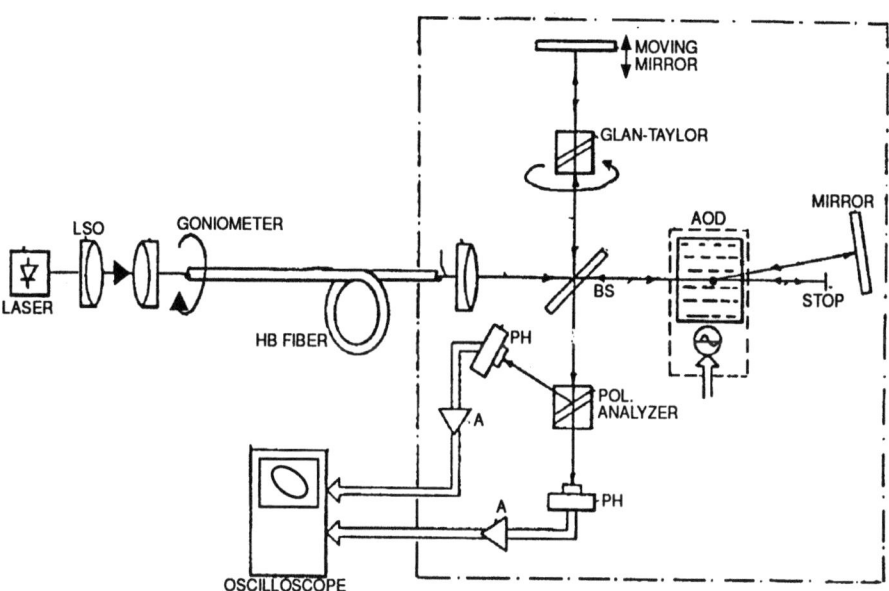

Figure 4. Two frequency Michelson interferometer for fibre polarimetry (AOD used as a frequency shift doubler).

The source is a semiconductor laser without particular spectral requirements, which emits the beam to be coupled to a SM or HB fibre. The radiation outgoing from fibre and containing the SOP to be determined is collimated and sent to a Michelson interferometer. A first beam propagating along the corresponding arm, enters an AOD, mounted with its optical axis oriented at the Bragg angle with respect to the incidence direction and driven by a suitable RF electrical signal. The shifted frequency output, impinges onto a mirror at right angles. The back-reflected beam is again deviated and shifted in frequency by the AOD. So, after a double deviation and shift, the outgoing radiation is exactely superimposed (through the recombination beam-splitter) to the reference beam (with the same frequency as the source and a linear polarization at 45° with respect to the system axes) coming from the other arm of the interferometer. The two recombined radiations are analyzed in polarization with a Glan-Taylor prism and separately detected. The resulting IF output is the double of the one driving the AOD.

Detected RF signals have amplitudes proportional to the two field components and the same relative phase as E_x and E_y. So they can be used to measure the SOP of the radiation at the output of the fibre through a vector voltmeter or to visualize the corresponding ellipse in real-time as previously described.

4.2. AOD USED AS FREQUENCY SHIFT DOUBLER AND BEAM-SPLITTER

In the setup of Fig. 5, the beam coming from the fibre is sent, after collimation, to the AOD which performs in this case the double task of splitting and recombining the incident radiations. The beam entering the AOD at the Bragg angle with respect to the optical axis of the device is splitted into two radiations: whose paths are shaped to build a Michelson configuration similar to the previous scheme of Fig. 4. In the present implementation the non deviated beam outgoing is collected by a total-reflection prism, sent onto a mirror and then reflected towards the prism and the device. The second beam, which is deviated and frequency shifted by the AOD is on the contrary sent to a Glan-Taylor prism, then to a mirror, and finally backreflected.

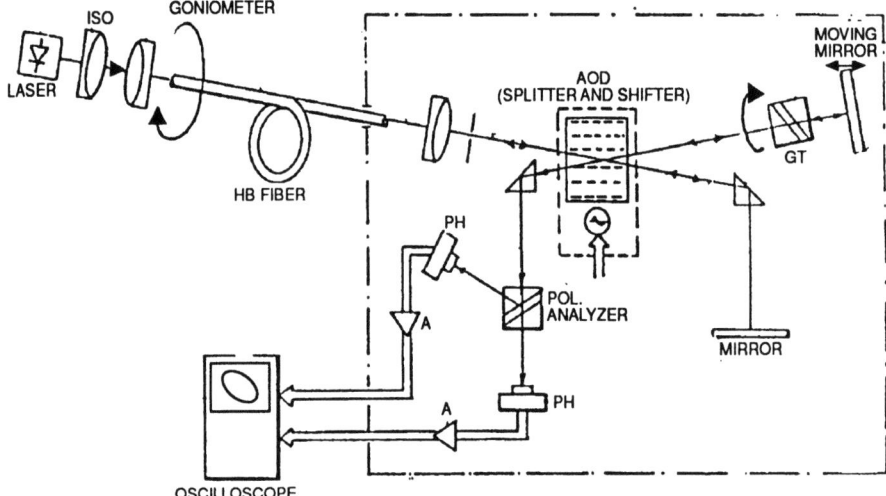

Figure 5. Two frequency Michelson interferometer for fibre polarimetry (AOD used as frequency shift doubler and beam-splitter).

The two aforesaid waves undergo two different frequency shifts ($\omega - \Omega$ and $\omega + \Omega$ respectively) corresponding to orders -1 and +1 of the Bragg diffraction. So after superposition, provided again by the AOD their beating is obtained at the same RF 2Ω as in the previous case. The following electronic treatment of the two S_x, S_y signals as well as the techniques for the visualization and measurement of the output SOP remain unaltered with respect to the schemes already described.

5. Heterodyne Stokes polarimetry

An alternative definition of the SOP with respect to the one based on the relative quantities R and ϕ (which represent the actual output from any of the Jones polarimeters) has already be outlined at the end of paragraph 2. In this case the SOP measurement of a radiation emerging from an optical fibre can be achieved using the Stokes parameters, as defined in eqs. (3). The more effective technique to realize a heterodyne system for the Stokes vector measurement is based on the electical mixing and low-pass filtering of the two RF signals S_x, S_y which can be obtained from the two outputs of whichever of the previous interferometric setups.

In Fig. 6 the practical implementation of a heterodyne Stokes polarimeter is shown. To simplify the setup a source operating in the visible light with a high coherence length was chosen. A typical dual frequency Mach-Zehnder interferometer, containing an AOD in a branch, produces the two fields $\vec{E}_1, \vec{E}_R$ described in the previous principle scheme (with Ω=120 MHz). A half wave-plate and a polarizer generate the required linear reference at 45°. In the present configuration, the two interferometric branches are balanced within the source coherence length, but the system would work also in the unbalanced condition (with a long fiber span in one arm) because of the aforesaid phase subtractivity property of the SOP outputs.

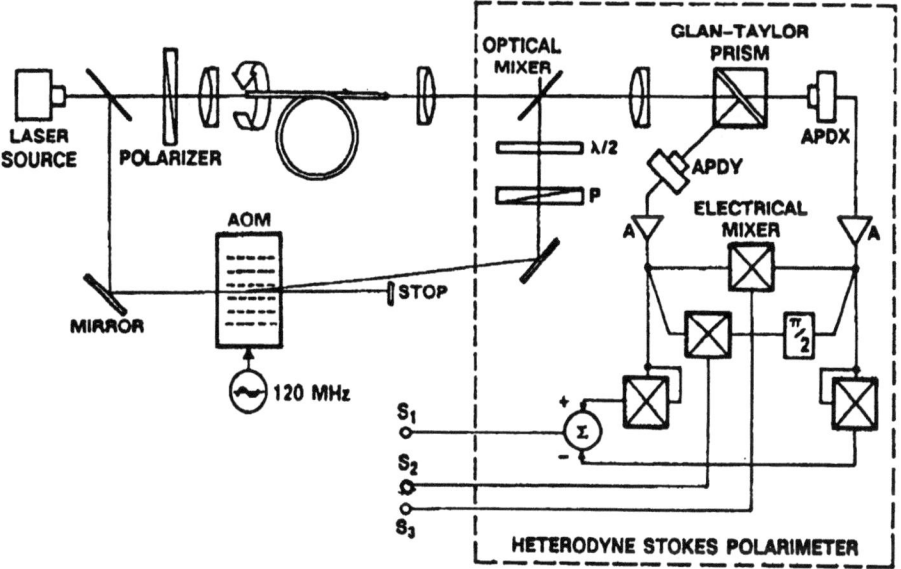

Figure 6. Heterodyne Stokes polarimeter. Experimental setup.

The detected signals S_x and S_y are properly processed through four electrical mixers, a 90° phase shifter and an analog subtractor to obtain finally the three Stokes parameters of the completely polarized wave to be measured.

In principle the amplitude and phase deviations, of the relevant SOP, due to imperfections of both the recombination beam-splitter and the analyzing Glan-Taylor prism, can be reduced to a product of a birefringence and a dichroism transformation with axes parallel to the analysis frame x, y (as the interferometer paths and the two detected output beams can be made parallel to the common setup plane with excellent precision using standard alignment techniques). Therefore these anomalies can be compensated introducing a constant relative phase and a fixed selective attenuation for the two field components. The simplest procedure to generate such effects relies on a differential amplification and a variable dephasing of the two RF signals. The corresponding electrical adjustments should be performed at the beginning of any measurement stage, using a known optical device (for istance a quarter wave-plate) on the measurement arm instead of the object under test, in order to calibrate the system output and to verify eq. (4).

6. Heterodyne polarimetry for fiber birefringence measurement

The performance of all the polarimetric instruments previously described can be checked by replacing the fibre or anything else which is to be characterized in polarization with a different object of known SOP properties, for example a quarter wave-plate which allows the polarimeter calibration. The plate (after insertion in the measurement arm) is oriented at different angles to produce a family of SOPs ranging from a linear to a circular state.

The experimental readings of the amplitude ratio R and the phase ϕ of the two output field components are taken from a vector voltmeter changing plate azimuths. On the other hand, the same output SOPs can be computed using the Jones formalism.

The comparison between theoretical and experimental results (in the case of Mach-Zehnder interferometers) is shown in Fig. 7 where the discrepancy of the measured points with respect to the exact curve is limited to 0.05 dB in terms of amplitude ratio and 0.1° in the relative phase.

Examples of polarization displays on the screen of the sampling oscilloscope are given in Fig. 8, where the linear polarization at the output of the quarter wave-plate corresponds to an imput SOP linear and aligned with the birefringence axes, while the circular polarization is obtained for a 45° rotation of the plate with respect to the previous orientation [3].

To verify the applicability of the heterodyne polarimetry to optical fibres beat-length measurements have been performed on a HB "bow-tie" fibre (manufactured by York Technologies) using a 1;55 µm DFB laser diode in a setup of the type shown in Fig. 2 (all the other schemes of the Mach-Zehnder class are suitable to the same characterization purpose).

In a polarization maintaining fibre the beat-length parameter L is inversely proportional to the modal birefringence B through the well known relation

$$L = \lambda/B \qquad (6)$$

where $B = n_x - n_y$ (n_x and n_y being the effective propagation indexes of the x, y linearly polarized modes) and λ is the operation wavelength. So, since the SOP repetition length L is simply measurable with an external perturbation technique, relying on the elasto-optic properties of the fibre material, birefringence B is generally obtained from L through eq. (6).

A photo of the "bow-tie" fibre end-face taken at the microscope is shown in Fig. 9. The cutoff wavelength is 1.4 µm and the fabrication value for L is 1 mm at 1.55 µm.

In the experiment a short length of fibre (1 m) was inserted in the measurement arm of the interferometer, whose path difference is approximately balanced in this condition. To produce the elasto-optic effect required for the measurement of L, a small fibre section (20 cm of the 1 m span) was placed inside a mechanical device which applies a nearly poit-like stress (the size of the contact area is ~40 µm maximum) through a roller bearing connected to a mobile equipment. The position of this mechanism is controlled by a long-run micrometric movement for the longitudinal scanning of the fibre. This produces a variation of the SOP at the fibre output with a spatial period equal to L. The maximum SOP effect is achieved when the input light has equal component amplitudes along the fibre birefringence axes, and the applied stress is oriented at ±45° with respect to them.

To determine the fibre beat-length an output mixer on the S_x, S_y signals was used, to obtain the second parameter s_2, according to the scheme previously discussed of the Stokes polarimeter. The relevant low frequency signal proportional to s_2 is then conveyed to an x-y recorder, which displays the variation of its amplitude versus time.

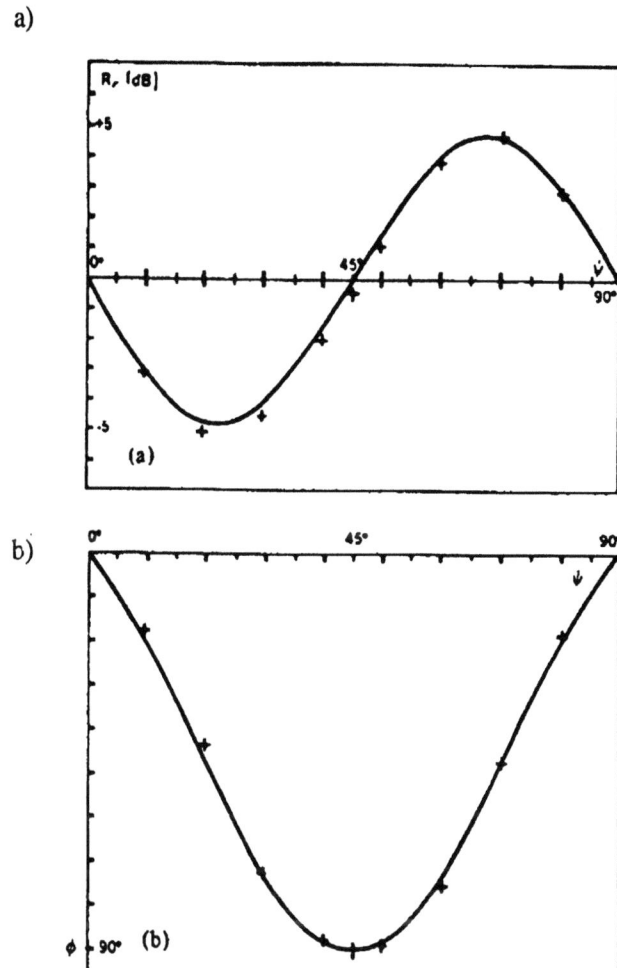

Figure 7. a) Amplitude ratio R (electrical dB) versus wave-plate azimuth ψ. The experimental and computed results are represented by cross marks and a full line respectively. b) Phase lag φ versus wave-plate azimuth ψ. The experimental and computed results are represented by cross marks and a full line respectively.

As the other s_1, s_3 parameters were not detected a preliminary calibration was performed at the beginning of the experiment to optimize the input and output SOPs with respect to the x, y analysis frame, defined by the Glan-Taylor orientation. A further improvement of the polarization signal could be obtained inserting a Soleil-Babinet compensator at the fibre output. Furthermore the fibre was manually rotated in the V-groove of the stressing device to achieve the optimal orientation for SOP change, as described above.

Fig. 10 shows a typical trace resulting from a 15 mm scanning length. The spatial period of this waveform (derived for example from the mutual distance of the highest peaks) provides an L value very close to the fabrication figure (0.997 mm at 1.55 μm). The regular presence of three minor spikes in each period can be attributed to a possible fibre torsion (with a pitch comparable with L) in the drawing stage, and to the rectangular core shape (as apparent from Fig. 10).

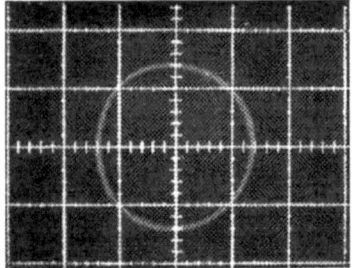 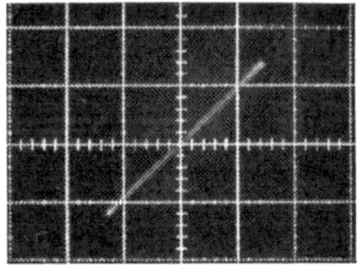

Figure 8. Linear and circular polarizations at the output of an quarter wave-plate.

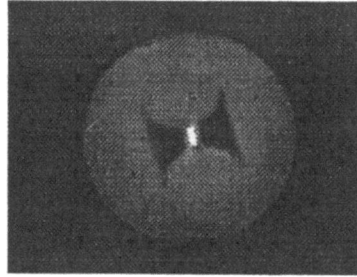

Figure 9. Cross section of a "bow-tie" polarization maintaining fibre.

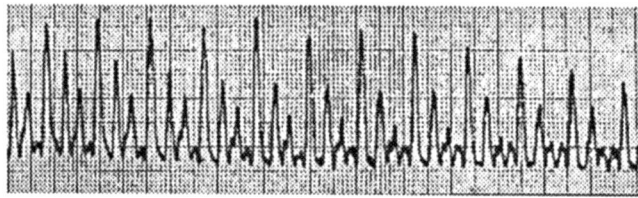

Figure 10. Recording of s_2 Stokes parameter at the mixer output.

7. Conclusions

Real-time heterodyne polarimetry has been presented as an approach to visualize and measure the SOP at the output of an optical fibre or, in general, of a birefringent object. Two experimental Mach-Zehnader schemes suitable for operation with narrow- and broad-band sources have been proposed and the relevant experiment reviewed. Two Michelson configuration using the AOD not only as frequency shift doubler, but also as a beam-splitter have been described, together with the corresponding laboratory demonstrations. Moreover the possibility of SOP measurement using the Stokes parameters has been discussed.

Finally the application of the heterodyne polarimetry has been presented to beat-length (or identically birefringence) measurement on HB fibres. The resulting values of the measurand, which can be defined with high resolution up to a submillimetric size, are in good agreement with the fabricant figure and exhibit a high repeatability.

8. References

1. R. Calvani, R. Caponi and F. Cisternino, (1986) Real-time heterodyne fiber polarimetry with narrow- and broad-band sources, *Journal of Lightwave Technology* **LT-4**, 877-883.
2. R. Calvani, R. Caponi, F. Cisternino and G. Coppa, (1987) Fiber birefringence measurements with an external stress method and heterodyne polarization detection, *Journal of Lightwave Technology* **LT-5**, 1176-1182.
3. R. Calvani, R. Caponi and F. Cisternino, (1985) A heterodyne Mach-Zehnder polarimeter for real-time polarization measurements, *Optics Communications* **54**, 63-67.
4. R. Calvani, R. Caponi and F. Cisternino, (1990) A heterodyne interferometric polarimeter for the detection of Jones and Stokes polarization vectors, *Fiber and Integrated Optics* **9**, 153-161.
5. R. Calvani, R. Caponi and R. Piglia, (1991) High speed polarimetric measurements for fiber optic communications, Fiber Optic Metrology and Standards, ECO4, The Hague (NL), 12-14 March, pp. 258-263.

APPLICATION OF ER-DOPED FIBER AMPLIFIERS TO OPTICAL MEASUREMENT TECHNIQUES

HARUO OKAMURA and KATSUMI IWATSUKI
NTT Transmission Systems Laboratories
1-2356, Take, Yokosuka-shi, Kanagawa 238-03,
Japan

1. INTRODUCTION

Optical-fiber-based measurements directly translate the physical quantities being detected into optical quantities such as optical intensity, phase, frequency or polarization by utilizing the characteristics of optical fiber. Intrinsic measurements, those that use optical fibers directly for measurement purposes, make the best use of the unique mechanical and optical properties of optical fibers and can capture very minute physical forces. Interferometric measurements, in particular, are very sensitive. The typical system consists of a light source, sensor head, interferometer and signal processing circuits. Although such components and related technologies have been intensively studied, further improvement is still desired in many respects. Toward this end, recently-developed optical-amplification techniques are expected to play an important role.

One of the first reports on optical amplifiers appeared in 1964. The first step was rare-earth-doped glass lasers. Semiconductor laser amplifiers have been studied and significant progress has been achieved. As to optical-fiber amplifier technologies, the initial fiber lasers were constructed with Nd- or Er-doped fibers. In the late 80's, the semiconductor-pumped Er-doped fiber amplifier(EDFA) was created and its high gain was confirmed.

Er-doped fiber amplifiers(EDFAs) are suitable for signal amplification in the 1550 nm transmission window. Their advantages over other optical amplifiers include high efficiency, high output power, wide band, low noise, high coupling efficiency to fibers, negligible non-linearity, and applicability to high-speed and/or multi-wavelength signal amplification. Although EDFAs are thus potentially very attractive, they have so far been studied mainly as transmission amplifiers.

This paper discusses the application of EDFAs to optical-fiber-based measurement techniques. Section 2 discusses light sources; EDF-ring lasers and its characteristic of multi-wavelength oscillation. Section 3 presents two new types of interferometers: a single-arm interferometer and an active ring-resonant interferometer. Section 4 discusses the introduction of EDF-laser oscillation to the signal processing functions of optical loss compensation and wavelength conversion. Section 6 provides general conclusions.

2. SOURCES

This section presents a unidirectional Er-doped fiber ring laser based on polarization-preserving fiber. A narrow optical bandpass filter and an optical polarizer are also used. The narrowest linewidth yet reported is observed with wide lasing-wavelength tunability. The first experiment involving a multi-wavelength EDF-RL is then discussed.

2.1. A SINGLE-FREQUENCY ER-DOPED FIBER RING LASER

To date, several narrow-linewidth EDF-ring lasers have been reported[1]-[3]. The left trace of Fig. 1 shows an early-stage EDF-ring laser and its observed spectral output[1]; single-mode oscillation was confirmed as shown in the middle and right photos.
Polarization maintaining fiber was used throughout the ring. The 15-m Er-doped PANDA fiber was pumped with a 1.48-mm LD. The lasing wavelength was adjusted with a 1-nm optical bandpass filter. The wavelength tuning ranged from 1.5493 nm to 1.5521 nm. A polarizer plus optical isolator was used for unidirectional and single-polarization lasing. The FSR of the ring cavity was 9.2 MHz.

Single frequency operation of the laser was confirmed. The ring-laser output, phase modulated at 3 MHz for scaling purposes, was examined with a scanning Fabry-Perot interferometer. Only one sharp peak existed over the FSR as shown in the upper middle photo. Shown in the lower photo is the peak observed at increased resolution. The spectral linewidth of the laser output was then measured with delayed self-heterodyne interferometer[4] with a 72 km delay fiber(resolution;1.4 kHz). Shown in the right photo is the spectrum analyzer display indicating that FWHM < 1 kHz. The linewidth strongly depends on acoustic noises, which fluctuate both the laser output power and phases thus broadening the linewidth. Elimination of acoustic noises would narrow and stabilize the linewidth.

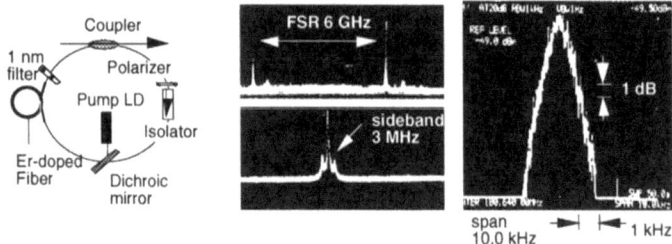

Fig. 1 An EDF ring laser and its output spectral observation

The lasing frequency of this ring laser hopped every 2-5 s, with some jumps exceeding several GHz. Reflections within the optical filter or outside the ring cavity might have established unwanted cavities and caused frequency hopping together with the effect of ASE. The complete elimination of mode-hopping demands that a bandpass filter narrower than 0.01 nm be placed in the ring cavity. Two cascaded Fabry-Perot filters were actively controlled and the mode hopping was suppressed for many hours[5]. The remaining issue is the complete stabilization of the oscillation frequency.

2. 2 SIMULTANEOUS OSCILLATION OF ER-DOPED FIBER RING LASER

Although EDF-RLs typically lase just at one wavelength, the wavelength at which the gain matches the cavity loss, a multi-wavelength EDF-RL has recently been reported[6]-[8]. This section describes its first demonstration[6].

The proposed multi-wavelength EDF-RL uses several cavities that share one EDFA, but each includes its own bandpass filter and attenuator. Each filter passband is adjusted to the appropriate lasing wavelength, λ_1, ..., λ_N (N: integral). Each attenuator is then precisely set such that the round-trip optical losses at λ_1, ..., λ_N match the EDFA gains at λ_1, ..., λ_N, respectively, thus making the effective gains equal. Simultaneous lasing at λ_1, ..., λ_N is now possible from one EDFA. If the gain is highly homogeneous, however, the acceptable cavity-loss variation that allows simultaneous lasing may be reduced.

Independent intensity or frequency modulation is possible, the IM crosstalk[9] between wavelengths can be kept within 100 kHz, because of the long fluorescence lifetime of Er^{3+}:glass. This level of crosstalk causes no problems in practical high-bit-rate transmission systems.

Fig. 2 shows the experimental set-up. Two ring cavities share an Al-codoped-EDF amplifier(Er:~1000 ppm and Al:~5000 ppm). Each cavity had a 1-nm tunable filter and an adjustable attenuator with an accuracy of 0.01-dB. Fig. 2 shows an example of the two lasing spectra. The wavelengths were tunable and their separation could be freely chosen from 0.5 nm to 14 nm.

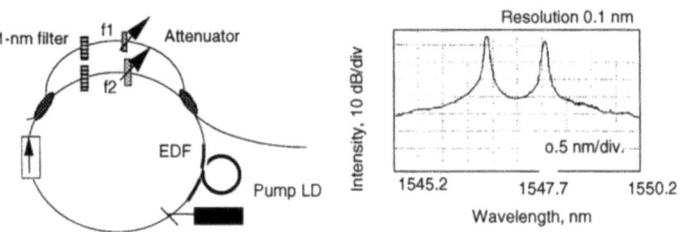

Fig. 2 Simultaneous oscillation at two wavelengths

In the oscilloscope display, two traces were independently observed that represent the lasing power obtained through a 1-nm filter. Crosstalk above several kilohertz was not observed because of the slow gain response of the EDFA, thus confirming that the independent intensity modulation of each channel is possible.

It was found that the simultaneous lasing is best maintained with higher pump powers against fluctuations in cavity loss. This is attributed to slight inhomogeneity[10] in the Er^{3+} gain. Although mode hopping was inevitable, each lasing light was single-moded around its lasing wavelength as confirmed with a scanning Fabry-Perot interferometer(FSR:6 GHz, finesse>6000).

Additional lasing cavities can be simply added by equipping each additional fiber with its own filter and attenuator. Frequency-superimposed laser-light sources are very useful, particularly if they support intensity or frequency modulation.

3. INTERFEROMETERS

Various interferometers have been developed for measurement applications. Single-pass interferometers such as Mach-Zehnder or Michelson interferometers suffer from the complexity of the two-arm configuration. Resonator-type interferometers have difficulty in reducing the losses typical of long-resonators to achieve a high finesse. To overcome such difficulties, this chapter discusses novel EDFA-supported interferometers for each category.

3.1 SINGLE-ARM INTERFEROMETER

Interferometers that use high power local lasers for homodyning can reduce the effect of receiver noise and halve the fiber length. They require that the LL be optically phase matched to the signal carrier. For this purpose, the optical PLL technique[11] demands a very wide band electric circuit. Another possibility is the use of the injection locking technique[12]. It can theoretically provide amplified signal carriers with minimized phase noises because the optical phase of the LL is locked to the signal carrier. However, no experimental studies have so far been reported.

This section addresses a novel configuration[13] based on the injection locking technique, wherein an EDF-ring laser is used as the LL.

3.1.2. Principle

Fig. 3 shows an interferometer using a ring laser built on a fiber coupler as the LL. The ring laser can be constructed with an EDFA or LD amplifier. With signals that are out of the LL's injection-locking range, the LL is injection locked to only the signal carrier, and the phase-modulated signal components simply pass through the coupler. Therefore, the LL output can be combined with the signal laser light at the coupler output. Phase modulated signals are thus converted into intensity modulated signals. The demodulated signals are theoretically free from additional phase noise if the center frequency and spectral linewidth of the original carrier and, environmental phase fluctuations do not cause the locking range to be exceeded.

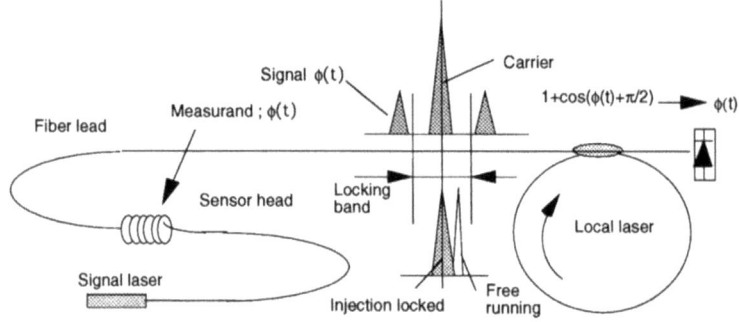

Fig. 3 One-arm interferometer using an injection-locked local laser

For achieving high performance, the demodulation condition must be optimized such that the operating point is stable at quadrature[14]. Moreover, the injection locking bandwidth and the ring cavity's FSR must be properly chosen depending on measurand frequency ranges.

(1)Acceptable signal frequency range Injection locking range, $\delta\theta$, can be written as[15]

$$\delta\theta = 2\sqrt{\frac{P_0}{P_{osc}}} \frac{(1-r^2)}{r} \qquad (1)$$

where q is the cavity tuning range, P_0, P_{osc} are master laser(signal laser light) power and self-oscillating power of the slave laser(LL), respectively. r^2 is the power transmitting ratio of the coupler. As optical phases of the LL are locked to those of the frequency components of the injected signal laser light within the locking range, signal components within the locking range are quenched by homodyne demodulation.

When the LL is injection locked to the signal carrier, frequency components at integer multiples of the ring cavity's FSR away from the carrier can be also injected in the ring cavity. They emerge with the injection-locked and regenerated signal carrier. The remaining signal frequency components bypass the ring. components within the locking range, homodyne-demodulated signals are quenched at frequencies equal to integer multiples of the FSR.

(2)Sensitivity and linearity The phase error between the signal carrier and the LL light is proportional to the frequency difference between the signal carrier and the free-running LL[16]. The phase difference is 90 degrees at just the edge of the locking range. This problem can be avoided with 3×3[17] or 4×4[18] couplers, which yield 3 or 4 optical outputs with 120-degree or 90-degree phase differences, respectively. Once the quadrature condition is maintained, demodulation sensitivity and linearity can be theoretically identical to those provided by 2-arm interferometers.

3.1.3 Experiments
(1) Set up Fig. 4 shows the experimental set-up. A single-mode LD with a 157 kHz linewidth, center-frequency stability within 25 kHz, and a maximum 5 mW output was used as the signal laser. Its output light was led to a phase modulator and a 5 km-long single mode fiber connected to the LL by a 2×2 PANDA coupler, on which the ring laser was constructed.

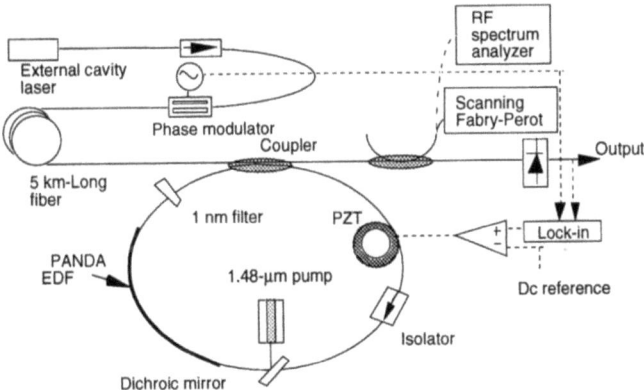

Fig. 4 Experimental set-up

(2) Verification of injection locking The ring cavity length of the LL was ramped with a PZT fiber stretcher and the RF spectrum analyzer at the photo detector was synchronously swept. As shown in the lower trace of Fig. 5, a beat-free region was clearly demonstrated which verifies the existence of injection locking. By contrast, when the ring laser frequency was detuned from the signal laser frequency with the 1-nm optical filter, the beat-free region disappeared as shown in the upper trace of this figure.

(3) Demodulation of the signal The ring cavity length fluctuated due to environmental disturbances thereby changing the free-running frequency of the ring laser and yielding a proportional phase error between the LL light and the signal light. Therefore, the ring cavity length was feed-back stabilized with the PZT fiber stretcher. Fig. 6 shows the effectiveness of the feed-back loop. This result confirmed that the oscillation of Er-doped fiber ring laser was controlled

to a single longitudinal mode at a single frequency under feed-back-stabilized injection-locking.

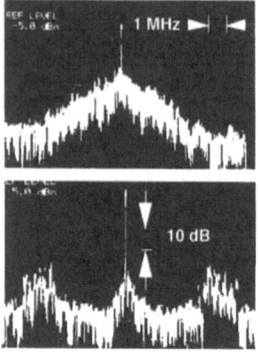

Fig. 5 Beat spectrum with/without injection locking

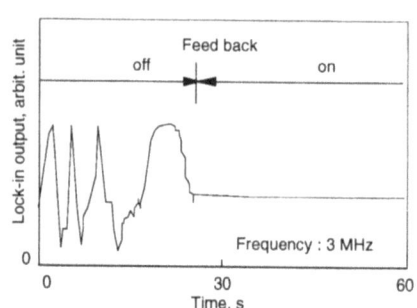

Fig. 6 Lock-in amplifier outputs with and without the feed back.

Fig. 7 shows demodulated signal amplitudes measured with a lock-in amplifier. Under the experimented condition, the injection locking half-bandwidth is calculated as ~0.22 MHz for FSR= 9.2 MHz, which roughly parallels the measured values. The demodulated signals below 1 MHz and around 9 MHz were quenched. It was then confirmed that as the injected signal light power increases, the locking bandwidth increases. Accordingly, high injection power suppresses demodulated signal amplitudes at higher frequencies than is true at low injection powers. This parallels Eq.(1).

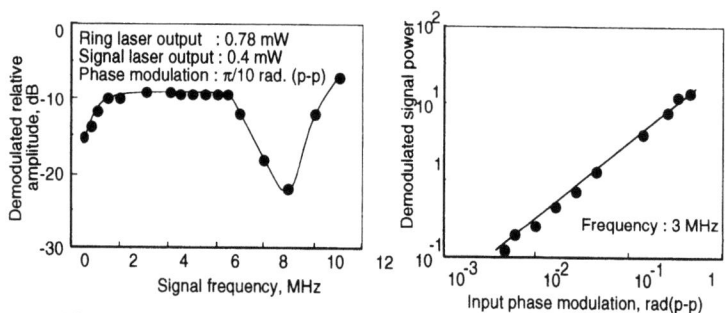

Fig. 7 Demodulation frequency response

Fig. 8 Demodulation linearity

Fig. 8 shows the depth of sinusoidal phase modulation vs. demodulated signal amplitude. The linearity exceeds 2 decades and the minimum detected phase shift is of the order of 10^{-3} rad.(p-p).

3.1.4 Conclusion

A new category of interferometric fiber-optic sensor was created based on an EDFA that uses a local laser. This configuration is suitable for detecting physical quantities far from the observer. Application of this technique to communications needs further development because of the much wider bandwidths common in such applications.

3.2 ER-DOPED FIBER RING RESONATOR

The narrowest linewidth reported for a fiber resonator is 270 kHz[19] for a one-coupler fiber ring at 633 nm with a finesse of 1260, and 500 kHz[20] for a Fabry-Perot resonator at 1520 nm with a finesse of 500. Such resonators demand extremely low loss couplers or carefully fabricated fiber-cavity endfaces to minimize the round-trip optical loss.

A more practical fiber resonator suitable for high-resolution spectrum analyzers and narrow-bandwidth filters would be a two-coupler-type[21] fiber-ring that accepts conventional fiber couplers plus other optical components, and yet still attains a high finesse. Such a fiber resonator would be possible if the round-trip optical loss is offset by an optical amplifier. Although an active resonator has been suggested[22] based on degenerate two-wave mixing, no experiment has so far been reported.

This section presents a high-finesse, fiber ring resonator that offers the narrowest linewidth and is the first to use a loss-compensating EDFA[23]. Also discussed is a novel resonator-linewidth measurement method[24] that does not demand a narrow-linewidth light source.

3.2.1 Resonator finesse enhancement

Fig. 9 shows the fiber ring resonator which incorporates two couplers and an optical amplifier with intensity gain G. E_1 is the input amplitude to the coupler(A) and E_2 is the recirculating amplitude in the ring. The resonator recirculating power is extracted from the coupler(B), which provides the output proportional to $|E_2|^2$. $|E_2/E_1|^2$ is given by using
$A = e^{-\alpha L} \sqrt{G(1-\eta)(1-\gamma)(1-K)}$, as

$$\left|\frac{E2}{E1}\right|^2 = \frac{(1-\gamma)K}{1+A^2-2A\cos\beta L}, \quad (2)$$

where g and K are the fractional intensity loss and intensity coupling coefficient of the coupler(A). Fiber attenuation, propagation constant and the fiber ring length are represented as α, β and L, respectively. The fractional over-all loss of optical components in the ring is η. This includes the power extracted by coupler(B). Under sharp resonance, resonator finesse F is approximated as

$$F = \frac{\pi}{(\beta L)_{HM}} \approx \frac{\pi\sqrt{A}}{1-A}, \quad (3)$$

where $2(\beta L)_{HM}$ is FWHM of the resonance peak. Full resonance, in particular, is attained when the round-trip intensity loss through the ring is just compensated for by the fractional intensity coupling at coupler(A). Thus, full resonance is achieved when

$$G = \frac{1-K}{e^{-2\alpha L}(1-\gamma)(1-\eta)} \quad (4)$$

From Eq. (3),
$$F = \frac{\pi\sqrt{1-K}}{k} \quad (5)$$

If G is further increased to
$$G = \frac{1}{e^{-2\alpha L}(1-\gamma)(1-\eta)(1-K)}, \quad (6)$$

"A" becomes unity and the finesse approaches infinity regardless of α, K, η and/or γ.

Fig. 10 shows the finesse value against round trip intensity transmittance, A^2. Fig. 11 plots the optical amplifier gain required to theoretically attain the ultimate finesse for various coupling coefficients of coupler(A) against $e^{-2\alpha L}(1-\gamma)(1-\eta)$. Figs. 10 and 11 indicate that, although just a small gain can theoretically raise the finesse even to infinity, small gain-variations at A≈1 would greatly fluctuate the finesse value.

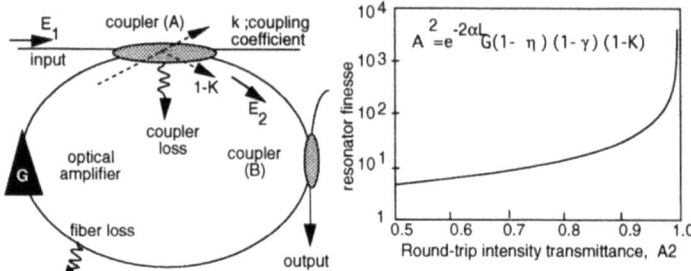

Fig. 9 The proposed resonator

Fig. 10 Finess enhancement vs intensity transmittance

3.2.2 Mode-filtered heterodyne method

The concept of this measurement method is outlined in Fig. 12, where two resonator-passbands are simultaneously illuminated by two uncorrelated lights whose spectral linewidths surpass the resonant linewidth. The beat noise appearing at the photo detector conveys the resonant linewidth information. If the resonator finesse is sufficiently high, the RF spectral-linewidth of the beat noise should be twice the resonant linewidth. This is because[25] Eq.(2) has a Lorentzian form under sharp resonance, and $\sin\beta L$ can be approximated as βL.

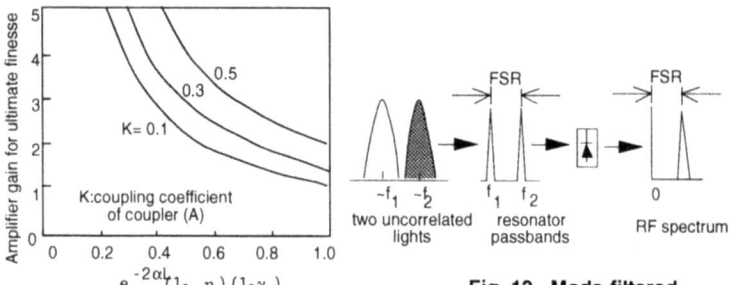

Fig. 11 Required gain for ultimate finess

Fig. 12 Mode-filtered heterodyne method

The uncorrelated lights can be obtained in two different ways: Sideband method and Broadband-light method.

Sideband method provides two uncorrelated lights by splitting one light into two and inducing a frequency shift, which corresponds to the FSR of the resonator, and time delay between them. Spectral linewidth of the source light should be narrower than the resonator's FSR but wider than the resonant linewidth.

Broadband-light method simply uses a light whose linewidth surpasses the resonator's FSR such that at least two resonator passbands are simultaneously excited. The broadened laser spectrum is a combination of the line spectra each of whose spectral power represents their existence probabilities. Just one line spectrum is excited at a time and its phase is uncorrelated to that of the next line spectrum to be excited. Therefore, the spectra extracted from the resonator are uncorrelated. They are, however, simultaneously received by the detector because the light recirculates in the resonator. They are thus heterodyned to yield at least one RF spectrum.

For both methods, resonator finesse f is calculated by using the FWHM value, δv, of the RF spectrum and the FSR, F, as

$$f = \frac{F}{\delta v} \qquad (7)$$

F is simply known from the RF beat frequency.

3.2.3 Experiments
(1) Finesse enhanced EDF-ring resonator Fig. 13 shows the experimental ring resonator with two 2×2 fiber couplers. The ring length was ~24 m. The FSR is ~8.5 MHz. A 1-nm bandpass filter was used to tune the resonator and prevent it from spuriously lasing. Polarization-maintaining fibers and components were used throughout the ring to attain single-polarization-state resonance. A length of the fiber was wound on to a PZT cylinder to sweep the ring length.

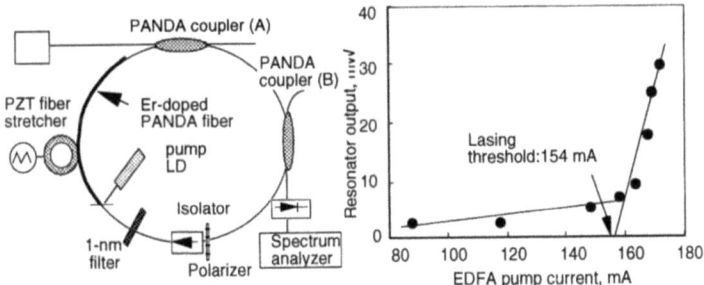

Fig. 13 Experimental set-up

Fig. 14 Pump current vs resonator output power

The ring resonator was excited with a 80-kHz-linewidth LD operating at 1551.0 nm. The optical amplifier was then appropriately pumped and the PZT fiber stretcher was scanned at 80 Hz with a phase excursion of ~6π (p-p). The 1-nm optical filter was tuned to 1551.0 nm such that resonant-peak signals were maximized on the oscilloscope. The resonant peak increased with pump power until the commencement of lasing. Fig. 14 shows resonator output power against the current injected into the EDFA's pump LD.

In most cases, the resonator started lasing at the frequency equal to that of the input light because of injection locking[26]. From this figure, the lasing threshold was determined as ~154 mA. This confirmed that the finesse was enhanced with the EDFA.

(2)Mode-filtered heterodyne method A 35-MHz-linewidth LD light at ~1550.5 nm was directly input to the resonator. The RF-beat spectral linewidths of the resonator output light, ranging up to about the 16th-order, were observed to be identical.

Shown in Fig. 15 is a spectral display of the 16-th order beat with an expanded horizontal scale; the EDFA was pumped to just under the lasing threshold. The HWHM value was 13~14 kHz so the finesse was determined as 650~600.

Fig. 16 shows the finesse measured with the mode-filtered heterodyne method. Variation in measured finesse is also shown. It was found that the higher the finesse, the more effectively the finesse was enhanced against the pumping power. The highest finesse, 560~500, corresponds to a round-trip intensity loss of less than 0.05 dB ($A^2 \geq 0.988$).

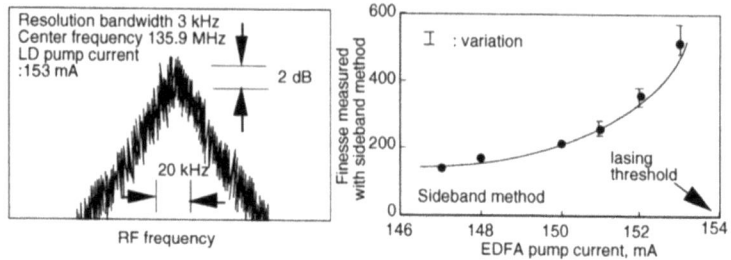

Fig. 15 RF spectrum of the 16-th beat   Fig. 16 Finess enhancement

(3)Application to optical spectrum analysis The experiment used a LD oscillating at 1551.0 nm with a 80-kHz linewidth that was sinusoidally phase modulated at 100 kHz and 1 MHz. The EDFA was pumped to just below the lasing threshold such that the resonator finesse was increased to the maximum possible value of 500~600.

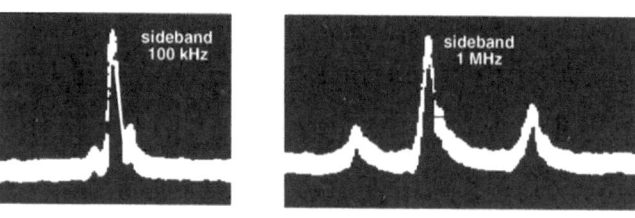

carrier linewidth 80 kHz

Fig. 17 Spectrum analyzer display resolving phase modulated optical spectra

The resonator was then triangularly scanned by the PZT fiber stretcher at 80 Hz. The spectral resolution of this scanning-ring spectrum analyzer should be 14~17 kHz: possibly the best resolution yet reported for direct optical-spectrum analysis. A 20-kHz resolution has recently been reported[27] elsewhere.
Fig. 17 shows typical outputs of the spectrum analyzer, where the 100 kHz and 1 MHz sidebands were resolved. A simple calculation reveals that a resonator-length change of ~0.02 μm results in a displayed frequency fluctuation of ~100 kHz.

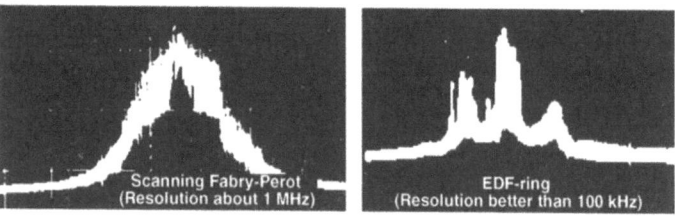

Fig. 18 Comparison of spectral observation (sideband 1 MHz)

Fig. 18 shows examples of direct spectral observation of another LD with a 350 kHz linewidth phase modulated at 1 MHz. The scanning Fabry-Perot interferometer used(FSR 6 GHz, finesse ~6000) offers one of the best spectral resolutions(~1 MHz).

3.2.4 Bi-directional EDF- ring resonator

The active fiber ring resonator discussed above has a narrow FS because of the length requirement of the EDF. Reported vernier structures can be used to expand the FSR.[28]-[30] However, the structures require that two or more independent resonators be concatenated so they suffer from instability and complexity.

To overcome this difficulty, a simple bi-directional ring resonator is presented here. Analytical treatment of this configuration is identical to what has been reported[28] for cascaded Fabry-Perot interferometers linked by an isolator.

Assume each resonator has a similar finesse $F_{cw} \sim F_{ccw}$ but a different length Lccw>Lcw, then $FSR_{cw+ccw}=mFSR_{cw}=(m+1)FSR_{ccw}$ where m(integer)= $(L_{cw}/L_{ccw})/(1-L_{cw}/L_{ccw})$. Here, L_{cw} and L_{ccw} have to be chosen such that the value of m is an integer. The effective finesse is raised to $0.32 F_{cw} F_{ccw}$ under the condition that the next-higher modes of the overlapped passband are reduced by 20 dB. Also indicated is that the overlapped passband narrows to 64%[29].

Fig. 19 shows the bi-directional ring resonator. The ring length is ~24 m[FSR~8.5 MHz). The circulator provides different path lengths and different FSRs in two counter-propagating directions. Input and output light spectra observed with a scanning Fabry-Perot interferometer (FSR: 6 GHz, finesse:~6000) are also shown in the figure. The uni-directional output spectra had a ~8.5-MHz spacing corresponding to the unidirectional FSR. The uni-directional resonant linewidth was as high as ~560 as determined by the "mode-filtered heterodyne method".(See Section 3.2.2). Bi-directional output spectrum is also shown in which only one dominant spectrum with negligibly small side-spectra was observed.

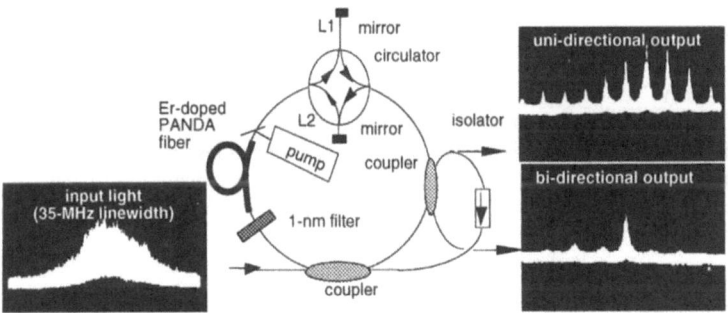

The resonator lengths 24 m and 26 m.

Fig. 19 Experimental set-up and input/output spectra

Though the adjacent overlapped passband was not simultaneously observable with the scanning F-P because of the narrow linewidth of the illuminating light, this figure confirms the finesse enhancement due to the vernier effect. Here, FSR_{ccw}=7.85 MHz and FSR_{cw}=8.5 MHz. Therefore, FSRcw+ccw=102 MHz and the effective finesse F_{cw+ccw}= 10625 when $F_{cw}=F_{ccw}$=560. To further expand the FSR, the path difference in both directions must be reduced. For stable operation, EDFA gain stabilization is necessary. The all-optical feedback scheme[31] would be applicable to automatically offset the loss fluctuation[32] thereby clamping the highest finesse.

3.2.4 Conclusion

A novel Er-doped fiber ring resonator, with an enhanced finesse ≥500, and a linewidth≤17 kHz, was operated as an optical spectrum analyzer and demonstrated a better-than-100-kHz resolution. The very narrow resonant-linewidth was measured with the novel mode-filtered heterodyne method.

4. SIGNAL PROCESSING APPLICATIONS

Previous sections dealt with applying the EDF-ring cavity to high-performance lasers and resonators. This section describes the application of the Er-doped fiber ring-cavity configuration to various optical signal processing functions such as optical-loss compensation, EDFA gain control and wavelength conversion in which laser oscillation is used as the basic principle.

4.1 LOSS COMPENSATION AND GAIN CONTROL

Transient gain saturation in EDFAs yields intensity noise in multi-channel transmission systems[33][34]. To overcome this problem, an all-optical feedback scheme[35] has recently been used to stabilize the gain at all channel wavelengths in WDM systems. This scheme has also been used to remotely switch EDFA gain[36]. Unlike digital transmission systems, fiber-optic measurements often deal with micro-radian phase information and analog signals at a speed slower than the EDFA gain's saturation. Therefore, care must be taken concerning phase noise and signal wave-form distortion when using EDFAs directly to amplify the sensor's signal.

This section discusses the use of the laser-oscillated feedback scheme with EDFAs to realize not only signal amplification, but also automatic compensation of fluctuating optical loss.

4.1.1 Theory

In the proposed loss compensating circuit, an EDFA with fractional intensity gain G is connected to the output of the targeted loss α. The EDFA output power is fed back to the input path via another loss β which is located out of the signal path. The two fiber couplers used have fractional coupling coefficients of κ_1 and κ_2. The output power Pout is expressed as

$$P_{out}=G(1-\kappa_1)(1-\kappa_2)(1-\alpha)P_{in}. \qquad (8)$$

With the laser oscillated feedback loop, the EDFA gain G at the lasing wavelength just compensates for the loop loss. Thus,

$$G = \frac{1}{\kappa_1\kappa_2(1-\alpha)(1-\beta)}. \qquad (9)$$

From (8) and (9), $P_{out} = \frac{(1-\kappa_1)(1-\kappa_2)}{\kappa_1\kappa_2(1-\beta)} P_{in}. \qquad (10)$

Thus P_{in} is linearly amplified to yield P_{out} irrespective of α. P_{out}/P_{in} is controlled by changing β.

When wavelength-superimposed signals with wavelengths of λ_κ (κ=1, 2, ..., N) are input to the EDFA with the feedback loop lasing at wavelength λ_f, the gain G_κ for each signal is expressed[35] under the assumption of homogeneous broadening as

$$G_\kappa = \exp[-\eta_k L + \frac{P^{IS}_f}{P^{IS}_k}(\ln\phi + \eta_f L)] \quad \text{for } \kappa=1, 2, ..., N, \qquad (11)$$

where η_k and η_f are the absorption constants at λ_f and λ_κ. P^{IS}_f and P^{IS}_k are the intrinsic saturation powers at λ_f and λ_κ, and, $\frac{1}{\phi}$ is the feedback loop attenuation. L is the EDF length. Eq. (11) uses $^4I_{15/2}$ and $^4I_{13/2}$ levels of the erbium atom and indicates that G_κ is independent of the signal power and also of the pump power, and is controllable through the loop attenuation. However, G_κ is wavelength dependent because $\alpha_{k,f}$ and $P^{IS}_{k,f}$ vary with wavelength. Accuracy of the proposed loss compensation method is therefore optimum in the wavelength ranges where $\eta_k=\eta_f$ and $P^{IS}_k = P^{IS}_f$, or $G\kappa=\phi$.

Concerning the dynamic performance of this system, since the EDFA yields a saturable(non-linear) gain and is used in a laser-oscillated feedback loop, system dynamics are complex. However, the system response is basically limited by the slow gain dynamics of the EDFA. This is because the loss compensation is achieved via gain saturation and recovery, and the feedback speed is much faster than EDFA dynamics.

4.1.2 Experiments and results

Figs. 20 and 21 shows the systems tested for loss compensation and gain control, respectively.

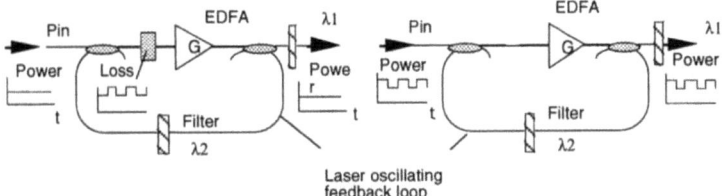

Fig. 20 Loss compensation   **Fig. 21 Gain control**

A feedback loop with a 1-nm filter was constructed by placing 3-dB couplers before and after the EDFA. A 35-MHz-linewidth LD provided input signals at 1548.5 nm from -19 dBm to +1.5 dBm. The loss was given by an AOM. A 70-m-long alminosilicate Er:glass fiber(Al:~5000 ppm., Er:~1000 ppm. wt.) was pumped at 1.48 µm. Al-codoped EDFs are known to be highly homogeneous with relatively flat gain spectrum[37],[38]. However, since some inhomogeneity sill exists[39], available wavelength range was estimated by changing the lasing wavelength and signal wavelength.

Fig. 22 shows typical spectral profiles of the amplified spontaneous emission(ASE) observed at coupler(B) output with lasing feedback. The ASE profile was almost clamped over the full ASE range. With the lasing wavelength around the first gain peak, no profile changes were observed. With the lasing wavelength at or around the second gain peak, the profile slightly increased in the first gain-peak area, which, however, was well within 1 dB for all six lasing wavelengths tested between 1534 nm and 1553 nm.

Fig. 23 shows the measured signal output power vs. EDFA pump power with the loop-loss and the signal input power fixed. The lasing wavelength was 1553 nm. Signal output was clamped with less than 0.4 % variation when the feedback loop lased. These results confirm the near homogeneous behavior of the tested EDFA across the full-wavelength range. Within the range wherein the gain matches the loop loss, this configuration can be operated as a complete loss compensator.

Fig. 24 shows ASE-spectral-profile changes, or small-signal-gain changes when the loss of attenuator(B) changed by 5 dB and 10 dB. With the lasing wavelength in the first gain peak area, the small-signal-gain change can be estimated to be identical to the loop loss change only at or very close to the lasing wavelength. With the lasing wavelength in the second gain peak area, the ASE level change matched the loop loss change from ~1540 to ~1553 nm. This suggests that the loss can be almost ideally recovered over this range for small signals if the lasing wavelength is chosen at or around the second gain peak.

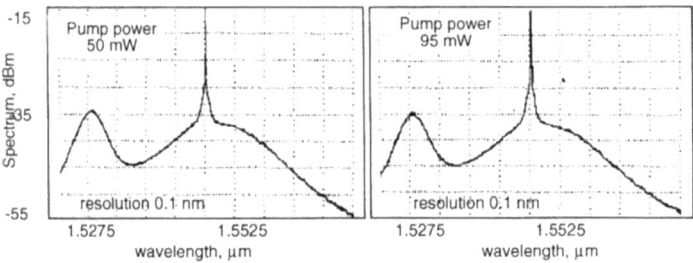

Fig. 22 Spectral profiles of ASE observed at coupler(B)

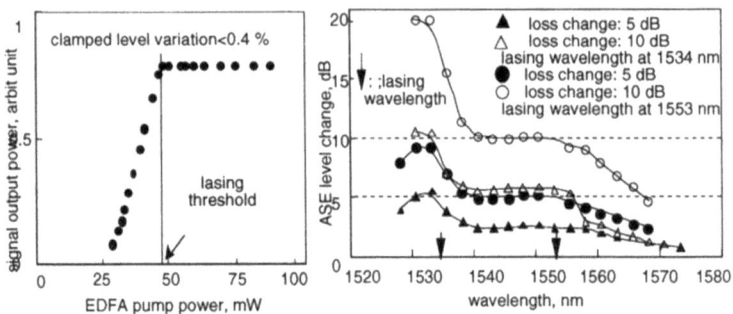

Fig. 23 Signal output power vs. pump power

Fig. 24 ASE-spectral profile changes against loss changes of ATT(B)

Shown in Fig. 25 and 26 are oscilloscope traces of output signals for loss compensation and for AGC, respectively, both with and without feedback. The signal and the feedback lasing wavelengths were 1548.5 nm and 1553 nm, respectively. Two cascaded 1-nm tunable optical bandpass filters were used to extract the signal power from coupler(B) output. The feedback lasing power was monitored at coupler(A) output via another 1-nm filter.

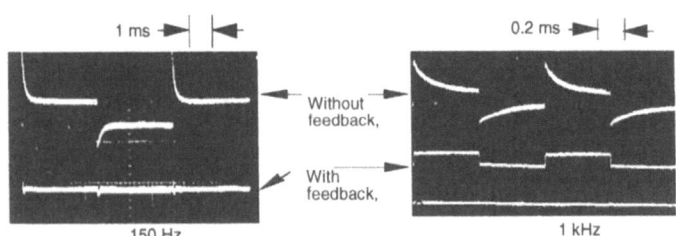

Fig. 25 Loss compemsation

Fig. 26 Gain control

In Fig. 25, without optical feedback, the output signal reflected the rectangular changes in optical loss. Moreover, optical surge was observed at the front end of the step due to the slow dynamics of the EDFA gain. With feedback, by contrast, the output signal was almost flat irrespective of the loss changes. For AGC, the output signal without feedback was distorted due to the EDFA's gain non-linearity, which almost completely disappeared when the feedback was employed. With a strongly changing input power, the EDFA had to be sufficiently pumped to ensure continuous AGC operation.

It was confirmed that AGC and loss compensation are very effective if the frequencies of input intensity fluctuations and local loss fluctuations are both lower than a few kilohertz. In such low frequency ranges, these two functions can be obtained simultaneously.

4.1.3 Conclusion

The effect of unwanted and varying optical loss was automatically offset with an EDFA used in an optical feedback configuration. Signal amplification is unaffected by such unwanted loss, and system gain can be controlled via an attenuator situated in the feedback loop. Optical loss of more than 15 dB was automatically recovered for signal wavelengths over a 13 nm range centered at about 1550 nm. The system dynamic range for loss fluctuation frequencies was 0 ~ several kilohertz. The proposed loss compensation scheme simultaneously provides an AGC function which well stabilizes EDFA gain is against slow changes in input signal power or pump power.

5.3 WAVELENGTH CONVERSION WITH AN IMPROVED EXTINCTION RATIO

Optical wavelength conversion techniques are essential in future broad band networks. They are applicable to wavelength routing[40] which greatly improves system capacity and flexibility in WDM networks. Reported techniques have used four-wave mixing[41], and devices include DFB lasers[42], DBR lasers[43],[44] and laser-diode amplifiers (LDAs)[45],[46] which were mostly based on saturable absorption or gain saturation. Those approaches, however, suffer from signal extinction ratios that are degraded by about ~3 dB or more.

To overcome this difficulty, this section proposes a novel wavelength converter[47] which converts optical signal frequency over a wide wavelength range, and yet achieves signal amplification with improved extinction ratios. Feasibility of operation at gigahertz speeds is also indicated.

5.3.1 Theory

(1) Working principle The proposed system is based on a ring-laser cavity incorporating a traveling-wave-type laser-diode amplifier as a fast-response optically-driven gain modulator, and an Er-doped fiber as a slow-response gain medium. Fig. 27 outlines the system, in which input signal wavelength is λ_1 and lasing wavelength is λ_2.

An intensity-modulated optical signal is injected into the LDA at λ_1 via the coupler at IM frequency f. If f<f_R (f_R is the relaxation resonance frequency of the LDA and is typically in the gigahertz range), the lasing-power at λ_2 increases/decreases with decreases/increases in the input signal power due to the depletion and recovery of the carrier thus realizing wavelength conversion. Since the LDA's gain is homogeneously broadened, wavelengths of the input signal and output signal, λ_1 and λ_2, can be widely chosen. Here, if the LDA is sufficiently pumped, the laser oscillation continues even in the presence of IM input signals thereby keeping the gain constant; the gain is clamped at the point where it balances the cavity losses.

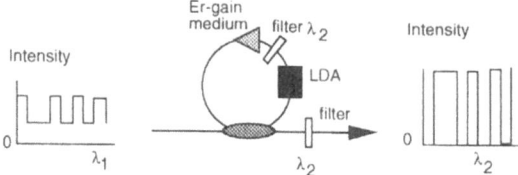

Fig. 27 Outline of the proposed system

When input signal power exceeds a certain limit P_{th}, however, the LDA gain can no longer balance the cavity loss because the carrier is excessively depleted. If the peak power P and frequency f of the IM input signal meet the following requirements, the lasing oscillation becomes intermittent corresponding to the input IM waveform. The requirements are $f<1/\tau$ (τ is the photon lifetime of the cavity in seconds, where, typically, $1/\tau<<f_R$), and $P>P_{th}$. If the wavelength conversion is based on this intermittent oscillation, the output-signal extinction ratio R approaches unity, where $R=(P_{max}-P_{min})/(P_{max}+P_{min})$ and P_{max} and P_{min} are the maximum and minimum signal powers, respectively. However, if a high peak power must be achieved with intermittent oscillation, the LDA must be sufficiently pumped and accordingly, Pth must increase which demands a large IM amplitude for the input signal.

To reduce the minimum amplitude required to initiate high-power intermittent oscillation, the present proposal adds an Er-doped gain medium into the cavity. Since this medium's gain response is slower than several kilohertz[47], its gain saturation follows the time-average of the rapid IM input signals and of the circulating laser intensity. Thus, the gain remains almost unchanged against the IM signal. Therefore, the Er-doped gain medium can be sufficiently pumped and contributes to providing high peak power under intermittent lasing oscillation; high-power input IM signals are not needed. The signal bandwidth, limited by the photon lifetime of the cavity, can be expanded by reducing the cavity length.

(2) LDA gain modulation The above wavelength-conversion effect depends on the LDA-gain modulation behavior against input IM signals, which is analyzed based on the carrier rate equation and propagation equation of the optical signal as follows.

Signal intensity I in the LDA is written as

$$\frac{dI}{dz} = \left(\frac{\Gamma G_0}{1+I/I_s} - \alpha\right)I, \qquad (12)$$

where z is the propagation direction of the optical signal, Γ is the optical-mode confinement factor for the active region, I_s is the saturation intensity, G_0 is the unsaturated gain coefficient, and α is the modal absorption coefficient. By integrating (12) over the entire length of the LDA, the gain saturation behavior against input signal power is expressed as[46]

$$\frac{I_{in}}{I_s} = \frac{\Gamma G_0 L - \ln(G)}{(G-1)}, \qquad (13)$$

where G is the amplifier gain defined as the ratio of the input intensity(I_{in}) to the output intensity expressed with linear scale, L is the amplifier length. α was neglected for simplicity. From Eq. (13),

$$\frac{d}{dG}\left(\frac{I_{in}}{I_s}\right) = -\frac{1}{G(G-1)} - \frac{\Gamma G_0 L - \ln(G)}{(G-1)^2}. \qquad (14)$$

This indicates that $|\frac{d}{dG}(\frac{I_{in}}{I_s})|$ monotonously decreases with G and, changes in the LDA-gain against input-intensity changes monotonously decrease with input signal power. Therefore, the gain-modulation amplitude decreases with increases in the DC bias level of the input power if the IM amplitude is unchanged.

(3) Operating frequency range The lowest conversion frequency is determined by the EDFA's gain response. This is because, if input IM frequency f is slower than the EDFA's gain response i.e. f<several kilohertz, LDA-gain changes are recovered by the Er-doped gain via the laser-oscillating feedback loop. Therefore, the lasing oscillation is no longer intermittent, but simply intensity modulated and signal extinction ratios start to rapidly decrease.

The highest conversion frequency depends on the total intensity-buildup time T_b from the initial noise level I_0 to the final steady-state oscillation level I_{ss}. T_b is written as

$$T_b \cong \frac{\tau_c}{r-1} \ln(\frac{I_{ss}}{I_0}), \qquad (15)$$

where τ_c is the cavity photon lifetime (or exponential decay time of optical signals in the cavity in the absence of laser gain) and r is the normalized inversion ratio or the ratio of the initial unsaturated gain coefficient to the cold-cavity loss coefficient[49]. This indicates that a short-cavity, high-gain laser pumped well above its threshold provides a short buildup time thereby attaining quick response.

5.3.2 Experiment and considerations

The proof-of-concept device used an 89-m-loop fiber-ring cavity(FSR;2.51 MHz), and a traveling-wave-type polarization-independent, 1.5-μm-GaInAsP LD amplifier followed by the EDFA and a 1-nm optical band-pass filter. An 80 kHz-linewidth LD provided the input signal at λ_1=1548.9 nm. The signal was intensity-modulated by a Mach-Zehnder device. The LDA was placed before the EDFA with a narrow band filter between them.

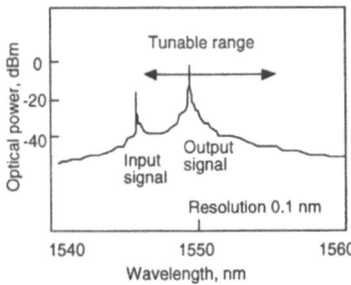

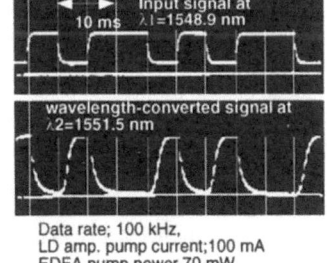

Fig. 28 Input/Output spectra   **Fig. 29 Input/Output waveform**

Fig. 28 shows an example of the output optical spectra observed at the output port without the 1-nm filter. Lasing wavelength was widely varied from 1546.5 to 1555.2 nm with the input wavelength fixed at 1548.9 nm.

Fig. 29 shows an example of waveform conversion for a data sequence at 0.2 Mb/s. Vertical scales for the input and output signals are different. It is clearly seen that the extinction ratio of the output waveform was enhanced.

Small-signal-gain of the LDA was then measured at signal wavelength λ_s=1548.9 nm against control-signal power at λ_c=1551.5 nm. The result agreed well with Eq.(13) thus confirming that the gain-modulation sensitivity increases with LDA pump power, but monotonously decreases with increases in the input signal power.

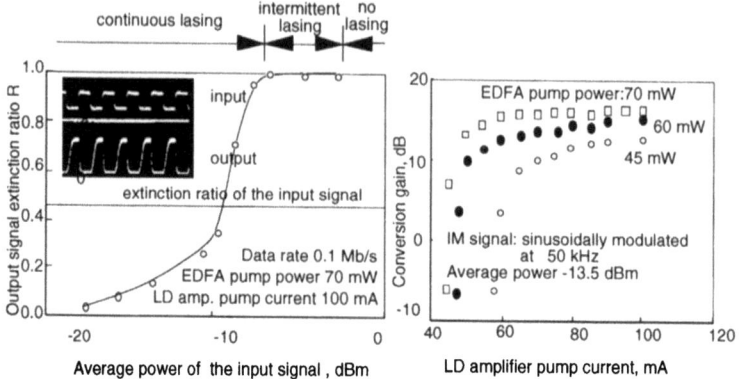

Fig. 30 Extinction ratio improvement   Fig. 31 Conversion gain

Fig. 30 shows the improvement in signal extinction ratio against input signal power. Although the wave form degraded due to the non-zero rise time of lasing, R improved with average input powers of ~-10 dBm to ~-1 dBm. For input signals<~-10 dBm, each lasing oscillation did not decay to zero because the carrier of the LDA was not sufficiently depleted due to the instantaneous recovery through the reduction in lasing power in the cavity. For input signals around -1 dBm, by contrast, carriers were deeply depleted such that lasing oscillation was possible only near the lasing threshold, thus the signal build-up was insufficient. When the input power further increased, lasing oscillation was totally suppressed and wavelength conversion was correspondingly blocked.

Complete lasing build-up and decay with intermittent laser oscillations was achieved at data rates of up to ~0.5 Mb/s with LDA pump=100 mA and EDFA pump=70 mW. At data rates over 0.5 Mb/s, the output waveform did not return to zero nor was the maximum possible value achieved so the converted signal extinction ratio degraded. This is because the data rate exceeded the maximum speed determined by the total intensity-buildup time T_b in Eq. (15). The highest possible data rate for the proposed wavelength conversion technique can be estimated considering the Er-doped silica-based planar ring laser that has been very recently reported. The laser has a cavity length of 92 mm(FSR=2.5 GHz)[49]. Another silica-based wave guide includes an LDA and forms a ring laser with a cavity length of 102 mm[50]. By simply integrating these two proposals, the feasibility of a high-speed wavelength converter is suggested. The cavity length can be reduced to about 100 mm and, accordingly, the operation speed can approach the gigahertz range.

Fig. 31 shows the conversion gain G, the output/input signal amplitude, measured with a 50 kHz sinusoidal IM input at -13.5 dBm and extinction ratio R=~0.8. R improved to ~1 and, the maximum gain was 18 dB.

5.3.3 Conclusion

The proposed device converted signal wavelengths over ~10 nm, and amplified signals by 18 dB while improving the extinction ratios. Although the operation speed was rather slow, up to gigahertz response is feasible simply by decreasing the cavity length based on the planer optical-waveguide technique.

The proposed device can be seen as an waveform-regenerator, which, though providing no re-timing function, discriminates signals and regenerates amplified signals. It does, however, yield much less ASE noise than conventional optical amplifiers. Therefore, if one or more of the cascaded EDFAs used in long-length transmission systems are replaced with such noise-reduced, regenerative amplifiers, unwanted ASE-noise accumulation would be effectively avoided.

6 GENERAL CONSIDERATIONS

This chapter clarifies the presented research field by indicating which issues have been confirmed and which issues remain unresolved. The relationship between the proposed techniques and existing techniques is clarified.

6.1 LIGHT SOURCES

Fig. 32 overviews the various light sources used in I-OFSs. The technical issues and research status concerning EDF-Ring lasers are also shown. The feasibility of EDF-RLs for measurement use has been confirmed by achieving narrow-linewidth single-mode oscillation and multi-frequency oscillation. Although mode-hop elimination has been achieved for "many hours"[5], the complete elimination of mode-hopping and perfect lasing-frequency stabilization have not been achieved so far.

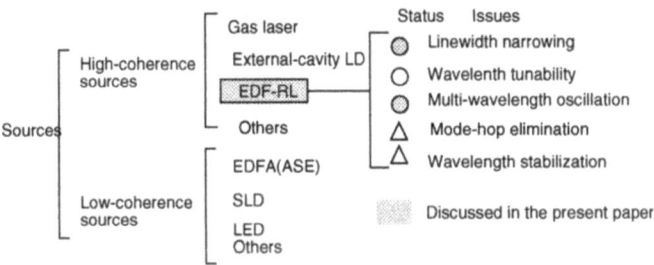

Fig. 32 Sources for optical measurements and research status of EDF-Ring lasers

6.2 INTERFEROMETERS

Figure 33 outlines several varieties of interferometers including the proposed one-arm interferometer and the active resonant interferometer. Technical problems and their study status are indicated.

The injection locking technique is applicable not only to one-arm interferometers, but also to homodyne detection for high bit-rate signals. Unfortunately, signal bandwidths for coherent communications are generally much wider than those used in measurement systems. Moreover, time-averaging techniques can not be used to improve S/N ratios as is usually possible in measurements. Therefore, when using injection locking for homodyne detection in communications, the locking range and the signal quenching effect at integer multiples of the ring cavity FSR must be carefully considered to demodulate wide band signals.

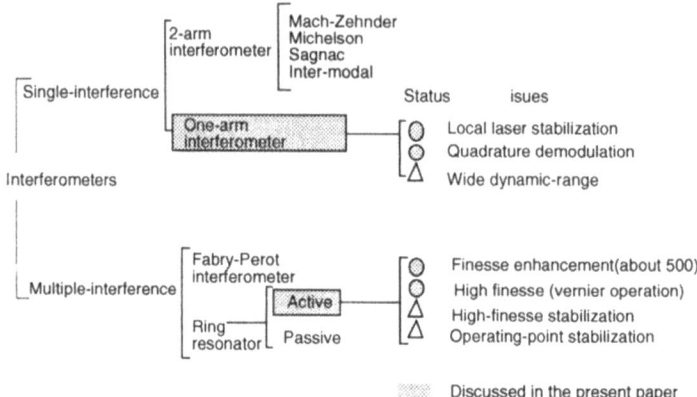

Fig. 33 Proposed interferometers and their research status

High stable finesse is achieved by dynamically optimizing the EDFA gain in the presence of input power fluctuations and round-trip optical-loss fluctuations. One solution to this would be to use the laser-oscillating optical feedback loop for automatic gain control and optical loss compensation as discussed in chapter 4.

6.3 SIGNAL PROCESSING

Fig. 34 shows the signal-processing schemes discussed and the study status of related technical issues.

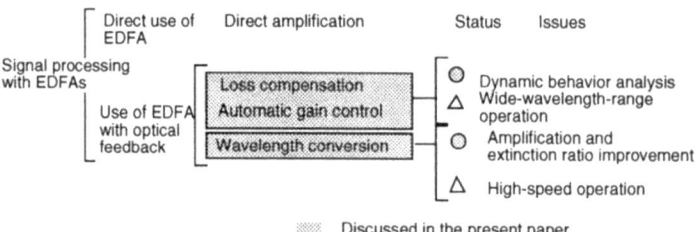

Fig. 34 Signal processing with EDFAs and research accomplishments

Concerning extinction-ratio-enhanced wavelength conversion based on the EDFA and LDA, just the theoretical basis was confirmed. One prime issue with this technique is to improve operation speed by reducing the cavity length. One solution to this would be to use an Er-doped silica-based planar ring resonator and to integrate into the resonator a semiconductor laser amplifier and other optical components.

7. GENERAL CONCLUSION

This paper has discussed a number of key techniques for interferometric optical measurements. Heavy emphasis was placed on using Er-doped fiber amplifiers. Based on the presented techniques, not only measurement performance but also the design flexibility of optical measurement equipment can be substantially improved. Important results and knowledge are summarized as follows.

(1) An Er-doped fiber ring laser(EDF-RL) oscillated in the single-longitudinal-mode with a less-than-1.4 kHz linewidth. Simultaneous oscillation of this laser at two wavelengths was attained from one gain media with a compound cavity structure.

(2) A new concept for single-arm interferometers was presented, wherein an EDF-ring laser is used as a local laser with injection locking. A 5-km-long single-arm sensor demodulated 10^{-3} rad phase modulation in the signal frequency range of 500 kHz-3 MHz.

(3) An EDF-ring resonator attained a finesse>500. The resonator operated as an optical spectrum analyzer and permitted direct optical spectrum observations with a better-than-100-kHz resolution. A resonant passband with a linewidth of only 17 kHz was measured with a relatively broad-band light. The resonator attains a finesse >10000 based on the vernier effect.

(4) An all-optical feedback scheme based on an EDFA was proposed that eliminates the effect of unwanted and varying optical loss over 14 dB at DC to several kilohertz.

(5) A wavelength converter using an Er-doped laser cavity and LD amplifier was proposed. The system not only converted signals over 10 nm wavelength range but also enhanced the signal extinction ratio and provided a signal gain of ~18 dB.

Many of the above techniques are applicable not only to optical measurements but also optical transmission systems and optical signal processing functions.

ACKNOWLEDGMENTS

The authors express their sincere gratitude to Dr. Tetsuya Miki, Executive Manager of NTT Transmission Systems Laboratories and Dr. Hideki Ishio, Executive Manager of Lightwave Communications Laboratory, Transmission Systems Laboratories for their instructions and encouragement. The authors are much indebted to Drs. Takeshi Ito, Kiyoshi Nakagawa, Masatoshi Saruwatari for their suggestions, supports. Also thanked are the co-workers in the Lightwave Communications Laboratory, particularly Drs. Shigeru Saito, Kyo Inoue, Osamu Ishida, Akira Naka and R. Esman for helpful discussions and a stimulating environment.

REFERENCES

(1) H. Schmuck, TH. Pfeiffer and G. Veith, "Widely tunable narrow linewidth erbium doped fiber ring laser", Electron. Lett., Vol. 27, No. 23, pp. 2117-2119, 1991

(2) J. L. Zyskind, J. W. Sulhoff, Y. Sun, J. Stone, L. W. Stulz, G. T. Harvey, D. J. Digiovanni, H. M. Presby, A. Piccirilli, U. Koren and R. M. Jopson, "Single mode diode-pumped tunable erbium-doped fiber laser with linewidth less than 5.5 kHz", Electron. Lett., Vol. 27, No. 23, pp. 2148-2149, 1991

(3) K. Iwatsuki, H. Okamura and M. Saruwatari, "Wavelength tunable single-frequency and single polarization Er-doped fiber ring laser with 1.4 kHz linewidth", Electron. Lett., Vol. 26, No. 24, pp. 2033-2035, 1990

(4) T. Okoshi, K. Kikuchi and A. Nakayama, "Novel method for high resolution measurement of laser output spectrum", Electron. Lett., Vol. 16, pp. 630-631, 1980

(5) H. Sabert., "Suppression of mode jumps in a single mode fiber laser", Opt. Lett., Vol. 19, NO. 2, pp. 111-113, 1994

(6) H. Okamura and K. Iwatsuki, "Simultaneous oscillation of wavelength-tunable, single-mode lasers using an Er-doped fiber amplifier", Electron. Lett., Vol. 28, No. 5, pp. 461-463, 1992

(7) J. W. Dawson, N. Park and K. J. Vahala, "Co-lasing in an electrically tunable erbium-doped fiber laser", Appl. Phys. Lett., Vol. 60, No. 25, pp. 3090-3092, 1992

(8) N. Park, J. W. Dawson, and K. J. Vahala, "Multiple wavelength operation of an Erbium-doped fiber laser", IEEE Photon. Technol. Lett., Vol. 4 No. 6, pp. 540-541, 1992

(9) M. J. Pettitt, A. Hadjifotiou and R. A. Baker, "Crosstalk in erbium doped fiber amplifiers", Electron. Lett., Vol. 25, No. 6, pp. 416-417, 1989

(10) E. Desurvier, J. W. Sulhoff, J. L. Zyskind and J. R. Simpson, "Study of spectral dependence of gain saturation and effect of inhomogeneous broadening in Erbium-doped Aluminosilicate fiber amplifiers", IEEE Photon. Technol. Lett., Vol. 2, No. 9, pp. 653-655, 1990

(11) S. Norimatsu and K. Iwashita, "10 Gbit/s optical PSK homodyne transmission experiment using external cavity DFB LDs", Electron. Lett., Vol. 26, No. 10, pp. 648-649, 1990

(12) S. Kobayashi and T. Kimura, "Coherence of injection phase-locked AlGaAs semiconductor laser", Electron. Lett., vol. 16, No. 17, pp. 668-670, 1980)

(13) H. Okamura and K. Iwatsuki, "Fiber optic sensor using an injection-locked local laser", IEEE J. Lightwave Technol., Vol. 9, No. 4, pp. 552-557, 1991

(14) D. A. Jackson, A. Preiest and A. Dandridge, "Elimination of drift in a single-mode optical fiber interferometer using a piezoelectrically stretched coiled fiber", Appl. Opt., vol. 19, pp. 2926-2929, 1980.

(15) C. J. Buczek, R. J. Freiberg, "Hybrid injection locking of high power CO_2 lasers", IEEE J. QE., Proc. IEEE, Vol. QE-8, No. 7, pp. 641-650, 1972

(16) S. Kobayashi, Y. Yamamoto and K. Kimura, "Optical FM signal amplification and FM noise reduction in an injection locked AlGaAs semiconductor laser", Electron. Lett., vol. 17, No. 22, pp. 849-851, 1981

(17) S. K. Sheem, "Fiber-optic gyroscope with (3×3) directional coupler", Appl. Phys. Lett., Vol. 37, No. 10, pp. 869-871, 1980

(18) Th. Niemeier and R. Ulrich, "Quadrature outputs from fiber interferometer with 4×4 coupler", Opt. Lett., Vol. 11. No. 10, pp. 677-679, 1986

(19) J. Stone and D. Marcuse, "Ultrahigh finesse fiber Fabry-Perot interferometer", IEEE J. Lightwave Technol., Vol. LT-4, No. 4, pp. 382-385, 1986

(20) C. Y. Yue, J. D. Peng, Y. B. Liao and B. K. Zhou, "Fiber ring resonator with finesse of 1260", Electron. Lett., Vol. 24, No. 10, pp. 622-623, 1988

(21) J. E. Bowers, S. A. Newton, W. V. Sorin and H. J. Shaw, "Filter response of single-mode fiber recirculating delay lines", Electron. Lett., Vol. 18, No. 3, pp. 110- 111, 1982

(22) Y. H. Ja, "Single-mode optical fiber and loop resonators using degenerate two-wave mixing", Proc. 7-th OFS, pp. 195-198, 1990

(23) H. Okamura and K. Iwatsuki, "Er-doped fiber ring resonator applied to optical spectrum analyzer with less than 100 kHz resolution", Electron. Lett., Vol. 27, No. 12, pp.1047-1049, 1991

(24) H. Okamura and K. Iwatsuki, "A finesse enhanced Er-doped fiber ring resonator", IEEE J. Lightwave Technol., Vol. 9, No.11, pp. 1991

(25) T. Okoshi, K. Kikuchi and A. Nakayama, "Novel method for high resolution measurement of laser output spectrum", Electron. Lett., Vol. 16, pp. 630-631, 1980
(26) H. Okamura and K. Iwatsuki, "Fiber optic sensor using an injection-locked local laser", IEEE J. Lightwave Technol., Vol. 9, No. 4, pp. 552-557, 1991
(27) K. Kalli and D. A. Jackson, "Ring resonator optical spectrum analyzer with 20-kHz resolution", Opt. Lett., Vol. 17, No. 15, pp.1090-1092, 1992
(28) P. Urquhart, "Compound optical-fiber-based resonators", J. Opt. Soc. Am. A, Vol. 5, No. 6, pp. 803-812, 1988
(29) I. P. Kaminow, P. P. Iannone, J. Stone and L. W. Stulz, " A tunable vernier fiber Fabry-Perot filter for FDM demultiplexing and detection", Photon. Technol. Lett., Vol. 1, No. 1, pp. 24-26, 1989
(30) K. Oda, N. Takato and H. Toba, "A wide-FSR waveguide double-ring resonator for optical FDM transmission systems", IEEE J. Lightwave Technol., Vol. 9, No. 6, pp. 728-736, 1991
(31) M. Zirngibl, "Gain control in Erbium-doped fiber amplifiers by an all-optical feedback loop", Electron. Lett., Vol. 27, No. 7, pp.560-561, 1991
(32) H. Okamura, "Automatic optical-loss compensation with an Er-doped fiber amplifier", Electron. Lett. Vol. 27, No. 23, pp. 2155-2156
(33) R. I. Laming, P. R. Morkel and D. N. Payne, "Multichannel crosstalk and pump noise characterization of Er^{3+}-doped fiber amplifier pumped at 980 nm", Electron. Lett., Vol. 25, No. 7, pp. 455-456, 1989
(34) K. Inoue, H. Toba, N. Shibata, K. Iwatsuki and A. Takada, "Mutual signal gain saturation in Er^{3+}-doped fiber amplifier around 1.54 μm wavelength", Electron. Lett., Vol. 25, No. 9, pp. 594-595, 1989
(35) M. Zirngibl, "Gain control in erbium-doped fiber amplifiers by an all optical feed back loop", Electron. Lett., Vol. 27, No. 7, pp. 560-561, 1991
(36) M. Zirngibl, "All optical remote gain switching in Er-doped fiber amplifiers", Electron. Lett., Vol. 27, No. 13, pp. 1164-1166, 1991
(37) E. Desurvire, C. R. Giles and J. R. Simpson, "Gain saturation effects in high-speed, multichannel Erbium-doped fiber amplifiers at λ=1.53 μm", IEEE J. Lightwave Technol., Vol. 7, No. 12, pp. 2095-2104, 1989
(38) E. Desurvire and J. R. Simpson, "Evaluation of $^4I_{15/2}$ and $^4I_{13/2}$ Stark-level energies in erbium-doped aluminosilicate glass fibers", Opt. Lett., Vol. 15, No. 10, pp. 547-549, 1990
(39) J. L. Zyskind, E. Desurvire, J. W. Sulhoff and D. J. DiGiovanni, "Determination of homogeneous linewidth by spectral gain hole-burning in an Erbium-doped fiber amplifier with GeO_2:SiO_2 core", IEEE Photon. Technol. Lett., Vol. 2, No. 12, pp. 869-871, 1990
(40) A. A. M. Saleh, "Optical WDM technology for networking and switching applications", OFC'92, ThC1, 1992
(41) G. Grosskopf, R. Ludwig and H. G. Weber, "140 Mbit/s DPSK transmission using an all-optical frequency converter with a 4000 GHz conversion range", Electron. Lett., Vol. 24, pp. 1106-1107, 1988
(42) P. Pottier, M. J. Chawki, R. Auffret, G. Klaveau and A. Tromeur, "1.5 Gb/s transmission system using all optical wavelength converter based on tunable two-electrode DFB laser", Electron. Lett., Vol. 27, No. 23, pp. 2183-2185, 1991
(43) B. Mikkelsen, T. Durhuns, R. J. Pedersen and K. E. Stubkjaer, "Penalty free wavelength conversion of 2.5 Gbit/s signals using a tuneable DBR-laser", ECOC'92, WeA10.4, 1992

(44) M. Oberg, S. Nilsson, T. Klinga and P. Ojala, "A Three-Electrode Distributed Bragg Reflector Laser with 22 nm Wavelength Tuning Range", IEEE Photon. Technol. Lett., Vol. 3, No. 4, pp. 299-301, 1991

(45) T. Durhuus, B. Fernier, P. Garabedian, F. Lebrond, J. L. Lafragette, B. Mikkelsen, C. G. Joergensen and K. E. Stubkjaer, "High-speed all-optical gating using a two-section semiconductor optical amplifier structure", CLEO' 92, CThS4

(46) M. J. Adams, J. V. Collins and I. D. Henning, "Analysis of semiconductor laser optical amplifiers", IEE Proceedings, Vol. 132, Pt. J, No. 1, pp. 58-63, 1985

(47) E. Desurvire, "Analysis of transient gain saturation and recovery in Erbium-doped fiber amplifiers", Photon. Technol. Lett., Vol. 1, No. 8, pp. 196-199, 1989

(48) Siegman, 'LASERS' University Science Book, Mill Valley, California(1986).

(49) T. Kitagawa, K. Hattori, Y. Hibino, Y. Ohmori and M. Horiguchi, "Laser oscillation in Er-doped silica-based planar ring resonator", ECOC'92, PDP. II-5, 1992

(50) Y. Hibino, H. Terui, A. Sugita and Y. Ohmori, "Silica-based optical waveguide ring laser integrated with semiconductor laser amplifier on Si substrata"' Electron. Lett., Vol. 28, No. 20, pp. 1932-1933, 1992

VIII Optical Communication Systems

OPTICAL FIBER COMMUNICATION TECHNOLOGY AND SYSTEM OVERVIEW

IRA JACOBS
Fiber and Electro-Optics Research Center
The Bradley Department of Electrical Engineering
Virginia Polytechnic Institute and State University
Blacksburg, Virginia, U.S.A. 24061

ABSTRACT

Basic elements of an optical fiber communication system include the transmitter (laser or LED), fiber (multimode, single mode, dispersion-shifted) and the receiver (PIN and APD detectors, coherent detectors, optical preamplifiers, receiver electronics). Receiver sensitivities of digital systems are compared on the basis of the number of photons-per-bit required to achieve a given bit-error-probability, and eye-degradation and error-floor phenomena are described. Laser relative-intensity-noise and nonlinearities are shown to limit the performance of analog systems. Networking applications of optical amplifiers and wavelength division multiplexing are considered, and future directions discussed.

1. Introduction

Although the light guiding property of optical fibers has been known and used for many years, it is only relatively recently that optical fiber communications has become both a possibility and a reality. Following the first prediction in 1966 [1] that fibers might have sufficiently low attenuation for telecommunications, the first low loss fiber (20 dB/km) was achieved in 1970 [2]. The first semiconductor laser diode to radiate continuously at room temperature was also achieved in 1970 [3]. The 1970's were then a period of intense technology and system development, with the first systems coming into service at the end of the decade. This is a relatively short time period for the development and introduction of a radically new communications technology into telecommunications networks worldwide. The 1980's saw both the growth of applications (service on the first trans-Atlantic cable in 1988), and continued advances in technology. Whereas present systems are primarily point-to-point, applications emphasis is now on optical networks.

This paper presents a tutorial overview of the basic technology, systems and applications of optical fiber communication. It is an update and expansion of a more descriptive and less technical earlier paper on these topics [4].

2. Basic Technology

This section considers the basic technology components of an optical fiber communications link; namely the fiber, the transmitter, and the receiver; and discusses the principal parameters that determine communications performance.

2.1 FIBER

An optical fiber is a thin filament of glass with a central core having a slightly higher index of refraction than the surrounding cladding. From a physical optics standpoint, light is guided by total internal reflection at the core-cladding boundary. More precisely, the fiber is a dielectric waveguide in which there are a discrete number of propagating modes [5]. If the core diameter and the index difference are sufficiently small, only a single mode will propagate. The condition for single mode propagation is that the normalized frequency V be less than 2.405, where

$$V = \frac{2\pi a}{\lambda}\sqrt{n_1^2 - n_2^2} \qquad (1)$$

and a is the core radius, λ the free space wavelength, and n_1 and n_2 are the indices of refraction of the core and cladding, respectively. Multimode fibers typically have a fractional index difference (Δ) between core and cladding between 1 and 1.5 % and a core diameter between 50 and 100 µm. Single mode fibers typically have $\Delta \approx 0.3\%$, and a core diameter between 8 and 10 µm.

The fiber numerical aperture (NA), which is the sine of the half-angle of the cone of acceptance, is given by

$$NA = \sqrt{n_1^2 - n_2^2} = n_1\sqrt{2\Delta} \qquad (2)$$

Single mode fibers typically have an NA about 0.1, whereas the NA of multimode fibers is in the range 0.2-0.3.

From a transmission system standpoint, the two most fiber parameters are attenuation and bandwidth.

2.1.1 *Attenuation*

There are three principal attenuation mechanisms in fiber: absorption, scattering, and radiative loss. Silicon dioxide has resonance absorption peaks in the ultraviolet (electronic transitions) and in the infrared beyond 1.6 µm (atomic vibrational transitions), but is highly transparent in the visible and near infrared.

Radiative losses are generally kept small by using a sufficiently thick cladding (communication fibers have an outer diameter of 125 µm), a compressible coating to buffer the fiber from external forces, and a cable structure that prevents sharp bends.

In the absence of impurities and radiation losses, the fundamental attenuation mechanism is Rayleigh scattering from the irregular glass structure which results in index of refraction fluctuations over distances small compared to the wavelength. This leads to a scattering loss

$$\alpha = \frac{B}{\lambda^4}, \text{ with } B \approx 0.9 \frac{dB}{km}\mu m^4 \qquad (3)$$

for "best" fibers. Attenuation as a function of wavelength is shown in Figure 1. The attenuation peak at $\lambda = 1.4$ μm is a resonance absorption due to small amounts of water in the fiber. Present long-distance fiber optic systems generally operate at wavelengths of 1.3 or 1.55 μm. The former, in addition to being low attenuation (about 0.32 dB/km for best fibers), is the wavelength of minimum intramodal dispersion (see Section 2.1.2). Operation at 1.55 μm allows even lower attenuation (minimum is about 0.16 dB/km).

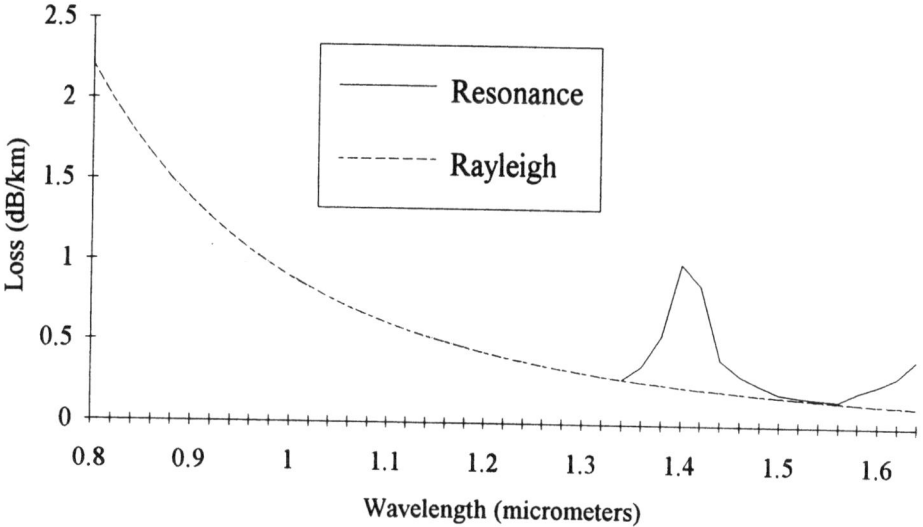

Figure 1. Fiber attenuation as a function of wavelength. Dashed curve shows Rayleigh scattering. Solid curve indicates resonance absorption at 1.4 μm from water and tail of infrared atomic resonances.

2.1.2 *Dispersion*

Pulse spreading (dispersion) limits the maximum modulation bandwidth (or maximum pulse rate) that may be used with fibers. There are two principal forms of dispersion: intermodal dispersion and intramodal dispersion. In multimode fiber, the different modes experience different propagation delays resulting in pulse spreading. For graded-index fiber, the lowest dispersion per unit length is given approximately by [6]

$$\frac{\delta\tau}{L} = \frac{n_1 \Delta^2}{10c} \quad \text{(intermodal)} \qquad (4)$$

(Grading of the index of refraction of the core in a nearly parabolic function results in an approximate equalization of the propagation delays. For a step-index fiber, the dispersion per unit length is $\delta\tau/L = n_1\Delta/c$, which for $\Delta=0.01$, is 1000 times larger than that given by Eq. 4.)

Bandwidth is inversely proportional to dispersion, with the proportionality constant dependent on pulse shape and how bandwidth is defined. If the dispersed pulse is approximated by a Gaussian pulse with $\delta\tau$ being the full-width at the half-power point, then the -3 dB bandwidth B is given by
$$B = 0.44/\delta\tau \qquad (5)$$
Multimode fibers are generally specified by their bandwidth in a 1 km length. Typical specifications are in the range from 200 MHz to 1 GHz. Fiber bandwidth is a sensitive function of the index profile, is wavelength dependent, and the scaling with length depends on whether there is mode mixing [7]. Multimode fibers are now generally used only when the bit rates and distances are sufficiently small that accurate characterization of dispersion is not of concern.

Although there is no intermodal dispersion in single mode fibers[1], there is still dispersion within the single mode (intramodal dispersion) resulting from the finite spectral width of the source and the dependence of group velocity on wavelength. The intramodal dispersion per unit length is given by

$$\delta\tau/L = D\,\delta\lambda \qquad \text{for } D \neq 0$$
$$= 0.2 S_o (\delta\lambda)^2 \qquad \text{for } D = 0 \qquad (6)$$

where D is the dispersion coefficient of the fiber, $\delta\lambda$ is the spectral width of the source, and

$$S_o = \frac{dD}{d\lambda} \text{ at } \lambda = \lambda_0, \text{ where } D(\lambda_0) = 0 \qquad (7)$$

If both intermodal and intramodal dispersion are present, the square of the total dispersion is the sum of the squares of the intermodal and intramodal dispersions. For typical digital systems, the total dispersion should be less than half the interpulse period T. From Eq. 5, this corresponds to an effective fiber bandwidth that is at least $0.88/T$.

There are two sources of intramodal dispersion: material dispersion which is a consequence of the index of refraction being a function of wavelength, and waveguide dispersion which is a consequence of the propagation constant of the fiber waveguide being a function of wavelength.

For a material with index of refraction $n(\lambda)$, the material dispersion coefficient is given by

---

[1] A single mode fiber actually has two degenerate modes corresponding to the two principal polarizations. Any assymetry in the transmission path removes this degeneracy and results in polarization dispersion. This is typically very small, and is generally only of concern in long distance undersea systems with linear repeaters [8]

$$D_{mat} = -\frac{\lambda}{c}\frac{d^2n}{d\lambda^2} \tag{8}$$

For silica based glasses, D_{mat} has the general characteristic shown in Figure 2. It is about -100 ps/km-nm at a wavelength of 820 nm, goes through zero at a wavelength near 1300 nm, and is about 20 ps/km.nm at 1550 nm.

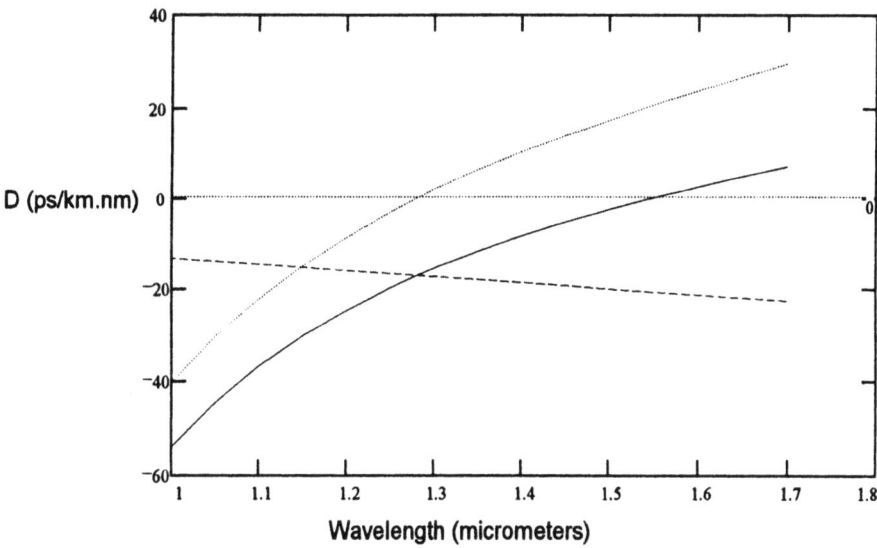

Figure 2. Intramodal dispersion coefficient as a function of wavelength. Dotted curve shows D_{mat}; dashed curve shows D_{wg} to achieve D (solid curve) with dispersion zero at 1.55 μm.

For step-index single mode fibers, waveguide dispersion is given approximately by [8]

$$D_{wg} \approx -\frac{0.025\lambda}{a^2 c n_2} \tag{9}$$

For conventional single mode fiber, waveguide dispersion is small (about -5 ps/km.nm at 1300 nm). The resultant D(λ) is then slightly shifted (relative to the material dispersion curve) to longer wavelengths, but the zero dispersion wavelength (λ_0) remains in the vicinity of 1300 nm. However, if the waveguide dispersion is made larger negative, by decreasing a or equivalently by tapering the index of refraction in the core, the zero dispersion wavelength may be shifted to 1550 nm. Such dispersion-shifted-fibers are now available commercially, and are advantageous because of the lower fiber attenuation at this wavelength, and the advent of erbium-doped-fiber amplifiers (see Section 5). Note that dispersion-shifted-fibers have a smaller slope at the dispersion minimum, $S_0 \approx 0.05 \text{ ps/km.nm}^2$ compared to $S_0 \approx 0.09 \text{ ps/km.nm}^2$ for conventional single mode fiber [10].

With more complicated index of refraction profiles, it is possible, at least theoretically, to control the shape of the waveguide dispersion such that the total dispersion is small in both the 1300 and 1550 nm bands [11]. Such dispersion-flattened-fibers are difficult to manufacture and (as yet) do not appear to offer significant system advantages.

2.2 TRANSMITTING SOURCES

Semiconductor light emitting diodes (LED) or lasers or the primary light sources used in fiber optic transmission systems. The principal parameters of concern are the power coupled into the fiber, the modulation bandwidth, and (because of intramodal dispersion) the spectral width.

2.2.1 *Light Emitting Diodes (LED)*

LEDs are forward biased PN junctions in which carrier recombination results in spontaneous emission at a wavelength corresponding to the energy gap. Although several milliwatts may be radiated from high radiance LEDs, the radiation is over a wide angular range, and consequently there is a large coupling loss from an LED to a fiber. Coupling efficiency (η = ratio of power coupled to power radiated) from an LED to a fiber is given approximately by [12]

$$\eta \approx (NA)^2 \quad \text{for } r_s < a$$
$$\eta \approx (a/r_s)^2 (NA)^2 \quad \text{for } r_s > a$$
(10)

where r_s is the radius of the LED. Use of large diameter, high NA multimode fiber improves the coupling from LEDs to fiber. Typical coupling losses are 10 to 20 dB to multimode fibers, and more than 30 dB to single mode fibers.

In addition to radiating over a large angle, LED radiation has a large spectral width (about 50 nm at λ = 850 nm, and 100 nm at λ = 1300 nm) determined by thermal effects. Systems employing LEDs at 850 nm tend to be intramodal dispersion limited, whereas those at 1300 nm are intermodal dispersion limited.

Owing to the relatively long time constant for spontaneous emission (typically several nanoseconds), the modulation bandwidth of LEDs are generally limited to several hundred MHz. Thus, LEDs are generally limited to relatively short-distance low-bit-rate applications.

2.2 *Lasers*
In a laser, population inversion between the ground and excited states results in stimulated emission. In semiconductor lasers, this radiation is guided within the active

region of the laser, and is reflected at the end faces. The combination of feedback and gain results in oscillation when the gain exceeds a threshold value. The spectral range over which the gain exceeds threshold (typically about 10 nm) is much narrower than the spectral width of an LED. Discrete wavelengths within this range, for which the optical length of the laser is an integer number of half-wavelengths, are radiated. Such a laser is termed a multi-longitudinal mode Fabry-Perot laser. Radiation is confined to a much narrower angular range than an LED, and consequently may be efficiently coupled into a small NA fiber. Coupled power is typically about 1 mw.

The modulation bandwidth of lasers is determined by a resonance frequency caused by the interaction of the photon and electron concentrations [13]. Although this resonance frequency was less than 1 GHz in early semiconductor lasers, improvements in materials have led to semiconductor lasers with resonance frequencies (and consequently modulation bandwidths) in excess of 10 GHz. This is important not only for very high speed digital systems, but now also allows semiconductor lasers to be directly modulated with microwave signals. Such applications are considered in Section 7.2.

Although multilongitudinal mode Fabry-Perot lasers have a narrower spectral spread than LEDs, this spread still limits the high-speed and long-distance capability of such lasers. For such applications, single longitudinal mode (SLM) lasers are used. SLM lasers may be achieved by having a sufficiently short laser (less than 50 μm), using coupled cavities (either external mirrors, or by cleaved coupled cavities [14]) or by incorporating a diffraction grating within the laser structure to select a specific wavelength. The latter has proven to be most practical for commercial application, and includes the distributed feedback (DFB) laser in which the grating is within the laser active region, and the distributed Bragg reflector (DBR) laser where the grating is external to the active region [15].

There is still a finite linewidth of SLM lasers. For lasers without special stabilization the linewidth is of the order of 0.1 nm. Expressed in terms of frequency, this corresponds to a frequency width of 12.5 GHz at a wavelength of 1550 nm. (Wavelength and frequency spread are related by $\delta f / f = \delta \lambda / \lambda$ from which it follows that $\delta f = c\, \delta \lambda / \lambda^2$.) Thus, unlike electrical communication systems, optical systems generally use sources with spectral widths large compared to the modulation bandwidth.

The finite linewidth (phase noise) of a laser results from fluctuations of the phase of the optical field resulting from spontaneous emission. In addition to the phase noise contributed directly by the spontaneous emission, the interaction between the photon and electron concentrations in semiconductor lasers leads to a conversion of amplitude fluctuations to phase fluctuations which increases the linewidth [13]. If the intensity of a laser is changed, this same phenomenon gives rise to a change in the frequency of the laser ("chirp"). Uncontrolled, this causes a substantial increase in linewidth when the laser is modulated, which may cause difficulties in some system applications. However, the phenomenon can also be used to advantage. For appropriate lasers under small signal modulation, a change in frequency proportional to the input signal can be used to frequency modulate and/or to tune the laser. Tunable lasers are of particular importance in networking applications employing WDM.

2.3 PHOTODETECTORS

Fiber optic systems generally use PIN or APD photodetectors. In a reverse biased PIN diode, absorption of light in the intrinsic region generates carriers which are swept out by the reverse bias field. This results in a photocurrent (I_p) that is proportional to the incident optical power (P_R), where the proportionality constant is the responsivity ($\mathfrak{R}$) of the photodetector; i.e., $\mathfrak{R} = I_p / P_R$. Since the number of photons per second incident on the detector is power divided by the photon energy, and the number of electrons per second flowing in the external circuit is the photocurrent divided by the charge of the electron, it follows that the quantum efficiency (η = electrons/photons) is related to the responsivity by

$$\eta = \frac{hc}{q\lambda} \frac{I_p}{P_R} = \frac{1.24 \, (\mu m. volt)}{\lambda} \mathfrak{R} \qquad (11)$$

For wavelengths shorter than 900 nm, silicon is an excellent photodetector with quantum efficiencies of about 90%. For longer wavelengths, InGaAs is generally used with quantum efficiencies typically around 60%. Very high bandwidths may be achieved with PIN photodetectors. Consequently, the photodetector does not generally limit the overall system bandwidth.

In an avalanche photodetector (APD), a larger reverse voltage accelerates carriers causing additional carriers by impact ionization resulting in a current $I_{APD} = MI_p$, where M is the current gain of the APD. As noted in Section 3 this can result in an improvement in receiver sensitivity.

3. Receiver Sensitivity

The receiver in a direct-detection fiber optic communication system consists of a photodetector followed by electrical amplification and signal processing circuits intended to recover the communications signal. Receiver sensitivity is defined as the average received optical power to achieve a given communication rate and performance. For analog communications the communication rate is measured by the bandwidth of the electrical signal to be transmitted (B), and performance is given by the signal-to-noise ratio (SNR) of the recovered signal. For digital systems, the communication rate is measured by the bit rate (R_b), and performance is measured by the bit error probability (P_e).

For a constant optical power transmitted, there are fluctuations of the received photocurrent about the average given by Eq. 11. The principal sources of these fluctuations are: signal shot noise (quantum noise, resulting from random arrival times of photons at the detector), receiver thermal noise, APD excess noise, and laser relative intensity noise (RIN, associated with fluctuations in intensity of the source[16].

3.1 DIGITAL ON-OFF-KEYING RECEIVER

It is instructive to define a normalized sensitivity as the average number of photons per bit $(\overline{N}_p)$ to achieve a given error probability which we take here to be $P_e = 10^{-9}$. Given $\overline{N}_p$, the received power when a 1 is transmitted is obtained from

$$P_R = 2\overline{N}_p R_b \frac{hc}{\lambda} \tag{12}$$

where the factor of two in Eq. (12) is because P_R is the peak power, and $\overline{N}_p$ is the average number of photons per bit.

3.1.1 *Ideal Receiver*
In an ideal receiver individual photons may be counted, and the only source of noise is that the number of photons counted when a 1 is transmitted is a Poisson random variable with mean $2\overline{N}_p$. No photons are received when a 0 is transmitted. Consequently, an error is made only when a 1 is transmitted and no photons are received. This leads to the following expression for the error probability

$$P_e = \frac{1}{2}\exp(-2\overline{N}_p) \tag{13}$$

from which it follows that $\overline{N}_p = 10$ for $P_e = 10^{-9}$. This is termed the quantum limit.

3.1.2 *PIN Receiver*
In a PIN receiver, the photodetector output is amplified, filtered, sampled, and the sample is compared with a threshold to decide whether a 1 or 0 was transmitted. Let I be the sampled current at the input to the decision circuit scaled back to the corresponding value at the output of the photodetector. (It is convenient to refer all signal and noise levels to their equivalent values at the output of the photodetector.) I is then a random variable with means and variances given by

$$\mu_1 = I_p \qquad\qquad \mu_0 = 0 \tag{14a}$$

$$\sigma_1^2 = 2qI_p B + \frac{4kTB}{R_e} \qquad\qquad \sigma_0^2 = \frac{4kTB}{R_e} \tag{14b}$$

where the subscripts 1 and 0 refer to the bit transmitted, kT is the thermal noise energy, and R_e is the effective input noise resistance of the amplifier. Note that the noise in the 1 and 0 states are different owing to the shot noise in the 1 state.

Calculation of error probability requires knowledge of the distribution of I under the two hypotheses. Under the assumption that these distributions may be approximated by Gaussian distributions with means and variances given by Eqs. (14), the error probability may be shown to be given by [17]

$$P_e = Q\left(\frac{\mu_1 - \mu_0}{\sigma_1 + \sigma_0}\right) \tag{15}$$

where

$$Q(\kappa) = \frac{1}{\sqrt{2\pi}} \int_\kappa^\infty dx\, \exp(-x^2/2) \tag{16}$$

It can be shown from Eqs. (11), (12), (14), and (15) that

$$\overline{N}_p = \frac{B}{\eta R_b}\kappa^2\left[1 + \frac{1}{\kappa}\sqrt{\frac{8\pi kTC_e}{q^2}}\right] \tag{17}$$

where

$$C_e = \frac{1}{2\pi R_e B} \tag{18}$$

is the effective noise capacitance of the receiver, and from Eq. (16), $\kappa = 6$ for $P_e = 10^{-9}$. The minimum bandwidth of the receiver is half the bit rate, but in practice B/R_b is generally about 0.7.

The Gaussian approximation is expected to be good when the thermal noise is large compared to the shot noise. It is interesting, however, to note that Eq. (17) gives $\overline{N}_p = 18$ when $C_e = 0$, $B/R_b = 0.5$, $\eta = 1$ and $\kappa = 6$. Thus, even in the shot noise limit, the Gaussian approximation gives a surprisingly close result to the value calculated from the correct Poisson distribution. It must be pointed out, however, that the location of the threshold calculated by the Gaussian approximation is far from correct in this case. In general, the Gaussian approximation is much better in estimating receiver sensitivity than in establishing where to set receiver thresholds.

Low input impedance amplifiers are generally required to achieve the high bandwidths required for high bit rate systems. However, a low input impedance results in high thermal noise and poor sensitivity. High input impedance amplifiers may be used, but this narrows the bandwidth which must compensated by equalization following the first stage amplifier. Although this may result in a highly sensitive receiver, the receiver will have a poor dynamic range owing to the high gains required in the equalizer [18]. Receivers for digital systems are generally implemented with transimpedance amplifiers having a large feedback resistance. This reduces the effective input noise capacitance to below the capacitance of the photodiode, and practical receivers can be built with $C_e \approx 0.1$ pF. Using this value of capacitance and $B/R_b = 0.7$, $\eta = 0.7$ and $\kappa = 6$, Eq. (17) gives $\overline{N}_p \approx 2600$. Note that this is about 34 dB greater than that given by the quantum limit.

3.1.3 APD Receiver

In an APD receiver, there is additional shot noise owing to the excess noise factor F of the avalanche gain process. However, thermal noise is reduced because of the current multiplication gain M before thermal noise is introduced. This results in a receiver sensitivity given by

$$\overline{N}_p = \frac{B}{\eta R_b} \kappa^2 \left[F + \frac{1}{\kappa} \sqrt{\frac{8\pi kTC_e}{q^2 M^2}} \right] \qquad (19)$$

The excess noise factor is an increasing function of M which results in an optimum M to minimize $\overline{N}_p$ [19]. Good APD receivers at 1300 and 1550 nm typically have sensitivities of the order of 1000 photons per bit. Owing to the lower excess noise of silicon APDs, sensitivity of about 500 photons per bit can be achieved at 850 nm.

3.1.4 Impairments

There are several sources of impairments that may degrade the sensitivity of receivers from that given by Eqs. (17) and (19). These may be grouped into two general classes: eye degradations and signal dependent noise.

An eye diagram is the superposition of all possible received sequences. At the sampling point, there is a spread of the values of a received 1 and a received 0. The difference between the minimum value of a received 1 and the maximum value of the received 0 is known as the eye opening. This is given by $(1-\varepsilon)I_p$ where ε is the eye degradation. The two major sources of eye degradation are intersymbol interference and finite laser extinction ratio. Intersymbol interference results from dispersion, deviations from ideal shaping of the receiver filter, and from low frequency cut-off effects which result in DC off-sets.

Signal dependent noise are phenomena that give a variance of the received photocurrent that is proportional to I_p^2 and consequently lead to a maximum signal-to-noise ratio at the output of the receiver. Principal sources of signal dependent noise are laser relative intensity noise (RIN), reflection induced noise, mode partition noise, and modal noise. RIN are a consequence of inherent fluctuations in laser intensity resulting from spontaneous emission [16]. This is generally sufficiently small that it is not of concern in digital systems, but is an important limitation in analog systems requiring high signal to noise ratio (see Section 7). Reflection induced noise is the conversion of laser phase noise to intensity noise by multiple reflections from discontinuities (such as at imperfect connectors.) This may result in a substantial RIN enhancement that can seriously affect digital as well as analog systems [20]. Mode partition noise occurs when Fabry Perot lasers are used with dispersive fiber. Fiber dispersion results in changing phase relation between the various laser modes which result in intensity fluctuations. The effect of mode partition noise is more serious than that of dispersion alone [21]. Modal noise is a similar phenomenon that results in multimode fiber when relatively few modes are excited and these interfere.

We account for eye degradations by replacing Eq. (14a) by

$$\mu_1 - \mu_0 = (1-\varepsilon)I_p \tag{20a}$$

and account for signal dependent noise by replacing Eq. (14b) by

$$\sigma_1^2 = 2qI_pB + \frac{4kTB}{R_e} + \alpha^2 I_p^2 B \qquad \sigma_0^2 = \frac{4kTB}{R_e} + \alpha^2 I_p^2 B \tag{20b}$$

and α^2 is the relative spectral density of the signal dependent noise. (It is assumed that the signal dependent noise has a bandwidth large compared to the signal bandwidth B.) With these modifications, the sensitivity of an APD receiver becomes

$$\overline{N}_p = \frac{\frac{B}{\eta R_b}\left(\frac{\kappa}{1-\varepsilon}\right)^2\left[F+\left(\frac{1-\varepsilon}{\kappa}\right)\sqrt{\frac{8\pi kTC_e}{q^2M^2}}\right]}{1-\alpha^2 B\left(\frac{\kappa}{1-\varepsilon}\right)^2} \tag{21}$$

where the PIN expression is obtained by setting F=1 and M=1. It follows from Eq. (21) that there is a minimum error probability ("error floor") given by

$$P_{e,min} = Q(\kappa_{max}) \quad \text{where} \quad \kappa_{max} = \frac{1-\varepsilon}{\alpha\sqrt{B}} \tag{22}$$

The existence of eye degradations and signal dependent noise cause an increase in the receiver power (called "power penalty") required to achieve a given error probability. This is illustrated in Figure 3 for a PIN receiver where the lower curve is for no degradation. The remaining curves are for eye degradation alone, signal dependent noise alone, and both present. The parameters were chosen such that eye degradation and signal dependent noise individually give a 2 dB power penalty at $P_e = 10^{-6}$. Note that when both are present, the power penalty is 13 dB at $P_e = 10^{-6}$ and the error floor is at $P_e = 4.5(10)^{-7}$.

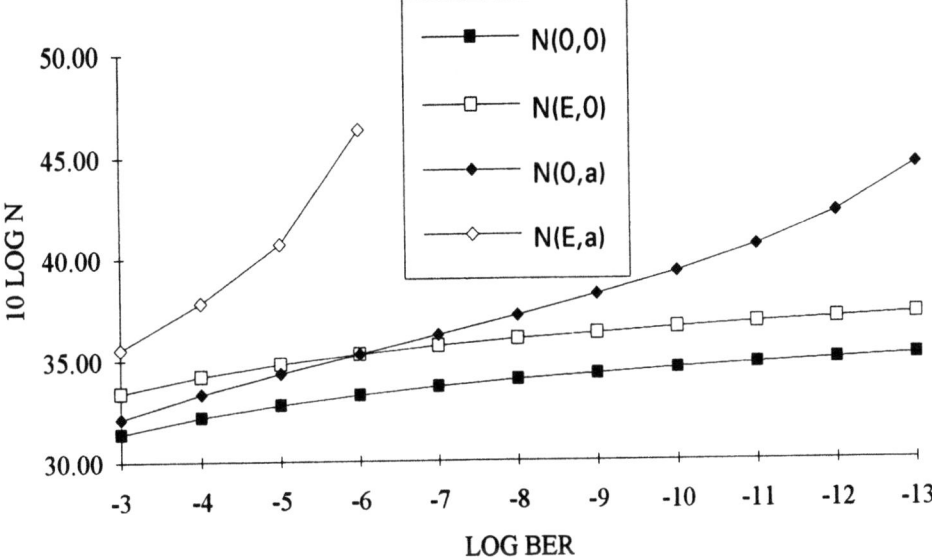

Figure 3. Sensitivity of PIN receiver in number of photons per bit (N) required for a given bit error rate (BER). N(0,0) denotes sensitivity without impairments. "E" indicates eye degradation impairment, and "a" signal-dependent noise.

4. Bit-Rate and Distance Limits

Bit-rate and distance limitations of digital links are determined by loss and dispersion limitations. The following example is used to illustrate the calculation of the maximum distance for a given bit rate. Consider a 2.5 Gbits/s system at a wavelength of 1550 nm. Assume an average transmitter power of 0 dBm coupled into the fiber. Receiver sensitivity is taken to be 3000 photons/bit, which from Eq. (12) corresponds to an average receiver power of -30.2 dBm. Allowing a total of 8 dB for margin, and for connector and cabling losses at the two ends gives a loss allowance of 22.2 dB. If the cabled fiber loss, including splices, is 0.25 dB/km, this leads to a loss-limited transmission distance of 89 km.

Assuming the fiber dispersion is D = 15 ps/km.nm and a source spectral width of 0.1 nm gives a dispersion per unit length of 1.5 ps/km. Taking the maximum allowed dispersion to be half the interpulse period, this gives a maximum dispersion of 200 ps, which then gives a maximum dispersion-limited distance of 133 km. Thus, the loss-limited distance is controlling.

Consider what happens if the bit rate is increased to 10 Gb/s. For the same number of photons per bit at the receiver, the receiver power must be 6 dB greater than that in the above example. This reduces the loss allowance by 6 dB corresponding to a reduction of 24 km in the loss-limited distance. The loss-limited distance is now 65 km (assuming all other parameters are unchanged). However, dispersion-limited distance scales inversely with bit rate, and is now 22 km. The system is now dispersion limited. Dispersion-shifted fiber would be required to be able to operate at the loss limit.

4.1 INCREASING BIT RATE

There are two general approaches for increasing the bit rate transmitted on a fiber: time division multiplexing (TDM) in which the serial transmission rate is increased, and wavelength division multiplexing (WDM) in which separate wavelengths are used to transmit independent serial bit streams in parallel. TDM has the advantage of minimizing the quantity of active devices but requires higher speed electronics as the bit rate is increased. Also, as indicated by the above example, dispersion limitations will be more severe.

WDM allows use of existing lower speed electronics, but requires multiple lasers and detectors, and optical filters for combining and separating the wavelengths, and the insertion loss of these filters needs to be included in the loss budget. Presently TDM is generally more economic, but TDM is approaching electronic speed limitations. Technology advances, including tunable lasers, transmitter and detector arrays, high resolution optical filters, and optical amplifiers (Section 5) are making WDM more attractive, particularly for networking applications (Section 6).

4.2 LONGER REPEATER SPACING

In principal, there are three approaches for achieving longer repeater spacing than that calculated above: lower fiber loss, higher transmitter powers, and improved receiver sensitivity (smaller $\overline{N}_p$). Silica based fiber is already essentially at the theoretical Rayleigh scattering loss limit. There has been research on new fiber materials that would allow operation at wavelengths longer than 1.6 µm, with consequent lower theoretical loss values [22]. There are many reasons, however, why achieving such losses will be difficult, and progress in this area has been slow.

Higher transmitter powers are possible but there are both nonlinearity and reliability issues that limit transmitter power. Since present receivers are more than 30 dB above the quantum limit, improved receiver sensitivity would appear to offer the greatest possibility. To improve the receiver sensitivity, it is necessary to increase the photocurrent at the output of the detector without introducing significant excess loss. There are two main approaches for doing so: optical amplification and optical mixing. Optical preamplifiers result in a theoretical sensitivity of 38 photons/bit [23] (6 dB above the quantum limit), and experimental systems have been constructed with sensitivities of about 100 photons/bit [24]. This will be discussed further in Section 5.2.1. Optical mixing (coherent receivers) will be discussed briefly below.

4.2.1 Coherent Systems

A photodetector provides an output current that is proportional to the magnitude square of the electric field that is incident on the detector. If a strong optical signal ("local oscillator"), coherent in phase with the incoming optical signal, were to be added prior to the photodetector, then the photocurrent will contain a component at the difference frequency between the incoming and local oscillator signals. The magnitude of this photocurrent, relative to the direct detection case, is increased by the ratio of the local oscillator to the incoming field strengths. Such a coherent receiver offers considerable improvement in receiver sensitivity. With on-off-keying, a heterodyne receiver (signal and local oscillator frequencies different) has a theoretical sensitivity of 36 photons/bit, and a homodyne receiver (signal and local oscillator frequencies the same) has a sensitivity of 18 photons/bit. Phase-shift keying (possible with coherent systems) provides a further 3 dB improvement. Coherent systems, however, require very stable signal and local oscillator sources (spectral linewidths small compared to the modulation bandwidth) and matching of the polarization of the signal and local oscillator fields [25].

An advantage of coherent systems, more so than improved receiver sensitivity, is that because the output of the photodetector is linear in the signal field, filtering for WDM demultiplexing may be done at the difference frequency (typically in the microwave range). This allows considerably greater selectivity than is obtainable with optical filtering techniques, but frequency selectivity and efficient use of the optical spectrum is not critical at optical frequencies [26]. The advent of optical amplifiers has slowed the interest in coherent systems.

5. Optical Amplifiers

There are two types of optical amplifiers: laser amplifiers based on stimulated emission, and parametric amplifiers based on nonlinear effects [27]. The former are currently of most interest for fiber optic communications. A laser without reflecting end faces is an amplifier, but it is more difficult to obtain sufficient gain for amplification than it is (with feedback) to obtain oscillation. Thus, laser oscillators were available much earlier than laser amplifiers.

Laser amplifiers are now available with gains in excess of 30 dB over a spectral range of more than 10 nm. Output saturation powers in excess of 10 dBm are achievable. The amplified spontaneous emission (ASE) noise power at the output of the amplifier, in each of two orthogonal polarizations, is given by

$$P_{ASE} = n_{sp} \frac{hc}{\lambda} B(G-1) \qquad (23)$$

where the spontaneous emission factor, n_{sp}, is equal to one for ideal amplifiers with complete population inversion.

The noise figure of an amplifier is generally defined as the ratio of the input to the output signal-to-noise ratio (SNR). This is confused for an optical amplifier because

SNR is defined electrically at the output of the photodetector. Since this is a square-law detector (output current proportional to the magnitude square of the input field), there are output noise terms involving the product of the signal and the ASE (signal-spontaneous beat noise) and the square of the ASE (spontaneous-spontaneous beat noise). Note that the latter involves both polarizations of ASE (unless a polarization filter is used) whereas the former involves only a single polarization. In the definition of noise figure, the input SNR is taken to be the SNR that would result without an amplifier in an ideal receiver in which there is only signal shot noise. If the gain of the optical amplifier is sufficiently large, this definition leads to a noise figure given by $2 n_{sp}$, but other factors may cause the electrically measured noise figure to differ from this [28,29].

5.1 COMPARISON OF SEMICONDUCTOR AND FIBER AMPLIFIERS

There are two principal types of laser amplifiers: semiconductor laser amplifiers (SLA), and doped fiber amplifiers, with the erbium doped fiber amplifier (EDFA) operating at a wavelength of 1.55 μm being of most current interest.

The advantage of the SLA, similar to laser oscillators, is that it is pumped by a DC current, it may be designed for any wavelength of interest, and that it can be integrated with electrooptic semiconductor components.

The advantages of the EDFA are that there is no coupling loss to the transmission fiber, it is polarization insensitive, it has lower noise than SLAs, it can be operated at saturation with no intermodulation owing to the long time constant of the gain dynamics, and it can be integrated with fiber devices. However, it does require optical pumping, with the principal pump wavelengths for the EDFA being either at 980 or 1480 nm.

5.2 COMMUNICATIONS APPLICATION OF OPTICAL AMPLIFIERS

There are four principal applications of optical amplifiers in communication systems [30]:
    a. Transmitter power amplifier
    b. To compensate for splitting loss in distribution networks
    c. Receiver preamplifier
    d. Linear repeater in long distance systems

The last application is of particular interest for undersea systems, where a bit rate independent linear repeater would allow subsequent upgrade of system capacity (either TDM or WDM) with changes only at the system terminals. To find the maximum number of amplifiers that may be placed in tandem, we first consider the sensitivity of a receiver with an optical preamplifier.

5.2.1 Optical Preamplifier

If an optical preamplifier with a sufficiently high gain is used, then receiver shot and thermal noise are negligible compared to the signal-spontaneous and spontaneous-spontaneous beat noise. Applying the same methodology as in Section 3.1.2, the receiver sensitivity (using the Gaussian approximation) is given by [31]

$$N_p = n_{sp}\left(\kappa^2 + \kappa\sqrt{\xi B_o / R_b}\right) \qquad (24)$$

where $\xi = 1$ or 2 dependent on whether or not a polarization filter is used, B_o is the optical bandwidth, R_b the bit rate, and $\kappa = 6$ for $P_e = 10^{-9}$. Note that for $n_{sp} = 1$, $B_o / R_b = 1$, and $\xi = 1$, which are the minimum possible values for these parameters, Eq. (24) gives $N_p = 42$ which compares quite well with the value of 38 calculated from the exact distribution of the noise [23]. The Gaussian approximation should be even better when B_o / R_b is larger, although (as noted previously), the Gaussian approximation is better for predicting N_p than it is for predicting the optimum receiver threshold.

The optical bandwidth dependence of N_p is a consequence of the spontaneous-spontaneous beat noise, within the electrical receiver bandwidth, being dependent on B_o. The signal-spontaneous beat noise is independent of the optical bandwidth, since this term is linear in the signal, and noise outside the electrical receiver bandwidth is removed by the receiver. It should also be noted that if the optical bandwidth is sufficiently large that ASE noise saturates the amplifier, then the sensitivity may be degraded beyond that given by Eq. (24).

5.2.2 Limitation on Maximum Length of Linearly Repeatered Systems

The above results may be used to calculate the maximum length of linearly repeatered systems using the following simple model. Assume M equally spaced optical amplifiers, each with gain G compensating for the loss between amplifiers. The total system gain, which is then equal to the total transmission loss that may be accommodated, is given by

$$\Gamma = G^M \qquad (25)$$

Let N_T be the average number of photons per bit at the output of each amplifier

$$N_T = \frac{P_T}{R_b} \frac{\lambda}{hc} \qquad (26)$$

Since the ASE noise at the end of the system is M times that from a single amplifier, the value of N_p required at the input to each amplifier is M times that given by Eq. (24). Consequently, it follows that

$$N_T = GMn_{sp}\left(\kappa^2 + \kappa\sqrt{\xi B_o / R_b}\right) \qquad (27)$$

Using Eq. (25) to express M in terms of Γ and G, and solving Eq. (24) for Γ gives

$$10\log\Gamma \approx \frac{10\log G}{G-1} \frac{N_T/n_{sp}}{\kappa^2 + \kappa\sqrt{\xi B_o/R_b}} \quad (28)$$

Note that the overall system gain is maximized when G→1 and M→∞.

As an example, it follows from Eqs. (26) and (28) that for $P_T = 1$ mW, $R_b = 5$ Gbits/s, $B_o/R_b = 10$, $n_{sp} = 2$, $\kappa = 6$, and $\xi = 2$, that $10 \log \Gamma = 1.38(10)^4$ dB. Thus, with low loss fibers, exceptionally long distances may be achieved in linearly repeatered systems, considering only loss and noise limitations.

However, in addition to the accumulation of ASE, there are other factors limiting the distance of linearly amplified systems, namely dispersion and the interaction of dispersion and nonlinearity [32]. There are two alternatives for achieving very long distance very high bit rate systems with linear repeaters: solitons which are pulses which maintain their shape in a dispersive medium [33], and dispersion compensation [34]. This is an active area of current research.

6. Fiber Optic Networks

A network is a communication system to *interconnect* a number of terminals within a defined geographic area, e.g. local area networks (LAN), metropolitan area networks (MAN), and wide area networks (WAN). In addition to the transmission function discussed throughout the earlier portions of this paper, networks deal also with the routing and switching aspects of communications.

Much of the current research on fiber optic networks is focused on passive star networks in which N transmitting and N receiving terminals are interconnected through an N x N star coupler. In an ideal star coupler, the signal on each input port is uniformly distributed among all output ports. If an average power P_T is transmitted at a transmitting port, the power received at a receiving port (neglecting transmission losses) is

$$P_R = \frac{P_T}{N}(1-\delta_N) \quad (29)$$

where δ_N is the excess loss of the coupler. Since an N x N star coupler, with N a power of 2, may be implemented by $\log_2 N$ stages of 2 x 2 couplers, we may conservatively assume that

$$1-\delta_N = (1-\delta_2)^{\log_2 N} = N^{\log_2(1-\delta_2)} \quad (30)$$

The maximum bit rate per user is given by the average received power divided by the product of the photon energy and the required number of photons per bit (N_p). The throughput Υ is the product of the number of users and the bit rate per user, and from Eqs. (29) and (30) is therefore given by

$$\Upsilon = \frac{P_T}{N_p} \frac{\lambda}{hc} N^{\log_2(1-\delta_2)} \tag{31}$$

Thus, the throughput (based on power considerations) is independent of N for ideal couplers ($\delta_2 = 0$), and decreases slowly with N ($\sim N^{-0.17}$ for $10\log(1-\delta_2) = 0.5\,\text{dB}$). It follows from Eq. (31) that for a power of 1 mW at $\lambda = 1.55$ μm and with $N_p = 3000$, the maximum throughput is 2.6 Tbits/s.

This may be contrasted with a tapped bus, where it may be shown that optimum tap weight to maximize throughput is given by 1/N, leading to a throughput given by [35]

$$\Upsilon = \frac{P_T}{N_p} \frac{\lambda}{hc} \frac{1}{Ne^2} \exp(-2N\delta) \tag{32}$$

Thus, even for ideal ($\delta = 0$) couplers, the throughput decreases inversely with the number of users. If there is excess coupler loss, the throughput decrease exponentially with the number of users, and is considerably less than that given by Eq. 31. Consequently, for a power limited transmission medium, the star architecture is much more suitable than a tapped bus. The same conclusion does not apply to metallic media where bandwidth rather than power limits the maximum throughput.

Although the above the indicates the large throughput that may achieved in principle with a passive star network, it doesn't indicate how this will be realized. Most interest is in WDM networks [36]. The simplest protocols are those for which fixed wavelength receivers and tunable transmitters are used. However, the technology is simpler when fixed wavelength transmitters and tunable receivers are used, since a tunable receiver may be implemented with a tunable optical filter preceding a wideband photodetector. Fixed wavelength transmitters and receivers, involving multiple passes through the network is also possible, but this requires utilization of terminals as relay points. Protocol, technology, and application considerations for "gigabit networks" (networks having access at gigabit rates and throughputs at terabit rates) is an extensive area of current research [37].

7. Analog Transmission on Fiber

Most interest in fiber optic communications is for digital transmission since fiber is generally a power rather than a bandwidth limited medium. There are applications, however, where it is desirable to transmit analog signals directly on fiber without converting them to digital. Examples are CATV distribution, and microwave links such as entrance links to antennas and interconnection of base stations in portable radio systems

7.1 CARRIER TO NOISE RATIO (CNR)

Optical intensity modulation is generally the only practical modulation technique for incoherent detection fiber optic systems. Let f(t) be the carrier signal that intensity modulates the optical source. For convenience, assume that the average value of f(t) is equal to zero, and that the magnitude of f(t) is normalized to be less than or equal to one. The received optical power may then be expressed as

$$P(t) = P_o[1 + mf(t)] \tag{33}$$

where m is the optical modulation index

$$m = \frac{P_{max} - P_{min}}{P_{max} + P_{min}} \tag{34}$$

The carrier-to-noise ratio is then given by

$$CNR = \frac{\frac{1}{2}m^2 \Re^2 P_o^2}{RIN \, \Re^2 P_o^2 B + 2q\Re P_o B + <i_{th}^2> B} \tag{35}$$

where $\Re$ is the photodetector responsivity, RIN is the relative-intensity-noise spectral density (denoted by α^2 in Section 3.1.3) and $<i_{th}^2>$ is the thermal noise spectral density (expressed as $4kT/R_e$ in Section 3.1.2). CNR is plotted in Figure 4 as a function of received optical power for a bandwidth of B = 4 MHz (single video channel), optical modulation index m = 0.05, $\Re$ = 0.8 A/W, RIN = -155 dB/Hz, and $\sqrt{<i_{th}^2>} = 7 \, pA/\sqrt{Hz}$. At low received powers (typical of digital systems) the CNR is limited by thermal noise. However, to obtain the higher CNR generally needed by analog systems, shot noise and then ultimately laser RIN become limiting.

7.2 ANALOG VIDEO TRANSMISSION ON FIBER [38]

It is helpful to distinguish between single channel and multiple channel applications. For the single channel case, the video signal may directly modulate the laser intensity ("AM" system), or the video signal may be used to frequency modulate an electrical subcarrier, and this subcarrier then intensity modulates the optical source ("FM" system). Eq. (34) then gives the CNR of the recovered subcarrier. Subsequent demodulation of the FM signal gives an additional increase in signal-to-noise ratio. In addition to this FM improvement factor, larger optical modulation indices may be used than in AM systems.
Thus FM systems allow higher signal-to-noise ratio and longer transmission spans than AM systems.

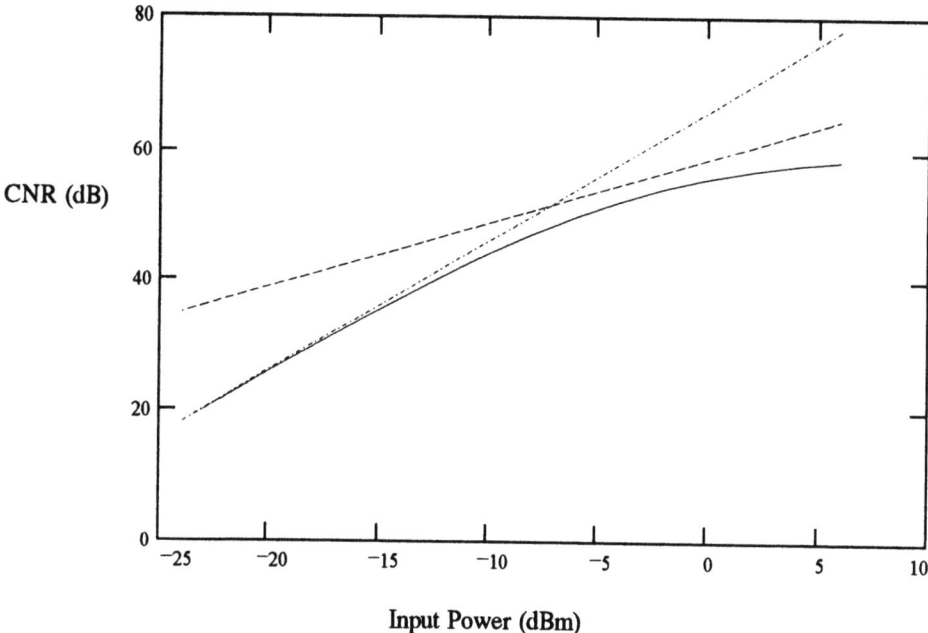

Figure 4. CNR as a function of input power. Straight lines indicate thermal noise (-.-.-.), shot noise (----) and RIN (.....) limits.

Two approaches have been used to transmit multi-channel video signals on fiber. In the first ("AM systems"), the video signals are electrically frequency division multiplexed (FDM), and this combined FDM system intensity modulates the optical source. This is conceptually the simplest system, since existing CATV multiplexing formats may be used.

In "FM systems," the individual video channels frequency modulate separate microwave carriers (as in satellite systems), these carriers are linearly combined and the combined signal intensity modulates a laser. Although FM systems are more tolerant than AM systems to intermodulation distortion and noise, the added electronics costs have made such systems less attractive than AM systems for CATV application.

Multichannel AM systems are of interest not only for CATV application but also for mobile radio applications to connect signals from a microcellular base station to a central processing station. Relative to CATV applications, the mobile radio application has the additional complication of being required to accommodate signals over a wide dynamic power range.

7.3 NONLINEAR DISTORTION

In addition to CNR requirements, multichannel analog communication systems are subject to intermodulation distortion. If the input to the system consists of a number of

frequencies given by all sums and differences of the input frequencies. Second-order intermodulation gives intermodulation products at frequencies $\omega_i \pm \omega_j$, whereas third-order intermodulation gives frequencies $\omega_i \pm \omega_j \pm \omega_k$. If the signal frequency band is such that the maximum frequency is less than twice the minimum frequency, then all second-order intermodulation products fall outside the signal band, and third-order intermodulation is the dominant nonlinearity. This condition is satisfied for the transport of microwave signals (e.g., mobile radio signals) on fiber, but is not satisfied for wideband CATV systems where there are requirements on composite-second-order (CSO) and composite-triple-beat (CTB) distortion.

The principal causes of intermodulation in multichannel fiber optic systems are laser threshold nonlinearity, [39] inherent gain nonlinearity, and the interaction of chirp and dispersion.

A multichannel FDM signal has essentially Gaussian amplitude statistics owing to the random phase of the individual channels. Consequently, if this signal is used as the input drive current to a laser transmitter, there is the possibility that the negative peaks will reduce the total current to the laser below threshold which clips these peaks and thereby causes distortion. To meet CSO and CTB requirements, the rms modulation index of the laser (μ) must be greater than 0.3 [40]. If a single constant amplitude input channel results in an optical modulation index m, and there are N such channels, then

$$\mu = \sqrt{\frac{Nm^2}{2}} \tag{36}$$

For a given CNR, Eqs. (35) and (36) may be used to calculate the maximum number of channels as a function of received optical power. Results indicate that for CNR=55 dB, about 100 channels may be obtained with a received optical power of 1 mW [40].

For "good" lasers, the effects of inherent laser nonlinearities should be less than the effects of threshold clipping at modulation indices of interest. An additional source of nonlinearity results from the interaction of chirp and fiber dispersion. Chirp imposes a nonlinear phase modulation on the signal, and the presence of dispersion converts phase to intensity modulation. This effect may be kept small by using wavelengths close to the dispersion minimum, Thus, threshold clipping is the fundamental nonlinearity limiting system performance.

8. Technology and Applications Directions

Fiber optic communications applications began with metropolitan and short distance intercity trunking . Technological advances, primarily higher-capacity transmission and longer repeater spacings extended the application to long-distance intercity transmission, both terrestrial and undersea. All of the signal processing in these systems (multiplexing, switching, performance monitoring) was done electrically, with optics serving solely to provide point-to-point links.

Current emphasis is on extending applications closer to the end user and to optical networks in which more of the functions are performed optically rather than electronically. Although there is interest in "all-optical networks," it is more likely that future networks will involve hybrid combinations of optical and electronic technologies and transmission media [41].

The huge bandwidth capability of fiber optics (measured in tens of terahertz) is not likely to be utilized by time-division-techniques alone, and WDM technology and systems will continue to receive increasing emphasis. Advances in optical filter technology, and in tunable lasers (including multiple individually addressable lasers on a single chip) will affect the architecture and the economics of such systems.

Optical amplifiers will play an important role in both long distance "transparent systems," and as a building block in optical networks. Amplifiers may be integrated with high insertion loss optical signal processing elements to allow architectures previously considered impractical.

Nonlinear phenomena, when uncontrolled, generally lead to system impairments. However, controlled nonlinearities are the basis of devices such as parametric amplifiers and switching and logic elements. Nonlinear optics will consequently continue to receive increased emphasis.

In many respects, the technological capability of fiber optic communications has expanded more rapidly than the communications applications that might utilize this capability. Consequently, continued effort needs to be directed towards applications research and to trials to explore and promote integrated wideband communications applications.

References

1. Kao, C.K. and Hockham, G.A. (1966) Dielectric-fiber surface waveguides for optical frequencies, *Proc. IEE,* **113**, 1151-1158.
2. Kapron, F.P., et.al.(1970) Radiation losses in glass optical waveguides, *Appl. Phys. Lett.,* **17**, 423.
3. Hayashi, I., Panish, M.B. and Foy, P.W. (1970) Junction lasers which operate continuously at room temperature, *Appl. Phys. Lett.,* **17**, 109.
4. Jacobs, Ira (1986) Fiber-optic transmission and system evolution, in T.C. Bartee (ed), *Digital Communications,* Howard W. Sams & Co., Indianapolis.
5. Gloge, D. (1971) Weakly guiding fibers, *Applied Optics,* **10**, 2252-2258.
6. Olshansky, R. and Keck, D. (1976) Pulse broadening in graded index fibers, *Appl. Optics,* **15**, 483-491.
7. Gloge, D, Marcatili, E.A.J., Marcuse, D. and Personick, S.D. (1979) Dispersion properties of fibers, in S.E. Miller and A.G. Chynoweth, (eds), *Optical Fiber Telecommunications,* Academic Press, San Diego.

8. Namihira, Y. and Wakabayashi, H.,(1991) Fiber length dependence of polarization mode dispersion measurements in long-length optical fibers and installed optical submarine cables, *J. Opt. Commun.*, 2,2.
9. Jones, W.B. Jr., (1988) *Introduction to Optical Fiber Communication Systems*, Holt, Rinehart and Winston, New York, 90-92.
10. Bellcore, (1986) *Digital Fiber Optic Systems Requirements and Objectives*, Technical Advisory TA-TSY-000038, Issue 3.
11. Cohen, L.G., Mammel, W.L. and Jang, S.J. (1982) Low-loss quadruple-clad single-mode lightguides with dispersion below 2 ps/km.nm over the 1.28 µm - 1.65 µm wavelength range, *Electron. Lett.*, **18**,1023-1024.
12. Keiser, G. (1991) *Optical Fiber Communications*, 2nd Edition, Ch. 5, McGraw Hill, New York.
13. Bowers, J.E. and Pollack, M.A. (1988) Semiconductor lasers for telecommunications, in S.E. Miller and I.P. Kaminow (eds), *Optical Fiber Telecommunications II*, Academic Press, San Diego.
14. Tsang, W.T. (1985) The cleaved-coupled-cavity (C^3) laser, *Semiconductors and Semimetals*, **22 part B**, 257-373.
15. Kobayashi, K. and Mito, I. (1988) Single frequency and tunable laser diodes, *J. Lightwave Technol.*, **6**, 1623-1633.
16. Mukai, T. and Yamamoto, Y. (1984) AM quantum noise in 1.3 µm InGaAsP lasers, *Electron. Lett.*, **20**, 29-30.
17. Green, P.E. Jr., (1993) *Fiber Optic Networks*, Ch. 8, Prentice Hall, Englewood Cliffs, New Jersey.
18. Personick, S.D. (1973) Receiver design for digital fiber optic communication systems I, *Bell Syst. Tech. J.*, **52**, 843-874.
19. Forrest, S.R. (1988) Optical detectors for lightwave communication, Ch. 14 in S.E. Miller and I.P. Kaminow (eds), *Optical Fiber Telecommunications II*, Academic Press, San Diego.
20. Gimlett, J.L. and Cheung, N.K. (1989) Effects of phase-to-intensity noise conversion by multiple reflections on gigabit-per-second DFB laser transmisssion systems, *J. Lightwave Technol.*, **LT-7**, 888-895.
21. Ogawa, K. (1982) Analysis of mode partition noise in laser transmission systems," *IEEE J. Quantum Electron.*, **QE-18**, 849-855.
22. Tran, D.C., Sigel, G.H. and Bendow, B. (1984) Heavy metal fluoride fibers: A review, *J. Lightwave Technol.*,**LT-2**, 566-586.
23. Henry, P.S. (1989) Error-rate performance of optical amplifiers, *Optical Fiber Communications Conference (OFC'89) Tech. Dig.*,THK3.
24. Gautheron, O., Grandpierre, G., Pierre, L., Thiery, J.-P and Kretzmeyer, P. (1993) 252 km repeaterless 10 Gbits/s transmission demonstration," *Optical Fiber Communications Conference (OFC'93) Postdeadline Papers.*, PD11.
25. Stanley, I.W. (1985) A tutorial review of techniques for coherent optical fiber transmission systems, *IEEE Commun. Mag.*, **23**, 37-53.
26. Green, P.E. and Ramaswami, R. (1990) Direct detection lightwave systems: why pay more? *IEEE LCS*, **1**, 36-49.
27. Agrawal, G.P. (1992) *Fiber Optic Communication Systems*, Ch. 8, John Wiley & Sons, New York.

28. Bellcore, (1992) *Generic Requirements for Optical Fiber Amplifier Performance*, Technical Advisory TA-NWT-001312, Issue 1.
29. Jacobs, Ira (1994) Dependence of optical amplifier noise figure on relative-intensity-noise," submitted to *J. Lightwave Technol.*
30. Li, T. (1993) The impact of optical amplifiers on long-distance lightwave telecommunications, *Proc. IEEE,* **81**, 1568-1579.
31. Jacobs, Ira (1990) Effect of optical amplifier bandwidth on receiver sensitivity, *IEEE Trans. Commun.,* **38**, 1863-1864.
32. Naka, A. and Saito, S. (1994) In-line amplifier transmission distance determined by self-phase modulation and group-velocity dispersion," *J. Lightwave Technol.* **12**, 280-287.
33. Agrawal, G.P. (1989) *Nonlinear Fiber Optics,* Ch. 5, Academic Press, Boston.
34. Kikuchi, K.and Lorattanasane, C. (1994) Compensation for pulse waveform distortion in ultra-long distance optical communication systems by using midway optical phase conjugator, *IEEE Photon. Technol. Lett.,* **6**, 104-105.
35. Green, P.E. Jr. (1990) *Fiber Optic Networks,* Ch. 11, Prentice Hall.
36. Goodman, M. (1989) Multiwavelength networks and new approaches to packet switching, *IEEE Commun. Mag.,* **27**, 27-35.
37. Smith, P.J., Faulkner, D.W. and Hill, G.R. (1993) Evolution scenarios for optical telecommunication networks using multiwavelength transmission, *Proc. IEEE,* **81**, 1580-1587.
38. Darcie, T.E., Nawata, K. and Glabb, J.B. (1993) Special issue on broad-band lightwave video transmission, *J. Lightwave Technol.,* **11**, no.1.
39. Saleh, A.A.M. (1989) Fundamental limit on number of channels in SCM lightwave CATV system, *Electron. Lett.* **25**, 776-777.
40. Chung, C.J., and Jacobs, Ira (1992) Practical TV channel capacity of lightwave multichannel AM SCM systems limited by the threshold nonlinearity of laser diodes, *IEEE Photon. Technol. Lett.,* **4**, 289-292.
41. Baack, C. and Walf, G. (1993) Photonics in future telecommunications, *Proc. IEEE,* **81**, 1624-1632.

MODERN OPTICAL COMMUNICATION SYSTEMS

Govind P. Agrawal

The Institute of Optics, University of Rochester
Rochester, New York 14627, USA

During the last few years the design and the performance of fiber-optic communication systems have been revolutionized by several major developments. Foremost among them are (i) the advent of the fiber amplifier, (ii) the use of optical solitons, and (iii) advances in multiplexing techniques so that multiple channels can be transmitted over the same fiber. The performance of an optical communication system operating at the bit rate B and transmitting information over a distance L is often quantified by the bit rate-distance product BL. In practice, optical loss and chromatic dispersion in optical fibers set the ultimate limit on the BL values. Before 1990 the BL product of most optical communication systems was limited to about 0.2 (Tb/s)-km if we use B = 2.5 Gb/s and L = 80 km as typical values. By using optical solitons together with fiber amplifiers this value for modern optical communication systems has increased by more than 1000-fold. This paper provides an overview of the design, the techniques, and the components used to realize such extraordinary advances over a relatively short duration of less than five years. It begins with an historical perspective as an introduction to the field of fiber-optic communication systems and then discusses in separate sections how fiber loss and dispersion are managed in modern lightwave systems. The last section is devoted to soliton communication systems.

1. Historical Perspective

The advent of telegraphy in the 1830s began the era of electrical communications. The BL product of early telegraphic systems was ~10 (b/s)-km. Figure 1 shows how the BL product has increased super-exponentially since 1850 through various technological advances. By 1970 the use of coaxial cables and microwaves resulted in electrical communication systems operating with BL values ~100 (Mb/s)-km. It was clear at that time that the existing technology was unable to boost the performance significantly without a fundamental change in the system design.

It was realized during the 1950s that an increase of several orders of magnitude in the BL product would be possible if optical waves were used as the carrier of information. However, neither a coherent optical source nor a suitable transmission medium was available during the 1950s. The invention of the laser and its demonstration in 1960 solved the first problem. Attention then focused on finding ways for using the laser light for optical communications. Many ideas were advanced during the 1960s, the most noteworthy being the idea of light confinement by using a sequence of gas lenses. It was suggested in 1966 that optical fibers might be the best choice, as they are capable of guiding the light in a way similar to the guiding of electrons in copper wires. The main problem was the high loss of optical fibers; fibers available during the 1960s had losses in excess of 1000 dB/km. A breakthrough occurred in 1970 when the fiber loss could be reduced to about 20 dB/km in the wavelength region near 1 μm [1]. At about the same time, GaAs semiconductor lasers capable of operating continuously at room temperature were demonstrated [2]. The simultaneous availability of a compact optical source and a low-loss optical fiber led to a worldwide effort for developing fiber-optic communication systems. Figure 2 shows the progress in the performance of lightwave systems realized after 1974 through five generations of development stages. The progress has indeed been rapid as evidenced by the million-fold increase in the BL product over a period of less than 20 years.

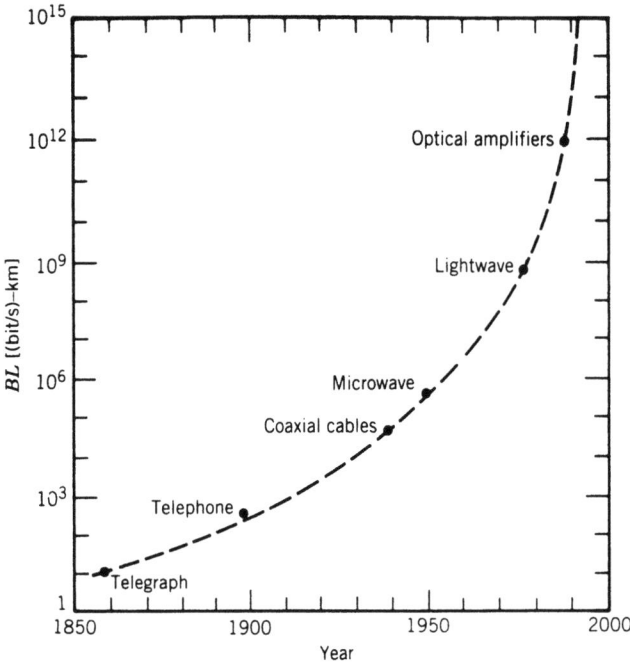

Figure 1. Increase in the Bit rate-distance product during 1850-2000 because of technological evolution.

The commercial deployment of lightwave systems followed the research and development closely. After many field trials, the first-generation lightwave systems operating near 0.8 μm began to be deployed in 1978. They operated at bit rates in the 50-100 Mb/s range and allowed a repeater spacing of about 10 km [BL < 1 (Gb/s)-km]. It was clear during the 1970s that the repeater spacing could be increased considerably by operating the lightwave system in the wavelength region near 1.3 μm where fiber loss is generally below 1 dB/km. Furthermore, optical fibers exhibit minimum dispersion in this wavelength region. This realization led to a worldwide effort for the development of InGaAsP semiconductor lasers and detectors operating near 1.3 μm. Such a laser was demonstrated [3] in 1977. The second generation of fiber-optic communication systems became available in the early 1980s and allowed a repeater spacing in excess of 20 km.

However, the bit rate of early systems was limited to below 100 Mb/s because of modal dispersion in multimode fibers [4]. This limitation was overcome by the use of single-mode fibers. A laboratory experiment in 1981 demonstrated 2-Gb/s transmission over 44 km of single-mode fiber [5]. The introduction of commercial systems soon followed. By 1987, second-generation 1.3-μm lightwave systems operating at bit rates up to 1.7 Gb/s with a repeater spacing of about 50 km were commercially available [BL ~80 (Gb/s)-km].

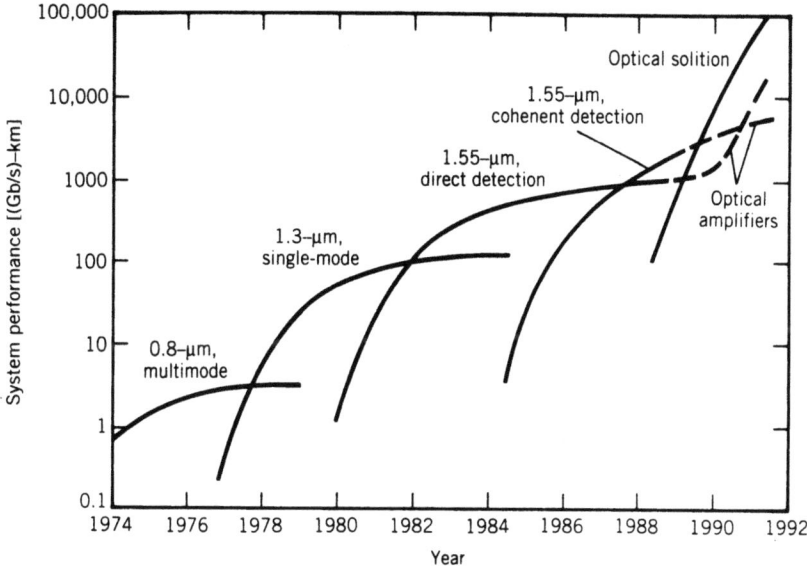

Figure 2. Increase in the Bit rate-distance product through five generations of fiber-optic communication systems.

The repeater spacing of the second-generation lightwave systems was limited by the fiber loss at the operating wavelength near 1.3 μm (typically 0.5 dB/km). The loss of silica fibers is minimum near 1.55 μm. Indeed, a loss of 0.2 dB/km in this spectral region was realized [6] in 1979. However, the introduction of third-generation lightwave systems operating at 1.55 μm was considerably delayed by large fiber dispersion near 1.55 μm. Conventional InGaAsP semiconductor lasers could not be used because of pulse spreading occurring as a result of simultaneous oscillation of several longitudinal modes. The dispersion problem can be overcome either by using dispersion-shifted fibers designed to have minimum dispersion near 1.55 μm or by limiting the laser spectrum to a single longitudinal mode. Both approaches were followed during the 1980s. By 1985 the laboratory transmission experiments showed the possibility of communicating information at bit rates up to 4 Gb/s over distances of 100 km. The third-generation 1.55-μm systems operating at 2.5 Gb/s with BL ~ 200 (Gb/s)-km became available commercially in 1989.

The fourth generation of lightwave systems is concerned with an increase in the bit rate through frequency-division multiplexing and an increase in the repeater spacing through optical amplification. Most of the attention initially focused on the compensation of fiber loss through optical amplification By 1991 erbium-doped fiber amplifiers (EDFAs) were developed to the extent that data transmission over 4500 km at 2.5 Gb/s and over 1500 km at 10 Gb/s [BL ~ 15 (Tb/s)-km] could be demonstrated in the laboratory experiments by using multiple EDFAs [7,8]. Such systems are limited by fiber dispersion rather than by fiber loss. Several approaches have been adopted for reducing the dispersive effects. The simplest among them consists of choosing the operating wavelength close to the so-called zero-dispersion wavelength at which the second-order dispersive effects vanish. Indeed, this approach has shown the possibility of data transmission over 21,000 km at 2.5 Gb/s and over 14,300 km at 5 Gb/s [BL ~ 70 (Tb/s)-km] by using a recirculating-loop configuration [9].

The fifth generation of fiber-optic communication systems is already at the stage of research and development. It is based on the novel concept of fiber solitons, optical pulses that preserve their shape during propagation in a fiber by counteracting the effect of dispersion through the fiber nonlinearity. Although the basic idea was proposed [10] as early as 1973, it was only after the advent of the EDFA that this approach became practical. Since 1990, several system experiments have demonstrated the eventual potential of soliton communication systems. Such systems can transmit signal over 24,000 km at 15 Gb/s resulting in a BL product ~360 (Tb/s)-km. Even larger values can be realized by employing wavelength-division multiplexing such that several soliton channels are transmitted simultaneously over the same fiber. This paper describes how such performance improvements are realized in modern communication systems. The reader is referred to recent texts for further details [11,12].

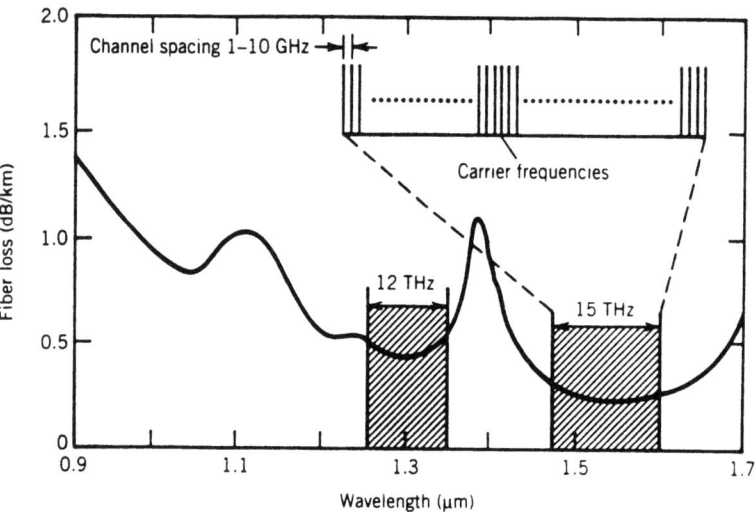

Figure 3. Low-loss transmission windows of silica fibers. The inset shows the possibility of multichannel operation.

2. Fiber-Loss Management

When an optical signal travels through an optical fiber, its power is attenuated because of several unavoidable effects such as material absorption, waveguide imperfections, and Rayleigh scattering. In the case of silica fibers, material absorption is negligible in the wavelength range 1.0-1.6 μm, and the loss is mainly due to Rayleigh scattering and residual water vapors. Figure 3 shows the loss spectrum of a state-of-the-art optical fiber.

Modern communication systems are designed to operate in the wavelength regions near 1.3 and 1.55 μm since the loss is minimum there. The low-loss "transmission windows" (hatched regions) are quite wide (~12-15 THz) and can support hundreds of channels with a suitable choice of carrier frequencies. In the transmission window located near 1.55 μm, the fiber loss is only 0.2 dB/km.

2.1 LOSS-LIMITED TRANSMISSION DISTANCE

In spite of a relatively low fiber loss, the transmission distance L is still limited by it. This can be understood by noting that optical receivers need a minimum amount of power P_{rec} in order to recover the transmitted information with high accuracy. Typically P_{rec} ~0.1 μW for bit rates ~1-10 GHz. If we use a typical value of 1 mW for the transmitter power P_{tr} together with the 0.2 dB/km fiber loss, the transmission distance L is limited to less than 200 km even under the best operating conditions. In practice, L is <150 km because of other sources of power loss such as fusion splices and fiber connectors. It can be increased by increasing the transmitter power P_{tr}. However, P_{tr} is limited to values ~10 mW because of the nonlinear effects such as stimulated Brillouin scattering (SBS), making it difficult to make L significantly larger than 150 km.

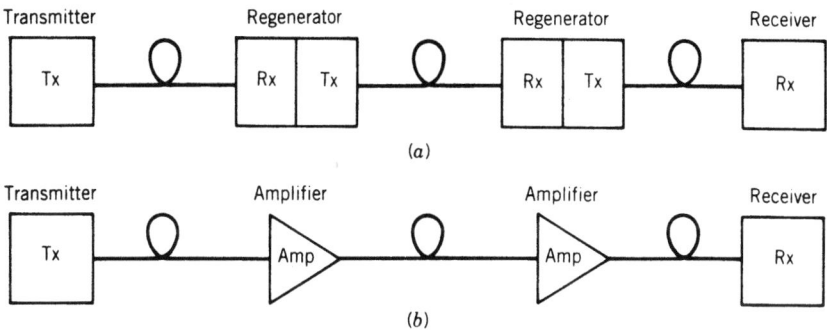

Figure 4. Compensation of fiber loss through (a) regenerators and (b) optical amplifiers.

For long-haul and undersea lightwave systems, it becomes necessary to compensate for fiber loss, as the signal would otherwise become too weak to be detected reliably. Such a compensation is traditionally carried out by using repeaters, often called regenerators because they regenerate the optical signal. Figure. 4(a) shows the schematic of such a system. A regenerator is nothing but a receiver-transmitter pair that detects the incoming optical signal, recovers the electrical bit stream, and then converts it back into an optical bit stream by modulating the transmitter. As shown in Fig. 4(b), loss compensation can also be carried out by using optical amplifiers which amplify the optical bit stream directly without having to convert it to the electrical domain. Although several type of optical amplifiers, such as semiconductor laser amplifiers and fiber Raman amplifiers, were considered during the 1980s, their use did not become practical until the advent of the EDFA in the late 1980s. Among the advantages offered by EDFAs are diode-laser pumping, high gain (30-40 dB) at moderate pump powers, relatively low noise, large bandwidth, and immunity to interchannel crosstalk when several channels are amplified simultaneously.

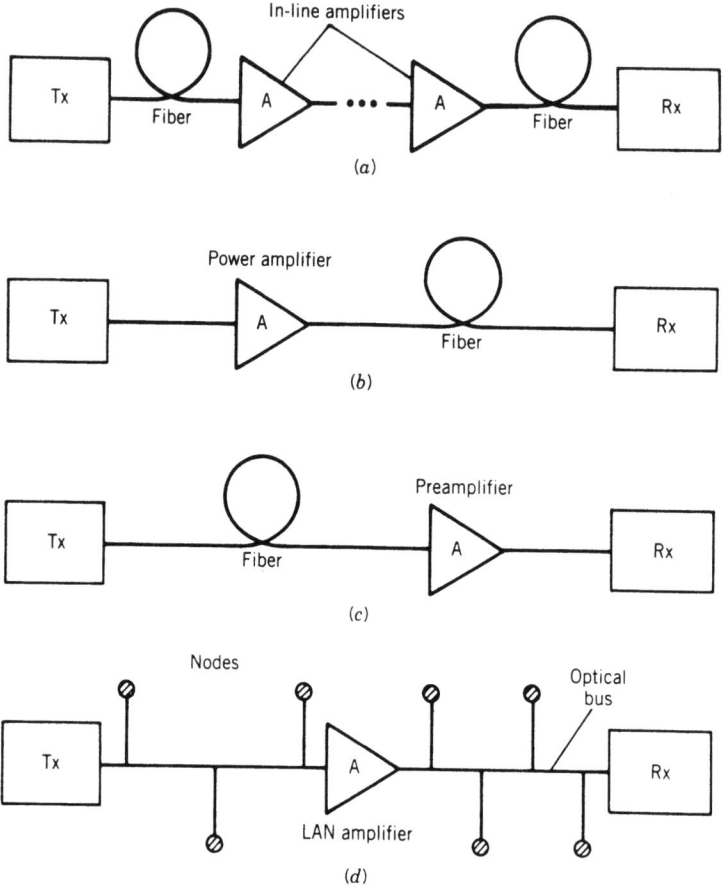

Figure 5. Four types of applications of optical amplifiers in lightwave systems.

2.2 ERBIUM-DOPED FIBER AMPLIFIERS

Fiber amplifiers makes use of rare-earth ions as a gain medium. These ions are doped inside the fiber core during the manufacturing process and pumped optically to provide the gain. Although doped-fiber amplifiers were studied [13] as early as 1964, their use became practical only in 1988, after the techniques for fabrication and characterization of low-loss doped optical fibers were perfected [14]. The amplifier characteristics such as the operating wavelength are determined by the dopants rather than by the silica fiber, which plays the role of a host medium. Many different rare-earth ions, such as erbium, holmium, neodymium, samarium, thulium, and ytterbium, can be used to realize fiber amplifiers operating at different wavelengths, covering visible to infrared region (up to 2.8 μm). EDFAs have attracted the most attention among them simply because they operate near 1.55 μm, the wavelength region in which the fiber loss is minimum. Their development has revolutionized the design of modern fiber-optic communication systems. EDFAs became available commercially in 1990. An undersea transpacific fiber-cable system is planned to make use of EDFAs by 1995. EDFAs are expected to play an important role in the design of modern fiber-optic communication systems. Figure 5 shows the four possible applications of optical amplifiers in lightwave systems. They can be used (a) as in-line amplifiers in place of regenerators, (b) as booster of transmitter power, (c) as a preamplifier at the receiver, and (d) for compensation of distribution losses in local-area networks.

2.2.1 Optical Pumping

All amplifiers need to be pumped so that the energy needed to achieve population inversion can be supplied. EDFAs are pumped optically. The amorphous nature of silica broadens the energy levels of erbium ions (Er^{3+}) into bands. Figure 6(a) shows the relevant energy levels of Er^{3+} in silica glasses. Many different transitions can be used to pump the EDFA. Initial experiments used the visible wavelengths emitted by high-power lasers, such as argon-ion, Nd:YAG, and dye lasers, for pumping even though such pumping schemes are relatively inefficient. Efficient pumping is possible by using semiconductor lasers operating near 0.98-μm and 1.48-μm wavelengths. Indeed, the development of such lasers was fueled by the need for suitable pump lasers for EDFAs. It is possible to obtain high amplifier gains in the range 30-40 dB with only a few milliwatts of pump power when EDFAs are pumped by using 0.98-μm or 1.48-μm semiconductor lasers. Efficiencies as high as 11 dB/mW have been achieved with 0.98-μm pumping [15].

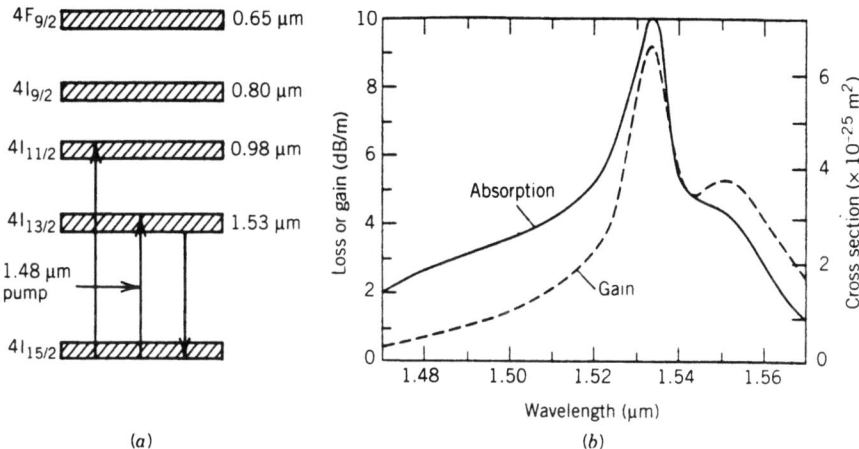

Figure 6. (a) Energy-level diagram and (b) absorption and gain spectra of erbium ions in silica fibers [20].

2.2.2 Gain Spectrum

The gain spectrum of EDFAs is affected considerably by the amorphous nature of silica and by the presence of other co-dopants such as germania and alumina within the fiber core [16-20]. The gain profile of isolated erbium ions is homogeneously broadened and its bandwidth is determined by the dipole relaxation time T_2. However, it is considerably broadened by the presence of silica glass. Structural disorders lead to inhomogeneous broadening of the gain profile, whereas Stark splitting of various energy levels is responsible for homogeneous broadening. Mathematically, the gain $g(\omega, \omega_0)$ should be averaged over the distribution of atomic transition frequencies ω_0 so that

$$g_{eff}(\omega) = \int_{-\infty}^{\infty} g(\omega, \omega_0) f(\omega_0) d\omega_0 , \qquad (1)$$

where $f(\omega_0)$ is the distribution function whose form depends on the fiber-core composition. Figure 6(b) shows the gain and absorption spectra of an EDFA whose core was doped with germania [20]. The gain spectrum is quite broad (FWHM > 10 nm) with a double-peak structure. The addition of alumina to the core broadens the gain spectrum even more. Attempts have been made to isolate the contributions of homogeneous and inhomogeneous broadening through measurements of spectral hole-burning. For germania-doped EDFAs the relative contributions of homogeneous and inhomogeneous broadening are found to be 4 and 8 nm respectively [17]. By contrast, the gain profile of aluminosilicate glasses is dominated by homogeneous broadening.

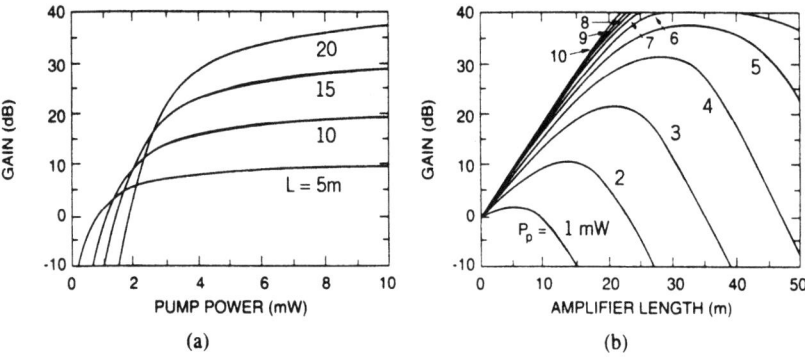

Figure 7. Calculated small-signal gain of an EDFA as a function of (a) pump power and (b) amplifier length [20].

2.2.3 Small-Signal Gain
The gain of EDFAs depends on a large number of device parameters such as erbium-ion concentration, amplifier length, core radius, and pump power. Considerable efforts have been made to develop an understanding of the gain characteristics through theoretical modeling. A three-level rate-equation model commonly used for some lasers can be adapted for EDFAs for all pumping wavelengths. It is sometimes necessary to add a fourth level to include the excited-state absorption. Figure 7 shows the calculated small-signal gain at 1.55 µm as a function of the pump power and the amplifier length by using typical parameter values [20]. For a given amplifier length L, the amplifier gain initially increases exponentially with the pump power, but the increase becomes much smaller when the pump power exceeds a certain value [corresponding to the "knee" in Fig. 7(a)]. For a given pump power, the amplifier gain becomes maximum at an optimum value of L and drops sharply when L exceeds this optimum value. The reason is that the latter portion of the amplifier remains unpumped and absorbs the amplified signal. Since the optimum value of L depends on the pump power P_p, it is necessary to choose both L and P_p appropriately. Figure 7(b) shows that a 35-dB gain can be realized at a pump power of 5 mW for L = 30 m. It is possible to design amplifiers such that high gain is obtained for amplifier lengths as short as a few meters. The qualitative features shown in Fig. 7 are observed in all EDFAs. The output saturation power can vary over a wide

range depending on the EDFA design, with typical values ~10 mW. EDFAs can therefore be used to boost the power of an optical transmitter.

2.2.4 Amplifier Noise

Since amplifier noise is the ultimate limiting factor for system applications, it has been studied extensively. Amplifier noise is generally quantified through the noise figure F_n that is related to the spontaneous emission factor by $F_n = 2n_{sp}$. Since EDFAs operate on the basis of a three-level pumping scheme, $n_{sp} > 1$. Thus, the noise figure of EDFAs is expected to be larger than the ideal value of 3 dB. Moreover, it is expected to depend both on the amplifier length L and the pump power P_p, just as the amplifier gain does. Numerical results show that the noise figure close to 3 dB can be obtained for a high-gain amplifier pumped strongly. The experimental results confirm that F_n close to 3 dB can be realized in EDFAs. A noise figure of 3.2 dB was measured [21] in a 30-m long EDFA pumped at 0.98 µm with 11 mW of power. A similar value was measured in another experiment [22] with only 5.8 mW of pump power at 0.98 µm. The measured values of F_n are generally larger for EDFAs pumped at 1.48 µm.

2.2.5 Multichannel Amplification

An advantage of optical amplifiers is that they can be used to amplify several communication channels simultaneously as long as the bandwidth of the multichannel signal is smaller than the amplifier bandwidth. The bandwidth of EDFAs is quite large (~2-4 THz or 15-30 nm), making them suitable for multichannel amplification. The relatively long (~10 ms) lifetime of the excited state offers an additional benefit. Many amplifiers, especially semiconductor laser amplifiers, suffer from interchannel crosstalk induced by gain modulation occurring at the frequency associated with the channel spacing. Such crosstalk does not occur only when the gain medium is unable to respond to such fast variations. Because of the long lifetime of erbium ions, gain-modulation effects are negligible when EDFAs are used for amplification of a multichannel signal. This property makes them quite useful for multichannel communication systems.

A second source of interchannel crosstalk is cross-saturation occurring because the gain of a specific channel is saturated not only by its own power (self-saturation) but also by the power of neighboring channels. This mechanism of crosstalk is common to all optical amplifiers, and EDFAs are no exception. It can be avoided by operating the amplifier in the unsaturated region. Experimental results support this conclusion. In one experiment [23] negligible power penalty was observed when an EDFA was used to amplify two channels operating at 2 Gb/s and separated by 2 nm as long as the channel powers were low enough to avoid gain saturation.

2.2.6 Ultrashort Pulse Amplification

The relatively large bandwidth (15-30 nm) of EDFAs suggests that they can be used to amplify optical pulses as short as a picosecond without distortion. In one experiment [24] 9-ps optical pulses were amplified by about 30 dB without significant changes in the pulse shape or width. Gain saturation was negligible at low repetition rates (<100 MHz) but became important when the repetition rate exceeded 100 MHz. In several other experiments EDFAs were used to amplify femtosecond optical pulses. The spectral bandwidth of such ultrashort pulses is generally comparable to the amplifier bandwidth such that the gain is reduced for spectral wings. This effect is known as gain dispersion and should be included for an understanding of the amplification process [25].

Furthermore, it is necessary to include intrapulse stimulated Raman scattering [26] a phenomenon through which high-frequency components of an ultrashort pulse can pump the low-frequency components of the same pulse. The amplification of ultrashort optical pulses is also influenced by the group-velocity dispersion (GVD) and the intensity dependence of the refractive index of the silica fiber acting as a host to the erbium ions which provide gain. Numerical simulations show that an input pulse can be amplified and simultaneously compressed by an EDFA when its peak power exceeds a certain value [25,26]. It can also split into several subpulses. From the standpoint of lightwave system applications, the potential of EDFAs lies in their ability to amplify picosecond pulses without distortion.

2.3 SYSTEM PERFORMANCE

EDFAs have found many applications in modern communication systems because of their excellent amplification characteristics, such as low insertion loss, high gain, large bandwidth, low noise, and low crosstalk. Indeed, they have been used for all four purposes shown in Fig. 5. This section discusses the applications of EDFAs as optical preamplifiers, as in-line amplifiers, and in local area networks.

2.3.1 Optical Preamplifiers

The low-noise property (the noise figure close to the ideal 3-dB value) makes EDFAs quite suitable as an optical preamplifier at the receiver. The receiver sensitivity can be improved 10-20 dB by using the EDFA as a preamplifier. In one experiment [27] only 152 photons/bit were needed for a lightwave system operating at bit rates in the range 0.6-2.5 Gb/s. In another experiment [28], a receiver sensitivity of -37.2 dBm (147 photons/bit) was achieved at the high bit rate of 10 Gb/s. It is even possible to use two preamplifiers in series; the receiver sensitivity was improved 18.8 dB by using this technique [29]. The basic idea is similar to coherent detection: Preamplification of the optical signal makes it strong enough that thermal noise becomes negligible compared with shot noise. Indeed, an experiment in 1992 demonstrated a record sensitivity of -38.8 dBm (102 photons/bit) at 10 Gb/s by using two-stage amplification in two EDFAs [30]. The sensitivity degradation was limited to less than 1.2 dB when the signal was transmitted over 45 km of dispersion-shifted fiber.

The most important performance issue in the design of lightwave systems making use of optical amplifiers is the contamination of the amplified signal by spontaneous emission. As a result of the incoherent nature of spontaneous emission, the amplified signal is noisier than the input signal. The signal-to-noise ratio (SNR) of the received signal at the decision circuit is degraded because of spontaneous-emission noise added by the optical amplifier. A lower SNR results in a higher bit-error rate (BER). The BER can be improved by increasing the received power; the relative increase required to offset the effect of spontaneous emission is referred to as the amplifier-induced power penalty.

The calculation of receiver sensitivity requires consideration of various noise sources such as shot noise, thermal noise, and spontaneous-emission noise. In general, both shot and thermal noise can be neglected since the receiver performance is limited by the spontaneous-emission noise. A simple model yields the following expression [11] for the receiver sensitivity expressed in terms of he average number of photons/bit:

$$N_p = \tfrac{1}{2} F_n [Q^2 + Q(2\Delta v_{opt}/B)^{1/2}], \tag{2}$$

where F_n is the amplifier noise figure, B is the bit rate, and Δv_{opt} is the optical bandwidth of the spontaneous-emission spectrum or of the optical filter sometimes used to improve the sensitivity. The parameter Q represents the SNR and is related the BER as [11]

$$BER = \frac{1}{2}\text{erfc}(Q/\sqrt{2}). \qquad (3)$$

Equation (2) is a remarkably simple expression for the receiver sensitivity. It clearly shows why amplifiers with a small noise figure must be used; the receiver sensitivity degrades as F_n increases. It also shows how optical filters can improve the receiver sensitivity by decreasing Δv_{opt}. The minimum optical bandwidth is equal to the bit rate ($\Delta v_{opt} = B$). Figure 8 shows N_p as a function of $\Delta v_{opt}/B$ for several values of the noise figure F_n by using Q = 6, a value required to achieve a BER of 10^{-9}. The minimum value of F_n is 2 for an ideal amplifier. Thus, by using Q = 6, the absolute minimum value from Eq. (2) is N_p = 44.5 photons/bit. This should be compared with N_p = 10 for an ideal receiver operating in the quantum-noise limit. Of course, N_p = 10 is never realized in practice because of thermal noise; typically, N_p exceeds 1000 for p-i-n receivers without optical amplifiers. Equation (2) shows that $N_p < 100$ can be realized when optical amplifiers are used to preamplify the received signal in spite of the degradation caused by spontaneous-emission noise.

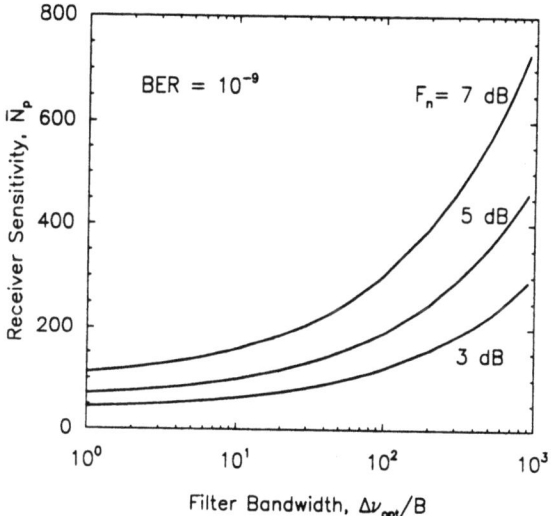

Figure 8. Receiver sensitivity as a function of filter bandwidth for several values of noise figures.

The improvement in receiver sensitivity can be used to increase the transmission distance of point-to-point fiber links used for intracity and island-to-island communication since intermediate repeaters can be eliminated. In an early experiment [31] the 1.8-Gb/s signal was transmitted over 250 km (without regeneration or in-line amplification) by using a combination of two EDFAs acting as a power amplifier at the transmitter and as a preamplifier at the receiver. The performance was improved in a 1992 transmission experiment in which a 2.5-Gb/s signal was transmitted over 318 km [32]. The bit rate was further increased to 5 Gb/s in another experiment that transmitted signal over 226 km of conventional fibers with a total dispersion of 4100 ps/nm [33]. In this experiment two EDFAs were used to boost the signal power from -8 to 15.5 dBm (about 35 mW). This power level is large enough that the nonlinear phenomenon of SBS becomes a problem. The experiment suppressed SBS by increasing its threshold through broadening of the laser linewidth from 35 to 160 MHz. Most experiments have used unidirectional transmission. In a novel scheme unrepeatered bidirectional transmission was realized by using two EDFAs on each side serving in a dual role [34]. Each EDFA acted as a power booster in the forward direction and as an optical preamplifier in the reverse direction. Even though the wavelength of two transmitters differed by 11 nm, the same EDFA can act as a power booster and preamplifier because of its wide gain bandwidth.

2.3.2 In-Line Fiber Amplifiers

The transmission distance of fiber-optic communication systems can be increased to thousands of kilometers if EDFAs are used as in-line amplifiers along the fiber link. An early experiment demonstrated transmission over 904 km at 1.2 Gb/s by using 12 EDFAs [35]. In a record-breaking coherent transmission experiment [36] the 2.5-Gb/s signal was transmitted over 2223 km by using 25 EDFAs placed at approximately 80-km intervals, offering a total gain of more than 440 dB. The effective BL product for this experiment exceeds 5.5 (Tb/s)-km. As expected, fiber dispersion is the limiting factor for lightwave systems with in-line amplifiers. Indeed, the experiment was possible only because dispersion-shifted fibers were used throughout the link. Even then, a 4.2-dB power penalty was observed in the experiment. In several transmission experiments performed in 1992 the total transmission distance could be increased beyond 10,000 km by using hundreds of EDFAs as in-line amplifiers. In one experiment [37] the 2.5-Gb/s signal was transmitted over 10,073 km by using 199 EDFAs. The bit rate was extended to 10 Gb/s in a similar experiment but the total transmission distance was limited to 6,000 km. The BL product of 60 (Tb/s)-km is nonetheless a record achievement made possible by the advent of the EDFA.

The performance level can be further improved if care is taken to decrease the effect of fiber dispersion by operating very close to the zero-dispersion length. Indeed, an effective transmission distance of 21,000 km at the bit rate of 2.4 Gb/s and of 14,300 km at 5 Gb/s was demonstrated in a laboratory experiment by using the recirculating fiber-loop configuration [9]. Such performance may be difficult to realize under field conditions, but the experiment clearly demonstrates the potential capacity of lightwave systems with cascaded optical amplifiers. Such systems are ideal for undersea transmission where the total distances are ~5-10,000 km. Indeed, undersea lightwave systems employing EDFAs are under development. One such transpacific system is scheduled to operate in 1995.

For undersea lightwave systems optical pulses propagate over thousands of kilometers. Even though power levels are relatively modest, the nonlinear effects can become important because of their accumulation over long distances. Numerical simulations show that fiber nonlinearities (specifically self-phase modulation) can limit the system performance unless the launched power is suitably optimized. One way to avoid the detrimental effects of fiber dispersion and nonlinearity is to make use of solitons for optical communications. Solitons balance the two potentially damaging effects in such a way that the pulse shape is maintained as the optical signal propagates through the fiber link. The use of EDFAs in soliton communication systems is discussed in Sec. 4.

2.3.3 Local Area Networks

EDFAs have also been used to overcome distribution and other kinds of losses in LANs and broadcast networks. An important application is for analog subcarrier-multiplexed multichannel video distribution networks whose power budget is inherently limited. In one experiment [38] an EDFA amplified 100 FM (frequency-modulated) video channels and six 622-Mb/s digital baseband channels simultaneously, covering a 34-nm spectral range. The amplified multichannel signal could be distributed to 4096 subscribers. The concept can be generalized to millions of subscribers by using multiple amplifiers. A broadcast network capable of serving 39.5 million subscribers was demonstrated [39] in 1990 by using two stages of fused-fiber couplers. An EDFA acting as a power amplifier boosted the multichannel signal during each stage. The distribution network was capable of broadcasting 384 digital video channels over 27.7 km. This experiment demonstrates the usefulness of EDFAs for distribution and broadcast networks.

The applications of EDFAs for analog video distribution has attracted most attention because of its practical use by the telephone and cable- television industry. In one 30-channel experiment [40] the power budget was extended to 45 dB by using three EDFAs. In a later experiment [41] the power budget was increased to 73 dB by using just three EDFAs. The system was capable of transmitting the composite video signal to 0.52 million subscribers and is suitable for local area networks such as fiber-to-the-home systems. In another experiment the broadcast network was capable of transmitting 35 video channels to 4.2 million subscribers [42]. The number of channels can be increased to several hundreds by using the technique of subcarrier multiplexing. Indeed, fiber amplifiers are being used by the cable-television industry since 1992.

Experiments have also been performed to show that EDFAs can enhance the performance of local-area networks where star couplers are used for establishing two-way communication between users. In one experiment [43] an EDFA was used as a preamplifier between the star coupler and the receiver. The 14-dB improvement in receiver sensitivity increases the number of users by about a factor of 20. Another scheme designs star couplers with gain [44]. Several star couplers and EDFAs are combined in such a way that the signal can be distributed without much loss. In addition, a single pump laser can be used to pump all EDFAs, making the scheme economically attractive. Recently a photonic dual-bus architecture with optical amplifiers has been proposed [45]. The use of fiber amplifiers permits multigigabit operation with 100 nodes over hundreds of kilometers.

3. Fiber-Dispersion Management

As mentioned in Sec. 2, with the availability of EDFAs, fiber dispersion sets the ultimate limit on the performance of modern lightwave systems. This section discusses the limitations imposed by fiber dispersion and the techniques used to circumvent them.

3.1 DISPERSION LIMITATIONS

Chromatic dispersion in optical fibers broadens optical pulses such that they spread beyond the alloted bit slot. Dispersion broadening can be studied by solving the Maxwell wave equation and accounting for the frequency dependence of the propagation constant $\beta(\omega)$. In the slowly varying envelope approximation, the amplitude $A(z, t)$ of the pulse envelope evolves according to the simple equation [11]

$$\frac{\partial A}{\partial z} + \frac{i}{2}\beta_2 \frac{\partial^2 A}{\partial t^2} - \frac{1}{6}\beta_3 \frac{\partial^3 A}{\partial t^3} = 0, \tag{4}$$

where $\beta_m = d^m\beta/d\omega^m$ is evaluated at $\omega = \omega_0$. The parameter β_2 and β_3 govern the effects of second- and third-oder dispersion respectively. In general, the GVD parameter β_2 dominates unless it vanishes because the operating wavelength coincides with the zero-dispersion wavelength λ_{zd} of the fiber. For standard telecommunication fibers, λ_{zd} occurs near 1.31 µm; it can be moved in the vicinity of 1.55 µm in especially designed dispersion-shifted fibers.

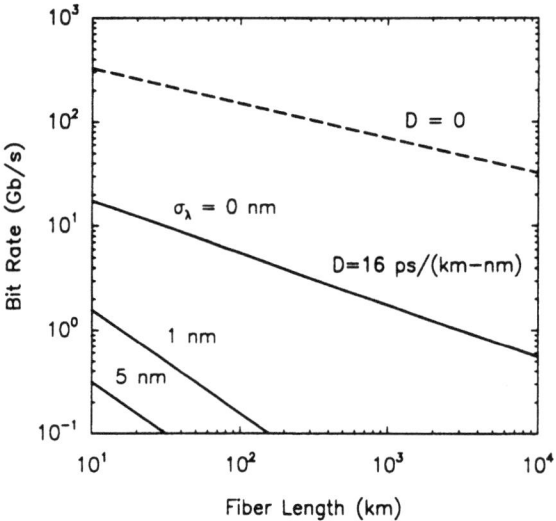

Figure 9. Dispersion limitations for single-mode fibers under various operating conditions.

Equation (4) can be easily solved by using the Fourier-transform method, and the result is

$$A(z, t) = \frac{1}{2\pi} \int_{-\infty}^{\infty} \tilde{A}(0, \omega) \exp\left(\frac{i}{2}\beta_2\omega^2 z + \frac{i}{6}\beta_3\omega^3 z - i\omega t\right) d\omega, \tag{5}$$

where $\tilde{A}(0, \omega)$ is the Fourier spectrum of the input pulse. Physically, each spectral component experiences a phase shift that results in pulse broadening in the time domain. The integral in Eq. (5) can be performed analytically for a Gaussian input pulse. For this reason, Gaussian pulses are often used in calculating the dispersion limitations [11]. The results depend on not only on the dispersion parameters but also on the spectral width of the laser source used for data transmission. For a multimode semiconductor laser the dispersion limitation is governed by $4BL|D|\sigma_\lambda < 1$, where σ_λ is the source spectral width and $D = -(2\pi c/\lambda^2)\beta_2$. Figure 9 shows the limiting bit rate B as a function of L for σ_λ = 1 and 5 nm and D = 16 ps/(km-nm). For σ_λ = 1 nm, GVD limits the BL product to BL < 16 (Gb/s)-km. The use of narrow-linewidth semiconductor lasers (such as distributed feedback lasers) in combination with an external modulator (to avoid frequency chirping) can improve the performance dramatically. In fact, the dispersion limitation in that case is simply given by $4B(|\beta_2|L)^{1/2} < 1$. The $\sigma_\lambda = 0$ line in Fig. 9 shows this condition. As an example, L could be as large as 120 km for B = 5 Gb/s, resulting in BL = 0.6 (Tb/s)-km. Further improvement in the BL product is possible by using dispersion-shifted fibers so that the GVD is parameter is small. If we use D = 1 ps/(km-nm) as a representative value, L could be as large as 2000 km for B = 5 Gb/s, resulting in BL = 10 (Tb/s)-km. Several different approaches discussed below are used to counteract the effects of fiber dispersion in modern lightwave systems.

3.2 OPERATION AT THE ZERO-DISPERSION WAVELENGTH

Conceptually, the simplest solution to the fiber-dispersion problem is to operate the lightwave system exactly at the zero-dispersion wavelength so that $\beta_2 = 0$ in Eq. (5). The third-order dispersion β_3 then limits the system performance [11] such that $B(|\beta_3|L)^{1/3} <$ 0.324. Dashed line in Fig. 9 shows this limit by using $\beta_3 = 0.1$ ps^3/km. The transmission distance under such operating conditions can exceed 10,000 km even at a relatively high bit rate of 30 Gb/s, resulting in BL = 300 (Tb/s)-km. From a practical standpoint, it is quite difficult to design such a lightwave system. The reason is that both the zero-dispersion wavelength of the fiber and the operating wavelength of the semiconductor laser are rarely known with enough precision to make an exact match possible. Moreover, both can change during system operation because of the aging-related changes in the operating conditions. The best one can hope is that the two wavelengths remain matched to a precision ~0.1 nm. However, even such small deviations are large enough that the system is performance is limited by the residual β_2 rather than by β_3. Nonetheless, β_2 can be made small enough that BL > 50 (Tb/s)-km can be realized in practice. In one transmission experiment BL = 71 (Tb/s)-km was obtained by transmitting the 5-Gb/s signal over 14,300 km of fiber in a recirculating-loop configuration [9]. It is also important to note that this approach for dispersion control cannot be extended for multichannel operation since only one channel can operate close to the zero-dispersion wavelength.

3.3 DISPERSION COMPENSATION

An interesting approach for dispersion management consists of minimizing the total dispersion by adding a dispersion compensator at the receiver even though the transmitted signal may experience large dispersion over the fiber link. This scheme is based on the observation that in a linear dispersive medium the effect of GVD is determined by the net dispersion since the frequency-dependent phase shifts introduced in different parts of the link are additive as is evident from Eq. (5). The technique is very useful for upgrading the installed fiber base from 1.3 to 1.55 µm where $\beta_2 \sim -20$ ps^2/km for standard telecommunication fibers. Without dispersion compensation such systems are dispersion limited to distances <100 km at a bit rate of a few Gb/s. However, the effect of fiber dispersion can be considerably reduced if the transmitted signal is propagated through a short length of fiber with large positive value of β_2 before it is amplified or detected [46,47]. Many techniques can be used to make fibers with large β_2. In one approach, values of β_2 ~600 ps^2/km can be obtained by using the LP$_{11}$ mode of a dual-mode fiber [46]. Of course, one does not have to use fibers for dispersion compensation; any device that effectively introduces large GVD can be used. Among the several possibilities are

Gires-Tournois interferometer [48], a micro-strip delay equalizer [49], a Bragg-grating filter [50], and a planar lightwave circuit [51].

3.3 PHASE CONJUGATION

A novel technique for dispersion compensation makes use of phase conjugation. Although the basic idea was proposed [52] in 1979, it was only in 1993 that its usefulness for practical lightwave systems was demonstrated [53-56]. Phase conjugation is achieved through the nonlinear phenomenon of four-wave mixing occurring in optical fibers when the transmitted signal is mixed with a pump wave [57]. The effect of phase conjugation is to convert the signal amplitude A(t) into its complex conjugate A*(t). Since the spectrum of A*(t) is the mirror image of that of A(t), this process is also referred to as spectral inversion.

Equation (4) can be used to see under what condition phase conjugation can lead to dispersion compensation. The conjugated field A*(t) satisfies an equation that is complex conjugate of Eq. (4). Since the β_2 term changes sign, the net effect is the same as if the sign of β_2 has been reversed. Thus, the effects of β_2 can be eliminated if phase conjugation or spectral inversion is performed in the middle of the fiber link. Note that the β_3 term does not change sign for propagation of A*(t) indicating that the effects of third-order dispersion are not eliminated by this technique. In fact, it is easy to show that all even-order dispersion terms cancel out while the odd-orders do not.

Several experiments have realized dispersion compensation through mid-way phase conjugation [53-56]. In one experiment [54], the 10-Gb/s signal could not be transmitted over more than 100 km because of large GVD ($\beta_2 = -20$ ps^2/km). However, the transmission distance increased to 360 km with mid-way phase conjugation. It is important to note that nondegenerate four-wave mixing used to generate the phase-conjugated signal changes the carrier frequency from ω_0 to $2\omega_p - \omega_0$, where ω_p is the pump frequency. As a result, the GVD parameter is generally different before and after phase conjugation. Thus, strictly speaking, complete cancellation requires a careful placement of the phase conjugator. Also, dispersion in the fiber used for four-wave mixing should be nearly zero to meet the phase-matching requirement [57]. This requires the use of a fiber whose zero-dispersion wavelength nearly coincides with the pump wavelength.

An added benefit of phase conjugation is that it can also eliminate the detrimental effects of Kerr nonlinearity that is responsible for self-phase modulation (SPM) in optical fibers [57]. The refractive index of silica fibers is slightly nonlinear such that $n = n_0 + n_2|A|^2$, where $n_2 \approx 3 \times 10^{-20}$ m^2/W. Even though the nonlinear term is quite small (~10^{-9}) compared to the linear modal index n_0, its effects can accumulate in long-haul lightwave systems where the signal is transmitted over thousands of kilometers by using optical amplifiers. SPM can be included by adding a nonlinear term to Eq. (4). If the β_3 term is assumed to be relatively small, Eq. (4) takes the form [57]

$$\frac{\partial A}{\partial z} + \frac{i}{2}\beta_2 \frac{\partial^2 A}{\partial t^2} - i\gamma |A|^2 A = 0, \qquad (6)$$

where $\gamma = (2\pi/\lambda)n_2$. Since both the dispersive and nonlinear terms change sign on taking the complex conjugate of Eq. (6), the effects of both can be eliminated simultaneously through mid-way phase conjugation [55]. Indeed, in one experiment [56] the transmission distance of 6000 km was realized at 10 Gb/s even when the input power was large enough for the SPM effects to become important. The experiment was performed in the normal-GVD regime ($\beta_2 > 0$) of the fiber. In the case of anomalous dispersion ($\beta_2 > 0$), GVD can be used to advantage by using optical solitons, as discussed next.

4. Soliton Communication Systems

Optical solitons form as a result of a balance between GVD and SPM, both of which limit the performance of fiber-optic communication systems when acting independently on optical pulses propagating inside the fiber. One can develop an intuitive understanding of how such a balance is possible by noting that both SPM and GVD impose frequency chirp on optical pulses during their propagation inside the fiber. In the case of anomalous GVD, the two chirps have opposite signs. Since the SPM-induced chirp is power dependent, at a certain power level, SPM and GVD can cooperate in such a way that the two chirps cancel each other perfectly. The optical pulse would then propagate undistorted in the form of a soliton even in a dispersive system. This section discusses the properties of optical solitons and their applications in modern lightwave systems [11].

4.1 FUNDAMENTAL AND HIGHER-ORDER SOLITONS

Pulse propagation in optical fibers in the presence of both GVD and SPM is governed by Eq. (6) which can be written in the normalized form

$$i\frac{\partial U}{\partial \xi} + \frac{1}{2}\frac{\partial^2 U}{\partial \tau^2} + N^2|U|^2 U = 0 \qquad (7)$$

by using the transformation

$$\tau = \frac{t}{T_0}, \quad \xi = \frac{z}{L_D}, \quad U = \frac{A}{\sqrt{P_0}}, \qquad (8)$$

where T_0 is the width and P_0 is the peak power of the input pulse, $L_D = T_0^2/|\beta_2|$ is the dispersion length ($\beta_2 < 0$), and $N^2 = \gamma P_0 L_D$. Equation (7) is known as the nonlinear Schrödinger equation [11] and can be solved exactly by using the inverse scattering method. The results indicate that solitons exist only for integer values of N. More specifically, when an input pulse of initial amplitude

$$U(0, \tau) = \text{sech}(\tau) \qquad (9)$$

is launched into the fiber, its shape remains unchanged during propagation when $N = 1$, but follows a periodic pattern for integer values of $N > 1$ such that the input shape is recovered at $\xi = m\pi/2$, where m is an integer. The soliton corresponding to $N = 1$ is called the fundamental soliton. Solitons corresponding to other integer values of N are called higher-order solitons. The parameter N describes the soliton order. By noting that $\xi = z/L_D$, the soliton period z_0, defined as the distance over which the N-th order soliton recovers its original shape, is given by

$$z_0 = \frac{\pi}{2} L_D = \frac{\pi}{2} \frac{T_0^2}{|\beta_2|}. \tag{10}$$

The soliton period z_0 and the soliton order N play an important role in the theory of optical solitons.

4.2 SYSTEM DESIGN

Since the fundamental soliton propagates undistorted even in the presence of GVD, soliton communication systems use solitons as "1" bits as shown schematically in Fig. 10. Note that soliton communication requires the use of the return-to-zero (RZ) format while most non-soliton systems operate with the non-return-to-zero (NRZ) format. It also requires a mode-locked laser capable of emitting picosecond (~10 ps) pulses at a repetition rate equal to the bit rate. Pulse shape should be as close to the hyperbolic secant as possible [see Eq. (9)]. Moreover, the peak power of input pulses should be such that the parameter N = 1 in Eq. (7). This condition relates the peak power to the pulse width and other fiber parameters as

$$P_0 = |\beta_2|/(\gamma T_0^2). \tag{11}$$

At first sight, such strict requirements on the input pulse appear to impose severe limitations on the system design. Fortunately, solitons are inherently stable to small perturbations. In essence, any input pulse with N in the range 0.5-1.5 evolves into a fundamental soliton as it propagates down the fiber. Since it sheds some energy in the form of undesirable dispersive waves during the adaption phase, one should try to match the ideal requirements as closely as possible. This is not difficult in practice. Mode-locked semiconductor or fiber lasers often emit pulses whose shape is quite close to the hyperbolic secant profile required for a soliton. An EDFA is used a power booster; its gain can be adjusted to satisfy the condition (11) for the peak power, making it relatively easy to launch solitons in fibers.

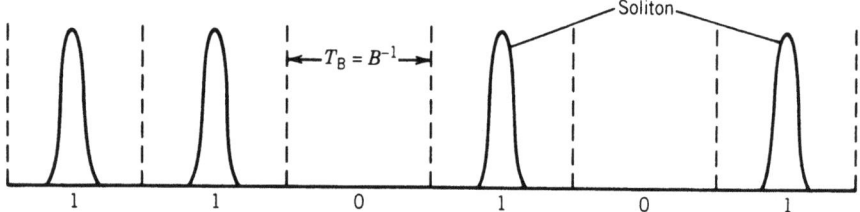

Figure 10. Soliton bit stream with the RZ format.

What is the right pulse width for soliton communication systems? It should be chosen to ensure that neighboring solitons are far enough apart that they do not interact with each other. This condition typically requires T_0 to be a small fraction of the bit slot (~5-10%). By using $T_0 = 0.05/B$ as a conservative design criterion together with typical values $\beta_2 = -2$ ps^2/km and $\gamma = 2$W^{-1}/km, $T_0 = 5$ ps and $P_0 = 40$ mW for a soliton lightwave system operating at 10 Gb/s. These values correspond to an average power of only 2 mW for the coded soliton bit stream.

4.3 SOLITON AMPLIFICATION

Optical solitons propagate undistorted only in the absence of fiber loss. Since the peak power decreases because of loss, the nonlinear effects weaken with propagation, destroying the delicate balance between GVD and SPM. However, as seen from Eq. (11), the condition $N = 1$ can be maintained if the pulse width increases. The net effect is that solitons begin to broaden as they propagate down the fiber. The simplest solution is to compensate for the fiber loss through periodic amplification of the soliton bit stream in a way similar to that shown in Fig. 4(b). The amplifier spacing L is an important design parameter for soliton communication systems as it plays a role analogous to the repeater spacing in non-soliton lightwave systems. From a practical standpoint L should be as large as possible in order to minimize the overall system cost. The amplifier spacing can be as large as 100 km for non-soliton 1.55-µm systems making use of EDFAs. For soliton communication systems the optimum value of L is generally smaller and depends on many system parameters.

An approach for estimating the amplifier spacing L is based on simulating the performance of soliton communication systems on a computer. Equation (11) is solved numerically after including the fiber loss α. The soliton amplitude U is increased at each amplifier to compensate for the fiber loss by using

$$U'(L, \tau) = \sqrt{G_0}\, U(L, \tau), \qquad (12)$$

where G_0 is the amplifier gain chosen such that

$$G_0 \exp(-\alpha L) = 1. \qquad (13)$$

Thus the soliton energy is boosted to its initial value at each amplifier. The system performance is gauged by the number of cascaded amplifiers that can be used before optical pulses degrade so much that the system becomes inoperable (as judged by the extent of eye degradation). The results show [58] that solitons can be propagated over a few thousand kilometers provided the amplifier spacing L is kept considerably below the soliton period z_0 given by Eq. (10). If we adopt $L < L_D$ as a design rule and relate T_0 to the bit rate B by using $T_0 = 0.05/B$, we obtain the criterion $B^2 L < (400|\beta_2|)^{-1}$. By using $\beta_2 = -20$ ps^2/km as a representative value near 1.55 µm for conventional fibers, $B^2L <$ 125 (Gb/s)^2-km. Thus, L must be below 20 km for B = 2.5 Gb/s and below 5 km for B = 5 Gb/s. These estimates improve by up to a factor of 10 for dispersion-shifted fibers because of their reduced dispersion. The amplifier spacing then becomes large enough (~30 km) that practical systems can be designed. The use of dispersion-shifted fibers appears to be necessary for soliton communication systems.

What is the ultimate limit on the number of amplifiers that can be cascaded? The answer is related to the amplifier noise that leads to jitter in the arrival time of solitons at the photoreceiver. If timing jitter is so large that solitons do not remain confined to the bit slot, the BER can increase drastically, making the system inoperable.

The origin of timing jitter can be understood by noting that the addition of amplified spontaneous emission to the signal introduces amplitude and phase fluctuations at the amplifier output. Amplitude fluctuations lead to fluctuations in the soliton width, while phase fluctuations change the carrier frequency of the soliton in a random manner. Width fluctuations are small enough that they do not affect the system performance significantly. However, this is not the case for fluctuations in the carrier frequency. The reason is that the group velocity depends on the carrier frequency ω_0. A change in ω_0 changes the group velocity of the soliton randomly after each amplifier. As a result, the arrival time of the soliton at the far end of the amplifier chain varies from bit to bit in a random manner. It is possible to calculate the variance of timing jitter by assuming that each amplifier produces velocity shifts independently. The ultimate BL product for dispersion-shifted fibers ($\beta_2 = -2$ ps^2/km) is found to be limited to $BL_T < 35$ (Tb/s)-km, where L_T is the total transmission distance. This limit is often referred to as the Gordon-Haus limit [59]. It can be relaxed considerably by using optical filters after each amplifier. This aspect is discussed next.

4.4 EXPERIMENTAL PROGRESS

The first experiment [60] that demonstrated the possibility of soliton transmission over transoceanic distances was performed in 1988 by using a recirculating fiber loop whose loss was compensated through the Raman amplification scheme. The main drawback from a practical standpoint was that the experiment used two color-center lasers for soliton generation and amplification. The availability of diode-pumped EDFAs by 1989 provided an opportunity to use them for amplification of picosecond pulses emitted by mode-locked or gain-switched semiconductor lasers so that the peak power is high enough for launching of fundamental solitons in optical fibers. Furthermore, EDFAs could also be used as in-line amplifiers to compensate for fiber loss. Many experimental demonstrations [61-80] of soliton communications were carried out worldwide starting in 1990.

The soliton communication experiments can be divided into two categories, depending on whether a direct fiber link or a recirculating fiber loop is used in the experiment. The experiments using fiber link are more realistic, as they mimic the actual field conditions. Early experiments [61-63] demonstrated soliton transmission over fiber lengths of roughly 100 km at bit rates 3-5 Gb/s. Input pulses were generated by using gain-switched or mode-locked semiconductor lasers and amplified by EDFAs to increase their peak power up to levels required for launching fundamental solitons into the dispersion-shifted fiber. An external modulator blocked the soliton pulse for "0" bits. The coded soliton-bit stream was transmitted through several fiber sections, and loss of each section was compensated by using an EDFA. The amplifier spacing was chosen to satisfy the criterion $L \ll z_0$ and was typically in the range 25-40 km.

In a variation of this technique both EDFAs and Raman amplifiers were used for soliton transmission [64]. Two EDFAs were used to boost the power of the input pulse train while the transmission fiber was used as a Raman amplifier by pumping it with high-power semiconductor lasers. The bit rate was as high as 20 Gb/s, and the signal was transmitted over 70 km by using the Raman amplification scheme. The soliton period z_0 was only 3.7 km. However, the amplifier spacing L could be made larger than z_0 (L = 23 km) because of the distributed nature of Raman amplification. Another experiment [65] demonstrated that soliton transmission over 200 km was possible at 20 Gb/s by using EDFAs when the pre-emphasis technique in which the input peak power exceeds the fundamental-soliton power was used to launch solitons with N > 1 (N was about 1.4 in the experiment). It was necessary to use relatively broad solitons (soliton width ~20 ps) to increase the soliton period beyond 100 km; the amplifier spacing was then 25 km. Since the bit slot is only 50-ps wide at 20 Gb/s, solitons in this experiment were quite closely spaced. However, soliton interaction did not limit the system performance because of a relatively small transmission distance. The maximum distance realized by this technique at B = 20 Gb/s was about 350 km [71]. In another experiment [69] solitons could be transmitted over 1000 km at 10 Gb/s. The soliton width in the experiment was 45 ps to increase the soliton period z_0 to a value beyond 400 km. An amplifier spacing of 25 km was chosen to satisfy L << z_0. Another experiment showed that the amplifier spacing can be as large as 71 km and the fiber can exhibit large dispersion variations provided the soliton period (z_0 = 350 km) is large enough that the average-soliton model remains valid [68].

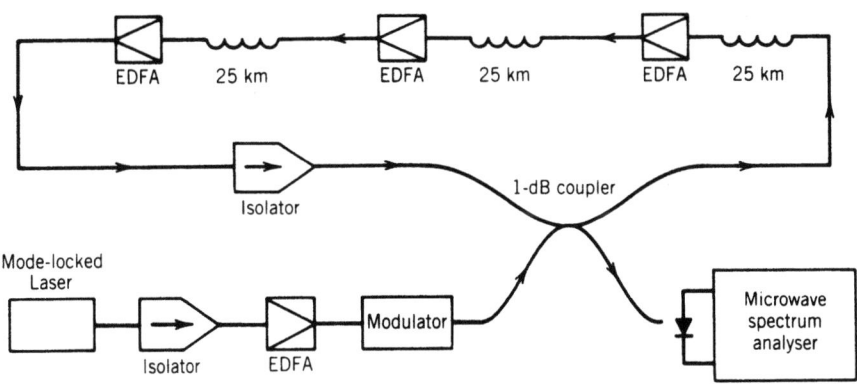

Figure 11. Experimental setup for soliton transmission in the recirculating-loop configuration.

Several experiments used a recirculating fiber loop to demonstrate the ultimate potential of soliton communication systems. Figure 12 shows the experimental setup for the recirculating-loop configuration. In an early experiment [66] 50-ps solitons were transmitted over 10,000 km by using a 75-km fiber loop containing three EDFAs spaced 25 km apart. The soliton period in this experiment was 680 km so that the condition L << z_0 was well satisfied. The effective bit rate was low (B = 200 Mb/s) since the experiment used a mode-locked color-center laser for launching solitons into the fiber loop. The use of a mode-locked semiconductor laser permitted an increase in the bit rate to 2.5 Gb/s [67]. In this experiment solitons could be transmitted over 12,000 km. The resulting bit rate-distance product BL = 30 (Tb/s)-km was limited mainly by the noise-induced timing jitter. The timing-jitter problem was solved partially in several later experiments by placing optical filters after each EDFA to reduce the wideband spontaneous emission noise. In one experiment [72] the 2.5-Gb/s signal was transmitted over 14,000 km.

Even higher bit rates and longer transmission distances are possible with further refinements. In a 1992 experiment [74] the use of an erbium-doped fiber laser for input pulses together with sliding-frequency guiding filters resulted in transmission over 15,000 km of a 5-Gb/s signal. In this scheme the center frequency of successive optical filters along the fiber link is shifted at a sliding rate ~5-10 MHz/km. Solitons adjust easily to such a weak perturbation and change their carrier frequency accordingly. By contrast, spontaneous emission noise passing through earlier filters is blocked by the later filters because of the frequency shift, resulting in a much smaller noise and reduced timing jitter. In a 1993 experiment, the use of sliding-frequency guiding filters provided error-free soliton transmission over 20,000 km at 10 Gb/s in a single channel and over 13,000 km at 20 Gb/s in a two-channel experiment [79]. Further improvements in the single-channel transmission experiments have resulted in soliton transmission over 35,000 km at 10 Gb/s and over 24,000 km at 15 Gb/s [80]. The BL product of 360 (Tb/s)-km clearly demonstrates the potential of soliton communication systems.

In a different approach, the timing-jitter problem was solved by using a $LiNbO_3$ modulator within the fiber loop [70]. The role of the modulator was to shape and re-time the soliton stream synchronously. In a 1991 experiment, solitons at 10 Gb/s could be maintained over one million kilometers when a modulator was used within the 510-km loop incorporating EDFAs with 50-km spacing [70]. In a later experiment solitons were controlled in both the time and frequency domain by using a modulator and a bandpass filter inside the fiber loop. No performance degradation was observed even after one million kilometers, suggesting that such an approach can maintain solitons over unlimited distances [78]. Several attempts have been made to increase the bit rate of soliton communication systems beyond 20 Gb/s. In a 1992 experiment [73] the 32-Gb/s solitons were transmitted over 90 km by using 16-ps pulses obtained from a mode-locked semiconductor laser. Since the bit slot is only about 31-ps wide at the bit rate of 32 Gb/s, neighboring 16-ps solitons were so close to each other that they interacted strongly, limiting the transmission distance to only 90 km. Another experiment [75] that transmitted 40-Gb/s solitons over 65 km suffered from the same problem. Nonetheless, in a 1993 experiment [77] the bit rate could be doubled to 80 Gb/s by using polarization-division multiplexing such that the neighboring bit slots carried orthogonally polarized soliton pulses. The 80-Gb/s signal was transmitted over 80 km by this technique. More recently, bit rates have been extended to 100 Gb/s by using ultrashort pulses generated from mode-locked fiber lasers. At such high bit rates the bit slot is so small that typically soliton width is ~1 ps or less. For such ultrashort solitons several higher-order nonlinear effects become quite important and are a subject of active investigation.

5. References

1. F. P. Kapron, D. B. Keck, and R. D. Maurer, *Appl. Phys. Lett.* **17**, 423 (1970).
2. I. Hayashi, M. B. Panish, P. W. Foy, and S. Sumski, *Appl. Phys. Lett.* **17**, 109 (1970).
3. K. Oe, S. Ando, and K. Sugiyama, *Jpn. J. Appl. Phys.* **16**, 1273 (1977).
4. D. Gloge, A. Albanese, C. A. Burrus, E. L. Chinnock, J. A. Copeland, A. G. Dentai, T. P. Lee, T. Li, and K. Ogawa, *Bell Syst. Tech. J.* **59**, 1365 (1980).
5. J. I. Yamada, S. Machida, and T. Kimura, *Electron. Lett.* **17**, 479 (1981).
6. T. Miya, Y. Terunuma, T. Hosaka, and T. Miyoshita, *Electron. Lett.* **15**, 106 (1979).
7. S. Saito, M. Murakami, A. Naka, Y. Fukuda, T. Imai, M. Aiki, and T. Ito, Paper PDP5, *Europ. Conf. Opt. Commun.*, Paris, Sept. 9-12 (1991).
8. N. Edagawa, H. Taga, Y. Yoshida, M. Suzuki, S. Yamamoto, and H. Wakabayashi, Paper PDP3, *Europ. Conf. Opt. Commun.*, Paris, Sept. 9-12 (1991).
9. N. S. Bergano, J. Aspell, C. R. Davidson, P. R. Trischitta, B. M. Nyman, and F. W. Kerfoot, *Electron. Lett.* **27**, 1889 (1991).
10. A. Hasegawa and F. Tappert, *Appl. Phys. Lett.* **23**, 142 (1973).
11. G. P. Agrawal, *Fiber-Optic Communication Systems* (Wiley, New York, 1992).
12. P. E. Green, Jr., *Fiber Optic Networks* (Prentice Hall, Englewood Cliffs, New Jersey, 1993).
13. C. J. Koester and E. Snitzer, *Appl. Opt.* **3**, 1182 (1964).

14. S. B. Poole, D. N. Payne, R. J. Mears, M. E. Fermann, and R. E. Laming, *J. Lightwave Technol.* **LT-4**, 870 (1986).
15. M. Shimizu, M. Yamada, H. Horiguchi, T. Takeshita, and M. Okayasu, *Electron. Lett.* **26**, 1641 (1990).
16. W. J. Miniscalco, *J. Lightwave Technol.* **9**, 234 (1991).
17. J. L. Zyskind, E. Desurvire, J. W. Sulhoff, and D. J. DiGiovanni, *IEEE Photon. Technol. Lett.* **2**, 869 (1990).
18. J. R. Armitage, *Appl. Opt.* **27**, 4831 (1988); *IEEE J. Quantum Electron.* **26**, 423 (1990).
19. C. R. Giles and D. DiGiovanni, *IEEE Photon. Technol. Lett.* **2**, 797 (1990).
20. C. R. Giles and E. Desurvire, *J. Lightwave Technol.* **9**, 271 (1991).
21. M. Yamada, M. Shimizu, M. Okayasu, T. Takeshita, M. Horiguchi, Y. Tachikawa, and E. Sugita, *IEEE Photon. Technol. Lett.* **2**, 205 (1990).
22. R. I. Laming and D. N. Payne, *IEEE Photon. Technol. Lett.* **2**, 418 (1990).
23. E. Desurvire, C. R. Giles, and J. R. Simpson, *J. Lightwave Technol.* **7**, 2095 (1989).
24. K. Inoue, H. Toba, N. Shibata, K. Iwatsuki, A. Takada, and M. Shimizu, *Electron. Lett.* **25**, 594 (1989).
25. G. P. Agrawal, *IEEE Photon. Technol. Lett.* **2**, 875 (1990).
26. G. P. Agrawal, *Opt. Lett.* **16**, 226 (1991); *Phys. Rev. A* **44**, 7493 (1991).
27. P. P. Smyth, R. Wyatt, A. Fidler, P. Eardley, A. Sayles, and S. Graig-Ryan, *Electron. Lett.* **26**, 1604 (1990).
28. T. Saito, Y. Sunohara, K. Fukagai, S. Ishikawa, N. Henmi, S. Fujita, and Y. Aoki, *IEEE Photon. Technol. Lett.* **3**, 551 (1991).
29. L. T. Blair and H. Nakano, *Electron. Lett.* **27**, 835 (1991).
30. A. H. Gnauck and C. R. Giles, *IEEE Photon. Technol. Lett.* **4**, 80 (1992).
31. K. Hagimoto, K. Iwatsuki, A. Takada, M. Nakazawa, M. Sarawatari, K. Aida, and K. Nakagawa, *Electron. Lett.* **25**, 662 (1989).
32. Y. K. Park, S. W. Granlund, T. W. Cline, L. D. Tzeng, J. S. French, J.-M. P. Delavaux, R. E. Tench, S. K. Korotsky, J. J. Veselka, and D. J. DiGiovanni, *IEEE Photon. Technol. Lett.* **4**, 179 (1992).
33. Y. K. Park, O. Mizuhara, L. D. Tzeng, J.-M. P. Delavaux, T. V. Nguyen, M. L. Kao, P. D. Yates, and J. Stone, *IEEE Photon. Technol. Lett.* **5**, 79 (1993).
34. E. Kannan and S. Frisken, *IEEE Photon. Technol. Lett.* **5**, 76 (1993).
35. N. Edagawa, Y. Toshida, H. Taga, S. Yamamoto, K. Mochizuchi, and H. Wakabayashi, *Electron. Lett.* **26**, 66 (1990).
36. S. Saito, T. Imai, and T. Ito, *J. Lightwave Technol.* **9**, 161 (1991).
37. T. Imai, M. Murakami, T. Fukuda, M. Aiki, and T. Ito, *Electron. Lett.* **28**, 1484 (1992).
38. W. I. Way, S. S. Wagner, M. M. Choy, C. Lin, R. C. Mendez, H. Tohme, A. Yi-Yan, A. C. Von Lehman, R. E. Spicer, M. Andrejco, M. A. Saifi, and H. L. Lemberg, *IEEE Photon. Technol. Lett.* **2**, 665 (1990).
39. A. M. Hill, R. Wyatt, J. F. Massicott, K. J. Blyth, D. S. Forrester, R. A. Lobbett, P. J. Smith, and D. B. Payne, *Electron. Lett.* **26**, 1882 (1990).
40. P. M. Gabla, C. Bastide, Y. Cretin, P. Bousselet, A. Pitel, and J. P. Blondel, *IEEE Photon. Technol. Lett.* **4**, 510 (1992).
41. H. Bülow, R. Fritschi, R. Heidemann, B. Junginger, H. G. Krimmel, and J. Otterbach, *IEEE Photon. Technol. Lett.* **4**, 1287 (1992).
42. H. Bülow, R. Fritschi, R. Heidemann, B. Junginger, H. G. Krimmel, and J. Otterbach, *Electron. Lett.* **28**, 1836 (1992).
43. A. E. Willner, E. Desurvire, H. M. Presby, C. A. Edwards, and J. R. Simpson, *IEEE Photon. Technol. Lett.* **2**, 669 (1990).
44. Y. K. Chen, S. Chi, and J. Y. Liaw, *IEEE Photon. Technol. Lett.* **5**, 230 (1993).
45. K. T. Koai and R. Olshansky, *IEEE Photon. Technol. Lett.* **5**, 482 (1993).
46. C. D. Poole, J. M. Wiesenfeld, and D. J. DiGiovanni, *IEEE Photon. Technol. Lett.* **5**, 194 (1993).

47. Y. Sorel, J. F. Kerdiles, C. Kazmierski, M. Blez, D. Mathoorasing, and A. Ougazzaden, *Electron. Lett.* **29**, 973 (1993).
48. A. H. Gnauck, C. R. Giles, L. J. Cimini, Jr., J. Stone, L. W. Stulz, S. K. Korotky, and J. J. Veselka, *IEEE Photon. Technol. Lett.* **3**, 1147 (1991).
49. K. Yonenaga and N. Takachio, *IEEE Photon. Technol. Lett.* **5**, 949 (1993).
50. B. Eggleton, P. A. Krug, L. Poladian, *OFC Conf.*, Paper ThK3, San Jose, California, Feb. 20-25, 1994.
51. K. Takiguchi, K. Okamoto, and K. Moriwaki, *IEEE Photon. Technol. Lett.* **5**, 561 (1993).
52. A. Yariv, D. Fekete, and D. M. Pepper, *Opt. Lett.* **4**, 52 (1979).
53. S. Watanabe, T. Naito, and T. Chikama, *IEEE Photon. Technol. Lett.* **5**, 92 (1993).
54. A. H. Gnauck, R. M. Jopson, and R. M. Derosier, *IEEE Photon. Technol. Lett.* **5**, 663 (1993).
55. S. Watanabe, T. Chikama, G. Ishikawa, T. Terahara, and H. Kuwahara, *IEEE Photon. Technol. Lett.* **5**, 1241 (1993).
56. K. Kikuchi and C. Lorattanasane, *IEEE Photon. Technol. Lett.* **6**, 104 (1994).
57. G. P. Agrawal, *Nonlinear Fiber Optics* (Academic Press, San Diego, California, 1989).
58. Y. Kodama and A. Hasegawa, *Opt. Lett.* **7**, 339 (1982); **8**, 342 (1983).
59. J. P. Gordon and H. A. Haus, *Opt. Lett.* **11**, 665 (1986).
60. L. F. Mollenauer and K. Smith, *Opt. Lett.* **13**, 675 (1988).
61. M. Nakazawa, K. Suzuki, and Y. Kimura, *IEEE Photon. Technol. Lett.* **2**, 216 (1990).
62. N. A. Olsson, P. A. Andrekson, P. C. Becker, J. R. Simpson, T. Tanbun-Ek, R. A. Logan, H. Presby, and K. Wecht, *IEEE Photon. Technol. Lett.* **2**, 358 (1990).
63. K. Iwatsuki, S. Nishi, and K. Nakagawa, *IEEE Photon. Technol. Lett.* **2**, 355 (1990).
64. K. Iwatsuki, K. Suzuki, S. Nishi, M. Saruwatari, and K. Nakagawa, *IEEE Photon. Technol. Lett.* **2**, 905 (1990).
65. M. Nakazawa, K. Suzuki, E. Yamada, and Y. Kimura, *Electron. Lett.* **26**, 1592 (1990).
66. L. F. Mollenauer, M. J. Neubelt, S. G. Evangelides, J. P. Gordon, J. R. Simpson, and L. G. Cohen, *Opt. Lett.* **15**, 1203 (1990).
67. L. F. Mollenauer, B. M. Nyman, M. J. Neubelt, G. Raybon, and S. G. Evangelides, *Electron. Lett.* **27**, 178 (1991).
68. A. D. Ellis, J. D. Cox, D. Bird, J. Regnault, J. V. Wright, and W. A. Stallard, *Electron. Lett.* **27**, 878 (1991).
69. E. Yamada, K. Suzuki, and M. Nakazawa, *Electron. Lett.* **27**, 1289 (1991).
70. M. Nakazawa, E. Yamada, H. Kubota, and K. Suzuki, *Electron. Lett.* **27**, 1270 (1991).
71. M. Nakazawa, K. Suzuki, E. Yamada, and H. Kubota, *Electron. Lett.* **27**, 1662 (1991).
72. L. F. Mollenauer, M. J. Neubelt, M. Haner, E. Lichtman, S. G. Evangelides, and B. M. Nyman, *Electron. Lett.* **27**, 2055 (1991).
73. P. A. Andrekson, N. A. Olsson, M. Haner, J. R. Simpson, T. Tanbun-Ek, R. A. Logan, D. Coblentz, H. M. Presby, and K. W. Wecht, *IEEE Photon. Technol. Lett.* **4**, 76 (1992).
74. L. F. Mollenauer, E. Lichtman, G. T. Harvey, M. J. Neubelt, and B. M. Nyman, *Electron. Lett.* **28**, 792 (1992).
75. K. Iwatsuki, K. Suzuki, S. Nishi, and M. Saruwatari, *Electron. Lett.* **28**, 1821 (1992).
76. H. Taga, M. Suzuki, N. Idagawa, Y. Yoshida, S. Yamamoto, S. Akiba, and H. Wakabayashi, *Electron. Lett.* **28**, 2247 (1992).
77. K. Iwatsuki, K. Suzuki, S. Nishi, and M. Saruwatari, *IEEE Photon. Technol. Lett.* **5**, 245 (1993).
78. M. Nakazawa, K. Suzuki, E. Yamada, H. Kubota, Y. Kimura, and M. Takaya, *Electron. Lett.* **29**, 729 (1993).
79. L. F. Mollenauer, E. Lichtman, M. J. Neubelt, and G. T. Harvey, *Electron. Lett.* **29**, 910 (1993).
80. L. F. Mollenauer, *Opt. Photon. News*, **5** (4), 15 (1994).

AN OPTICAL COHERENT TRANSMISSION SYSTEM BASED ON POLARIZATION MODULATION AND HETERODYNE DETECTION AT 155 Mbit/s BITRATE AND 1.55 μm WAVELENGTH

R. CALVANI, R. CAPONI, F. DELPIANO and G. MARONE
CSELT - Centro Studi e Laboratori Telecomunicazioni S.p.A.
Via G. Reiss Romoli, 274 - 10148 Torino - Italy -
Fax: +39.11.2285.085

1. Introduction

The digital modulation schemes of optical coherent transmissions which show the highest efficiency are based on angle modulation techniques, such as frequency and phase shift keying (FSK, PSK). Their optimum performance are due to the advantage of a complete exploitation of the source emission, since sending an equal amount of energy for each bit, the maximum transmitted power remains unchanged at any time.

In the case of a binary FSK modulation, the aforesaid property is related to the signal structure, whose spectrum can be compared with a double ASK (amplitude shift keying) system (one for each tone) carrying a complementary information on the two superimposed channels. The pulse pattern of one ASK channel (at angular frequency ω_1) exactly compensates the sequence from the other (at ω_2), so the total power remains invariant as required. From this point of view the two frequency channels carry a bidimensional message in the modulation space and the information content can be represented with a redundant coding [1].

Such interpretation can be extended to the vector space of the electromagnetic field, where the state of polarization (SOP) of the light is defined. In this novel approach the two channels carrying the signal, correspond to different directions of the electric field vector and in particular to a pair of linear orthogonal SOPs of the same optical carrier. Using this modulation scheme, the binary information is transmitted by switching the lightwave SOP between the aforesaid two states at any bit change. Therefore, according to other classic modulation formats, this unconventional technique exploiting the vector properties of the electromagnetic field has been called polarization shift keying or POLSK [2, 3, 4].

As regards a typical drawback of optical coherent transmission systems with respect to radiowave communication apparata, it is known that the laser shows a phase noise which affects both the transmitter source and the local oscillator. A recent solution of this problem employed in heterodyne receivers in the case of angle modulation is called non-coherent postdetection and consists in an envelope, square-law or delay demodulation.

The first alternative has been applied on both detected channels of a double ASK (POLSK) experiment at 560 Mbit/s bitrate and 1.3 μm wavelength [3]. Another possibility realized in a POLSK experiment, is to detect the relevant phase shift $(0, \pi)$ corresponding to the information bits (1, 0 respectively) through a cross product of the two output channels at intermediate frequency (IF), which provides a synchronous subtraction of the unwanted phase noise. Other receiver configurations experimentally demonstrated or simply proposed include delay schemes [5] or square-law devices [6]. Nevertheless the fundamental property which unifies all coherent POLSK versions is just the phase noise cancellation which purifies the baseband data improving the signal to noise ratio.

The main advantage of the POLSK technique, from the viewpoint of the bandwidth occupancy, is that the IF signal shows the spectrum of a single ASK channel, while the corresponding sensitivity is in principle the same as in the FSK system (as will be discussed in the following).

In the present paper an experimental POLSK system with square-law detection on two orthogonal polarization channels is demonstrated.

Transmission tests at bitrates up to 155 Mbit/s have been performed using two DFB lasers at 1.548 μm as source and local oscillator.

The system performance has been reviewed by taking into account the double-ASK cofiguration of the POLSK technique and by illustrating its analogy with FSK on the basis of the error probability function.

Moreover the optical setup is accurately described, in particular the polarization modulator (simply implemented modifying an OGW 2x2 switch) and the heterodyne receiver configuration.

As regards experimental results, specific tests on the phase noise subtractivity show an excellent ratio (about 10^7) between the IF and the demodulated linewidth and an output eye-diagram at 155 Mbit/s which compares favourably with the simulated one.

2. Performance of the POLSK system

In the present experiment the digital polarization modulation has been accomplished, at the transmitter stage, by using a modified scheme of an OGW 2x2 matrix based on an integrated Mach-Zehnder interferometer. In this special application the four ports are equipped with polarization maintaining fibres and the two outputs provide linear orthogonal SOPs (corresponding to bits 0, 1). At the receiver stage after the fibre line the two input SOPs are completely recovered with manual compensation.

According to the vector diagram of Fig. 1, the received signal can be represented in the x, y frame of the output polarization analyzer (parallel to the 0, 1 SOP directions respectively) as follows

$$\vec{E}_s = \sqrt{P_s} \exp\{i[\omega_s t + \phi_s(t)]\}\{m(t)\vec{i} + [1 - m(t)]\vec{j}\} \qquad (1)$$

where P_S is the signal power, ω_S the corresponding angular frequency, $\phi_S(t)$ the source phase noise, m(t)=0, 1 the information message, and $\vec{i}, \vec{j}$ the unit vectors of the chosen x, y axes. Owing to this orientation of the SOP splitter, the data flow is divided between the two detectors, receiving the sequence m(t) and [1-m(t)] respectively at the same bitrate.

In a similar way the local oscillator (LO) beam at angular frequency ω_L, linearly polarized at 45° in the same reference as the signal, can be represented in the form

$$\vec{E}_L = \sqrt{P_L/2} \exp\{i[\omega_L t + \phi_L(t)]\}(\vec{i} + \vec{j}) \qquad (2)$$

where P_L is the LO power and $\phi_L(t)$ its phase noise, totally uncorrelated with respect to $\phi_S(t)$. To fulfill the condition of quantum limited detection, P_L is assumed sufficiently high, which overcomes all the receiver noises.

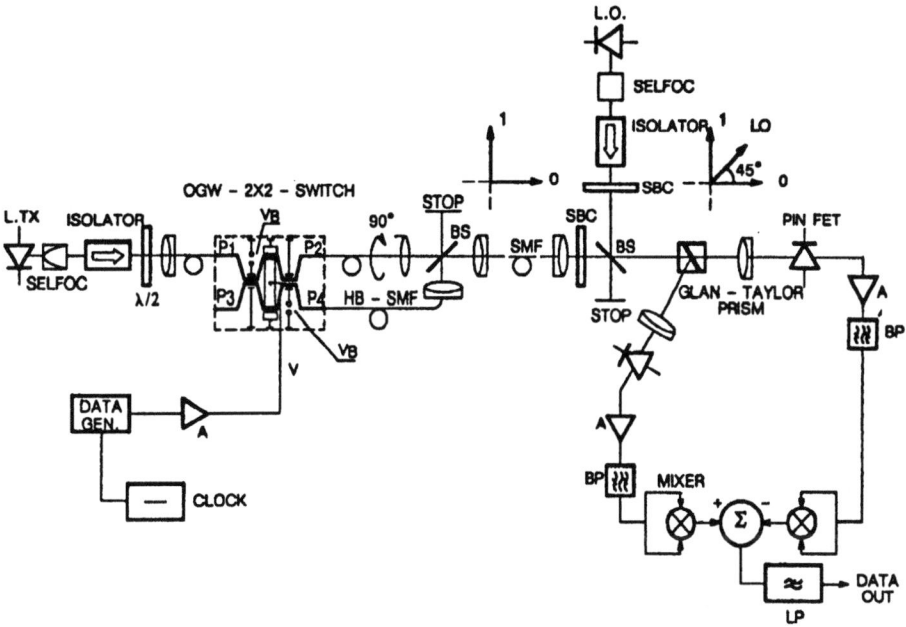

Figure 1. Experimental setup with transmitted and received SOP diagrams.

After superposition of the two beams (1), (2), and SOP analysis along the x, y axes of the given frame, the two photodetected current terms oscillating at IF $\Omega = \omega_S - \omega_L$ are

$$I_x(t) = R_c\{2m(t)\sqrt{P_S P_L/2} \cos[\Omega t + \phi(t)] + n_x(t)\}$$

$$I_y(t) = R_c\{2[1 - m(t)]\sqrt{P_S P_L/2} \cos[\Omega t + \phi(t)] + n_y(t)\}$$

(3)

where $\phi(t) = \phi_S(t) - \phi_L(t)$ is the IF phase noise, R_c is the responsivity term of both channels (=detector responsivity·amplification) assumed as equal. Furthermore, $n_x(t)$ and $n_y(t)$ are zero-mean Gaussian noises with white spectrum and $\sigma^2/2$ variance (being generated by $P_L/2$ power each). This parameter can be expressed in terms of photon energy hv, as follows

$$\sigma^2 = 2 < n_i^2(t) > = 2h\nu P_L W \qquad \text{(i=x, y)} \qquad (4)$$

where W is the IF filter bandwidth and is assumed sufficiently large to leave undistorted the whole phase noise spectrum [7].

Using the quadrature expansion with argument $\Omega t + \phi(t)$ for the IF currents (according to [8]), equations (3) become

$$I_x(t) = R_c\{[m(t)\sqrt{2P_sP_L} + p_x(t)]\cos\Phi(t) - q_x(t)\sin\Phi(t)$$
$$I_y(t) = R_c\{[1-m(t)]\sqrt{2P_sP_L} + p_y(t)\}\cos\Phi(t) - R_c q_y(t)\sin\Phi(t) \quad (5)$$
$$\text{with} \quad \Phi(t) = \Omega t + \phi(t)$$

where $p_i(t)$, $q_i(t)$ (for i=x, y) are phase and quadrature components with $\sigma^2/2$ variance.

At the decision stage, the square-law detector (with K mixing efficiency) is followed by a subtractor and a low-pass filter (Fig. 1), so the resulting baseband signal is

$$D(t) = \frac{KR_c^2}{2}\{[m(t)\sqrt{2P_sP_L} + p_x(t)]^2 + q_x^2(t) - \\ -\{[1-m(t)]\sqrt{2P_sP_L} + p_y(t)\}^2 - q_y^2(t)\} \quad (6)$$

This expression does not include the phase noise contributions which are cancelled in the square-law demodulation.

Equation (6) can be transformed in a symmetric form corresponding to a bipolar signal which is described by the Rice and Rayleigh probability function [8]. Finally, by a suitable integration of these distributions the error probability P_e of the POLSK system becomes

$$P_e = \frac{1}{2}\exp(-P_sP_L/\sigma^2) \quad (7)$$

This is just the standard result of FSK, which proves the perfect equivalence with the present POLSK technique from the viewpoint of the system sensitivity.

3. System description: transmitter

In the present experiment, two main parts compose the transmitter: the source module and the digital SOP modulator (Fig. 1). The laser source is an InGaAsP DFB diode manufactured by Fujitsu (type FLD150F2RH) operating at 1.548 μm with 20 MHz linewidth, included in a module suitable to provide an excellent thermal and mechanical stability. A feedback control, based on a Peltier cell and a thermistor, allows a precise temperature adjustment, while the bias current is supplied by a highly stable driver. The aforesaid regulations give a fine tuning of the lasing wavelength over a range of about 2 nm, so permitting an excellent superposition of the source and LO center frequencies.

The beam collimation is obtained in the laser module through a SELFOC lens with a convex input surface and anti-reflection coating at the output, to avoid backward reflection into the laser cavity together with a 70 dB isolator. A half wave-plate rotates the linear SOP of the beam to maximize the polarization coupling with the external modulator.

As regards the digital polarization modulator (DPM), a very stable configuration has been selected, based on an integrated spatial switch. A Mach-Zehnder OGW matrix of commercial type with two input and two output ports has been considered to realize the 2x2 switch. This scheme is obtained with two directional couplers connected by two parallel guides, where the phase shift of $\pm\pi/2$ (for the upper and lower path respectively) is generated via the electro-optic effect. When the voltage V (applied to high speed switching electrodes: 3 GHz bandwidth) is off, the beam entering in P_1 will exit in P_4 (cross state), while the straigth-through state with output in P_2 occurs for V=12.5 Volts, corresponding to the half-wave voltage (which gives a phase shift of π between the two interfering fields).

In Fig. 2 are shown the experimental curves of the behaviour of the output normalized power (at ports P2 and P4) versus V. The sinusoidal shape interpolates the experimental points according to the theory of the device and the measured cross-talk at each output is of the order of 25 dB. An independent dc voltage V_B provides a fine tuning of the 3 dB couplers over 30 nm wavelength range. All the four ports are pig-tailed with polarization maintaining fibres having birefringence axes aligned to the TE field direction of the integrated waveguide (i.e. parallel to the plane of Fig. 1). Using this facility, the spatial modulation produced by the 2x2 matrix can be easily transformed into an SOP switching function through a 90° rotation of either output fibre and successive recombination of the two orthogonal SOPs on an output beam-splitter (BS).

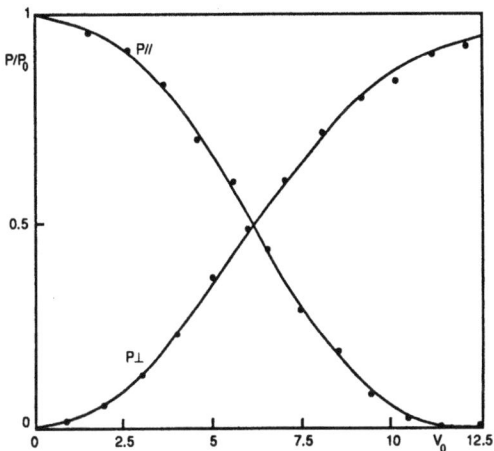

Figure 2. Response curves of the SOP modulator (P// comes from P_4 port, P⊥ from P_2).

The integrated modulator has a fibre-to-fibre insertion loss of 6 dB. Moreover, taking into account the coupling efficiency and the 3 dB penalty of the beam-splitter, the overall losses of the DPM can be estimated to exceed 10 dB. The two polarized outputs often exhibit a slight selective attenuation, due to the different losses of the two channels. These can be electrically equalized, through the splitting ratio of the two directional couplers, by changing the bias voltage V_B.

The information signal at the transmitter output is launched into 1 km of conventional single mode fibre (λ_{co} = 1.23 μm and 0.23 dB/km attenuation at 1.55 μm). Of course the two linear SOPs change during fibre propagation, nevertheless their mutual orthogonality is maintained to a high degree.

4. System description: receiver

At the fibre output a complete recovery of the two linear orthogonal SOPs (with the same orientation as the transmitted ones) has been achieved. The thermal effects on both the modulator pig-tails and the fibre channel induce quite slow SOP fluctuations, so manual readjustments are needed at intervals of several minutes.

As regards the receiver structure the LO laser is a DFB diode of the same kind as the transmitter one. Temperature and current controls are set to provide a stable IF beating at $\Omega/2\pi$=450 MHz. The mixing between the signal and the LO (having a power level of 0.6 mW) is achieved through a BS with 50% efficiency at the operation wavelength. The transmitted and reflected beams at the output of the recombination BS are affected by polarization effects which can be compensated inserting a SBC also in the LO arm to obtain at the BS output a 45° linear SOP. After beam recombination, polarization splitting is performed through a Glan-Taylor prism with axes parallel to the directions of the transmitted SOPs. In this way the two outputs (orthogonally polarized) contain complementary information patterns (ASK coded) as pointed out previously (3).

The IF signals are separately photodetected with a pair of PIN-FET diodes, amplified, filtered and square-law demodulated using two electrical mixers. Taking into account that the bitrate is 155 Mbit/s, the bandpass filters are centered at 450 MHz IF with a 3 dB bandwidth of W=590 MHz. The resulting baseband channels are finally subtracted to obtain the bipolar form of the signal, which is finally lowpass filtered with a cutoff frequency of 100 MHz and read out with a decision threshold placed on the zero level.

5. Experiment and simulation results

All the experimental results on polarization modulation are obtained with bitrates ranging from 64 Mbit/s up to 1 Gbit/s in NRZ (Non-Return-Zero) code. The pattern used are either a 24 bit word or a pseudo-random sequence. Some preliminary tests have been performed at the transmitter and receiver stage.

As regards the DPM, the two SOP signals associated with the 0, 1 bits have been found orthogonal within 2°-3°. The traces of the two polarization channels with a 24 bit word at a bitrate of 140 Mbit/s are shown in Fig. 3, where the expected complementarity of the two information flows is displayed. The amplitude difference between the two sequence are due to residual dichroism inside the modulator unit, as previously mentioned. Same tests up to 1 Gbit/s have given similar results as regards the response in the time domain.

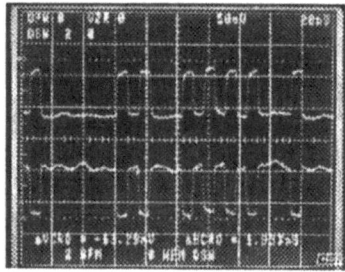

Figure 3. Orthogonally polarized POLSK signals (with a 24 bit word, NRZ code, 140 Mbit/s bitrate) at the transmitter output.

The previously mentioned property of the phase noise subtractivity has been simply checked by comparing the IF spectrum of one channel with the corresponding baseband output [4,6]. In the present experiment the IF linewidth is about 60 MHz (Fig. 4a). Conversely, in Fig. 4b is shown the first harmonic line of the demodulated signal (with a 50 MHz square wave input), which does not exceed 10 Hz FWHM (limited by the instrument resolution). The high degree phase noise cancellation can be pointed out by the ratio between the two aforesaid linewidth, which gives a spectral contraction factor of

$6 \cdot 10^6$. This characteristic has been confirmed also in a system test with a 32 MHz square wave modulation, well inside the IF linewidth. The received signals after demodulation are of the same quality as the ones presented in the following.

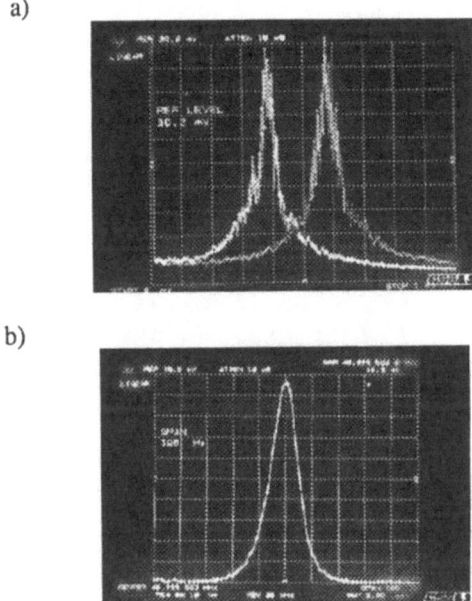

Figure 4. Phase noise subtractivity test: a) IF spectra of the two received ASK channels (corresponding to different recording times and frequencies only for illustration purposes); b) first harmonic peak of the demodulated signal at 50 MHz.

The heterodyne detection experiments have been performed on each ASK channel with the same bit sequence at 140 Mbit/s used to check the transmitter. The corresponding traces (Fig. 5a), exhibit an asymmetrical noise distribution between the 0 and 1 levels of the two complementary outputs (as expected in an ASK signal), while the up and down transitions are well resolved. The dichroism is higher than at the DPM output, which is due to a certain unbalance in the electronic processing of the two SOP channels. To show the full dynamic of the system, the two recorded signals (after averaging over 10 sample) are finally subtracted, as shown in Fig. 5b.

A complete test of operation of the POLSK system has been performed using a pseudo-random sequence at 155 Mbit/s. The resulting eye diagram is shown in Fig. 6, where the trace exhibits a symmetric noise distribution on the 0, 1 levels, as expected from the theory. In fact, the noise asymmetry of the two ASK signals is removed due to the final subtraction of the two demodulated channels typical of the POLSK system output.

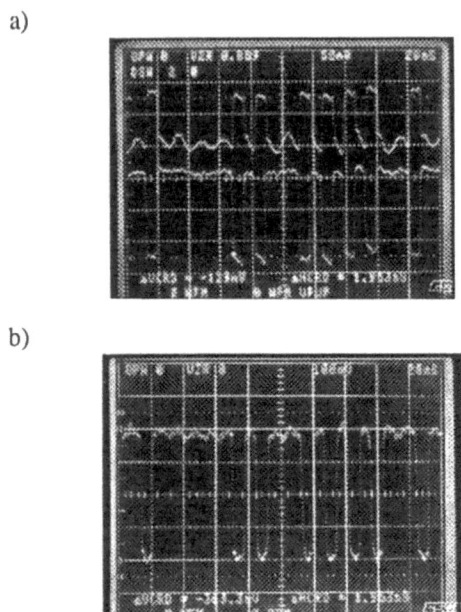

Figure 5. Received POLSK signals (with a 24 bit word, NRZ code, 140 Mbit/s bitrate); a) demodulated ASK signals on the two orthogonal channels; b) combined baseband output.

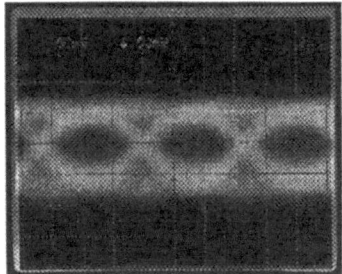

Figure 6. Experimental eye-diagrams of the POLSK signal at 155 Mbit/s.

To study the performance of the POLSK technique, numerical simulations at different bitrates have been performed taking into account the same system parameters used in the experiment. A software package suited to the analysis of coherent transmission has been utilized, with a procedure especially developed for reproducing the phase noise of a 20 MHz linewidth laser [8]. The DPM has been implemented following the input/output response curves of the experimental device, but with a bandwidth limited to 1 GHz. The fibre channel has been considered only from the viewpoint of the attenuation, neglecting the chromatic and polarization dispersion effects.

The simulated PIN-FET receiver exhibits a 4 GHz bandwidth and takes into account the thermal noise sources and the shot noise terms according to the signal and LO levels with their proper spectral densities [8, 9]. The two ASK channels include a pair of identical bandpass filters (10 poles Butterworth type) centered at 450 MHz IF with 600 MHz 3 dB bandwidth. The mixers and the following subtractor used in the receiver scheme have been assumed with ideal characteristics, while the final lowpass filter (6 poles Bessel type) has 90 MHz cutoff frequency.

In Fig. 7 is shown the received eye diagram of a pseudo-random data flow at 155 Mbit/s using NRZ code and 45 dB fibre attenuation. The symmetry of the pattern around the zero threshold is complete as in the case of the experiment (Fig. 6). Moreover the eye opening compares favourably with the previous diagram.

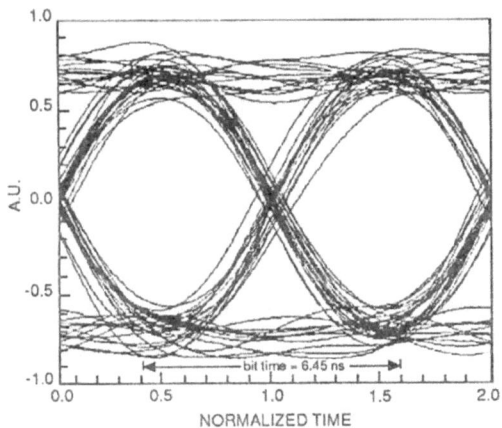

Figure 7. Eye diagram of the simulated dual ASK POLSK system at 155 Mbit/s.

6. Conclusions

The first experiment of coherent transmission with polarization modulation and heterodyne detection at 155 Mbit/s bitrate and 1.55 µm wavelength is presented. The theory of the POLSK system has been reviewed taking into account the separate square-law demodulation of the two detected channels.

The experimental setup includes two DFB lasers inserted in high thermal and mechanical stability modules. The DPM is obtained using a 2x2 switch of commercial type with output polarization pig-tails adjusted in such a configuration to convert the space modulation into the polarization one.

Pseudo-random transmission tests at 155 Mbit/s over a short fiber length have given good results in terms of the eye opening. The phase noise subtractivity at the system detection stage has been verified to a high degree, giving a reduction ratio (IF beating width/demodulated linewidth) of the order of 10^7.

The comparison of the experiment and the corresponding simulation, on the basis of the eye diagram at 155 Mbit/s bitrate, is in good agreement.

7. References

1. T.Okoshi, K. Kikuchi, (1988) Coherent optical fibre communications, Tokyo, *KTK Scientific Publ.*.
2. R. Calvani, R. Caponi, F. Cisternino, U.S. Patent No. 4,817,306, applied priority data: April 1986.
3. E. Dietrich, B. Enning, R. Gross, H. Knupke, (1987) Heterodyne transmission of a 560 Mbit/s optical signal by means of polarization shift keying, *Electronics Letters* **23**, 421-422.
4. R. Calvani, R. Caponi, F. Cisternino, (1988) Polarization phase shift keying: a coherent transmission technique with differential heterodyne detection, *Electronics Letters* **24**, 642-643.
5. S. Benedetto, P. Poggiolini, R. Calvani, R. Caponi, (1990) Performance evaluation of polarization shift keying modulation systems, OFC '90, San Francisco (CA), 22-26 January, Tech. Dig. Vol. 1, WWM33, 134.
6. S. Betti, F. Curti, G. DeMarchis, E. Iannone, A. Trifiletti, (1990) Experimental antipodal Stokes parametzrs shift keying transmission system at laser linewidth exceeding the bitrate, ECOC'90, Amsterdam (NL), 16-20 September, Proc. Vol. 1, 353-356.
7. L. G. Kazowsky, (1986) Impact of laser phase noise on optical heterodyne communication systems, Journal of Optical Communications, **7**, 66-78.
8. R. Caponi, R. Calvani, G. Marone, P. Poggiolini, (1989) Optical heterodyne communications with polarization modulation, OE/FIBERS'89, Boston (MA), 5-8 September, SPIE Proc. Vol. 1175, 136-153.
9. R. Calvani, R. Caponi, F. Delpiano, G. Marone, (1991) An experiment of optical heterodyne transmission with polarization modulation at 140 Mbit/s bitrate and 1.55 µm wavelength, GLOBECOM'91, Phoenix (AZ), 2-5 December, 1587-1591.

IX Optical Sensors

FIBER SENSOR REVIEW[†]

D.A. JACKSON
Physics Laboratory,
The University of Kent, Canterbury, Kent, CT2 7NR, U.K.

1. Introduction

It is well known that optical techniques offer extremely high precision in metrological applications, particularly if the technique is based upon optical interferometry. Unfortunately, it has been very difficult to apply these high precision techniques outside the laboratory as environmental perturbations tend to cause rapid misalignment of the optical components. Another factor that has also limited the application of optics outside the laboratory is the requirement of a 'line of sight' between all the optical components making it difficult to achieve remote sensing. This situation has changed dramatically during recent years with the incorporation of fibre optic waveguides into sensors, where the fibre optic waveguide can act as the sensor or simply serve as a transceiver link to transfer light to a remote sensor; the remote sensor being fabricated either from the fibre or from conventional optical components [1].

2.1. FIBRE OPTIC INTERFEROMETERS (FOI)

Fibre optic variants of most of the classical optical interferometers have been demonstrated. Of these configurations, the Mach Zehnder, Michelson, Fizeau/Fabry-Perot and Sagnac[*], (Figure 1), show the most promise as basic sensing elements. The transduction mechanism common to all these FOI when used as sensors is that the measurand induces an optical phase change which is detected as a change in intensity in the interferometer's output signal. Of the possible measurands that can induce a phase change in the fibre, strain and temperature produce by far the largest effects. For a typical fibre illuminated with a source at ~800nm - typical of a laser diode - the strain and temperature sensitivities are 10^7 rad m^{-1} and 10^2 rad K^{-1} m^{-1} respectively. With appropriate signal processing, phase resolutions in excess of 10^{-6} rads/$\sqrt{Hz}$ can be detected. [3] Lower sensitivity differential (or polarimetric) interferometers based upon the bimodal properties of the optical waveguide have also been developed, the sensitivity of these interferometers is typically two orders of magnitude less.

[†] Some of the material presented here has been taken from "Recent Progress in Monomode Fibre Optic Sensors" to be published in Measurement & Science Technology 1994

[*] The fibre optic gyroscope is not discussed in this review - a complete review of this sensor is given in reference [2]

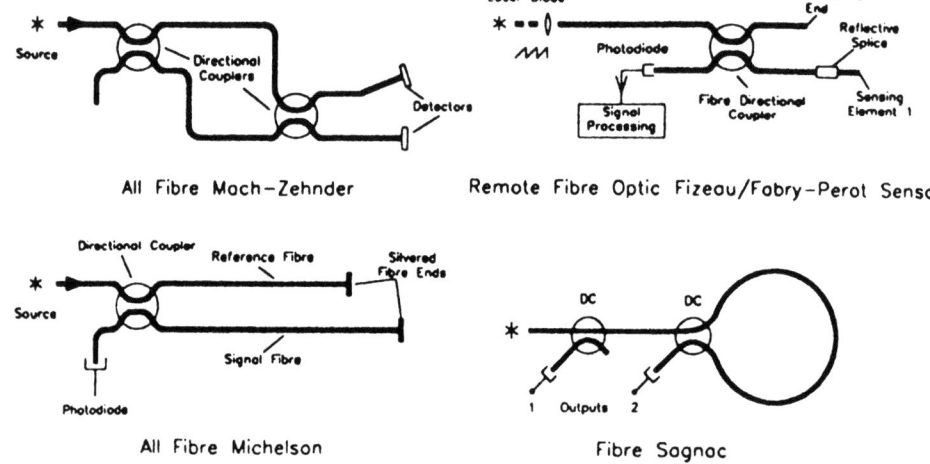

Figure 1.
Fibre Optic Interferometers, local and remote configurations. In the Michelson interferometer Faraday rotator mirrors [4] can be used to replace the mirrored fibre ends to prevent signal fading, output port 1 of the fibre Sagnac interferometer provides the reciprocal output.

2.2. SIGNAL PROCESSING

The output signal I from a 2 beam interferometer can be written
$$I = C(1 + V \cos \phi(t)) \qquad (1)$$
where C is a constant including such factors as the conversion efficiency of the photodetector, V is the fringe visibility and $\phi(t)$ is the time dependent differential phase between the two arms of the interferometer. The periodic nature of this transfer function tends to make signal recovery more complex than from a conventional electronic sensor with a linear transfer function, such as a strain gauge for example. For displacement resolutions of the order of the wavelength of the source, (i.e 0.3µm for a source wavelength of 0.6µm) simple 'fringe counting' will suffice - providing the fringe direction is known - if this is not the case then additional optical components are required in order to generate a second output from the interferometer which is out of phase by 90° (i.e. a quadrature signal). To obtain higher sensitivity it is necessary to determine the instantaneous phase of the interferometer. In a conventional interferometer this is usually accomplished by incorporating some form of high bandwidth frequency shifter such as an acousto-optic modulator in one of the arms of the interferometer, such that heterodyne processing may be used. Phase resolutions of $\sim 10^{-6}$ rads/$\sqrt{\text{Hz}}$ can be achieved, with the additional advantage that the 'fringe direction' can also be determined, eliminating the directional ambiguity associated with equation (1).

2.3. APPLICATIONS OF FOI FOR PERIODIC MEASURANDS

2.3.1. *Acoustic Pressure Sensing - Hydrophone*

The main interest in acoustic sensors has been in the development of novel types of hydrophone. [5] [6] Most hydrophones have been based on the Mach-Zehnder type of interferometer in which the sensing arm is mounted on a mandral fabricated from metal or plastic and 'potted' in an elastomeric compound such that a good acoustic impedance match between the aqueous medium and the fibre sensor is achieved, whilst at the same time being sufficiently robust to survive the large hydrostatic pressure encountered when deployed. The reference arm is deployed inside the body of the hydrophone such that it is not subject to the pressure fluctuations. Special coatings have been applied to the reference fibre to minimise its pressure sensitivity. [7] Sensitivities of 10 dB below sea state zero (acoustic noise floor in the sea) with a dynamic range of 60 db have been reported for these new sensors. The flexible nature of the fibres allows a large variety of geometrical configurations such that the optical fibre hydrophone can be designed to act as a point, gradient, or directional device and readily incorporated into towed arrays. [8]

2.3.2. *Accelerometers*

a) Compliant cylinder design

Apart from inertial guidance applications accelerometers tend to be mainly used in vibrational studies where the relative amplitudes of the spectral components in the signal must be measured.

As the fibre exhibits very high strain sensitivity [9] there has been considerable interest in designing new forms of accelerometers based upon a fibre sensing element, where the fibre forms part of the spring in a 'mass loaded spring' accelerometer. In the first accelerometer of this type the mass was simply supported by the fibre which also formed part of a fibre interferometer. Although the sensitivity of this arrangement was high it is not very practical because of the high sensitivity to orthogonal accelerations. A more practical approach is shown in Figure 2(a) where the fibre is wrapped under tension around a compliant mandral loaded by the seismic mass. In the original design [10] only one mandrel was used, an improved design by Gardner [11] used two mandrels in a Push-Pull arrangement. A major advantage of the compliant cylinder fibre combination is that it effectively operates as a mechanical amplifier with the amplification factor being directly dependent on the number of fibre turns. Sensitivities of $> 10^{-9} g \sqrt{Hz}$ have been demonstrated by this type of optical accelerometer, which are difficult to achieve with any conventional accelerometer, a typical response [12] is shown in fig 2(b). Possible applications for such high resolution accelerometers are in geophysical surveying, earthquake studies and microgravity studies on space shuttles.

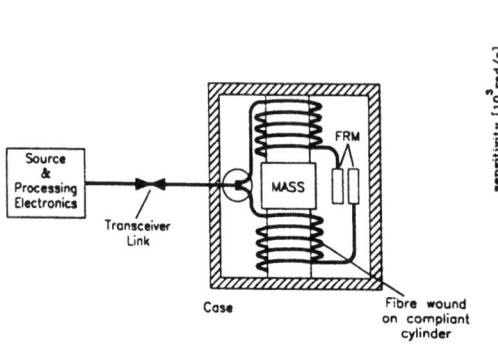
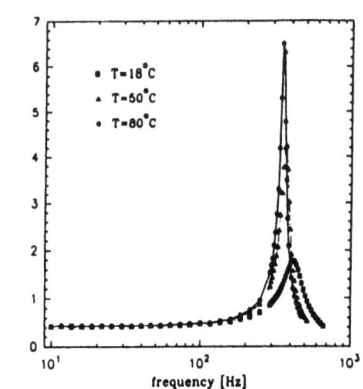

Figure 2a.
Compliant cylinder accelerometer, based upon a fibre Michelson interferometer incorporating FRM's to prevent polarization fading, operational ranges > 10Km are possible, with $10^{-9} g \sqrt{Hz}$ sensitivities.

Figure 2b.
Shows the variation of the sensitivity of a prototype high temperature compliant cylinder accelerometer as a function of frequency at various temperatures.

b) Diaphragm based accelerometers

A rather different design for an optical accelerometer [13] is based upon a taut diaphragm loaded with a central mass, this also acts as a mass loaded spring accelerometer where the spring is the taut diaphragm. This accelerometer is shown in Figure 3 and is based on a hemispherical Fabry-Perot cavity formed between the distal end of the fibre transceiver link and a very low cost metal mirror mounted centrally on the diaphragm acting as the mass. The sensitivity of this accelerometer is $\sim 10^{-7} g/\sqrt{Hz}$. One of the major advantages of this accelerometer is that the all metal construction makes it both extremely rugged and well suited for high temperature operations.

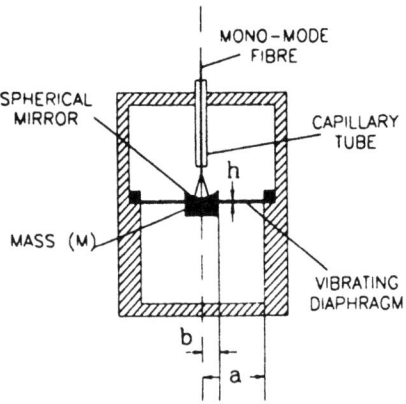

Figure 3.
Simple mass 'loaded spring' accelerometer based upon a hemispherical Fabry Perot type cavity capable of operating at very high temperatures.

2.3.3. *Temperature*

The very large thermo-optic coefficient of a typical optical fibre (induced phase change/°C makes it an ideal thermometer or heat flux sensor when the induced phase change is measured interferometrically. Given a modest optical resolution of only 10^{-9}m, temperature changes of less than a millidegree can be determined using a 5mm long fibre Fabry Perot type sensor. Most applications for temperature sensors require that the output is always unambiguous, hence low coherence processing appears to be the optimum method of for signal recovery.

There are several applications for fibre optic temperature sensors in which it is unnecessary to determine the absolute phase, for example in the study of transient phenomena. Miniature Fibre Fabry Perot/Fizeau (FFP/F) probes have been demonstrated with bandwidths in excess of 200 KHz [14] which makes it ideally suited for the study of fast phenomena such as very fast chemical reactions and heat diffusion studies. Sensors of this type have been used for heat diffusion studies in turbine blades [15]. Other periodic measurands such as **strain, displacement** and **force** can also be measured using simple Fabry Perot/Fizeau probes.

2.3.4. *Flow*
Contact - Vortex Shedding

Vortex shedding occurs in obstructed turbulent flows; the well know phenomenon of 'whining' telephone lines in strong winds is an example of Vortex shedding - the vibration frequency of the obstructing body is directly related to the flow velocity. In a conventional vortex shedding flow meter, the movement of this body is measured by a piezo-electric element. As the movement of the body is relatively small, the resulting electrical signal is similarly small. It is possible to use a fibre optic strain sensor to measure the vortex shedding frequency - either directly by using the fibre as the obstruction (in gas flows) [16] or indirectly by interrogating the movement of a conventional vortex shedder, where the motion of the shedder is measured externally to the flow. The performance obtainable with optical interrogation is comparable to that achievable with a conventional system, however, the optical system has the advantage that it is not affected by electromagnetic interference and can also operate at much higher temperatures than is possible with vortex shedders using a piezoelectric element [17].

2.4. LOW COHERENCE OR WHITE LIGHT INTERFEROMETRY

Low coherence interferometry is a very powerful technique for the measurement of absolute optical path difference. In a typical fibre optic sensing system two interferometers are operated in tandem [1] where the sensor is deployed remotely and the receiving interferometer (RI) is located in a benign environment. The system is illuminated with light from a source with a very short coherence length (C_L) ideally $C_L < 10\mu m$. The sensor is then designed such that the OPD is greater than the source coherence length hence no optical interference will be seen at the sensor output. When this optical signal is injected into the receiving interferometer optical interference is observed for the condition

$$|\text{OPD sensor} - \text{OPD}_{RI}| \leq C_L.$$

For a typical low coherence source the envelope of the autocorrelation function has a Gaussian or Lorentzian profile hence the visibility of the interference signal decreases very rapidly. This rapid reduction in fringe visibility can be exploited to locate a fiducial point in the transfer function of the tandem interferometer. The OPD of the sensor can then be determined, for example by identifying this fiducial point by scanning the OPD of the RI, accuracies better than 10nm have been achieved by these techniques [18]. Closed loop operation is also possible and similar accuracies have been achieved [19]. Two wavelength techniques [20] can also be applied to this problem, if the maximum OPD change of the sensor is less than C_L, the advantage of this being that the scanning range of the RI is only ~ $\lambda/2$. In many ways this approach represents a universal measurement concept for any measurand which can be transduced to a displacement in the range 0 ~50 μm.

The general advantages of low coherence interferometry when applied to fibre optic sensor systems are:

(i) Capability of determining the absolute phase of the sensor when the system is switched on, enabling the system to be used for quasi static measurands such as temperature, pressure, etc.

(ii) Sensor systems can be operated such that :
   (a) the measurement accuracy is virtually independent of the source stability, or
   (b) the effects of wavelength instabilities are greatly reduced

(iii) The optical sensors can be fabricated with very short optical cavities ~50μm - which tends to greatly reduce the effects of measurand cross-sensitivity, this is particularly important for instruments such as precision pressure sensors for example.

2.4.1. *Applications for Low Coherence Interferometry (LCI)*
LCI is very well suited for the recovery of quasi-static measurands detected using short FFP/F cavities, particularly when combined with dual wavelength sensing techniques as bandwidths up to ~500 Hz are possible.

2.4.2. *Temperature Sensor*
As discussed in section 2.3.3. FFP/F cavities are ideally suited for temperature measurement, the sensing element, basically a short length of monomode fibre is formed by simply cleaving both ends of the fibre. For low temperature applications the short fibre sensing cavity can be bonded to the transceiver fibre with a suitable adhesive. However a better approach is to fusion splice the sensing fibre, to the transceiver fibre, in this case it is necessary that the ends of the cavity are coated to ensure reflection occurs at the interface. Metals cannot be used because the arc will 'track down' the coating causing it to evaporate, ordinary dielectric coatings are also unsuitable as they are damaged by the high temperature. Successful reflective splices have been reported for fibres coated with evaporated TiO_2 film [21,22]

2.4.3. *Pressure Sensor*
Pressure sensors based upon Fabry Perot/Fizeau cavities formed between the distal face of the fibre and a diaphragm or similar pressure have also been designed which offer very high performance. Miniature optical pressure sensor have recently been demonstrated for medical [23], process control [24] and very high pressure applications [25]. In

Figure 4 is shown a pressure sensor suitable for process control applications based on low coherence interferometry, a fibre optic temperature sensor is included to enable errors in the measurement of the position of the diaphragm due to thermal expansion to be minimised.

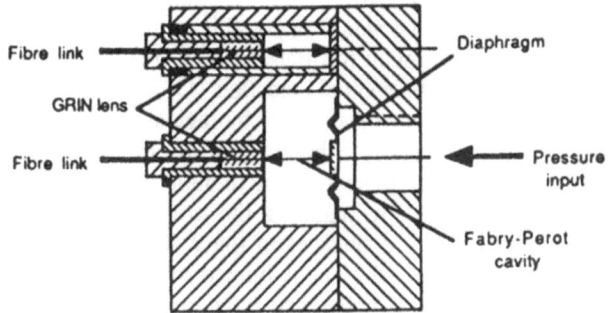

Figure 4.
Prototype optical pressure sensor based on a diaphragm and an external Fizeau type cavity formed between the diaphragm and the distal face of the fibre. Processing is via low coherence interferometry, a fibre optic temperature sensor is included to enable temperature compensation.

2.4.4. *Optical Coherence Domain Reflectometry* (OCDR)

The positions of very weak reflecting surfaces in a (semi) transparent sample can be located by using the sample as one of the mirrors of a Michelson interferometer, the other mirror driven at a frequency of f_r is mounted on a motor driven translation stage. The interferometer is illuminated by a low coherence source. The positions of the

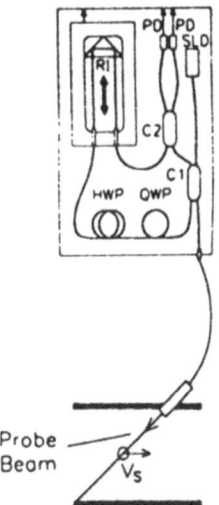

Figure 5.
Schematic diagram of a distributed fibre optic laser Doppler veolicmeter (DLDV), based upon low coherence interferometry. The sampling point in the flow V_s is controlled by the OPD of the RI.

reflecting surfaces are then located by varying the OPD of the Michelson whilst monitoring the output signal at f_r. The spatial resolution reported was ~10μm with a dynamic range of 100dB. [26] Remote measurements are possible by incorporating optical fibres into the interferometer. A distributed LDV system [27] has been developed based on this concept, Figure 5; here the fibre is placed in the flow and the position where the velocity is measured in the flow along the path of the probe beam, is governed by the OPD in the RI. A possible application for this system is the measurement of blood flows where it would be possible to move the measurement point away from the stagnation area around the fibre tip. A very large dynamic range is achieved with this system using differential detection. OCDR is also being developed for medical applications, for example the depth of biological tissues [28] and the length and positions of the lens and cornea of the eye [29,30] are being examined by this technique. It is also possible to use this technique to build up images of objects buried in highly scattered materials [31] by scanning the sample in the x-y plane whilst measuring the position of reflecting surfaces in the Z direction (Figure 6). At the present time the application of the technique is limited by the relatively long scanning time; if this can be reduced, optical coherence reflectometry (or tomography) could replace X-ray techniques in the diagnosis of cancer in soft tissues.

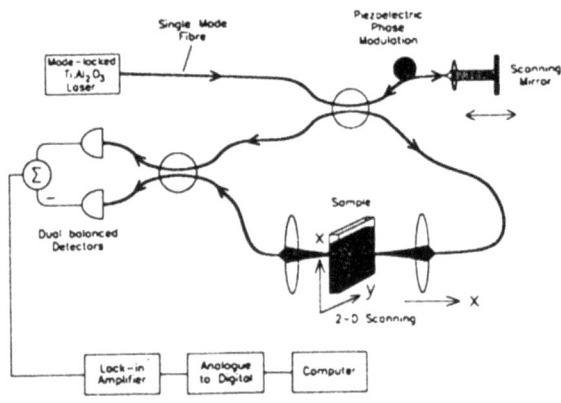

Figure 6.
Schematic diagram of a transillumination tomographic system based upon low coherence interferometry.
The sampling position along the Z direction is determined by the position of the scanning mirror.

3. Current and Magnetic Field Sensors Exploiting the Faraday Effect

3.1. CURRENT SENSORS

The possibility of using the Faraday effect to measure large currents at high potential was identified very early in the development of fibre optic sensors [32]. The Faraday effect is the rotation of the polarization azimuth ϕ_F which occurs when an optical beam propagates through a medium subject to a magnetic field. Despite the nearly 20 years of research virtually no commercial exploitation has occurred. The major advantages of using optical techniques in current measurement, compared with conventional current transformers, may be summarised as follows:

1. Passive dielectric sensor which can be addressed via optical fibre links giving highly effective isolation from high line potentials

2. Freedom from the saturation effects which can occur in a conventional transformer

3. Optical signal not corrupted by electro-magnetic interference

4. Highly linear response over a wide frequency range

5. Remote, high speed measurements for monitoring or metering purposes, and

6. Compact and light weight measuring devices at low cost.

In order to use the Faraday effect for current measurement it is necessary that the action of the sensor is to perform the following integral equation,

$$\phi_F = \int_{L_F} V_{(\lambda T)} \bar{H} \, dL_F \quad \text{along a closed optical path,} \quad (2)$$

Where $V_{(\lambda T)}$ is the material dependent Verdet constant, which is both dispersive and temperature dependent and L_F the interaction length.

If the integral is performed along a closed optical loop subject to the condition that the state of polarization of the light remains constant then by Ampéres law, ϕ_F is directly related to the enclosed current and will not be affected by the relative position of the current with respect to the optical sensor or magnetic fields outside the optical sensing loop [33].

For optical current sensors (OCS) that use optical fibre as the sensing element, this condition is generally not met due to the presence of the intrinsic birefringence induced by core ellipticity and asymmetric stress, or the extrinsic birefringence caused in deploying the fibre sensing element. These problems can be ameliorated to some extent by annealing the fibres to reduce the level of bend induced linear birefringence[34]. An alternative approach is to fabricate the sensor from high Verdet constant birefringence free bulk glass. As the light is not guided the sensor incorporates reflecting surfaces in order to force the light to propagate in a closed loop without changing its state of polarization. This has been achieved using dual quadrature reflections [35] or critical angle reflection [36]. Sensors exploiting critical angle reflection can be fabricated in an operable format allowing it to be deployed without interrupting the current [37].

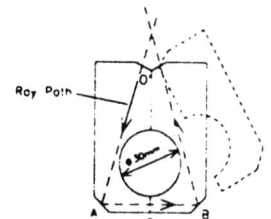

Figure 7.
Optical current clamp based upon the Faraday effect, optical fibre links are used to enable remote operation.

3.2. FARADAY EFFECT MAGNETIC PROBE SENSORS

Miniature high speed magnetic field probes have been developed based upon the Faraday effect. The simplest method of processing is to implement the system as a polarimeter, however if any mechanical drifts occur then the sensitivity will become time dependent. An alternative approach is to use heterodyne processing. The basic concept of the magnetic probe, utilizing heterodyne signal processing, is shown in Figure (8). The probe is illuminated via a laser source oscillating at two different frequencies, ω_1 and ω_2, in orthogonally polarized modes. These two independent outputs from the laser are then transferred to the probe via a high-bi fibre with the light at frequencies ω_1 and ω_2 being guided in different orthogonal modes of the fibre. At the probe head a reference frequency ($\omega_1 - \omega_2$) is generated by combining the two beams at polarizer (1) with its transmission axis at 45° to the eigenmodes of the hi-bi fibre. The remaining part of the input signal is coupled into the sensing element via a $\lambda/4$ plate where the two linear states are transformed into circular states. These states then transverse the sensor and are reflected at the mirror and retrace their paths back through the sensor. After passing through the $\lambda/4$ plate and polarizer(2), these states are converted to a linear state at ($\omega_1-\omega_2$) where the phase depends on the instantaneous value of ϕ_F. If the applied magnetic field is periodic then the amplitude of the phase deviation is proportional to the amplitude of the field and the rate of deviation, its frequency., i.e. the resultant signal is a phase modulated carrier at $\omega_1-\omega_2$ and can be detected with a spectrum analyzer. If the hi-bi link is subject to environmental noise the effects of this noise may be eliminated, to first order, by comparing the modulated signal with the reference signal using a phase meter. Using Hoya Fr-5 glass sensitivities of 7×10^2 nT have been achieved [38]. This heterodyne technique is particularly effective for measuring small amplitude high frequency magnetic fields and could be used for probes based upon Ga:YIG films which have demonstrated $100\text{pT}\sqrt{\text{Hz}}$ sensitivities at 500Hz with bandwidths up to 1GHz [39].

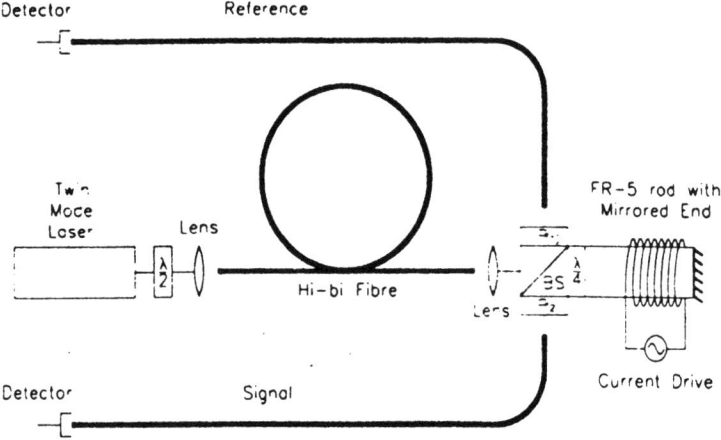

Figure 8.
Remote magnetic field probe based upon Faraday rotation and a common path optical fibre heterodyne interferometer.

4. Distributed Sensors

Amongst the hierarchy of sensors of all types i.e. conventional and optical, is the distributed fibre sensor as it offers continuous sensing over tens of kilometers with good accuracy and high spatial resolution. Measurands which have been addressed to date are temperature and strain.

In the last ten years, several different schemes have been demonstrated for realising a distributed temperature sensor (DTS) with an optical fibre as the sensing element. The evolution of the state of polarisation along the fibre [40] the temperature dependences of the Rayleigh back-scattering coefficient [41] and the ratio of Stokes to anti-Stokes intensities for Raman scattering [42] have all been used as the sensing mechanism. Commercial systems are available based upon Raman scattering, using a multimode fibre for the sensor in conjunction with a high power pulsed laser source, a temperature resolution of 1-2°C with a spatial resolution of 1-2m over distances of up to 8km have been reported [43].

4.1. RAMAN DTS

When a laser beam is injected into a fibre it is scattered by spontaneous fluctuations in the dielectric constant of the medium, these fluctuations are caused by the fundamental transport mechanism in the medium arising from molecular vibrations, (Raman scattering), propagating density fluctuations, (Brillouin scattering) and non propagating entropy fluctuations, (Rayleigh scattering). The Raman shift frequency is relatively large such that a simple monochrometer can be used to separate the back scattering Raman signal from the strong laser signal which arises from either the Brillouin and Rayleigh scattered signals or from impurities or reflections occurring in the system, the latter usually gives rise to much stronger signals. In Raman scattering the frequency of the light is both up and down shifted by the molecular vibrations. The ratio of these intensity components, known as Stokes and Antistokes lines is directly related to the absolute temperature of the scattering region. In order to measure the temperature of the fibre at a specific location the laser must be pulsed. The location in the fibre is then determined by the time taken from injection of the pulse to detection of the signal, which equals $2n\,l/c$ where n is the refractive index l is the physical position in the fibre and c the velocity of light. The spatial resolution is equal to $\tau/2$ where τ is the pulse width. A Raman DTS system is shown in Figure 9.

4.2. BRILLOUIN BASED DISTRIBUTED SENSORS

The Brillouin frequency shift depends upon temperature and strain thus provided the effects of these measurands can be separated, it offers advantages when compared with Raman scattering. In principle spontaneous or stimulated Brillouin scattering could be used as the basic interaction mechanism, however as the frequency shift is relatively small (~12GHz at 1.3 μm), the high resolution instrumentation required for such a system would be both expensive and insufficiently robust for commercial applications.

Recently, a related technique based on the variation with temperature of the centre frequency of the Brillouin gain curve, has been demonstrated by Kurashima et al.

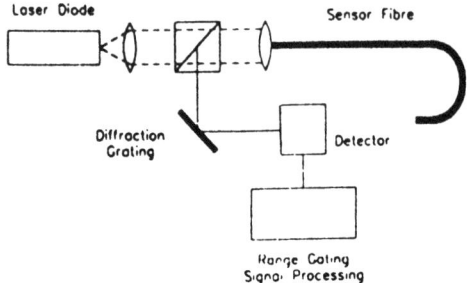
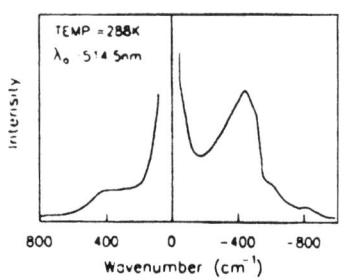

Basic System   Raman Backscattering Signal

Intensity ratio of Stokes to AntiStokes $= \left(\frac{\lambda_s}{\lambda_0}\right) e^{-\frac{h\nu}{\kappa T}}$ Range 1 - 8km

Accuracy claimed $\pm 1°C$ Spatial Resolution ~ 1- 5m

Figure 9.
Distributed temperature sensor baseng in multimode fibre. The inset shows the Raman spectrum of quartz.

[1990]. They reported a temperature resolution of 3°C, combined with a spatial resolution of 100m over 1.2km of fibre.

The technique operates as follows: Light from a cw laser is launched into one end of the monomode sensing fibre, while short pulses of light from a pulsed laser are launched into the other end. When the pulsed laser frequency is greater than that of the cw laser by the Brillouin frequency shift, the cw laser will be amplified, experiencing Brillouin gain. The Brillouin frequency shift depends on the fibre temperature; so, when the fibre temperature is non-uniform, the probe will experience gain for a given laser frequency difference only in those parts of the fibre at a specific temperature.

If the intensity of the cw beam emerging from the fibre is monitored following the launch of a pump pulse, an increase in the intensity will be observed whenever Brillouin gain occurs. The time delays between the launch of the pump pulse and these increases in the received probe signal, correspond to round-trip times for light travelling to and from the regions of gain. These times provide the positional information. If the laser frequency difference is adjusted, then the probe will experience gain in parts of the fibre at a different temperature. Therefore by slowly scanning one of the laser frequencies the temperature distribution along the fibre length can be mapped out.

Using this technique a temperature resolution of 1°C with 5m spatial resolution for a sensing length of more than 22km. [45] has been demonstrated.

Very recently it has been shown that by monitoring the Brillouin loss rather than the gain the range can be further increased; here the roles of the lasers are reversed such that when their frequency difference equals the Brillouin frequency power is coupled into the pulsed beam[46]. A range of 51km has been achieved with 1°C temperature resolution and 5m spatial resolution. This system is shown in Figure 10. Brillouin scattering is sensitive to temperature and strain, hence it is feasible to make simultaneous measurements of temperature and strain [47].

Applications for distributed temperature sensors include monitoring the temperature of tunnels, natural gas pipe lines, fire alarms for ships, museums etc., distributed strain sensors will find application in monitoring the structural integrity of dams, buildings in areas subject to earthquakes and general environmental monitoring.

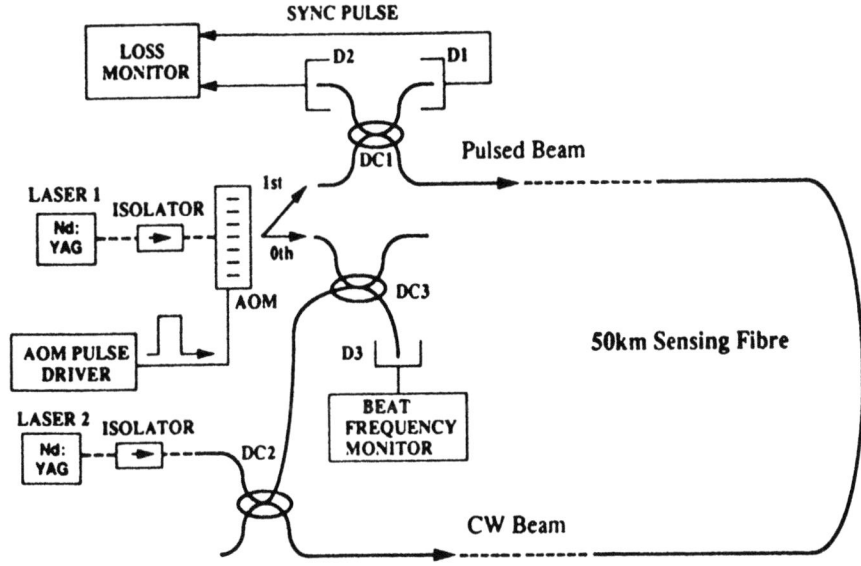

Figure 10.
Schematic diagram of a distributed temperature sensor based upon Brillouin loss in monomode fibre at 1.3μm. Sensing lengths >50Km have been achieved.

5. Quasi Distributed Fibre Sensors

Quasi distributed sensors are also being developed in which a large number of discrete sensors can be implemented at arbitrary locations on a single fibre.

5.1. HIGHLY BIREFRINGENT FIBRE SYSTEMS

Quasi distributed sensors have been developed which exploit the bimodal nature of the optical fibre. The sensor operates as follows light is injected into the fibre such that only one of its eigenmodes is populated. At a desired sensing point a mechanical device, is used to apply a force to the fibre which causes light to be coupled into another mode of the fibre. Coupling between orthogonal LP_{01} modes [48] of highly birefringent fibre and the LP_{01} & LP_{11} modes [49] of standard fibres has been exploited. As the propagation constants of the fibre modes are different there is a phase difference between the original beam and the coupled beam at the output face of the fibre. Hence if these beams are combined via a polarizer interference effects will be observed where the amplitude of this signal is proportional to the coupling between the eigenmodes and hence the force (or pressure) applied to the coupling unit. Signals from other arbitrary locations can be generated in the same manner. When there are several coupling units along the fibre it is necessary to be able to separate the resultant signals. This may be achieved using a receiving interferometer (RI) as indicated in Figure 11 where the sensing fibre is now illuminated with a low coherence source. If the induced optical path difference between the uncoupled and coupled beams at location N is different by at least the coherence length of the source from locations N-1 and N+1 (and similarly for all coupling points), then separate interference patterns will be observed at the output of the RI when the OPD is equal to OPD_{N-1}, OPD_N, OPD_{N+1} etc. Although this technique is promising the number of sensors that can be deployed is limited by spurious signals arising due to multiple cross coupling.

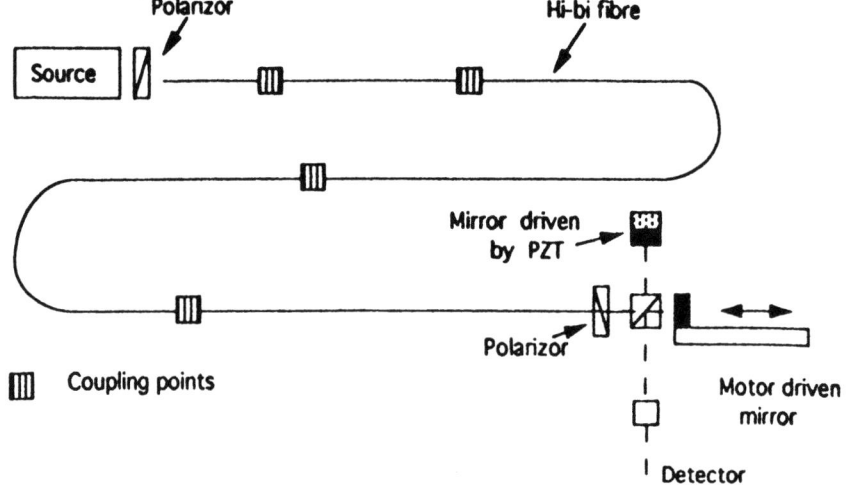

Figure 11.
Schematic diagram showing a quasi distributed sensor based upon coupling between the orthogonal modes of 'Hi-bi' fibre using a RI to demodulate the output signal.

5.2. BRAGG GRATING SENSORS

One of the most exciting developments in the field of fibre optics has been the introduction of the Bragg grating reflector. These reflectors can be 'written' in the fibre using a holographic system (illuminated with an Eximer laser operating at ~260nm) to generate a periodic intensity profile [50]. The fibre is illuminated from the side and a permanent grating is induced by the photorefractive effect. Dependent parameters such as the central reflecting wavelength, the reflectivity and the bandwidth of the grating can be selected. These new components can be used for numerous applications in optical communications and in optical sensors. Although these grating can be used as point sensors they are attracting considerable interest for application as sensing elements in quasi-distributed sensors, as the grating sensors can be written point by point in series at any arbitrary location along a fibre without inducing any changes in the fibre dimensions. The sensitized fibre can be readily incorporated into complex structures [51], such as composites, without compromising the structural integrity. In this type of application the gratings will be written at specific positions along the fibre such that when it is incorporated into the structure each sensing element will be located at a critical monitoring point.

For capital items such as passenger aircraft, a large number of sensors will be required. In order to interrogate and demultiplex the output signals from a serial array of grating sensors, it is necessary that each sensor can be uniquely identified from the wavelength at which it has maximum reflectivity. This can be achieved in principle if each grating sensor has a specified working range $\Delta\lambda_{1s}$, $\Delta\lambda_{2s}$, ...$\Delta\lambda_{ns}$, where $\Delta\lambda_{1s} = (\lambda_1 - \lambda_2)$, $\Delta\lambda_{2s} = (\lambda_2 - \lambda_3)$, and $\Delta\lambda_{ns} = (\lambda_{n-1} - \lambda_n)$, the instantaneous central reflecting wavelength of each grating sensor is within the specified range $\Delta\lambda_{1s}$, $\Delta\lambda_{2s}$, ...$\Delta\lambda_{ns}$,etc. If now the system is illuminated with a broadband source with a bandwidth $\geq(\lambda_1 - \lambda_n)$, then the backreflected signal will consist of (n-1) frequency components, where the central wavelength of each component is directly related to the number of lines per millimeter of each grating reflector and hence to the current value of the measurand that it is designed to monitor.

Measurement of a large number of wavelengths with high precision is possible with costly instruments such as an optical spectrum analyzer; however, this is not feasible in a practical application owing to its size and weight and the frequent need to recalibrate the instrument. An interrogation scheme based on a Mach-Zehnder interferometer acting as a wavelength discriminator was reported by Kersey [52]. This approach offers extremely high performance for a single sensor but is not easy to use for multiplexing a large number of sensors without incorporating time-division multiplexing. Recently a novel approach for multiplexing the output signals from a network of grating sensors, that permits virtually simultaneous interrogation of all the sensors, that is based on matching a receiving grating to a corresponding sensor has been reported [53]. This system is shown in Figure 12. A somewhat similar approach has been adopted where a high finesse Fabry Perot [54] is used sequentially to measure the centre wavelength of each sensor. Microstrain resolutions have been achieved by both approaches.

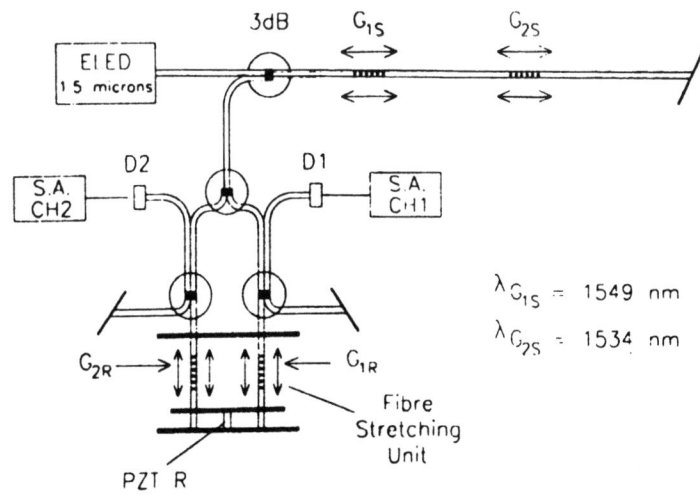

Figure 12.
Simple system to multiplex and demultiplex Bragg grating sensors based upon matched sensor-receiver grating pairs.

6. Conclusions

Fibre optic based measurement technologies have become very diverse enabling them to be applied to a variety of disciplines such as basic science, engineering, aerospace, medicine etc. The simplest implementation in which the fibre serves as a transceiver link between a remote external sensor and a central processing unit is likely to grow in importance because the power of a wide range of conventional optical laboratory instruments can be exploited away from the benign laboratory environment. The ability to implement miniature external sensors for pressure and temperature measurements clearly offers sensing opportunities ranging from minimally invasive diagnostic procedures in the body to monitoring heat transfer in a rocket motor at 1000°C. Although commercial exploitation is relatively slow for most optical fibre sensors the fibre optic gyroscope will undoubtedly go into commercial production in the near future; fully developed sensors such as the hydrophone are also in this category despite being developed for military purposes. The unique sensing capabilities of very long range distribution sensors are likely to be exploited in fire protection and structural monitoring applications. Possibly the most significant innovation in the whole field of fibre optic based sensors is the invention of the Bragg grating which can now be written into the fibre before the primary coating is applied, heralding the availability of very low cost, customized sensor networks.

REFERENCES

1. Jackson, D.A. and Jones, J.D.C. (1989) Interferometers, in Dakin and Culshaw (eds), *Optical Fibre Sensors*, Artech House Publishing, pp. 329-376.
2. Lefevre, H.C. (1993) *The Fibre Optic Gyroscope*, Artech House, Boston.
3. Jackson, D.A. (1985) Monomode optical fibre interferometers for precision measurement, *J.Phys.E: Sci Instrum, Instrument Science and Technology* 18, 981-1001.
4. Martinelli, M. (1989) A Universal Compensator for Polarization changes induced by birefringence on a retracing beam, *Optics. Comms.* 72, 341-344.
5. Bucaro, J.A., Dardy, H.D. and Carome, E.F. (1977) Fiber-Optic Hydrophone, *J. Acoust. Soc.* 62, 1302.
6. Cole, J.H., Johnson, R.L. and Bhuta, P.G. (1977) Fiber optic detection of sound, *J. Acoust. Soc. Am.* 62, 1136.
7. Bucaro, J.A. (1986) Optical fiber sensor coating, in Chester, A.N. Martellucci and Scheggi, A.M. (eds), *Proceedings of the Nato Advanced Study Institute on Optic Fiber Sensor Eric*, Martinus Nyhoff Dordrecht, Netherlands, pp.321.
8. Yurek, A.M. (1992) Status of fiber optic acoustic sensing, *Proc. 8th OFS Conference Monterey, Ca.*, 338-341.
9. Hocker, G.B. (1979) Fiber-optic sensing of pressure and temperature, *Applied Optics* 18, 1445.
10. Kersey, A.D., Jackson, D.A. and Corke, M. (1982) High-sensitivity fibre optic accelerometer, *Electron. Letts.* 18, 559-561.
11. Gardner, D.L., Hofler, T., Baker, S.R., Yarber, R.K. and Garrett, S.L. (1987) A fiber-optic interferometric seismometer, *JLWT* LT5, 953-960.
12. Pechstedt, R-D, Webb, D.J. and Jackson, D.A. Optical fibre accelerometers for high temperature applications, Proc. SPIE Conf. Boston, 2070,352-359.
13. Gerges, A.S., Newson, T.P., Jones, J.D.C. and Jackson, D.A. (1989) High-sensitivity fiber-optic accelerometer, *Opt. Letts.* 14, 251-253.
14. Farahi, F., Jones, J.D.C., and Jackson, D.A. (1991) High-Speed Fiber-Optic Temperature Sensor, *Opt. Letts.*, 16 1800-1802.
15. Kidd, S.R., Sinha, P.G., Barton, J.S. and Jones, J.D.C. (1992) Utilization of Fibre Fabry-Perot Interferometers in the Determination of Heat Transfer Transients in Wind Tunnels, *Proc. 8th OFS, Monterey, Ca.*, pp.73-76.
16. Akhavan Leilabady, P., Jones, J.D.C., Kersey, A.D., Corke, M. and Jackson, D.A., (1984) Monomode fibre optic vortex shedding flowmeter, *Electron. Letts.* 20, 664.
17. Chu, B.C.B., Newson, T.P. and Jackson, D.A. (1991), Fibre Optic Based Vortex Shedder Flow Meter, ECO'4, The Hague, 1504, pp.251-257.

18. Bosselman, T. (1987) Optical Fiber Sensors in Chester, A.N. Martellucci and Scheggi, A.M. (eds), *Proceedings of the Nato Advanced Study Institute on Optic Fiber Sensor Eric*, Martinus Nyhoff Dordrect, Netherlands, pp.429-432.
19. Gerges, A.S., Farahi, F., Newson, T.P., Jones, J.D.C. and Jackson, D.a. 91987) Interferometric fibre optic sensor using a short coherence length source, *Electron. Letts* **23**, 1110-1111.
20. Kersey, A.D. and Dandridge, A. (1987) Dual wavelength approach to interferometric sensing, in Scheggi, A.M. (ed), *Fiber Optic Sensors, II*, SPIE, Hague **798**, pp.176-181.
21. Lee, C.E., Taylor, H.F., Markus, A.m. and Udd, E., (1989) Optical Fiber Fabry-Perot Embedded Sensor, *Opt. Letts* **14**, 1225-1227.
22. Inci, M.N., Kidd, S.R., Barton, J.S. and Jones, J.D.C., (1993) High temperature miniature fibre optic interferometric thermal sensors, *Meas. Science and Tech* **4**, 382-387.
23. Rao, Y.J. and Jackson, D.A. (1993) Prototype Fibre-Optic based Fizeau medical pressure sensor using coherence reading, *Opt. Letts.* **18**, 2153-2155.
24. Rao, Y.J. and Jackson, D.A. (1993) Prototype fibre-optic based pressure probe with built-in temperature compensation with signal recovery by coherence reading, *Appl. Opt.* **32**, 7111-7113.
25. Rao, Y.J. and Jackson, D.A., (1993) Prototype fiber-optic-based ultrahigh pressure remote sensor using dual-wavelength coherence reading, *Electron. Letts.* **29**, 2142-2143.
26. Youngquist, R.C., Carr, S. and Davies, D.N. (1987) Optical coherence domain reflectometry: a new optical evaluation technique, *Opt. Letts.* **12** 158-160.
27. Gusmeroli, V. and Martinelli, M. (1991) Distributed Laser Doppler Velocimeter, *Opt. Letts* **16** 1358-1360.
28. Chivaz, X., Marques-Weddle, F., Salathe, R.P., Novak, R.P. and Gilgen, H.H. (1992) High resolution reflectometry in biological tissues, *Opt. Letts* **17**, 4-6.
29. Swanson, E.A., Huang, D., Hee, M.R., Fujimoto, J.G., Lin, G.P. and Puliafito (1992) High speed optical coherence domain reflectometry, *Opt. Letts.* **17** 151-153.
30. Chen, S., Gratton, K.T.V. and Palmer, A.W., (1993), A compact optical device for eye length measurement, *Photonic Tech. Letts.* **5**, 729-731.
31. Hee, M.R., Izatt, J., Jacobson, M.J., Fujimoto, J.G. and Swanson, E.A. (1993) Femtosecond transillumination optical coherence tomography, *Opt. Letts.* **18** 950-952.
32. Rogers, A.J. (1977) Optical methods for measurement of voltage and current on power systems, *Optics and Laser Technology*, 273-283.
33. Bush, S.P. and Jackson, D.A., (1992) Numerical Investigation of the effects of Birefringence and Total Internal Reflection on Faraday Effect Current Sensors, *Applied Optics*, **31** 5366-5374.
34. Day, G.W., (1989), Recent advances in Faraday effect Sensors in Arditty, H.J., Dakin, J.P.and Kersten, R.Th. (eds) Proc. of 6th Int. Conf. OFS'89, Paris France, Springer Verlag, Berlin, *Proc. in Physics* **44**, 250-254.
35. Sato, T., Takahashi, G. and Inui, Y. (1983) European Patent Application 0088419
36. Chu, B.C.B., Ning, Y.N. and Jackson, D.A. (1992) Faraday current sensor that uses a triangular-shaped bulk-optic sensing element, *Opt. Letts.* **17**, 1167-1169.
37. Ning, Y.N. and Jackson, D.A. (1993) Faraday effect optical current clamp using a bulk-glass sensing element, *Opt. Letts.* **18**, 835-837.
38. Bartlett, S.C., Farahi, F. and Jackson, D.A. (1990) Current Sensing using Faraday rotation and a common path optical fiber heterodyne interferometer, *Rev. Sci. Instrum*, **61** 2433-2435.
39. Day, G.W., Deeter, M.N. and Rise, A.h. (1991) Faraday effect sensors: a review of recent progress in Culshaw, B., Moore, E.L. and Zhipeng, Z. (eds) *Advances in Optical Fiber Sensors*, Wuhan, China. SPIE Optical Engineering Press, 11-26.
40. Rogers, A.J. (1980) Polarisation optical time domain reflectometry, *Electron. Letts* **16**, 489-490.
41. Hartog, A.H. (1983) A distributed temperature sensor based on liquid-core optical fibres, *I.E.E.E. JLWT*, **LT-1** 498-509.
42. Dakin, J.P., Pratt, D.J., Bibby, G.W. and Ross, J.N. (1985) Distributed optical fibre Raman temperature sensor using a semi-conductor light source and detector, *Electron. Letts.* **21** 569-570.
43. Dakin, J.P. Distributed Optical Fibre Sensors in Dakin, J.P. and Kersey, A.D. (eds) *Distributed and Multiplexed Fiber Optic Sensors II*, Proc. SPIE 1797, Boston, pp.76-108.
44. Kurashima, T., Horiguchi, T. and Tateda, M. (1990) Distributed-temperature sensing using stimulated Brillouin scattering in optical silica fibers, IOpt. Letts. **15**, 1038-1040.
45. Bao, X., Webb, D.J. and Jackson, D.A. (1993) 22-km distributed temperature sensor using Brillouin gain in an optical fiber, *Opt. Letts*, **18**, 552-554.
46. Bao, X., Webb, D.J. and Jackson, D.A. (1993), 32-km distributed temperature sensor based on Brillouin loss in an optical fiber, *Optics Letts*. **18**, 1561-1563.
47. Bao, X., Webb, D.J. and Jackson, D.A. (1994), Combined Distributed Temperature and Strain Sensor using Brillouin Loss in an Optical Fibre, *Optics Letts*.**104**, 298-302.
48. Chen, S. and Giles, I.P. (1990) Optical coherence domain polarimetry: intensity and interferometric type for quasi-distributed optical fibre sensors, *Proc SPIE Conference 1370* p.217. Fibre Optic Sensors Structures and Skins III(1989) San Jose Ca.
49. Kotrotsios, G. and Parriaux, O. White light interferometry for distributed sensing on dual mode fiures. *Proc. 6th Int Conference of Optical Fibre Sensors* p.568, Paris France.

50. Meltz, G., Morey, W.W. and Glenn, W.H. (1989) Formation of Bragg gratings in optical fibers by a transverse holographic method, *opt. Letts.* **14**, 823-825.
51. Dunphy, J.R., Meltz, G., Lamm, F.P. and Morey, W.W. (1990) Multifunction distributed optical fiber sensor for composite cure and response monitoring, in Udd, E and Claus, R.O. (eds) *Fiber Optic Smart Structures and Skins III*, **1370**, pp.116-118.
52. Kersey, A.D., Berkoff, T.A. and Morey, W.W. (1992) Fiber-grating based strain sensor with phase sensitive detection, *Proc. of 1st European Conf. on Smark Structures and Materials*, Glasgow, pp.61-67.
53. Jackson, D.A., Lobo Ribeiro, A.B., Reekie, L. and Archambault, J.L. (1993) Simple Multiplexing Scheme for a fibre-optic grating sensor network, *Opt.Letts.* **18** 1192-1194.
54. Kersey, A.D., Berkoff, T.A. and Morey, W.W., (1993) Multiplexed fiber Bragg grating strain sensor system with a fiber Fabry-Perot wavelength filter, *Opt. Letts.* **18**, 1370-1372.

FIBRE OPTIC SENSORS: Potential, applications and state of the art of the technology

M.R.H. Voet, GLÖTZL, Rheinstetten, Germany and IDENTITY, Mol, Belgium, Y. Verbandt, L. Boschmans, IDENTITY, H. Thienpont, Free Brussels University, F. Berghmans, SCK/CEN, Mol/ Free Brussels University

1. Abstract:

This paper briefly introduces of the early years of optical fibre communications and the evolution of fibre optic technology. A short overview of the fibre optic principles is presented and the events leading to the commitment to optical fibres sensors are highlighted. The advantages of fibre optic sensors are discussed in some specific applications, namely Long-term Observation of physical parameters in the Civil Engineering Industry, such as an underground laboratory, a copper mine or a large hydroelectric power station. Finally, the subsequent developments in European R&D programmes leading to the realisation of fibre optic sensors as well as distributed sensor systems are summarised.

2. Introduction :

For several years, fibre optic sensing devices have been only used for straightforward on/off functions such as presence and position. More recently, they gained interest because fibre optic sensors represent a novel exciting technological challenge for a multitude of sensing applications. Most physical properties can be sensed optically with fibres, e.g. light intensity, displacement, temperature, pressure, rotation, sound, strain, magnetic field, electric field, flow, liquid level, chemical analysis, and vibration.

Glass fibre seems to be the logical alternative carrier to copper in plant monitoring and process control because of its high data rates, immunity to electromagnetic interference, safety in flammable or explosive atmospheres, resistance to high temperature and chemically induced corrosion and hence endurance in hostile environments. The potential of using glass for long term sensing purposes is obvious. Has glass not already survived for centuries, even without significant composition and property changes ?

The potential benefits of the use of fibre optic sensing systems have, however, not yet led to a major breakthrough and greater acceptance of this technology in the present industrialised world. The principle obstacle to the increased use of fibre optics for sensing purposes is clearly identified as being the cost. At present, fibre based sensing and monitoring systems still remain considerable more expensive than conventional sensing, and acquisition systems. Furthermore, there is still no technical standard for fibres, connectors and related equipment. But the most important parameter is that most engineers are unaware of the fibre optic opportunities and the many industrial applications. The new functionalities of optical fibre sensors and their integration in large networks are the two main reasons why this class of sensors have their place in the industrial world.

If fibre optic sensing is to establish itself as a fundamental tool for environmental and industrial measurement, the technology has to be transfered from the laboratory prototype to the real applications. The outline of this contributions is as follows. We give a brief overview of optical communications and fibre optics basics in chapters 3 and 4. Three examples which are representative for the broad classes of optical fibre sensors, are explained in chapter 5. We will discuss an intensity modulated sensor (microbend optical fibre piezometer), an interferometric sensor (polarimetric pressure sensor) and a spectrally modulated sensor (extrinsic Fabry-Perot sensor) in detail. In chapter 6, some concepts and systems for the multiplexing of these sensors are presented.

In conclusion, we present some examples of practical applications 'in the field'. The diversity of applications in civil engineering is emphasized and some aspects of the practical installation are explained.

3. In the beginning there was light [1]

The beginnings of optical communications go back to antiquity: Various realisations, such as heliograph or a semaphore, or even smoke signals are embodiments of optical communications. However these early ideas had, as we know, some very serious limitations, were very unreliable, time consuming and had an inadequate capacity for carrying information.

There are accounts of studies of light propagation along a jet of water as far back as 1820. In 1870 Tyndall proved that light could be catched into cylindrical tubes filled with water. Also Graham Bell reported on experiments with transmission of speech signals along a light beam in 1880.

The theory of propagation of electromagnetic waves along dielectric cylinders, the principle behind modern fibre-optic communication, was well established at the beginning of this century. Especially the work of Hondros and Debeye should be mentioned. After the invention of the laser, the optical fibre communications developed rapidly. The availability of various sources (Light Emitting Diodes, Superluminescent Diodes, Diode Lasers, ...) and detectors for various wavelenghts as well as the development of the interconnection of fibres, sources and detectors were important steps to enable us to install and maintain fiber optic installations on practically the same level as copper wiring techniques.

Historically, sensing applications using fibre optics were oriented toward the very simple sensors, such as card readers for computers and outage indicators to determine blown out of light. Technology has taken fibre optic sensors to the other extreme. Very sophisticated sensors using interferometric techniques are in development. Compared to conventional displacement sensors these devices have a 4 to 5 orders of magnitude higher resolution. In general such high sensitivities are not required for most applications except for military use. Most industrial sensors fall somewhere in between these extremes.

Finally, multiplexed, networked and distributed optical fibre sensors are far more attractive to develop, for both logistic and economic reasons, than implementing a large number of separate point sensors.

4. Basic Fibre Optic Principle [3],[4],[9],[10]

The propagation of light into optical fibres is explained very simply by the phenomenon of total reflection of electromagnetic rays when falling on a layer with a lower value of the refractive index. If the incidence angle is larger then a certain critical value, the light is totally reflected and stays confined in the core of the fibre (see also figure 1 [2]).

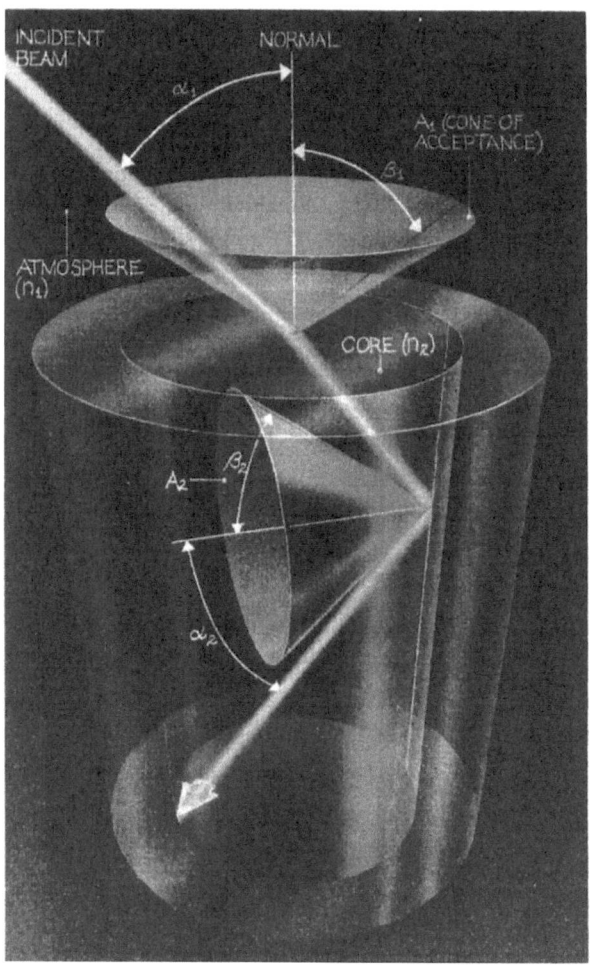

Fig. 1 : Propagation of light into optical fibres

In practice, optical fibres are made of two concentric cylinders of glass with a slightly higher index of refraction for the central part or core. Because of the different indexes, light transmitted in the fibre bounces off the interface between the different glasses and is retained in the core.

Figure 2a is showing a step-index multimode fiber, with a typical core diameter of 50 micron and a cladding (outside diameter) of 125 micron. Because the index difference between the core and the cladding glasses is a few percent and the core diameter is large compared to the optical wavelength, different modes can propagate at different angles and velocities through the fibre. This step-index fibre (the index of refraction differs in a abrupt step from core to cladding) exhibits strong mode dispersion which will limit the bandwidth of the signal being carried through the fibre.

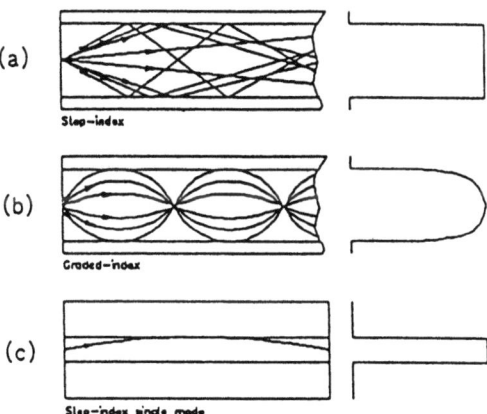

Fig. 2 : Three basic types of optical fibre

To avoid the mode dispersion the graded-index fibre (the index difference between the core and cladding glasses has a parabolic profile) has been developed (see also fig. 2b)

The high order modes moves through a longer path than the low angle rays but the glass in the path for the high order modes has a lower index of refraction than for the axial rays. The net effect is an actual optical path of about the same length for both high- and low order modes. We therefore have very low dispersion and this fibre may be used to carry very high bandwidth data over considerable distances.

Finally figure 2c shows a step index single mode fibre which has a very small core diameter (about 5-7 micron, at near infrared optical wavelength) and will therefore support only one mode.

This fibre has practically zero mode dispersion. In reality this fibre supports two orthogonally polarized modes which are degenerated and propagated with the same phase velocity. Environmental effects, such as temperature and mechanical distortions of the fibre, will influence the output polarization state. The polarization state of the light will change randomly in time due to external perturbations to the fibre. This badly affects the signal-to-noise ratio of coherent transmission systems. In order to enhance this signal-to-noise ratio, highly-birefringent fibres were developed to maintain the linear polarization state of the light during propagation, even in the presence of perturbations. Fig. 3 shows a typical HiBi fibre, called Panda fibre together with 2 other types of HiBi fibres nl. Bow-tie and Elliptical optical Jacket. The two rods apply a thermal stress to the core, which makes the latter anisotropic.

Due to this indexanisotropy, the two orthogonally polarised guided modes are no longer degenerated and propagated with a different phase velocity. This difference, called modal birefringence is sensitive to external perturbations and can be used for sensing mechanical perturbations to the fibre (see section 5).

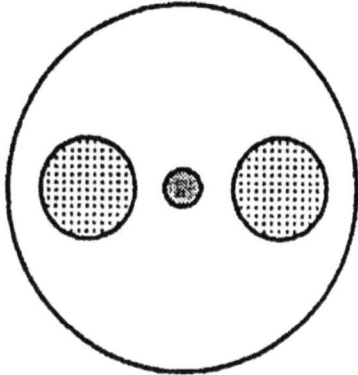

Fig. 3a : Panda Fiber

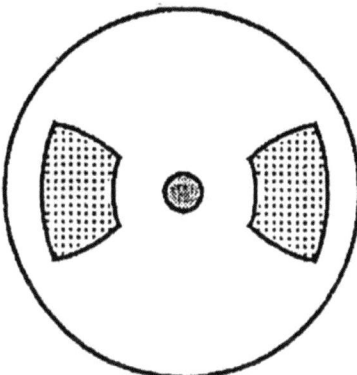

Fig. 3b : Bow-tie

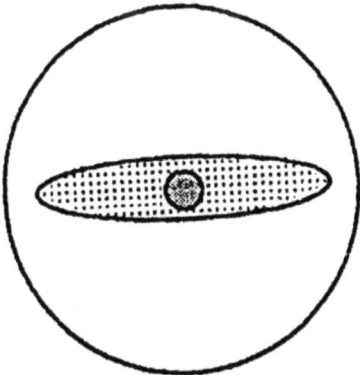

Fig. 3c : Elliptical optical Jacket

5. Fibre Optic Sensing

5.1. INTRODUCTION

Optical fibre sensors can be divided into two major categories : intrinsic and extrinsic sensors. Intrinsic fibre sensors are optical fibre devices where the measurand through a transducer mechanism induces a modulation of the light propagation characteristics in the fibre. Thus, the light does never leave the fibre. The signal is a measurement of the changes in fibre geometry, material properties, etc... induced by the external perturbation to be monitored. Extrinsic sensors use the optical fibre only as lead-in and lead-out of the optical signal. The light is coupled out of the fibre to be modulated in free space by the measurand. It is then coupled back into a lead-out fibre. This type of design is mostly used for chemical spectral analysis.

In this chapter, we will discuss two types of intrinsic sensors, i.e. a microbend and a polarimetric sensor and one type of extrinsic spectral modulation sensor.

5.2. INTENSITY MODULATED FIBRE OPTIC SENSORS

To explain in a more practical way the fundamental working principles of the fibre optic sensing technology, we know that an optical fibre can sense material and environmental characteristics by modulating light propagated in it.

For sake of simplicity, a fibre optic sensing system can be schematised as follows: a light source, a guidance fibre and a detector. (see also figure 4)

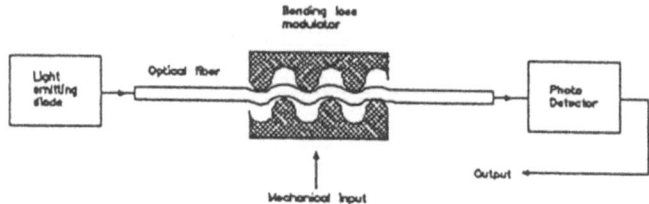

Fig. 4 : Microbend Optical Fibre Sensor

In between a fibre optic intensity sensor can be placed where measurand causes a change in light propagation characteristics. Once modulated, the light is transmitted to detectors, which interpret the information carried. Modulation may alter the intensity, frequency, phase or polarization of the light.

For our purposes we will focus on the intensity modulated fibre optic sensor principle, e.g. the microbend sensor of figure 4 and the GLÖTZL microbend fibre optic piezometer in figure 5.

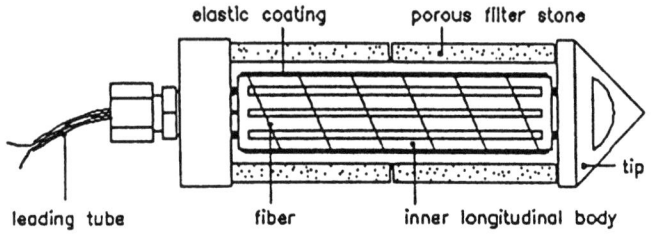

 Glötzl

Fig. 5 : Microbend Fibre Optic Piezometer

As we know that for every fiber there exists a critical angle at which the mode will fail to be guided in the core, but will radiate into the cladding. External mechanical forces will change the local radius of fiber curvature, consequently changing the amount of light reaching the optical detector. This microbending technique [6] used to monitor the loss in optical fibres has been proposed to monitor the deformation of an elastic coating or membrane into the recesses of an inner longitudinal body fixed in a pressurized housing.

Furthermore when we consider a idealized microbend sensor where the sensing fibre is sandwiched between deformer plates or a membrane , the transmission coefficient for light propagating through the bent fiber will change following Lagakos, Cole and Bucaro [7],

$$(1) \quad \Delta T = \frac{\Delta T}{\Delta X} \cdot A_p \left(K_f + \frac{A_s.Y_s}{l_s} \right)^{-1} \cdot \Delta P$$

where ΔP is the change in pressure... cf. p.14

Therefore when requiring a high sensitivity pressure sensor the ratio of $A_s.Y_s/l_s$ should be very small so that the compliance between brackets will be determined by the bent fibre force constant K_f which is in fact quite large.

In this case the equation becomes

$$(2) \quad \Delta T = \frac{\Delta T}{\Delta X} \cdot A_p K_f^{-1} \Delta P$$

On the same hand the above equations could be used to sense temperature as for a temperature sensor a coefficient alpha will be introduced for the thermal expansion of the spacer or membrane.

Therefore equation (1) becomes

$$(3) \quad \Delta T = \frac{\Delta T}{\Delta X} \cdot A_s \alpha_s Y_s \left(K_f + \frac{A_s.Y_s}{l_s} \right)^{-1} \cdot \Delta \theta$$

where $\Delta \theta$ is the temperature change.
As in equation (2) to fulfil our requirements for a sensitive temperature sensor one should have $K_f l_s << A_s Y_s$.

In this condition equation (3) becomes simply

$$(4) \quad \Delta T = \frac{\Delta T}{\Delta X} \cdot \alpha_s l_s \Delta \theta$$

We therefore notice that in our transduction mechanism the transmission coefficient ΔT will be influenced, in relation of course to the mechanical construction of the sensor, with:

for pressure:
* the surface of the membrane A_p
* the bent fibre force constant K_f
* the pressure ΔP

for temperature:
* the surface of the membrane A_s
* the thermal expansion coefficient alpha
* the Young's modulus
* the change in temperature ΔQ

So finally the change in optical transmission through the bent fiber results in a change in optical energy incident on the photodetector. After linearity and hysteresis measurements, the change in the output detector signal can be memorized and the environmental parameter can in principle be presented directly in mbar or in hPa on a display unit.

In practice the above mentioned single sensor system consist of " a cascade coupling " [5], where the real sensor is connected with a fiber delay coil or reference coil, all in the same housing.(see also figure 6).

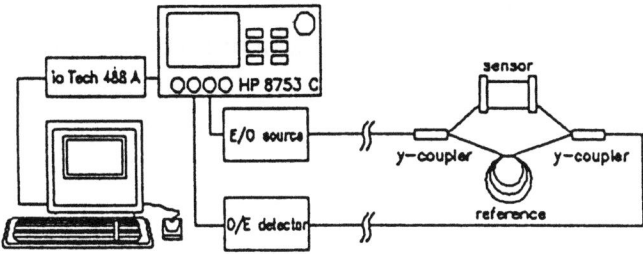

Fig. 6 : Test set-up

The reference coil can be used to compensate for temperature and other environmental parameters, such as LED ageing, lead fibre influences, etc. Also as we will see further on in the network system, this reference coil can act as a delay line for passive multiplexing.

5.3. POLARIMETRIC FIBRE OPTIC SENSORS

It has been established that polarisation maintaining (PM) highly birefringent (HiBi) single mode (SM) optical fibres can be used to measure high hydrostatic pressures up to 200 MPa. The measuring principle is based on the influence of the hydrostatic pressure on the birefringence of the HiBi fibre. The application of axial strain instead of hydrostatic pressure influences the birefringence in exactly the same way.

If the two modes of the HiBi fibre are equally excited, i.e. by coupling linearly polarised light in the fibre at an angle of 45° with respect to the two polarisation axes of the HiBi fibre, the phase difference $\Delta\Phi$ between the two polarisation modes (vacuum wavelength λ) after a propagation length L is given by

$$\Delta\Phi(p,T) = \frac{2\pi L}{\lambda} \cdot \Delta n(p,T)$$

where the difference between the refractive indices of the fast and slow axis Δn is the only parameter which is effectively influenced by the applied hydrostatic pressure p, but also by the temperature T. If we succeed in eliminating the influence of temperature on the response of the sensor, it is clear that the simple measurement of the polarisation state of the light coming out of the device can be related to the hydrostatic pressure of the environment. In order to achieve temperature desensitisation of the device an equal length of the same fibre is added to the sensing part, but spliced at 90° with respect to the polarisation axes of the sensing element. If this second segment is kept at the same temperature of the sensor but at atmospheric pressure the influence of the temperature is cancelled out. This can be seen as follows : the phase differences depending of the temperature in both segments are the same in magnitude but have opposite signs because the 90° splice has exchanged the fast and slow axes of the two HiBi fibre segments (fig.7).

Polarimetric Sensor Prototype

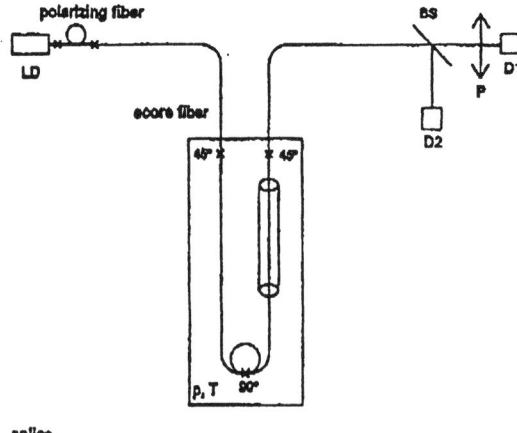

Fig. 7 : Polarimetric Sensor Prototype

The fibre used in this prototype is Corning elliptical core HiBi fibre. This type was chosen for its good polarization maintaining properties (Beat length < 1.5 mm) and its favorable price compared to other commercially available HiBi fibres. Figure 8a show a typical response of this prototype. We are currently investigating the long-term stability of the prototype.

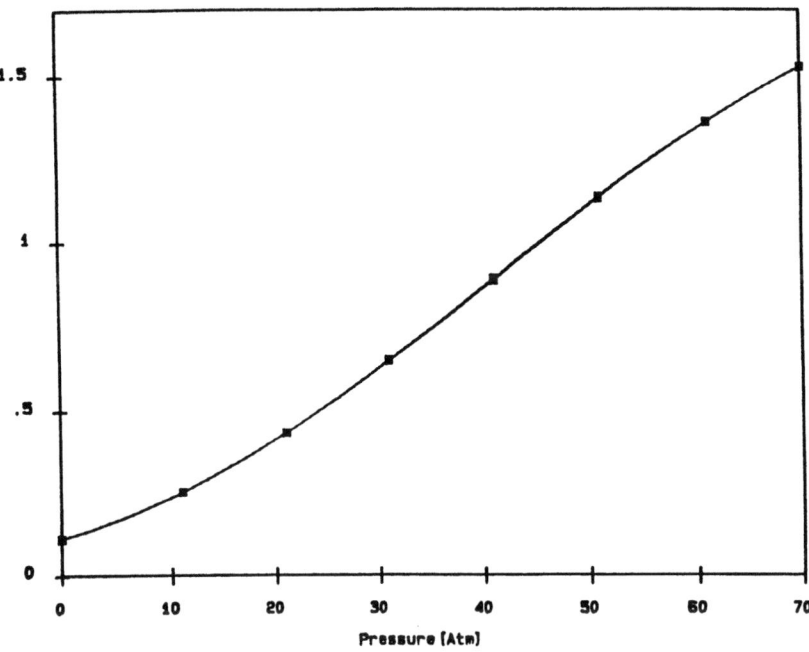

Fig. 8a : Response of the prototype

It can be seen from numerical finite-element modelisation that the application of a suitable coating can enhance the pressure sensitivity (+ 10%) while decreasing the temperature sensitivity. The technological design considerations which have to be studied very carefully are the precision of the splicing angles, the protection of the splices without influencing the polarization properties of the fibre and the insertion of the sensor in its housing. This insertion has to be pressure-tight for very high pressures and may not induce a temperature gradient between the sensing and the compensating part of the HiBi fibre. The very careful balancing of the two fibre lengths which is needed for this temperature compensation scheme remains a technological challenge because of the inherent features of the splicing procedure.

5.4. SPECTRAL MODULATION FIBRE OPTIC SENSORS

Figures 8b and c show the principle of a spectrally modulated optical fibre pressure cell. Light from a broad-band LED is guided through a standard multimode fibre to a Fabry-Perot cavity at the end. The rear mirror of the cavity is a Si diaphragm (fig. 8c) which deflects under the ambiant pressure. The spectral characteristics of the cavity change with a variation of optical path length (n.d, where n is the refractive index and d the physical thickness). This change is monitored by a set-up of two detectors and a dichroic beamsplitter.

This system has numerous advantages. It is composed of low cost optical fibre and components and it has a large linear measurement range as can be seen in fig.8d. The latter is due to the careful choice of design parameters :

- dichroic filter sprectral response
- LED spectral asymmetry
- resonator characteristics

This sensing principle is very versatile. It can easily be adapted for measurements of temperature, index of refraction and density of a gas. For the first, one introduces in the cavity a material with a very high refractive index change with temperature. A temperature change hence induces a modification of the optical path length. The latter two measurements are made possible by allowing the cavity to communicate with the ambient environment (fluid or gas flow).

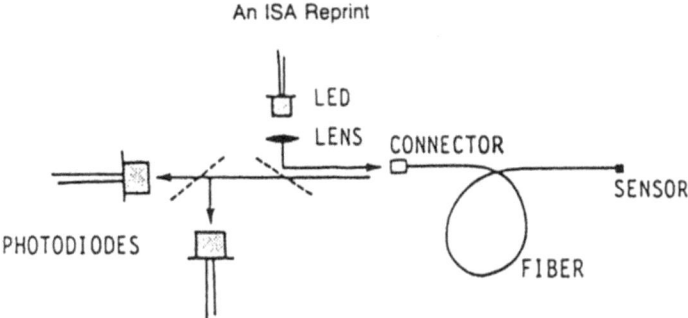

Fig. 8b : Spectral Microshift System Design

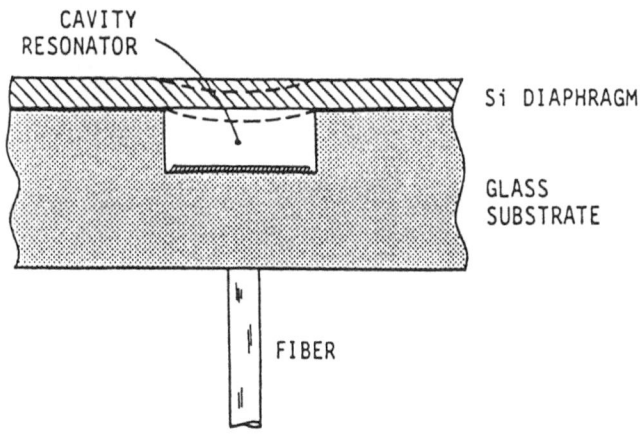

Fig. 8c : Resonant Cavity Pressure Sensor

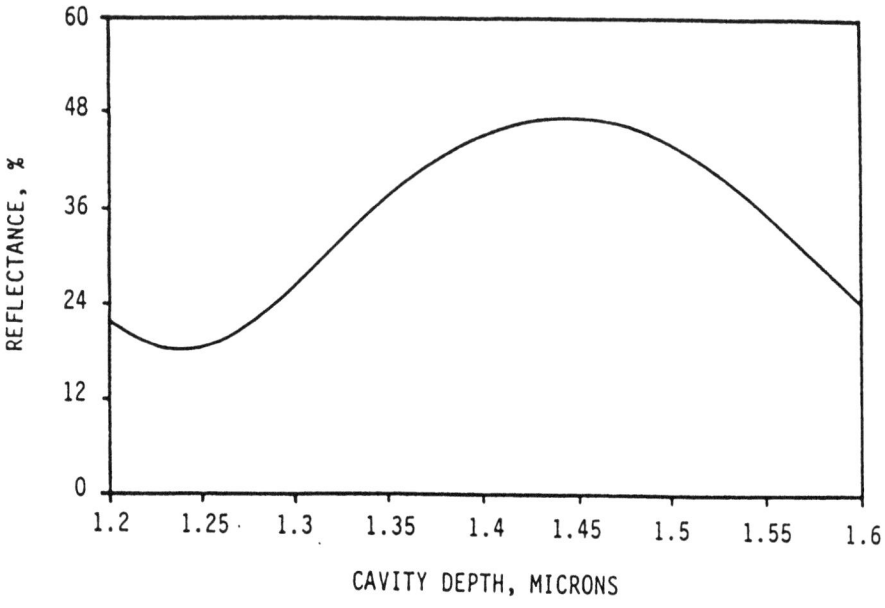

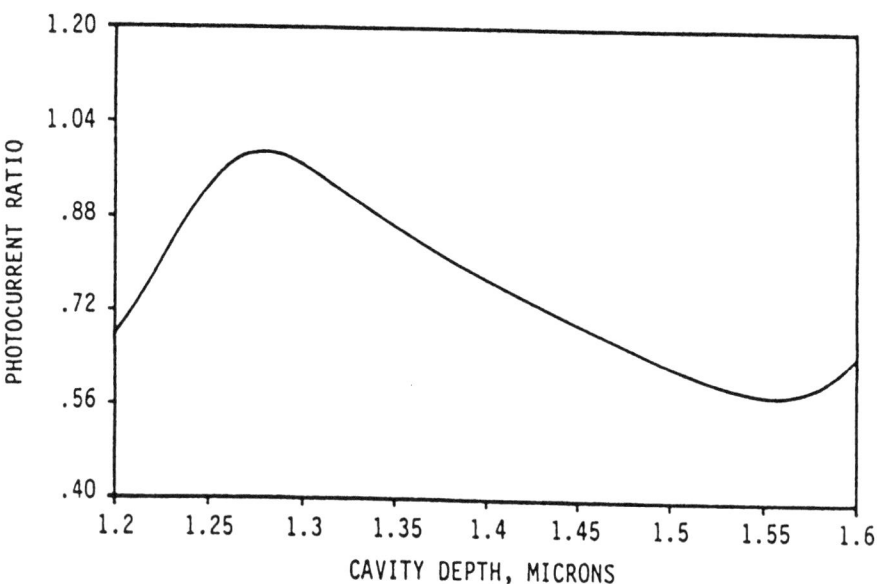

Fig. 8d : Comparison of reflectance (i.e. intensity modulated) and photocurrent ratio signals from a linearized spectral microshift pressure sensor

6. Fibre Optic Sensor Networking [5],[8]

6.1. INTRODUCTION

As many of the applications in process control require a large number of sensors spatially distributed within the area concerned, it would be very promising to integrate the above mentioned single fiber optic intensity modulated sensors into a multiple sensor system. Several multiplexing techniques have recently been developed.

As a multiplexing technique was demonstrated [5] (see also fig. 9), combining the interrogation of time division multiplexed sensors with signal processing in the frequency domain, 4 to 8 of the above microbend pressure- or temperature sensors could be implemented into the network system (see also fig. 11).

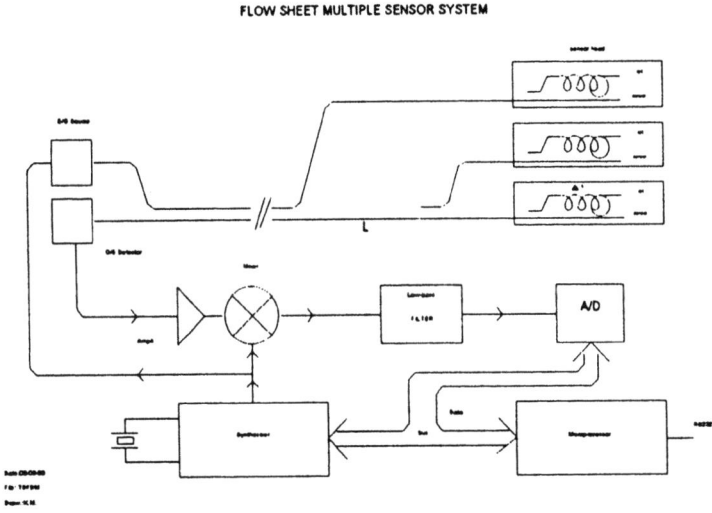

Fig. 9 : Flow Sheet Multiple Sensor System

The most straightforward way of implementing a sensor network with polarimetric sensors is in the form of a star network. For this purpose an optical switch with suitable polarization properties has been designed. See figure 10.

Polarimetric Sensor : Multiplexing

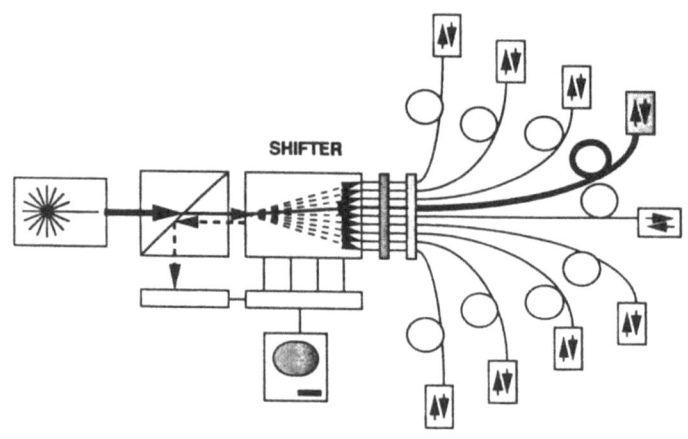

Fig. 10 : Polarimetric Sensor : Multiplexing

Multiplexing spectrally modulated fibre sensors requires, in general terms, a spectrophotometer, i.e. a wavelength sensitive device. If several sensors can be designed with different spectral characteristics (see for instance, Bragg gratings with different periods) a simple Fabry-Perot timable cavity can be used to distinguish the signals comming from them.

6.2. MULTIPLEXING SET-UP FOR INTENSITY MODULATED SENSORS

This principle of the line neutrality has already been explained in detail in ref. 5 and 6, but in a simple approach it allows us to detect and to eliminate the influence of disturbances on the lead fiber such as bendings, connector losses, temperature variations and response variations of the LED and detector. Furthermore in using a reference fiber coil in the sensor head and different fiber lengths in between several sensors in one sensor network, the system allows us to multiplex up to 8 sensors, combining the interrogation of time division multiplexed sensors with signal processing in the frequency domain. The sensor network system is therefore an all fiber set-up where LED, detector, interrogation and signal processing are all together located in one housing in the control room area. Several of these sensor control units can than be combined in one ring control unit where on their turn these ring control units can be interconnected in a fibre optic LAN (See also fig. 11).

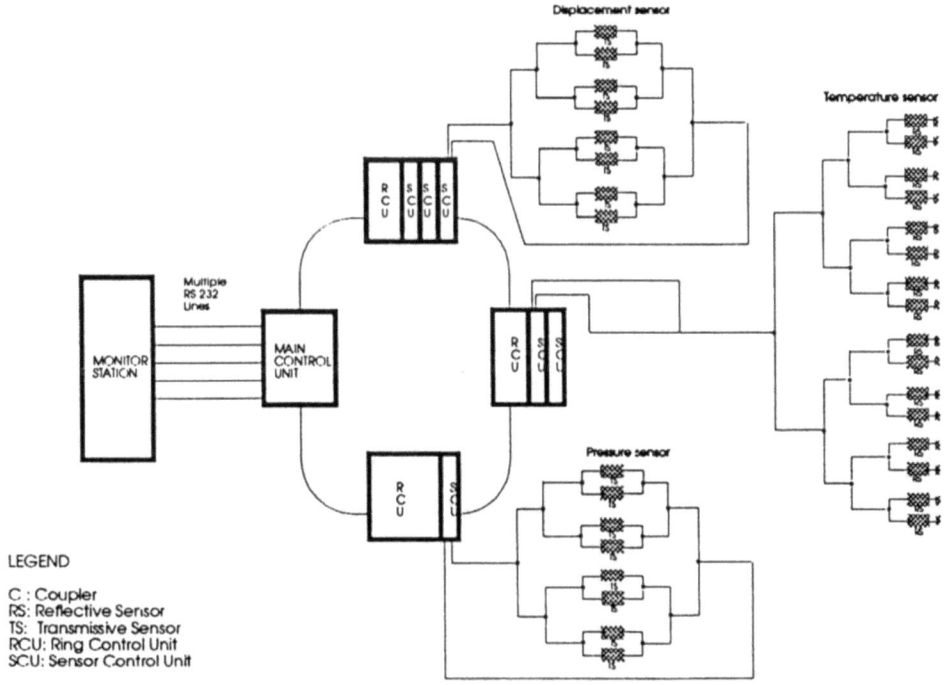

LEGEND

C : Coupler
RS: Reflective Sensor
TS: Transmissive Sensor
RCU: Ring Control Unit
SCU: Sensor Control Unit

FIBER OPTIC MULTIPLE SENSOR RING ARCHITECTURE SYSTEM

Fig. 11 : Fiber Optic Multiple sensor ring architecture system

Fibers coming out of this area can interrogate optically the different sensors at distances of up to 2 km in a passive redundant way (couplers, sensors and reference coil). The reference coil allows furthermore to detect and to check at any time in the future the initial state of the sensor even if the control unit will be changed or improved.

6.2.1. Setup of an intensity modulated network :

A typical set up of an intensity modulated network is depicted on figure 12. As can be seen, the network consists out of the following components :

- an optical source to launch the power into the network,
- star couplers to divide the power over the different sensors on the one hand, and to combine the power again at the detector side,
- mechanical connections (mechanical splices) to obtain a flexible set-up of the network,
- optical fiber delay lines,
- an optical detector (PIN, PINFET, APD) converting the optical signal into an electrical signal.

The sensor or sensor array itself consists out of (ref. box on fig. 12) :

- again star couplers, now dividing and combining the optical power to the intensity modulated sensor fibers (in this particular case microbend loops),
- the intensity modulated loops or branches which are now fusion spliced on the star couplers.

It has to be mentioned that in order to make a distinction possible between, all the branches of the whole network must have different lengths. Therefore, different delay loops are inserted at different places in the sensor network.

Optical Fiber Sensor Arrays detecting Impact and Damage.

Set up:

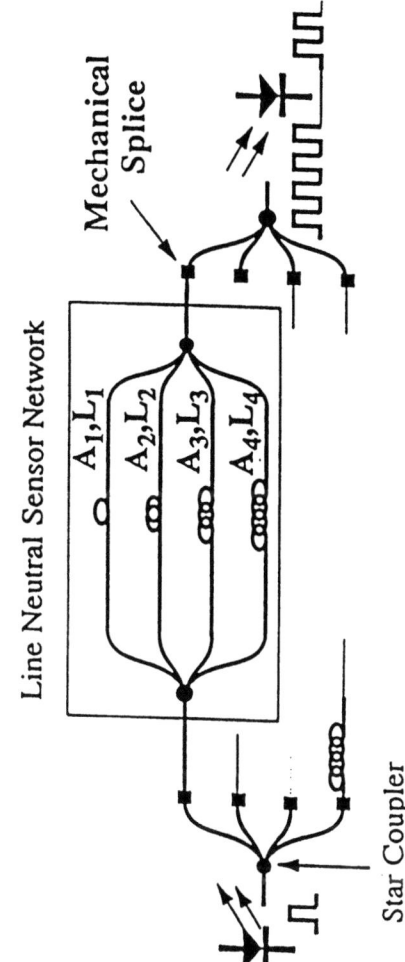

Identity EEIG

Fig. 12 : Optical Fibre Sensor Arrays detecting Impact and Dammage

6.2.2. Intuitive working principle (time domain) :

When launching a sharp pulse at the source side, a series of pulses will be obtained at the detector side due to the different lengths (indicated by L1...L4) of all the branches in the network. The height of the received pulses will be influenced by :

- the height of the launched pulse (aging...),
- the losses in the star coupler (power is divided over the branches + extra losses in the coupler).
- the losses in the mechanical connections (these losses are not constant when disconnecting and connecting again!),
- losses in the leading fibers,
- losses in the fusion splices (remain constant),
- the physical quantity acting on the sensor path (for example : the influence of the pressure on the transmission coefficient in a microbend pressure cell).

The multiplication of all these attenuations are represented on the figure by the capital letter
A1... A4.

By isolating one of the branches (let's take branche L4, A4) of the physical quantity, the height of the pulse, coming from this brance, is only determined by the attenuation coefficients of : the aging of the source, losses in the couplers, losses in the leading fibers and losses of the mechanical splices. This branche therefore serves as a branch to measure these specific attenuations, and is called from now on the reference branche or reference path. In every housing of a sensor, measuring one or multiple physical quantities, one reference path is included.

Now, when taking the ratios of the heights of the pulses form the sensor branches with repect to the height of the pulse of the reference path, a figure will be obtained which is only depending on the physical quantity acting on the sensor path; a "Line Neutral Optical Fibre Sensor (LINOFS)" is born.

On fig. 13, a measurement result is given of a sensor network, consisting out of 4 sensors (4 housings, with in every housing one sensor fiber, one reference fiber and two Y couplers) or eight branches. Pressure is applied at one of the four sensors, and the ratio "pulse height sensor path/pulse height regerence path" or s/r ratio is presented as a function of the applied pressure (in reality, not the height of the pulses is measured, as will be declared later).

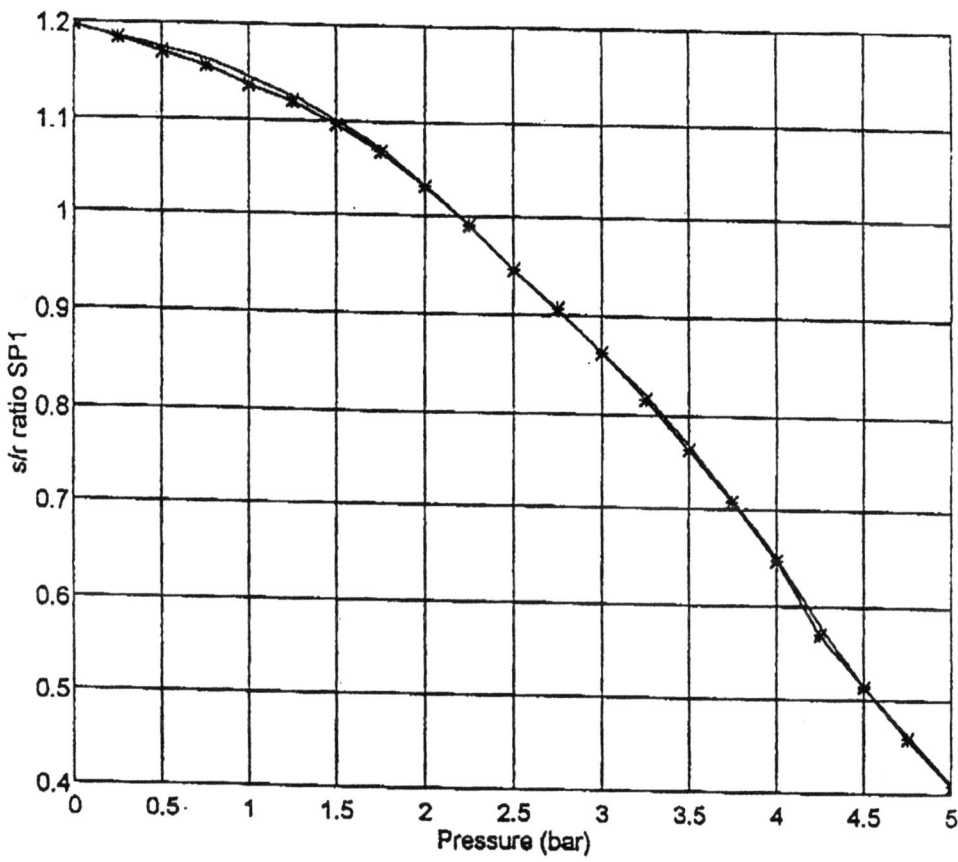

Fig. 13 : Measurement of a sensor network

6.2.3. Time domain versus frequency domain :

The previous explanation shows that the working principle of such a network is very simple, although some problems will have to be faced, as there are :

- very short pulses have to be applied, with as much energy in them as possible,
- the pulse width must be very narrow to make a distinction at the receiver side
 possible between the pulses of the all the branches.
- at the receiver side, very high speed acquisition electronics are necessary to acquire the pulses.
- due to the intrinsic very large energetic spectrum of a pulse, the receiver should possess a very large bandwidth, resulting in a very poor signal to noise ratio of the detector.
- the shape of the pulse will be spreaded out while the pulse is travelling along the fiber, due to a phenomenon called 'phase dispersion', making distinction of the pulses at the receiver side very difficult.

Based on these considerations, and the fact that the network is nearly an academic example of a 'time invariant linear system', we go over to the frequency domain.

By appluing the rule that ' the inverse Fourrier transform of the transfer function of a time invariant linear system is the impulse response of the system', we know that it is possible to make our measurements in the frequency domain and obtaining the specific information of the network (lengths and attenuation coefficients of the branches) by using a mathematical tool.

The advantages of doing the measurements in the frequency domain are :

- since the network is excited at well known frequencies, at the receiver side only these components have to be filtered out, resulting in a very high signal to noise ratio.
- to accomplish these measurements, affordable electronic components are available.

On figure 14, the influence on the transfer function of the same four sensor network as in the previous example is shown, when applying pressure at only one sensor. In the vertical plane, the amplitude relation between input and output is shown, with respect to the frequency; the third axis shows the change of transfer function when changing the pressure at one sensor.

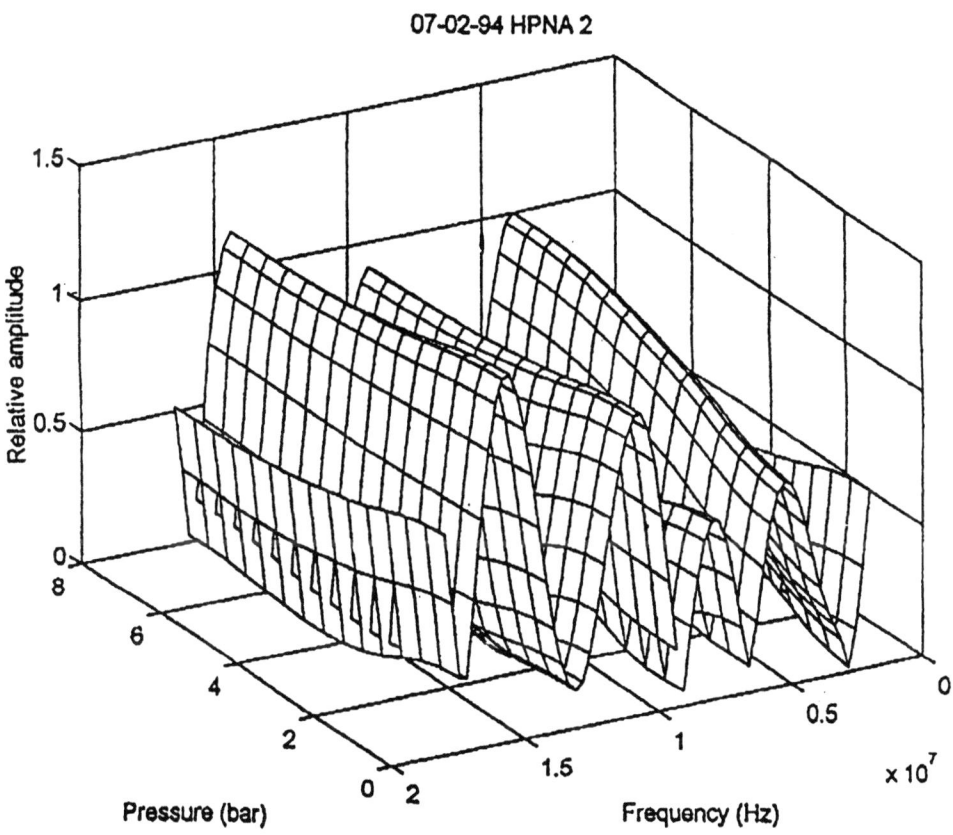

Fig. 14 : Influence of transfer function

6.2.4. Optimizing the number of frequency components

So far, it has been seen that it is possible to make the measurements in the frequency domain, and then obtain the same coefficients as would be measured in the time domain (attenuation coefficients and lengths of the branches) by making Fourrier transforms.

This method, however, has the disadvantage that more excitation frequencies are used than strictly necessary, as will be seen in the following paragraphs.

In order to determine and explain how many frequency components have to be used as a function of the number of sensors, let's look how a one sensor network (two branches) behaves when exciting at a certain frequency.

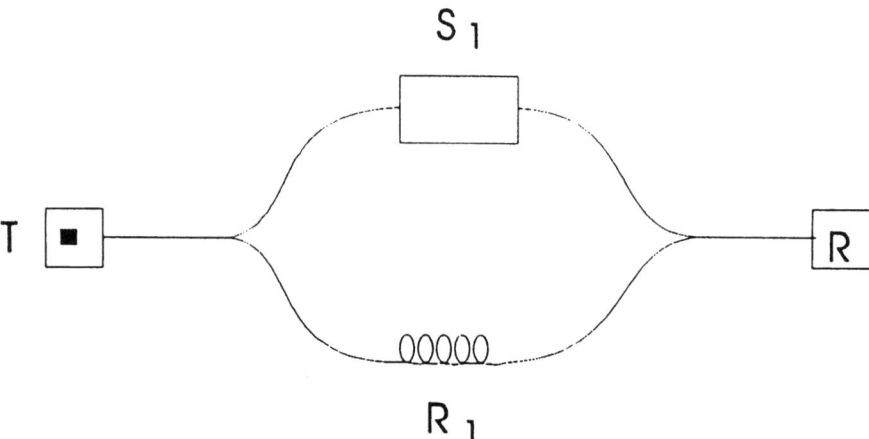

Fig. 15 : One sensor Network

At the frequency f_1 we have the following receiving signal

$$A_1 \cos(w_1 t + \psi_1) = T_1 S_1 \cos(w_1 t + \varphi_1) + T_1 R_1 \cos(w_1 t + \varphi'_1)$$

Identical at f_2

$$A_2 \cos(w_2 t + \psi_2) = T_1 S_1 \cos(w_2 t + \varphi_2) + T_1 R_1 \cos(w_2 + \varphi'_2)$$

with
$$\varphi_1 = \frac{2\pi f_1 n L_1}{C_0}$$

and
$$\varphi'_1 = \frac{2\pi f_1 (L_1 + \Delta L)}{C_0}$$

Here,
L_1 is the length of the fiber between the control unit and sensor 1
ΔL is the length of the roll of reference
n is the index of refraction of the fibre
C_0 is the speed of the light in the vacuum
T_1 is the influence of the line

With the help of a complex parameter estimation algorithm and a measurement with more frequencies, we can determine the parameters l_1, l_2 et Δl. We will suppose that those parameters are known.

Going to the imaginary notation and writing the formulas in a matrix allows to write :

$$\begin{pmatrix} A_1 e^{i\psi_1} \\ A_2 e^{i\psi_2} \end{pmatrix} = \begin{pmatrix} e^{i\varphi_1} & e^{i\varphi_1'} \\ e^{i\varphi_2} & e^{i\varphi_1'} \end{pmatrix} \cdot \begin{pmatrix} S_1 T_1 \\ R_1 T_1 \end{pmatrix}$$

This represents a linear acquisition system which is very easy to resolve. The result that we are looking for, is the ratio between $T_1 S_1$ and $R_1 T_1$. The result is independent of the influences of the line (T_1).

During the installation and with the help of the same formulas, they will also determine the whole length of the fibres between the control unit and the sensors. These formulas can easily be extended to 2 or more sensors.

So, it is clearly that for a one sensor network, only two frequencies have to be used; in general : for a N sensor network (with 2N branches) 2N excitation frequencies are necessary.

The formulas, however, suppose that the lengths of the fibers are known, what is not the case at installation time.

To obtain these lengths, a special estimation algorithm is used (Newton Gauss) at regular timestamps. In figure 16, the result is given of the length estimation of one branch in a four sensor network. The measurement is taken over a period of one week, and the uncertainty of length is about 4 cm at a length of 115.10 m

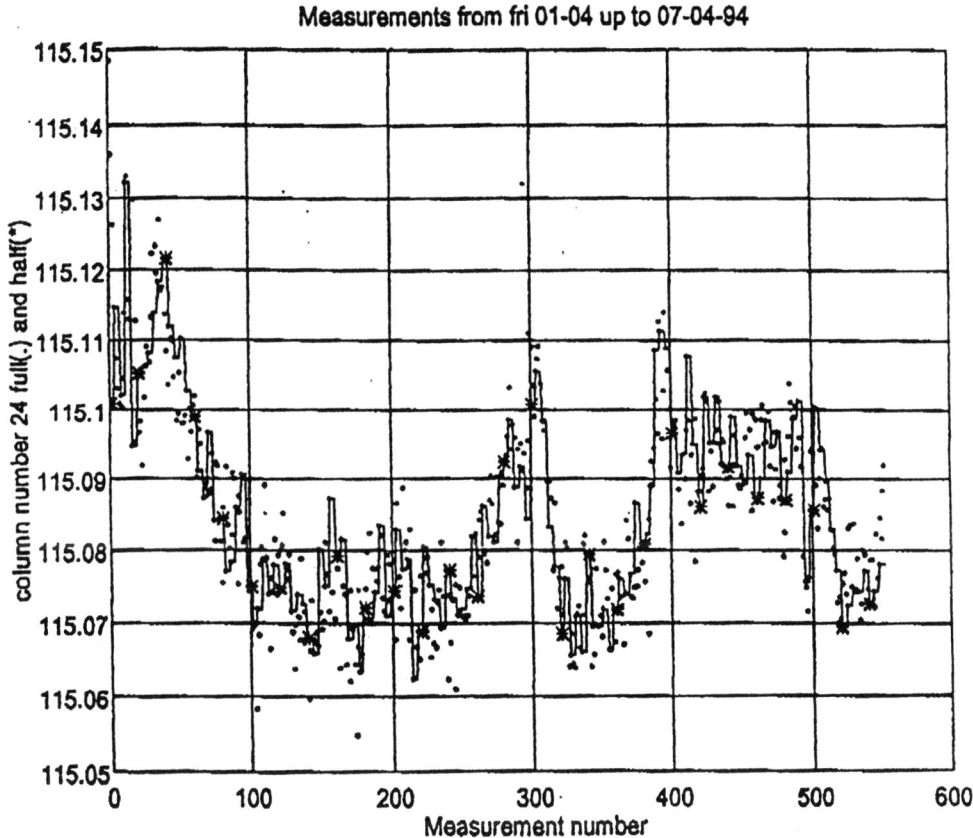

Fig. 16 : Length estimation of one branch in a four sensor network

6.3. MULTIPLEXING WITH THE OPTICAL SHIFTER [11]

An electrically addressed all-optical fibre switch is shown in fig.17a. The working is as follows.

The LCR's Liquid Crystal Retarders act as polarization modulators and provide for electrically driven 0 or $\lambda/2$-retardation in order to switch from TE to TM polarization and back (see fig. 17b). The switching times reach 30 ms. The B6s (Calcite, splitting angle α about 6°) perform the spatial shift of the linearly polarised (TE or TM) incoming beams, respectively in the up, right, down and left direction (see fig. 17c). This configuration allows a laser beam to be shifted in 9 different symmetrically spaced directions originating from the incoming beam. The will-chosen length d of the crystals makes it possible to displace the beam from one core of an optical fibre to another in an array of optical fibres.

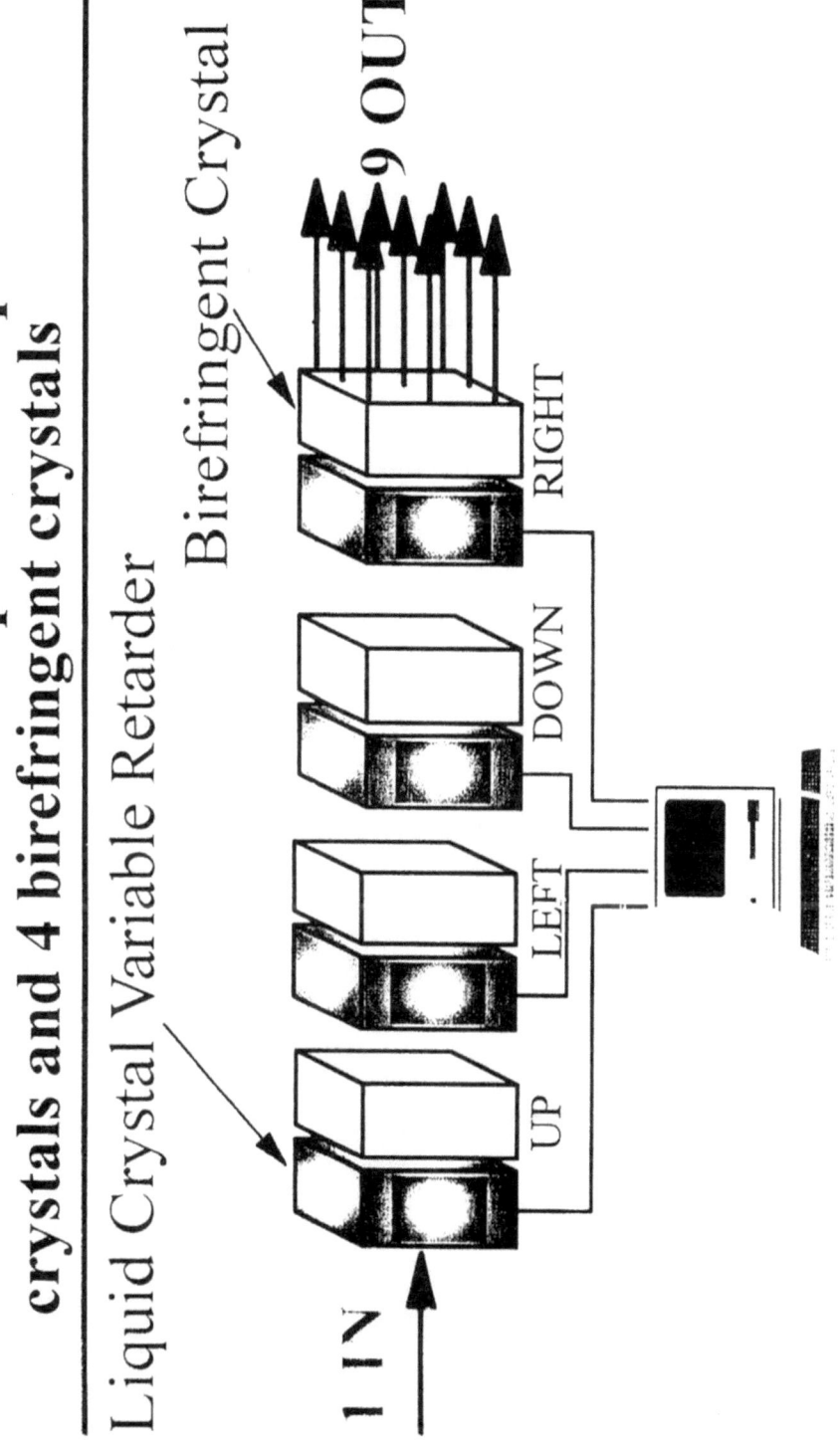

Fig. 17a : An electrically addressed all-optical fibre switch

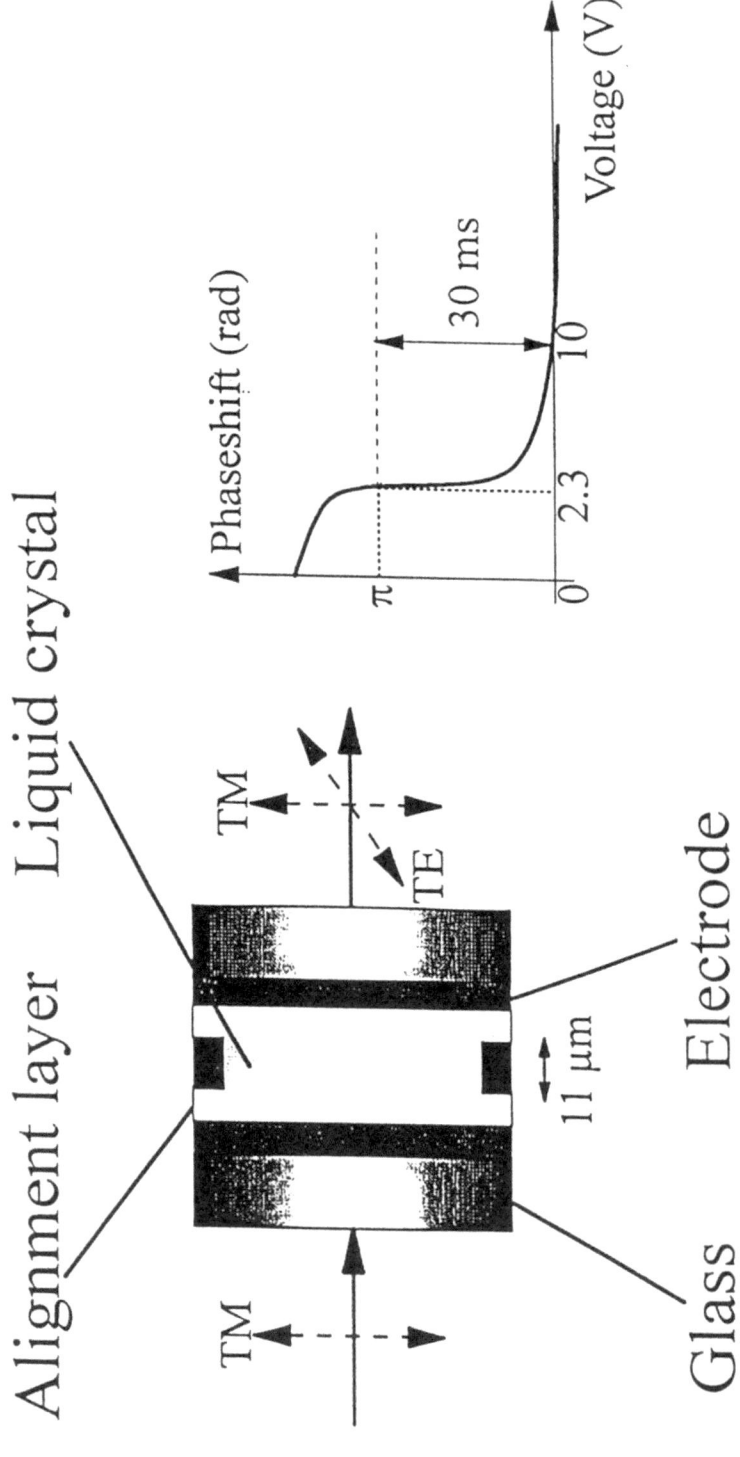

Fig. 17b : The liquid crystal variable retarder acts as a polarization modulator

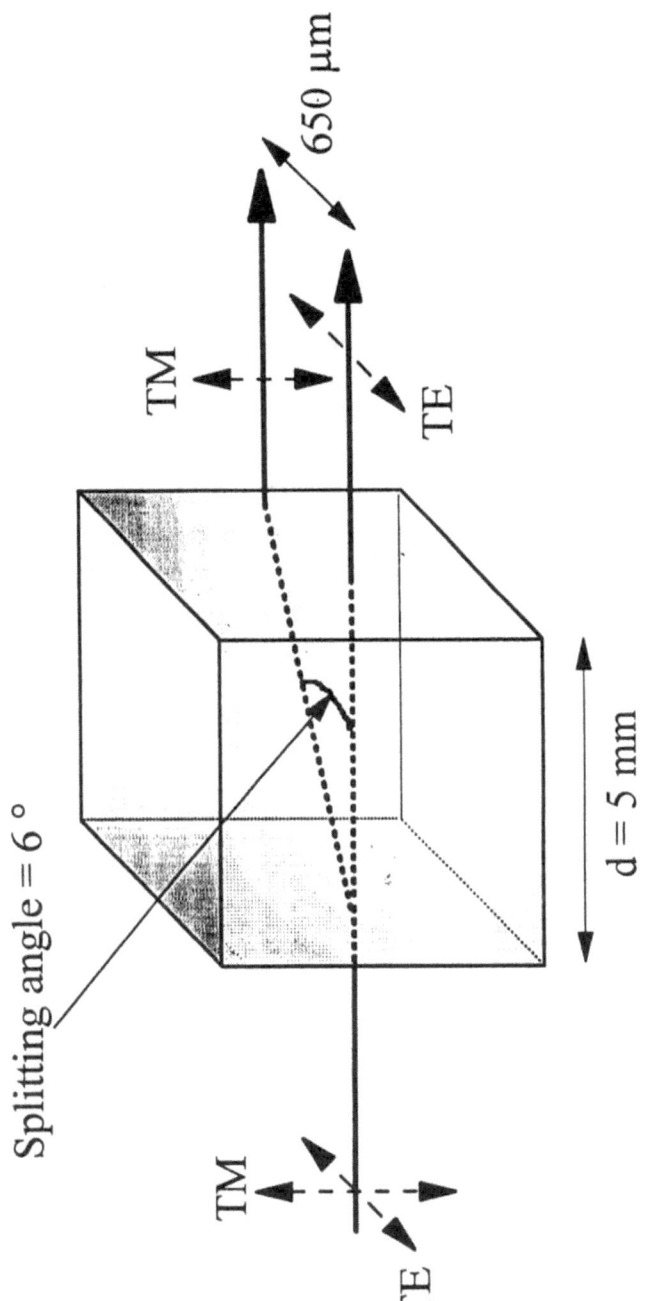

Fig. 17c : The birefringent crystal shifts the laser beam in a particular direction

7. Field Test and Practical Applications

7.1. APPLICATION 1

The use of "**fibre optic pressure sensors**" in the underground facility "**HADES**".

The HADES Underground Research Facility (URF) is an experimental installation at the site of the SCK/CEN, the Nuclear Research Center at Mol in the north-eastern part of Belgium. The facility, located in a deep clay layer (the Boom clay), is designed and built to accomodate experiments related to the disposal of high-level and long-lived radioactive wastes. HADES is unique in that sense that it is, of the small number of facilities of this kind in the world, the only one in a clay stratum built-on purpose. The HADES-URF (see also fig. 18) is made up of four parts: the surface facilities and services area, the access shaft, an underground laboratory, and a test drift . It has been continuously in full operation for various experimental programs since more than 10 years.

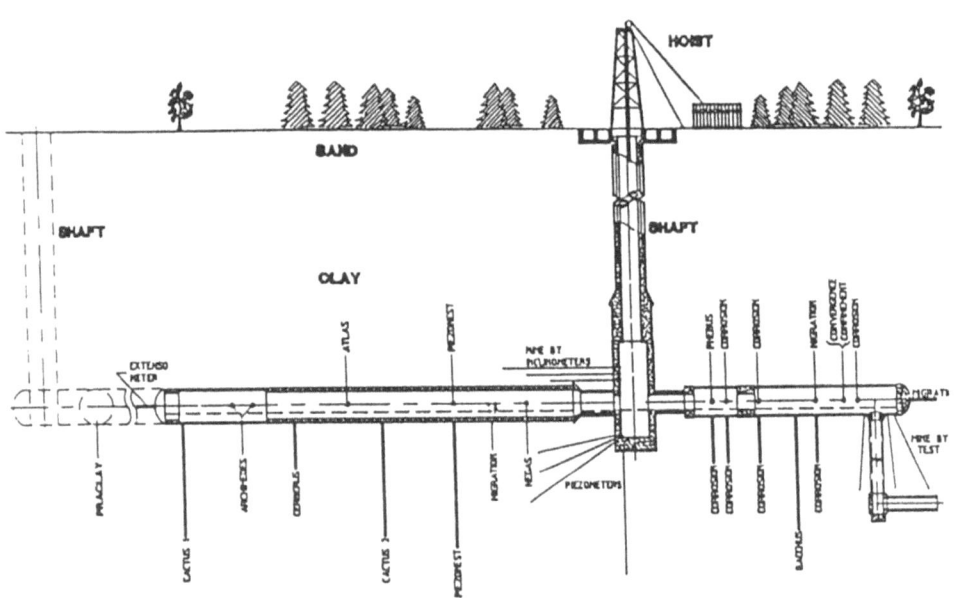

Fig. 18 : In-situ test programme HADES

When assessing different candidate materials to be applied in a radioactive waste repository, it is necessary to investigate their long-term individual performances, as well as their functioning in the overall system. Theoretical, laboratory and in-situ tests are developed in order to select the most suitable materials and to investigate their performances in view of the design of a multi component system. This requires extensive testing and experimentation over several years, since it is necessary to extrapolate the performances over extremely long periods of time. The design and construction of underground structures in deep clay (e.g. for radioactive waste disposal) require expertise in design, forecasting, implementation and assessment of tests. In the HADES facility, experiments can be designed and performed at different scales, supported by in-house modelling and analysis.

Typical experiments

Tunneling, Boring, Sealing, Backfilling and Plugging Experiments - These experiments deal with the specific problems encountered in shaft sinking, tunnelling and boring in a clay medium, and with the materials to be used for backfilling or sealing of the openings. Problems dealt with include the selection, placement and long-term behaviour of various candidate materials (e.g. concrete and clay based mixtures) under conditions of stress, heat, hydration, irradiation, chemical interaction and so on. Experiments on all these factors can be carried out at various scales. They require monitoring and survey of the test conditions and results.

Geotechnical Instrumentation - Given the novel and highly specific nature of the tests, it has been necessary to develop new, high-performance measuring tools and methods, with long-term performance and durability w.r.t. external factors such as a large temperature range, radiation and chemical corrosion, etc.. The facility makes it possible to test specific devices and set-ups under realistic conditions of combined effects, and thus to develop, demonstrate and improve new techniques. One example of a new technology being tested in the HADES-URF is a **"distributed environmental monitoring network using optical sensors"** for the survey of the local hydrological conditions.

Observation network

The more local hydrological conditions around the URF are investigated in greater detail by a series of piezometers buried in the clay host rock together with a serie of fibre optic piezometers.
The data from the conventional piezometers, build in as a redundant system, is used with experimental models to determine various hydraulic and hydrological conditions and parameters (such as local pressure gradients and local permeability).

Fibre optic piezometer data together with the conventional ones are indicated in figure 19. A typical set-up can be seen in figure 20.

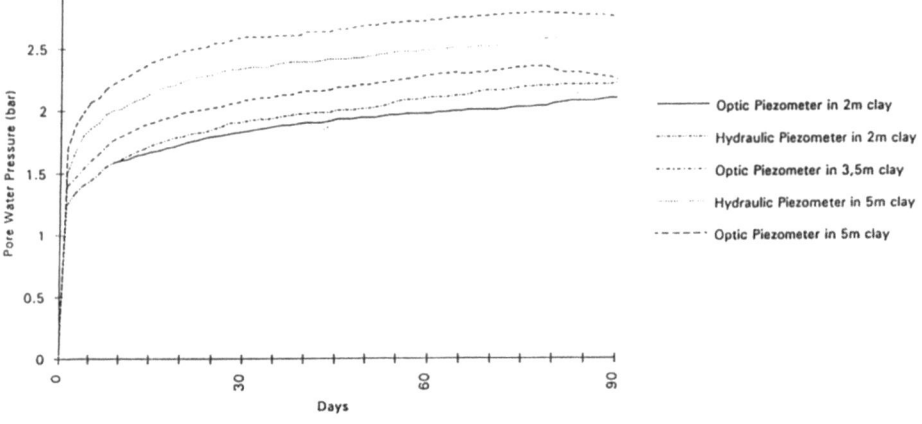

Fig. 19: Pressure Buildup

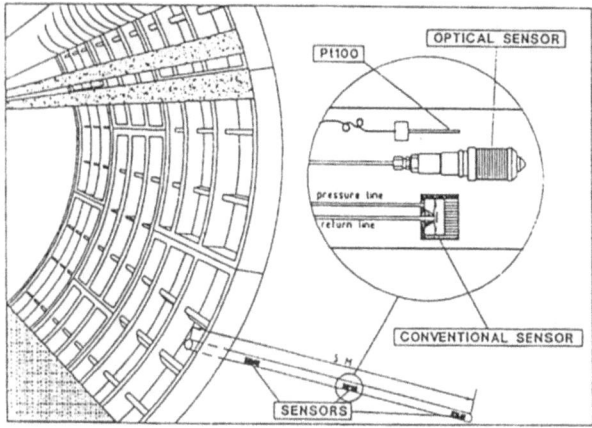

Fig. 20: DEMOS Test

7.2. APPLICATION 2

The use of **"fibre optic pressure sensors"** for the measurement of pressure in the foundation of the **"L'Eau d'Heure"** dam.

The **"L'Eau d'Heure"** dam at Cerfontaine (South Belgium) is a rock surrounded dam for holding the 17 million cubic meter surface water reservoir of an hydroelectric power station. The dam reservoir was filled in 1977 and is fully operational in the hydroelectric power system since 1990.

A scheme of the dam structure is given in figure 21 and from the cross section shown, it can be learned that the surface water tightness is provided an asphalt screen, whereas an in the underground, upstream of the dam, a watertight screen is made by a cement grouted curtain. Into the impervious core of that curtain a network of drain holes has been placed from a inspection gallery. As part of the overall survey programme of the dam and of its stability the monitoring of the drains in the foundations has been undertaken since its filling. Indeed the stability of the dam structure depends to a large extent on the stress conditions of the foundation. The conditions can in many cases only be derived from the effects the stresses induce. When these stresses pass certain thresholds important consequences might follow, such as volume changes, saturation of the geomedium, loss of cohesion, erosion of the filling materials of the dam and its embankment. In order to keep the stresses in the dam foundation to acceptable levels as well injections in the impervious core as drainage or both together can

DAM OF EAU D'HEURE

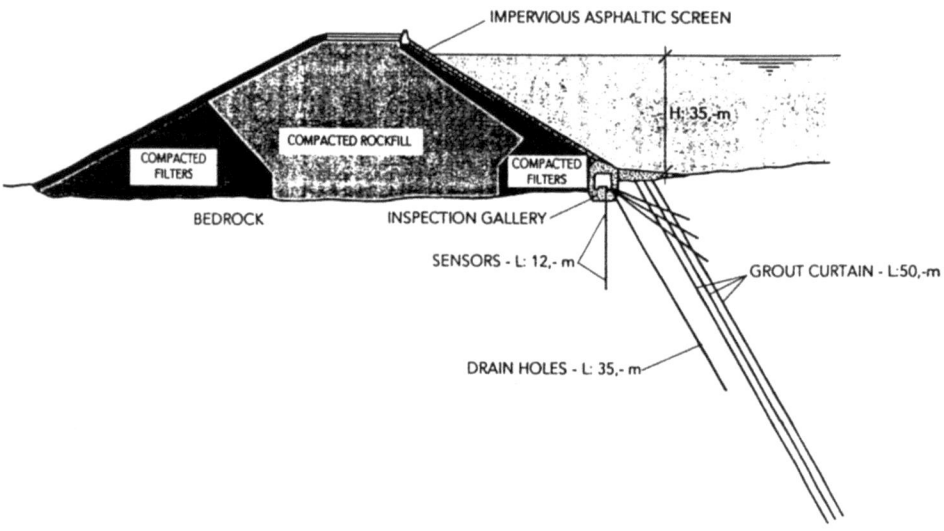

Fig. 21 : Dam of Eau d'Heure

Typical observations

The monitoring of the performances of the drainage network is thus very important for security reasons; it allows an on-line control and maintenance. In first instance is it important to observe the evolution in time of the drainage and to try to identify the reasons for these (e.g. saturation of the medium, clogging of the voids, etc.). Observation made since the filling of the L'Eau d'Heure dam point towards an increase in the interstitial pressures and a decrease in the drain rates in the last ten years. Complementary tests allow to conclude that these changes are very probably due to a clogging in the draining network. It was therefore decided to develop a new drain network downstream of the previously poured impervious screen. In the configuration of the network one had to take into account the geometry of the dam structure, the nature and geometry of its components and the assumed influences. The distance between the new drain holes is 5 meter.

Instrumentation and Observation Network

The direct observation of the interstitial pressures is performed by conventional pneumatic point piezometers and fibre optic piezometers installed at different depths in bore holes of the drain network. A first piezometer is placed between the bedrock and the extrados of the floor inspection gallery and a second is located 10 meter below the interface between the bedrock and the inspection gallery floor (see fig. 22). The piezometers are mounted on a 10 mm diameter stainless steel stem allowing to retrieve the piezometers in case of necessity. The boreholes are sealed off with a mixture of bentonite to assure appropriate hydraulic isolation.

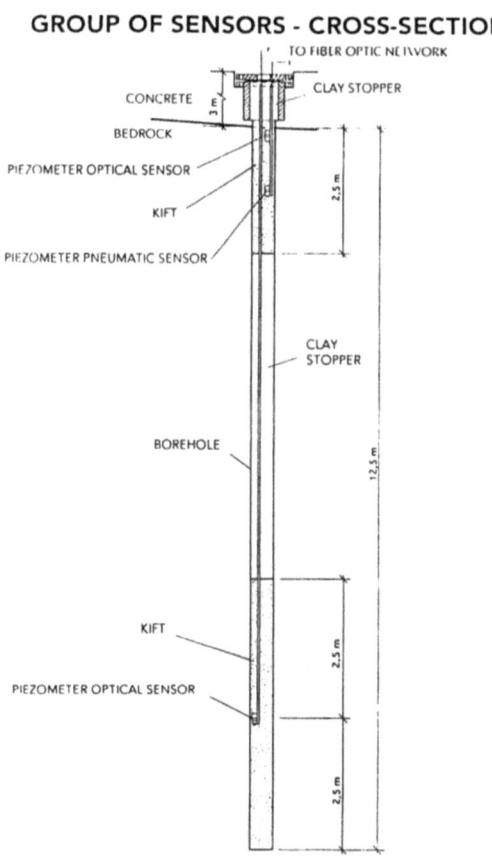

Fig. 22 : Group of sensors - Cross section

8. Acknowledgements

Part of this work was supported by the Commission of the European Communities nl. in the Brite Euram projects DEMOS BE 3077, STABILOS BE 5553 and a Feasibility Study BRE2-CT93-0712.
The authors are much indebted for the help of the Photonetics team nl. Fr. Xavier-Desforges, A. Quelquejay, H. Lefèvre.

9. References

[1]
Karbowiak, A.E. : "In the beginning there was light", Proc. 7th Optical Fibre Sensors Conference, OFS7, Sydney, New South Wales, 1990, 1-6.

[2]
Galileo Electro optics. Cover art Work.

[3]
Krohn, D.A. : "Fiber Optic Sensors, Fundamentals and Applications", Instrument Society of America, ISBN 0-87664-997-5, 1988.

[4]
Kingsley, S.A., Harmer, A.L. : "Fibre Optic Sensors and their Potential in Industry", Batelle Technical Inputs to planning/ Report N°48, 1986.

[5]
Voet, M.R.H., Barel, A.R.F. : "Line-Neutral fibre optic sensing, combining time division- and frequency division multiplexing techniques", Proc. ECO3, The Hague, March 1990.

[6]
Voet, M.R.H., Barel, A.R.F. : " Performances of the Glötzl fibre optic pore water pressure transducer in a line neutral sensing system", Proc. 7th Optical Fibre Sensors Conference, OFS7, Sydney, New South Wales, 1990, 159-162.

[7]
Lagakos, N., Cole, J.H., Bucaro, J.A. : "Microbend fiber-optic sensor", Applied Optics, Vol 26, N°11, 1 june 1987.

[8]
Schrever, K., De Vilder, J., Rolain, Y., Voet, M.R.H., Barel, A.R.F., : "Designing enhanced maintainability fiber optic networks", Proc. OFS(C), Wuhan, China, 1991.

[9]
G.P. Agrawal : "Fibre Optic Communication Systems", Hohn Wiley & Sons, Inc. 1992, ISBN 0-471-54286-5

[10]
Gowar : "Optical Communication Systems", Prentice Hall, 1984, ISBN 0-13-638056-5

[11]
F. Berghmans et Al. " Optical computing subsystems for smart sensing : a reconfigurable interconnection module adapted to multiplexing optical fibre sensor arrays", Proposed for presentation at the 10th International Conference on Optical Fibre Sensors, Glasgow, Scotland, 1994

FIBER OPTIC SENSORS
FOR ENVIRONMENTAL MONITORING

An Overview of Technologies and Materials

A.G. MIGNANI, M. BRENCI, A. MENCAGLIA
IROE-CNR
Via Panciatichi, 64, I-50127 Firenze, Italy

Summary

The major types of fiber optic sensors for environmental monitoring are presented. Two categories of sensors are described: those for monitoring pollutants (hydrocarbons, herbicides, ammonia) and those for monitoring environmental conditions (humidity, pH, and carbon dioxide partial pressure of sea- and ocean water).

1. Introduction

Industrialization and the consequent pollution of the environment mandate urgent revision of numerous industrial processes and rapid implementation of effective strategies for environmental monitoring and control. Environmental monitoring encompasses two conceptually different tasks: *diagnostics*, or measurements performed to determine whether pollutants are present and, if so, to what extent, and *surveillance*, or measurements of preset parameters performed on a continual basis. To achieve successful environmental monitoring, it is necessary to proceed from preventive diagnostics to non-stop surveillance.

The market for surveillance instrumentation is currently expanding in pace with research efforts. The present European market of 209 MECU is forecast to expand to 336 MECU by the year 2000. Optical

---

(∗) This paper was presented at a poster session.

instrumentation for air, water, and biological analyses is expected to occupy approximately 30% of the market (Figure 1).

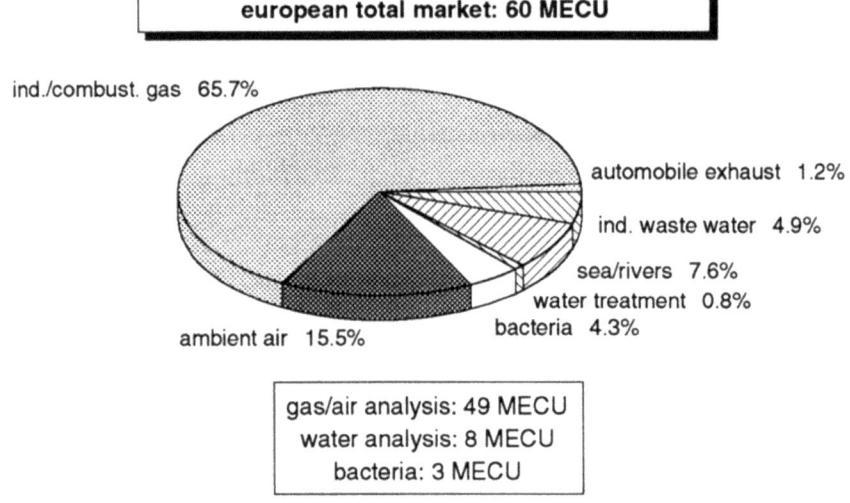

Figure 1. Optically-monitored environmental-sectors: European market share.

Environmental surveillance with optical instrumentation entails two types of measurements: *remote* and *on-site*. Remote measurements are performed with systems such as LIDAR to analyze the atmosphere and vegetation and to map water surface pollution. On-site measurements mainly involve sampling of elements for subsequent analysis via various types of spectrophotometry.

Fiber optic sensors (FOSs) offer several advantages over conventional optical instrumentation for on-site measurements:

- Their main advantage is that they can be permanently positioned in hard-to-access areas, including underground sites, providing the possibility of carrying out highly localized measurements without sampling for non-stop monitoring in real-time.

- The sensor's working principle enables direct analysis of

the parameter.

- The electro-optical power supply and control units can be located remotely from the measuring area.

- Fiber with low attenuation allows long-haul networks with the possibility of multiplexing several sensors controlled by a single electro-optical unit.

In this overview, the basic criteria for implementing FOSs for environmental monitoring will be described. Examples will be given, with a special focus on instrumentation that has demonstrated feasibility and successful operation [1] [2] [3] [4] [5] [6].

2. Fiber Optic Sensors

2.1. WORKING PRINCIPLE

The working principle of FOSs is based on the modulation of one of the parameters of the fiber-guided light such as phase, intensity, wavelength or polarization state, by the parameter under investigation. The complexity of the electro-optical system, the type of components selected, and thus cost of the sensor are related to the operating principle.

The ideal FOS for environmental monitoring should possess the following characteristics:

- reliability

- automatic or semiautomatic operation for use by operators who have little or no technical background.

- low cost installation and maintenance.

These requirements limit the selection of the sensor's operating principle and impose limitations on the complexity of the electro-optical system. FOSs for environmental applications are mostly of the intensity modulation type, owing to the low cost of their components and the simplicity of their architectures with respect to that of conventional systems. They can be either *intrinsic* or *extrinsic*, according to whether the intensity modulation is produced directly in the fiber, which is sometimes modified, or by an external transducer connected to the fiber (Figure 2).

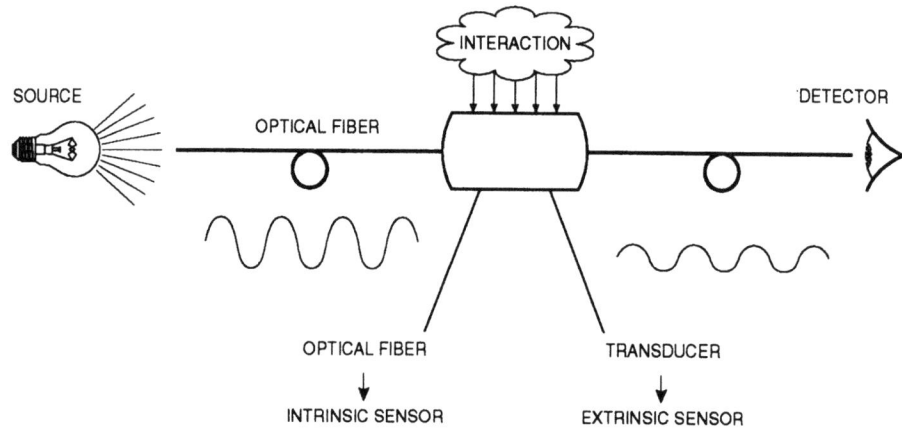

Figure 2. Working principle of intensity-modulated FOSs.

2.2. ARCHITECTURE

A FOS of the intensity modulation type can be viewed as a compact electro-optical module connected to the measuring probe by a multimode optical fiber.

The source can be some form of conventional light bulb (lamp): generally, additional optical components are required to couple the optimal optical power into the fiber. If a halogen lamp is used, interference or dichroic filters may also be necessary when the sensor has a specific operating wavelength. LEDs and laser diodes are the most compact sources and may be readily "pigtailed" to the fiber. These sources offer a wide variety of wavelengths; the laser diodes offer the highest output power. The detectors are normally PIN-type photodiodes, sometimes with filters to provide spectral response.

The optical fiber connection carries the light intensity from the source to the probe and returns the intensity modulated by the measuring parameter to the detector. Generally, fibers having diameters larger than

100 μm are used to maximize the source's coupling efficiency. The fibers can be all silica (AS) or plastic silica (PCS), either bare or cabled, according to application.

The connection can be either *single fiber*, in which case the fiber serves for both illumination and detection, i. e. it acts as a transceiver link, or *two fibers*, in which case one fiber serves to illuminate the sample (or measurand volume) and one for detection. In the case of transceiver link directional coupler can be used to separate the returning and illuminating beams.

Depending on the specific application the system may be fabricated in an all fiber format or as a hybrid, where some of the functions are performed with conventional miniature optics.

2.3. MAIN PROBLEMS

The main problems regarding intensity modulation FOSs are [7]:

Sensitivity to Light Propagation. A problem common to all intensity modulation FOSs is sensitivity to propagation conditions. Since the information is contained in the intensity of the guided light, any modulation not correlated to the state of the measuring parameter upsets the measurement. For example, an incorrect interpretation can be caused by fluctuations in the source or attenuations introduced by fiber curvatures. To solve this problem, a reference system must be used to compensate for the undesired intensity fluctuations, despite the increase in system complexity and thus cost it entails. Source intensity fluctuations can be offset by normalizing the measurement signal S with a reference signal R proportional to the source intensity. Other spurious fluctuations such as those produced by propagation events can be offset by illuminating the sensor at two wavelengths, one of which is modulated in intensity by the investigated parameter and the other of constant intensity. Since they both travel on the same fiber, their ratio provides a measurement devoid of propagation errors.

Signal to Noise Ratio. The S/N depends critically on the source power and the overall optical system.

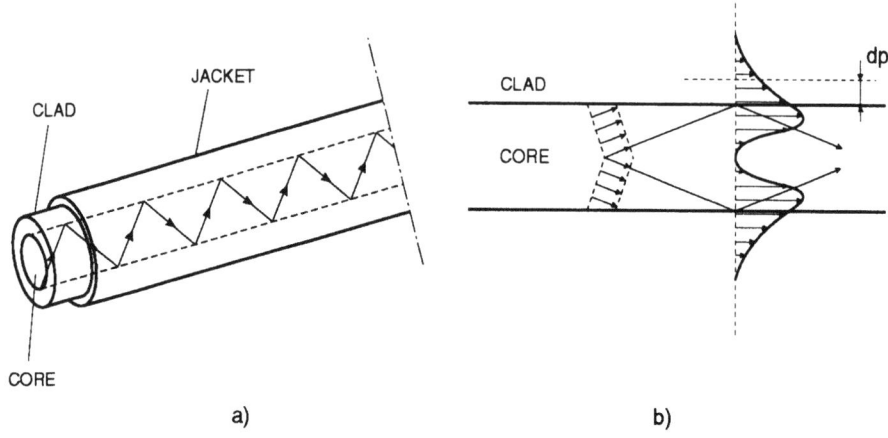

Figure 3. Light propagation inside an optical fiber: a) ray-tracing; b) wave optics.

3. Intrinsic Sensors: Treatments for Sensing Fibers

An optical fiber is composed of two concentrically layered glass or plastic cylinders [8]. The inner cylinder, termed *core*, has a refractive index of n_1; the outer cylinder, termed *cladding*, has a refractive index of $n_2 < n_1$. The incident rays reaching the core-cladding interface at angles $\theta > \theta_c$ ($\sin\theta_c = n_2/n_1$) are guided inside the core by total reflection at the core-cladding interface (Figure 3a). The amplitude of the electromagnetic field decays by 1/e at a distance of $d_p = \lambda/[2\pi/(n_1^2 \sin^2\theta - n_2^2]^{1/2}$ (Figure 3b).

The optical fiber can be sensitized by modifying a cladding or a core section [9] [10]. If one or more discrete fiber sections are modified only point measurements can be made, while treatment of the whole fiber allows for distributed measurements.

3.1. CLADDING TREATMENTS

Cladding treatments are the following:

Simple Cladding Removal. The simplest treatment is to remove the cladding. A variation in the refractive index of the medium surrounding the core produces a variation in angle θ_c and thus in the intensity of the fiber-guided light. Variations in the refractive index can be detected with very high precision.

Cladding Removal with Evanescent Wave Modulation. Cladding removal with evanescent wave modulation is conceptually different from refractometry. Any phenomenon that disturbs the intensity of the evanescent wave also produces modulation in the intensity of the fiber-guided light. The more the evanescent wave penetrates the cladding (i.e., rays penetrating the interface n_1-n_2 at angle $\theta \approx \theta_c$), the greater the interaction. The simplest form of interaction is absorption. This can be accomplished either by removing the cladding and having the parameter to be analyzed interact directly with the evanescent wave or treating the core-cladding interface with a material whose absorption is modulated by the external parameter in the evanescent zone.

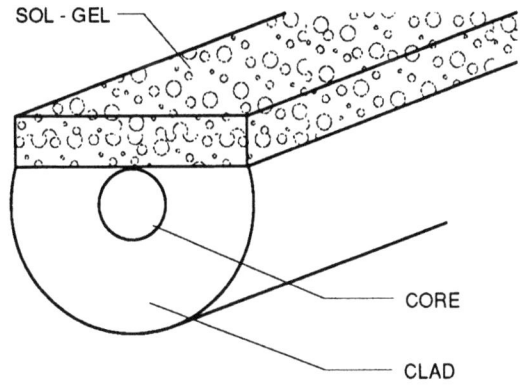

Figure 4. Sol-gel coating for fiber optic intrinsic sensing.

Sol-Gel Claddings. A layer of sol-gel makes an extremely versatile cladding, as shown in Figure 4 [11]. The sol-gel process, is the same as that used in making glass and ceramics at low temperatures via hydrolysis and polymerization of organic precursors. A combination of metal alkoxides, solvent, catalyst and water are mixed to obtain homogeneity. The hydrolysis and condensation polymerization produce a viscous gel, which is an amorphous porous material containing liquid solvent in its pores. The gel is heated at less than 100°C so that most of the liquid is expelled, leaving the porous oxide, and then densified at a higher temperature. The materials and their concentrations can be modified to vary both the material's refractive index and porosity. This process is especially suited to producing a thin layer deposited on the fiber core.

An uncladded PCS fiber covered by a layer of sol-gel can either be used directly or else a reagent with absorption or fluorescence varying with an external parameter can be trapped in the sol-gel.

Claddings with Color Variations. The color, and thus absorption spectrum, of the cladding can be modulated by an external parameter.

3.2. CORE TREATMENTS

Porous core sections, either silica or plastic, are commonly used [12]. Two sensing techniques are:

1) Soaking porous sections of core in a coloring agent to produce modulated absorption by an external parameter. Optical fibers with several sections sensitive to the same parameter constitute a near-distributed sensor that can be easily interrogated by OTDR.

2) Soaking porous sections of the same fiber with different coloring agents to create a single fiber sensitive to several parameters, obviously using multiwavelength illumination and detection systems.

4. Extrinsic FOSs: Optical Fiber Transducers

The extrinsic FOS uses the optical fiber to guide the light to some form of conventional analytical chemical sensors. As shown in Figure 5, one or more fibers are connected to a reagent-transducer to measure the modulation of the reagent's optical properties by interaction with the parameter under investigation. The probe is termed "*optrode*", from *op*tical + elec*trode*. The most frequently used reagents feature absorption or fluorescence modulation.

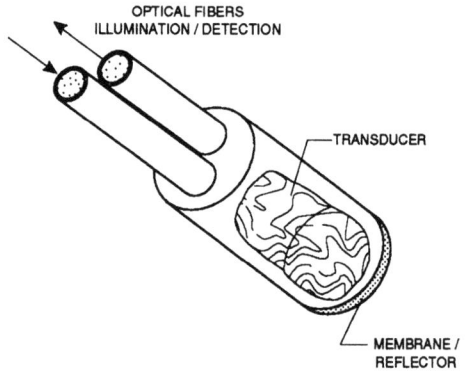

Figure 5. Fiber optic optrode.

Absorption modulation, as described by the Lambert-Beer law, is the linear relation between the absorbance of the transducer A and its concentration C.

$$A = \log(I_{in}/I_{out}) = \varepsilon Cl$$

where:

I_{in} = intensity of the light before transiting the transducer
I_{out} = intensity of the light after transiting the transducer
ε = molar extinction coefficient of the transducer
l = optical path of the light inside the transducer.

Fluorescence modulation is, in many cases, described by the simplified Parker relation between the transducer's fluorescence intensity F and its concentration C:

$$F = 2.3 I_{in} \Phi_f \varepsilon Cl \, k$$

where:

Φ_f = fluorescence efficiency
k = constant accounting for the fact that only a part of the emitted light is actually observed.

Many FOSs of this type have been proposed, since many reagents are suitable [13]. However, the construction of the probe housings, which must contain reagent at the fiber end to initiate the interaction with the test parameter and at the same time provide a high signal-to-noise ratio, is proving exceedingly difficult.

5. Optical Fiber Spectroscopy

Almost all manufacturers of classical spectroscopic instrumentation offer the possibility of substituting the quartz cell module with an optical fiber connection for on-site measurements. In addition to instrumentation for general use, dedicated fiber optic spectrophotometers (spectroFOS) have been developed that operate only on specific wavelengths of parameter sensitivity and sometimes with measuring cells dedicated to the specific problem.

AS fibers and semiconductor components can be used in compact spectrophotometers to analyze gases such as methane, propane, carbon dioxide, and nitrogen oxides. The absorption bands of the gases or combinations and the upper harmonics can lie in a spectral range (near infrared) in which the transmission of the AS fibers is optimal, and for which active semiconductor components are readily available (Figure 6).

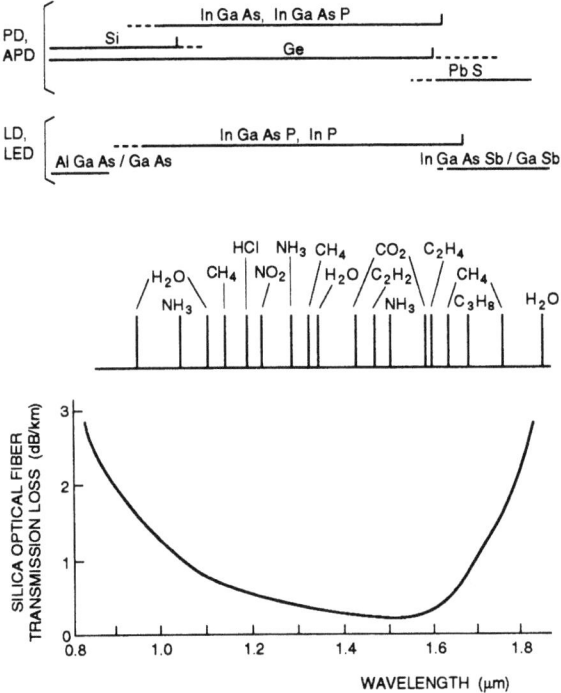

Figure 6. Absorption of some gases compared with optical fibers and semiconductor characteristics.

6. Environmental Monitoring with FOSs

Different types of FOSs are used to monitor pollutants and environmental conditions such as the following:

Methane and Propane. Monitoring methane and propane is essential in numerous industrial plants and mining operations, as well as in urban areas, where methane is used as a heating fuel and propane as a vehicle fuel. FOSs developed for methane and propane sensing are based on the techniques summarized in Table 1.

TABLE 1. Summary of FOSs for methane and propane monitoring

	METHANE and PROPANE	
intrinsic	absorption analysis at $\lambda=3.39$ µm	tapered fiber [14]
spectroFOS	absorption analysis at $\lambda=1.33$ or 1.66 µm • reduced sensitivity • more active/passive components available	sol-gel cladding [15] bulk cell coupled to fibers [16] [17] [18]

TABLE 2. Summary of FOSs for monitoring different types of hydrocarbons

	HYDROCARBONS		
intrinsic	sea- and ocean water (crude, diesel oils)	leakage (ground pollution)	vapors (gasoline)
	silylated fiber-core • frustrated total reflection at core-water interface [19] (Figure 7)	striated plastic fiber • pollutant as a new cladding inducing intensity modulation [20]	bare PCS fiber • evanescent wave modulation [21] [22]
spectroFOS	fiber optic gas chromatography for CCl_4 [23] [24] [25] • sample excited to its emission point by a RF discharge in a low pressure inert atmosphere • fiber optic cable to monitor the emission		
	time resolved fluorescence spectroscopy for aromatic hydrocarbons [26] [27] [28]		

Hydrocarbons. Oil rigs, refineries, and tankers that discharge hydrocarbons make the oil industry a major pollutant. In addition, leakage from storage tanks and pipelines can penetrate to the water table, thereby endangering drinking water sources. Hence, accurate monitoring of these pollutants, which are toxic and cancerogenic, is a prime concern. FOSs developed for monitoring different types of hydrocarbons are based on the techniques summarized in Table 2.

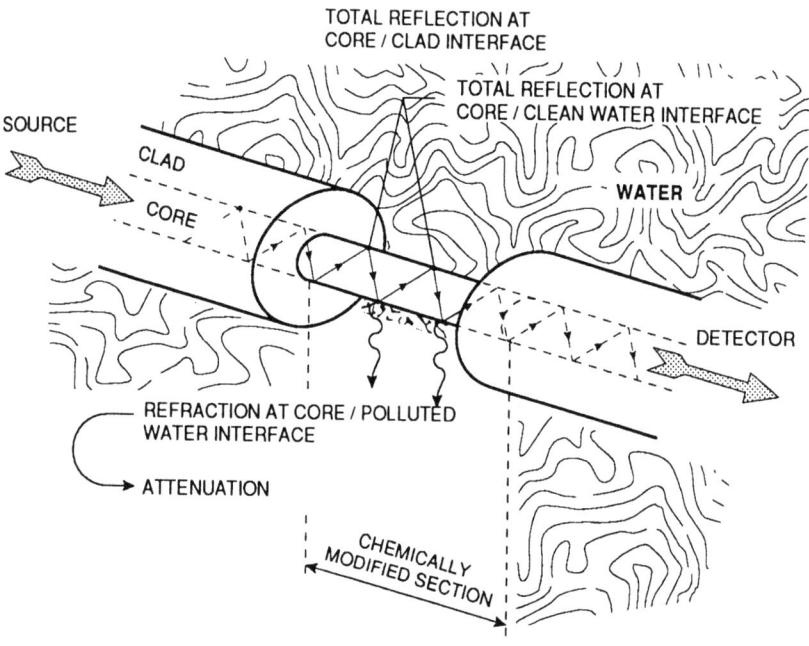

Figure 7. Monitoring of hydrocarbons by a silylated fiber.

Organochlorides. Toxic substances in drinking water distribution plants, primary plants, wells, and the water table must be monitored non-stop. Some common pollutants are trichloroethylene (ClCH=CCl$_2$) (TCE) and chloroform (CHCl$_3$). TCE, a probable cancerogenic whose tolerability is only a few micrograms per liter, is widely used as a degreaser and in drycleaning processes, and is one of the most common pollutants of drinking water sources. The gas-chromatographic techniques normally used to quantify these pollutants can give rise to evaluation errors owing to the compounds' volatility. Ideally, a water monitoring system should be continuous operating directly in the source without sampling, and provide real-time response that allows speedy intervention.

TCE and chloroform are monitored by optrodes made of solutions of pyridine, which changes the color and thus the absorption when exposed to various types of organochlorides [29] [30].

Herbicides and Pesticides. Highly polluting herbicides and pesticides are used in agriculture and in clearing construction sites. The current practice is to reduce usage and monitor the treated areas [31].

Optical fiber fluoro-immunosensors have been developed for herbicide detection, by using an intrinsic sensor with antibodies immobilized on the core surface. When the fiber is dipped in complex solutions, the immobilized molecules bond with the complementary antibody molecules. This bond, which occurs in the evanescent zone, can be measured. Interaction of the evanescent wave occurs with fluorescence excited at the core-cladding interface. Two measuring situations are possible: 1) the antibodies and their complementary antigens are intrinsically fluorescent (Figure 8a), or 2) the antibodies are marked by a fluorescent molecule (Figure 8b). The sensor response consists of a fluorescence signal whose intensity increases or decreases, respectively, with the antigen concentration [32]. Atrazine derivative herbicides has been quantified by this technique silylating the core with the isotiocianate fluoresceine used as a marker. The minimum detection limit for atrazine is 0.1 ng/ml [33] [34].

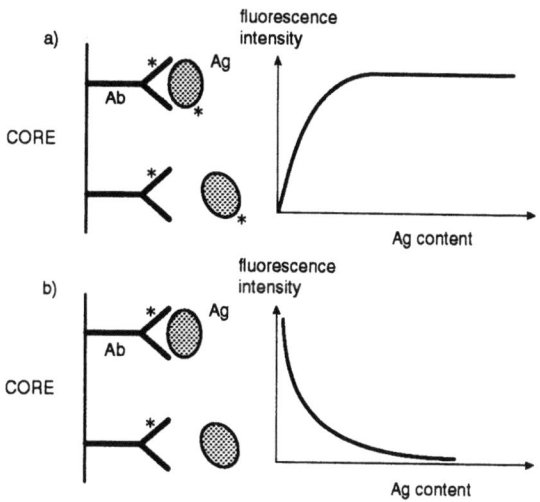

Figure 8. Fiber optic fluoro-immunoassay.

Ammonia and Humidity. Nitrogen in the form of ammonia (NH_3) is the most widely used, and costly, fertilizer. It is responsible for the presence of nitrates in the subsoil. However, small amounts of ammonia dissipated into the atmosphere is more an economic waste than an environmental concern. Relative humidity (%RH), an important meteorological parameter, is defined as %RH=100P/PS, i.e., the percentage of water vapor pressure P in a gas mixture, for example, in air, in relation to the saturated water vapor pressure PS, considered at the same temperature [35]. Table 3 summarizes the techniques used by FOSs for ammonia or humidity detection.

TABLE 3. Summary of FOSs for ammonia or humidity monitoring

	AMMONIA or HUMIDITY	
intrinsic	microporous cladding [36]	
	cladding with color variations	PCS fiber [37] [38]
		fluorescent fiber [39] [40]
	porous core soaked with colored reagents	silica [41] [42]
		plastic [43]
extrinsic	fluorescent optrode [44]	

pH and Carbon Dioxide Partial Pressure (pCO_2) of sea- and ocean water. Monitoring the pH of bodies of water is essential in ensuring the proper implementation of the chemical and biological processes such as the exchange of carbon dioxide between the atmosphere and the water. Since the pH of sea- and ocean water is typically 8, a sensor whose sensitivity ranges between 6 and 9 and whose accuracy exceeds 0.1 pH units should be used. The oceans intake or dissipate CO_2 according to region, season, and other episodic events. Being able to measure the carbon dioxide partial pressure and its atmosphere-water surface variations is of primary importance in determining the CO_2 cycle. This measurement should be carried out non-stop over periods varying from six months to a year in a range of 200-600 µatm, with an accuracy of at least 1%. At this writing, no electric or optical instrument is capable of performing this task automatically and at a reasonable cost. pH and pCO_2

FOSs have been developed by the techniques summarized in Table 4.

TABLE 4. Summary of FOSs for monitoring pH or pCO_2 of sea- and ocean water

	pH or pCO_2
intrinsic	sol-gel cladding + fluorescent or colored reagents [45] [46]
extrinsic	optrode [47] [48]
	micro-flow cell [49] [50] [51]

7. Conclusions

Despite some problems, fiber optic sensors are being increasingly used for environmental monitoring. Forecasts for the European market in the decade 1990-2000 indicate an annual increase of 20%, with expected sales topping 7.5 MECU by 1995 and 19 MECU by the year 2000. These optimistic forecasts are mainly based on the excellent prospects for networking sensors and eventually incorporating them into fiber optic communication networks. Another promising application which, owing to fiber compactness and intrinsic crosstalk, should not be long in coming, is the integration of groups of sensors measuring different parameters in a single probe housing.

8. References

1. B. Culshaw and J. Dakin (eds) (1988) *Optical Fiber Sensors. vol. I: Principles and Components, vol. II: Systems and Applications*, Artech House, Boston and London.
2. R.A. Lieberman and M.T. Wlodarczyk (eds) (1988) *Chemical, Biochemical and Environmental Applications of Fibers*, SPIE Proc. **990**, Bellingham WA.
3. R.A. Lieberman and M.T. Wlodarczyk (eds) (1989) *Chemical, Biochemical and Environmental Fiber Sensors*, SPIE Proc. **1172**, Bellingham WA.
4. R.A. Lieberman and M.T. Wlodarczyk (eds) (1991) *Chemical, Biochemical, and Environmental Fiber Sensors II*, SPIE Proc. **1368**, Bellingham WA.

5. R.A. Lieberman (ed) (1992) *Chemical, Biochemical, and Environmental Fiber Sensors III*, SPIE Proc. **1587**, Bellingham WA.
6. Moslehi, B., Shahriari, M.R., Schmidlin, E.M., Anderson, M. and Lukasiewicz, M.A. (1992) Optical fiber simplifies gas-sensing systems: sniffing out gases with fiberoptic chemical sensors offers many advantages over more traditional methods, *Laser Focus World* **4**, 161-167.
7. Brenci, M., Mencaglia, A. and Mignani, A.G. (1991) Problems and solutions in fiberoptic amplitude-modulated sensors, in O.D.D. Soares (ed), *Fiber-Optic Metrology and Standards*, SPIE Proc. **1504**, Bellingham WA, 212-220.
8. Kapany, N.S. (1967) *Fiber Optics*, Academic Press, New York-S. Francisco-London.
9. Lieberman, R.A., Recent progress in intrinsic fiber optic chemical sensing, in Ref. 4, 15-24.
10. Lieberman, R.A. (1993) Recent progress in intrinsic fiber-optic chemical sensing II, *Sensors and Actuators B* **11**, 43-55.
11. MacCraith, B.D. (1993) Enhanced evanescent wave sensors based on sol-gel-derived porous glass coatings, *Sensors and Acuators B* **11**, 29-34.
12. Tabacco, M., Zhou, Q. and Nelson, B., Chemical sensors for environmental monitoring, in Ref. 5, 271-277.
13. Scheggi, A.M. and Baldini, F. (1993) Chemical sensing with optical fibres, *International J. of Optoelectronics* **8**, 133-156.
14. Tai, H., Tanaka, H. and Yoshino, T. (1987) Fiber-optic evanescent-wave methane-gas sensor using optical absorption for the 3.392-μm line of a He-Ne laser, *Optics Letters* **12**, 437-439.
15. Stewart, G., Muhammad, F.A. and Culshaw, B. (1993) Sensitivity improvement for evanescent-wave gas sensors, *Sensors and Actuators B* **11**, 521-524.
16. Chan, K., Ito, H., Inaba, H. and Furuya, T. (1985) 10-Km-Long fiber-optic remote sensing of CH_4 gas by near infrared absorption, *Applied Physics B* **38**, 11-15.
17. Chan, K., Ito, H. and Inaba, H. (1984) An optical-fiber-based gas sensor for remote absorption measurement of low-level CH_4 gas in the near-infrared region, *IEEE J. of Lightwave Technology* **LT2**, 234-237.
18. Inaba, H. (1986) Remote sensing of environmental pollution and gas dispersal using low-loss optical fiber network system, in W. Waidelich (ed) *Laser-Optoelectronics in Engineering*, Springer-Verlag, Berlin, 619-629.
19. Kawahara, F.K., Fiutem, R.A., Silvus, H.S., Newman, F.M. and Frazar, J.H. (1983) Development of a novel method for monitoring oils in water, *Analytica Chimica Acta* **151**, 315-327.
20. Brossia, C.E. and Wu, S.C., Low-cost in-soil organic contaminant sensor, in Ref. 4, 115-120.
21. Carome, E.F., Fischer, G. and Kubulins, V. (1993) Fiber-optic sensor system for hydrocarbon vapors, *Sensors and Actuators B* **13-14**, 305-308.
22. Fischer, G., Carome, E.F., Kubulins, V.E. and Burgess, L.W., Fiberoptic hydrocarbon sensor system, in Ref. 5, 258-270.

23. Griffin, J.W., Olsen, K.B., Matson, B.S., Nelson, D.A. and Eschbach, P.A., Fiber optic spectrochemical emission sensors, in Ref. 2, 55-68.
24. Griffin, J.W., Matson, B.S., Olsen, K.B., Kiefer, T.C. and Flynn, C.J., Fiber optic spectrochemical emission sensors: a detector for chlorinated and fluorinated compounds, in Ref. 3, 99-107.
25. Anheirer, N.C., Olsen, K.B. and Griffin, J.W. (1993) Fiber-optic spectrochemical emission sensor: a detector for volatile chlorinated compounds, *Sensors and Actuators B* **13-14**, 447-453.
26. Inman, S.M., Thibado, P., Theriault, G.A. and Lieberman, S.H. (1990) Development of a pulsed-laser fiber-optic-based fluorimeter: determination of fluorescence decay times of polycyclic aromatic hydrocarbons in sea water, *Analytica Chimica Acta* **239**, 45-51.
27. Panne, U. and Niessner, R. (1993) A fiber-optical sensor for polynuclear aromatic hydrocarbons based on multidimensional fluorescence, *Sensors and Actuators B* **13-14**, 288-292.
28. Zung, J.B., Woodleee, R.L., Fuh, M.R.S. and Warner, I.M., Fiber optic based multidimensional fluorometer for studies of marine pollutants, in Ref. 2, 49-54.
29. Angel, S.M. and Ridley, M.N. (1989) Dual-wavelength absorption optrode for trace level measurements of trichloroethylene and chloroform, in Ref. 3, 115-122.
30. Angel, S.M., Langry, K., Roe, J., Colston, B.W., Daley, P.F. and Milanovich, F.P., Preliminary field demonstration of a fiber-optic TCE sensor, in Ref. 4, 98-104.
31. Shropshire, G.J. and DeShazer, J.A. (1993) Optical sensors aid agriculture, *Laser Focus World* **5**, 79-84.
32. Andrade, J.D., VanWagenen, R.A., Gregonis, D.E., Newby, K. and Lin, J.N. (1985) Remote fiber-optic biosensors based on evanescent-excited fluoro-immunoassay: concept and progress, *IEEE Transactions on Electron Devices* **ED32**, 1175-1179.
33. Bier, F.F., Stocklein, W., Bocher, M., Bilitewski, U. and Schmid, R.D. (1992) Use of a fibre optic immunosensor for the detection of pesticides, *Sensors and Actuators B* **7**, 509-512.
34. Oroszlan, P., Duveneck, G.L., Ehrat, M. and Widmer, H.M. (1993) Fiber-optic atrazine immunosensor, *Sensors and Actuators B* **11**, 301-305.
35. Loughlin, C. (1990) Tutorial: moisture content and humidity, *Sensor Review* **7**, 137-140.
36. Ogawa, K., Tsuchiya, S., Kawakami, H. and Tsutsui, T. (1988) Humidity-sensing effects of optical fibres with microporous SiO_2 cladding, *Electronics Letters* **24**, 42-43.
37. Blyler Jr., L.L., Ferrara, J.A. and MacChesney, J.B. (1988) A plastic-clad silica fiber chemical sensor for ammonia, in *Optical Fiber Sensors*, Tech. Dig. vol. 2, Optical Society of America, Washington D.C., 369-372.
38. Russell, A.P. and Fletcher, K.S. (1985) Optical sensor for the determination of moisture, *Analytica Chimica Acta* **170**, 209-216.

39. Muto, S., Ando, A., Ochiai, T., Ito, H., Sawada, H. and Tanaka, A. (1989) Simple gas sensor using dye-doped plastic fibers, *Japanese J. of Applied Physics* **28**, 125-127.
40. Muto, S., Fukasawa, A., Kamimura, M., Shinmura, F. and Ito, H. (1989) Fiber humidity sensor using fluorescent dye-doped plastics, *Japanese J. of Applied Physics* **28**, L1065-L1066..
41. Zhou, Q., Shahriari, M.R., Kritz, D. and Sigel Jr., G.H. (1988) Porous fiber-optic sensor for high-sensitivity humidity measurements, *Analytical Chemistry* **60**, 2317-2320.
42. Shahriari, M.R., Zhou, Q. and Sigel Jr., G.H. (1988) Porous optical fibers for high-sensitivity ammonia-vapor sensors, *Optics Letters* **13**, 407-409.
43. Zhou, Q., Kritz, D., Bonnell, L. and Sigel Jr., G.H. (1989) Porous plastic optical fiber sensor for ammonia measurements, *Applied Optics* **28**, 2022-2025.
44. Posch, H.E. and Wolfbeis, O.S. (1988) Optical sensors, 13: fibre-optic humidity sensor based on fluorescence quenching, *Sensors and Actuators* **15**, 77-83.
45. MacCraith, B.D., Ruddy, V., Potter, C., O'Kelly, B. and McGilp, J.F. (1991) Optical waveguide sensor using evanescent wave excitation of fluorescent dye in sol-gel glass, *Electronics Letters* **27**, 1247-1248.
46. Ding, J.Y., Shahriari, M.R. and Sigel Jr., G.H. (1991) Fibre optic pH sensors prepared by sol-gel immobilization technique, *Electronics Letters* **27**, 1560-1562.
47. Serra, G., Schirone, A. and Boniforti, R. (1990) Fibre-optic pH sensor for seawater monitoring using a single dye, *Analytica Chimica Acta* **232**, 337-344.
48. Boisdé, G. and Pérez, J.J. (1987) Miniature chemical optical fiber sensors for pH measurements, in A.M. Scheggi (ed), *Fiber Optic Sensors II*, SPIE Proc. **798**, Bellingham WA, 238-245.
49. DeGrandpre, M.D. (1993) Measurement of seawater pCO_2 using a renewable-reagent fiber optic sensor with colorimetric detection, *Analytical Chemistry* **65**, 331-337.
50. DeGrandpre, M.D., A renewable-reagent fiber optic sensor for ocean pCO_2, in Ref. 5, 60-66.
51. Woods, B.A., Ruzicka, J., Christian, G.D., Rose, N.J. and Charlson, R.J. (1988) Measurement of rainwater pH by optosensing flow injection analysis, *Analyst* **113**, 301-306.

X Future Trends

NONLINEAR OPTICS: THEORY AND APPLICATIONS

U.L. ÖSTERBERG
Thayer School of Engineering
Dartmouth College
Hanover, N.H. 03755
U.S.A.

1. Introduction

The field of nonlinear optics has become extremely important for ultrafast signal processing in modern fiber optic communication systems. For the light propagating in the optical fiber, nonlinear effects have both positive and negative side-effects. One positive effect is the formation of solitons due to the Kerr effect in the glass, however, the same nonlinearity also causes crosstalk between different communication channels. Similarly, Raman scattering is advantageous in that it can be used to generate new wavelengths and produce tunable fiber lasers, at the same time Raman scattering also produces crosstalk in a wavelength division multiplexed system. The conclusion from these two simple examples is that it is important to understand nonlinear effects to optimize the usage of an optical fiber.

Two other chapters in this book discusses nonlinear effects in optical fibers, more specifically general theory of nonlinear optics and solitons. This chapter intends to give a brief introduction to the following very important nonlinear effects occuring in an optical fiber: Stimulated Raman scattering (SRS), Stimulated Brillouin scattering (SBS), Four-Photon Mixing (FPM), and Two-Photon Absorption (TPA). Common for all of these nonlinearities is that they are due to the third-order nonlinear susceptibility, $\chi^{(3)}$.

The outline of this chapter is as follows: first we give an introduction to nonlinear susceptibilities, spontaneous versus stimulated processes and resonant versus non-resonant transitions. Next we discuss phonons in glasses as an introduction to Raman and Brillouin scattering followed by a discussion of Stoke's and anti-Stoke's pulse propagation in an optical fiber. Four-photon mixing in optical fibers is then described followed by an introduction to two-photon absorption and how it effects pulse propagation in fibers.

2. Nonlinear susceptibilities

When strong electromagnetic fields are interacting with matter the induced polarization field **P** is most commonly described through a Taylor series of the field amplitudes,

$$\mathbf{P}(\mathbf{r},\omega_s) = \chi^{(1)}(\omega_s;-\omega_s)\mathbf{E}(\mathbf{r}) + \chi^{(2)}(\omega_s;\omega_1,\omega_2)\mathbf{E}_1(\mathbf{r})\mathbf{E}_2(\mathbf{r}) + \chi^{(3)}(\omega_s;\omega_1,\omega_2,\omega_3)\mathbf{E}_1(\mathbf{r})\mathbf{E}_2(\mathbf{r})\mathbf{E}_3(\mathbf{r}) + \ldots \quad (1)$$

where ω_s is the frequency of the generated polarization, $\chi^{(n)}$ is the electric susceptibility of first, second, and third order for n=1, 2, 3, respectively, $\mathbf{E}(\mathbf{r})$ are the electric field amplitudes and ω_1, ω_2, ω_3 etc. are the carrier frequencies for the different electric fields. Inherent in this description is that the nonlinear terms are small, and this is the reason why we can use a Taylor series approximation. Small nonlinear terms indicate that the light frequencies used to excite the material do not coincide with an electronic resonance frequency in the material. We will only be concerned with non-resonant phenomena in this chapter. To describe resonant nonlinear effects, equation (1) above is not suitable [1].

The electronic susceptibilities are technically referred to as tensors and they have all the information needed to describe the interaction between the material and the light. A susceptibility tensor can be described by (for a mathematically correct definition see [2]),

$$\chi^{(n)}_{i,j,k,\ldots}(\omega;\omega_1,\omega_2\ldots) = \frac{\text{spatial dispersion}}{\text{frequency dispersion}} = \sum \frac{\langle g|\mathbf{r}|f\rangle}{\omega_0^2 - \omega^2 - i2\omega\gamma}. \quad (2)$$

The subscript indices; i,j,k..., are connected with the structural symmetry of the material (spatial dispersion) and the particular polarization of the electromagnetic waves. The denominator describes the frequency dispersion with ω being the frequency of an electromagnetic wave, ω_0 being a resonant frequency in the material and γ being the width of the resonance. The summation is over all possible states that can occur in the material *while* it is interacting with the electromagnetic fields. As the rank (=n+1) increases, the numerator and denominator becomes more complicated as more terms like $\langle g|r|f \rangle$ and $\omega_0^2 - \omega^2 - i2\omega\gamma$ have to be multiplied together and we refrain from giving the complete expression for $\chi^{(3)}$ due to its complexity [3]. As can be seen from equation (2), the electronic susceptibilities are complex quantities. It is common to separate the susceptibilities into a real and imaginary part, for the third-order nonlinear susceptibility this could look like,

$$\chi_{ijkl}^{(3)}(\omega_s;\omega_1,\omega_2,\omega_3) = \chi_{Real}^{(3)} + i\chi_{Imaginary}^{(3)}. \tag{3}$$

In general, the real part describes light-matter interactions that leaves the material in the original energy state, while the imaginary part describes interactions that transfers energy between the electromagnetic wave and the material in such a way as to leave the material in a different energy state than the original state. Processes described by the real part are commonly refererred to as parametric processes and an example of such a process is four-photon mixing. It is interesting to note that nonlinear processes governed by the real part of the electric susceptibility requires phasematching while processes due to the imaginary part do not [4]. Examples of processes described by the imaginary part are Raman and Brillouin scattering, and two-photon absorption.

Important for the discussion of Raman and Brillouin scattering is the distinction between spontaneous versus stimulated processes. In simple terms, spontaneous Raman and Brillouin scattering are due to fluctuations in one or more optical properties caused by the internal energy of the material. Stimulated scattering is driven by the light field itself, actively increasing the internal fluctuations of the material.

Finally, we would like to mention that all the electronic susceptibilities in this chapter are made up of electric-dipole transitions. When these transitions are between real energy levels of the material we talk about resonant processes. In general, resonant nonlinear processes are strong and slow; strong because the susceptibility gets large at resonances and slow because the electrons have to physically get relocated. In this chapter we will only be concerned with non-resonant nonlinear processes. These nonlinearities are distinguished by their small susceptibilities but very fast response. This is in part because the electrons are making only virtual transitions. A virtual energy level can be thought of as an energy level that exist for the combined system, matter+light.

3. Raman and Brillouin scattering

1. Phonons in glasses

From basic solid state physics we know that describing vibrational excitations in a crystal is usually done with the concept of phonons. Phonons are described with a wavefunction, just like the electron or any other -on particle. The equation of motion for this wavefunction describes how e.g. temperature and sound propagates through the solid as vibrational excitations. If the atoms move in opposite directions to each other the propagation is done by optical phonons, if on the other hand the atoms move in unison the propagation is done by acoustical phonons. Additionally, since the crystal lattice is three-dimensional there exists both longitudinal and transversal optical and acoustical phonons. A generic ω-k diagram for a crystal is shown in Figure 1. The ordinate is unlabeled on purpose, to be able to use the same diagram for glass. However, in the case of a crystal the ordinate should be labeled in fractions of the first Brillouin zone.

So, what about glasses. From a crystal structure point of view, glasses are different from their crystal counterparts in that they lack long-range order, Figure 2. How does this difference affect phonon activity in the glass? To answer that question we quote the following paragraph from S.R.Elliott [6]. "At the outset one must ask the question whether, in the absence of long-range order, propagating vibrational excitations exist. The answer of course is in the affirmative, since it is common experience that window glass transmits both sound and heat. The difference between amorphous and crystalline materials essentially lies in the fact that periodicity cannot be used to simplify the dynamical equations by the introduction of Bloch states." It is therefore possible to use most of the concepts that we have learned from crystalline solid state physics when we are describing what is happening in an amorphous media such as glass.

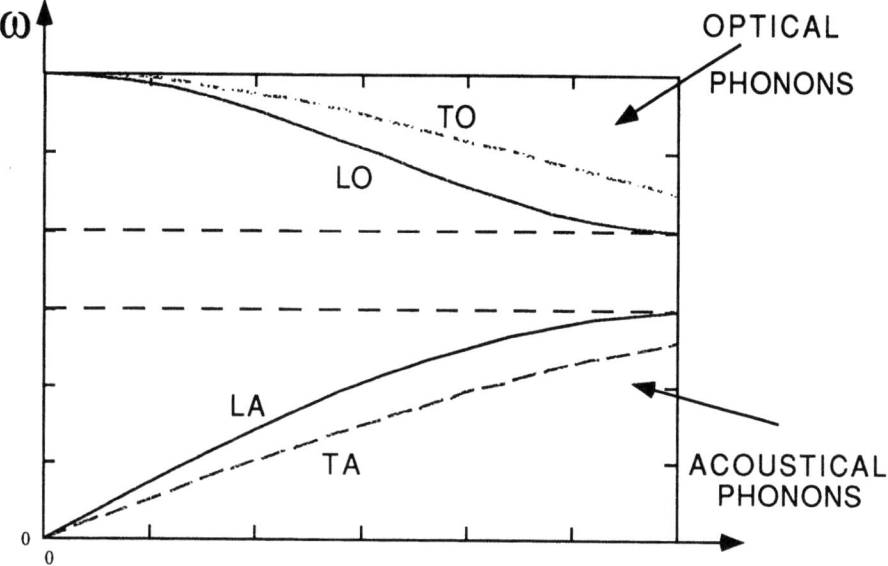

Figure 1. Generic ω-k diagram for phonons.

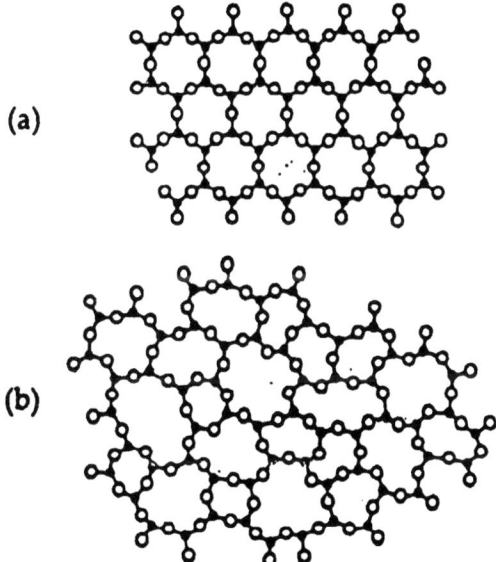

Figure 2. Network structure of SiO_4^{4+} tetrahedrons: (a) quartz crystal, (b) quartz glass. Adapted from [5].

One can describe Raman scattering as due to optical phonons and Brillouin scattering as due to acoustical phonons. Despite the many differencies between SRS and SBS, they can both be described by propagating vibrational exitations and therefore we can derive the governing equations for SRS and SBS simultaneously.

Both SRS and SBS start from spontaneously emitted photons at the Stoke wavelength. The Stoke wavelength is one vibrational phonon away on the "red" side of the pump photon. These photons are ususally referred to as noise, and in the case of optical fibers one can show that there is approximately one photon at every Stoke wavelength for each allowed mode of the light [7]. The physical origin of this noise is of course the fact that the glass molecules, e.g. at room temperature, have enough energy to vibrate around an equilibrium and scatter the incoming laser light.

2. Density waves, electromagnetic waves and Raman scattering

The purpose of this section is to introduce the various approximations that are made in order to derive the coupled equations that describe propagation of SRS. These equations can also be used for SBS by just changing the gain constant. In order to comply with the vast literature on stimulated scattering we will continue to use cgs units. The two equations we start from are the wave equation for the electromagnetic waves and the equation of motion for the molecular vibrations. The electromagnetic waves are $\mathbf{E}_p$ (pump wave) and $\mathbf{E}_s$ (Stoke's wave). The molecular vibration is described through the vibrational coordinate Q, Figure 3.

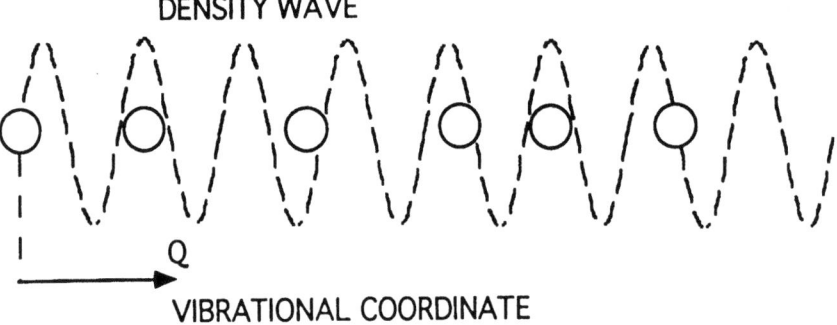

Figure 3. Illustration of a density wave propagating through a glass network.

$$\nabla^2 \mathbf{E} - \frac{\varepsilon}{c^2}\frac{\partial^2 \mathbf{E}}{\partial t^2} = \frac{4\pi}{c^2}\frac{\partial^2 \mathbf{P}_{NL}}{\partial t^2}$$

$$\frac{\partial^2 Q}{\partial t^2} + \omega_{ph}^2 Q + 2\Gamma\frac{\partial Q}{\partial t} = \mathbf{P}_{NL}\cdot\mathbf{E}$$

(4)

The total field $\mathbf{E}=\mathbf{E}_p+\mathbf{E}_s$, ω_{ph} is the phonon vibration frequency, Γ is related to the phonon lifetime and $\mathbf{P}_{NL}$ is the nonlinear polarization. The linear portion of $\mathbf{P}$ is already incorporated into ε on the left hand side of the wave equation. We have already noted that the nonlinear polarization for Raman scattering is proportional to the imaginary part of the third-order susceptibility. We will use the susceptibility for the final equations, but to gain more insight into how the density wave Q is coupled to the electromagnetic wave $\mathbf{E}$ we will write the nonlinear polarization as a function of the optical polarizability for the individual molecule,

$$\mathbf{P}_{NL} = N\left(\frac{\partial\alpha}{\partial Q}\right)Q\mathbf{E}$$

(5)

where $\alpha = \alpha_0 + \left(\frac{\partial \alpha}{\partial Q}\right) Q$ is the optical polarizability and N is the total number of molecules contributing to the scattering light in a unit volume. The static term α_0 (if it has a spatial dependence) is responsible for Rayleigh scattering. The dynamic variations of the optical polarizability, which is driven by the electric field, causes the Raman scattered light. To proceed we start by assuming the following functional forms of the various fields;

$$E_p = \frac{1}{2}\left\{e_p(z,t)e^{i(k_p z - \omega_p t)} + c.c\right\}$$

$$E_s = \frac{1}{2}\left\{e_s(z,t)e^{i(k_s z - \omega_s t)} + c.c\right\} \quad (6)$$

$$Q = \frac{1}{2}\left\{q(z,t)e^{i(k_q z - \Delta \omega t)} + c.c\right\}$$

these are plane waves propagating in the z-direction, for which $k_q = k_p - k_s$ and $\Delta\omega = \omega_p - \omega_s$. If the fields in equation (6) are inserted into equations (4) and (5) we obtain three coupled differential equations of second order. For the two electromagnetic equations we will adopt the slowly-varying envelope approximation (SVEA) which will reduce those equations to first order and for the density wave equation we will assume that the phonon lifetime is so short that Q can be considered a constant during the duration of the pump pulse. The approximation for the density wave means that we are ignoring the frequency response of the nonlinear polarization. For Raman scattering in glasses this is an excellent approximation since the phonon lifetime is $\approx 10^{-13} - 10^{-14}$ seconds. The lifetime is estimated from the inverse of the Raman gain spectra, Figure 4.

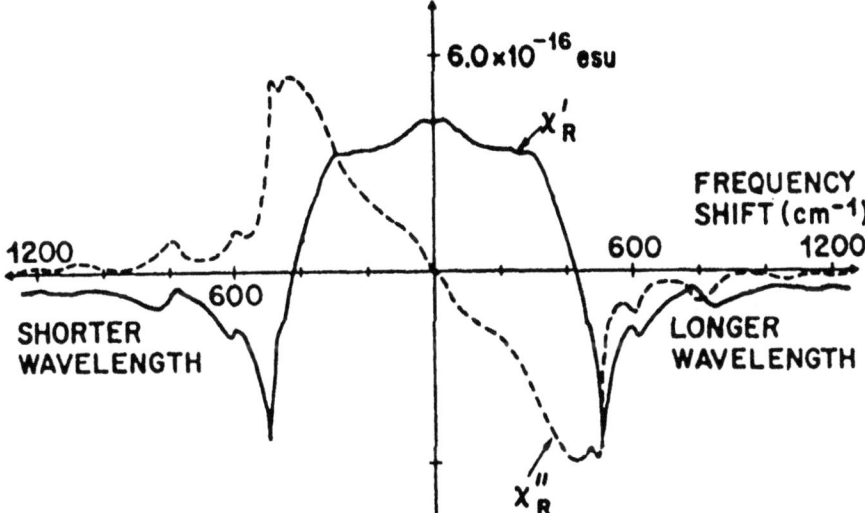

Figure 4. Real part of the third order susceptibility (solid line) and imaginary part (dashed line). From [8].

Using the above approximations we arrive at the follwing set of coupled first-order differential equations

$$\frac{\partial e_p}{\partial z} + \frac{1}{v_p}\frac{\partial e_p}{\partial t} = -g_p |e_s|^2 e_p$$

$$\frac{\partial e_s}{\partial z} + \frac{1}{v_s}\frac{\partial e_s}{\partial t} = -g_s |e_p|^2 e_s \qquad (7)$$

where $g_i = \dfrac{3\omega_i^2 \mu_0 \chi^{(3)}_{Imaginary}}{4k_i}$, $\chi^{(3)}_{Imaginary} \propto N\left(\dfrac{\partial \alpha}{\partial Q}\right)^2$ and v_p and v_s are the phase velocities for the pump and Stoke light pulses, respectively. It is clear from equation (7) that our earlier statement is correct, stimulated Raman scattering is self phasematched. Absorption can be incorporated by adding $-\dfrac{\alpha_p}{2}e_p$ and $-\dfrac{\alpha_s}{2}e_s$ to the right hand sides, respectively.

For the steady-state case it is easily verified that the Stokes signal grows exponentially with distance. The reason for this is that the imaginary part of $\chi^{(3)}$ is negative. For the same reason the anti-Stokes signal is exponentially decreasing with distance. For those who have actually recorded a spectrum from an optical glass fiber in which a high intensity lightwave is propagating it is well known that a peak is occurring on the shorter wavelength side where the anti-Stokes peak is expected. The reason for this peak is four-photon mixing between the pump light and the Stokes signal. Four-photon mixing will be discussed later in this chapter. It should therefore be stressed that the ratio between the Stokes and anti-Stokes signals in stimulated Raman scattering experiments do not carry any temperature information about the system the way the ratio does for spontaneous Raman scattering.

There are a number of issues that we have ignored so far in our brief description of SRS. The purpose of the next few paragraphs is to bring to the attention some of the more important properties of Raman propagation in optical fibers.

2.1 Raman gain and frequency spectrum

If we convert equation (7) from amplitude to intensity ($I = \dfrac{c\,n}{8\pi}|E|^2$) the solution for the Stokes' signal can be written as $I_s(z) = I_s(0) e^{g_s I_p z}$. For the Si-O vibration in an optical glass fiber the gain factor $g_s \approx 10^{-11}$ cm/W at $\lambda \approx$ 1 μm. Because of the exponential gain most of the stimulated Raman scattered light is shifted 440 cm^{-1} (1cm^{-1} ≈ 30 GHz), which corresponds to the peak of the Raman cross section, Figure 4. The width of the Raman gain is approximately 40 THz. For temporally short optical pulses (≈100 picoseconds and less) the frequency dependence of the Raman gain has to be incoporated to fully describe the evolution of the Raman scattered light. For pulses that are less than a few hundred femtosecond the approximation of Q being constant during the pulse duration is no longer valid and the differential equation for Q has to be incorporated into equation (7).

2.2 Walk-off

When one numerically solves equation (7) it is common to use a coordinate system that moves with either the pump or the Stokes pulse. If we somewhat arbitrarily choose to move along with the pump pulse ($t' = t - \tau_p z$; $z' = z$) equation (7) changes into,

$$\begin{aligned}\frac{\partial e_p}{\partial z'} &= -g_p|e_s|^2 e_p \\ \frac{\partial e_s}{\partial z'} + (\tau_s - \tau_p)\frac{\partial e_s}{\partial t'} &= -g_s|e_p|^2 e_s\end{aligned} \qquad (8)$$

The new factor ($\tau_s - \tau_p$) takes into account the fact that due to dispersion the Stokes and the pump pulse do not travel with the same velocity. For a pump pulse at 1.06 μm and a Stokes pulse at 1.12 μm the "walk-off" is approximately 2-3 ps/m with the Stokes pulse moving faster than the pump pulse. Some consequences of this walk-off are; the

Raman conversion is not as large for short pulses as it is for (the same peak power) long pulses, the frequency spectrum gets severely distorted for the pump pulse and the Stokes pulse aqcuires a substantial chirp. This latter effect is of importance for optical fiber-grating pulse compressors and for Raman amplifiers [9].

2.3 Polarization

If a polarization preserving fiber is used, it can be shown that a pump wave polarized perpendicularly to the Stokes wave does not transfer energy. This can be seen from the symmetry properties of the third-order susceptibility tensor describing an amorphous material such as glass. In the case of a standard non-polarization preserving optical fiber it becomes more difficult. Unless special precautions are taken both the pump and the Stokes pulse exit the fiber randomly polarized.

2.4 Overlap integral

When dealing with nonlinear effects in waveguides one has to remember that light can propagate in different modes and that the same mode "occupies" a different volume depending on the wavelength. In the case of Raman scattering in a waveguide the electromagnetic fields in equation (6) should be modified from being plane waves to having an arbitrary transverse profile,

$$\mathbf{E} = \frac{1}{2}\left\{ e_p(z,t) \cdot F(R,\varphi) \cdot e^{i(k_p z - \omega_p t)} + c.c \right\}. \tag{9}$$

The normalized projection of the nonlinear contribution to the Stokes wave is then given by the following factor [9],

$$\langle s|p \rangle = \frac{\int_0^{2\pi}\int_0^{\infty} |F_s(R,\varphi)|^2 |F_p(R,\varphi)|^2 R\, dR\, d\varphi}{\int_0^{2\pi}\int_0^{\infty} |F_s(R,\varphi)|^2 R\, dR\, d\varphi} \tag{10}$$

the factor $\langle s|p \rangle$ should be multiplied by the factors on the right hand side of equation (7). It is worthwhile noticing, that to a surprisingly large degree it is sufficient to use the plane wave analysis for nonlinear interactions in optical fibers. This is partly because a mode has the same propagation constant across its tranverse profile. If indeed the modal nature of the light has to be taken into account, it is often enough to approximate the modes using a Gaussian profile or derivatives thereof.

3. Brillouin scattering

In the case of Raman scattering the optical phonon wave originates from the vibrational mode in the Si-O bond. For Brillouin scattering the acoustical phonon wave can arise from either optical absorption or electrostriction. For optical fibers, absorption is negligible in the visible and near-infrared so the main reason for the scattering is from electrostriction. Therefore, the phonon wave in Brillouin scattering is a pressure wave originating from the density variations in the glass where the light intensity is high.

The coupled equations in (7) are still valid but the gain constant g is obviously different. First of all, the gain for Brillouin scattering is approximately two orders of magnitude larger than for Raman scattering, i.e. $g \approx 10^{-9}$ cm/W. The acoustic wave is also a lot slower than the optical wave so the frequency downshift for the Brillouin scattered Stokes wave is only 1 cm^{-1}. Another big difference between the Brillouin and Raman gain, is the frequency width. As can be seen from Figure 5, the width is only in the order of 10^8 Hz for the Brillouin gain. As with any gain process, if the pump bandwidth is larger than the gain bandwidth the effective gain is reduced, similarly for the Brillouin gain constant. The gain therefore scales as $g \propto \frac{\Delta v_{Brillouin}}{\Delta v_{pump}}$. This means that for short optical pulses (i.e.

large bandwidth) propagating in an optical fiber, Brillouin scattering is negligible. For coherent systems, where narrow bandwidth cw lasers are being used, Brillouin scattering is a very noticeable effect.

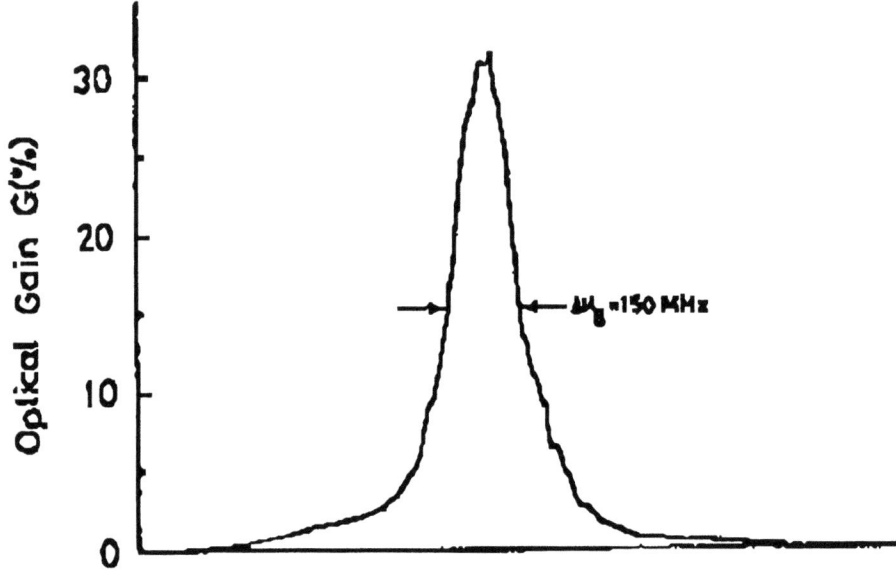

Figure 5. Brillouin gain in an optical fiber. Adapted from [10].

Brillouin scattering differs from Raman scattering in a number of other ways also. For example, the optical phonon wave in Raman scattering consists of both longitudinal and transverse waves, while the acoustical phonon wave in Brillouin scattering only consists of a transvere wave, unless very special experimental conditions have been set-up. Another difference is that the Raman scattered light propagates in both the forward and backward directions, while the Brillouin scattered light only propagates in the backward direction. The reason for this can be seen from the simple phasematching diagram in Figure 6.

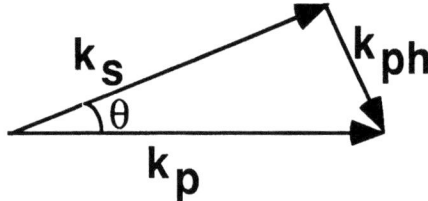

Figure 6. Wavevector mismatch.

It is easy to verify that $k_{ph}^2 = k_p^2 + k_s^2 - 2k_p k_s \cos\theta$, where k_{ph} is the wavevector of the phonon and k_p and k_s are the wavevectors for the pump laser and the Stokes wave, respectively. θ is the angle between the electromagnetic wavevectors. In the case of Brillouin scattering, $k_p \approx k_s$ which leads to the following modification of the above equation, $k_{ph} \cong 2 \cdot k_p sin\left(\frac{\theta}{2}\right)$. It is evident from this last equation that the forward scattered light is zero and the backward scattered light is maximum.

4. Four-photon mixing

Four-photon mixing is a nonlinear parametric process, and as such it is due to the real part of the third-order nonlinear susceptibility tensor, Figure 4. The two main four-photon mixing schemes are illustrated in Figure 7. In the degenerate case, two pump photons are generating one Stoke and one anti-Stoke wave. The referral to Stoke and anti-Stoke is just to illustrate that one light wave is red shifted and the other is blue shifted, relative to the pump frequency. In the non-degenerate case two pump photons at different wavelengths are generating a Stoke and an anti-Stoke wave. The generation of these new wavelengths is due to third-order nonlinear mixing between the pump photons and (originally) noise at the Stoke and anti-Stoke wavelengths. For a given fiber, only the noise at wavelengths that will fulfill the energy and momentum equations in Figure 7, will grow. Due to the constraints of phasematching it is unusual to have optimized FPM fibers longer than 5-10 metres. An exception to this is when the pump wavelength is in the vicinity of zero dispersion.

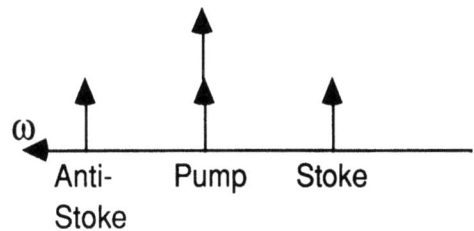

$$\omega_s - \omega_p = \omega_p - \omega_a$$

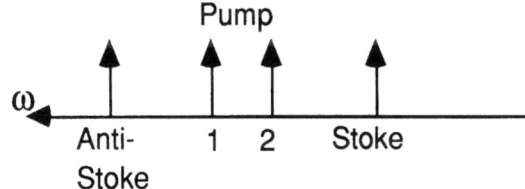

$$\omega_s - \omega_{p2} = \omega_{p1} - \omega_a$$

$$k_s - k_2 = k_1 - k_a \quad \text{PHASEMATCHING}$$

Figure 7. Energy and momentum conservation for four photon mixing.

To derive the equations of motion for the Stoke and anti-Stoke waves we, again, start from the wave equation in equation (4). For four-photon mixing the nonlinear polarization is $\mathbf{P} = \chi^{(3)}_{Real}\mathbf{E}\cdot\mathbf{E}\cdot\mathbf{E}$ (in cgs-units). The total electric field $\mathbf{E}$ is $\mathbf{E} = \mathbf{E}_p + \mathbf{E}_s + \mathbf{E}_a$, where p,s, and a refer to pump, Stoke and anti-Stoke wave, respectively. If we again adopt SVEA and assume a non-depletable pump we obtain the following two coupled set of equations [11],

$$\frac{de_s}{dz} = i\cdot c_1\cdot \chi^{(3)}_{Real}\cdot e_p^2 e_a^* \cdot e^{i\kappa z}$$

$$\frac{de_a^*}{dz} = i\cdot c_2\cdot \chi^{(3)}_{Real}\cdot e_p^2 e_s \cdot e^{-i\kappa z}$$

(11)

where the * refers to the complex conjugate and $\kappa = \Delta k - \delta k_s - \delta k_a + \delta k_{p1} + \delta k_{p2}$ (for the non-degenerate case). It is clear from equation (11) that if phasematching is achieved, the Stoke and anti-Stoke signals grow exponentially. For phasematching to occur κ has to be zero. The different contributions to the phasemismatch are; material and waveguide dispersion that are both incorporated into Δk, and the influence of the intensity dependent refractive-index that is incorporated through the terms δk_i. In most cases it is sufficient to balance the material and waveguide dispersion to get "close enough" to the right phasematching conditions. If we subsequently concentrate on the Δk term we find that it can be written as,

$$\Delta k = \Delta k_{mat} + \Delta k_{wav} \tag{12}$$

where

$$\Delta k_{mat} = 2\pi\cdot\left[\frac{n_s}{\lambda_s} + \frac{n_a}{\lambda_a} - \frac{n_1}{\lambda_1} - \frac{n_2}{\lambda_2}\right]$$

$$\Delta k_{wav} = 2\pi\cdot(\Delta n)\cdot\left[\frac{b_s^{lm}}{\lambda_s} + \frac{b_a^{lm}}{\lambda_a} - \frac{b_1^{lm}}{\lambda_1} - \frac{b_2^{lm}}{\lambda_2}\right]$$

(13)

n is the refractive index, Δn is the refractive index difference between core and cladding and b is the normalized propagation constant. It should be emphasized that the superscripts lm refer to different mode combinations (LP-modes) and that all four waves can be, and sometimes are, in different modes. If we use typical fiber parameters we can plot Δk_{mat} and Δk_{wav} as functions of the shift in the same diagram, Figure 8. First of all, note that it is the negative waveguide dispersion that is plotted, so where the two curves are crossing each other Δk is zero. The largest frequency shifts are typically obtained for the degenerate case with the two pump photons in the highest order mode (LP01) and the Stoke and anti-Stoke in the next higher order mode (LP11). The next size frequency shift is typically obtained by using the birefringence within the optical fiber and the smallest frequency shift is obtained for the non-degenerate case with the two pump photons in different modes [11].

5. Two-photon absorption

Two-photon absorption (TPA) has been known for many years as an excellent, complimentary tool to one-photon absorption (OPA) for probing microscopic properties of matter since it obeys different selection rules than OPA [12]. Recently, it was also shown that TPA is very important for determining the figure-of-merit for a material to be used for optical switching [13].

TPA in optical fibers have received a lot of attention lately, primarily due to its connection to photoinduced effects in glasses [14]. It has been observed that a periodic $\chi^{(1)}$ can be induced in an optical fiber from absorption of either two green (514 nm) or two blue (488 nm) photons. Additionally, interference between two-photon absorbed IR (1.06 μm) photons and one-photon absorbed green (0.532 μm) photons can produce a periodic $\chi^{(2)}$ in the glass of the optical fiber. A simplified energy level diagram of germanium doped SiO_2 is shown in Figure 9.

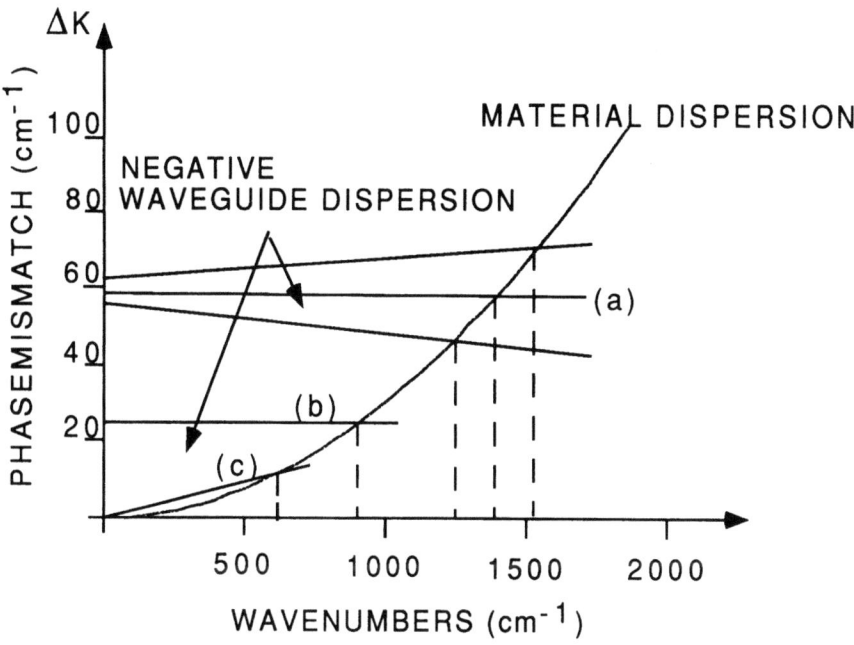

Figure 8. Phasematching curves for four-photon mixing in optical fibers: (a) degenerate pump waves, (b) birefringence, and (c) non-degenerate pump waves. Adapted from [11].

Many, but not all, of the defects that are believed to be of importance for photoinduced effects in optical fibers are shown in the energy diagram [15]. It is evident from this energy diagram that higher order absorption is needed to alter the properties of the glass with visible or infrared light.

Two-photon absorption can schematically be thought of as in Figure 10. In TPA the summation has to be performed over all possible real and virtual states. This includes virtual states that are at an energy level higher than the final state. What may seem as a violation of conservation of energy is actually not, the reason being that the lifetime of the virtual state is so short that the energy uncertainty places the energy level within reach of just one photon.

One method for measuring the TPA coefficient is to plot the inverse transmission versus input power, Figure 11. In this diagram the linear absorption can be obtained at the intercept between the measured data and the abscissa and the TPA coefficient is obtained from the slope of the measured data. To realize that this is the case, we can start from the following equation, that describes how the light intesity is absorbed due to OPA and TPA,

$$\frac{dI}{dz} = -\alpha I - \beta I^2 \tag{14}$$

where I is the light intensity and α is the linear absorption coefficient and β is the two-photon absorption coefficient. If we integrate equation (14) and use the power instead of the intensity we obtain the following equation [13],

$$\frac{1}{T} = \frac{1}{T_0} + \frac{\beta L_{eff} P_{in}}{C_2 e^{-\alpha L} A_{eff}} \tag{15}$$

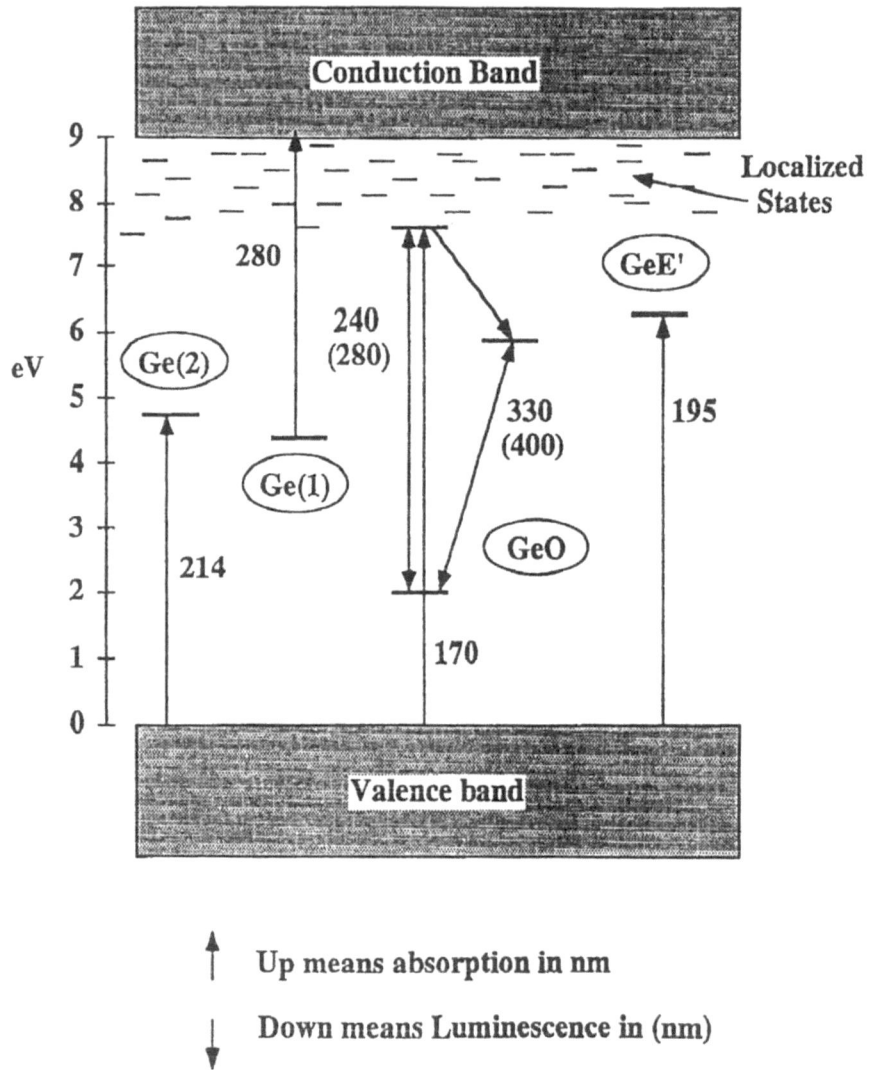

Figure 9. Schematic energylevel diagram for germanium doped silica glass.

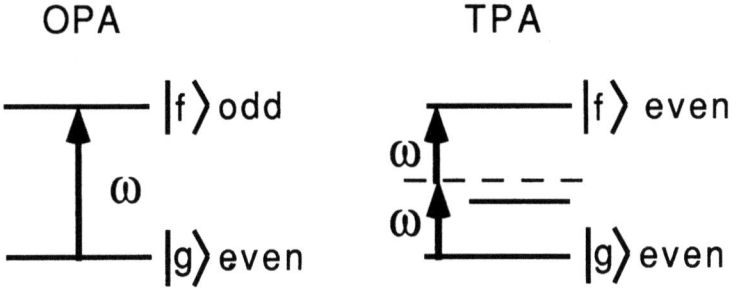

$$W \propto \sum_i \frac{\langle g|p|i\rangle\langle i|p|f\rangle}{\omega_{ig} - \omega_1} + \sum_i \frac{\langle g|p|i\rangle\langle i|p|f\rangle}{\omega_{ig} - \omega_2}$$

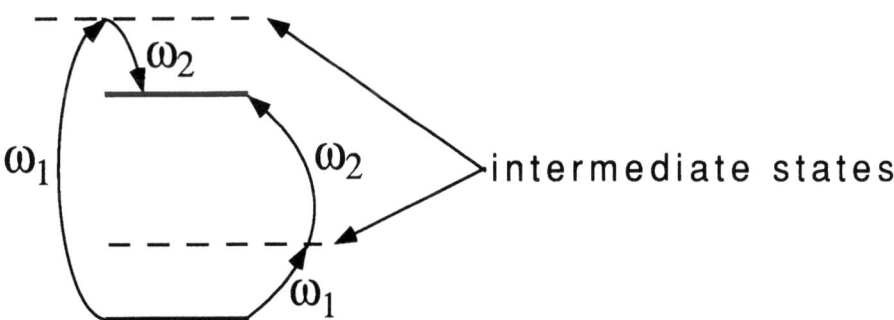

Figure 10. (Top) Difference between OPA and TPA, (middle) transition probability for TPA, and (bottom) illustration of virtual states in TPA transitions.

where P_{in} is the icident laser power, $L_{eff} = \int_0^L e^{-\alpha z} dz$, A_{eff} is the effective core area (for light close to cut-off this area is approximately the geometric area), C_2 is the fraction of light coupled out of the fiber core, and T_0 is the linear transmission.

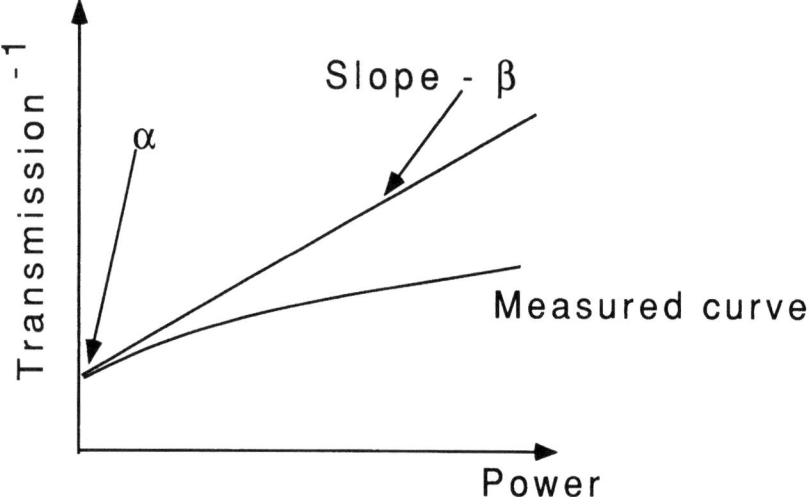

Figure 11. Illustration of result obtained from a power dependent absorption measurement.

In optical fibers the TPA coefficient is so small that intense laser pulses have to be used in order to obtain a signal. In such experiments the measured curve seems to saturate, Figure 11. In these cases the time dependence has to be incorporated. For gaussian temporal pulses this is easily done and a linearized curve is obtained from which β can be extracted.

Some typical values for β in a germanium-doped silica glass fiber are; $\beta < 10^{-3}$ cm/W for $\lambda = 0.527$ µm and $\beta \approx 10^{-6}$ cm/W for $\lambda = 1.053$ µm [16].

6. Conclusion

Even though the third-order nonlinearity in glasses is very small compared to most other nonlinear materials, third-order nonlinearities in optical fibers are very important. The reason is of course that the light in an optical fiber is confined to a small area over long distances.

The conclusion from this and the other chapters on nonlinear effects should be that it is essential for anyone working with optical fibers to understand the fundamentals of nonlinear optics.

7. References

1. Butylkin, V.S., Kaplan, A.E., Khronopulo, Y.G., and Yakubovich, E.I. (1989) *Resonant Nonlinear Interactions with Matter*, Springer-Verlag, Heidelberg.
2. Boyd, R.W. (1992) *Nonlinear Optics*, Academic Press, Boston.
3. Flytzanis, C. (1975) Theory of nonlinear optical susceptibilities, in H. Rabin and C.L. Tang (eds), *Quantum Electronics:A Treatise, volume I*, Academic Press, New York.
4. Marcuse, D. (1980) *Principles of Quantum Electronics*, Academic Press, New York.
5. Izumitani, T. (1984) Optical glass, UCRL-TRANS-12065, Lawrence Livermore National Laboratory.

6. Elliott, S.R. (1990) *Physics of Amorphous Materials*, Longman Scientific & Technical, Harlow.
7. Smith, R.G. (1972) Optical power handling capacity of low loss optical fibers as determined by stimulated Raman and Brillouin scattering, *Applied Optics* **11**, 2489-2494.
8. Chraplyvy, A.R., Marcuse, D. and Henry, P.S. (1984) Carrier induced phase noise in angle-modulated optical fiber systems, *IEEE J. Lightwave Technology* **LT-4**, 6-12.
9. Schadt, D. (1989) Contributions to nonlinear pulse propagation in single mode optical fibers, Ph.D. Thesis, TRITA-FYS 2081, Stockholm.
10. Shibata, N., Waarts, R.G. and Braun, R.P. (1987) Brillouin-gain spectra for single-mode fibers having pure silica, GeO_2-doped, and P_2O_5-doped cores, *Optics Letters* **12**, 269-271.
11. Stolen, R.H. and Bjorkholm, J.E. (1982) Parametric amplification and frequency conversion in optical fibers, *IEEE J. Quantum Electronics* **QE-18**, 1062-1072.
12. Mahr, H. (1975) Two-photon absorption spectroscopy, in H. Rabin and C.L. Tang (eds), *Quantum Electronics:A Treatise, volume I*, Academic Press, New York.
13. Mizrahi, V., Delong, K.W., Stegeman, G.I., Saifi, M.A. and Andrejco, M.J. (1989) Two-photon absorption as a limitation to all optical switching, *Optics Letters* **14**, 1140-1142.
14. Agrawal, G.P. (1989) *Nonlinear Fiber Optics*, Academic Press, Boston.
15. Gallagher, M.D. (1993) Induced optical effects in germanium-doped silica glass, Ph.D. Thesis, Dartmouth College.
16. Weitzman, P.S. (1994) Modeling of photoinduced second harmonic generation in silica based glasses, Ph.D. Thesis, Dartmouth College.

ADVANCED OPTICAL FIBRE OPTOELECTRONICS

Professor Brian Culshaw
Electronic & Electrical Engineering Department
University of Strathclyde
204 George Street
Glasgow G1 1XW, United Kingdom

ABSTRACT

Fibre optics is over 30 years old and in its relatively short history has achieved astonishing technological maturity. The techniques which we currently see in the development laboratory and in real applications are, almost without exception, based upon concepts enunciated in the 1960s and 1970s. The suggestion that their could be "advanced" optical fibre optoelectronics implies that equally profound new ideas are currently emerging. This paper speculates on the somewhat unusual uses of fibre optics which are currently in the research laboratories. Whether these will lead to anything so profound as the fibre laser or the idea of zero dispersion is probably doubtful but the area is still buoyant and energetic embracing thoughts ranging from the scanning tunnelling microscope through time reversal systems.

INTRODUCTION

The suggestion that fibre optics could play a major role in telecommunication systems was first made in the mid 1960s. Whether it was English[1] or French[2] in origin could be debated endlessly but in practice its origins are now of little practical relevance. At the early stages a - now trivial - break-even point on fibre performance was set at 20dB/km attenuation and around one nanosecond/km dispersion to ensure that the fibre optic transmission medium could supplant the technical achievements for copper coaxial cable in which the fundamental limitation introduced by skin effect phenomena severely limit high frequency performance possibilities.

Within the same era the laser appeared and indeed the optical fibre amplifier[3] was first suggested. Consequently by 1970 the thought had emerged that an all-optical transmission system could be feasible though perhaps dispersion could be a problem. In the early 1970's it was first appreciated that the zero dispersion point within silica glass characteristics could be accessed thereby providing the opportunity for a linear low dispersion transmission medium[4]. In the same era the potential offered by solitons also became apparent in permitting zero dispersion transmission in a non-linear medium[5].

Indeed one could argue that the foundations of modern optical fibre systems were all laid well before 1975. Since then we have seen very significant technological advances in both the fabrication of fibres themselves and in the design and implementation of the surrounding infrastructure such as sources and detectors and associated electronics. The impact on transmission systems for communications has been dramatic. An excellent figure of merit for a communication system is its capacity in Gb/s multiplied by the distance in kilometres between repeaters. The best obtainable with copper cables is somewhat less than 1 Gb/s.km. Figure 1 graphically illustrates the numerous orders of magnitude improvement which have been achieved within the thirty years of fibre communication system history. Further since the cost of such communication systems is in real terms approximately inversely dependent upon this figure of merit, fibre optic transmission has enabled much of the communications revolution which we have witnessed over the last quarter of a century by providing extremely low cost high capacity digital links over immense distances.

We could then argue - and convincingly - that "advanced optical fibre optoelectronics" is already with us. The powerful economic benefits in the communications industry have exploited the conceptual foundation so imaginatively laid twenty years ago to excellent effect. We now see all-fibre transmission lines with guided wave optical fibre amplifiers involving little or no regenerative electronics and encompassing transmission distances in the thousands of kilometres.

The inevitable conclusion is that advances in optical fibre optoelectronics must in future impact upon other application sectors. This logic, whilst sound, has its black side in that only the communications industry has been able to identify the necessary economic benefits to stimulate the very significant technological advances which we have witnessed over the last quarter of a century. Without this economic stimulus, the science and technology would have languished as a curiosity. No other single application of optical fibre has anything approaching this market potential. Consequently the investment possible within any new applications sector is orders of magnitude lower.

However this should not detract from its interest and stimulus. The technical mind thrives on adversity so the lack of predictive substantial investment has often stimulated ingenuity and application which is creative and productive whilst requiring only the investment appropriate to the sector it serves.

Within this paper we shall discuss some of these rather unusual, often creative, often idiosyncratic concepts which have expanded the science and technology of fibre optics during the past decade or thereabouts. None of these ideas and concepts has yet become a really big issue though perhaps the ideas of all optical routing switches will someday make their mark. The remainder have made their presence felt in that other area of information science - instrumentation. Here we have a fragmented and specialised industry which historically has thrived on ingenuity rather than investment and in which the ingenious exploitation of guided wave optics will make a major impact.

The structure of a discussion on advanced optical fibre optoelectronics could run along either applications oriented or technology oriented pathways. In what follows we shall opt for the latter primarily because many of the application sectors have yet to emerge.

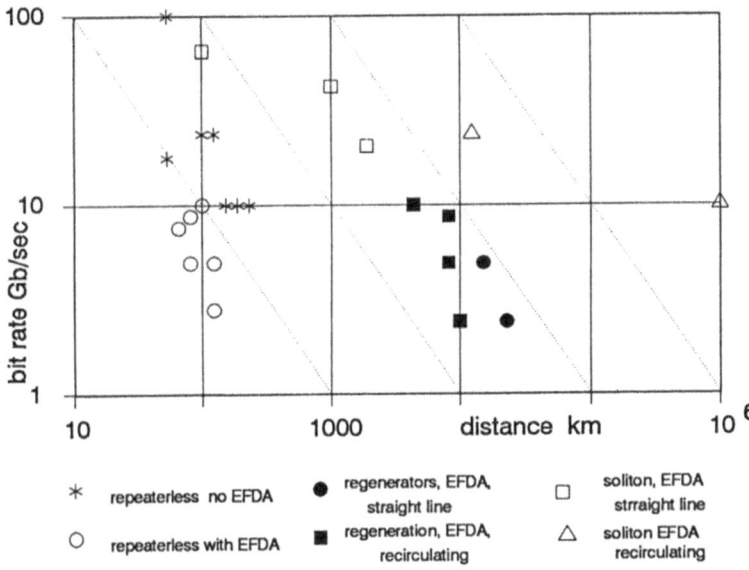

Figure 1: Achieved performance in Gb.s^{-1}km for experimental optical fibre communication systems at 1.5μm wavelength (EFDA Erbium Amplifier).

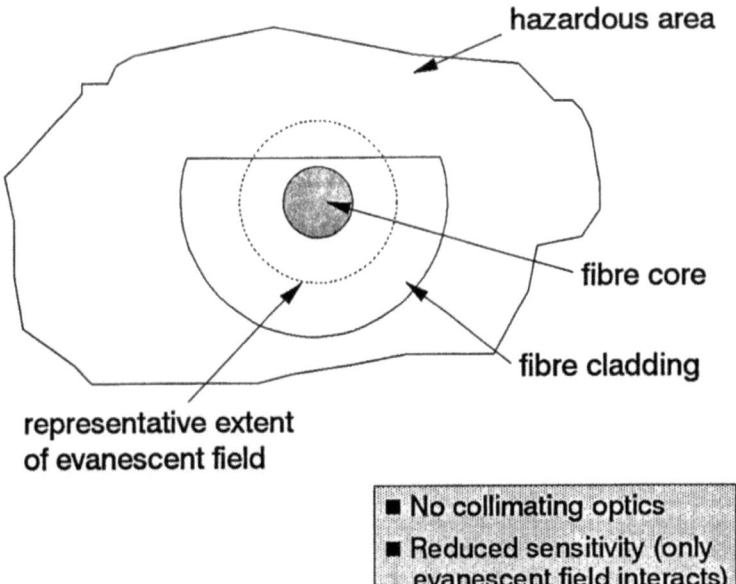

The discussion also will venture outside the narrow definition of optoelectronics but since any optical fibre application involves transmitting light and detecting it electronically I feel that we shall always be staying within the literal bound.

The discussion which follows will examine some of the trends in the following:

- the optical fibres themselves, their structure, their geometry and their materials

- coating and cabling technologies and novel means whereby these may be exploited

- optical effects occurring within optical fibres, their optimisation and their exploitation

- components fabricated within and upon optical fibres

- optical fibre sub-systems, their features and their applications.

It would be misleading to even hint that this treatment covers all the options. The essential observations to be drawn from what follows are that the general area of optical fibre technology remains scientifically and technically very much alive and that the ingenuity factor continues to pull optical fibre principles out of the tightly specified communications domain into the more fragmented and frequently more challenging disciplines imposed by instrumentation and signal processing.

FIBRE OPTICS - WHERE NEXT

Optical Fibres and Devices

By far the majority of the effort on optical fibre design, research and development has been devoted towards achieving the extremely high quality communications grade fibres which feature strongly elsewhere in this volume. These fibres address attenuation and dispersion issues in the low loss windows at 1.3 and 1.55 µm wavelengths. The emphasis is on single mode operation within these wavelength regions using nominally circular core geometries and largely ignoring the polarisation dispersion between the two quasi degenerate modes which such fibres support.

To support these fibres there is also significant effort on praseodymium and erbium all-fibre amplifiers. The latter is proved and working as a commercial entity, the former is rapidly emerging from the research laboratories and is targeted at the 1.3 µm band where the long recognised zero dispersion point greatly facilitates waveguide design. Low dispersion waveguides and the availability of an amplifier can compensate for the slightly greater losses which are fundamental to operation at the shorter wavelength. Again these amplifiers are described elsewhere in this volume.

There is however a growing subculture which has modified the basic fibre fabrication processes to slightly different ends with particular reference to sensing and

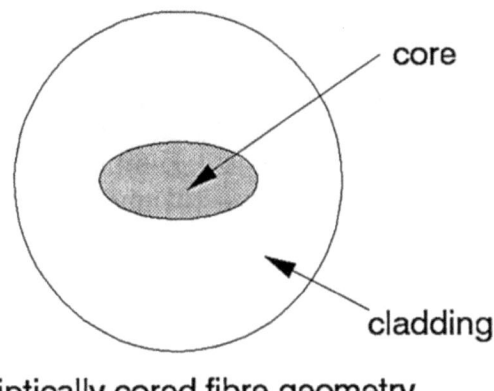

elliptically cored fibre geometry

Figure 3 Geometrical birefringence in optical fibres.

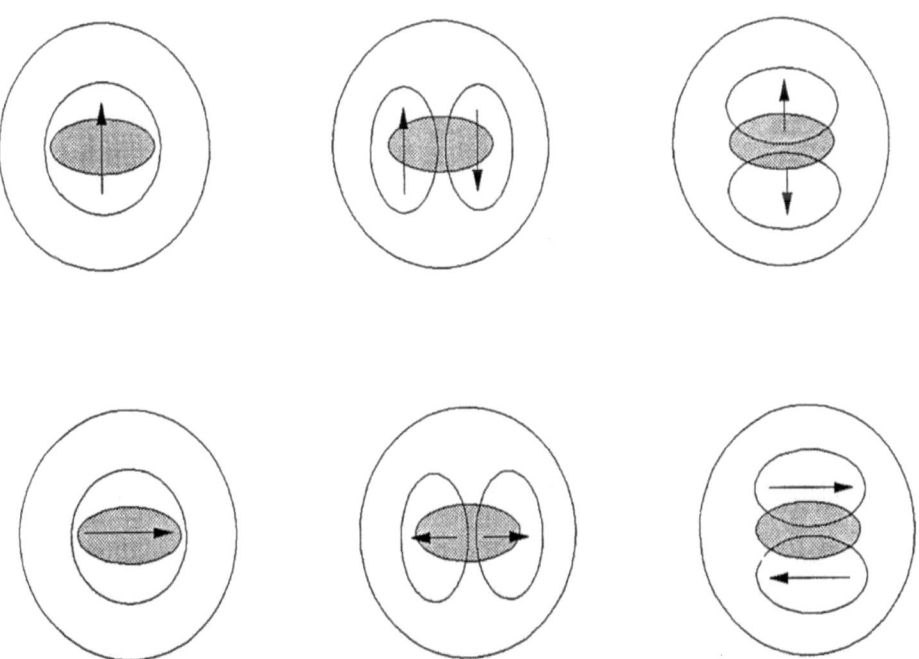

Representative modal fields for elliptical cored fibre suporting two lowest order mode sets

Figure 4 Modes in low-moded birefringence fibres.

instrumentation applications. One example of such a fibre is shown in Fig. 2. In this fibre almost half of the preform has been ground away prior to drawing resulting in a D shaped cross section in this case with a circular core. If the flat of the D is close to the core then some of the evanescent field of light travelling along the fibre will penetrate through the surface and address the medium outside. This configuration has been used effectively as a spectroscopic probe for methane gas. The illumination through the fibre has focused on the rotational overtone at 1.66 μm wavelength and the sensitivity achieved in an integrated measurement promises the later availability of distributed monitoring of methane gas concentration as a function of position along the fibre[6].

This represents a simple modification to a communications grade fibre. A slightly less simple variant involves forming an elliptical core[7] within the fibre (Fig. 3). The pronounced asymmetry of this geometry, especially if the core is fabricated from or doped to a high concentration with germania, introduces a pronounced asymmetry in the propagation constants of modes polarised along the major and minor axes of the ellipse. The beat length between these modes in such fibres is typically a few millimetres and is sufficient to ensure that the coupling between light launched in one mode and the orthogonal mode is very small (typically -50dB m). These fibres are relatively inexpensive to produce and are beginning to find their way into mass applications in fibre optic gyroscopes[8].

These fibres have also found their place in overmoded device design. Under normal circumstances the fibre will be operated in the single mode regime. However at somewhat shorter wavelengths the fibre becomes slightly overmoded and depending upon the aspect ratio of the ellipse[9] can be considered as a six moded guide (see Fig. 4). These modes all have different propagation constants and different dependencies of these propagation constants upon parameters such as temperature and strain. Cautious selection of pairs of modes from this set (to emphasise the distinction between strain and temperature sensitivities) can produce single fibre sensing systems which are capable of discriminating between strain and temperature effects by measuring propagation delays and their variation with mode number[10]. Temperature resolutions of 1° and strain resolutions of a few tens of microstrains have been demonstrated using this principle (Fig. 5).

The elliptical core has also produced a fascinating variance on the fibre laser principle. Neodymium doped elliptical core fibres have recently become available as experimental samples. These fibres - doped with the simplest of rare earth dopants - have already shown polarisation selective lasing operation (the effective gain is greater for one polarisation state than the other) and have also been shown to exhibit very sensitive optically driven switching characteristics. These switches have demonstrated "time gain". The lifetime of the metastable state in neodymium is known to be in the region of 400 microseconds. However by exploiting clustering effects[11] this can be cut by 3 orders of magnitude. Whilst the effect is clearly considerably slower than Kerr induced switches[12], which have also been demonstrated in elliptically cored (but not neodymium doped) fibre, the sensitivity in the doped fibre is 3 to 4 orders of magnitude higher. These effects are but recently demonstrated and in the very early stages of a

POLARIMETRIC AND MODE-MODE SENSORS

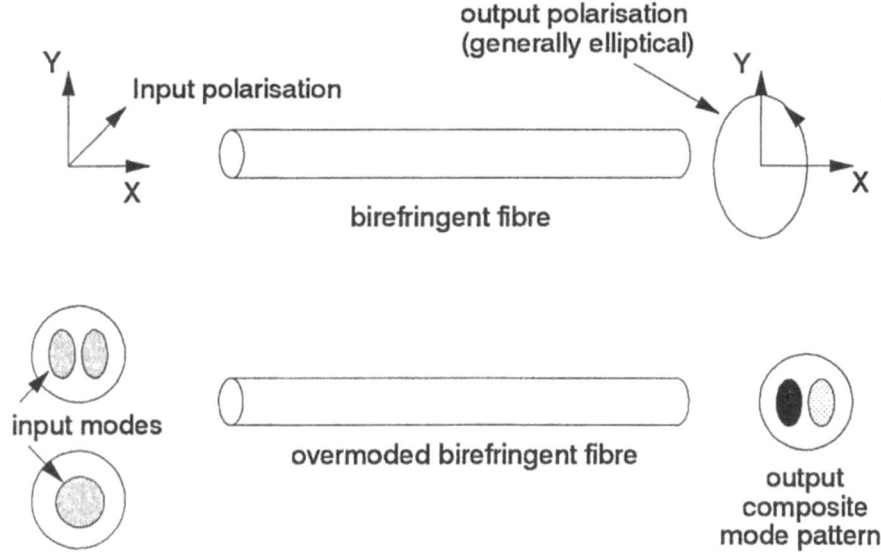

Figure 5 Principle of one form of dual measurement for combined temperature:strain monitoring. The birefringence measurement is performed at a longer wavelength beyond cut-off.

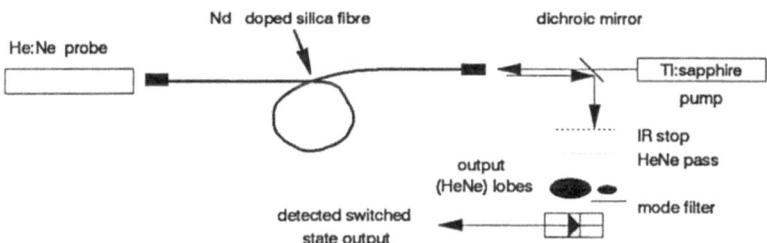

Experimental demonstration of two moded Nd fibre switch

Figure 6 Switching in overmoded fibre. The output, at He:Ne 633nm wavelength alternates between the two output lobes.

research programme but the prospects for all optical routing switching remain the dream of the communications engineer (Figure 6) The arguments for optical switching are very seductive but the observation that the power of semiconductor switching devices continues to increase whilst optical devices are often physically long and cumbersome must always bring a hint of caution into any discussion on this topic.

That said the role of fibres in interconnects, especially as the all-optical back plane, shows considerable promise[13, 14]. The motivation here stems primarily from the low radiation losses which guided wave optics offer. There are however also more subtle benefits associated with the minimisation of clock skew in very high bit array systems. There is also promise for the D fibre shown in figure 2 in this area. The fact that the evanescent field penetrates the surface of the flat of the D implies that in principle when two D fibres are placed together (Fig. 7) the coupling between them can be designed such that a predetermined optical power fraction is transmitted from one to the other. The coupling length can be effectively controlled by the angle at which the two D fibres intersect and the coupling strength by the distance of the flat surface from the core. It is therefore in principle quite feasible to design a fibre ribbon which cross couples to a second fibre ribbon simply by laying one on top of the other at an appropriate angle. The lateral and longitudinal tolerances are irrelevant provided that the fibres intersect at the appropriate angle and the propagation constants match.

In the interconnect applications especially for back plane use the optical losses of the waveguide are largely irrelevant and even though the data rates can be extremely high the transmission distances are so short (typically less than 1m) that specifically designed dispersion characteristics are not necessary. Consequently polymeric waveguides are beginning to have a significant impact in this area[15]. In this context simple fabrication processes which enable the cost effective production of complex waveguide arrays are the principal benefits.

Plastic waveguides have also been used for many years in sensing applications. Perhaps the most appealing is the florescent guide (Fig. 8a) which has been known for many years as a particularly visual means for demonstrating the presence of ultraviolet light[16]. The principal intellectual point about this waveguide is that it enables light coming in from the side (which therefore in principle could never be guided) to excite longer wavelength light with a different spatial distribution than the input light, some of which can be guided. Additionally the fact that this light is at longer wavelength implies (usually) that the losses which this fluorescent excitation experiences are significantly lower. The principle can also be used to provide a crude indication of the location of the ultraviolet source by measuring the relative amplitudes of the fluorescence at each end and assuming that the losses of the fibre are stable at the fluorescent wavelength.

More recently the general concept of this fluorescent light has been modified somewhat by using a hollow silica fibre[17] filled with lasing dye (rhodamine 6G) and using the stimulated emission from the dye. In this case the visible output stimulated by local ultraviolet radiation (Fig. 8b) is heavily attenuated by the dye but over lengths of a

Another use for D fibres: Back plane interconnect

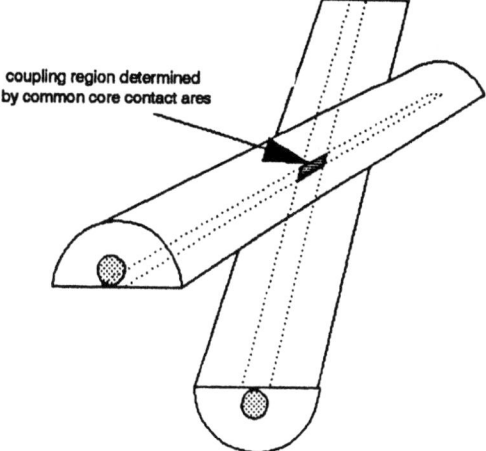

Figure 7 Contact coupling using D fibres, more usefully configured in multi-way flat cables.

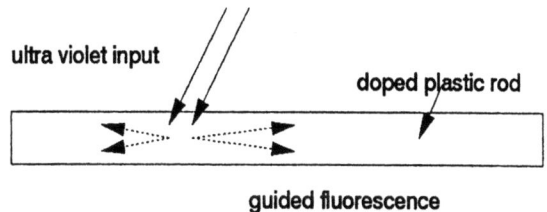

(a) ultra violet to visible conversion in doped plastic guide

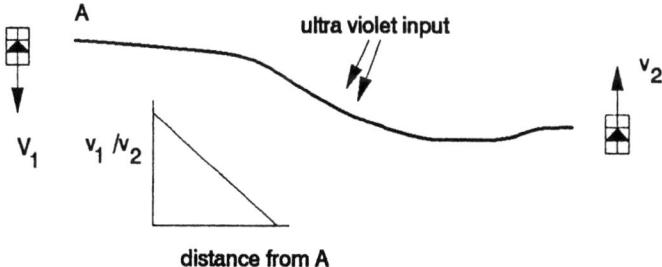

(b) rhodamine 6G filled fibre version

Figure 8 (a) Ultra-violet - visible conversion in plastic rod guides.
(b) Similar conversion using the more efficient fluorescence from a lasing dye.

metre or so the ratio of the outputs from the two ends of the fibre can give an accurate indication of the location of the localised ultraviolet pulse.

Plastic fibres have been used for many years in this mode, indeed such converting fibres have been used to ensure that the visible output from quasi passive displays exceeds the ambient light by adding the output from the ultraviolet in the ambient through converting this into the visible and guiding it to the illuminating fibre end. Plastic fibres have also been doped with photo chromic material with the same end in view.

Coating and Cabling Technologies

The customary function of coatings in cables is to ensure that the surrounding environment does not interfere with the properties of the optical fibre. Thus the coatings are generally designed to prevent moisture ingress stimulating the propagation of micro cracks and thereby destroying the mechanical integrity of the fibre and the cable structure is designed to both facilitate handling, especially pulling through confined ducts, and to provide mechanical and environmental protection. In the communications context the coatings are usually based upon acrylates and the cabling structure involves typically mounting the fibres in a loose tube to prevent mechanical contact with the world outside and encasing these tubes, together with a strength member along the axis of the cable, in a combination of plastic and/or metal coverings.

This accounts for the vast majority of fibre optic cable. Within the remaining few percent there is a rich diversity of geometries and objectives which reflect the desire and the need to adapt the basic optical fibre configuration to specialist applications where usually the coating is required to interact either mechanically or optically with the optical fibre to produce modulation of the light therein. Among the many coatings which have been evaluated the most prominent are:

- metals notably aluminium, gold and indium

- different plastic/polymer materials especially polyimides

- electro optic and piezoelectric materials

- carbon, as a specialised hermetic seal.

Metal coatings are usually applied to provide additional thermal or mechanical protection over and above that available from the standard acrylates. For example aluminium coatings have been used to protect fibres which are stored in a harsh environment with tight bend radii (for example in delay lines and payout spools). Gold coatings enable optical fibres to operate at very high temperatures. Most of the polymer based coatings are unsuitable for use over 150 to 200°C. Gold can be used at temperatures up to 1000°C. Indium coating is a relatively specialised procedure occasionally used to provide thermal and mechanical equalisation in fibre optic gyroscopes though the fact that indium melts at 113°C underlines the need for care in its application.

Piezoelectric and magnetostrictive coatings are used to change local variations in magnetic or electric fields into mechanical perturbations on the optical fibre usually as the basis of phase modulated sensor systems though occasionally (with piezoelectrics) in phase modulators. The most commonly used piezoelectric material is polyvinylidene flouride (PVDF) which can be readily applied and simply poled but again is limited in its range of operating temperatures. There are a variety of nichrome based magnetostrictives some of which are based upon loaded ceramics or glasses (metglas) which can act as magnetostrictive coatings.

The cabling function is occasionally totally inverted to ensure that the cable *is* sensitive to the environment typically through either direct coupling (Fig. 9a) or through micro bend induced variations in optical loss induced through mechanical changes (Fig. 9b). There is a considerable art in designing such cables to ensure that the perturbation which is induced in the characteristics of the fibre is repeatable, single valued and controlled. The direct coupled cable is typically used as a length transducer so that any changes in the length of the external cable assembly should be accurately transferred to the fibre within it *without* inducing any excess losses. For the micro bend cable mechanical changes induce changes of loss and this loss change must not be so great that it obscures considerable lengths of the cable beyond the input end whilst producing a detectable signal.

Moving in closer to the fibre itself there have been a great number of experiments with novel cladding materials including some which have realised practical products. The simplest is a cryogenic alarm probe which utilises differential temperature coefficients of refractive index between the core and the cladding designed to make the index difference zero at a temperature of the order of -50°C. This has been used as a distributed leak detector for liquefied natural gas storage systems. The cladding has also played its part as a thin film transduction system and host medium. For example many silicone resins are porous to some gases in solution so can be used as extraction membranes for spectroscopic measurements. In a similar vein cladding materials loaded with antibody:antigen variants have been used as biosensors exploiting the index change induced by selective binding.

Optical Effects Within Fibres

Optical effects in fibres may be divided into two broad categories:

- phenomena which are inherent within the usual (typically silica:germania) constituents of the fibre core and cladding. These include stimulated Brillouin scatter, stimulated Raman scatter and the optical Kerr effect

- optical phenomena induced by the presence of deliberately introduced dopants within the fibre of which the fibre laser is the most striking example.

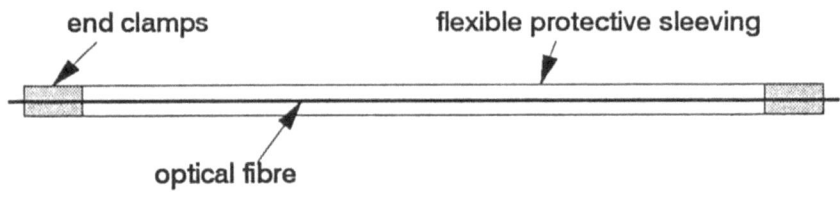

(a)

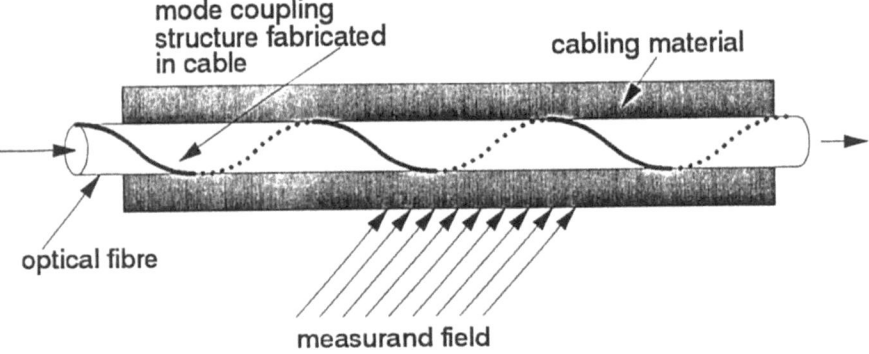

(b)

Figure 9 (a) Fibre cable for strain to length transfer in time delay measurement systems.
(b) Strain to attenuation transfer via microbend cables.

Fibre lasers have been covered in detail elsewhere in this volume. To date the emphasis has been upon the use of erbium and praseodymium as amplifiers and sources for the 1.5 and 1.3 µm bands respectively. Neodynium is the simplest lasing medium to incorporate within the fibre structure and we have already seen the use of this as a non-linear switching device with extremely high sensitivity. The neodymium fibre laser has an extremely wide gain curve (of the order of 150 nm - wider than any other lasing medium) so that neodymium doped fibres and potentially very useful as super luminescent sources. They have been put to particularly good effect in fibre optic gyroscopes where their high output power and extremely broad bandwidth are particularly useful. The principal difficulty lies in finding a suitable photo detector since their operating wavelength (~ 1 µm) is on the edge of the silicon detectivity curve.

The really exciting feature of rare earth doped all-fibre lasers is that with appropriate doping and frequency control suitable rare earths can be located to cover most of the wavelength band from about 400 nm to about 3 µm. Of particular current interest here are thulium doped lasers which offer promise in the 1.66 µm band which is extremely important for methane gas detection. The fibre laser promises to be an extremely versatile highly coherent light source with excellent frequency control. Whilst there is still further research to be completed in optimising both the dopants and the tuning mechanisms there is no doubt the fibre lasers offer very substantial opportunities as general purpose optical sources.

Stimulated Brillouin scattering and stimulated Raman scattering are phenomena (Fig.10) which occur in all optical fibre systems. Both SBS and SRS are threshold phenomena[18] implying that above particular power density there will be frequency conversion via these phenomena. SRB has been put to good use as the basis of a distributed temperature sensor which exploits the fact that the population in the Stokes and anti-Stokes levels depends in an "exp E/kt" way upon temperature and to a very good approximately upon nothing else[19].

The Raman spectrum is itself frequently exploited as a means for chemical analysis, particularly of surfaces. For some materials the position of the spectrum is also stress dependent. This is especially so in polymers where the effect has been used to detect and measure mechanical strains though to date this has not been used in polymer fibres.

SBS is a narrow band frequency shift the value of which depends upon the wavelength of the exciting light and the acoustic velocity within the fibre. The scattering occurs when the Kerr effect induced grating (see below) within the fibre coincides exactly in period with the acoustic wavelength, typically around the 10GHz mark. Measurements of the Brillouin shift will therefore give an indication of the acoustic velocity provided that the optical wavelength is well controlled. This in turn depends upon the material density, stiffness and the stress to which the material is subjected. The phenomenon can be put to good effect in a strain sensor[20] the concept of which is now well developed and for which commercial prototypes are available. The coherence requirements for SBS dictate that the resolution achievable using this method is in the region of a few tens of micro strain over gauge lengths of several metres. However the transduction mechanism is essentially loss free so that extremely long lengths of fibres may be addressed.

The most dramatic exploitation of the optical Kerr effect (where the index depends upon optical intensity) is in soliton propagation. However its importance has been well recognised in other areas, for example since silica is an almost loss free transmission medium the accumulated non-linear Kerr phase for a nominal loss, say, 3dB exceeds by orders of magnitude that achievable in any other more conventional non-linear optical medium for the same total loss. This observation has been incorporated into such laboratory novelties as non-linear switching loop mirrors[21] and is also critical to the operational of optical fibre gyroscopes[22]. The Kerr effect has also been used in strain sensing[23] where, in effect, stress induced changes in birefringence are addressed into the frequency domain by using non-linear Kerr effect induced mixing.

Advanced Optical Fibre Components
The evanescent field is a powerful tool. Its power lies in the ability to accurately locate the position of this field via the manipulation of the main guiding medium whilst at the same time being able to observe closely located interactions of this field outside the main guide with other media.

Perhaps the most dramatic exploitation of the evanescent field is in the photon tunnelling microscope (Fig. 11). Here the evanescent field is that emerging from the surface of a precisely polished prism for light incident at an angle exceeding the critical angle. The sample to be studied is placed on this polished surface within the evanescent field. This field is addressed using an optical fibre probe (Fig. 12)[24]. This probe which is piezoelectrically driven has dimensions which are typically of the order of manometers so that it samples the intensity of the evanescent field at sub-wavelength dimensions. Consequently "images" with resolutions of a very small fraction of a wavelength (determined by the tip size) can reveal detail at the molecular level[25]. This new probe has exciting possibilities in for example observing the progress of chemical reactions most of which are very much surface phenomena and in materials analysis.

The evanescent field has also been most efficiently exploited in a wide range of all-fibre overlay devices (Fig. 13). These use either D shaped fibre or conventional fibre polished down in order to access the evanescent field at the surface. The overlay can take a great may forms ranging from another fibre (thereby realising an adjustable fibre coupler - a well established principle[26]). In a more advanced form this overlay can be a dielectric whose index may be measured with extremely high precision (10^{-6}) or thermally or electro optically tunable filtering system or a modulator or a polarisation beam splitter, polariser or polarisation controller. The scope of these devices is only limited by the scope of all realistic and usable overlay materials[27, 28].

The use of non-reciprocal devices in optics opens up the intriguing possibilities - including the concept of time reversal[29]. The phase conjugate mirror[30] is one manifestation of this but this of course is in bulk optics. Another similar concept which

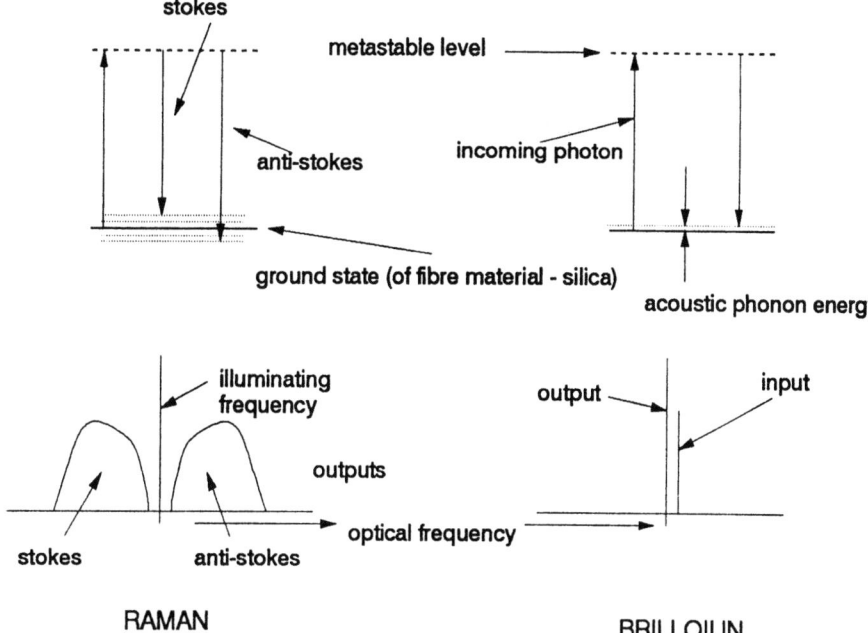

Non-linear scattering processes in optical fibre

Figure 10 Raman and Brillioun scattering processes. Both processes are dominated by the properties of silica but can be modified slightly be dopants. The processes also occur in all other fibre materials.

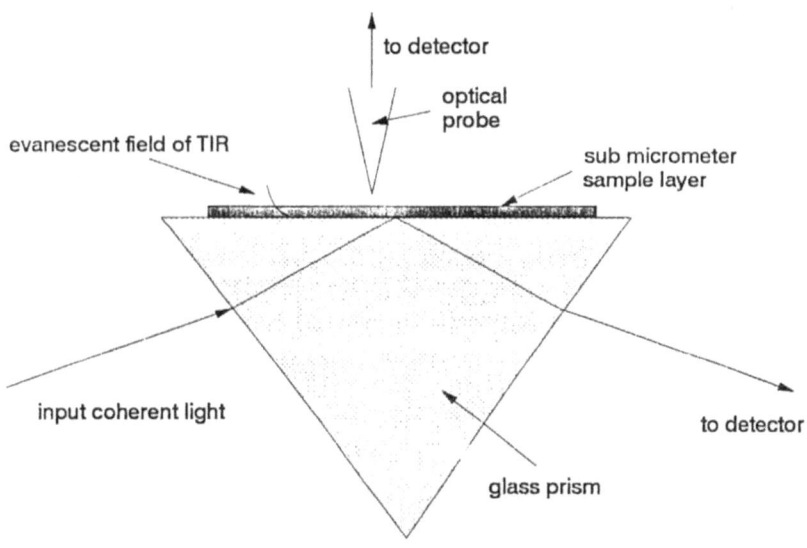

Principle of photon tunnelling microscope

Figure 11 Evanescent field interactions in the phonon tunnelling microscope. The

does have applications in guided wave optics is that of the orthogonal mirror shown schematically in Fig. 14. This particular device ensures that for any polarisation input state introduced to the system, the reflected output is orthogonal to the input regardless of the polarisation properties of the intervening elements. This basic idea has particular relevance in the generalised isolator which prevents reflected power from interfering with, for example, the operation of a semiconductor laser diode and in certain types of interferometric optical fibre sensing systems where detection in an orthogonal polarisation state can significantly enhance the signal to noise ratio.

These are but a few example of a buoyant research topic continually exploring new concepts in guided wave optical devices and in particular those utilising optical fibres as an essential feature within their structure. These examples will perform unique and useful functions within the guided wave format and all promise to make significant contributions to the future evolution of fibre optic technology.

SYSTEMS SUBSYSTEMS AND APPLICATIONS

Fibre optics currently finds applications in communication systems and in instrumentation with, especially in the former, gradually increasing contributions from guided wave signal processing.

From the component oriented ideas briefly reviewed in the previous section we can identify at least two subsystem concepts of significant promise to be important. These are:

- increasingly flexible variants on the theme of the fibre laser

- guided wave optical switching systems.

The fibre laser will be an extremely important general purpose optical source with wide ranging applications emerging within the next few years. Currently the emphasis lies upon the fibre as an amplifying medium in communications systems where unprecedented bandwidth and excellent noise performance feature strongly. The fibre laser produces an output with excellent spatial coherence and with temporal coherence which can be readily adjusted by incorporating carefully selected components within the fibre cavity. The lasing medium has an extremely broad tuning range and by the use of different rare earth dopants the range of accessible wavelengths can scan the visible and near to mid infrared almost continuously. Careful selection of overlay devices[28] can produce narrow band single longitudinal mode operation which is tunable throughout the entire lasing spectrum and which can through modifications to the cavity design also operate in Q-switched and mode locked configurations. The fibre laser is then a source with extremely high spatial and spectral power density which can be used as a uniquely flexible optical probe.

All optical switching has remained the objective for the more esoteric research in optical systems for probably quarter of a century. The basic concept is to control light

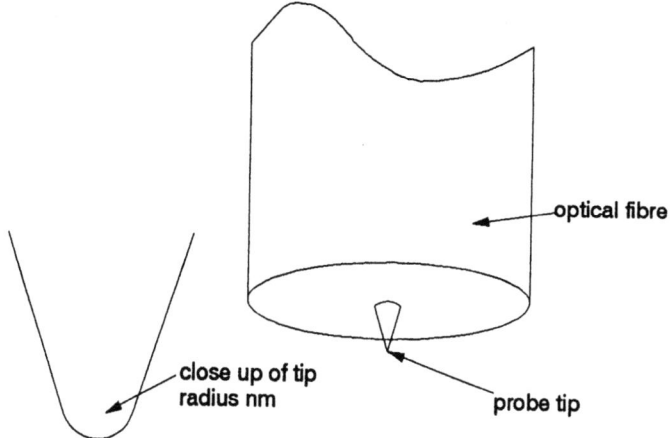

Figure 12 The PTM prove is usually formed by carefully controlled selective etching. Its tip radius determines the resolution.

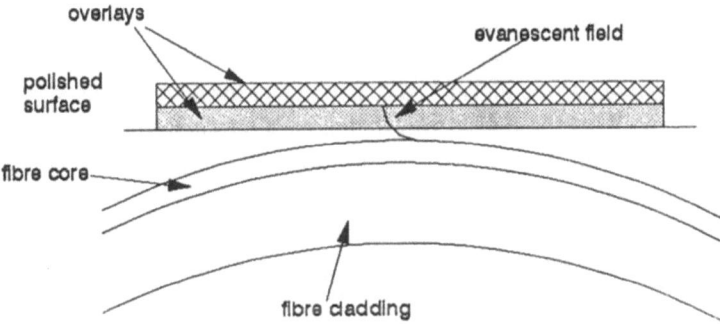

Figure 13 Optical fibre overlay devices exploit evanescent field interactions in carefully designed coupled layers.

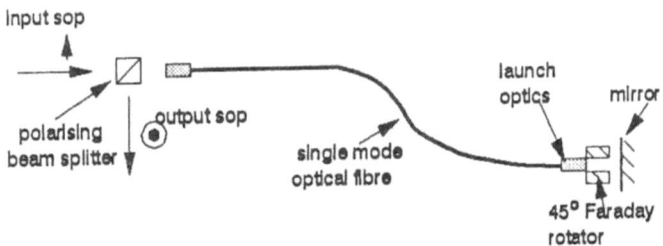

Figure 14 The orthogonal mirror in reflection mode.

by light. The actual realisation of all-optical systems has been thwarted by the invariably high power densities which are required in order to produce the switching. In some cases, power densities are low but need relatively complex often thermally sensitive bulk optic devices (with the consequent difficulties in interfacing to a guided wave optical system) or have an implicit requirement for long interaction lengths. For example the device in Fig. 6 which is about 1m in length is one of the shortest relatively high sensitivity all guided systems reported to date. The 1m of fibre has a 5 nanosecond transient time and obviously requires packaging and storing within any potential switching matrix. In comparison the electronic switching equivalent device occupies a chip area of a few microns square on a substrate which is a few hundred microns thick and which is readily switched. Consequently despite the intellectual appeal of all optical switching it seems likely that the electronic crosspoint will continue to dominate. That said, there has been some success in limited arrays of electronically controlled integrated optic switches though these have yet to break the thousand crosspoint barrier - a bench mark achieved in the semiconductor industry around 1970.

SOME CONCLUDING COMMENTS

This paper has dipped into a selection of the more unusual manifestations of optical fibre guided wave components, sub-systems and systems. Whether these concepts may be regarded as "advanced fibre optic electronics" is open to debate. However, they can certainly all be regarded as the unusual, speculative and the adventurous side of optical fibre technology. Many of these ideas will find applications and some, notably the fibre optic back plane interconnect and possibly all fibre switching systems, could make a significant contribution within the communications industry. However it is far more likely that the esoteric, the oddball and the specialist will find their homes in the specific niches within the more diverse techniques required for instrumentation systems including signal processing and display technologies. Regardless, I believe that fibre optics will continue to interest and challenge researchers and development engineers and will also continue to find ever more diverse applications and exploitation avenues.

BIBLIOGRAPHY

1. Kao, C.K. and Hockham G.A. (1966) Dielectric fibre surface waveguides for optical frequencies, *Proc. IEE* 133, p1151.
2. Werts, A. (Sept. 1966) Propagation de la lumiere coherente dans les fibre optiques, *L'Onde Electronique* 474, p967-980.
3. Koester, C.J. and Snitzer, E (1964) Amplification in a fiber laser, *Applied Optics* 3, p1182-1186.
4. Dyott, R.B. and Stern J.R. (Feb.1971) Group delay in glass fibre waveguides, *Electronics Letters* 7, 3, p82.
5. Zakharov, V.E. and Shabat, A.B. (1972) Soviet Physics, *JETP* 34, p62.
6. Jin, W., Stewart, G., Culshaw, B., Murray, S., Pinchbech, D. (1993) Absorption measurement of methane gas, *Optics Letters* 18, 16, p1364.
7. Dyott, R.B., Cozens, J.R. and Morris, D.G. (June 1979) Preservation of polarisation in optical fibres with elliptical cores, *Electronics Letters* 15, 13, p380.
8. Berkey, G.E. (Jan.1992) Developments in elliptical core polarisation maintaining fibre, *Proc OFS(8)*, p121, Monterey Ca (IEEE).
9. Dyott, R.B. (1994) Composition of LP modes in elliptically cored fibre, *Electronics Letters* 30, 9, p728.
10. Vengsarkar, A.M., Michie, W.C., Jankovic, I, Culshaw, B., Claus, R.O. (1994) Fibre optic dual technique for simultaneous measurement of strain and temperature, *IEEE J-LT* 12, 1, p170.
11. Sadowski, R.W., Digonnet, M.J.F., Pantell, R.H. and Shaw, H.J. (Sept.1993) Sub microsecond and optical switching in neodymium doped fibres, *Proc. SPIE* 2073, p166, Boston.
12. Park, H.G., Pohalski, C.C. and Kim, B.Y. (1988) Optical Kerr switch using elliptical core two mode fiber, *Optics Letters* 13, 9, p776.
13. Chen, R.T. (Jan. 1994) Guided wave optoelectronic interconnects: their potential and future trends, *Proc. SPIE* 2153, p196, Los Angeles.
14. Mackenzie, F., Payne, R. and Rush, J.D. (Jan.1994) D-fibre optical backplane interconnect, *Proc. SPIE* 2153, p218, Los Angeles.
15. Nakagaisa, K., Kowalewski, T., Phelps, C.W., Rook, D.L. and Krchnavck, R.R. (Jan.1994) Optical channel waveguides based upon photo-polymerizable di/tri acrylates, *Proc. SPIE* 2153, p208, Los Angeles.
16. Johnstone, J.S., Private communication ~1980.
17. Yoshino, T., Takahashi, Y., Tamura, H. and Ohde, N. (Sept.1993) Some special fibres for distributed sensing of uv light, electric field or strain, *Proc. SPIE* 2071, p242, Boston.
18. Agrawal, G.P. (1989) *Nonlinear fiber optics*, Academic Press, Boston.
19. Dakin, J.P., Pratt, D.J., Bibby, G.W. and Ross, J.N. (1985) Temperature distribution measurement using Raman ratio thermometry, *Proc. SPIE* 566, p249
20. Shimizu, K., Horiguchi, T., Koyamada, Y. and Kurashima, T. (to be published 1994) Measurement of distributed strain and temperature using Brillouin scatter, *IEEE JLT* (also in *Proc. OFS(10) Glasgow 1994 (SPIE)*).
21. Dyott, R.B., Bello, J. (Aug. 1983) Polarisation holding directional coupler made from elliptically cored fibre having a D section, *Electronic Letters* 19, p601.

22. Bergh, R.A., Culshaw, B., Cutler, C.C., Lefevre, H.C. and Shaw, H.J. (1982) Compensation of the optical Kerr effect in fibre optic gyroscopes, *Optics Letters* 7, p282.
23. Cokgor, I., Handerek, V.A. and Rogers, A.J. (May 1993) A practical Kerr effect distributed optical fibre sensor with polarisation diversity detection, *Proc.OFS(9)*, p79, Florence, (IROE, Firenze).
24. Ohtsu, N. (1994) Progress of high resolution photon tunnelling microscope with a nanometric fibre probe, *Proc. OFS(10)*, Glasgow (SPIE).
25. Collins, G.P. (May 1994) Near field optical microscopes take a close look at individual molecules, *Physics Today*, p17.
26. Bergh, R.A., Cutler, C.L. and Shaw, H.J. (1980) Single mode fibre optic directional coupler, *Electronic Letters* 18, p260.
27. Johnstone, W., McCallion, K., Langford, N. and Gloag, A. (1994) Electro-optically tunable fibre lasers for fibre optic sensing systems, *Proc. OFS(10)*, Glasgow (SPIE).
28. Moodie, D. and Johnstone, W. (1993) Wavelength tunability of components based on evanescent coupling, *Optics Letters* 18, 2, p1025.
29. Yariv, A. (1978) Phase conjugate optics and real time holography, *IEEE JQE*-14, 650.
30 Peffer, D.M. (Jan. 1986) Applications of optical phase conjugation, *Scientific American* 56, 254.
31. Martinelli, M. (1994) Time reversal for the polarisation state in optical fibre circuits, *Proc. OFS(10)*, Glasgow (SPIE).

PLASTIC OPTICAL FIBRES (POF)

Characteristics, metrology and applications

DEMETRI KALYMNIOS
School of Electronic and Communications Engineering
and Applied Physics
University of North London
166-220 Holloway Road
London N7 8DB

1. Characteristics

As the name implies plastic optical fibres (POF) are made entirely from plastic or polymeric materials. These fibres were first introduced about twenty years ago, but it is during the last five years or so that a new impetus in their further development and use is observed. The main reason for this is the fact that they can offer a cheaper and more cost-effective solution than glass optical fibres (GOF) in many applications where the basic advantages of fibre optics are essential but over relatively short distances. The attenuation of POF is typically two orders of magnitude ($\sim$ 200 dB/km) higher than that of GOF (0.4 - 10 dB/km) and this represents the most serious disadvantage of POF when compared with GOF. However POF offer a number of advantages such as greater flexibility and less susceptibility to transmission variations due to vibrations. They have a large numerical aperture (NA), typically 0.47, which makes them more suitable for use with low-cost light-emitting diodes, and they also have a large core diameter (0.5 - 1 mm) which makes alignment and interconnections easy with simple inexpensive injection moulded connectors.

The most widely used POF today is the PMMA-type using polymethylmethacrylate core, with a thin (10 - 20 μm) cladding layer of fluorinated polymer. Figure 1 shows the basic step-index (SI) structure, and Figure 2 shows the attenuation characteristics of a PMMA core. Lowest absorption of light is observed at around 570 nm in the green with the next lowest absorption window in the red at around 650 nm. Currently the latter is the most frequently used window, because of the wide availability of powerful and inexpensive red LEDs, mainly of the AlGaAs or GaAsP type. In general POF cables have a protective jacket of polyethylene or PVC, typically 1 mm

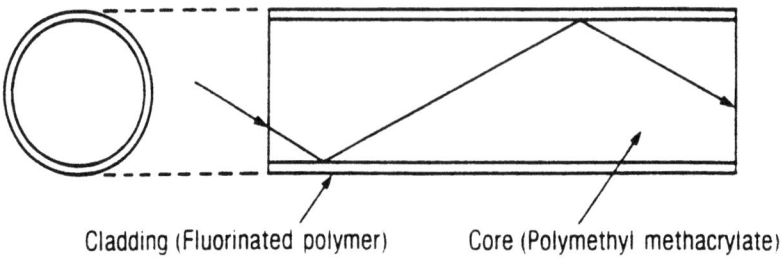

Figure 1. Structure of POF and light transmission.

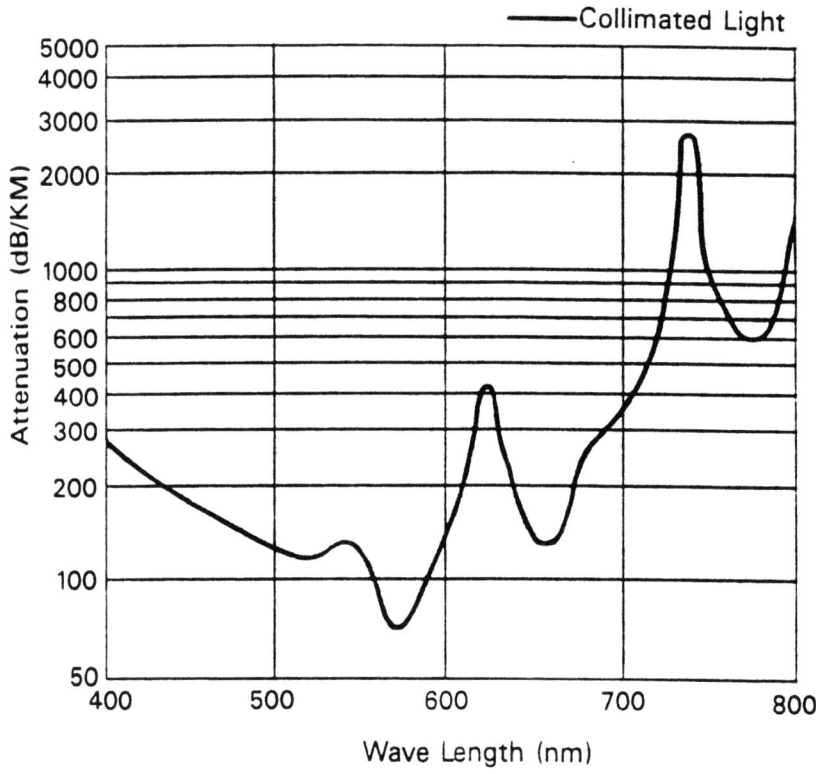

Figure 2. Spectral attenuation of PMMA, POF
(Eska Extra, Mitsubishi Rayon Co., Ltd.)

TABLE 1. Eska family POF (Ref 1)

	Eska	Super Eska	Extra	D	F
Core	PMMA	PMMA	PMMA	PMMA	Eng. Plastic
Cladding	F-polymer	F-polymer	F-polymer	F-polymer	F-polymer
Light Transmsn dB/km	<500	<400	<200	<200	<1000
N.A. (Number)	0.5	0.5	0.47	0.54	0.75
(Angle)	60	60	56	85	97
Heat Resistance					
Dry (°C)	70	70	85	115	125
Wet (°C)	60	60	85	80	85
Main Application	Display Inf panel	Light guide Sensor	Data-trans Sensor	Heat resist Data-trans	Heat resist Data-trans

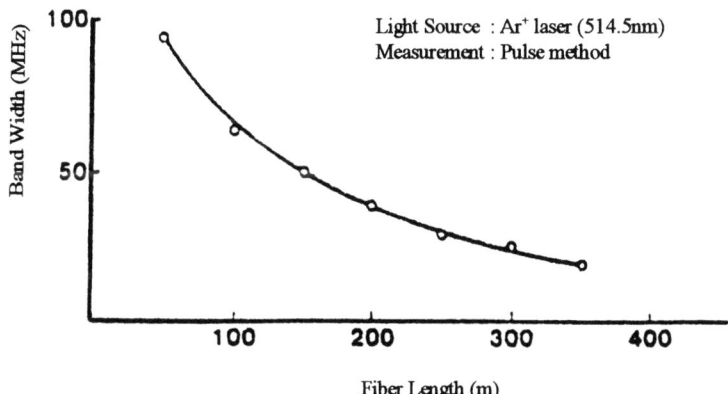

Figure 3. Eska Extra bandwidth (Ref. 1)

thick. Cables containing two or more fibres can also be obtained. Table 1 shows a well-known range of POF cables produced by one leading manufacturer, together with some typical application areas[1]. The relatively low maximum operating temperatures for POF is due to the low glass-transition temperature that plastics generally possess. Although this is another clear disadvantage of POF compared with GOF, nonetheless, serious consideration is given for its use in harsh and high temperature (125° C) environments such as the engines of automobiles[2]. POF for the higher temperature applications is mostly made of polycarbonate (PC) and exhibits 5 to 6 times higher losses than PMMA[3]. In spite of this it can still be useful for optical links up to 10 m or so.

Inherent to the step-index structure of POF is a limited bandwidth length product (~ 5MHz.km) due to intermodal (multimode) dispersion. Typical measured performance with a laser source is shown in Figure 3, but for launch conditions of higher NA, such as with LEDs, the performance is expected to be significantly below that shown. Achieving a modulation bandwidth of 10 Mbits/s (Ethernet) and 100 Mbits/s (Fibre Distributed Data Interface, FDDI) over a distance of 100 m or so is of particular interest in data communications with fibres for office Local Area Networks (LANs). It will be seen in a later section, that office LANs represent a huge proportion of the networks market, and it is not surprising that recent developments in POF have this particular market very much in mind.

2. POF production techniques

The basic steps in the production of step-index POF are essentially those used in the production of step-index GOF [4][5]. Following the production of a suitable core material preform, the fibre is formed by drawing from one end of the preform under controlled temperature conditions while the rotation speed of a fibre take-up drum generally determines and controls the diameter of the fibre produced. Cladding is added during the fibre-drawing process by passing the just formed fibre through a suitable molten polymer. Different preform production techniques may be more appropriate for different polymeric core materials. In one approach the core material preform is in the form of a polymerised fixed-size rod from which a certain length of fibre can be drawn. In a second, the core material is kept in the molten state and allowed to polymerise just before being ejected under gas pressure through a nozzle to the fibre-drawing equipment. With the latter approach much longer lengths of fibre are possible since the molten core material can be continually replenished. Purity of the core material is most important in minimising the scattering losses arising from density and composition variations. Detailed descriptions of the production techniques for step-index PMMA and Polystyrene (PS) POF are given by Ashpole et al [6].

In order to overcome the disadvantage of the small bandwidth length product with (SI) POF, graded index (GI) POF have also been developed. With (GI) POF, a bandwidth length product improvement of around 200 times that of (SI) POF is obtained when the core refractive index profile becomes parabolic. The parabolic index is obtained at the preform production stage through a random copolymerization technique, also known as 'interfacial-gel'. In this technique a pure PMMA tube is filled with two monomers M_1 (e.g. MMA, n = 1.492) and M_2 (e.g. benzylmethacrylate, BzMA, n = 1.562), and is placed in a furnace at 80 - 100° C. Monomer M_1 has a lower refractive index and smaller molecular size than M_2. In this arrangement the inner wall of the tube is initially swollen by the monomer mixture and then a gel phase is formed which accelerates polymerization. The copolymer formed at the inner wall of the tube gradually solidifies towards the centre of the tube. During this process the smaller size monomer M_1 diffuses faster into the copolymer gel state, leaving more monomer M_2 of higher refractive index in the centre of the tube. During the above process some PMMA molecules dissolve and also diffuse towards the centre of the tube. A (GI) POF fibre is drawn from this GI preform using the same fibre-pulling technique as with (SI) POF. The drawn fibre maintains essentially the same index profile as that of the GI preform. This technique has been developed by researchers at Keio University [7] who make reference to various monomers M_1 and M_2 that have been tried. As one example, (GI) POF of diameter 500 μm - 1000 μm, with attenuation of 150 - 200 dB/km at 652 nm, and bandwidth length products of 1 GHz.km have been produced with MMA as monomer M_1, and BzMA as monomer M_2, in a pure PMMA tube of 20 mm diameter.

3. POF components and interconnections

The prime benefit in seeking to use POF rather than GOF in applications where this is possible, relates to cost-saving for the whole fibre-optic system. But equally important may be the fact that as a result of using a larger size fibre with higher NA, interconnections and coupling to active devices is a lot easier, more efficient, and a lot more reliable than with the standard 50/125 or 62.5/125 GOF. This is particularly the case in industrial environments where the presence of vibrations may have an adverse effect on the stability of glass fibre-optic systems.

3.1 LIGHT SOURCES AND DETECTORS

The high numerical aperture of POF contributes to efficient coupling, since most surface emitting LEDs are near-Lambertian emitters [4]. Further increase in the coupling efficiency is possible through matching of the size of the active element of the LED (or photodetector) to that of the core of the fibre. Increasing the size of the active element, however, also increases the capacitance of the device and this leads to a slower risetime and hence a reduced bandwidth. A trade-off solution between the conflicting requirements for higher output and higher coupling efficiency on one hand, and higher speed of operation on the other, may be found through the use of active devices (both emitters and detectors) that employ either moulded plastic microlenses placed very close to the active elements, or a moulded hemispherical lens which forms part of the plastic encapsulation of the device. The lens arrangement increases coupling efficiency while still using a small-size active element. Such devices tend to be more expensive than their counterparts that use no lenses. In a more recent development a 3 dB improvement in the coupling efficiency with 1 mm POF has been obtained using an embossed reflector surrounding the semiconductor active chip [8]. In this arrangement, apart from the surface emission, light from the sides of the active region is also captured and coupled into the fibre.

Of the two lowest loss windows of PMMA (Figure 2) the red is the one most frequently used because the available red LEDs are more powerful and faster than the available green LEDs. The most powerful red LEDs may typically couple in excess of 600 μW, with a risetime of 50 ns - 150 ns, into 1 mm POF at full numerical aperture. As a rule, the more powerful the LED the slower the risetime. By comparison, the best green LEDs currently available can only couple up to 10 μW into 1 mm POF, with risetimes in the microseconds region. Of course, one could also use semiconductor lasers that could couple power in the mW region with subnanosecond risetimes, but these are not cheap and would therefore offset the cost benefits of using POF.

3.2 CONNECTION TECHNIQUES

The large core diameter, and large NA of POF make connection particularly easy and reliable. There is no need for connectors with tight mechanical tolerances as with GOF, and most connection techniques utilise simple plastic moulded connectors. Termination with most of these connectors can be done by crimping, while the end of the fibre may either be polished or be cut with a sharp blade, which, if heated, produces a smoother surface. The improvement in transmission between a rough end surface (produced with a 600 grit abrasive paper) and a polished surface (produced with 3 μm lapping film following the 600 grit stage) is about 2 dB [9]. This 2-stage polish can be done either wet (water), or dry, and is fast, taking no more than about 3 minutes. Continuing the polishing with a finer (0.3 μm) lapping film can remove most light scratches, but this stage would take significantly longer (10 - 15 minutes per surface) to complete, and the improvement in transmission, estimated to be less than 1 dB, may not be warranted for many applications.

The simplest and cheapest form of connection (both for emitters and detectors) is illustrated in Figure 4, where a heat shrinking sleeve over the fibre and device may also be used to secure the push-fit connection. These light-link components [10] are also offered with a plastic encapsulation suitable for board mounting, and have a split end tubular receptacle with a screw nut for securing the fibre, Figure 5. The variation of coupling into 1 mm POF using these devices is shown in Figures 6 and 7 as a function of axial and lateral misalignment. Similar performance can also be expected from alternative connection systems. Figures 8 and 9 show such alternatives which have a simple push latch-on/off action. Encapsulation of the active devices allows for close stacking, (i.e. transmitter next to receiver) so that a duplex connector and cable can be used [9].

Active components can also be obtained which are encapsulated in metallic SMA receptacles for use with metallic SMA connectors. These may be preferable for the

753

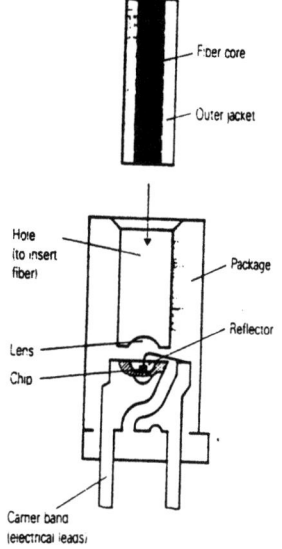

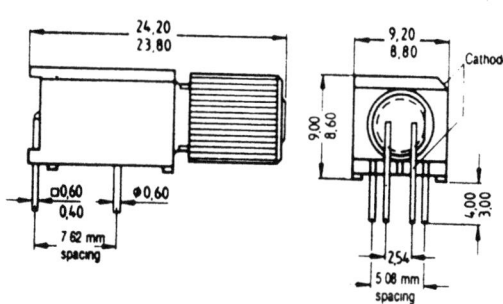

Figure 4. Cross section of light-link emitter diode (Siemens, Ref.10)

Figure 5. Light-link components with screw connection (Siemens)

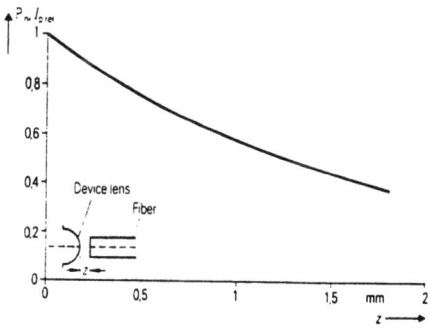

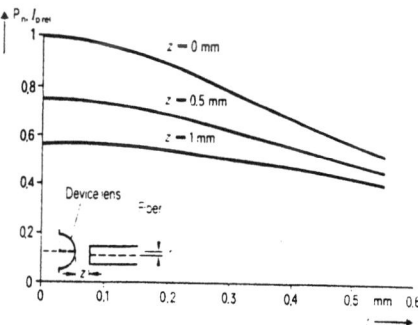

Figure 6. Relative change of coupled-in power Pin and photocurrent Ip measured at the receiver with distance z between device lens and fiber.(Ref.10)

Figure 7. Relative change of coupled-in power Pin and photocurrent Ip measured at the receiver with lateral misalignment r of fiber to device lens. (Ref.10).

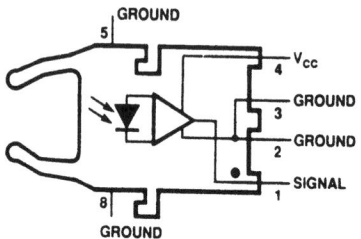

Figure 8. Versatile Link active components (Ref. 9)

HFBR-4501 (GRAY)/4511 (BLUE) SIMPLEX CONNECTOR

HFBR-4503 (GRAY)/4513 (BLUE)
SIMPLEX LATCHING CONNECTOR

HFBR-4506 (PARCHMENT) DUPLEX CONNECTOR

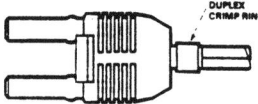

HFBR-4516 (PARCHMENT) DUPLEX
LATCHING CONNECTOR

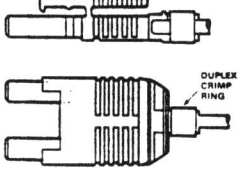

HFBR-4505 (GRAY)/4515 (BLUE) ADAPTER

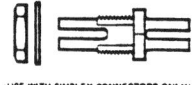

·USE WITH SIMPLEX CONNECTORS ONLY!

Figure 9. Versatile Link connectors (Ref.9)

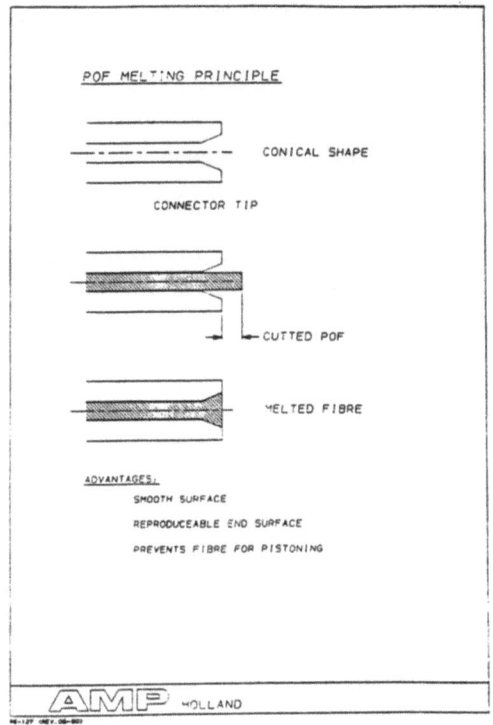

Figure 10. SMA metallic connector tip. (Ref. 11)

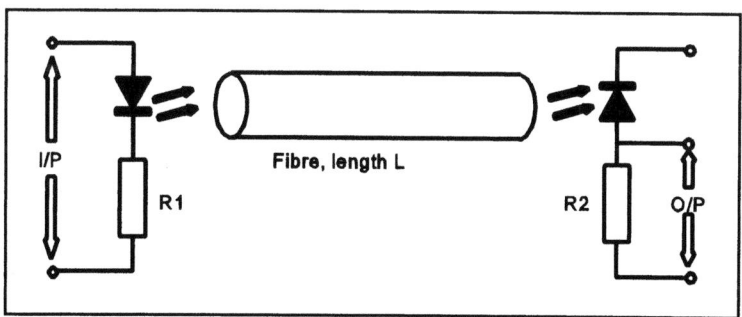

Figure 11. Basic fibre optic link circuit

harsher industrial environments. Figure 10 shows the tip of a specially developed metallic SMA connector, which is polished by melting the fibre under temperature control by pressing it against a heated polished metallic plate [11]. The formed conical part of the fibre may also stop 'pistoning' (movement of the fibre with respect to its jacket) in one direction. Pistoning occurs due to a difference in the linear thermal expansion coefficients of the fibre and its surrounding jacket. More intricate connectors in which the effects of pistoning in both directions is eliminated have also been developed [11]. Two fibres can also be joined using a simple short tube carrying the two fibre ends with either an index matching liquid between them or, for a more permanent joint, using a suitable adhesive. A permanent splice loss as low as 0.2 dB can be achieved in this way [12].

4. Typical performance

A typical fibre optic system consists of a transmitter incorporating an emitter which couples a light signal into a fibre of length L, the output of which is coupled into a photodetector. The simplest form of transmitter and receiver circuits is shown in Figure11. Depending on the application more sophisticated circuits may be used and the photodetector would be used with some type of preamplifier [4] [13], most commonly, a transimpedance amplifier.

When transmitting pulses, the 3-dB electrical bandwidth B of a fibre-optic system is taken to be that of a low pass filter to an ideal rectangular pulse stream. The bandwidth of the system is related to a system risetime t_{syst} (10% - 90%) and is given by [14]

$$B_{syst} = \frac{0.35}{t_{syst}}$$

The system rise time contains contributions from the source, the fibre-optic cable and the detector. For design purposes an acceptable approximation for their relationships is

$$t_{syst}^2 = t_{source}^2 + t_{cable}^2 + t_{det}^2$$

Source and detector rise/fall times are normally quoted by the manufacturers for typical driving and detecting conditions. The cable risetime is similarly related to all dispersion effects in the fibre and is

$$t_{cable}^2 = \Delta t^2 (\text{multimode}) + \Delta t^2 (\text{material})$$

where

$$\Delta t(\text{multimode}) = \frac{Ln_1}{c}\left(\frac{n_1}{n_2} - 1\right) = \frac{L(NA)^2}{2n_1 c}$$

and

$$\Delta t(\text{material}) = \frac{L\lambda\Delta\lambda}{c} \cdot \frac{d^2 n}{d\lambda^2}$$

Waveguide dispersion in multimode POF is negligible.
For a numerical aperture of 0.47, $n_1 = 1.492$, $d^2n/d\lambda^2 = 1.43 \times 10^{11}$ m^{-2} [15], $\lambda = 650$ nm, and $\Delta\lambda = 50$ nm, these equations lead to $\Delta t(\text{multimode}) = 0.263$ ns/m, $\Delta t(\text{material}) = 0.016$ ns/m. Thus for POF $\Delta(\text{multimode})$ dominates, and to a good approximation $t_{cable} = \Delta t(\text{multimode}) = 0.263$ ns/m. For 100 m of cable this corresponds to a cable bandwidth, $B_{cable} = 13.3$ MHz.
Using a typical LED risetime of 50 ns, detector risetime of 10 ns, and 100 m POF cable the expected system bandwidth is:

$$t_{syst} = \sqrt{50^2 + 26.3^2 + 10^2} = 57.4 \text{ ns}$$

$$B_{syst} = \frac{0.35}{57.4 \times 10^{-9}} = 6 \text{ MHz}$$

Selecting a faster LED with, say, 20ns rise time, the system bandwidth would increase to 10 MHz.

Whether or not such a link is feasible (assuming availability of all components), also depends on the amount of light that can be delivered to the detector and the desirable signal-to-noise ratio for a reliable link. The expected output from a fibre of length L is given by:

$$dBm_{out} = dBm_{in} + \alpha L$$

where α is the attenuation constant in dB/m.
This equation is plotted in Figure 12 for typical outputs from various sources (neglecting coupling losses).

This graph provides an indication of the link length available provided we know the sensitivity of the detector in dBm. For most digital systems, the detector is most likely to be 'thermal noise limited' in which case:

$$S/N = \frac{R_L(\rho P)^2}{4kT_e B}$$

where,

P(W) is the signal input to detector, R_L (Ω) the effective load resistance, T_e (K) the effective temperature, k the Boltzmann constant, ρ the detector responsitivity(taken to be 0.3 A/W at 660 nm).

Assuming an effective temperature of 300 K (which could be higher due to preamplifier noise) then for the signal-to-noise ratio of 21.5dB (142:1) required for a bit-error-rate (BER) of 10^{-9} the minimum signal strength can then be calculated for various frequencies. The results of such calculations are plotted in figure 13.

Considering a link at 10 MHz (Ethernet 10 Mbits/s) and allowing a probable coupling and connector loss of 4 dB at the detector end, and recognising that the output of the LED is more likely to be at the lower end (i.e. -10 dBm, or -14 dBm allowing for the 4 dB loss above) in view of the higher frequency, then a link length of over 120 m is indicated by line 5 on Figure 12. This is typical of what is possible with current LEDs and 1mm (SI) POF, the limit being imposed both by the frequency and output of LEDs as well as by the bandwidth of the cable at full NA. If a laser is used (of spectral width 0.2 nm) with a launch NA of 0.1 (only multimode dispersion is significant) at Ethernet frequencies, enough signal output would be obtained for only up to 250 m, line 2 Figure 12. The cable bandwidth in this case, as estimated from the cable rise time for 250 m, should be about 125 MHz, and would not limit the performance of the link. Similarly, from Figure 12 at the FDDI data rate, a laser would deliver sufficient output at 200 m, and the cable bandwidth (156 MHz) would not seem to be a limiting factor with such a link. However, in practice, the measured bandwidth is considerably less than the calculated one because mode dependent attenuation and mode coupling lead to an equilibrium NA value of ~ 0.3 after 150-200 m irrespective of the launch numerical apertures. Recent experimental results [16] recorded a bandwidth of 50 MHz for 100 m of fibre with launch NA of 0.65 and 170 MHz for the same length with a launch NA of 0.1.

One obvious way of removing the (SI) POF bandwidth limitations would be to use a graded index (GI) POF. Such fibres are currently being developed [7] [17] with a reported bandwidth length product of at least 1 GHz.km. Another way of extending the bandwidth of a (SI) POF link has been demonstrated at speeds of 265 Mbit/s and 532 Mbits/s over 100 m using a laser with low numerical aperture launch conditions and first-order pre-emphasis of the data signal to compensate for the low pass filtering characteristics of the fibre [18].

Another consideration in the design of any link relates to the anticipated variation in the performance due to temperature fluctuations and their effects on receiver sensitivity and emitter efficiency. Ensuring that the detector always receives a minimum signal $P_{D(min)}$ for reliable performance and at the same time ensuring that there is no case for receiving a signal greater than a $P_{D(max)}$ (a level which may lead to saturation and most likely undesirable pulse broadening), results in link lengths considerably less than those suggested by using the graphs of figures 12 and 13. In general the maximum coupled power, $P_{T(max)}$ less the minimum system loss must not exceed the maximum allowable signal at the detector, i.e.

$$P_{T(max)} - L_{min} < P_{D(max)}$$

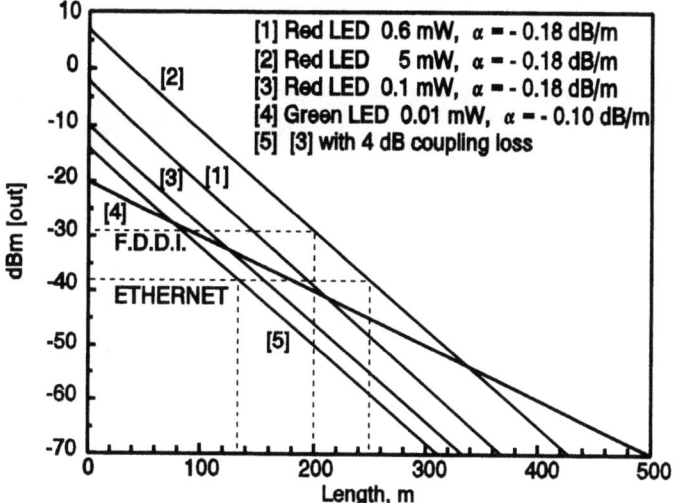

Figure 12. Fibre output vs length for a number of sources.

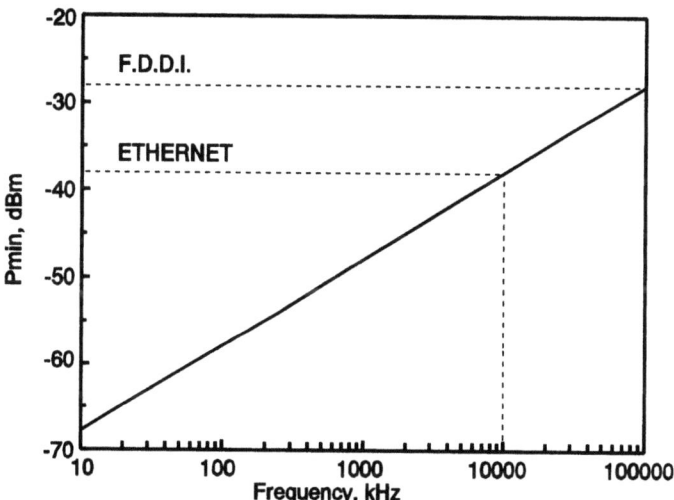

Figure 13. Minimum required signal strength vs frequency.

TABLE 2. Typical POF Link Performance, (Ref. 19)

TYPICAL POF Link Performance Low Loss POF, Latched Components	50 MBd Rate &		65m Length	
Temperature, Deg (C):	25		0	70
	Low	High	Low	High
SOURCE OUTPUT SPECIFICATIONS:				
Center Wavelength, Uc(um)	0.662	0.662	0.657	0.671
Spectral Width, Uw(nm)	25	25	23.8	27.1
Average Mod. Output Power, dBm	-12.5	-12.5	-13.70	-11.50
Extinction Ratio, (%)		50		50
Rise (Fall) Time Ts, 10-90% (ns)	5.0	6.0	3.0	8.0
Over Shoot, (%)		50		75
Duty Cycle Distortion (DCD), ns		2		2
RECEIVER INPUT SPECIFICATIONS:				
Average Mod. Input Power, dBm	-28.7	-8.4	-27.7	-9.50
Max Input 10-90 % Res. Time, (ns)		9.2		9.2
Input Eye Opening @ BER, (ns)	10	10	10	10
FIBER SPECIFICATIONS:				
Core/Cladding Diameter, (um)	960	1000	960	1000
Attenuation @ Ur (um), (dB/km)		170		170
Modal B.W. (MHz*km) @ L=0.1 km	4		4	
Dispersion Slope So(ps/nm^2*km)		1.4		1.4
Dispersion Minimum @ Uo(um)=		0.8		0.8
LINK SPECIFICATIONS:				
Signaling Rate, (mBd)		50		50
Link Length, (km)		0.065		0.046
Total Link DCD, (ns)	3	3	3	3
Bit Error Rate, BER*10^-9		1		1
Receiver Bandwidth, (MHz)	90		90	
Calculate @ Wavelength Uc(um)=	0.662	0.662	0.657	0.671
Fiber Attenuation @ Uc, (dB)=	13.38	13.38	8.41	12.41
Fiber Tests @ Reference Ur(um)=		0.65		0.65
LINK ANALYSIS RESULTS:				
Active Input Response, To(ns)	7.69	8.37	5.45	9.20
Channel Response Time, Tc(ns)	8.61	9.23	6.70	9.99
Fiber Spectral Attenuation, (dB)	13.61	13.61	9.12	11.54
Specified Aging Margin, (dB)	2.50	2.50	2.50	2.50
Allowed Passive Loss, (dB)	0.09	0.09	2.38	-0.04
Total Specified Budget, (dB)	16.2	16.2	14.00	14.00
Specified Dynamic Range, (dB)	20.3		18.20	
Detector Relative Loss, (dB)	-0.13	-0.13	-0.08	-0.22
Eye Loss @ Zero Offset, (dB)	0.00	0.00	0.00	0.00
Eye Opening Penalty, (dB)	0.68	0.90	0.20	1.23
Required Low Rate & Ctr Eye, dBm	-29.25	-29.47	-27.82	-28.71

and at the same time the minimum coupled power in the fibre $P_{T(min)}$ less the maximum system loss must be sufficient for reliable detection i.e.

$$P_{T(min)} - L_{max} > P_{D(min)}$$

The 'optical power budget' (OPB) which determines how much loss one can tolerate in the system for reliable operation under all conditions is determined from:

$$OPB = P_{T(min)} - P_{D(min)}$$

and the maximum system loss can be determined from:

$$L_{max} + \text{Power margin} < OPB$$

where a power margin of 3 dB is usual allowing for component ageing, and other environmental changes. L_{max} contains all loss contributions i.e. coupling losses, splice or interconnection losses and the cable loss. Table 2 provides detailed performance figures for a typical POF link offered by one manufacturer [9][19].

5. Applications of POF

5.1 SHORT DISTANCE DATA LINKS

From the typical performance analysis, and figures 12 and 13, it can be seen that at a more modest data rate of 100kbits/s or less, the link could extend to more than 200 m. This distance is more than adequate in many industrial, electrically noisy environments where various process control and other types of links may be required. These could be RS232-C data links, IEEE-488 connections to data buses [9] and others. Video transmission is possible over distances in excess of 50 m, and such links could be particularly attractive for short-distance closed TV applications such as remote inspection of machinery, security in factories and superstores, entrance surveillance of buildings, and similar.

There are many other applications, apart from in factories, where various transducer signals (analogue or digital) need to be transferred from their source through an electrically noisy environment (i,e, high voltage/current switching, lightning, heavy current power lines etc) to a well isolated area with sensitive instrumentation. One such environment is found in hospitals with electrocardiographs or x-ray equipment. As the European standards on electro-magnetic compatibility (EMC) for various equipment are being enforced, one will see increasingly the use of fibres inside equipment so as to reduce electro-magnetic interference (EMI) effects, originating from the same or other pieces of equipment. The low cost of POF links would be particularly attractive for such applications. Other areas where POF links may find wide use are in point-of-sale terminals where these are connected to a central computer, connections between dispensing machines and cash registers such as in petrol stations, and similarly in motorway toll control points when these become widely in use.

5.2 OFFICE LANs

Broadband ISDN and the growing interest in bringing this into the office and eventually to the home, creates a huge market for high-speed, short-distance communications links. It is estimated that the 100 m desk-top links will accommodate 95% of all commercial LAN users, while increasing the link length to 300m will increase LAN coverage to 99%. With this prospect in sight, POF technology has to compete very hard with the other competing technologies, copper wires and GOF, both of which are currently trying to develop strongholds in this market. The main interest in LAN broadband services is currently with FDDI. This is a dual token ring network operating at a data rate of 100Mbit/s, with Manchester encoding. This network is specified for multimode GOF and will form the backbone of LANs on to which 100m links can be attached for distribution in offices through active hubs or concentrators [19]. Relevant LAN standards such as IEEE 802.3 (10 Base-T Standard) IEEE 802.3 token ring LAN, ASC X3T9.5 FDDI allow for the use of copper wires in the form of unshielded twisted pair (UTP) or dual-shielded twisted pair (STP) for various link lengths and speeds, provided these comply to specified EMI emission standards. The only type of fibre attachment specified so far is the 62.5/125 GOF with 1.3μm opto-electronic technology. At this stage, none of the relevant LAN standards make reference to the equally cheap alternative, when compared with (UTP) or (STP), that of POF. This situation may well change once POF technology can deliver links of 100m at the FDDI rate. POF FDDI links might eventually win over copper wire because of the EMI immunity. Current POF technology produces FDDI links up to 20m operating at 125 Mbaud utilising 4B5B encoding, an alternative to Manchester encoding, that requires a fundamental frequency of 62.5 MHz for the fibre, as opposed to 100 MHz. These links are reported as loss limited [20]. At the same time a cheap GOF alternative is also being offered through the use of 200 μm core SI multimode hard clad silica (HCS) fibres. The main advantage offered by HCS over POF is the lower attenuation ($\sim$ 10 dB/km) at 650 nm, NA 0.37 making a 100m link at 125 Mbaud possible [20].

5.3 AUTOMOBILE

There is a possibility for the extensive use of POF in the car for a number of purposes, such as illumination, sensing, data transmission, etc [21] [22]. Cheapness for all components used, especially the connectors, is of paramount importance in view of the high density of connections involved in a car. Connector simplicity, ease of termination, and reliability are most essential. It is envisaged that the car of the future will have a computer controlling a number of functions within the car, and also receiving real-time traffic and navigation information, and displaying it on a colour CRT. Following the widespread use of CD technology or digital audio tapes at home, one will also want that technology in one's car. All these facilities will amount to having a LAN within the car, with data transmission rates varying from a few kilobits to tens of Megabits. Many of the standard topologies [23] have been considered for their applicability in the car, the most favoured one being a passive star network [22]. For such networks a (N x N) passive star coupler with as little an excess loss as possible is essential, together with other types of couplers such as (1 x N) for the distribution of data or illumination signals. Considerable effort is therefore expended in the development of cheap, very low excess loss couplers of all types. The market potential for such LAN components is clearly immense.

5.3.1 *Components for POF LANs*

The important parameters characterising the performance of fibre-optic couplers are listed in table 3. The desired features in all couplers are as low an excess loss as possible, which in turn leads to an insertion loss value very close to the theoretical (that with 0 dB excess loss), uniform distribution of a common signal to all channels, that is, 0 dB output channels power deviation, and very high directionality, that is, very high isolation between channels on the same side (input or output) of the coupler.

TABLE 3. Coupler parameters

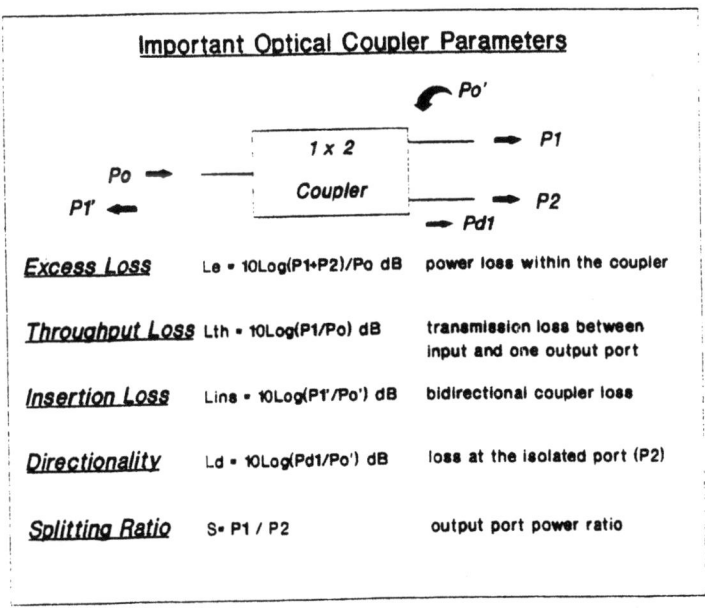

One type of POF star coupler utilises technology developed for producing GOF star couplers [24]. In this method pre-tapered POF are arranged in a bundle, and under controlled temperature conditions, are twisted and pulled until the tapered twisted region assumes the optimum specified length. With this method (2 x 2), (5 x 5) and (15 x 15) couplers have been fabricated, with excess losses consistently less than 2 dB. The coupler is put in a protective tube and secured at the ends with adhesive.

For unidirectional networks such as rings, loops and dual busses, passive taps may be used to distribute the signal at the various nodes. By bending the fibre a variable ratio-tap can be achieved, the ratio being determined by the bend radius [25], while the tap fibre is held close to the bend and the intervening region is filled with a suitable material of refractive index less than that of the fibre clad. Continuous variation of tap ratio between -11 dB and -16 dB can be achieved by varying the bend radius between 4 mm and 12 mm with screw adjustments. The excess loss varies with bend radius the worst being less than 1 dB. Fabrication of the taps without destroying the bus fibre is reported to be both easy and reliable.

In another method of producing star couplers ultrasonic welding is used to bond two fibres over a certain mixing length [26]. Rays from the input fibre propagate through the welded region into the other. The best split ratio that has been shown possible is 1:1.3 with a mixing length of 20 mm and the welded compressed region of the fibres about 10% of their diameter (12 mm^2 welded area). The (2 x 2) coupler thus produced can be made in (N x N) star couplers by crossnetting the mixing regions with other (2 x 2) couplers in series. An (8 x 8) coupler, for example, can be formed using 8 fibres and 12 welds forming 12 (2 x 2) couplers placed in series in sets of 4. Such a coupler has shown an excess loss of only 1.6 dB with a largest power deviation between any two channels of 2.6 dB. For a (16 x 16) coupler the excess loss becomes about 2.8 dB but the overall length of the mixing region must be increased to more than 100 mm.

An alternative (N x N) star coupler has been produced by interfacing the polished end surfaces of two bundles of fibres onto the polished ends of a mixing rod with the same diameter as the bundle [27]. To improve mixing of the rays in the rod and achieve good power deviation, a rod of length 500 mm is used and is bent through 180° in order to force low order modes into higher order ones. Using a PMMA rod of diameter 3 mm, bend radius 50 mm, and two bundles of 7, 1mm POF, an excess loss of 2.1 dB, with power deviation of 0.65 dB has been achieved. The mixing rod and polished interfaces are all encapsulated.

Until very recently POF couplers were not commercially available, and this may still be the case for small quantities. This was also true for (2 x 2) or (1 x 2) {50:50, 3 dB} couplers, also known as Y-couplers which are needed in many other applications apart from automotive. To overcome this problem Bougas and Kalymnios [28] adopted a low technology approach to produce (1 x 2) couplers, with no particular concern about mass production viability. Originally, couplers were produced in which one end of the three fibres was profiled through polishing in such a way that two polished surfaces, at an angle to the fibre axis, formed an apex, and two well-separated beams exited the fibre symmetrically on either side of the apex. Coupling normally onto the other two fibres was achieved by placing these close to the profiled fibre and at a specified angle. The coupler performance for various profiled apex angles was investigated, and an optimum arrangement resulted in an excess loss of 2.5 dB, splitting ratio 1.08 and directionality of -17 dB [28]. These couplers do not have index matching at the interfaces and their performance depends strongly upon the quality of the surfaces finish. Special polishing jigs had to be made and the assembled coupler was sandwiched between two acrylic plates for ruggedness. Couplers of this type have been performing reliably for a long time, but their production is fairly elaborate and not suitable for mass production.

In an even more straightforward 'do-it-yourself' approach (1 x 2) couplers have been produced by symmetrically overlapping the ends of three fibres whose ends are polished normal to their axis. With this configuration index matching or permanent fusion can enhance the coupler performance significantly. Typical values are 1.9 dB for excess loss with index matching, and 3.3 dB without index matching; a splitting ratio of 0.6 - 0.9 dB and directionality of -27 dB [29]. It has also been shown that by cascading these simple couplers, (1 x N) tree couplers can be formed, but the excess loss increases by that of the (1 x 2) coupler for every stage in the cascade. Easy assembly and protective encapsulation for these couplers is achieved using two acrylic plates, one of which has suitable grooves for positioning the fibres. The method of construction of these couplers easily lends itself to mass production techniques.

In another recent development [30] [31] (1 x 2) couplers, for 0.5 mm and 1 mm POF, are made by using a technique of micromoulding polymeric (PMMA) substrates with waveguides and alignment grooves. At the Y-junction the structure has an angle of 15° and a taper length of 1.5 mm. The three fibres are placed in the micromoulded grooves and the Y- region is filled with a material of refractive index consistent with the NA of the fibres. Essential to the production of these couplers is the fabrication of a metal mould insert for embossing the substrate. The metal mould is produced by a technique known as LIGA, based on deep-etch x-ray lithography in combination with precision electroforming and moulding processes. Both 3 dB and other splitting ratio couplers have been produced. All of these exhibit an excess loss of less than 1.7 dB, excellent power deviation (< 0.1 dB), and directionality of -31 dB. These couplers are reported to be commercially available.

5.4 SENSORS AND OTHER APPLICATIONS

The attenuation of fibres is increased when exposed to nuclear radiation, and this may limit their use in a nuclear environment. This increase for (GI) GOF is typically 1 dB/m (measured at 850 nm) for a radiation dose of 10^4 rads. Beyond this dose the attenuation becomes very high and the fibre becomes brittle. Attenuation measurements with both PMMA and PS POF have been recorded between 10^4 rads and 10^6 rads with a typical induced attenuation of about 2 dB/m measured at 550 nm [32], with no reported permanent damage. By doping the core of POF with various additives a large number of visible colours are emitted from the fibre when irradiated with γ-rays, x-rays, UV or even ambient light. The presence of a visible signal reaching the ends of a length of doped fibre serves as an indication of the incidence of the radiation. This principle of operation is widely used in high-energy physics with scintillation counters and many of these employ scintillating POF [33]. The characteristics of POF fluorescent fibres have been examined in detail [34] and a number of interesting applications using either scintillating or fluorescent fibres have been reported [35]-[41].

Many fibre optic sensing techniques for various physical parameters such as position, vibration, pressure, temperature etc., are based on intensity or wavelength modulation and can use multimode fibres [42]. If the sensor head and its control or signal processing electronics are separated by a distance that can be covered by POF, then this possibility becomes attractive, in view of the advantages that POF offer. This area of applications is expected to grow and some examples have been cited [43]-[45].

Medical and chemical sensor applications is a fast-growing field and, again, the low cost of POF may prove very attractive in many of these. The basic characteristics of POF for medical use have been examined [46] and some initial applications in this area have been reported [47] [48]. Chemical sensing utilising porous POF has also been demonstrated with pH, CO, and ammonia sensors [49].

Another interesting area with a potentially sizeable market for POF is that of illumination and displays. Illumination at airports, buildings, and theatres, has been reported feasible with very interesting and pleasing aesthetic possibilities. In all these applications 'cold light' is carried by bundles of POF served by one central lighting unit, thus simplifying maintenance and increasing reliability [50]. In yet another example, the illumination of gardens and swimming pools is being proposed [51]. Displays with POF, apart from simple spot indicators or ribbon displays already widely in use, may evolve into giant screens suitable for large public gatherings. A 'fibre-optic tube' equivalent to the CRT may be possible through the use of tapered and suitably transformed bundles of fibres [52]. The possibility for such an application is considered very exciting, with huge market potential.

6. Measurements and Standards

The most important measurement for a POF data link is the bit-error-rate (BER), usually required to be less than 10^{-9}. This refers to the number of bits in error in a sequence of correct bits, i.e. 1 error bit in 10^9 correct bits. BER rate is normally measured with a fairly expensive instrument known as 'BER test set' which generates a pseudo random bit sequence (PRBS) (but known at all times) driving the transmitter while a bit-by-bit error detector at the receiver end records the errors. The BER is simply the number of errors which occur over the number of bits transmitted during some interval of time. It is unlikely that most users will have such an instrument to qualify their POF links, and therefore other measurements which indicate an expected BER, such as overall link loss, power margin, and link bandwidth must be used.

There are no developed standards yet which are specific to POF applications. This situation, however, is bound to change as POF become more widely used.

Various measurement techniques that are used with GOF are also adapted for POF. The attenuation of POF is mostly measured using the cut-back method [4], and the bandwidth from time-domain measurements [15] [53] [54] with known launch conditions since these strongly affect the results. Interestingly, an optical time-domain reflectometer for POF has also been developed, and has been used to measure the attenuation, length and faults or splice losses for fibres up to 100 m long [55].

For industrial applications, in particular, knowledge of the environmental influence on the performance of POF is most important. A number of measurements on the performance of POF under variable conditions, such as mechanical (bending, tensile force, impact etc.), climatic (temperature, humidity etc.), chemical and biological (solvents, acids, salt-water etc.) have been undertaken [56]. Many of these measurements have been performed in accordance with techniques specified either for GOF or some other relevant national industrial standard, or international (CEI/IEC). An international POF interest group (POFIG) has been started in the USA (1992), and one of its important aims is that of seeking representation for POF on the various international standard committees, CCITT, ANSI, IEEE etc.

7. Future Developments and Markets

Currently, the limitations in POF data links arise from both the limited bandwidth of the fibre and the availability of fast and powerful inexpensive emitters. The bandwidth limitation has already been removed with the development of the graded-index POF [17], and very soon these fibres should become widely available. Attempts to develop more efficient and faster LEDs are already under way [57]. Utilizing the quaternary material, InGaAsP, and metal organic chemical vapour deposition (MOCVD) for the epitaxial growth on a GaAs substrate, both red and blue LEDs have been fabricated with superior characteristics to their current equivalents (AlGaAs or GaAsP) grown by liquid-phase epitaxy. The new red LED (670 nm) has a rise time of 7.9 ns and delivers -12 dBm average power into a 1mm POF. A data rate of 100 Mbits/s over 30 m was demonstrated with a wide open eye pattern. With the green LED (573 nm) the attenuation of the fibre is only about 80 dB/km and much longer links may become possible. Initial tests have shown that 10 Mbits/s over 100 m is now easily achievable. The search for faster and more efficient LED emitters is bound to continue, with promising prospects for the future.

Proponents of POF believe that the interest in these fibres has now been revitalized, and future applications in networks with 1 mm POF may well extend beyond FDDI to ATM, Fibre Channel, Escson, Sonet, and fibre-to-the-home. Marketing reports have been cited which estimate that by 1998 POF sales will surpass those of GOF. Major growth areas are thought to be the simple point-to-point links both for industrial and consumer applications while automobile networks and links will also develop into a major market. Sensors applications, especially with doped POF, will continue to increase and may well establish new and sizeable markets.

8. Conclusion

The basic advantages and characteristics of POF have been reviewed and typical performance figures for simple POF data links, utilising inexpensive optoelectronic components and connectors have been presented. There is a wide range of applications for which the use of POF is being considered and the market prospects for many of these applications are most promising. Interest in POF has grown rapidly during the last few years. Following the first international meeting on POF in Boston in 1991, two more international conferences wholly dedicated to POF and its applications have taken place. The proceedings of these have been cited extensively, and readers are referred to these for more detailed reading. This year's POF conference was held in Yokohama, Japan (October 1994), and at the time of preparation of this script, the proceedings were not available. It is almost certain that further new developments in POF and its applications will have been presented.

9. Acknowledgments

I would like to thank my colleagues N. Ioannides and I W Rogers for their help in the preparation of this paper.

10. References

1. Kitazawa, M. (1987) An overview of recent development of plastic optical fibres. *J. of Opt. Sensors*, **2**, 35-43.
2. Teshima, S., Munekuni, H., Katsuta, S. (1992) Plastic optical fibre for automotive applications. *Proc. 1st Int. Conf. POF and Applications, Paris, 44-48*.
3. Tanaka, A., Sawada, H., Takoshima, T., Wakatsuki, N. (1987) New Plastic Optical fibre Using Polycarbonate Core and Fluorescence-Doped Fiber for High Temperature Use. *Fiber and Integrated Optics, 7, 139-158*.
4. Senior, J.M. (1992) *Optical Fiber Communications, 2nd Ed.*, Prentice-Hall Inc.
5. Cheo, P.K. (1990) *Fiber Optics and Optoelectronics*, Prentice-Hall Inc.
6. Ashpole, R.S., Hall, S.R., Luker, P.A. (1993) Polymer Optical Fibres - a case study, in *POF Design Manual and Handbook and Buyers Guide*, Information Gatekeepers Inc., Boston, pp 6-15.
7. Koike, Y., Yukie, H., Eisuke, N. (1991) Graded-index polymer optical fiber by new random copolymerization technique, *Plastic Optical Fibers, SPIE* **1592**, 62-72.
8. Krumpholz, O., Pressmar, K., Schlosser, E. (1993) LED-carrier with reflector for plastic optical fibers, *Proc. 2nd Int. Conf. on POF and Applications*, The Hague, 125-126.
9. Hewlett-Packard Co. (1991) *Versatile Link*, Commercial Literature.
10. Hirschmann, G. (1987) Light-Link Optocouplers are Faster, *Siemens Components* **XXII**, 3, 96-99.
11. AMP Inc. (1993) *Connectors for POF and Applications*, Commercial Literature, The Hague, June 28-29.
12. Carson, S.D., and Salazar, R.A. (1991) Splicing Plastic Optical Fibers, *SPIE*, **1592**, 134-138.
13. Gowar, J. (1984) *Optical Communication Systems*, Prentice-Hall International.
14. Hoss R.J. (1990) *Fibre optic communications handbook*, Prentice-Hall Inc.
15. Meier, J., Lieber, W., Heinlein, W., Groh, W., Herbrechtsmeier, P., Theis, J. (1987) Time-Domain Bandwidth measurements of Step-index Plastic Optical Fibres. *Electron. Lett.*, **23**, 22, 1208-1209.
16. Takahashi, S. (1993) Experimental studies on launching conditions in evaluating transmission characteristics of POFs. *Proc. 2nd Int. Conf. POF and Applications*, The Hague, 83-85.
17. Koike, Y., Ishigure, T., Horibe, A., Nihei, E. (1993) High-band width polymer optical fiber and its applications. *Proc. 2nd Int. Conf. POF and Applications*, The Hague, 54-58.

18. Walker, S., Bates, R.J.S. (1993) Towards gigabit plastic optical fibre data links : present progress and future prospects, *Proc. 2nd Int. Conf. POF and Applications*, The Hague, 8-13.
19. Hanson, D. (1993) Wiring with Plastic, in *POF Design Manual and Handbook and Buyers Guide*, Information Gatekeepers Inc., Boston, pp 41-46
20. Hewlett Packard Co. (1994) *Fibre-optic Solutions for 125 MBd Data Communications at Copper Wire Prices*, Application Note 1066.
21. Harmer, A.L. (1984) Fibre-optics in automobiles, *SPIE*, **468**, 174-185.
22. Steele, R.E., Schmitt, H.J., (1987) Development of fiber optics for passenger car applications, *SPIE*, **840**, 2-9.
23. Stallings, W. (1990) *Local Networks*, 3rd Edition, Macmillan Publishing Co.
24. Imoto, K., Sano, H., Maeda, M. (1986) Plastic optical star coupler, *Appl. Opt.* **25**, 19, 3443-3447.
25. Kagami, M., Sakai, Y., Okada, H. (1991) Variable-ratio tap for plastic optical fibre, *Appl. Opt.* **30**, 6, 645-649.
26. Yuuki, H., Ito, T., Sugimoto, T. (1991) Plastic star coupler, *SPIE*, **1592**, 2-11.
27. Woesik, van, E., E.Th.C., Post, J. (1993) N*N Bi-directional transmissive star coupler, *2nd Int. Conf. POF and Applications*, The Hague, 127-131.
28. Bougas, V., Kalymnios, D. (1991) Plastic fibre couplers using simple polishing techniques, *SPIE*, **1504**, 298-302.
29. Kalymnios, D., (1992) Plastic optical fibre tree couplers using simple Y-couplers, *Proc. 1st Int. Conf. POF and Applications*, Paris, 115-118.
30. Rogner, A. (1992) Micromoulding of passive network components, *Proc. 1st Int. Conf. POF and Applications*, Paris, 102-104.
31. Rogner, A., Pannhoff, H. (1993) Characterisation and qualification of moulded couplers for POF-networks, *2nd Int. Conf. POF and Applications*, The Hague, 136-139.
32. Chiron, B. (1989) The behaviour of POF under nuclear radiation: applications for nuclear detector scintillators, data transmission and illumination. *IEE Colloquium on 'Plastics materials for optical transmission'* Digest No. 1989/140, 5/1-5/8.
33. Henschel, H., Kohn, O., Schmidt, H.U. (1993) Radiation sensitivity of plastic optical fibres, *2nd Int. Conf. POF and Applications*, The Hague, 99-104.
34. Bross, A.D. (1991) Scintillating plastic optical fibre radiation detectors in high energy particle physics, *SPIE*, **1592**, 122-132.
35. Grammatico, A. (1992) Manufacturing and applications of plastic optical scintillating fibres, *1st Int. Conf. POF and Applications*, Paris, 64-67.
36. Chiron, B. (1991) Highly efficient plastic optical fluorescent fibers and sensors, *SPIE*, **1592**, 86-95.
37. Ikhlef, A., Skowronek, M. (1993) One-dimension radiation position sensor using a plastic scintillating fiber, *2nd Int. Conf. POF and Applications*, The Hague, 158-161.
38. Fabini, P., Pazzi, G.P., Linari, R., Pelfer, P.G., Tartoni, N. (1993) Measurement of thickness and defects by a matrix of plastic scintillating fibres, *2nd Int. Conf. POF and Applications*, The Hague, 153-157.
39. Laguesse, M. (1992) Characterization of an x-ray detector with fluorescent plastic optical fibre readout, *1st Int. Conf. POF and Applications*, Paris, 68-73.
40. Laguesse, M. (1993) Sensor applications of fluorescent plastic optical fibres, *2nd Int. Conf. POF and Applications*, The Hague, 14-19.
41. Destruel, P., Farenc, J., Saad, A., Llop, X. (1992) Luminescent plastic optical fibers in the field of active sensors, *1st Int. Conf. POF and Applications*, Paris, 74-79.
42. Jackson, D.A., Jones, J.D.C. (1986) Fibre optic sensors, *Optica Acta*, **33**, 12, 1469-1503.
43. Kalymnios, D. (1993) Linear and scalable optical sensing technique for displacement, in K.T.V. Grattan and A.T. Augousti (eds), *Sensors VI*, IOP Publishing Ltd., Great Yarmouth, pp 291-294.
44. Ioannides, N., Kalymnios, D., Rogers, I.W. (1993) Experimental and theoretical investigation of a POF based displacement sensor, *2nd Int. Conf. POF and Applications*, The Hague, 162-165.
45. Felgenhauer, R., Haack, A., Streuf, M. (1993) Fibre-optic distance sensor with POF, *2nd Int. Conf. POF and Applications*, The Hague, 171-173.

46. Kosa, N.B. (1991) Key issues in selecting plastic optical fibers used in novel medical sensors, *SPIE*, **1592**, 114-121.
47. Suzuki, F. (1991) Novel plastic image-transmitting fiber, *SPIE*, **1592**, 150-157.
48. Falciai, R., Baldinni, F., Bechi, P. (1992) Bile enterogastric reflex sensor using plastic fibers, *1st Int. Conf. POF and Applications*, Paris, 128-131.
49. Zhou, Q. (1991) Development of Chemical Sensors Using Plastic Optical Fiber, *SPIE*, **1592**, 108-113.
50. Jaquet, P. (1991) Plastic optical fibre applications for lighting of airports and buildings, *SPIE*, **1592**, 165-172.
51. Ono, T. (1993) Illumination of architecture and landscape with plastic optical fibre, *2nd Int. Conf. POF and Applications*, The Hague, 71-74.
52. Loubeyre, F., Garcia, T., Dejeus, J.M., Truchot P. (1992) Optical anamorphosis, *1st Int. Conf. POF and Applications*, Paris, 132-133.
53. Takahashi, S., Ichimura, K., (1991) Time domain measurements of launching-condition-dependent bandwidth of all-plastic optical fibres, *Electron. Letts.*, **27**, 3, 217-218
54. Takahashi S. (1993) Experimental studies on launching conditions in evaluating transmission characteristics of POFs, *2nd Int. Conf. POF and Applications*, The Hague, 83-85
55. Mohr St. (1992) Optical time-domain reflectometry for plastic optical fibres, *1st Int. Conf POF and Applications*, Paris, 82-85
56. Daum W., Brockmeyer A., Goehlich L. (1992) Environmental qualification of polymer optical fibres for industrial applications, *1st Int. Conf POF and Applications*, Paris, 91-95
57. Fukuoka K., Iwakami T., Schumacher K. (1993) High speed and long-distance POF transmission systems based on LED transmitters, *2nd Int. Conf. POF and Applications*, The Hague, 43-46.

XI Further Posters

REFRACTIVE INDEX PROFILE OF PLANAR MICROLENSES

S. RÍOS (1), E. ACOSTA (1), M. OIKAWA (2) AND K. IGA (3)
(1) Laboratorio de Óptica, Escuela de Óptica, Universidad de Santiago de Compostela, Campus Sur, 15706 Santiago de Compostela, Spain. Phone and FAX: 34-81-521984
(2) Nippon Sheet Glass Co. Ltd. Shimbashi Sumimoto Bldg. 5-11-3 Shimbashi, Minato-ku, Tokyo 105, Japan. Phone: 81-3-3436-8721, FAX: 81-3-3436-8721
(3) Tokyo Institute of Technology, 4259 Nagatsuta, Midori-ku, Yokohama 227, Japan. Phone and FAX: 81-45-922-5358

1. Introduction

Knowledge of the index profile of any optical component is needed to characterize optical properties such as focal length, numerical aperture, aberrations... as well as to perform optical devices design. In this way, the determination of the index profile of planar microlenses (PML) is needed to characterize them and design devices containing these elements as mux/demux configurations, couplers, processors and so on. In a previous work, S. Mizawa, M. Oikawa and K. Iga proposed a 3-D representation of the index fitting to a matrix representation, which is related to the power series expansion with respect to the axial and radial distances [1]. The fit coefficients are the matrix elements. It has been also shown that 4x4 matrix leads to a good approximation. This representation means 16 fit coefficients, and to perform design in terms of such a number of parameters is not an easy task.

In this work we present a general method to determine functional expressions of the refractive index profile of PML reducing the number of coefficients.

2. Experimental Data

The experimental data of the index profile of PML is usually obtained from interferometric methods such as the total shearing one [2,3]. The total shearing interference pattern represents the equi-index fringes of the distributed index of the lens (see Fig.1). From each equi-index fringe we can determine the coordinates of the lens with the same refractive index. These are the only experimental data needed for the fit.

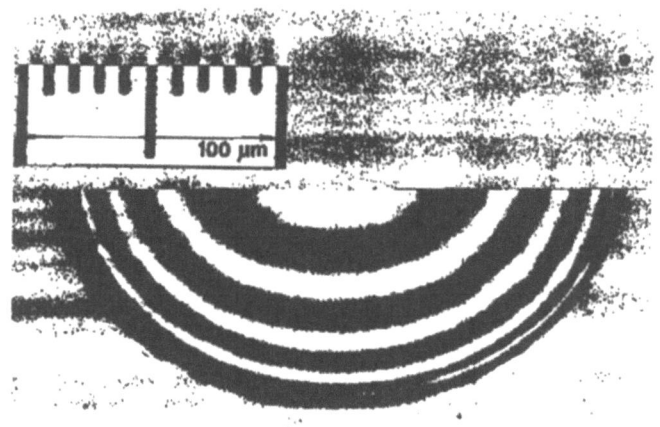

INDEX DISTRIBUTION OF PLANAR MICROLENS

Figure.1. Total shearing interference pattern of a PML.

3. Description of the Fit Method and Results

The experimental data of each equi-index fringe have been fitted to a polynomial expression in the form:

$$z^2 = \sum_{i=0}^{N} a_i r^i \qquad (1)$$

where N is the maximum number of coefficients needed for a good fit. Linear algorithms have been used to perform this fit, and by means of an F-test [4], the significance of each term in the fit has been determined. The conclusion obtained with this analysis is that ellipses give a good aproximation. In this case N=2 and i=1 presents no significance in the fit, meaning that all the ellipses are centered in the same coordinate system. Therefore each ellipse depends only on two parameters, namely m_1 and m_2, representing respectively the minor axis size and the ellipticity of each fringe. Both m_1 and m_2 depend on the refractive index profile of the fringe, $m_1=m_1(n)$, $m_2=m_2(n)$.

As each fringe can be represented by:

$$z^2 + m_2^2 r^2 = m_1^2 \qquad (2)$$

a dependence between m_1 and m_2 has been found in such a way that m_1 could be represented in terms of only r,z and the fit coefficients between m_1 and m_2. Therefore the relation between m_1 and m_2 can be expressed as:

$$m_2^2 = \frac{a + bm_1^2}{1 + cm_1^2} \qquad (3)$$

where the fit was performed by linear algorithms and the value of the coefficients are:

$$a = .2215 \text{(dimensionless)}$$
$$b = 2.107 \times 10^{-4} \, \mu m^{-2} \qquad (4)$$
$$c = 1.666 \times 10^{-4} \, \mu m^{-2}$$

Fig.2 represents this fit. The regression coefficient and the standard deviation of the fit are $r^2=0.991$ and $\sigma=0.011$ respectively. The standard deviation and the confidence limits for the coefficients are shown in Table.1. Fig.3 represents the residuals in %. It can be observed that the highest error from the fit to the experimental data of m_1 and m_2 is smaller than 8%.

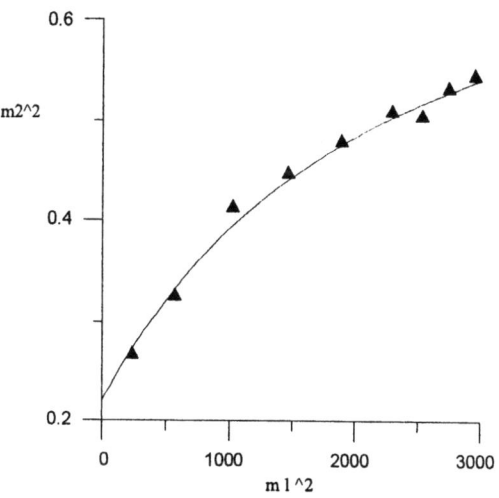

Figure.2. Representation of m2^2 as a function of m1^2 given by eq.3

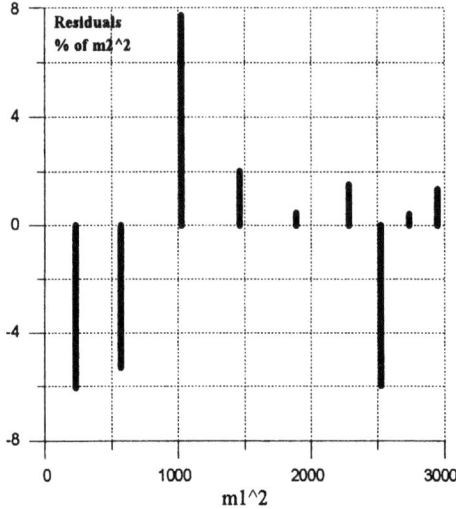

Figure.3. Residuals in % corresponding to the fit of eq.3.

Table1. Standard deviation and confidence limits of the coefficients for the fit of eq.3.

coefficients	Std. error	90% confidence limits
a=0.221	0.014	0.193...0.249
b=0.00021	4e-05	0.00012..0.00029
c=0.00016	2e-05	0.00011..0.00021

In this way the ellipses can be represented in terms of m_1 as follows:

$$m_1^2 = \frac{(bz^2 + cr^2 - 1) + \sqrt{(bz^2 + cr^2 - 1)^2 + 4b(z^2 + ar^2)}}{2b} \quad (5)$$

where the "-" sign of the solution has been discarted because it has not physical meaning.

To check the accuracy of the elliptical approximation, as well as the interrelation between m_1 and m_2, the ellipses defined now by eq.4 were plotted and overlaped on the fringe pattern of Fig.1 obtaining that the fringes can be accurately reproduced.

The last step is to fit m_1 versus n^2. This fit must have enough precission to avoid the transmission of errors due to two consecutive fits. In this case 4 parameters were needed and non-linear algorithms, as Marquard's, were used [5]. The relation between m_1 and n^2 is as follows:

$$\ln(n_0^2 - n^2) = \frac{D + Em_1^2 + Fm_1^4}{m_1^2} \qquad (6)$$

where the coefficients are:

$$\begin{aligned} n_0^2 &= 2.89 \\ D &= -375 \mu m^2 \\ E &= -0.89 \, (\text{dimenssionless}) \\ F &= 4.2 \times 10^{-4} \end{aligned} \qquad (7)$$

Fig.4 represents this fit and Fig.5 the residuals. The regression coefficient and the standard deviation of the fit are r^2=0.999 and σ=0.009 respectively. The standard deviation and the confidence limits for the coefficients are shown.in Table.2. It can be seen that the error from the fit to the experimental data is smaller than 8%. n_0 is a fit coefficient and represents the highest value of the distributed index.

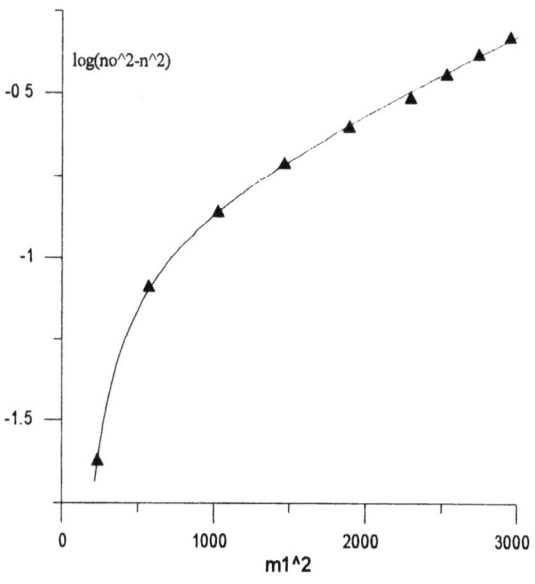

Figure.4. Representation of the fit given in eq.6

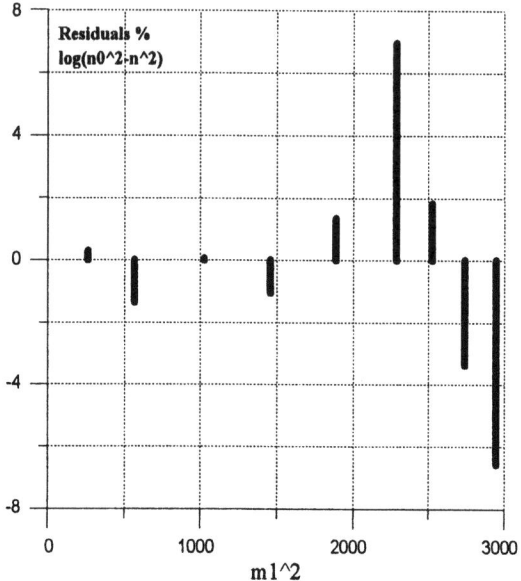

Figure5. Residuals corresponding to the fit of eq.6.

Table2. Standard deviation and confidence limits of the coefficients for the fit of eq.6.

coefficients	Std. error	90% confidence limits
a=-375	8	-392...-355
b=-0.895	0.015	-0.924...-0.865
c=0.000102	3e-06	9.6e-05...1.08e-04

So, the final expression for the index distribution of the PML becomes:

$$n^2 = n_0^2 - K\exp\left\{-D'\frac{g(r,z)-\sqrt{g^2(r,z)+4bf(r,z)}}{2f(r,z)}\right\}$$
$$\exp\left\{F'\frac{g(r,z)+\sqrt{g^2(r,z)+4bf(r,z)}}{2b}\right\} \quad (8)$$

where

$$g(r,z) = cr^2 + bz^2 - 1$$
$$f(r,z) = z^2 + ar^2 \qquad (9)$$

and the coefficients are

$$n_0^2 = 2.89, a = 0.22 \,(\text{dimenssionless})$$
$$b = 2.107 \times 10^{-4} \,\mu m^{-2}, c = 1.666 \times 10^{-4} \,\mu m^{-2}$$
$$D' = 4.319 \,\mu m^2, F = 1.1762 \times 10^{-4} \,\mu m^{-2} \qquad (10)$$
$$K = 0.1273 \,(\text{dimenssionless})$$

All together the fit has been made by means of 7 parameters. Fig.6 represent the equi-index firnges obtained from the fits of eqs.3,6 and Fig.7 shows a 3D representation of the index distribution given by eq.8. The function of n^2 is continuous as well as the derivative (with respect to both r or z). This is a very important condition in the trial functions since this implies the existence of solutions of the ray tracing equations.

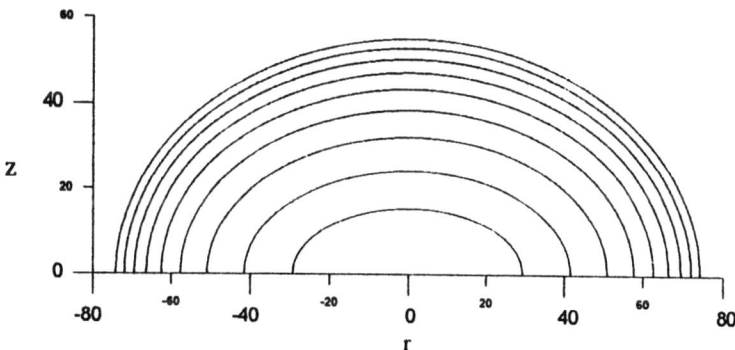

Figure 6. Equi-index fringes obtained using the fits of eqs. 3 and 6

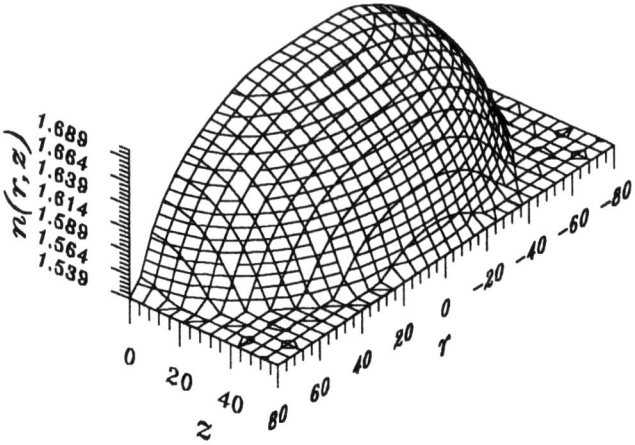

Figure.7. 3D representation of the index distribution given in eq.8.

4. Conclusions

We have presented a method to fit the refractive index profile of planar microlenses to analytical expressions from their total shearing interference patern. This method means a reduction in the number of fit coefficients compared to the most common fit to a matrix expression including 16 fit coefficients.

5. References

1. Mizawa, S., Oikawa, M. and Iga, K.(1982) *Jap. J. Appl. Phys.* **21**, L589.
2. Kokubun, Y. and Iga, K. (1982) Index profiling of distributed-index lenses by a shearing interference method, *Applied Optics* **21**, 1030-1034.
3. Kokubun, Y. and Iga, K. (1980) Refractive index profile measurement of preform rods by a transverse differential interferogram, *Applied Optics* **19**, 846-851.
4. Bevington, P.(1969) *Data Reduction and Error Analysis for the Physical Sciences*, Mc Graw-Hill, New York.
5. Marquard, D. W. (1963) *J. Sec. Ind. Appl. Math.* **2**, 431.

6. Acknowledgment

This work was supported by the CICYT, Ministerio de Transportes, Turismo y Comunicaciones (Contract TIC 93/0606).

FIBRE INTERFEROMETER FOR PHASE AND POLARISATION MEASUREMENT OF PASSIVE FIBRE ELEMENTS [*]

M. SZUSTAKOWSKI, L. R. JAROSZEWICZ
Institute of Technical Physics, Military University of Technology
01-849 Warsaw 49, ul. Kaliskiego 2, POLAND

An application of a measuring system based on fibre optic Sagnac interferometer for measurements of phase-amplitude characteristics of fibre elements is described. Due to the application of a set of phase-polarization detectors, such a system may be also applied to obtain phase characteristics of measuring heads used in fibre optic interferometer. Used instrumentation allows to study elements in the frequency range from 100 Hz to 1.4 MHz.

1. Introduction

Nowadays fibre optic analogues of all classic interferometric systems are known [1,2]. Besides characteristic features of fibre optic technology, e.g., light conducting in a closed waveguide, those systems inherit all disadvantages of their volume counterparts. The main disadvantage is an instability of action if environment parameters, especially temperature, are changed. Due to thermal instability, a testing of fibre optic elements and phase heads in two-beam interferometer, e.g., Mach-Zehnder or Michelson type is troublesome and difficult [3].

In this paper the measuring system of fibre elements and phase heads using fibre optic Sagnac interferometer is described. Due to stable action of this interferometer, reproducible results may be obtained.

2. Detection of Periodic Phase Disturbance in Fibre Optic Sagnac Interferometer

Phase modulator used in fibre optic technology [4] is a classic example of the phase fibre optic transducer which changes vibrations into phase modulation of light wave. This modulator is obtained by a rigid assembling of single-mode fibre with length of L on a piezoceramic element. In case of substratum elongation by ΔL, the phase shift of optical wave which propagates in the fibre optic takes place. This phase shift may be described as [5]:

$$\phi_o \approx \tfrac{2\pi}{\lambda} N_{ef} \cdot \Delta L \qquad (1)$$

---

[*] The paper was presented at a poster sesion

where λ stands for light wavelength and N_{ef} is an effective refractive index of the optic fibre.

An application of sine enforcement signal to such phase transducer causes harmonic phase shift of light wave described by an expression [6]:

$$\phi_m(t) = \phi_o \sin(\omega t + \beta), \qquad (2)$$

where β stands for an initial phase of disturbance.

In general, parameters β and ϕ_0 may be dependent on several factors. For instance, in case of piezoceramic element they depend on the applied voltage (V), the enforcement frequency (ω) and on the construction of the transducer (a geometry of assembling of the fibre to the substratum). The aim of this paper is to demonstrate a possibility of measurement of those quantities, describing frequency-voltage characteristics of given transducer, by the application of Sagnac interferometer.

2.1. CHARACTERISTICS OF OPERATION OF FIBRE OPTIC SAGNAC INTERFEROMETER

A reciprocal two-beam fibre optic Sagnac interferometer with schematically presented elements is shown in Fig. 1. Output laser radiation is divided by an X-type coupler in two beams cw and ccw, then it is passing by a sensor loop containing fibre optic element FOE and eventually it is put together by the same coupler. A destructive interference picture existing on the second coupler input, passing through the additional X-type coupler, is received by the set of two detectors; directly in case of D1 and after passing a polarizer P in case of D2.

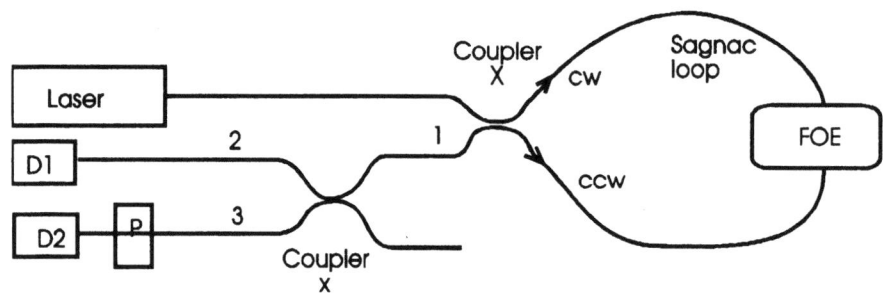

Figure 1. Fibre optic Sagnac interferometer with polarization-phase detection.

The description of the system action may be done by means of Jones matrix formalism [7]. Assuming suitable double helix coil of the sensor loop [8] one may compensate the Sagnac effect. For the perfect X-type coupler 3 dB and coherent interaction of both beams using above formalism one may obtain optical power on the output (1) in form of:

$$p = p_0 - 2|\tilde{p}| \cos \Delta\phi \qquad (3)$$

where:
$$p_0 = a_0^+(|P_{cw}|^2 + |P_{ccw}|^2)a_0; \quad |\tilde{p}| = |a_0^+ P_{ccw}^+ P_{cw} a_0|;$$

$$\Delta\phi = \arg(\tilde{p}) = arctg\left[\frac{Im(a_0^+ P_{ccw}^+ P_{cw} a_0)}{Re(a_0^+ P_{ccw}^+ P_{cw} a_0)}\right] \quad (4)$$

while $\mathbf{P}_{cw}$ and $\mathbf{P}_{ccw}$ are Jones' matrices for beams propagating through the loop clockwise and counter-clockwise, respectively, and $\mathbf{a}_0$ is Jones vector describing electric field on optic fibre input. For above system configuration we have:

$$\mathbf{P}_{ccw}^+ \mathbf{P}_{cw} = -\tfrac{1}{4}\overleftarrow{\mathbf{M}}^+ \mathbf{In} \ln \overrightarrow{\mathbf{M}} = -\tfrac{1}{4}\overleftarrow{\mathbf{M}}^+ \overrightarrow{\mathbf{M}} \quad (5)$$

while **In** matrix is an inversion matrix connected with an existence of a change of propagation direction of output beams in comparison with input beams [9]. In turns, **M** is Jones matrix of fibre optic element FOE (in general, the whole sensor loop). Upper arrows describe this direction of wave propagation through the loop for which given matrix has been calculated, while $^+$ is the operation of Hermite conjugation. Moreover, as it has been proved in [10], for the optical element independent on time in a linear medium, with the same modes at the input and output and no nonreciprocal (Sagnac and Faraday) effects the following relation is correct for waves passing in the opposite directions (cw and ccw waves):

$$\overleftarrow{\mathbf{M}} = \overrightarrow{\mathbf{M}}^T \quad (6)$$

where superscript T describes matrix transposition.

According to the theorem on the optical system equivalence [7], any non-dissipative optical element may be presented as a product of a rotator $\mathbf{R}(\alpha)$, where α stands for the rotation angle of the polarization plane due to rotator passing, and phase retarder $G(\zeta,\delta)$, where δ is a phase retardation between orthogonal field components, while ζ is the direction of normal axes of the optical element. Therefore one can describe an expression:

$$\mathbf{M} = \mathbf{G}(\zeta,\delta)\mathbf{R}(\alpha) \equiv \mathbf{R}(\zeta)\mathbf{G}(\delta)\mathbf{R}(-\zeta)\mathbf{R}(\alpha) = \mathbf{R}(\zeta)\mathbf{G}(\delta)\mathbf{R}(\zeta') \quad (7)$$

where: $\zeta' = \zeta - \alpha$.

The good example showing characteristic features of the Sagnac interferometer is polarization controller. This element consists of two fibre optic loops. The mutual rotation of those loops enables to obtain any modification of polarization state of light beam. Using the description of this element given in [11] and assuming its matching one may describe:

$$\overrightarrow{\mathbf{M}} = \mathbf{R}(\zeta)\mathbf{G}(\delta)\mathbf{R}(\alpha) \quad (8a)$$

where:
$$\zeta = 0.92\psi - \pi/4; \quad \delta = \pi/2 + 0.92(\varphi - \psi); \quad \alpha = -0.92\varphi - \pi/4 \quad (8b)$$

while ψ and φ are rotation angles of the first and the second loop, respectively.

Then, assuming that the input radiation is linearly polarized along 'x'-axis of coordinate system:

$$\mathbf{a}_0 = \begin{bmatrix} 1 \\ 0 \end{bmatrix} \qquad (9)$$

one obtain the following expression for the intensity measured on the detector

$$p = \tfrac{1}{2}\{1 - |\tilde{p}|\cos[Arg(\tilde{p})]\} \qquad (10)$$

where:

$$\tilde{p} = \tfrac{1}{4}\{\left[1 - (1+\cos 2\delta)\sin^2(\alpha+\zeta)\right] + i[\sin 2\delta(\cos^2\zeta - \cos^2\alpha)]\} \qquad (11)$$

As one can see, by changing the angle between the loops from which the polarization controller is built, an arbitrary modification of the polarization properties of the loop is possible.

The above situation changes dramatically if the optical element is time-dependent. This is exactly the case of phase transducer harmonically dependent on time, we are interested in. The example of such a transducer is phase modulator obtained by coiling N coils of single-mode optic fibre on a piezoceramic moulder with a radius of a [5]. In the case of ideal preparation, when the optic fibre is coiled without twist, the matrix description of this element is as follows [8]:

$$\overrightarrow{\mathbf{M}} = \mathbf{D}(2\pi N a \beta_0)\mathbf{G}(mN/a) \qquad (12)$$

where $\mathbf{D}(...)$ is the matrix of constant-phase retarder [11], β_0 stands for propagation constant in fibre core, m for a constant equal to 1.22×10^{-9} [m].

An application of sine voltage with a frequency of ω to the piezoceramics causes the following change of piezoceramics diameter:

$$a(t) = a[1 + \alpha \sin(\omega t + \beta)], \quad \alpha \ll 1 \qquad (13)$$

Then the Jones matrix for this element placed in the fibre optic loop with length of L close to one of the coupler branches takes the form for waves cw and ccw, respectively:

$$\overrightarrow{\mathbf{M}} = \mathbf{D}[2\pi N\beta_0 a\{1 + \sin(\omega t + \beta)\}]\mathbf{G}\left[\tfrac{mN/a}{1+\alpha\sin(\omega t+\beta)}\right] \qquad (14a)$$

$$\overleftarrow{\mathbf{M}} = \mathbf{D}[2\pi N\beta_0 a\{1 + \alpha\sin[\omega(t-\tau) + \beta]\}]\mathbf{G}\left[\tfrac{mN/a}{1+\alpha\sin[\omega(t-\tau)+\beta]}\right] \qquad (14b)$$

where $\tau = LN_{ef}/c$ stands for retardation time of ccw beam signal in relation to cw beam signal.

Consequently, this time retardation between interacting beams causes the change of system behaviour because, as it may be shown even for linear input polarization ($\mathbf{a}_0$ described by (9)), the obtained result is:

$$\tilde{p} = \tfrac{1}{4}\exp\{-i[(\phi_0 - b\alpha)[\sin(\omega t + \beta) + \sin[\omega(t-\tau) + \beta]] + 2b]\} \qquad (15)$$

where $\phi_0 = 2\pi\beta_0\alpha Na$ stands for the amplitude of phase modulation (1), $b = mN/a$ for a constant related to the birefringence enforced in the fibre by its bending.
As one can see, time-dependent phase transducer introduces a constant phase shift $2b$ and two kinds of modulation. The first one is phase modulation with amplitude of ϕ_0, the second one is polarization modulation with amplitude of $b\alpha$. The phase modulation is predominant, however:

$$\frac{\phi_0}{b\alpha} = \frac{2\pi\beta_0}{m}a^2 \approx 2 \qquad (16)$$

Nevertheless, if input radiation is linearly polarized and uniformly induces both modes the resulting expression takes the following form:

$$\tilde{p} = \tfrac{1}{4}\cos\{b\alpha[\sin(\omega t + \beta) + \sin[\omega(t-\tau) + \beta]] + 2b\}e^{-i\phi_0\{\sin(\omega t+\beta)+\sin[\omega(t-\tau)+\beta]\}} \qquad (17)$$

In other words, polarization modulation is predominant.
From this reason, studies of dynamically changed phase element require to take into consideration a possibility of polarization modulation. In the described system this requirement is fulfilled by an application of polarization detector (detector D2 preceded by a polarizer, see Fig 1). An occurrence of alternating signal on this detector is an evidence of the presence of polarization modulation.

2.2. MEASUREMENT OF DYNAMIC PHASE CHARACTERISTICS OF STUDIED ELEMENT

In order to measure parameters of studied phase element (2) the configuration of Sagnac interferometer shown in Fig. 2 has been applied. Laser light is introduced to the interferometer loop by 1 branch of X-type coupler. Polarization controller (PC1) placed before the coupler optimizes conditions of the induction of the sensor loop. The detection of output signal is performed in 4 branch of the coupler. An introduction of an additional X-type coupler in this branch allows to divide output power to two detectors D1 and D2. In the branch containing D2 detector preceded by the polarizer an additional polarization controller (PC4) is placed. This controller allows to match output polarization state and polarizer characteristic. In order to avoid the effect of Earth revolution and other disturbances on the measurement result, i.e., Sagnac effect, the interferometer loop is coiled to a shape of double helix. The studied element is placed close to one of the ends of the interferometer loop.
Two polarization controllers (PC2 and PC3) placed inside the loop, according (11) allow to adjust a working point of the system by an arbitrary modification of the polarization properties of the loop. This possibility is shown in the Fig. 3.
For this configuration of the measuring system to study phase element one should adjust conditions of the system action with the minimum polarization modulation (at the level of detection system noise). It may be obtained by turn of controllers PC3 and PC2 and the change of element supplying voltage. In case of this adjustment the signal on the D2 detector for an arbitrary adjustment of PC4 is assumed as $(P_2(t)=0)$. In other

words, polarization modulation is neglected ($b\alpha \sim 0$ in Eq. (15)). Then the power detected by the D1 detector may be described as:

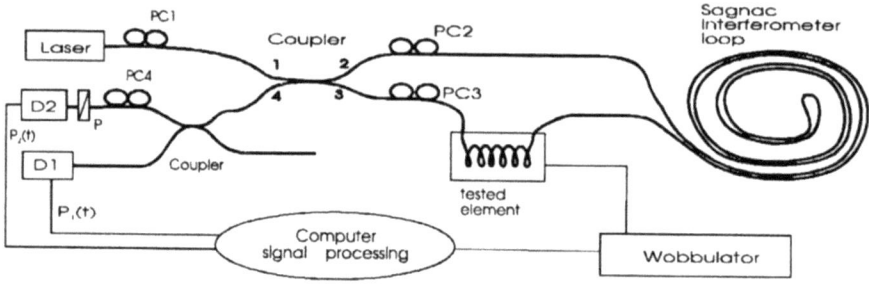

Figure 2. The scheme of Sagnac interferometer adapted to study of phase element.

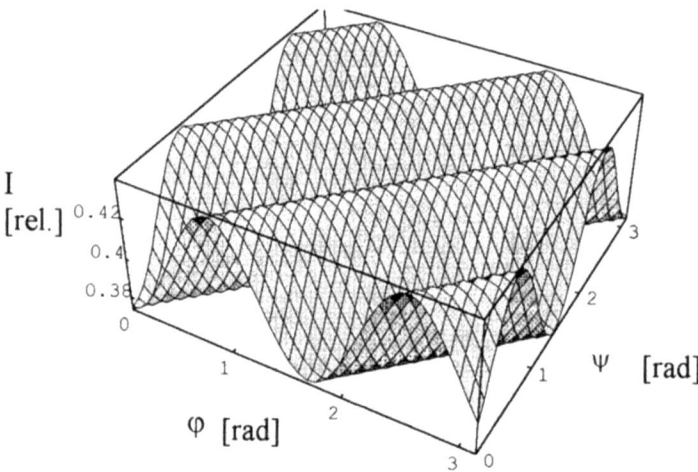

Figure 3. The modification of the output signal of the Sagnac interferometer by means of polarization controller PC2; ψ and φ are angles of arrangement of fibre optic loops forming PC2.

$$P_1(t) = P_o \left\{ \begin{array}{l} 1 + \cos\left\{\phi_o \sin\left(\frac{\omega\tau}{2}\right)\cos\left[\omega\left(t - \frac{\tau}{2}\right) + \beta\right]\right\}\cos\Delta\varphi - \\ \sin\left\{\phi_o \sin\left(\frac{\omega\tau}{2}\right)\cos\left[\omega\left(t - \frac{\tau}{2}\right) + \beta\right]\right\}\sin\Delta\varphi \end{array} \right\} \quad (18)$$

where P_0 stands for the power of semiconductor laser output signal and $\Delta\varphi$ for the total phase shift introduced by birefringence of the studied element, the sensor loop and polarization controllers PC2 and PC3.

For the working point adjusted by polarization controllers as $\Delta\varphi = -\pi/2$ one may obtain:

$$P_1(t) = P_o\left\{1 + \sin\left\{\phi_o \sin\left(\frac{\omega\tau}{2}\right)\cos\left[\omega\left(t - \frac{\tau}{2}\right) + \beta\right]\right\}\right\}. \tag{19}$$

This signal allows to obtain the amplitude of phase disturbance generated by phase transducer as:

$$\phi_o = \frac{(2n+1)\frac{\pi}{2} - (-1)^n \arccos\left[\left(\frac{P_1(t)}{P_o}\right)_{max}\right]}{2\sin\left(\frac{\omega\tau}{2}\right)} \tag{20}$$

where n stands for successive signal maxima for which $P_1(t)/P_0=1$. Those extrema occur in case of amplitude ϕ_0 increase. It is easy to notice that the maximum of signal alternating component $P_1(t)$ on the basic enforcement frequency ω (n=0) is equal P_0 for amplitude increase ϕ_0. Therefore by measurement of maximum value V_{maxpp} (just before sinusoid turn) one may normalize the signal $P_0=V_{maxpp}$. The above result with an additional assumption n=0 allows to obtain ϕ_0 as:

$$\phi_o = \frac{\pi}{4\sin\left(\frac{\omega\tau}{2}\right)} \arccos\left[\frac{P_1(t)}{V_{maxpp}}\right] \tag{21}$$

3. The Way to Measure of Studied Element Characteristics

In the measuring system (see Fig. 4) the working point is adjusted in quadrature ($\Delta\varphi=-\pi/2$) by polarization controllers (PC). The maximum voltage V applied to the studied element is adjusted by an oscillator in such way that the output remains linear. Then a chosen frequency range is studied by the oscillator, and the system output signal is registered by a A/D converter card on a IBM/386 computer. Due to possibility of the occurrence of mismatching between the oscillator and the studied element, additional the load impendance of the studied system is registered.

Figure 4. The installation for measurement of phase elements.

The output intensity from detector D1 and a magnitude of element loading vs. frequency is presented in Fig. 5. Fig. 5a concerns phase modulator while Fig. 5b small loudspeaker pzt type used in post cards (as example of phase head), respectively. The modulator has been prepared by rigid mounting single-mode optic fibre 12.5 meters long on the hollow piezoceramic cylinder made by K-2 material. The cylinder diameter, height and wall thickness were 50 mm, 30 mm and 5 mm, respectively. In case of loudspeaker studies (flat cylinder of pzt with diameter 15 mm and 1.25 mm thick) the modulation inside fibre has been obtained by mounting in its central part a straight segment of optic fibre 5.0 mm length.

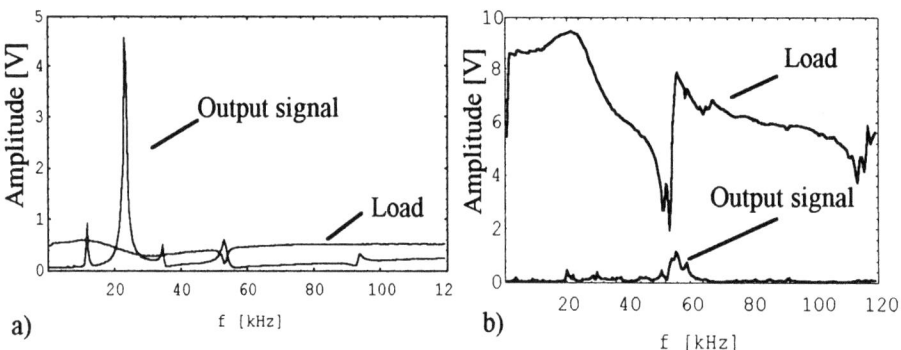

Figure 5. The registered values of the intensity and loading for
a) phase modulator, b) small phase head frequency.

The obtained data allow to calculate in a numerical way modulator parameters, i.e., ϕ_0 in [rad], initial phase β in [rad], modulation efficiency $\eta=\phi_0/(U\,l)$ in [rad/Vm], where l stands for the length of fibre in the modulator and U for applied voltage. The modulation efficiency for above studied elements is presented in Fig. 6.

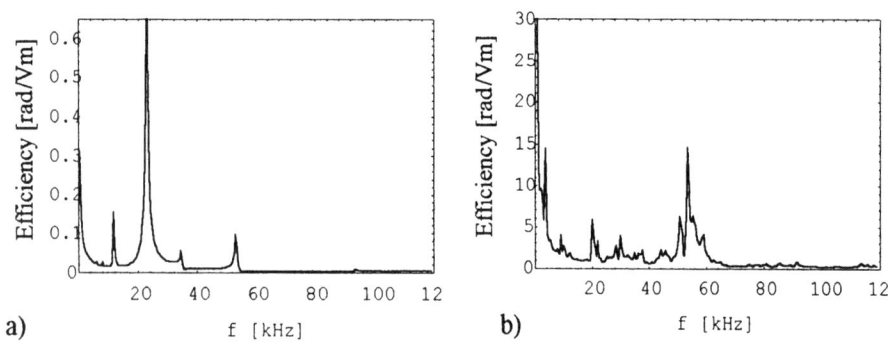

Figure 6.. The calculated modulation efficiency vs. enforcement
frequency: a) phase modulator, b) phase head.

4. Conclusion

The presented system together with the measurement methodics allows to precise measurement of frequency characteristics of phase modulators used in fibre optic systems. Moreover, it may be successfully applied for studies of measuring phase heads prepared in all-fibre technology, e.g., for phase refractometer, with high reproducibility of results.

Additional by using eq. (1) the proposed system may be successfully adopted for studies of vibrations with very small amplitude (nanometer order). The measurement of nanometer vibrations in a classical way is difficult and troublesome. It is caused by the necessity of using of interferometer system and depositing a reflective layer on the studied element in order to ensure the proper action of the interferometric system. The application of above described fibre optic system does not require depositing of reflective layers, but only the temporary mounting of the optic fibre segment.

5. References

1. Proceedings of the Fibre Optic Sensors Conference FOS(9), (1993) Firenze, Italy 3-6.05.93.
2. Optical Fibre Sensor Technology. Workshop notes (1989), *Workshop organised by Cranfield and Sira Communications*, London.
3. DePaul, R.P. (1991) Introduction to interferometric fibre optic sensors, *SPIE's Internat. Symp. on OE/Fibers*, Boston USA, 3 Sept.
4. DePaula, R.P., Moore, E.L. (1984) Review of all-fiber phase and polarisation modulators, *SPIE Proc.*, **478**, 3.
5. Berg, R.A. (1983) All-fiber gyroscope with optical-Kerr-effect compensation, *Doctor's Thesis*, Standford University.
6. Berg, R.A., Lefevre, H.C., Shaw H.J (1981) All single-mode fibre optic gyroscope with long-term stability, *Opt. Lett.*, **6**, 502.
7. Jones, R.G. (1991) A new calculus for treatment of optical systems, *J. Opt. Soc. Am.*, **31**, 488.
8. Jaroszewicz, L.R., Kieżun, A., Swiłło M. (1994) *Pat. Appl.*, **P-301.913**.
9. Jaroszewicz, L.R. (1988) Ploarization transfer analysis in fiber optic gyroscope, *Doctor's Thesis*, MUT,
10. Ulrich, R. (1982) Polarization and depolarization in the fibre optic gyroscope, *Fiber optic rotation sensors and related technologies*, Springer-Verlag, New York 22.
11. Jaroszewicz, L.R., Szustakowski, M. (1990) The matrix analysis of the polarisation transformer, *Opt. Appl.*, **20**, 81.

This work has been done under financial support of State Committee for Scientific Research Grant No. 2-0173-91-01.

FMCW-LIDAR WITH TUNABLE TWIN-GUIDE LASER DIODE

A. Dieckmann
Siemens AG, Corporate Research and Development
Otto-Hahn-Ring 6, 81739 Munich, Germany

Prof. Dr.-Ing. M.-C. Amann
University GhK-Kassel,
Heinrich-Plett Str. 40, 34109 Kassel, Germany

Abstract

The realisation of a coherent frequency modulated continuous wave LIDAR aimed for the accurate measurement of short distances employing a distributed feedback tunable twin-guide laser diode is demonstrated. Theoretical calculations on the ultimate limits in accuracy of distance measurements are carried out. A single shot relative accuracy of $8*10^{-5}$ (8µm) has been achieved at a distance of 10cm. The results presented here prove the predictions calculated from theory.

1. Introduction

Recent progress in the area of laser technology has rendered it possible to set up light detection and ranging (LIDAR) systems for distance measurement and scanning applications employing semiconductor laser diodes. In particular, by the realisation of electronically tunable laser diodes, the coherent frequency modulated continuous wave (FMCW) technique has gained particular interest [1-5]. This is due to the high accuracy in distance measurement applications, which is to be expected for widely tunable laser diodes.

Here we present for the first time a FMCW-LIDAR system which utilizes a distributed feedback tunable twin-guide (DFB-TTG)-laser diode. Among the electronically tunable laser diodes, the DFB-TTG-laser diode exhibits by far the largest continuous tuning range Δf up to about 1.4THz [6,7] for the emission frequency (Table 1). Owing to its inverse proportionality to Δf, the accuracy of distance measurements can be optimized using DFB-TTG-lasers as light sources. Further decisive advantages of the DFB-TTG-laser diode represent its inherent continuous tuning behaviour, the simple tuning scheme requiring only one single frequency control current [8] and its suitability for rapid scanning speeds (> 10MHz).

TABLE 1: Frequency tuning ranges of electronically tunable laser diodes
(DBR = distributed Bragg reflection / DFB = distributed feedback)

laser diode type	DFB-TTG	DBR	DFB
max. tuning range (in GHz)	1400	600	800
typical tuning range (in GHz)	600	400	300

This paper concentrates on the ultimate limits in accuracy of distance measurements as determined by the laser diode phase noise. The results presented here prove theoretical calculations on the phase noise limited accuracy of such a system [9].

2. Principles of FMCW-LIDAR systems for distance measurements

The principle setup of a FMCW-LIDAR system is depicted in Figure 1. The DFB-TTG-laser diode is frequency-modulated by applying a current ramp to the tuning section while the current into the active section is kept constant (see Figure 3). The laser output passing an optical isolator is fed into a Michelson interferometer and split into two beams. One beam (local oscillator) is fed directly onto the photodiode PD and the other one (object signal) is directed onto the object. The object-signal reflected from the object at a distance R is mixed with the local oscillator signal and then square-law detected on the photodetector.

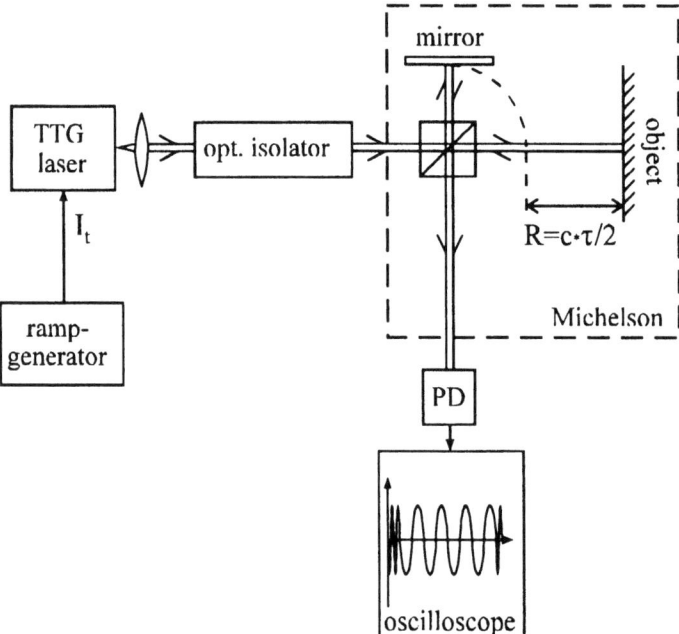

Figure 1: Principle setup of a FMCW-LIDAR system
(Michelson interferometer marked by broken line)

As illustrated in Figure 2 the object distance R causes a round trip time delay $\tau = 2R/c$ (c: light velocity) between both signals, which is converted in PD into an RF-signal with an intermediate frequency f_{if}. Assuming a linear frequency ramp the following equation holds for f_{if}:

$$f_{if} = \frac{\tau}{T_s} \Delta f \qquad (1)$$

where T_s denotes the sweep duration time of the current ramp.

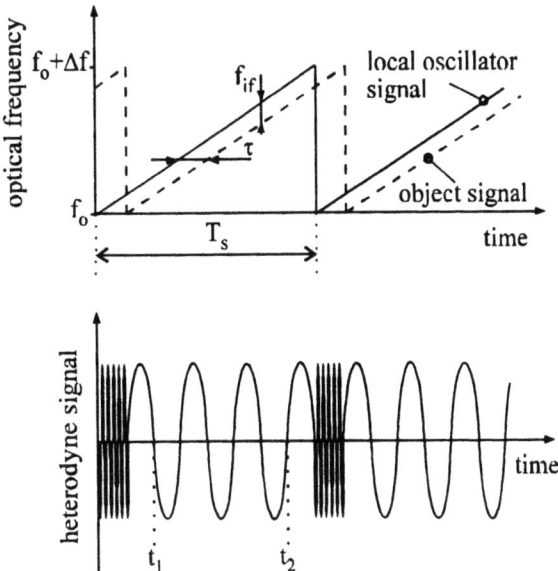

Figure 2: Instantaneous frequencies of the interferometer arms and heterodyne signal at the output of the photodetector

The corresponding photocurrent is proportional to

$$I_{Ph} \propto \cos(2\pi f_{if} t + \Theta) \qquad (2)$$

with the phase factor

$$\Theta = 2\pi\tau(f_o - \frac{\tau}{T_s}\Delta f) + \Delta\Phi(t,\tau) \qquad (3)$$

where the phase difference $\Delta\Phi(t,\tau) = \Phi(t) - \Phi(t-\tau)$ contains the phase-noise of the laser [8]. The resulting heterodyne signal is shown in Figure 2. With a given frequency tuning range Δf the time delay τ and therefore the distance R can be calculated by equation (1) after measuring the intermediate frequency f_{if} as

$$f_{if} = \frac{N_{if} - 1}{2(t_2 - t_1)} \qquad (4)$$

where N_{if} denotes the number of zerocrossings within measuring period $t_2 - t_1$ (c. f. Figure 2).

3. Distributed Feedback Tunable Twin-Guide (DFB-TTG) laser diode

3.1 LASER STRUCTURE AND WAVELENGTH TUNING MECHANISM

The limits in accuracy of the FMCW-LIDAR system are set by the employed DFB-TTG-laser diode and its phase noise. Therefore the basic principles of the TTG-laser diode are explained by means of its longitudinal cross section (Figure 3).

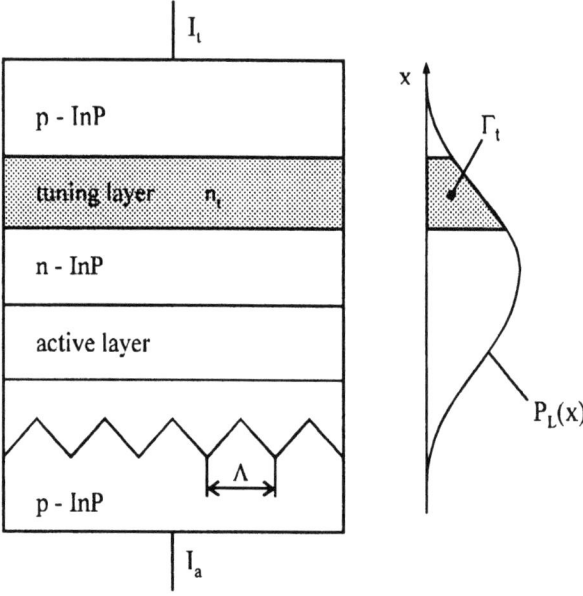

Figure 3: Schematical longitudinal section of the DFB-TTG-laser diode and distribution of the optical power in the laser structure

The tuning current I_t is applied via the top p-InP layer, while the active current I_a is applied via the substrate of the laser. $P_L(x)$ represents the distribution of the optical power in the laser diode and the confinement factor G_t describes the portion of the optical field which propagates in the tuning region. The grating with a grating period Λ provides the single mode emission of the laser output: the Bragg wavelength λ_o is related to Λ by

$$\lambda_o = 2\Lambda \cdot n_{eff} \qquad (5)$$

where n_{eff} denotes the effective refractive index of the laser structure.

The basic physical mechanism underlying the electronic wavelength tuning in the TTG-laser diode is the refractive index control by carrier injection. Exploiting the "plasma effect", the refractive index of the tuning region n_t is changed by the injection of carriers into the tuning region while applying I_t. The corresponding change of the effective refractive index Δn_{eff} leads to a wavelength change $\Delta\lambda$:

where

$$\Delta\lambda = \frac{\Delta n_{eff}}{n_{eff}} \lambda_o \qquad (6a)$$

$$\Delta n_{eff} = \Gamma_t \cdot \Delta n_t (I_t) \qquad (6b)$$

3.2 LASER CHARACTERISTICS

A typical tuning charateristic of a TTG-laser diode is shown schematically in Figure 4a. With tuning currents around 60mA one typically achieves optical frequency shifts in the regime of 600GHz. Due to the various recombination processes in the tuning region, the change in the optical frequency f_o is related to the tuning current I_t in a nonlinear manner. This is to be compensated by a suitable circuitry within a LIDAR system with a TTG-laser diode since the nonlinear tuning characteristic leads to a shift in the intermediate frequency f_{if}.

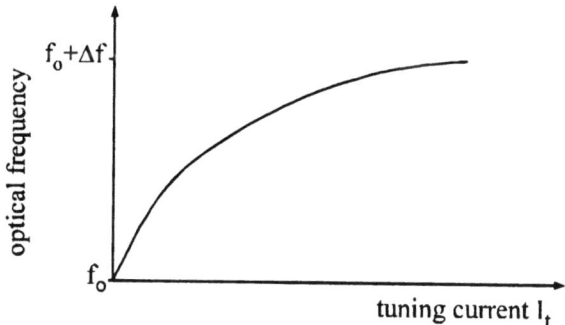

Figure 4a: Frequency tuning characteristic

Similar to other tunable lasers with a passive tuning region, the optical output power of the TTG-laser decreases with I_t (Figure 4b). This is because changes of the real part of the refractive index are accompanied by optical losses according to the Kramers Kronig-relation.

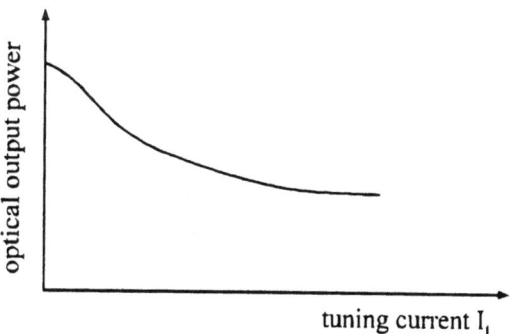

Figure 4b: Optical output power

The spectral linewidth $\Delta\nu$, which is an important parameter to assess the influence of the laser phase noise on the accuracy of distance measurements, is depicted in Figure

4c as a function of I_t. The observed excess linewidth broadening for $I_t > 0$ is caused mainly by shot noise of carriers injected into the tuning region. The resulting fluctuations in the refractive index lead to a broader linewidth [10].

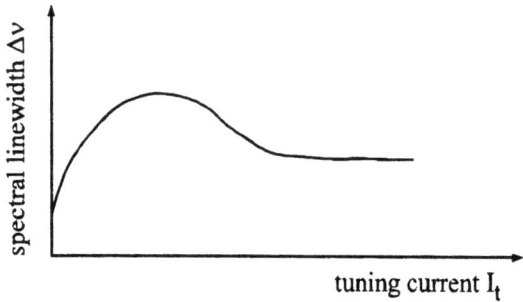

Figure 4c: Spectral linewidth

4. Phase noise limited accuracy of distance measurements in FMCW-LIDAR with TTG-laser diode

Assuming a precise measurement of the intermediate frequency f_{if}, the laser phase noise $\Phi(t)$ determines the limits in accuracy of distance measurements. To describe the phase noise the random-walk process [11] serves as a model. The corresponding nonstationary process is characterised by a zero mean and a variance increasing with time t:

$$\overline{\Phi(t)} = 0 \tag{7a}$$

$$\overline{\Phi^2(t)} = 2\pi\Delta vt \tag{7b}$$

Mixing the local oscillator signal and the object signal on the photodetector (Figure 1) leads to a phase noise

$$\Delta\Phi(t,\tau) = \Phi(t) - \Phi(t-\tau) \tag{8}$$

whose process becomes stationary with zero mean and Gaussian distribution [12]:

$$\overline{\Delta\Phi(t,\tau)} = 0 \tag{9a}$$

$$\overline{\Delta\Phi^2(t,\tau)} = 2\pi\Delta v\tau \tag{9b}$$

For an accurate description of the phase noise of tunable laser diodes, this model is to be modified. Besides instantaneous phase changes, the spontaneous emission of photons also causes delayed phase changes resulting from relaxation oscillations following changes in field intensity [11,13]:

$$\overline{\Delta\Phi_a^2(t,\tau)} = 2\pi\Delta v_a\left(\tau + \frac{\alpha^2}{1+\alpha^2}\frac{\cos 3\delta - \exp(-\Gamma\tau)\cos(\Omega\tau - 3\delta)}{2\Gamma\cos\delta}\right) \tag{10}$$

where Δv_a is the spectral linewidth due to spontaneous emission, α denotes the linewidth enhancement factor, Ω and Γ are the angular frequency and damping constant of the relaxation oscillation repectively, and $\delta = \arctan(\Gamma/\Omega)$.

A specific contribution $\Delta\Phi_t$ of electronically tunable laser diodes to the phase noise arises from shot noise due to carrier injection into the tuning region [9]:

$$\overline{\Delta\Phi_t^2(t,\tau)} = 2\pi\Delta\nu_t\left(\tau + \tau_n\left(\exp(-\tau/\tau_n) - 1\right)\right) \quad (11)$$

where $\Delta\nu_t$ denotes the excess linewidth broadening due to shot noise and τ_n is the carrier lifetime in the tuning region of about 3-50ns.

The variance of the total phase error $\Delta\Phi$, which is composed of the phase jitters occuring at t_1 and t_2, is given by

$$\overline{\Delta\Phi^2} = \overline{\Delta\Phi_a^2(t_1,\tau)} + \overline{\Delta\Phi_a^2(t_2,\tau)} + \overline{\Delta\Phi_t^2(t_1,\tau)} + \overline{\Delta\Phi_t^2(t_2,\tau)} \quad (12)$$

This phase error limits the accuracy in measuring the intermediate frequency f_{if} and therefore the relative accuracy of distance measurements $\Delta R/R$:

$$\frac{\Delta R}{R} = \sqrt{\frac{\overline{\Delta\Phi^2}}{\Psi_{if}^2}} \quad (13)$$

where $\Psi_{if} = (N_{if} - 1)\pi$ is the total phase of the heterodyne signal within the measuring period $t_2 - t_1$.

As long as the round trip time delay τ remains well below the carrier lifetime in the tuning region, the corresponding shot noise does not contribute to $\Delta\Phi$. As a consequence $\Delta R/R$ will be better than expected from equation (9), which is only valid for a white laser frequency noise. Figure 5 shows the relative accuracy $\Delta R/R$ as a function of the distance R calculated from $\Delta\Phi_a(t,\tau)$ (broken line), $\Delta\Phi_t(t,\tau)$ (broken and dotted line) and the total phase error $\Delta\Phi$ (solid line).

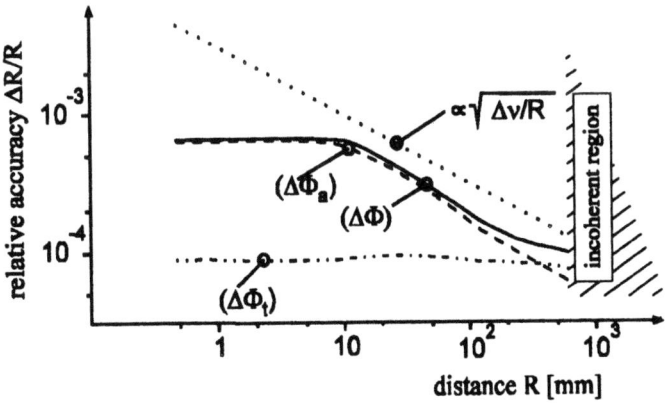

Figure 5: Phase noise limited relative accuracy $\Delta R/R$ of distance measurements

Obviously the relative accuracy at small distances up to 10cm is determined by

$\Delta\Phi_a(t,\tau)$, since $\Delta\Phi_t(t,\tau)$ contains no high frequency components. On the other hand, at larger distances the phase error from $\Delta\Phi_t(t,\tau)$ gains in importance. In this regime $\Delta R/R$ can be approximated by the following equation (white frequncy noise limit), which is obtained by inserting the square root law for the phase error (9b) into equation (13):

$$\frac{\Delta R}{R} = \sqrt{\frac{c\Delta v}{2\pi\Delta f^2 R}} \qquad (14)$$

with $\Delta v = \Delta v_a + \Delta v_t$. The resulting curve is displayed as dotted line in Figure 5.

5. Measurement of the relative accuracy of distance measurements

5.1 EXPERIMENTAL SETUP

As mentioned in section 3.2, the nonlinear tuning characteristic of the TTG-laser diode leads to a frequency shift with a resulting time dependent intermediate frequency $f_{if}(t)$, illustrated in Figure 6.

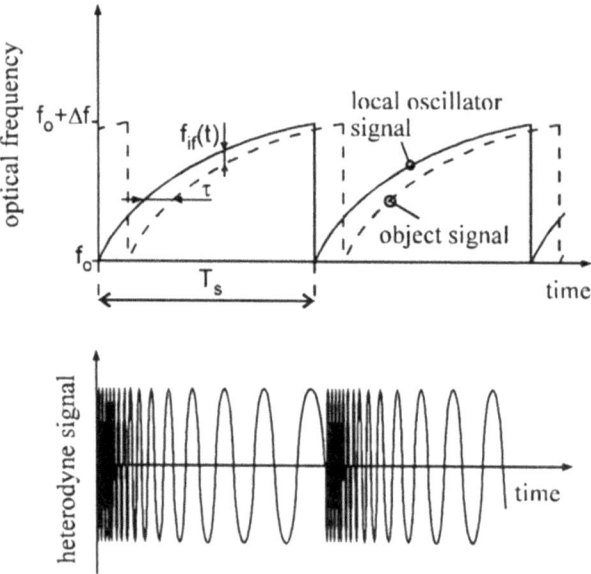

Figure 6: Nonlinear tuning characteristics and instantaneous frequencies in two interferometer arms with the resulting heterodyne signal

The quotient between the maximum and the minimum intermediate frequency is about 3-4. Therefore it is necessary to predistort the current ramp applied to the tuning section in order to obtain a time-linear-tuning characteristic. A Mach-Zehnder-interferometer integrated in the setup serves to compensate for the remaining nonlinearities of the frequency-against-time relationship. The entire setup of the FMCW-LIDAR system is shown in Figure 7.

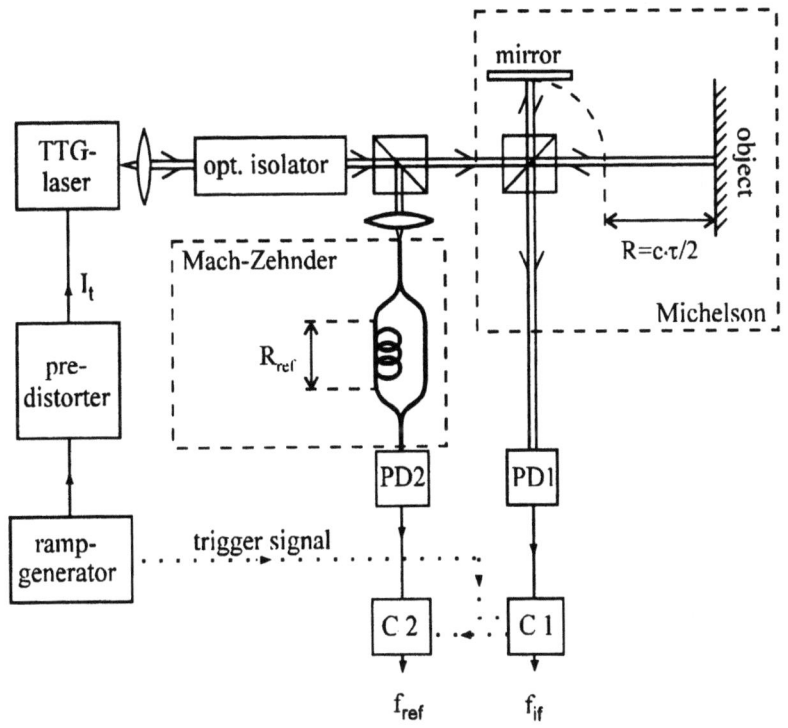

Figure 7: Experimental setup of FMCW-LIDAR system
(trigger signal marked by dotted lines, interferometers marked by broken lines)

The Mach-Zehnder-interferometer also eliminates the need to measure Δf directly. Instead the distance R can be easily related to the reference frequency f_{ref} by

$$R = \frac{f_{if}}{f_{ref}} \frac{R_{ref}}{2} \tag{15}$$

for $\Delta\Phi = 0$. The factor 2 accounts for the double path of the object signal through distance R.

The intermediate frequency f_{if} of the probe signal is measured with the frequency counter C1, which is triggered by the ramp generator. The second frequency counter C2 measuring f_{ref} is triggered by the gate-time signal of C1. The limit for the relative accuracy imposed by the setup itself is given by

$$\frac{\Delta R}{R} = \frac{2\delta}{N_{ref} - 1} \tag{16}$$

where δ denotes the normalized deviation of the strict time-linear-tuning characteristic and N_{ref} is the number of zerocrossings of the reference signal. With $δ < 1.8\%$ and N_{ref} of about 1500 the accuracy limit imposed by the setup is smaller than the phase noise limited accuracy of the entire system given by [5]

$$\frac{\Delta R}{R} = \sqrt{\frac{\overline{\Delta\Phi_{if}^2}}{\Psi_{if}^2} + \frac{\overline{\Delta\Phi_{ref}^2}}{\Psi_{ref}^2}} \qquad (17)$$

where $\Delta\Phi_{if}$ and $\Delta\Phi_{ref}$ are the total phase errors of the interferometers and $\Psi_{ref} = (N_{ref} - 1)\pi$.

5.2 RESULTS AND DISCUSSION

At each distance point the quotient of the intermediate frequencies f_{if} and f_{ref} was measured a hundert times. The measure of the single shot relative accuracy of distance measurements was obtained by calculating the mean and the relative mean square error of the quotients related to each distance point. The results are displayed in Figure 8 marked by the dots. The solid line represents $\Delta R/R$ for the employed TTG-laser diode calculated with our theory, while the broken lines shows the results obtained from the previous formulas assuming a white laser frequency noise.

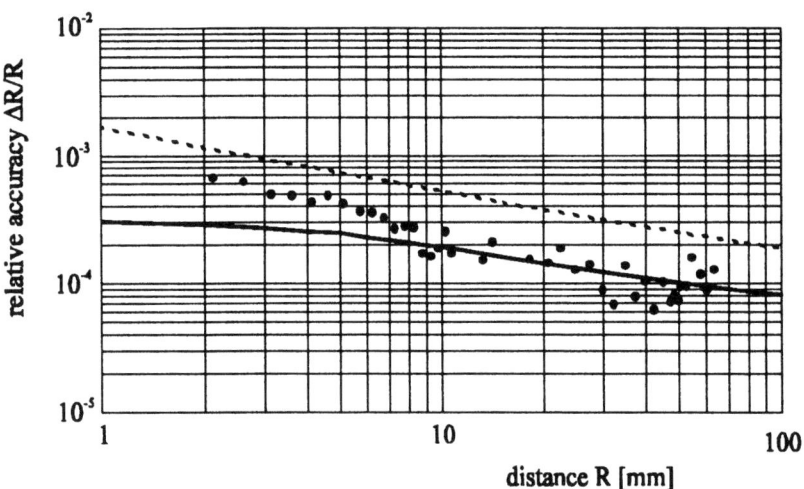

Figure 8: Measurement of the relative accuracy of distance measurements
(theory curve marked by solid line, measurements marked by •, previous
estimation from white frequency noise marked by dotted line)

The measurements of $\Delta R/R$ presented here prove for the first time the improved accuracy of small distance measurements ($R < 10$cm) with DFB-TTG laser diodes as predicted by recent analysis [9].

A further gain in accuracy is achievable by exploiting more information contained within the heterodyne signals. For example by sampling every zeocrossing of the object signal, an improvement in accuracy by an order of magnitude can be obtained at distances between 5 - 10cm.

6. Conclusion

A FMCW-LIDAR system with a TTG-laser diode for was demonstrated. Measurements with this system yielded a relative accuracy of distance measurements of $2 \cdot 10^{-4}$ and of $8 \cdot 10^{-5}$ for distances of 1cm and 10cm respectively. The results presented here prove for the first time the theoretical predictions concerning the relative accuracy. Together with its simple tuning scheme the TTG-laser diode proves to be a promising light source for highly accurate short distance measurement applications using the FMCW-LIDAR technique.

7. Acknowledgements

The authors gratefully acknowledge the fruitful discussions with M. Claassen, B. Borchert, T. Wolf and A. Ebberg.

8. References

1. Strzelecki, E.M., Cohen, D.A., and Coldren, L.A. (1988) Investigation of tunable single frequency diode lasers for sensor applications; *IEEE J. Lightwave Technology* **LT-6**, 1610-1618
2. Slotwinski, A.R., Goodwin, F.E., and Simonson, D.L. (1989) Utilizing AlGaAs laser diodes as a source for frequency modulated continuous wave (FMCW) coherent laser radars; *SPIE Proc. - Laser Diode Technology and Applications* **1043**, 245-251
3. Beheim, G., Fritsch, K. (1983) Remote displacement measurements using a laser diode; *Electronics Letters* **21**, 93-94
4. Burrows, E.C., Liou, K.-Y. (1990) High resolution laser LIDAR utilizing two-section distributed feedback semiconductor laser as a coherent source; *Electronics Letters* **26**, 577-579
5. Dieckmann, A. (1994) FMCW-LIDAR with tunable twin-guide laser diode; *Electronics Letters* **30**, 308-309
6. Wolf, T., Illek, S., Rieger, J., Borchert, B., Amann, M.-C. (1993) Tunable twin-guide (TTG) distributed feedback (DFB) laser with over 10 nm continuous tuning range; *Electronics Letters* **29**, 2124-2125
7. Wolf, T.; Illek, S.; Rieger, J.; Borchert, B.; Amann, M.-C. (1994) Extended

continuous tuning range (over 10nm) of tunable twin-guide lasers; *Proceedings of CLEO 94;* paper CWB 1
8. Schanen, C.F.J., Illek, S., Lang, H., Thulke, W., Amann, M.-C. (1990) Fabrication and lasing characteristics of $\lambda = 1.56\mu m$ tunable twin-guide (TTG) DFB lasers; *IEE Proceedings* **137, Pt. J**, 69-73
9. AMANN, M.-C. (1992) Phase noise limited resolution of coherent LIDAR using widely tunable laser diodes; *Electronics Letters* **28**, 1694-1696
10. AMANN, M.-C., SCHIMPE, R. (1990) Excess linewidth broadening in wavelength tunable laser diodes; *Electronics Letters* **26**, 279-280
11. HENRY, C.H. (1982) Theory of the linewidth of semiconductor lasers; *IEEE Journal of Quantum Electronics* **QE-18**, 259-264
12. MOSLEHI, B. (1986) Analysis of optical phase noise in fibre-optic systems employing a laser source with arbitrary coherence time; *IEEE Journal of Ligthwave Technology* **LT-4**, 1334-1351
13. HENRY, C.H. (1983) Theory of the phase noise and power spectrum of a single mode injection laser; *IEEE Journal of Quantum Electronics* **QE-19**, 1391-1397

FIBRE-OPTIC WHITE-LIGHT INTERFEROMETER INSENSITIVE TO POLARIZATION FADING

L. A. Ferreira[a], J. L. Santos[b,a], F. Farahi[c]

a) Centro de Optoelectrónica, INESC Porto, Rua José Falcão, 110, 4000 Porto, Portugal.
b) Laboratório de Física U. Porto, Praça Gomes Teixeira, 4000 Porto, Portugal.
c) Department of Physics, U. of North Carolina, Charlotte, North Carolina 28223, U.S.A.

Abstract

Interferometric systems consisting of two tandem interferometers with white-light illumination have been used as optical-fibre remote-sensing systems. Recently, it has been shown that both phase and amplitude of the output signal are significantly affected by externally induced birefringence in the fibre linking the interferometers; in some cases, this may cause total fading of the interference signal. In this poster, a new scheme is proposed to overcome this problem in white-light fibre-optic interferometric sensors. The approach is based on birefringence compensation using Faraday rotator mirror elements. The principle is experimentally verified, and the results confirm the theoretical expectation.

In interferometric optical-fibre sensors the noise (and scale error) caused by fluctuations in the polarization states of the interfering beams (polarization-induced fading) are well known as a limiting factor of the performance of these systems[1]. Many researchers have studied the implications of changes in the input state of polarization (SOP) on both the visibility and phase of fibre optic interferometers[2-4], and a few techniques to eliminate these effects have been considered[2,3,5-9]. Recently[10], these studies have been extended to fibre-optic white-light interferometry. Prior to this work, it was commonly believed that environmental perturbations on the fibre linking two tandem interferometers had no net effect on the system. But Gauthier et al. have shown[10] that both phase and amplitude of the output signals are significantly affected by externally induced birefringence in the downlead fibre and, in some cases, this may cause total fading of the interference signal.

In this poster, we propose a new scheme which eliminates this problem for reflective systems. This scheme utilizes Faraday rotators followed by mirrors which act as universal compensators for any kind of polarization change (and polarization induced phase changes) induced by (reciprocal) birefringence on a retracing beam[11,12]. The effectiveness of this scheme will be theoretically analyzed by Jones calculus, and some experimental results will be presented.

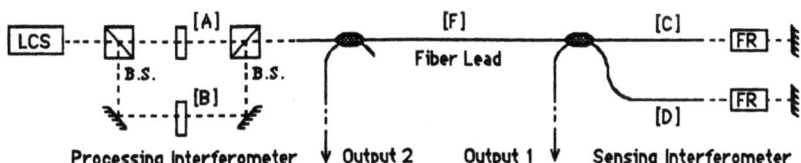

Fig.1 - White light tandem interferometric system, with the retardations of its components represented by Jones matrices [A], [B], [C], [D], [F]; FR: Faraday rotator.

Fig.1 shows a white-light interferometric system in which two unbalanced interferometers are linked by a length of single mode fibre. The sensing interferometer is a fibre Michelson and the processing interferometer is a conventional Mach-Zehnder. The system is considered balanced to within the coherence length of the input light. Optical elements are placed in each arm of the processing interferometer to represent general optical retarders, with matrices [A] and [B], respectively. The retardation of the fibre lead is represented by [F], and those of the fibre arms of the sensing interferometer by [C] and [D] (see Fig.1). The electric fields for outputs 1 and 2 can be written as:

$$\vec{E}_1 = [\vec{C}] \cdot [MFR] \cdot [\vec{C}] \cdot [\vec{F}] \cdot [\vec{A}] \cdot \vec{E}_{in1} \cdot e^{i(\phi_A + \phi_F + 2\phi_C)}$$

$$+ [\vec{D}] \cdot [MFR] \cdot [\vec{D}] \cdot [\vec{F}] \cdot [\vec{A}] \cdot \vec{E}_{in1} \cdot e^{i(\phi_A + \phi_F + 2\phi_D)}$$

$$+ [\vec{C}] \cdot [MFR] \cdot [\vec{C}] \cdot [\vec{F}] \cdot [\vec{B}] \cdot \vec{E}_{in1} \cdot e^{i(\phi_B + \phi_F + 2\phi_C)}$$

$$+ [\vec{D}] \cdot [MFR] \cdot [\vec{D}] \cdot [\vec{F}] \cdot [\vec{B}] \cdot \vec{E}_{in1} \cdot e^{i(\phi_B + \phi_F + 2\phi_D)} \quad (1)$$

$$\vec{E}_2 = [\vec{F}] \cdot [\vec{C}] \cdot [MFR] \cdot [\vec{C}] \cdot [\vec{F}] \cdot [\vec{A}] \cdot \vec{E}_{in2} \cdot e^{i(\phi_A + 2\phi_F + 2\phi_C)}$$

$$+ [\vec{F}] \cdot [\vec{D}] \cdot [MFR] \cdot [\vec{D}] \cdot [\vec{F}] \cdot [\vec{A}] \cdot \vec{E}_{in2} \cdot e^{i(\phi_A + 2\phi_F + 2\phi_D)}$$

$$+ [\vec{F}] \cdot [\vec{C}] \cdot [MFR] \cdot [\vec{C}] \cdot [\vec{F}] \cdot [\vec{B}] \cdot \vec{E}_{in2} \cdot e^{i(\phi_B + 2\phi_F + 2\phi_C)}$$

$$+ [\vec{F}] \cdot [\vec{D}] \cdot [MFR] \cdot [\vec{D}] \cdot [\vec{F}] \cdot [\vec{B}] \cdot \vec{E}_{in2} \cdot e^{i(\phi_B + 2\phi_F + 2\phi_D)} \quad (2)$$

In the above equations $\vec{E}_{in1} = k_1 \vec{E}_{in}$ and $\vec{E}_{in2} = k_2 \vec{E}_{in}$, $\vec{E}_{in}$ is the input electric field to the system, and constants k_1 and k_2 account for loss and beam splitting ($k_1 = 1/4\sqrt{2}$ and $k_2 = 1/8$ for equal beam splitting in a lossless system). The [MFR] matrix refers to the path FR+Mirror+FR; when the Faraday rotator is tuned to produce a 45° rotation, this matrix can be simplified as[11,12]

$$[MFR] = \begin{bmatrix} 0 & -1 \\ -1 & 0 \end{bmatrix} \quad (3)$$

In this case the polarization azimuth of the reflected wave is rotated by 90° relative to the polarization azimuth of the input wave.

For each optical element, we use two kinds of representations (different arrow directions in the matrix) depending on the direction of light propagation. For example, $[\vec{F}]$ is the Jones matrix describing the propagation in the fibre lead in the forward direction, and $[\vec{F}]$ is the Jones matrix describing the propagation in the same fibre, but in the backward direction. The two matrices are related by[13,14]

$$[\vec{F}] = [R] \cdot [\vec{F}]^t \cdot [R]^{-1} \quad (4)$$

where $[\vec{F}]^t$ is the transpose of $[\vec{F}]$ and [R] gives the coordinate rotation due to reflection. We now substitute equations (3) and (4) into equations (1) and (2) to obtain the intensity of the output signals:

$$I_1 = \vec{E}_1^\dagger \cdot \vec{E}_1 = 4\vec{E}_{in1}^2 + \vec{E}_{in1}^\dagger \cdot [\vec{A}]^\dagger \cdot [\vec{B}] \cdot \vec{E}_{in1} \cdot e^{-i\Delta\phi} + \vec{E}_{in1}^\dagger \cdot [\vec{B}]^\dagger \cdot [\vec{A}] \cdot \vec{E}_{in1} \cdot e^{+i\Delta\phi} \quad (5)$$

$$I_2 = \vec{E}_2^\dagger \cdot \vec{E}_2 = 4\vec{E}_{in2}^2 + \vec{E}_{in2}^\dagger \cdot [\vec{A}]^\dagger \cdot [\vec{B}] \cdot \vec{E}_{in2} \cdot e^{-i\Delta\phi} + \vec{E}_{in2}^\dagger \cdot [\vec{B}]^\dagger \cdot [\vec{A}] \cdot \vec{E}_{in2} \cdot e^{+i\Delta\phi} \quad (6)$$

where "†" denotes the complex conjugate transpose, $4\vec{E}_{in1}^2$ and $4\vec{E}_{in2}^2$ are the result of incoherent mixing of the waves, and the other terms represent coherent mixing which produce the interference signals. $\Delta\phi = (\phi_A + 2\phi_D) - (\phi_B + 2\phi_C)$ is the small phase difference for a nearly balanced system.

From equations (5) and (6) it turns out that the output signals are independent from the birefringence of the fibre lead ([F]). It is also evident, from these equations, that the visibility has its maximum value when [A] = [B]. In practice, it is possible to achieve this maximum value, since the polarization state of the optical beam propagating in the processing interferometer can always be controlled. In this particular case, equations (5) and (6) can be simplified to give

$$I_1 = 4\vec{E}_{in1}^2 \left[1 + \frac{1}{2}\cos\Delta\phi\right] \quad (7)$$

$$I_2 = 4\vec{E}_{in2}^2 \left[1 + \frac{1}{2}\cos\Delta\phi\right] \quad (8)$$

These results show that the visibility is independent of the system birefringence and has the maximum possible value in a white light system (0.5).

An experiment was implemented to test the concept outlined above. The set-up is shown in Fig.2. The optical source was a pigtailed SLD (Superlum, 361/A) which provided 2 mW of optical power at $\lambda=830$ nm. The coherence length was measured to be ≈ 25 μm. Since light from a SLD is unpolarized, a polarizer was utilized to produce linearly polarized light with a controllable azimuth. The path-imbalance of the processing interferometer was adjusted to match that of the sensing interferometer, within the coherence length of the light source, to observe an interference signal at the output of the whole system. A 50 cm length of the fibre lead was wrapped around PZT1; an electrical signal was applied to this PZT transducer to simulate the environmental perturbations. The arms of the sensing interferometer had an air path to allow the insertion of the Faraday rotators which were mounted on translation stages. The second fibre stretcher (PZT2) was utilized in the sensing interferometer to induce a periodic phase modulation, desirable for signal processing.

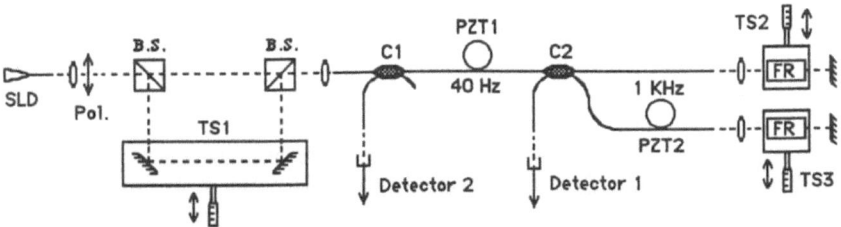

Fig.2 - Scheme of the experimental arrangement used to test the elimination of the lead sensitivity.

The system was illuminated with a vertically polarized beam. A sawtooth waveform with adequate amplitude was applied to PZT2 to generate a pseudo-heterodyne carrier with frequency f=1 kHz. A 40 Hz signal was applied to PZT1 to simulate the environmental perturbation in the fibre lead. Fig.3a shows the spectrum of the signal from output 2 for the tandem configuration without the Faraday rotators. Two strong sidebands appear in this figure, which are produced by modulation of the fibre lead. It is obvious that these sidebands can not be distinguished from a signal with the same frequency, if such a signal was applied to the sensing interferometer. Fig.3b shows the frequency spectrum of the signal obtained from output 2, but in this case the Faraday rotators were introduced to the system. The effect of Faraday rotators in suppressing the lead-induced signal is clear. In addition, the fringe visibility (which is proportional to the carrier amplitude) was substantially increased, approaching the theoretical limit (0.5).

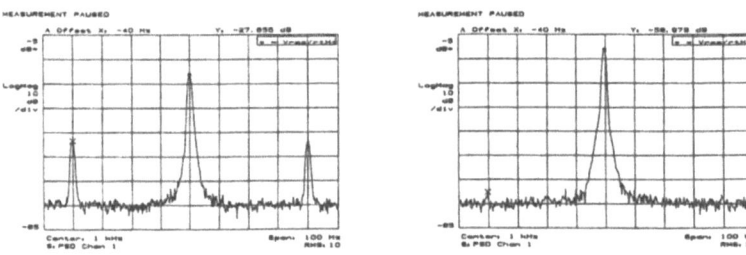

Fig.3 - Signal at output 2 with a vertically polarized input and a 40 Hz perturbation applied to the fibre lead: **a** - without Faraday rotators; **b** - with Faraday rotators.

The results shown in Fig.4 were obtained when the azimuth of the input light was rotated by 45° with respect to the previous case. All the other experimental conditions remained unchanged. Again, as expected, the Faraday rotators were very effective in eliminating the effect of induced birefringence in the fibre lead. This time, however, the fringe visibility was decreased. This was expected since, in the proposed system, the vertical direction coincides with one of the eigendirections of the matrices [A] and [B]. This is due to the fact that, in a conventional Mach-Zehnder interferometer, the polarizations of the two output beams coincide only for vertically polarized input light.

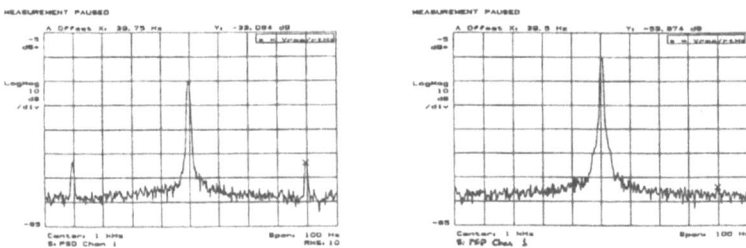

Fig.4 - Signal at output 2 with a linearly polarized input with an azimuth of 45° and a 40 Hz perturbation applied to the fibre lead: **a** - without Faraday rotators; **b** - with Faraday rotators.

It was also verified that the signal from output 1 exhibits fibre lead insensitivity, as can be seen in the results shown in Fig.5. These results confirm our predictions based on the analysis of equation (5).

Clearly, in this situation, the retracing condition is not satisfied. However, as the results of Fig.5 confirm, this condition is not strictly necessary for the Faraday rotator based scheme to work. In fact, if the birefringences of the two arms of the sensing interferometer are equal, then the fringe visibility is independent of the input polarization state to the sensing interferometer. Therefore, in the configuration of Fig.1, the Faraday rotators eliminate the effect of relative birefringence in the arms of the fibre interferometer. However, it is important to note that, in the analysis presented in this paper, we have assumed that the polarization state of the electric field does not change in the coupling region of the fibre directional coupler. In general, the characteristics of fibre directional couplers depend on the input state of polarization. This, in turn, limits the effectiveness of Faraday rotators as compensators for polarization change of a retracing beam. A quantitative assessment of this problem is currently under investigation.

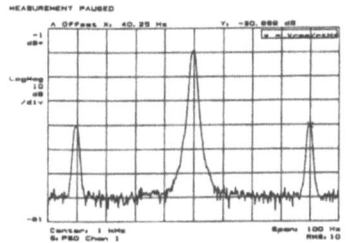 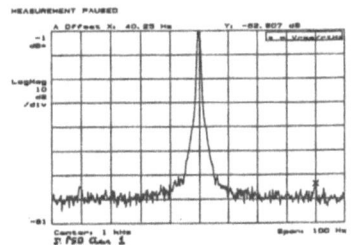

Fig.5 - Signal at output 1 with a vertically polarized input and a 40 Hz perturbation applied to the fibre lead: **a** - without Faraday rotators; **b** - with Faraday rotators.

In conclusion, a new scheme based on Faraday rotator mirror elements has been proposed in this poster to eliminate the polarization fading in reflective white light interferometry systems induced by fibre lead perturbations. The working principle of this scheme has been theoretically analyzed, and experimental results have been presented which confirm our theoretical predictions.

Acknowledgments: L. A. Ferreira acknowledges financial support from JNICT ("Junta Nacional de Investigação Científica e Tecnológica").

REFERENCES

[1] D. Jackson, J. Jones, Optica Acta, **33**, 1469 (1986).
[2] A. D. Kersey, A. Dandridge and A. B. Tveten, Opt. Lett., **13**, 288 (1988).
[3] A. D. Kersey, M. J. Marrone, A. Dandridge and A. B. Tveten,
 J. Lightwave Technol., **6**, 1599 (1988).
[4] A. D. Kersey, M. J. Marrone and A. Dandridge, J. Lightwave Technol., **8**, 838 (1990).
[5] H. C. Lefevre, Electron. Lett., **16**, 778 (1980).
[6] N. J. Frigo, A. Dandridge and A. B. Tveten, Electron. Lett., **20**, 319 (1984).
[7] K. H. Wanser and N. H. Safar, Opt. Lett., **12**, 217 (1987).
[8] A. D. Kersey and M. J. Marrone, Electron. Lett., **24**, 931 (1988).
[9] A. D. Kersey, M. J. Marrone and M. A. Davis, Electron. Lett., **27**, 518 (1991).
[10] R. R. Gauthier, F. Farahi and N. Dahi, Opt. Lett., **19**, 138 (1994).
[11] M. Martinelli, Opt. Commu., **72**, 341 (1989).
[12] N. C. Pistoni and M. Martinelli, Opt. Lett., **16**, 711 (1991).
[13] E. Brinkmeyer, Opt. Lett., **6**, 575 (1981).
[14] A. Yariv, Appl. Opt., **26**, 4538 (1987).

QND MEASUREMENT OF THE PHOTON NUMBER USING KERR EFFECT IN OPTICAL FIBRES

V. SOCHOR, I. PAULIČKA,
Czech Technical University, Faculty of Nuclear Sciences and Physical Engineering, Department of Physical Electronics, Břehová 7, 115 19 Prague 1, Czech Republic

1. Abstract

An experiment of quantum nondemolition (QND) measurement of the photon number using only one single-mode optical fibre is reported in this paper. Two beams with different wavelength counter-propagate in a single-mode optical fibre and interact with each other in an optical Kerr medium. The value of the optical Kerr constant is measured by means of this scheme. A relationship between quantum nondemolition measurement and direct detection of the photon number is obtained.

2. Introduction

In QND measurements the photon number is detected without disturbing its free motion state. The QND detection was first proposed to be used in overcoming the quantum limit in detecting gravitational waves [1, 2], but recently the relationship with quantum optics has been established.

Earlier experiments, using two fibres in a Mach-Zehnder interferometer configuration, had to use temperature compensation to overcome instabilities [3]. The experimental set-up, described in this paper, uses only one fibre and temperature compensation is therefore automatically obtained. This configuration can also be used in measuring the optical Kerr coefficient n_{2b} of silica fibres [4,5].

3. QND Measurement

A 1.06 μm cw YAG laser beam which serves as the signal beam propagates counter-directionally to a probe beam from a He-Ne laser at 0.6328 μm in an optical single-mode silica fibre (Fig. 1). The fibre is 300 m long and acts as the

Kerr medium. The signal beam is chopped with a frequency of 540 Hz so that the photon number can be measured by means of a lock-in amplifier. A prism is used to separate the two different wavelengths. The power of the YAG laser is 11.5 mW at the output of the fibre while the probe He-Ne laser power is about 100 μW. A polarizer and a quarter wave plate are positioned in front of one input end of the fibre to convert the probe beam into a circular polarization state. A phase retarder is placed at the opposite end of the fibre to convert the output probe beam into a linear polarization state. The prism does not change the polarization of the probe and signal beams. This linearly polarized infrared beam modulated by the chopper with a duty cycle of 0.5 is coupled into the fibre in the direction opposite to that of probe light. These two optical signals interact in the Kerr medium.

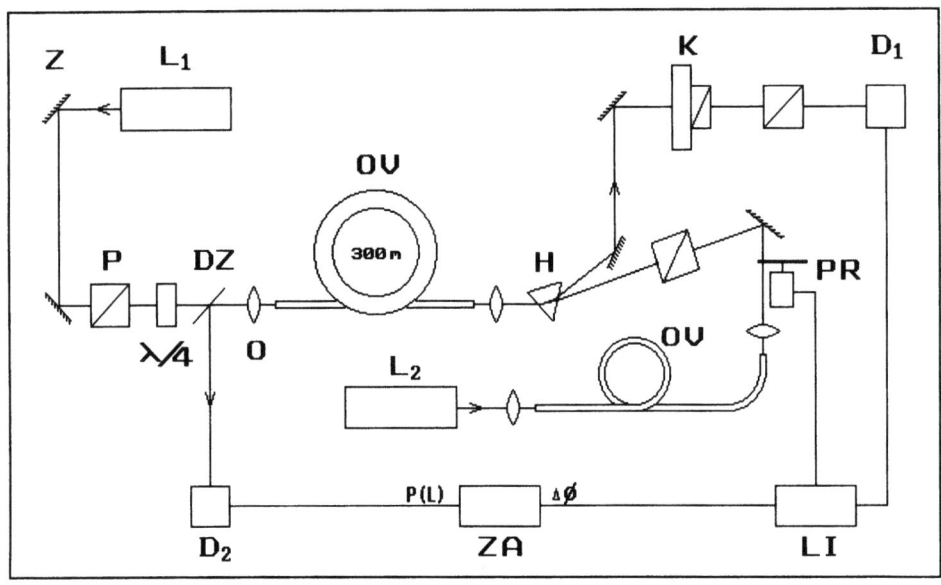

Fig. 1: The experimental scheme of QND measurement.
L_1 - He-Ne laser (λ = 632.8 nm), L_2 - Nd:YAG laser (λ = 1060 nm), K - phase compensator, Z - mirrors, P - polarizers, $\lambda/4$ - quarter-wave plate, DZ - beam-splitter, O - lenses, OV - single-mode optical fibre, H - prism, D_1 - He-Ne laser detector, D_2 - Nd:YAG laser detector, PR - chopper, LI - lock-in amplifier, ZA - plotter.

A silica optical fibre can be considered as a linear optical medium when transmitting low optical power. But when the high power is coupled into the fibre, the refractive index of the fibre is not a constant any more due to the optical Kerr effect. The refractive index is modulated by the intensity of the signal and it can be expressed as:

$$n = n_o + n_{2b} \frac{P_s}{A_e} \tag{1}$$

where n_o - refractive index of the fibre core, n_{2b} - Kerr constant of the fibre, P_s - intensity of the signal light, A_e - effective area of the fibre core.

The intensity dependent refractive index of the medium will cause a modulation of the phase of the probe wave and produce a phase shift. The phase shift of the probe beam can be obtained from the intensity changes of the He-Ne laser beam detected by a lock-in amplifier. When the YAG laser signal is blocked the phase change of the probe beam is zero. Obviously, this change is caused by the Kerr effect. The optical Kerr coefficient can be given by the relation:

$$n_{2b} = \frac{\lambda_p A_e \Delta\Theta}{2\pi L P_s} \quad (2)$$

where λ_p - vacuum wavelength of the probe wave, $\Delta\Theta$ - phase shift of the probe light, L - interaction length.

The value of n_{2b} calculated from (2) is 4.8×10^{-20} m²/W, which is close to results published by other authors [6, 7]. The photon number of the signal wave can also be obtained from this phase shift of the probe beam. The theoretical relationship between the signal photon number and the probe phase shift is given by:

$$N_s = \frac{\lambda_p \lambda_s A_e \Delta\Theta}{4\pi h f_m c L n_{2b}} \quad (3)$$

N_s - signal photon number per modulated pulse, λ_s - the vacuum wavelength of the signal wave, h - Planck constant, f_m - chopper modulation frequency, c - the vacuum velocity of light.

4. Experimental Results

In this experiment, the modulated signal has a power of 11.5 mW and produces a phase shift in the probe beam of 25.6 mrad. From this the photon number of the signal wave per pulse is calculated to be 5.63×10^{13}. This photon number is obtained nondemolitionally, without changing the free motion state of the signal beam. The output power of the signal beam was also detected using a power meter. By changing the power of the signal light the correlation was established between the QND measurement of the photon number and the direct detection of the photon number. The results obtained by these two methods are in good agreement.

5. Conclusion

In this experiment, the photon number of the signal wave is detected nondemolitionally. The classical relationship between the QND measurement and the direct detection of the signal photon number is established. This one-single-mode fibre scheme has a lower noise output than two fibre configuration proposed in [1] and there is no need for a the temperature compensation.

6. References

1. Caves, C.M., Thorne, K.S., Drever, R.W.P., Sandberg, V.D., Zimmerman, M. (1980) Rev. Mod. Phys. **52**, p. 341.
2. Milburn, G.J., Walls, D.F. (1983) Phys. Rev. **A 28**, p. 2065.
3. Imoto, N., Watkins, S., Sasak, Y. (1987) Opt. Comm. **61**, p. 159.
4. Paulička, I., Sochor, V. (1991) in Proc. Int. Conf. *Applied Optics '91"*, ČSVTS Prague, p. 191.
5. Zhang, J., Paulička, I., Sochor, V. (1992) in Proc. *18th Int. Quant. Electron. Conf. IQEC'92*, Vienna, Austria, p. 156.
6. Davison, A., White, I.H. (1989) Opt. Lett. **14**, p. 802.
7. Mikhailov, A. V., Wabnitz, S. (1990) Opt. Lett. **15**, p. 1055.

LASER TRANSITIONS CHARACTERIZATION BY SPECTRAL AND THERMAL DEPENDENCES OF THE TRANSIENT OSCILLATIONS IN THE FIBER LASERS

Oleg G. Okhotnikov [1], and José R. Salcedo [2]

Centro de Optoelectrónica, INESC,
R. José Falcão 110, 4000 Porto, Portugal

[1] On leave from the General Physics Institute, Russian Academy of Sciences, Moscow, Russia.
[2] Dept. Fisica, Universidade do Porto, Praça Gomes Teixeira, 4000 Porto, Portugal.

I. Introduction

One of the advantages of fiber lasers over their bulk counterparts is the possibility of low threshold operation due to the simplicity in assuring high pumping densities. In spite of the higher pumping thresholds that are characteristic of three-level lasers, as compared to four-level lasers, this feature of fiber lasers has allowed the demonstration of cw operation in rare-earth doped fiber lasers at room temperature on three-level transitions something not yet achieved in bulk lasers.

Rare-earth doped fibers exploiting the three-level transitions in Er^{3+}, Nd^{3+}, Pr^{3+} are now used in many applications, including fiber lasers and optical amplifiers. For these latter, it is important to know the process that governs the transient buildup of the emission, including the population inversion dynamics, the effect of spontaneous emission and the nature of the laser transition.

The temperature dependence of the performance of erbium-doped fiber amplifier is an important consideration for practical applications since inhomogeneously broadened fiber amplifiers inserted periodically in an amplifier cascade were shown to provide significant interchannel power equalization in wavelength-multiplexed systems [1]. For a complete characterization of the cooled amplifiers it is therefore important to study their parameters in order to clarify the mechanism of amplification. The most common tool is based on an optical spectroscopy, and employs fluorescence and absorption processes [2-4]. However, it is even more difficult to correctly measure all of the material parameters required for detailed device modeling. Moreover, methods commonly used are based on the principle of reciprocity, i.e. the absolute magnitude of the emission cross sections is derived from the absorption spectra, otherwise these values are interrelated through the spectral line profile of the spontaneous emission [2-5].

In this paper, we demonstrate that the dynamics of the transient buildup of the laser emission permits a direct experimental test of the gain medium parameters since the relaxation oscillation frequency depends directly on the transition cross section, the level degeneracies and partition function for the ground and excited electronic states. It is the purpose of this paper to expand the dynamic measurements to the study of the laser materials and relate them to the spectral properties to extract useful gain medium parameters, and compare them with the theoretical predictions of the transient laser output.

Substantial efforts have been devoted to determining the emission and absorption cross sections for the $^4I_{15/2}$-$^4I_{13/2}$ transition in Er^{3+}-doped fibers because of their importance in laser and amplifier development. Here we show that the transition mechanism can be easily characterized by the spectral dependence of the transient oscillations. We also present results on the properties of the transient oscillations in Nd^{3+}-doped fiber laser for the three-level $^4F_{3/2}$-$^4I_{9/2}$ transition at 900 nm in comparison with the well known four-level $^4F_{3/2}$-$^4I_{11/2}$ transition at 1060 nm. Experiments performed on the same Nd^{3+}-doped fiber laser clearly demonstrate the different behavior of the transient oscillation for these transitions.

II. Model

The following rate equations [3] of a typical 3-level system provide the conceptual basis of this method:

$$\frac{dN^u}{dt} = R - N^u \gamma_2 - \sigma c \eta' (f^u N^u - f^\ell N^\ell) q \quad (1)$$

$$\frac{dq}{dt} = \sigma c \eta (f^u N^u - f^\ell N^\ell) q - \gamma_c q . \quad (2)$$

Small-signal analytic solutions can be obtained, yielding an expression for the relaxation oscillation frequency and decay rate:

$$\omega^2_{relax} = \sigma c \eta (f^u + f^\ell)(R - R_{th}) \quad \text{or, equivalently} \quad (3)$$

$$\omega^2_{relax} = \gamma_c \gamma_2 (1 + \frac{\sigma c \eta f^\ell N}{\gamma_c})(r-1) \quad (4)$$

$$\gamma_{relax} = \frac{\sigma c \eta (f^u + f^\ell)}{2\gamma_c}(R - R_{th}) + \frac{\gamma_2}{2} . \quad (5)$$

The symbols used in this equations are defined as follows:
N^ℓ, N^u are the number of ions in lower and upper laser level within the volume of the active material.
N is total number of active ions.

R, R_{th} are pump rate and threshold pump rate, $r = R/R_{th}$.

γ_2, γ_c are decay rates for the population of the laser transition and for the power in the cold cavity.

σ is cross section of the laser transition.

q is photon density in the resonator.

L, ℓ are total cavity length and length of the gain medium and

$$\eta = \frac{\ell}{L+\ell(n-1)}, \quad \eta' = \frac{L}{L+\ell(n-1)}.$$

n is the refractive index and c is speed of light.

f^u and f^ℓ are the fractional thermal occupations of the i-th and j-th sub-level, within the upper (u) and lower (ℓ) levels.

From (4), it is clear that the wavelength dependent term in the bracket ($k(\lambda) = \frac{\sigma c \eta f^\ell N}{\gamma_c}$) disappears for lasers operating on four-level transitions with a negligible population on the terminal level ($f^\ell = 0$), that gives well-known relation [3]: $\omega^2_{relax} = \gamma_c \gamma_2 (r-1)$. An important consequence of this feature (not observed for four-level systems) is that the relaxation oscillation frequency depends directly on the absorption ($\sim \sigma f^\ell$) at the signal wavelength, due to the finite thermal population of the ground level. Thus, the wavelength dependence of the relaxation oscillations supplies a method to distinguish three- and four-level transitions, and this can be useful in spectroscopic studies as well as to determine the parameters of the laser transition. The "effective" cross sections for emission $\sigma_E^{eff} = f^u \sigma(\lambda)$, and absorption $\sigma_A^{eff} = f^\ell \sigma(\lambda)$ [4] may be determined directly from equations (3) and (4) using the experimentally measured dependencies $\omega_{relax}(\lambda)$ and $R_{th}(\lambda)$.

III. Experimental Setup

Fig.1 show the experimental setup of neodymium (a) and erbium (b) doped fiber lasers. Since mode hopping and mode competition leads to irregular or chaotic transient oscillation, a dispersion cavity was used to prevent these effects and to obtain spectral tunability. The linear cavity containing a fiber amplifier as a gain medium was defined by a dichroic (or HR) flat mirror and reflection grating in a Littrow configuration. Two intracavity AR coated lenses were used to collimate the beam from the single-mode fiber onto the flat mirror and diffraction grating. The pump was provided by coupling the output of a cw Ti:sapphire laser operating at 810 nm (or 980 nm) into the fiber with a dichroic mirror (fiber WDM). A 50:50 beam splitter (fiber coupler) provided the output signal. Index matching cells were used to eliminate reflections from the fiber/air interfaces which would otherwise cause unwanted lasing limiting the wavelength tuning range. A chopper placed in the

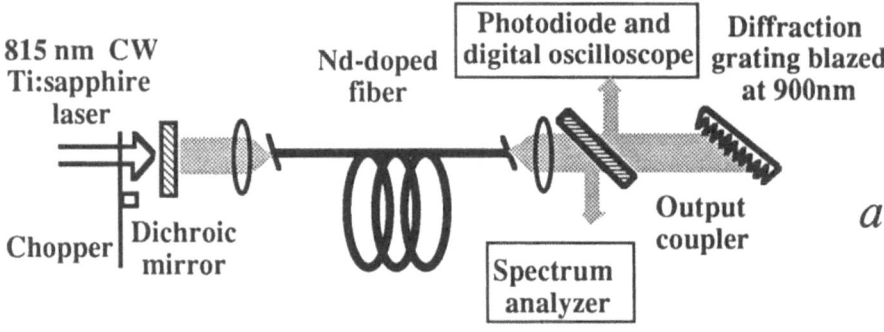

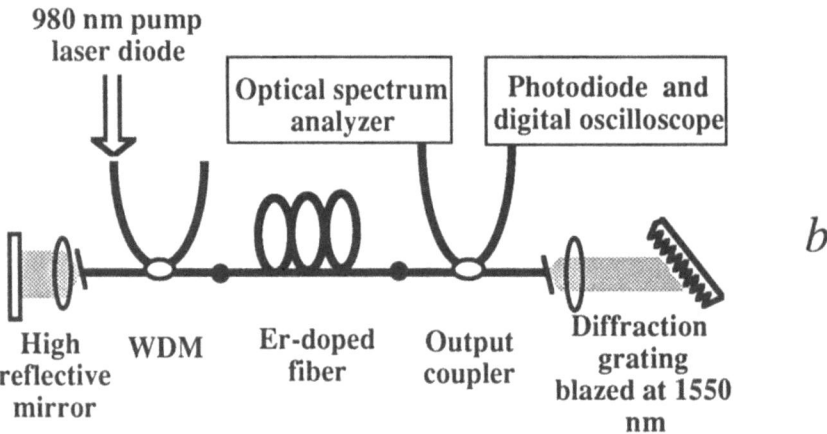

Figure 1. Experimental configurations of the tunable neodymium (a) and erbium (b) fiber laser used in relaxation oscillations measurements.

collimated pump beam was used to observe the transient evolution of the laser towards its stationary state. Fig.2 shows a typical relaxation oscillation pulse train with an exponential decay envelope. The relaxation oscillation frequency ω_{relax} was determined from the repetition period of damped small-amplitude nearly sinusoidal oscillation. The data was measured using an optical spectrum analyzer and a digital oscilloscope. For cooling, the fiber amplifier was immersed directly into a liquid-nitrogen bath.

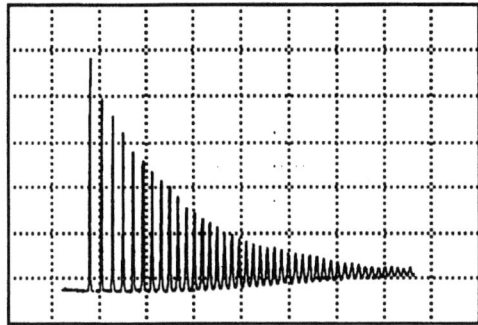

Figure 2. Typical transient oscillations from tunable fiber laser.

IV. Spectroscopy of the Transient Oscillations in Nd^{3+}-doped Fiber Laser for the Four-level $^4F_{3/2}$-$^4I_{11/2}$ (1060nm) and Three-level $^4F_{3/2}$-$^4I_{9/2}$ (900nm) Transitions.

The 10-m-long Nd^{3+}-doped fiber laser had a core diameter of 3.0 µm, a dopant concentration of 750 ppm-wt and showed an absorption of 3.1 dB/m at 810 nm. Figures 3(a) and 3(b) show photoluminescence spectra and the tuning curves around 1060 nm and 900 nm, respectively. For the 1060-nm region, we obtained a continuous tuning range of 55 nm, from 1050 to 1110 nm. For the 900-nm region, continuous tunability of 35 nm from 890 to 925 nm was available. Figures 4(a) and 4(b) show a plot of $(\omega_{relax}/2\pi)^2$ against the normalized pumping rate, around 1060 nm and 900 nm, respectively, taking the lasing wavelength as a parameter. Experimentally, we always observed linear dependencies for $(\omega_{relax})^2$ vs r-1, however, we observed that the relaxation oscillations originating at the four- and three-level transitions exhibited qualitatively different wavelength behavior. The slopes of the linear dependencies of $(\omega_{relax})^2$ vs r-1 were observed to change noticeably with wavelength for the three-level transition ($^4F_{3/2}$-$^4I_{9/2}$) relaxation oscillations; in contrast, the linear dependence was wavelength independent for the four-level transition ($^4F_{3/2}$-$^4I_{11/2}$) oscillations, as expected from the preceding analysis.

Thus, wavelength dependence of the relaxation oscillations observed supplies the method to distinguish three- and four-level transitions, and confirms the validity of the simplified rate equation approach.

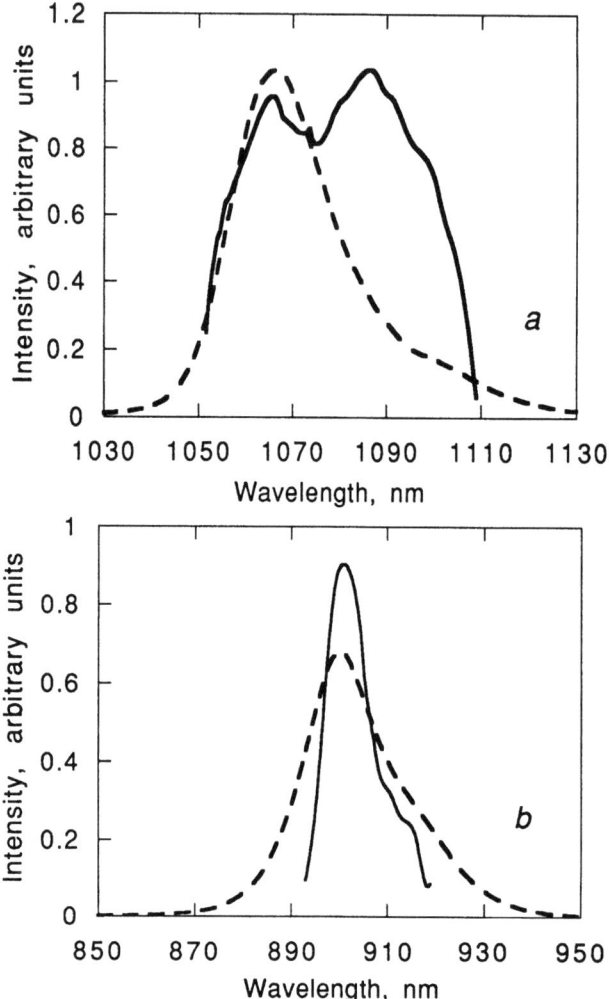

Figure 3. Photoluminescence spectra (dashed lines) and laser output power variations with wavelength (solid lines) for the four-level $^4F_{3/2}$-$^4I_{11/2}$ transition (a) and for the three-level $^4F_{3/2}$-$^4I_{9/2}$ transition (b) in Nd-doped fiber laser.

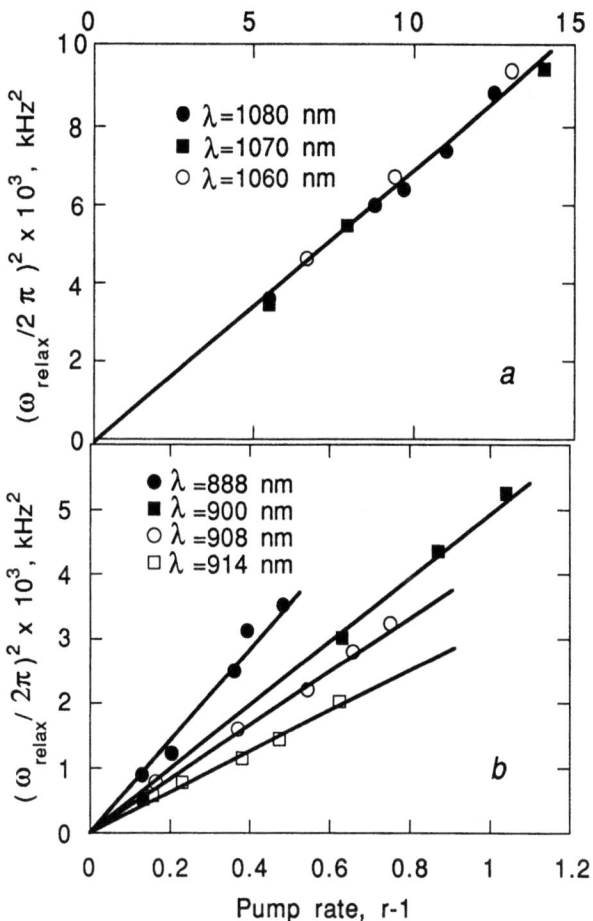

Figure 4. $(\omega_{relax}/2\pi)^2$ versus normalized pumping rate with the lasing wavelength as a parameter for the $^4F_{3/2}$-$^4I_{11/2}$ transition (a) and for $^4F_{3/2}$-$^4I_{9/2}$ transition (b) of the Nd^{3+}-doped fiber laser.

V. Transient Oscillations in Er-doped Fiber Laser

The dynamics of the transient buildup of the emission at 77 °K and room temperature was studied in Er^{3+}-doped fiber laser. The 5.2 m long fiber has a core diameter of 3.3 μm with a dopant concentration of 2000 ppm-wt and displays

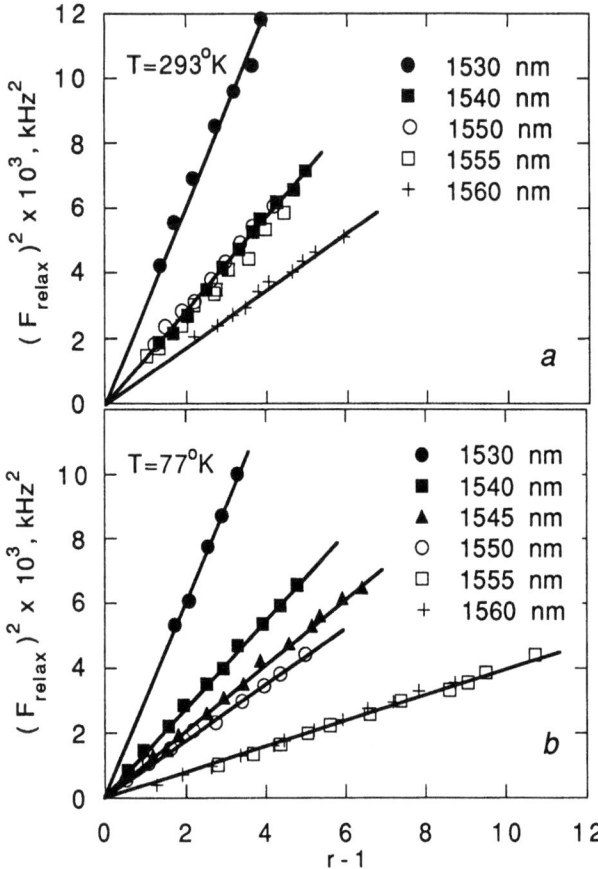

Figure 5. $(\omega_{relax}/2\pi)^2$ versus normalized pumping rate with the lasing wavelength as a parameter for T=297°K (a) and T=77°K (b) for the Er^{3+}-doped fiber laser.

absorption of 9 dB/m at 980 nm. The transient oscillations were investigated in the long-wavelength tail of the gain spectrum, since the occupation probability of the highest sublevels of $^4I_{15/2}$ manifold relative to that of the ground sublevel strongly decreases as the fiber is cooled. Figures 5(a) and 5(b) show a plots of $(\omega_{relax})^2$ against normalized pumping rate with the lasing wavelength as a parameter for T=297°K and T=77°K, respectively. The spectral change of the slopes with temperature is due to the variation of the population distribution between the components of the $^4I_{15/2}$ manifold. Wavelength dependence of the k(λ)-factor can be derived from these plots assuming that τ_s does not depend on temperature [7]. The temperature dependent

Boltzmann's distribution of electrons within the Stark-split manifold is clearly seen from k(λ) curves (Figures 6(a) and (b)). It is also evident that the temperature has a

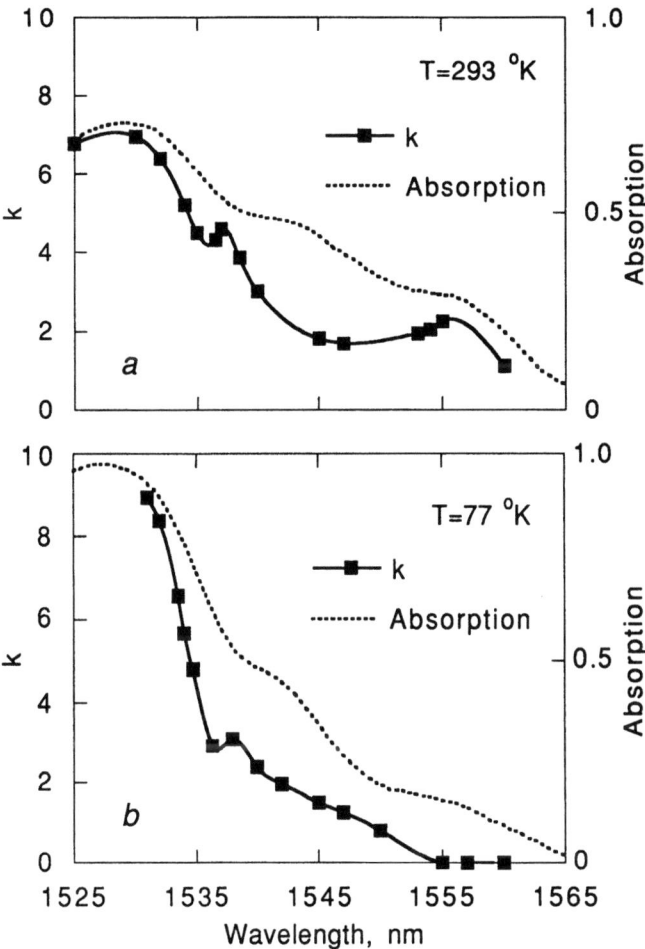

Figure 6. Wavelength dependence of the relaxation oscillation parameter k derived from the plots presented in Fig. 5 and absorption spectrum at T=297°K (a) and at T=77°K (b).

pronounced effect on the individual absorption transitions start from Stark sublevels with different equilibrium populations. In particular, it is deduced from these Figures that the lasing transitions at λ>1555 nm become four-level at liquid-nitrogen temperature (k=0). Moreover, the positions of the Stark components in the ground

$^4I_{15/2}$ manifold and hence, the individual transition wavelengths, can be determined more accurately from the k(λ) line shape, since the sublevels are resolved more clearly than in the absorption spectrum. The positions of the transitions at 1527 nm and 1557 nm are in agreement with those previously identified [8]. The difference between absorption spectrum and k(λ) in the 1538-1543 nm region can be attributed to the fact that k-factor is proportional to the *actual homogeneous cross-section*. At least qualitatively, k(λ)-lineshape is similar to that calculated for homogeneous cross-section by deconvolution from the experimental cross-section [9]. Variation of the homogeneous linewidth across the spectrum (and with temperature) may also be responsible for the observed feature.

The dependence of the relaxation oscillation frequency and threshold pump power on the wavelength and corresponding σ_E^{eff} and σ_A^{eff} obtained directly from the measurements of the relaxation oscillation frequency by the methodology discussed above are shown in Figures 7(a) and (b). The cold cavity lifetime $\tau_c = 1/\gamma_c$ =69 ns was obtained from the measurements of the relaxation oscillation frequency and oscillation decay time. The spontaneous lifetime of the laser transition was assumed to be of τ_s =10 ms [5]. The results compared well to experimental data obtained by conventional fluorescence/absorption spectroscopic measurements.

VI. Conclusions

In summary, we describe the spectral properties of the transient oscillations in Nd^{3+}- and Er^{3+}-doped fiber lasers. Experiments performed on the Nd^{3+}-doped fiber laser clearly demonstrate the different behavior of the transient oscillation for the three-level $^4F_{3/2}$-$^4I_{9/2}$ transition at 900 nm and for the four-level $^4F_{3/2}$-$^4I_{11/2}$ transition at 1060 nm. Therefore, the wavelength dependence of the relaxation oscillations supplies the method to distinguish three- and four-level transitions, and this can be useful in spectroscopic studies as well as to determine the parameters of the laser transition [10]. From the temperature measurements of the relaxation frequency in an Er^{3+}-doped fiber laser, it was confirmed that the laser transition mechanisms for λ>1555 nm were different at T=293 °K (quasi-three-level) and at T=77°K (four-level). This remarkable feature, that depends on the temperature distribution of electrons within the Stark-split manifold, should be taken into account for careful description of inhomogeneously broadened fiber amplifier. It was shown that the absorption cross-section exhibits the pronounced effect on the relaxation oscillation frequency and, therefore, plays important role in determining laser transient dynamics [11-12]. A new method to determine the laser transition cross sections for both emission and absorption based on the measurements of the relaxation oscillation frequency is proposed [13]. The magnitudes of the emission and absorption cross sections can be derived from these measurements independently, without the aid of the reciprocity principle, commonly involved for this determinations. In addition, the homogeneous cross sections for emission and absorption may be also determined *directly* using the experimentally measured dependencies of the relaxation oscillation frequency and threshold pump rate on wavelength.

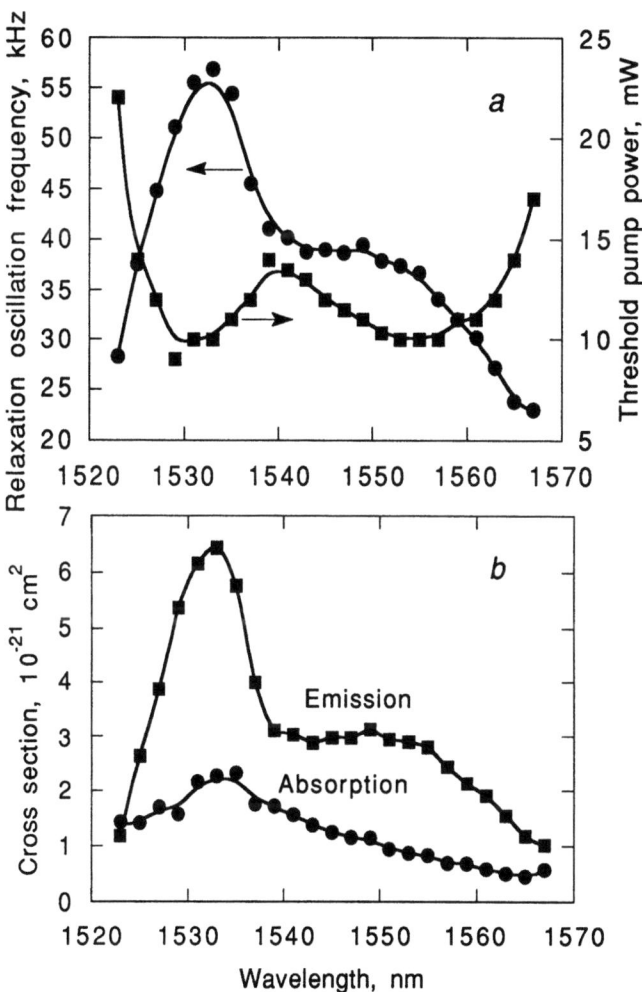

Figure 7. The relaxation oscillation frequency and threshold pump power versus lasing wavelength (a) and effective fluorescence and absorption cross sections σ_E^{eff} and σ_A^{eff} of the Er-doped fiber derived from wavelength dependence of the relaxation oscillation frequency.

Acknowledgment

The authors would like to thank V.V. Kuzmin for assistance. This research was partly supported by a NATO Fellowship.

References

[1] E.L. Goldenstein, V. da Silva, L. Eskildsen, M. Andrejco, and Y. Silberberg, IEEE Photon. Technol. Lett., **5**, p. 543, (1993).

[2] D.E. McCumber, Physical Review, **136**, pp. A954-A957, (1964).

[3] W. Koechner, "Solid-state laser engineering", Springer-Verlag New-York Inc., 1976.

[4] S.A. Payne, L.L. Chase, L.K. Smith, W.L. Kway, and W.F. Krupke, IEEE J. Quantum Electron., **QE-28**, pp. 2619-2630, (1992).

[5] W.L. Barnes, R.I. Laming, E.J. Tarbox, and P.R. Morkel, IEEE J. Quantum Electron., **QE-27**, pp. 1004-1009, (1991).

[6] C.R. Giles, C.A. Burrus, D.J. DiGiovanni, N.K. Dutta, and G. Raybon, IEEE Photon. Technol. Lett., **4**, pp. 363-365, (1991).

[7] N. Kagi, A. Oyobe, and K. Nakamura, J. Lightwave Technol., **9**, p. 261, (1991).

[8] E. Desurvire, and J.R. Simpson, Opt. Lett., **15**, p. 547, (1990).

[9] E. Desurvire, J.W. Sulhoff, J.L. Zyskind, and J.R. Simpson, IEEE Photon. Technol. Lett., **2**, p. 653, (1990).

[10] O.G. Okhotnikov, and J.R. Salcedo, Appl. Phys. Lett., **64** (20), pp. 2619-2621, (1994).

[11] O.G. Okhotnikov, and J.R. Salcedo, in *European Conference on Lasers and Electro-Optics/European Quantum Electronics Conference*, Amsterdam, 1994, paper QTuB6.

[12] O.G. Okhotnikov, and J.R. Salcedo, "Laser transitions characterization by spectral and thermal dependences of the transient oscillation", Optics Lett., accepted for publication.

[13] O.G. Okhotnikov, V.V. Kuzmin, and J.R. Salcedo, IEEE Photon. Technol. Lett., **6**, pp. 362-364, (1994).

NONLINEAR M-LINES SPECTROSCOPY OF DMOP-PPV POLYMER PLANAR OPTICAL WAVEGUIDES: QUASI-PERMANENT AND FAST ELECTRONIC REFRACTIVE INDEX CHANGES.

F.MICHELOTTI[1], T.GABLER[2], H.H.HÖRHOLD[2], R.WALDHÄUSL[2], A.BRÄUER[2]
[1]Università di Roma "La Sapienza"-Dipartimento di Energetica
INFM & GNEQP of CNR
Via A.Scarpa, 16 I-00161 Roma ITALY
[2]Friedrich Schiller Universitat - Jena - GERMANY

Abstract

Among conjugated polymers, the family of the poly-phenylene-vinylenes (PPV) has recently attracted much attention due to its high fluorescence quantum yield [1-4] and nonlinear optical response [5]. The first excited state is an excitonic $1B_u$ level [6], classifying the transition from the ground state ($1A_g$) as dipole allowed and giving rise to high luminescence quantum yield [4].

We report on prism coupling experiments in slab poly(1,4-phenylene-1,2-bis (4-methoxyphenyl) vinylene (DMOP-PPV) waveguides. In these measurements upconverted fluorescence at lambda=520 nm has been observed, using excitation wavelengths between 730nm<lambda<880nm.

In figure 1 the angular behaviour of the coupling efficiency eta(theta) versus the input angle theta for a one prism coupling experiment is shown for 5ps pulses at lambda=830nm. The three curves have been obtained for low (I_{gw}=13 MW/cm^2), high (I_{gw}=67 MW/cm^2) and low guided wave intensities respectively. A shift of the coupling angle is observed which is due to a permanent change of the refractive index of the film (Dn= $-5.7 \cdot 10^{-4}$). Illuminating the sample with an UV incoherent lamp below the prism we observed a return of the shift back towards the original resonance curves. The quasi-permanent change of n has been interpreted by means of a four level model taking into account trap levels due to the amorphous nature of the polymer film [7]. In such a model, excitation with wavelengths in the IR range, which cannot excite molecules from the ground state to the first excited state but which can cause untrapping from localized levels in the gap, gives rise to a depletion of the density of the population of trapped carriers with a consequent decrease of the refractive index at those wavelenghts. The model can explain the presence of upconverted fluorescence too.

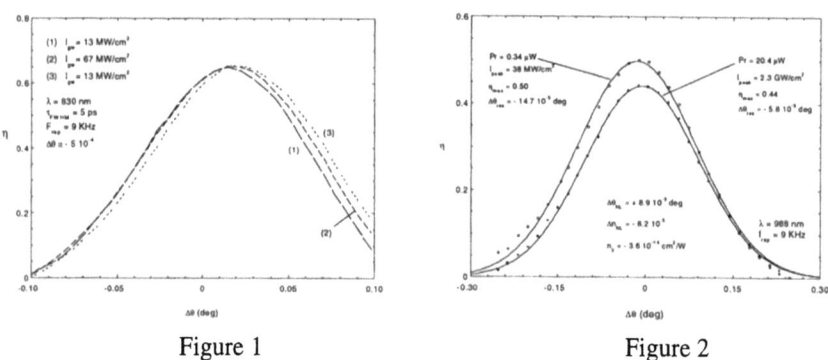

Figure 1   Figure 2

By exciting with radiation at a wavelength which is far from the absorption of such levels (lambda=988nm) and using shorter pulses (250fs) we have been able, in a nonlinear coupling experiment, to detect a fast and retrievable change of the couling angle. In figure 2 the coupling efficiency curves corresponding to two measurements at low (I_{gw}=38 MW/cm^2) and high (I_{gw}=2.3 GW/cm^2) guided wave intensities are shown. From the shift of the coupling angle and from the decrease of the coupling efficiency we exstimated a nonlinear refractive index coefficient n_2= -3.6 10^{-14} cm^2/W.

REFERENCES
1. N.F. Colaneri, D.D.C. Bradley, and R.H. Friend, P.L. Born and A.B. Holmes, C.W. Spanquer, Phys. Rew. B42, 670 (1990)
2. E.L. Frankevich and A.A. Lymasev; I. Sokdik, and F.E. Karasz; S. Blumstengel; R.H. Boughman; H.H. Hörhold, Phys. Rew. B46, 9320 (1992)
3. R. Kursting, U. Lemmer, R.F. Mahrt, K. Leo, H. Kurz, H. Bassler, and E.O. Göbel, Phys. Rew. Lett., 70, 3820 (1993)
4. Z.G. Soos, D.S. Galvão and S. Etemad, Adv. Mater., 6, 280 (1994)
5. U.Bartuch, A.Bräuer, P.Dannberg, H.H.Hörhold, D.Rahbe, Int.Jour.of Optoel., 7, 275 (1992).
6. D.D.C. Bradley, A. R. Brown, P.L. Burns, J.H. Burroughes, R.H. Friend, A.B. Holmes, K.D. Mackay, R.N. Marks, Synth.Met., 41-43, 3135 (1993).
7. F.Michelotti, T.Gabler, H.Hörhold, R.Waldhäusl and A.Bräuer, "Prism coupling in DMOP-PPV polymer waveguides", in press on Opt.Comm. (1994).

PLASTIC OPTICAL FIBRE (POF) DISPLACEMENT SENSOR

N. IOANNIDES, D. KALYMNIOS and I.W. ROGERS
*SECEAP, University of North London, Holloway Road
London N7 8DB, United Kingdom*

Abstract

In many industrial applications accurate measurements of linear displacements need to be made rapidly, the range of the displacement varying from the submicron region to many millimetres. Of the possible techniques used an optical, non contact method is often preferred.

The sensor shown schematically in figure 1 is based on the inverse square law and is suitable for the longer ranges.

This novel optical method for measuring displacement gives an output which varies linearly with displacement and with a simple adjustment to the sensor dimensions can be used to cover a number of different ranges from a few millimetres to many centimetres.

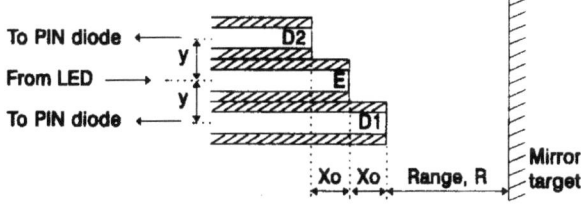

Figure 1

This arrangement was chosen since a simplified theoretical treatment showed that when the output signal S is defined by:

$$S = (S1+S2) / (S1-S2)$$

where S1 and S2 are the outputs from the two detectors D1 and D2 respectively then S varies linearly with the range R for $R > 4Xo$. Using a least squares fit routine over the useful measured S and by taking the inverse of this, the range R could be determined.

The sensor is using Plastic instead of Glass Optical Fibres for the reason that Plastic Optical Fibres have a fibre diameter of 1mm and so there is no need for any kind of lenses or any other optics to assist in collecting enough light.

Three 5 m long PMMA fibres, with their protective sheath are used one emitting (E) and two receiving (D1, D2), displaced from each other parallel to the axis of the detector by a scale factor Xo and perpendicular to this axis by an amount Y, both Xo and Y being of the order of millimetres. The three fibres are held side by side in the same plane using a simple holder. The emitting fibre is coupled to a red LED and emits about 400 μW on to a target. The detecting fibres are coupled to two 13 mm^2 active area PIN photodiodes with transimpedance amplifiers.

From the experimental results obtained by using a reflective target and by having the Xo and Y separations set at 3mm and 2.2mm respectively, acceptable linearity was confirmed over the range 10mm < R < 80mm for a temperature range of 20°C < T < 35°C. Figure 2 shows experimental results. Each experimental point shown represents the spread of S values from 9 independent runs. The sensor performance remained unchanged when the current through the LED was varied from 80 mA down to 8 mA and also when the op amp power supply was varied from ± 18 V down to ± 10 V.

Experimental results were also obtained by using diffuse targets (Al beadblasted plate and white matt paper). Acceptable linearity was confirmed for the range 10mm < R < 50mm (figure 3).

The slope of the experimental points taken by using diffuse targets was exactly half the slope of the points when using reflective targets for this particular Xo separation as was predicted theoretically.

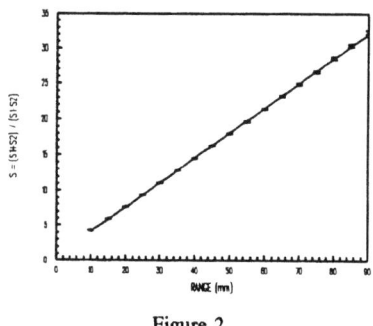

Figure 2   Figure 3

The conclusions from both experimental and theoretical results are that a displacement sensor based upon this principle could be constructed which would be both linear and scalable, the maximum measurable displacement being set by the available power from the emitter and the limiting signal to noise ratios of the detectors.

Work is continuing since it is believed that an exclusively POF sensor of this type, could find many useful applications.

References:

1. Kalymnios, D. (1993) Linear and scalable optical sensing technique for displacement, Sensors series, Sensor VI, KTV Grattan and AT Augousti, pp 291 - 3.

2. Ioannides, N., Kalymnios, D., and Rogers, I.W. (1993) Experimental and theoretical investigation of a POF based displacement sensor, 2nd international conference in plastic optical fibres and applications, The Hague, June 28 - 29.

A NEW REFLECTIVE PLASTIC OPTICAL FIBRE DISPLACEMENT TRANSDUCER

S. HADJILOUCAS, L.S. KARATZAS, D.A. KEATING, AND M.J. USHER.
Department of Cybernetics, School of Engineering and Information Sciences, University of Reading, P.O. Box 225, Whiteknights, Reading, Berks., RG6 2AY, U.K.

1. Abstract

A novel, dual plastic optical fibre amplitude modulating displacement transducer of improved sensitivity is presented. Results of different fibre optic configurations and the merits of cutting the fibres twice so as to improve coupling and make full use of the critical angle are shown.

2. Introduction

Plastic, dual, reflective type, amplitude modulating fibre optic displacement sensors are cheap, small in size, light, flexible, robust, and low in energy requirements. If coupling between emitter and receiver fibres could be improved, the signal-to-noise ratio would also be improved. Figure 1a shows a simple reflective optical fibre displacement transducer. The polymer cable has a core and cladding diameter of 1.00 mm, a core refractive index (n_1) of 1.49, and a clad refractive index (n_2) of 1.42, giving a numerical aperture (NA) of 0.47, an acceptance angle (θ_a) of 27.85°, and a critical angle (θ_c) of 71.76°. The fraction of light returning to the detector (for a given fibre radius and acceptance angle) is directly dependent on the sensor gap, that is the fibre-to-reflector separation as cited in [1] and [4]. Generally, for the uncut fibres there is a small non-linear region, then a very steep and linear part (region used for measurements) and finally a decreasing region according to $1/(distance)^2$.

In order to improve the responsivity of the transducer, a simple way is to cut the emitter (T_x) and receiver (R_x) fibres at an angle (α) as to make full use of the critical angle (Figure 1b). This ensures improved coupling and therefore better signal-to-noise ratio. The condition for (α) is the one for which the ray emerging from the fibre is parallel to its sloping face, and using Snell's law to cut the fibres gives $n_0 \sin 90 = n_1 \sin(18.24+\alpha)$ so $\alpha = 23.84°$. This value is used for the reflectance displacement transducer shown in Figure 1b, where the fibres are placed next to each other.

A further improvement in performance can be obtained by cutting the fibres twice, at angles of 24° and 66°, as shown in Figure 1c. To our knowledge, the only dual fibre-optic sensor which investigates the effect of varying the angle between the two fibres has been described by Powell [5], where an increase in sensitivity of only one order of magnitude is reported. The advantage of the double-cut configuration is that it maximises the amount of the reflected light that enters the receiving fibre, and minimises the reflected light that is lost in all other directions.

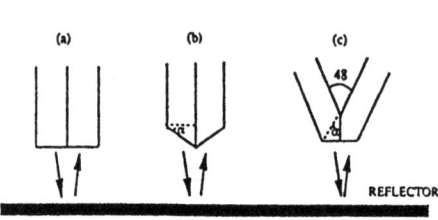

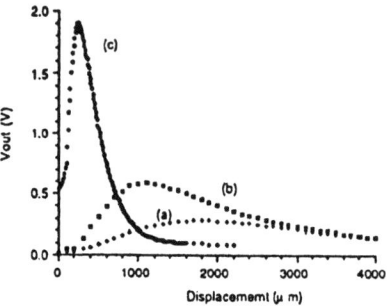

Figure 1. Different configurations of fibre-optic pairs. The angle of cutting for (b) is 24° and for (c) are 24° and 66° respectively.

Figure 2. Response curves for (a) uncut, (b) single-cut, and (c) double-cut fibres, using a mirror as a reflector.

3. Results and Discussion

The comparative responses of the three configurations, when a mirror is used as a reflector, are shown in Figure 2. It can be seen that a significantly better response is obtained by cutting the fibres twice (Figure 2c). The uncut fibre configuration gives a responsivity (front slope) of 183.3 V/m and a linear range of 1200 µm. The single-cut fibres give a responsivity of 769.2 V/m and a linear range of 650 µm. Finally, the double-cut fibre configuration give a responsivity of 11880 V/m and a linear range of 80 µm. It can be observed that the linear range of the double cut sensor is 15 times smaller than that of the uncut. Furthermore, the second linear part of the double-cut configuration (Figure 2c) allows operating the transducer in its back slope. Such operation has the advantage of greater stand-off distance but at the expense of responsivity. It was found necessary to use a thin opaque film (paint) between the two fibres (Figure 2c), to minimise light directly coupled from the emitter to the receiver. Further improvements could be made, resulting to a more sensitive transducer, when coupling the same amount of light into a plastic optical fibre of 100 µm. Finally, the advantages of cutting the fibres may be used to improve coupling in a scaleable displacement sensor as cited in [2] and [3].

4. Conclusions

The design of a new optical fibre displacement transducer has been presented. The merits of cutting the fibres twice at an angle to increase responsivity, have been demonstrated. The improved responsivity (11880 V/m) and the small working range (80 µm) of the double-cut configuration is an advantage when considering future integration of the sensor with optoelectronic devices.

5. References

1. Hochberg, R.C. (1986) Fiber-optic sensors,. *IEEE Transactions on Instrumentation and Measurement* 35, 447-450.
2. Ioannides, N., Kalymnios, D., and Rogers, I.W. (1993) Experimental and theoretical investigation of a POF-based displacement sensor, *Proc. 2nd Inter. POF and Applications Conference*, The Hague, 162-165.
3. Kalymnios, D. (1993) Linear scaleable optical scaling technique for measurement and displacement, in K.T.V. Grattan and A.T. Augousti (eds) *Sensors VI: Technology, Systems and Applications*, IOPP Sensors Series, Bristol and Philadelphia, pp. 291-294.
4. Krohn, D.A. (1986) Intensity modulated fiber optic sensors: An overview, *SPIE, Fiber Optic and Laser Sensors IV.* 718, 2-11.

WHITE LIGHT INTERFEROMETER USING LINEAR CCD ARRAY WITH SIGNAL PROCESSING

STEPHEN R. TAPLIN, ADRIAN GH. PODOLEANU, DAVID J. WEBB, DAVID A. JACKSON
Physics Laboratory, University of Kent at Canterbury, Kent CT2 7NR, United Kingdom.Fax: +44 227 475423

Many recent publications have demonstrated the advantages of analysing the spectral content of interferometric outputs[1,2]. They show clearly that by looking at the spectral characteristics of the output from an interferometer illuminated by a white light source, a sensor can be produced requiring no initialisation, and since the frequency response is highly linear, the system may be easily calibrated.

The experimental setup depicted in the figure is used to determine the cavity width of a Fabry-Perot interferometer. Low coherent light is obtained by using a Super Luminescent Diode (SLD). The effect of the interferometer in the spectral domain is to generate channelled spectra, produced by multiplying the source spectrum with the transfer function of the interferometer described by the Airy function in the case of the Fabry-Perot interferometer.

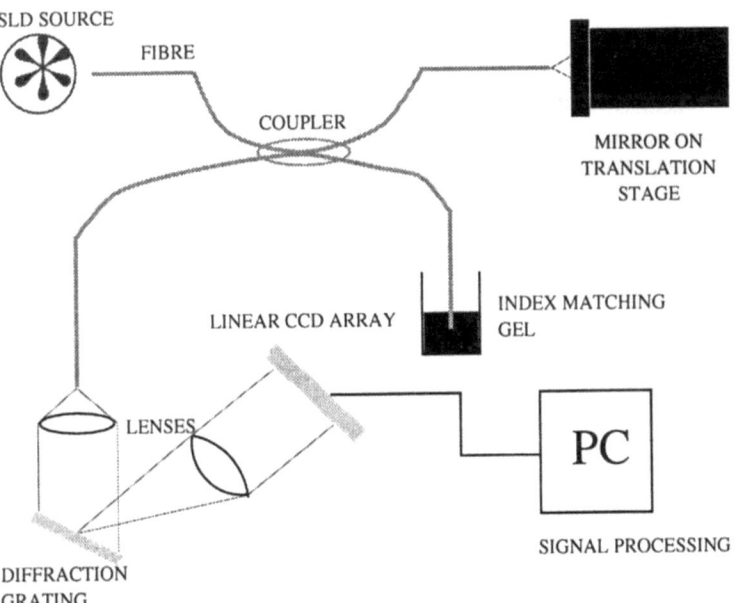

Figure: Experimental Arrangement

The technique described here monitors the spectral output from the interferometer directly with the aid of a diffraction grating and a linear CCD array (1728 pixels). The channelled spectrum is directed as a collimated beam toward a diffraction grating blazed at 800nm with 1200 lines per mm which filters the composite frequencies. The first or second order diffracted beam is then directed onto the CCD, through a collimating lens, whose output may be monitored on an oscilloscope or sampled into a computer for analysis.

For a given cavity width, the CCD signal is sampled into a computer. A waveform comprising the channelled spectrum filtered by the Gaussian profile of the source may be obtained. The signal is then passed through a Fast Fourier Transform (FFT) in order to determine the signal's frequency components. The maximum of the peak in the frequency domain is determined by the cavity width. A Gaussian fit is performed on the data points in the frequency domain since the FFT of the Gaussian source profile is Gaussian also. This permits a more accurate evaluation of the centre frequency should it lie between two points in the FFT.

The signal analysis is performed on a number of cavity widths calibrated by a micron resolution stepper motor which controls the mirror. A frequency, derived from the signal analysis, is associated with each cavity width and a plot is achieved demonstrating high linearity (0.25μm RMS deviation from linearity for the range 250-750μm).

Calibration of the system may be performed using the line of best fit through the series of data points indexed to the motor translation stage (1μm resolution over 100μm range). A more improved calibration may be achieved by interpolating between two known reference cavity widths or by characterising the grating/CCD system directly using known spectral lines.

A measurement system of sub-micron accuracy is described that is able to provide absolute measurements in the millimetre range. This range has traditionally been difficult to access with optical techniques, being too long for unambiguous interferometric sensing and too short for the time of flight or the sub-carrier modulation approaches. Since there is no scanning or tracking mirror in the optical system there is the potential for high bandwidth measurements.

1. Norton, D. A. (1992) System for Absolute Measurements by Interferometric Sensors, *SPIE Fiber Optic and Laser Sensors X*, **1975**, 371-376.

2. Taplin, S., Podoleanu, A. Gh., Webb, D. J. and Jackson, D. A. (1993) Displacement Sensor Using Channelled Spectrum Dispersed on a Linear CCD Array, *Electron. Lett.*, **29**, 896.

RADIATION-TO-GUIDED FIELD COUPLING IN Ti:LiNbO₃ INTENSITY MODULATORS

M. MARCINIAK
Academy of Telecommunications
05-131 Zegrze 300/11, Poland

M. SZUSTAKOWSKI
Military Technical Academy
00-908 Warsaw, Poland

Abstract

Experiments and BPM simulations of light propagation in Ti:LiNbO₃ Mach-Zehnder modulators reveal transverse light oscillations, described by a field divergence parameter developed in the paper.

Taking into account this parameter one can decrease power losses in the interferometer - typically from 4.26 dB to 0.29 dB.

WHITE LIGHT FIBER OPTIC INTERFEROMETRY- A TECHNIQUE FOR MONITORING TISSUE PROPERTIES

Scott Shukes, Faramarz Farahi and, M. Yasin Akhtar Raja
Physics Department, University of North Carolina at Charlotte, Charlotte, North Carolina, USA

Robert Splinter
Laser and Applied Technologies Laboratory, Carolinas Medical Center, Charlotte, North Carolina, USA

Abstract

Currently, no adequate non-invasive methods are available for in-situ analysis of laser-tissue interaction. This paper describes an integrated diagnostic and therapeutic laser procedure through in-situ monitoring of the optical tissue characteristics in a non-invasive manner. Biological tissue is a highly scattering medium. A proposed system uses a broadband (white-light) interferometer that addresses the signals obtained from different depths of the scattering medium. When the Michelson interferometer is illuminated by a broadband light source an interference signal can only be observed by a photodetector if and only if the two arms of the interferometer are identical in path length. In a scattering medium whereby the scattering optical signals are produced at different depths of the medium, the interference signal is produced by signal perturbed at a given depth. Thus the tissue may be scanned in depth by axial movement of the reference mirror. Identification of the local optical characteristics will provide discrimination between healthy and pathological conditions in addition to real time assessment of dosimetry.

HOLOGRAPHIC INTERFEROMETRY OF IMMERSED SYSTEMS IN ELECTROCHEMISTRY

K. HABIB
Materials Application Department
KISR, PO BOX 24885, SAFAT 13109
Kuwait

Abstract

In recent work conducted by the author a novel technique for monitoring the mechanochemical behaviour, i.e. stress corrosion cracking, corrosion fatigue, and hydrogen embrittlement, of metallic electrodes in aqueous solutions has been developed. The technique incorporates holographic interferometry for measuring microscopic deformation and electrochemical techniques for determining the corrosion current of metallic samples. In addition, the author has recently reported mathematical models describing the cathodic deposition and anodic dissolution of metals in aqueous solutions by holographic interferometry. As a result, one may suggest that the techniques of holographic interferometry have many useful applications in the field of electrochemistry yet to be explored. The objective of the present work is to demonstrate the advantages and limitations of holographic interferometry in the field of electrochemistry.

PRESSURE EFFECTS OF OPTICAL FREQUENCY IN EXTERNAL CAVITY DIODE LASERS

HANNU TALVITIE[1], JOHAN ÅMAN[2], HANNE LUDVIGSEN[1],
ANTTI PIETILÄINEN[1], LESLIE PENDRILL[3] AND ERKKI IKONEN[1]
[1]*Helsinky Univ. of Technology, Otakaari 5A, FI-02150 Espoo, Finland*
[2]*Chalmers Univ. Technology, S-41296 Gothenburg, Sweden*
[3]*Swedish Nat. Testing and Research Institute, S-50115 Borås, Sweden*

Abstract

A report is given on the construction of a passively stabilised external cavity diode laser operating at 780 nm. The relative frequency stability obtained is 10^{-8}. The sensitivity of laser frequency to changes in air pressure was studied and subsequently eliminated.

METHODS OF MEASUREMENT FOR ANGLE-POLISHED OPTICAL FIBER LAUNCHING EFFICIENTS

DARRIN P. CLEMENT and ULF ÖSTERBERG
Thayer School of Engineering
Dartmouth College
Hanover, NH 03755
USA

Abstract

It is well known that single-mode fiber technology requires strict alignment tolerances to maintain coupling integrity. With acceptable bit-error-rates (BERs) of 10^{-12} for telephony and as low as 10^{-15} for datacommunication applications, even small link losses can render a system inoperable. Before standards can be established for in-the-field matings and products, it must be well understood how to measure coupling changes in the lab, as these are more easily controlled so that small changes may be observed. This paper discusses experimental techniques to measure the influence of small changes of the overall coupling efficiency between a laser diode and a single-mode fiber. Particular emphasis is placed on fibers with non-zero endface angles (i.e. angle-polished fibers). It is increasingly important to model and measure the effects of out-of-plane misalignments (OPM) and this paper will present some of the primary methods and concerns involved.

THE APPLICATION OF A FIBER OPTIC GYROSCOPE FOR STUDIES OF EXTINCTION RATIO OF FIBER OPTIC POLARIZER

J. L. JAROSZEWICZ, M. SZUSTAKOWSKI
Institute of Technical Physics, Military University of Technology
PL-01-908 Warsaw, Kaliskiego Street 2, Poland

Abstract

The measurement method for an evaluation of extintion ratio of fiber optic polarizer is presented. The proposed method consists in an application of Sagnac interferometer in a configuration used in a fiber optic gyroscope. It allows to evaluate the extinction ratio unmeasurable by means of other known methods. Moreover, it enables to study of polarization properties of the polarizer without necessity of free propagation of the optical beam.

BLOCKS AND ARCHITECTURE FOR A MULTICHANNEL FIBRE OPTIC CORRELATOR

ADRIAN GH. PODOLEANU*, RYAN K. HARDING, DAVID A. JACKSON
Physics Laboratory, University of Kent at Canterbury
KENT CT2 /NR, United Kingdom

Abstract

There are two key operations implemented in a fibre optic correlator: the delay and the multiplication. This paper presents the choice of optical and microwave delays to optimise performance and cost and a comparative analysis of different ways to implement the multiplication blocks using the AND operation. The ANDing of 0.35 ns FWHM pulses was demonstrated experimentally using either very fast avalanche photodiodes with thresholding or a GaAs double gate MOSFET. Their relative performances are compared. For stationary signals, as an inexpensive way to expand the number of correlator channels by a factor of five, a five-step microwave switchable delay line was built and used. The ability to measure the correlation function of a 500 MHz sinusoidal signal is proved using double gate MOSFETS as AND gates, optical implementation for the delays longer than 5 ns and microwave implementation for the delays shorter than 5 ns.

CALIBRATION OF FIBRE OPTICAL MEASUREMENT EQUIPMENT AT SP

R. ANDERSSON, M. HOLMSTEN and L. LIEDQUIST
Swedish National Testing and Research Institute

Abstract

The Swedish National Testing and Research Institute, SP is the national institute in Sweden for technical evaluation, testing, certifying, measurement technique and research. SP is the largest centre among the Nordic countries for metrology technique. SP holds the national standards for most of the physical quantities and their traceability to international standards. SP supplies the Swedish industry and society with measurement support, competence and traceable measurements. SP offers a broad and nation-wide calibration service for measurements in most of the industrial sectors. Considerable research activities are of great importance to develop competence within each measurement area. We develop improved national standards, methods and measurement instruments continuously. SP collaborates with technical universities and participates in the international metrology collaboration, especially within the International Bureau for Weight and Measures, BIPM and the European metrological federation, EUROMET. SP was established because "rational testings and scientific research are of great importance in a stronger international competition", as cited from the proposal to a new National Testing Establishment, in 1903.

FIBRE APPLICATION IN ABSOLUTE RANGING INTERFEROMETRY WITH COMPUTER GENERATED HOLOGRAMS (CGH)

L.L. WANG, M. DEININGER, K. GERSTNER, T. TSCHUDI
Institute of Applied Physics, Technische Hochschule Darmstadt
Hochschulstr. 6, D-64289 Darmstadt, Germany

Abstract

One main problem of an interferometric measurement is to evaluate the object distance from the interference function. One of the known methods which delivers the object phase, is the phase step method. Here we introduce CGH's to realise the phase steps parallel without a mechanical change of the path and it is used fibre in the absolute ranging interferometry.

MONITORING THROUGH-THE-THICKNESS BEHAVIOUR OF COMPOSITES USING FIBER OPTIC SENSORS

W.E. WOLFE
The Ohio State University
Columbus, Ohio 43210, USA

B.C. FOOS
Wright-Laboratory
Wright-Patterson Air Force Base, Ohio 45433, Usa

Abstract

A dynamic stress based theory for composite laminates that assumes a piecewise linear distribution of in-plane stresses across each lamina is developed. Unlike displacement based approaches this theory accurately predicts transverse stresses through the lamina thickness even in regions of high stress gradients such as free edge delamination coupons. To validate the accuracy of the transverse stresses predicted by the theoretical model, a novel experimental technique is being developed. The experimental program utilizes unintrusive extrinsic Fabry-Perot interferometric strain sensors (EFPI-SS) embedded through the lamina thickness to monitor the transverse normal stress / strain. This experimental technique is unique in that to the authors' knowledge it is the first study of its kind to directly monitor through-the-thickness internal behaviour in multiply composite materials.

FIBRE-OPTIC FABRY-PEROT SENSOR FOR VIBRATION AND PROFILE MEASUREMENTS

I. PAULICKA, V. SOCHOR, J. STULPA
Czech Technical University, Faculty of Nuclear Sciences and Physical Engineering, Department of Physical Electronics
Brehová 7, 115 19 Prague 1, Czech Republic

Abstract

In this paper we demonstrate the operation of a feedback stabilised quadrature phase-shifted extrinsic Fabry-Perot (FP) fibre-optic sensor for the remote detection of the amplitude and the relative phase of subnanometer vibrations and displacements. The surface of a measured object facing at a short distance the end of single-mode fibre, acts as a reflector, thereby creating an air gap of Fabry-Perot cavity. To provide the surface profile measurements the fibre end is vibrating laterally at a constant distance from the plane of object surface. A brief description of the device is given, and its operation is described. A resolution in the order of 10^{-11}m for the vibration amplitude measurement was obtained. In the surface scanning regime a vertical deviations around 1 nm can be distinguished.

APPLICATION OF OTDR IN THE MINING INDUSTRY

V. KUMAR and DINESH CHANDRA
Department of Electronics and Instrumentation
Indian School of Mines, Dhanbad-826004, India

Abstract

This paper describes various possible applications of optical time domain reflectometer in the mining industry. Emerging trends are presented to show how the principle of OTDR can be used as distributed temperature sensor. A new concept of this technique is enunciated for the measurement of displacement of rock mass which may be used for the remote monitoring of strata movement above a longwall coal mine, collapse of a longwall panel, opencast mine, mine subsidence, underground nuclear tests.

PARAMETRIC AMPLIFICATION BY FOUR-WAVE MIXING FOR DISTRIBUTED OPTICAL FIBRE SENSORS

J. ZHANG, V.A. HANDEREK and A.J. ROGERS
*Department of Electronic and Electrical Engineering, Kings College London
Strand, London WC2R 2LS, United Kingdom*

Abstract

The principle of Four-Wave-Mixing parametric gain and its application in Pump-Probe architecture distributed optical fibre sensor systems is described. Signal level and bandwidth are estimated. Finally an experimental arrangement is proposed.

INTRACAVITY LASER DIODE PUMPED AND MODULATED ER^{3+} DOPED FIBER LASERS

OLEG G. OKHOTNIKOV and JOSÉ R. SALCEDO
*Centro de Optoelectrónica, INESC
Rua José Falcão 110, 4000 Porto, Portugal*

Abstract

We report on a pulsed compound Er^{3+}-doped laser, that simultaneously uses an intracavity laser diode for optical pumping, gain or phase modulation and as a bandwidth control etalon. We demonstrate a simple method to produce high-energy pulses with peak powers up to 100 times the average power and with durations on the order of 1 µs using the synchronization of relaxation oscillations by drive current modulation of the intracavity pump laser diode. The forced sustained relaxation oscillations is shown to depend on wavelength, and the frequency locking curve exhibits strong asymmetry with hysteresis behavior. Tunable locked spiking operation provides interesting tool for characterization of the laser dynamics.

Mode-locked pulses with durations adjustable from 200 ps to less than 50 ps were generated by current modulation of a 1.48 μm pump laser diode incorporated in an Er-doped fiber laser cavity. The laser diode exhibits sufficient phase modulation without degradation in pump ability, and thus demonstrates a very practical and simple approach to optical pulse generation.

We propose and demonstrate a novel mechanism for Q-switching an Er-doped fiber laser. The fiber laser cavity includes a current modulated laser diode that serves both as a pumping source and nonlinear etalon, and is terminated by a diffraction grating that serves as a frequency dependent reflector. Current modulation of the intracavity laser diode produces strong frequency modulation effects in the pumping diode laser modes. The laser is effectively Q-switched when diode mode frequency is swept through the reflection window of the diffraction grating in the fibre laser cavity acted as frequency-dependent reflector. The use of pump laser diode switch has the advantage of maintaining the light in a guided waveform and permits the extraction of pulse average power close to cw average power. High switch hold-off with switch time <1ns can be easily obtained by frequency modulation of the laser diode modes in dispersive resonator.

MODE-LOCKING OF A 2.7 μm ERBIUM-DOPED FLUORIDE FIBER LASER

CHRISTIAN FRERICHS

Czech Technical University, Faculty of Nuclear Sciences and Physical
Technische Universität Braunschweig, Institut für Hochfrequenztechnik
P.O. Box 3329, D- 38023 Braunschweig, Germany

Abstract

Pulsed lasers around 3 μm have potential applications in the field of medicine, e.g. laser ablation of tissue, and in sensing and spectroscopy, e.g. for investigation of ultrashort processes. Fluoride fiber lasers in this range have the advantages of low thresholds and high efficiencies and can be cw pumped by laser diodes.

A mode-locked fiber laser at 2.7 μm was realized for the first time, utilizing the 'flying-mirror' or kinematic mode-locking technique. The laser transition was $^4I_{11/2}$ -> $^4I_{13/2}$ in Er^{3+}, laser-diode pumped at 792 nm. The ZBLAN fiber doped with 1000 ppm Er^{3+} had a core diameter of 30 μm, a numerical aperture of 0.15 and a length of 134 cm. An input mirror with R @ 2.7 μm = 100% and T @ 792 nm = 80% and an output mirror with R @ 2.7 μm = 80% were butted to the fiber endfaces, forming the active resonator. The radiation diverging from the output endface of the fiber was collimated by an IRGN6 lens and led to an external mirror with R @ 2.7 μm > 98%, reflecting it into the active cavity. Power was coupled out of this external resonator by a 50% beamsplitter. The external mirror was mounted on a loudspeaker driven by a sine generator. The output radiation was focussed onto a fast InSb detector, the signal of which was preamplified and displayed on a 400 MHz oscilloscope. The response time of the detection setup was about 2.5 ns.

Moving the external mirror either manually or by the loudspeaker caused trains of pulses. Within each loudspeaker period several pulse trains occurred. No pulses were observed in the vicinity of the return points of the vibrating mirror. Each pulse train (typical length 5 - 10 μs) consisted of several hundreds of pulses. The pulse lengths were measured down to 3.5 ns and their fall times down to 2.5 ns, indicating that these measured values were detector resolution limited. The pulse repetition rate was 81 MHz, this representing approximately the roundtrip time of the active resonator. The pulses could be generated over a large range of loudspeaker frequencies, amplitudes and pump powers. For pulse generation it was favourable to have the greatest possible velocity of the external mirror. Average output powers exceeding 1 mW at a launched power of 190 mW could be measured.

The mechanism of pulse generation is not yet clear. A possible explanation could be that the mirror translation causes frequency shifts due to the doppler effect, hence exchanges energy between adjacent modes and locks them after several roundtrips.

AUTHOR INDEX

Agrawal, Prof. Govind P., 593
Almeida, Prof. Silvério P., 3
Amann, Prof. Markus-Christian, 217
Anacleto, Dr. Joaquim, 343
Anderson, Dr. R., 836
Andonovic, Prof. Ivan, 263

Barlow, Dr. Arthur, 429
Berghmans, Dr. Francis, 131

Calvani, Dr. Ricardo, 527, 617
Clement, Dr. Darrin P., 835
Corredera, Dr. Pedro G., 453
Corrons, Prof. Antonio, 441
Culshaw, Prof. Brian, 727

Dieckman, Dr. Andreas, 791
Dubard, Dr. J., 157, 399, 489

Ferreira, Dr. Luis A., 803
Frerichs, Dr. Christian, 840

Gauthier, Dr. F., 103, 469

Habib, Dr. Khalid, 834
Hadjiloucas, Dr. Sillas, 829

Ioannides, Dr. Nicos, 827

Jackson, Prof. David A., 629
Jacobs, Prof. Ira, 567
Jaroszewicz, Dr. L.R., 835

Kalymnios, Dr. I., 749
Kashyap, Dr. Raman, 69
Kumar, Dr. Virendra, 838

Laude, Dr. Jean Pierre, 193

Malzahn, Dr. Eric, 511
Marciniak, Dr. M., 833
Men, Dr. C. Le, 175, 241, 353
Michelotti, Dr. Francesco, 825
Mignani, Dr. Anna G., 691

Okamura, Dr. Haruo, 541
Okhotnikov, Dr. Oleg G., 813, 840
Oksanen, Dr. Lauri, 415
Österberg, Prof. Ulf, 711

Paulicak, Dr. I., 838
Pelayo, Prof. J., 47
Podoleanu, Dr. Adrian Gh., 836

Rebolledo, Prof. J., 305
Rios, Dr. Susana, 773

Shukes, Dr. Scott, 833
Soares, Prof. Olivério D.D., 19
Sochor, Dr. Vaclav, 809
Szustakowski, Prof. M., 781

Talvitie, Dr. Hannu, 834
Taplin, Dr. Stephen R., 831

Unrau, Prof. Ing. U., 113

Vita, Dr. Pietro Di, 327
Voet, Dr. Marc R.H., 647

Wang, Dr. Lingli, 837
Wolfe, Prof. William E., 837

Zhang, Dr. Jian, 839

SUBJECT INDEX

Absorption bands, 142
Absorption cross section measurement, 313
Absorption losses, 366
Accreditation, 20
Active components, 241
Active fibre, 335
Active silica fibres, 327
Adjustable fibre coupler, 738
Ageing, 457
All optical routing switching, 732
All-optical network, 40
Alternative test mtehod, 335
Ammonia and humidity, 703
Amplified spontaneous emission (ASE), 335, 518
Amplifier noise, 601
Analog transmission on fiber, 585
Annealing, 138
Applications of FOI for periodic measurands, 631
Applications of plastic optical fibres, 747
Artefacts, 429
ASE interpolation method, 519
Attenuation coefficient α, 366
Attenuation, 68

B(V) curves, 390
Backscattering technique, 372
Bandwidth^2× distance, 382
Bending losses, 368
Bit-rate and distance limits, 579
Booster amplifiers, 512
Bragg grating sensors, 642
Brillouin amplification, 287
Brillouin based distributed sensors, 639
Brillouin scattering, 717

Calibration, 429
Carbon dioxide partial pressure, 704
Chalcogenide glasses, 118
Characterisation of OFAs, 328
Characteristics of plastic optical fibres, 747
Chromatic dispersion, 338, 374, 432
Cladding embrittlement, 137
Cladding, 353, 696
Coating and Cabling Technologies, 734
Coherent detection, 218
Coherent systems, 581
Coherent transmission, 220, 617
Colour centres, 132, 137, 151
Communication systems, 593
Concentricity errors, 356
Connectors, 159
Continuous wavelength tuning, 222
Core, 353, 696
Couplers, 175
Cross phase modulation, 86
Crush, 423
Cryogenic radiometer, 461
Current sensors, 636
Cut-back technique, 369
Cut-off wavelength, 209, 353, 390

D shaped cross section, 715
Decontamination, 147
Density waves, 714
Detectors calibration, 450
Diffusion losses, 367
Digital on-off-keying receiver, 575
Dispersion compensation, 37, 607
Dispersion curve D(λ), 384
Dispersion equation β(k), 390
Dispersion measurement, 37
Dispersion shifted (DS) fibres, 383

Dispersion, 33, 569
Distributed Bragg reflector (DBR), 218
Distributed feedback (DBF), 218
Distributed sensors, 639
Dose rates, 148

EDFA test system, 521
EDFAs, 332
Effective spectral bandwidth, 382
Electrical spectrum analyser, 334
Electrical test method, 518, 519
Electron-spin resonance, 132
Elliptical core, 730
Emission and absorption cross sections, 344
Environmental conditions, 338
Environmental monitoring, 691
Environmental properties, 334
Environmental testing of connectors, 405
Equilibrium mode distribution, 33
Er-doped fiber ring laser, 542
Erbium doped fiber amplifiers (EDFAs), 512, 541
Erbium, 327
Erbium-doped fiber, 127, 332
Erbium-doped silica fibre modeling, 309
ETSI, 328, 331, 332
EUROMET, 29
Evanescent sensors, 15
Exotic fibres, 103
Extrinsic FOSs, 696
Extrinsic sensors, 15

Far field power distribution, 394
Far field scanning, 394
Faraday effect, 636
Fiber drawing, 119
Fiber lasers, 113
Fiber optic Networks, 584
Fiber optic sensors, 12
Fiber optic spectrophotometers, 699

Fiber optic test set, 432
Fiber sensor, 133, 629, 647, 655, 691, 692
Fiber test set, 429
Fiber-dispersion management, 605
Fiber-loss management, 596
Fibre amplifier, 343
Fibre attenuating, 353
Fibre curl, 416
Fibre dispersion, 353
Fibre geometry, 353
Fibre gratings, 37
Fibre lasers, 737
Fibre Metrology, 19
Fibre optic back plane interconnect, 742
Fibre optic interferometers, 629
Fibre optic pressure sensors, 683
Fibre optic principle, 651
Fibre optic sensing, 655
Fibre optic sensor networking, 666
Fibre probes, 35
Fibre spectral attenuation curve, 368
Fibre tale, 103
Fibre-optic instrumentation trends, 39
Filtered power meter TM, 334
Flat dispersion, 383
Fluorescence measurement, 317
Fluoride glasses, 117, 124
Four-photon mixing, 719
Frequency response, 380
Fundamental mode, 389
Fusion reactors, 147

Gain, 333, 334, 516
Gain coefficient, 348
Gain measurement, 319
Gain spectrum, 599
γ-radiation, 134, 136
Geometrical characterisation, 356
Glasses, 114
Graded-index, 354
Group index, 429

Group velocity, 374

Guided propagation, 352
Hard-clad silica fibres, 110
HE11 mode, 390
Herbicides and pesticides, 703
Heterodyne fibre polarimetry, 527
Heterodyne mixing, 518
Heterodyne stokes polarimetry, 534
Highly birefringent fibre systems, 641
Historical survey, 3
Holium-doped fiber, 129
Hollow silica fibre, 732
Hydrocarbons, 702

IEC, 328, 330
Impulse response method, 379
In-line fiber amplifiers, 604
In-line techniques, 354
Induced attenuation, 131
Input and output signal power, 333
Insertion loss technique, 372
Insertion loss, 169
Instrumentation trends, 38
Intelligent measurements, 40
Intensity modulated fibre optic sensors, 655
Interferometry technique, 387
International Standardisation Bodies, 328
Intramodal dispersion, 375, 376
Intrinsic sensors, 13, 696
Ionising radiation, 131
Irradiation facility, 148
ITU-T, 328, 329, 330

Kerr effect, 78, 738

Laser diodes, 253
Laser sources, 38
Laser, 218
Lifetime measurement, 324
Light emission, 232

Light emitting diodes (LED), 245, 572
Line amplifier, 333
Linearly polarizes (LP) modes, 389
Local area networks, 605
Loss calibration, 479
Loss compensation and gain control, 552
Low coherence or white light interferometry, 633
LP 01 & LP11 filtering, 391
LP m, μ, 389
Luminescence, 137

Macrobending, 423
Magnetostrictive coatings, 729
Material dispersion, 374
Measurement procedures, 31
Mechanical and environmental testing, 400
Mechanical diameter measurement, 358
Mechanical properties, 334
Mechanical testing of connectors, 402
Methane and propane, 700
Metrology, 747
Metrological strategy, 23
Microbending losses, 368
Microbending, 32
Military environment, 145
Modal bandwidth, 381
Modal noise, 34
Mode distribution, 33
Mode equilibrium, 370
Mode field diameter, 209, 353
Mode filtering, 34
Multichannel amplification, 601
Multimode fibres bandwidth, 377

National Metrological Institutes, 29
Near field intensity pattern, 395
Near field light distribution, 357
Neutron γ-ray, 145
Noise, 333, 334

Noise accumulation, 338
Noise figure, 517
Noise in semiconductors, 447
Noise level, 516
Nominal values and tolerances, 357
Non-circularities, 356
Non-invasive measurements, 36
Non-linear (Soliton) transmission, 291
Non-linear diffusion processes, 367
Non-linear polarisation, 72
Non-linear response, 455
Non-linear distortion, 587
Nonlinear effect, 11
Nonlinear optical fibres, 69
Nonlinear optics, 711
Nonlinear refractive index, 78
Nonlinear susceptibilities, 711
Nonlinear wave propagation, 76
Normalized frequency parameter V, 355
Normalized propagation constant B, 374
Nuclear fusion, 134
Nuclear installations, 132
Nuclear power plants, 134, 147
Nuclear radiation, 131
Numerical Aperture NA, 354

OFA characteristics, 336
OFA devices and sub-systems, 332
OFA parameters, 332
Operation at the zero-dispersion, 607
Operation, administration and maintenance, 338
Optical access network, 339
Optical amplifier metrology, 39
Optical amplifiers, 263, 327, 343, 512, 581
Optical attenuators, 175
Optical connectors, 335
Optical demultiplexer, 334
Optical distribution network, 339
Optical fiber communication technology, 567

Optical isolator, 335
Optical non-linearities, 338
Optical power meter, 334
Optical power, 334
Optical preamplifiers, 602
Optical pumping, 599
Optical safety aspects, 338
Optical spectrum analyser, 334, 515
Optical time domain reflectometry, 275, 373
Optically amplified interoffice line systems, 339
Optically amplified reveiver, 333
Optically amplified submarine line systems, 339
Optically amplified transmitter, 333
Optrode, 698
Organic glasses, 106
Organochlorides, 702
OTDR calibration, 469
OTDR measurement curve, 471
Out-of-band insertion loss, 333, 334
Overlay devices, 740

Passive components, 37
PDFA, 524
Performance testing, 32
Petermann definitions, 394
pH, 704
Phase conjugation, 608
Phase shift technique, 384
Photo bleaching, 137
Photodetector calibration, 441
Photodetectors, 574
Photons in glasses, 712
Photon tunneling microscope, 738
Photophone, 7
Plastic optical fibres (POF), 747
Plesiochronous digital hierarchy, 336
PMD measurement, 64
Polarimetric fibre optic sensors, 660
Polarisation instabilities, 97

Polarisation loss, 33
Polarisation mode dispersion, 429
Polarisation response, 70
Polarisation, 338
Polarization controller, 521
Polarization dependence, 181
Polarization dependent gain, 522
Polarization extinction method, 519
Polarization mode dispersion, 61, 522
Polarized light, 47
Polsk system, 618
Polycrystalline fibres, 107
Polymer optical fibres, 38
Polymeric waveguides, 732
Power (Booster) amplifiers, 333
Power meter calibration, 453
Pre-Amplifier (PA), 333
Predictive models, 145
Preform fabrication, 117
Propagation constant β, 374
Pulse delay technique, 385
Pump filter, 335
Pump laser, 335
Pump leakage, 333, 334
Pumping efficiency, 348
Pure silica core fibres, 132

Quasi distributed fibre sensors, 641

Radiation conditions, 144
Radiation damage, 136
Radiation effects, 131, 136, 137
Radiation hardening, 145
Radiation hardness, 132
Radiation induced attenuation, 132
Raman amplification, 284
Raman and Brillouin scattering, 712
Raman and Brillouin, 367
Raman DTS, 639
Raman scattering, 712
Rare-earth-doped fibre amplifiers, 305
Rayleight diffusion, 367

Receiver sensitivity, 574
Recirculating delay line, 475
Reference test method, 335
Reflectance calibration, 484
Reflectance, 333, 334
Refracted near field, 358
Refractive index changes, 137
Refractive index profile, 355
Reliability properties, 334
Remote fibre testing systems, 36
Repeater spacing, 580
Repeater spacing enlargement, 38
Return loss, 168
Ribbon fibre measurements, 415
Ribbon geometry, 418
Robotics, 135

Safety properties, 333
Sampled grating, 231
Saturated output power, 348
Second harmonic generation, 90
Self-focusing damage, 92
Self-phase-modulation, 80
Sensors, 35
Separability, 413
Shot noise, 518
Signal level, 516
Signal-to-noise ratio, 517
Single mode laser diodes, 217
Single-mode propagation, 390
Small-signal gain, 600
Software accreditation, 40
Sol-gel, 697
Soliton amplification, 611
Soliton communication systems, 609
SOP measurement, 527
Source spontaneous emission, 519
Spectral attenuation, 209
Spectral linewidth, 226
Spectral modulation fibre optic sensors, 663
Spectral responsivity, 443

Splices, 159
Standardisation bodies, 327
Standardisation, 327
Standardisation specifications, 34
Standards, 19, 131
Step-index, 395
Stimulated brilloin, 737
Stimulated raman scattering, 88
Stimulated raman, 737
Strain, 429
Strippability, 419
Superstructure grating, 231
Synchronous digital hierarchy, 336

Telemanipulation, 134
Time gain, 730
Time reversal, 738
Torsion, 422
Traceability, 26, 29
Transfer standards, 40
Transit time, 374

TTG (Tuneable Twin-Guide) laser, 223
Tunable laser source, 515
Twice degenerated, 390
Two-photon abosorption, 720

Ultrashort pulse amplification, 601
Uncertainty, 29
Uniformity, 36

Wavelength conversion, 555
Wavelength dependence, 183
Wavelength division multiplexing
    (WDM), 339, 513
Wavelength tunable laser diodes, 217
WDM long distance line systems, 339
Working principle of FOSs, 693
World market of fibre optics, 26

Y-laser, 230

Zero dispersion, 728

MIX
Papier aus verantwortungsvollen Quellen
Paper from responsible sources
FSC® C105338

If you have any concerns about our products,
you can contact us on
ProductSafety@springernature.com

In case Publisher is established outside the EU,
the EU authorized representative is:
**Springer Nature Customer Service Center GmbH
Europaplatz 3, 69115 Heidelberg, Germany**

Printed by Libri Plureos GmbH
in Hamburg, Germany